P9-EDP-937

FLORA
OF THE
PRAIRIES AND PLAINS
OF
CENTRAL NORTH AMERICA

BY

PER AXEL RYDBERG, PH.D.

Late Curator of the Herbarium of
The New York Botanical Garden

IN TWO VOLUMES

Volume
One

DOVER PUBLICATIONS, INC.
NEW YORK

Published in Canada by General Publishing Company, Ltd., 30 Lesmill Road, Don Mills, Toronto, Ontario.

Published in the United Kingdom by Constable and Company, Ltd., 10 Orange Street, London WC 2.

This Dover edition, first published in 1971, is an unabridged and corrected republication of the work originally published in one volume by The New York Botanical Garden in 1932.

International Standard Book Number: 0-486-22584-4
Library of Congress Catalog Card Number: 79-166434

Manufactured in the United States of America
Dover Publications, Inc.
180 Varick Street
New York, N.Y. 10014

CONTENTS

[For this Dover reprint edition, the text has been divided into two volumes. Volume One consists of pages 1 through 503 (Family 87: Oxalidaceae); Volume Two begins with Family 88: Linaceae at the bottom of page 503 and continues through page 969.]

PREFACE

A few days before his lamented death, on July 25, 1931, the author of this Flora, feeling that his strength was no longer adequate to the task of proof-reading, asked the undersigned to take over and complete the editorial work. The author had sent to the printers all of the manuscript for the main text and had seen the paged proofs of nearly half of the book. Manuscript for the Summary, Glossary, Abbreviations of Authors' Names, and Index was still to be prepared. The Glossary, with slight changes, has been reprinted from the author's Flora of the Rocky Mountains and Adjacent Plains. Dr. John Hendley Barnhart has supplied the Abbreviations of Authors' Names, as he did also for the previous volume. The author, had his life been prolonged a few weeks, would doubtless have prepared an Introduction, discussing the area covered by the Flora, the plant ecology of the region, etc. This ground is, fortunately, in part covered by his paper on "A Short Phytogeography of the Prairies and Great Plains of Central North America," published in Brittonia (1: 57–66 + *plate*. F 1931). From a brief prospectus left by Dr. Rydberg, one gathers that this Flora aims to be a complete manual of the Spermatophyta and Pteridophyta of the states of Kansas, Nebraska, Iowa, Minnesota, South Dakota, and North Dakota, and of southern Manitoba and southeastern Saskatchewan. It includes also most of the species occurring in the prairie regions of Illinois, southern Wisconsin, and northern Missouri, and on the plains of eastern Colorado, eastern Montana, and southern Saskatchewan. No Flora of just this area has previously been published, although its borders have been invaded by other well-known manuals. Extra-limital ranges are freely cited in the text. A posthumous paper by the author, entitled "Taxonomic Notes on the Flora of the Prairies and Plains of Central North America," published in October, 1931 (Brittonia 1: 79–111), included Latin diagnoses of the new species and one new genus (*Denslovia*) recognized in the present volume—also new combinations of names and discussions of changes in nomenclature, in respect to which the author has now endeavored to follow the International Rules. In footnotes Dr. Rydberg acknowledges assistance from Mr. Kenneth K. Mackenzie in the treatment of *Carex* and from Mr. Willard W. Eggleston in the account of *Crataegus*. The author made extensive personal collections in the area covered by the Flora. In 1926 and 1929 he visited the principal herbaria of that region and at various times he received for study collections from particular localities or was favored with loans of specimens of certain groups from herbaria of state universities and other institutions.

A biographical sketch of Dr. Rydberg, written by Dr. John Hendley Barnhart, appears in the *Journal of The New York Botanical Garden* for October, 1931.

The author and editor are much indebted to Mr. R. I. Cratty of the Iowa State College for reading the first proofs of the main text. Professor F. C. Gates of the Kansas State Agricultural College has kindly read proof of many of the later pages. For assistance in proof-reading and in preparing the index, the editor is grateful also to several others, especially to Miss Carol H. Woodward and Miss Rosalie Weikert.

MARSHALL A. HOWE.

THE NEW YORK BOTANICAL GARDEN, MARCH, 1932.

ABBREVIATIONS

The well-known abbreviations of the states of the United States are here omitted.

Adv. = adventive
Alp. = Alpine Zone
Alta. = Alberta*
Am. = America or American
Ap = April
Arctic = Arctic Zone.
Au = August
Auth. = Authors†
[B] = Britton's Manual
[BB] = Britton & Brown's Illustrated Flora
B. C. = British Columbia
Boreal = Boreal Zone
c = central
C. Am. = Central America
cm. = centimeter
D = December
dm. = decimeter
e = eastern
Eur. = Europe
Eurasia = Europe and northern Asia
F = February
(Fl. Colo.) = The Author's Flora of Colorado
(Fl. Mont.) = The Author's Catalogue of the Flora of Montana and Yellowstone Park
[G] = Gray's New Manual
Greenl. = Greenland
Ja = January
Je = June
Jl = July
Labr. = Labrador
L. Calif. = Lower California
L. Son. = Lower Sonoran Zone
m. = meter
Mack. = Mackenzie Territory
Man. = Manitoba
Mex. = Mexico‡
mm. = millimeter
Mont. = Montane Zone
Mont. = Montana
Mr = March
My = May
n = northern
N = November
N. Am. = North America
Nat. = Naturalized
N. B. = New Brunswick
ne = northeastern
Newf. = Newfoundland
N. S. = Nova Scotia
nw = northwestern
O = October
Ont. = Ontario
P. E. I. = Prince Edward's Island
Plain = Subboreal Plains Zone
Que. = Quebec
[R] = The Author's Flora of the Rocky Mountains
s = southern
S = September
S. Am. = South America
Sask. = Saskatchewan
se = southeastern
Son. = Sonoran Zone
St.Plains = Staked Plains Zone
Subalp. = Subalpine Zone
Subarctic = Subarctic Zone
Subboreal = Subboreal Zone
Submont. = Submontane Zone
sw = southwestern
Trop. = Tropics
w = western
W. Ind. = West Indies

SIGNS

- (short dash) between figures or words means that the two figures or two words denote the extremes of variation

§ Subgenus or section of a genus.

— (long dash) between the names of two or more states denotes the extent of distribution.

× denotes a hybrid between the two species mentioned.

MEASUREMENTS

1 mm. = 1/25 inch
3 mm. = ⅛ inch
1 cm. = ⅖ inches
5 cm. = 2 inches
1 dm. = 4 inches
1 m. = 40 inches or 3⅓ feet
300 m. = 1000 feet

---

* Many people use the abbreviation "Alb.", but as far as the author has been able to ascertain, the official one is "Alta."

† Used in cases of misapplications of names where the author first using it in such a sense has not been ascertained.

‡ Observe the difference between N. M. (New Mexico) and n Mex. (northern Mexico).

PTERIDOPHYTA

FERNWORTS

Plants of two distinct generations, in one consisting of a plant-body (sporophyte), which has stems containing vascular tissue and produces spores asexually, in the next, developing from the spore, consisting of a thalloid-body (gametophyte or prothallium), which bears the sexual reproductive organs (archegones and antherids). The plant-body develops from the oosphere within the archegone, after being fertilized by spirally coiled motile bodies (spermatozoids), produced by the antherids.

KEY TO THE FAMILIES.

PAGE

Leaves usually broad, entire or dissected, not scale-like; fern-like plants.
 Spores of one kind, minute, borne in sporanges.
  Vernation not spirally coiled; sporangia ringless, leathery, opening by a transverse slit, arranged in spikes or panicles. 1. OPHIOGLOSSACEAE. 1
  Vernation spirally coiled; sporangia membranous.
   Sporangia opening by a longitudinal slit, without a ring. 2. OSMUNDACEAE. 3
   Sporangia provided with a ring, which opens elastically. 3. POLYPODIACEAE. 4
 Spores of two kinds, minute microspores (male) and larger macrospores (female), borne in sporocarps.
  Plant rooting in the mud; leaves 4-foliate, petioled. 4. MARSILEACEAE. 16
  Plant minute, floating; leaves entire or 2-lobed. 5. SALVINIACEAE. 17
Leaves scale-like or awl-like; moss-like or rush-like plants.
 Sporanges in an apical cone, borne under peltate bracts; stem usually hollow, rush-like. 6. EQUISETACEAE. 17
 Sporanges in the axils of small leaf-like bracts; stem solid.
  Leaves awl-like, elongate, borne on a short thick corm-like caudex; water plants. 7. ISOETACEAE. 20
  Leaves scale-like, flat, borne on a distinct stem; land plants.
   Spores uniform, minute. 8. LYCOPODIACEAE. 21
   Spores of two kinds, microspores and macrospores. 9. SELAGINELLACEAE. 23

Family 1. **OPHIOGLOSSACEAE.** ADDER-TONGUE FAMILY.

Succulent plants with short fleshy rootstocks and fibrous roots. Frond consisting of a single sterile leafblade and a narrow, simple or compound fertile sporophyll, borne on a common stalk. Sporanges formed from the inner tissues, naked, opening by a transverse slit. Spores yellow, all alike. Prothallia subterranean, usually without chlorophyll.

Sporanges in a distichous spike, coherent; veins of the sterile blade reticulate. 1. OPHIOGLOSSUM.
Sporanges in ours in panicles, distinct; veins of the lobed or dissected sterile blade free. 2. BOTRYCHIUM.

1. **OPHIOGLÓSSUM** (Tourn.) L. ADDER'S-TONGUE.

Plant small with an erect fleshy rootstock. Sterile blade entire, lanceolate to reniform, sessile or short-stalked, the veins freely anastomosing. Sporophyll narrow, cylindric, formed of two rows of coalescent sporanges. Spores yellow. Bud for the following season's growth borne at the apex of the rootstock, exposed, free.

f. 1.

1. **O. vulgàtum** L. *Fig. 1.* Frond 5–40 cm. high; sterile blade sessile near the middle of the plant, ovate to elliptic, 5–10 cm. long, obtuse or rounded at the apex; midvein indistinct. Meadows: N.S. — Fla. — Mo. — Neb. — Mont. — Alaska; Eur. *Temp.* Je–Au.

2. BOTRÝCHIUM Sw. MOONWORT. GRAPE-FERN.

Plants with short erect rootstocks. Stalk of the frond sheathing at the base, containing the bud for the next year. Sterile blade simple and lobed, or pinnately compound or decompound, the veins free. Sporophyll 1–3-pinnate, the branches bearing 2 rows of naked sporanges, these globular, without a ring, opening transversely into valves. Spores numerous, yellow.

Bud for the following season wholly enclosed in base of the stalk; sterile blade more or less fleshy; cells of the epidermis straight.
  Sterile blade pinnately lobed or dissected, inserted usually near or above the middle of the plant, sessile or short-petioled, the primary segments sessile; buds glabrous.
    Primary segments of the sterile blades entire or crenulate.
      Sterile blade petioled, erect in the bud; lobes, if any, obovate, few. — 1. *B. simplex.*
      Sterile blade sessile or nearly so, slightly bent over in the bud; lobes several, fan-shaped. — 2. *B. Lunaria.*
    Primary divisions of the sterile blade distinctly toothed or lobed; sporophyll and sterile blade both bent over in the bud.
      Sterile blade and the sporophyll both sessile or nearly so at the end of the long common stalk. — 3. *B. lanceolatum.*
      Sterile blade short-stalked, the sporophyll long-stalked. — 4. *B. neglectum.*
  Sterile blade subternately decompound, the primary segments stalked; buds pilose.
    Segments of the sterile blade acute; leaves thin in drying.
      Segments of the sterile blades serrate-dentate. — 5. *B. obliquum.*
      Segments of the sterile blades laciniate. — 6. *B. dissectum.*
    Segments of the sterile blade obtuse; leaves thick and leathery in drying. — 7. *B. silaifolium.*
Bud for the following season exposed on one side; sterile blade sessile, thrice ternate, membranous; epidermal cells flexuose. — 8. *B. virginianum.*

1. **B. símplex** E. Hitchcock. Frond 4–20 cm. high; sterile blade short-stalked, ovate, obovate, or oblong in outline, simple, or pinnately lobed, with 3–7 cuneate to lunate divisions, rounded at the apex; sporophyll long-stalked, simple or pinnate. Meadows or open woods: N.S.—Md.—Minn.—Colo.—Ore.; Eurasia. *Boreal.* My–Je.

f. 2.

2. **B. Lunària** (L.) Sw. *Fig. 2.* Frond 3–28 cm. long; sterile leaf-blade oblong or rarely ovate or deltoid-ovate, rounded above, once pinnately divided; segments fan-shaped or lunulate or reniform, often imbricate, entire, or rarely radially incised or cleft into cuneiform lobes; sporophyll 1–3 times divided. Woods and grassy places: Greenl.–N.Y.—n Ohio—Mont.—Colo.—Calif.—B.C.—Alaska; Calif.; Old World. *Boreal—Mont.—Submont.* The

var. *onondagense* (*B. onondagense* Underw.), which is distinguished by its rather distant fan-shaped segments, is the usual form in the United States.

3. **B. lanceolàtum** (S. G. Gmel.) Ångstr. Frond 5–30 cm. long; common stalk long; sterile leaf-blade broadly deltoid, acute, 1–2 times pinnately or subternately divided, the primary divisions oblong to ovate or oblong-lanceolate, the segments ovate, ovate-lanceolate or suboval, entire or incised; stalk of sporophyll mostly shorter than the rather diffuse panicle; sporanges large, crowded, sessile or broadly short-pedicelled. Woods and damp hillsides; Alaska —B.C.—Wash.—Wyo.; N.D.—N.S—Greenl; Old World. *Boreal.*

4. **B. neglèctum** Wood. Frond 5.5–32 cm. long; common stalk usually long; sterile leaf with a stalk 0.2–1.7 cm. long, oblong or sometimes deltoid, subacute, the primary divisions oblong or ovate, the segments usually oblong, rounded at apex, entire or incised; sporophyll commonly diffusely branched. Wooded places: Que.—Md.—Ohio—Wis.—Colo.—S.D.; Eur. *Mont.—Boreal.*

5. **B. oblìquum** Muhl. Frond 1.5–5 dm. high; sterile blade long-stalked, inserted near the base, 5–7 cm. broad, somewhat pentagonal in outline, ternately pinnatifid, the primary divisions stalked, twice pinnatifid, the ultimate divisions obliquely ovate or lanceolate, acute; sporophyll long-stalked, 3-4-pinnate. Moist woods and thickets: N.B.—Fla.—Mo.—Minn.—Man. *Temp.* My–Je.

6. **B. dissèctum** Spreng. Fronds 2–4 dm. long; sterile blade long-stalked, inserted near the base, triangular or subpentagonal in outline, up to 1.5 dm. wide; sporophyll long-stalked, 3–4-pinnate. Woods and thickets: Va.—Iowa—Minn. *Canad.—Allegh.*

7. **B. silaifòlium** Presl. Frond 9.5–60 cm. long; sterile leaf-blade broadly deltoid to pentagonal, acute or obtuse, 3–4 times pinnately or subternately divided, the primary divisions oblong or oblong-lanceolate to deltoid or pentagonal, the subdivisions next to the last oblong-lanceolate, pinnately divided into 2–5 pairs of oblique, oblong, narrowly elliptic, or sometimes ovate, decurrent lateral segments, and with broader rhomboid tips, the segments crenulate; panicle of the sporophyll diffuse. *B. ternatum intermedium* D. C. Eat. [G]. Fields and open woods: Ont.—Que.—N.J.—Pa.—Neb.—Wis.; Alaska—B.C.—Calif. *Boreal.*

8. **B. virginiànum** (L.) Sw. Frond 6.5–60 cm. long; sterile leaf-blade thin, 3–5 times divided, broadly deltoid, the subdivisions next to the last narrowly oblong or lanceolate to ovate or deltoid, pinnately divided, the segments oblong, blunt, incised; sporophyll mostly long-stalked. Wooded places: Que.–Fla.—Tex.—Colo.—Ore.—B.C.; trop. Am. and the Old World. *Temp.—Mont.*

Family 2. **OSMUNDACEAE.** Royal Fern Family.

Large ferns with stout rootstocks. Fronds once to twice pinnate, the stipes winged at the base, the free veins extending to the margins. Sporanges naked, large, globose, mostly stalked, borne on modified pinnae, in ours covering them, or in some genera of the Old World borne in clusters on the lower side, opening by longitudinal slits; ring wanting.

1. **OSMÚNDA** (Tourn.) L. Royal Fern.

Tall ferns of swamps and low-lands, with large crowns of many fronds. Fronds long-stalked, pinnate or bipinnate, with regularly forked veins, the fertile portion much contracted and devoid of chlorophyll. Sporanges short-stalked, thin, reticulate, opening in halves.

Blades bipinnate, some of them fertile at the apex; ultimate pinnules oblong, finely serrulate. 1. *O. regalis.*
Blades pinnate, woolly when young; pinnae pinnatifid into oblong entire segments.
Fronds dimorphous, the sterile ones longer than the fertile ones; pinnae with a tuft of tomentum at their bases. 2. *O. cinnamomea.*
Fronds either wholly sterile, or some of the larger ones with some of the middle pinnae fertile; tufts of tomentum wanting. 3. *O. Claytoniana.*

1. **O. regàlis** L. Fronds 5–20 dm. high, bipinnate; pinnae 1.5–3 dm. long, the upper ones fertile, forming a terminal panicle; pinnules oblong, 2–5 cm. long, serrulate, rounded or cordate at the base. Royal Fern. Wet woods: Newf.—Fla.—Miss.—Neb.—Man. *E. Temp.* My–Je.

2. **O. cinnamòmea** L. *Fig. 3.* Sterile fronds 8–15 dm. high; stipe clothed with ferruginous tomentum when young, 3 dm. long or more; pinnae linear-lanceolate, deeply pinnatifid into oblong, obtuse, entire segments; fertile fronds smaller, appearing earlier, 2–9 dm. high, bipinnatifid; sporanges cinnamon-colored. Cinnamon Fern. Swamps: Newf.—Fla.—N.M.—Man.; Mex.—Braz.—W. Ind. *E. Temp.* My–Je.

f.3.

3. **O. Claytoniàna** L. Fronds 3–13 dm. high, loosely tomentose when young, the inner and larger ones fertile in the middle; pinnae oblong-lanceolate, deeply cleft into ovate-oblong, entire segments; fertile pinnae in the middle 2–5 pairs, smaller than the sterile ones, fully pinnate, at first greenish, later dark-brown. Clayton's Fern. Swamps: Newf.—N.C.—Mo.—Neb.—Man. *E. Temp.* My–Jl.

Family 3. POLYPODIACEAE. Fern Family.

Sporophytes consisting of a rhizome and leaves (fronds) coiled in the bud. Sterile fronds leaf-like. Fertile fronds (sporophylls) either leaf-like or partially or completely non-foliaceous, bearing the sporanges on their lower surface or at their margins, commonly in clusters (sori). Sori naked or furnished with a special covering (indusium). Sporanges stalked, furnished with an incomplete vertical ring of thickened cells (annulus), opening transversely. Prothallia green, above ground.

Fertile fronds with contracted berry-like or necklace-like subdivisions, not foliaceous.
Veins of the sterile fronds netted. 1. Onoclea.
Veins of the sterile fronds free. 2. Pteretis.
Fertile and sterile fronds foliaceous, alike or differing; veins free.
Sori on the under surface of the fronds, each provided with a special indusium not connected with the margin of the frond.
Sori roundish.
Indusia inferior or attached at base at one side of the sorus.
Indusia inferior, stellate or split into spreading hairs. 3. Woodsia.
Indusia attached at base at one side of the sorus, at first arched over it, finally thrown back or evanescent. 4. Filix.
Indusia superior, stellate or split into spreading hairs.
Indusia peltate. 5. Polystichum.
Indusia orbicular-reniform, adherent at the sinus. 7. Dryopteris.
Sori oblong or linear, or shaped like a horseshoe or shepherd's crook.
Veins reticulate; fronds simple, rooting at the tip. 8. Camptosorus.
Veins free; fronds pinnate, not rooting at the tip.
Sori all straight or rarely slightly curved, single on the sides of the veins. 9. Asplenium.

Sori, at least in part, shaped like a horseshoe or shepherd's crook, crossing the fertile vein and more or less recurved upon it. 10. ATHYRIUM.
Sori naked, or protected, at least at first, by the revolute or reflexed margins or portions of the margins of the frond.
Margin of the frond reflexed over the sori, more or less modified.
Sori borne on the under side of reflexed lobes of the frond. 11. ADIANTUM.
Sori not borne on the under side of the reflexed portions of the margin of the frond.
Sori borne on a continuous vein-like receptacle connecting the ends of the veinlets. 12. PTERIDIUM.
Sori borne on the veins at or near their tips.
Sori extending down the veins; edges of the fertile fronds finally opening out flat; sterile and fertile fronds markedly dissimilar. 13. CRYPTOGRAMMA.
Sori marginal or submarginal; sterile and fertile fronds alike or somewhat dissimilar.
Sori confluent, forming a submarginal band; segments of the fronds glabrous or nearly so. 14. PELLAEA.
Sori distinct or contiguous; segments usually pubescent, tomentose or scaly. 15. CHEILANTHES.
Margin of the frond flat or merely revolute, not modified.
Margin of the frond revolute; sori more or less confluent, forming a marginal band. 16. NOTHOLAENA.
Margin of the frond flat; sori dot-like on the veins.
Stipes jointed to the rhizome; frond simply pinnate. 17. POLYPODIUM.
Stipes not jointed to the rhizome; frond 2–3-pinnatifid. 6. PHEGOPTERIS.

1. **ONOCLÈA** L. SENSITIVE FERN.

Ferns, with coarse dimorphous fronds from creeping rhizomes. Sterile fronds foliaceous. Fertile fronds with rigid, contracted rounded subdivisions. Sori roundish, on elevated receptacles, partly covered by delicate hood-like indusia attached to the bases of the receptacles.

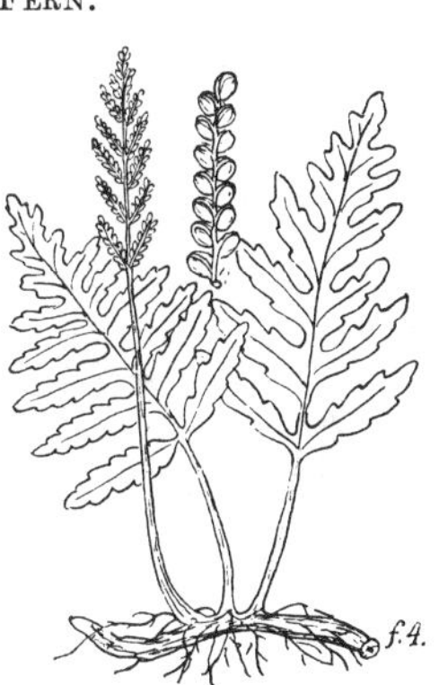

1. O. sensíbilis L. *Fig. 4.* Sterile fronds 3–13.7 dm. high; blade deltoid-ovate, pinnatifid; segments lanceolate-oblong, entire, undulate or sinuate-pinnatifid; rachis winged; veins anastomosing; fertile fronds 3–7.6 dm. high, bipinnate; pinnules rolled into closed balls, finally dehiscent; veins free; sporangia dorsal on the simple or forked veinlets. Damp places: Newf.—Fla.—Tex.—S.D.—Sask. *E. Temp.—Submont.*

2. **PTERÈTIS** Raf. OSTRICH FERN.

Plants with coarse dimorphous fronds, growing in crowns from stoloniferous rhizomes. Sterile fronds foliaceous. Fertile fronds contracted, with revolute margins covering the sori. Sori roundish, on elevated cylindrical receptacles partly covered by delicate lacerate fugacious indusia inferiorly attached. *Struthiopteris* Mett., *Matteuccia* Todaro.

1. P. nodulòsa (Michx.) Nieuwland. Scales of base of stipe pale brown to cinnamon-colored, thin, membranous; fronds abruptly acuminate at apex, gradually reduced towards base; sterile fronds 0.6–3 m. long, broadly oblanceolate or spatulate, with lanceolate or linear, pinnatifid pinnae 5–18 cm. long; segments oblong, obtuse or acute; fertile fronds shorter, with rigid, upcurved, commonly pinnatifid necklace-shaped pinnae; veins free, pinnate; veinlets simple. *Onoclea nodulosa* Michx. Related to but distinct from the European *P. Struthiopteris* (L.) Nieuwland. *Matteuccia Struthiopteris* (L.) Todaro. Wet places: Newf.—Va.—S.D.—B.C. *E. Temp.*

3. WOÓDSIA R. Br.

Small tufted ferns. Fronds 1–3-pinnatifid. Sori rounded, borne on the simply branched free veins. Indusia basal, often evanescent, thin, attached by their bases under the sporanges, either small and open, or irregularly bursting or cleft.

Indusia divided to near the center into slender, hairlike divisions, curled over the sporanges.
Stalk of the frond articulate below; fronds never glandular.
Blades and stalks distinctly chaffy beneath; lower pinnae 9–13-lobed. 1. *W. ilvensis.*
Blades and stalks glabrous or nearly so; lower pinnae 3–7-lobed.
Blades lance-oblong, sparingly chaffy-hirsute beneath; divisions of the indusium numerous. 2. *W. alpina.*
Blades linear or lance-linear, perfectly glabrous; divisions of the indusium few. 3. *W. glabella.*
Stalk of the frond not articulate; fronds usually somewhat glandular.
Frond pubescent with flattened articulate hairs and stalked glands. 4. *W. scopulina.*
Frond without articulate hairs, minutely glandular-puberulent or glabrate.
Frond glabrous; indusia cleft almost to the center. 5. *W. oregana.*
Frond glandular-puberulent; indusia cleft halfway down or slightly more. 6. *W. mexicana.*
Indusia with a few broad segments, merely jagged, at first covering the sori. 7. *W. obtusa.*

1. **W. ilvénsis** (L.) R. Br. Fronds 5–15 cm. long; blades oblong-lanceolate, glabrate and green above, rusty-chaffy beneath as well as the stalk; pinnae crowded, oblong, obtuse, their segments crowded; sori submarginal, confluent when old. Rocks: Greenl.—Me.—N.C.—Iowa—Minn.—Alaska; Eurasia. *Arctic.—Canad.* Je–Au.

2. **W. alpìna** (Bolton) S. F. Gray. Fronds 4–13 cm. long, 6–35 mm. wide; stipe chestnut-colored, shining; pinnae cordate-ovate, the lower 5–7-lobed, the lobes sub-entire; sori near the margins, the divisions of the indusium thread-like, numerous. Moist rocks: Greenl.—Vt.—Man.—Alaska—Eurasia. *Arct.—Boreal.*

3. **W. glabélla** R. Br. Fronds tufted, 2.5–15.5 cm. long; stipes usually straw-colored; blades linear or narrowly lanceolate, somewhat narrowed toward base, smooth, pinnate; pinnae deltoid to roundish-ovate, crenately 3–5-lobed, the lobes also crenate; indusia divided into narrow, jointed, hair-like, curving divisions. Moist rocks: Alaska—Greenl.—N.Y.—Minn.; Eu. *Boreal—Subarct.*

4. **W. scopulìna** D.C. Eat. *Fig. 5.* Fronds 5.5–35 cm. long; blades lanceolate, pinnate; pinnae mostly oblong-ovate, deeply pinnatifid; segments short, ovate or oblong, crenate-serrulate; indusia delicate, deeply cleft into laciniae which terminate in short hairs. On rocks: Minn.—Iowa—Neb.—Ariz.—Calif.—B.C.; Que. and N.C. *W. Temp. Submont.—Mont.*

5. **W. oregàna** D.C. Eat. Fronds 5–25.6 cm. long; blades lanceolate-oblong, pinnate; pinnae triangular-oblong, obtuse or subacute, pinnatifid; segments oblong or ovate, obtuse, toothed or crenate; teeth often reflexed over the sori; sori submarginal. On rocks: B.C.—Calif.—Ariz.—Colo.—Minn.—Sask.—Que. *W. Temp.—Submont.—Subalp.*

6. **W. mexicàna** Fée. Fronds 5–30 cm. long; blades lanceolate, pinnate, minutely glandular; pinnae triangular-lanceolate or

rarely suboblong, pinnately divided; segments finely toothed, the teeth ending in delicate semitransparent tips which are ciliated in young fronds; sori submarginal. *W. Cathcartiana* B. L. Robins. [G.] Rocks: Minn.—S.D.—Colo.—N.M.—Ariz.—Mex. *Allegh.—Son.—Mont.*

7. **W. obtùsa** Torr. Fronds 2–5 dm. long; blades broadly lanceolate, slightly reduced at base, pinnate to bipinnate, minutely glandular; pinnae oblong or triangular-ovate, pinnately parted or pinnate below; segments oblong, obtuse, crenate-dentate, or the lower pinnatifid with toothed lobes. N.S.—N.H.—Ga.—Ariz.—Tex.—Minn.—B.C. *Temp.—Mont.*

4. FÍLIX Adans. BLADDER FERN.

Ferns, with pinnately compound fronds of rather thin texture. Sori roundish, indusiate, borne on the veins. Indusia delicate, hood-like or flattish, attached at one side of and partly under the sorus, at first arched over it, finally thrown back or withering. Veins free. *Cystopteris* Bernh.

Blades of the fronds deltoid-lanceolate or broadly lanceolate, twice or thrice pinnate.
 Blades elongate-lanceolate, 3–12 dm. long; basal pair of pinnae the largest; rachis not margined. 1. *F. bulbifera.*
 Blades broadly lanceolate, 2–4.5 dm. long; basal pair of pinnae usually slightly shortened; rachis margined. 2. *F. fragilis.*
Blades of the fronds deltoid-ovate, three to four times pinnate. 3. *F. montana.*

1. **F. bulbífera** (L.) Underw. Rhizome short; fronds clustered; blades 3–8 cm. high, minutely more or less glandular beneath, especially on the rachises and midribs; pinnae oblong-ovate to lanceolate-oblong, pinnate; pinnules unequally oblong-ovate, obtuse, variously incised to deeply pinnatifid; rachis and pinnae commonly bulbiferous beneath, the bulblets producing new plants after falling to the ground; indusia short, convex, truncate at the attachment. *C. bulbifera* Bernh. [G]. Damp places, especially about rocks: Newf.—N.S.—Man.—Iowa—Ark.—Ga.; Utah—Ariz. *Boreal—Temp.—Son.*

2. **F. frágilis** (L.) Gilib. *Fig. 6.* Rhizome creeping; fronds clustered or slightly scattered, glabrous, 1–5 dm. high; stipe and primary rachis slender, brittle, stramineous or brownish below; pinnae deltoid-lanceolate or deltoid-ovate; segments decurrent, variable, roundish-oval, ovate, rhomboid-ovate, or ovate-lanceolate, dentate, with short obtuse teeth, or deeply toothed or cleft, with narrower teeth, or cleft or sometimes pinnatifid, with toothed segments; indusia delicate, convex, roundish or commonly pointed at the attachment, often toothed or laciniate at apex. *Cystopteris fragilis* Bernh. [G]. Rocky places; Greenl.—Ga.—Okla.—Calif.—Alaska; trop. Am., and the Old World. *Plain—Alp.—Arct.—Temp.—Trop.*

f. 6.

3. **F. montàna** (Lam.) Underw. Rhizome slender, creeping; fronds scattered, 12–45 cm. long; stipes slender; blades often subternate; basal pair of pinnae much the largest, unequally deltoid-ovate; pinnules deeply divided; segments oblong, deeply toothed or divided; indusia convex, ovate, soon thrown back or evanescent. Greenl.—Que.—B.C.—Alaska; Colo.; Eurasia. *Mont.—Subalp.—Boreal—Arctic.*

5. POLÝSTICHUM Roth. HOLLY FERN, CHRISTMAS FERN.

Ferns with stout rootstocks. Fronds firm, evergreen, pinnate or pinnately decompound, usually with sharply toothed or spinulose margins, the stalk and rachis chaffy, the veins free. Sori round, borne on the veins, with a peltate orbicular indusium.

Fronds normally simply pinnate; pinnae auricled at the base on the upper margin.
  Upper spore-bearing pinnae of the fertile fronds much narrower than the sterile ones. 1. *P. acrostichoides.*
  Upper spore-bearing pinnae similar to the sterile ones. 2. *P. Lonchitis.*
Fronds more or less bipinnatifid.
  Pinnae serrate in the distal part, pinnately lobed at the base. 3. *P. scopulinum.*
  Pinnae pinnatifid throughout, i.e., frond truly bipinnatifid. 4. *P. Braunii.*

1. **P. acrostichoìdes** (Michx.) Schott. Fronds in a crown, 2–6 dm. high, the stalks and rachis chaffy; blades lanceolate, once pinnate; pinnae 2–8 cm. long, oblong-lanceolate, acutish, serrulate with spinulose teeth, usually with a lobe on the upper margin at the base; upper part of the fertile frond contracted, the smaller pinnae in age wholly covered by the confluent sori; indusium entire. Rocky places: N.S.—Fla.—Tex.—Kans.—Wis. *E. Temp.* Jl–Au.

2. **P. Lonchìtis** (L.) Roth. *Fig. 7.* Fronds growing in a crown, 8–22 cm. long; stipes and rachises chaffy with light brown scales; blades linear-lanceolate, gradually tapering toward base; pinnae 1–4.5 cm. long, densely spinulose-toothed, the teeth mostly spreading; sori medial or supramedial. Woods: Greenl. — N.S. — Minn. — Alta. — Colo.—Calif.—Alaska. *Arct.—Mont.*

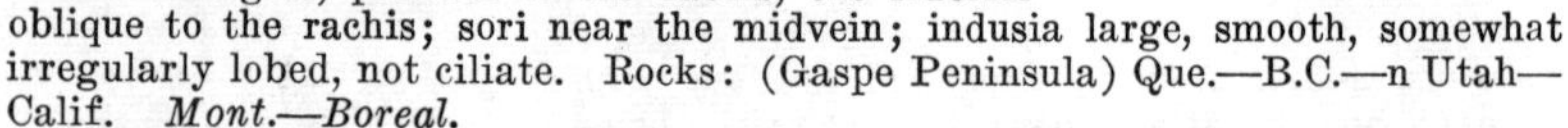

3. **P. scopulìnum** Maxon. Fronds 6.5–43 cm. long; stipes and rachis conspicuously chaffy, with light brown scales; blades lanceolate or linear, commonly somewhat narrowed toward base; pinnae ovate or ovate-oblong, mostly obtuse, serrate in the outer part, with pointed or aculeate, not spinescent teeth, pinnately lobed at base, the superior basal lobe the largest, parallel to the rachis, the inferior oblique to the rachis; sori near the midvein; indusia large, smooth, somewhat irregularly lobed, not ciliate. Rocks: (Gaspe Peninsula) Que.—B.C.—n Utah—Calif. *Mont.—Boreal.*

4. **P. Braùnii** (Spenner) Fée. Fronds in a crown, 1.4–8.7 dm. long; stipe and rachis chaffy, with bright brown scales; blades lanceolate, gradually narrowed toward base; pinnae oblong-lanceolate from a slightly broader base; segments ovate-oblong, produced on the upper side, obliquely cut away on the lower, commonly acute, sharply toothed, scaly; indusia small, entire. Rocky woods: Newf.—Mass.—Pa.—Mich. *Boreal.*

6. PHEGÓPTERIS (Presl) Fée. OAK FERN, BEECH FERN.

Ferns with creeping rootstocks. Fronds 1–3-pinnatifid, the stipe not articulate to the rootstock. Midveins and midribs uniting at a broad angle, the veinlets simple or forked. Sori borne on the veins, without indusium, Hairs long and soft, 1-celled, simple, often hooked at the apex. *Thelypteris* Slosson; not Schmidel.

Fronds twice pinnatifid; pinnae sessile, adnate to the winged rachis.
  Blade longer than broad; lowest pair of pinnae deflexed. 1. *P. polypodioides.*
  Blade as broad as long or broader; lowest pair of pinnae largest, spreading. 2. *P. hexagonoptera.*
Fronds ternate, the pinnae stalked; rachis not winged.
  Blade glabrous, the primary 3 pinnae subequal and spreading. 3. *P. Dryopteris.*
  Blades minutely glandular, the lower pinnae smaller than the terminal one. 4. *P. Robertiana.*

1. **P. polypodioìdes** Fée. Rhizome creeping; fronds scattered, 11–55 dm. long; blades triangular or nearly so, mostly longer than broad, sparingly

hairy on both surfaces, especially on the veins; hairs unicellular; pinnae mostly linear-lanceolate, the basal pair deflexed and advanced; segments oblong, obtuse, entire or slightly crenate, the basal decurrent and adnate to the main rachis; rachis scaly beneath; sori submarginal. *Dryopteris Phegopteris* C. Chr. [BB). *P. Phegopteris* Underw. *T. Phegopteris* Slosson [R]. BEECH FERN. Wooded places: Greenl.—Newf.—Va.—Minn.—Ore.—Alaska; Eu. *Mont.—Subalp.—Arct.—Boreal.*

2. **P. hexagonóptera** (Michx.) Fée. Rootstock creeping, chaffy; frond 4–7 dm. high; blade broadly triangular, slightly pubescent and glandular; uppermost pinnae oblong, obtuse, the middle ones lanceolate, acuminate, the lowest ones broadest at the middle, pinnately divided; sori near the margins. *D. hexagonoptera* C. Chr. [BB]. Dry woods: Que.—Fla.—La.—Minn. *E. Temp.*

3. **P. Dryópteris** (L.) Fée. *Fig. 8.* Rhizome slender, creeping; fronds scattered, the three primary divisions stalked, 1–2-pinnate; segments oblong, obtuse, entire or toothed; rachis not winged; sori submarginal. *T. Dryopteris* Slosson [R]. *D. Dryopteris* Britton [BB]. OAK FERN. Woody places: Greenl.—Newf.—Va.—Iowa—S.D.—Colo.—Ariz.—Ore.—Alaska; Eur. *Arctic—Boreal.—Allegh.*

4. **P. Robertiàna** (Hoffm.) A. Br. Rhizome slender, creeping; fronds scattered, 13–62 cm. long somewhat rigid; blades triangular-ovate, the three primary divisions stalked, the lateral smaller in proportion than in the preceding species; sori submarginal. *T. Robertiana* Slosson. *D. Robertiana* C. Chr. [BB]. Woods: Que.—Ia.—Mo.(?)—Ida.—Alaska.—Eur. *Arct.—Boreal.*

7. DRYÓPTERIS Adans. SHIELD FERN, MALE FERN.

Ferns with stipes not articulated to the rhizomes and mostly with bipinnate to decompound fronds. Midribs and midveins of the pinnae's subdivisions attached at a very acute angle. Sori borne on the veins, round, usually indusiate. Indusia reniform or occasionally resembling those in *Filix*. Fronds furnished with scales, at least on the stipes and rachises, often glandular, without true hairs, consisting of a single cell or row of cells. Scales thin, entire or fimbriate, always consisting of 2 or more rows of cells, which are mostly long and narrow, with flexuose walls. *Aspidium* Sw.

Fronds membranous, not evergreen; veins simple or once forked.
  Lower pinnae gradually decreasing in size, the lowest one very small. — 1. ***D. noveboracensis.***
  Lower pinnae not reduced, almost equaling the middle ones. — 2. ***D. Thelypteris.***
Fronds firmer, mostly evergreen; veins, at least the lower ones, more than once forked.
  Segments of the leaves not spinulose-tipped.
    Fronds small; pinnae 40–60, 4–10 mm. broad; rachis chaffy throughout. — 3. ***D. fragrans.***
    Fronds large, the pinnae fewer; 10–90 mm. broad; naked or with deciduous scales.
      Sori marginal. — 4. ***D. marginalis.***
      Sori not marginal.
        Indusia concave, firm; basal scales lance-linear, caudate-attenuate. — 5. ***D. Filix-mas.***
        Indusia flat, thin; basal scales ovate-oblong to deltoid.
          Blades ovate to oblong, not conspicuously narrowed below; basal scales firm, dark, chestnut-brown. — 6. ***D. Goldiana.***
          Blades linear-oblong or lanceolate, conspicuously narrowed at the base; basal scales thin, dull, light-brown. — 7. ***D. cristata.***

Segments of the leaves spinulose-tipped.
 Indusia glabrous or nearly so.
  Pinnules decurrent on the winged midrib; basal scales light-brown. 8. *D. spinulosa.*
  Larger pinnules not decurrent; basal scales dark brown. 10. *D. dilatata.*
 Indusia distinctly glandular.
  Blades ovate-lanceolate, scarcely narrowed below.
   Scales light brown. 9. *D. intermedia.*
   Scales dark brown. 10. *D. dilatata.*
  Blades elongate-lanceolate, narrowed at the base. 11. *D. Boottii.*

1. **D. noveboracénsis** (L.) A. Gray. Rootstock slender, creeping; fronds 3–6 dm. high; stipe short; blade lanceolate, tapering both ways, once pinnate, long-acuminate; pinnae 5–10 cm. long, lanceolate, long-acuminate, deeply pinnatifid, pilose along the midrib and veins; segments oblong, obtuse; sori near the margins; indusia small, glandular, withering. *A. noveboracense* Sw. [G]. Woods and thickets: Newf.—Ga.—Ark.—(?)Minn. *Canad.—Allegh.* Jl–S.

2. **D. Thelýpteris** (L.) A. Gray. Rootstock slender, creeping; 3–7 dm. high; blades lanceolate, scarsely narrowed at the base, once pinnate, short-acuminate; pinnae 4–7 cm. long, linear-lanceolate, pubescent beneath, deeply pinnatifid; segments oblong, obtuse; sori nearly median, crowded; indusia small, glabrous. *A. Thelypteris* Sw. [G]. Marshes: N.B.—Fla.—Tex.—Neb.—Man. *E. Temp.* Je–Au.

3. **D. fràgrans** (L.) Schott. Fronds borne in a crown, 4.5–40 cm. long, aromatic; stipe and rachis chaffy with bright brown scales; blades lanceolate to narrowly oblanceolate, somewhat narrowed toward the base, nearly or quite bipinnate; pinnae oblong-lanceolate to deltoid-lanceolate; segments oblong, obtuse, adnate-decurrent, subentire to deeply incised; indusia very large, ragged and somewhat glandular at the margin. *A. fragrans* Sw. [G]. On rocks: Alaska—Ellesmereland—Me.—Minn; Eur. *Arct.—Boreal.*

4. **D. marginàlis** (L.) A. Gray. Rhizome with golden brown scales; fronds in a crown, 1.2–10 dm. long; blades evergreen, coriaceous, ovate-oblong or ovate-lanceolate, usually bipinnate or nearly so; lower pinnae unequally deltoid-lanceolate, those above lanceolate to broadly oblong-lanceolate; segments oblong or lanceolate, mostly obtuse or subacute, falcate or subfalcate, subentire to crenately lobed; indusia smooth. *A. marginale* Sw. [G]. Rocky places and old wood in swamps: Que.—N.S.—Ga.—Okla.—Sask. *E. Temp.*

5. **D. Fílix-más** (L.) Sw. *Fig. 9.* Fronds borne in a crown 1.5–10 dm. long, 6–30 cm. broad; blades broadly oblong-lanceolate, somewhat narrowed toward the base, bipinnatifid or bipinnate; pinnules oblong, obtuse, serrate at apex and obscurely so on the sides, the larger incisely lobed; sori nearer the midvein than the margin; indusia glabrous or glandular, often some with one or more glands and others smooth on the same frond. *A. Filix-mas* Sw. [G]. MALE FERN. Damp woods: B.C.—Ore.—Ariz.—N.M.—S.D.—N.S.—Greenl; Eur. *Arct.—Boreal.*

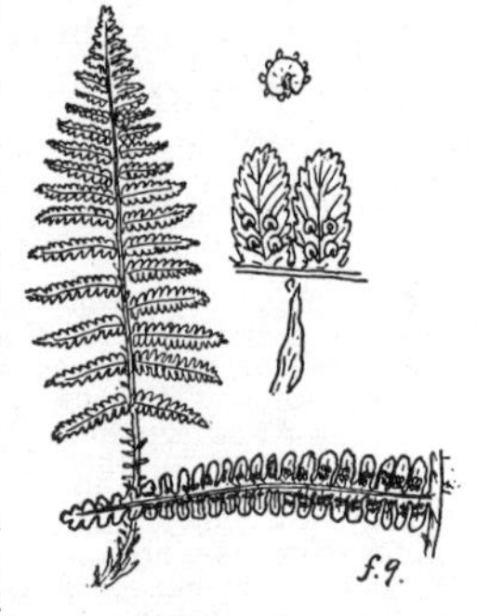

6. **D. Goldiàna** (Hook.) A. Gray. Rootstock stout; fronds many in a crown, 5–10 dm. high; stipe 3–5 dm. long, densely covered below by dark lustrous scales; blade ovate or oblong, short-acuminate, mostly twice pinnate; pinnae 1–2.5 dm. long, broadly lanceolate, pinnatifid almost to the midrib; segments oblong, acute, subfalcate, serrate; sori near the midrib, distinct; indusia glabrous, nearly orbicular. *A. Goldianum* Hook. [G]. Rich woods: N.B.—N.C.—Tenn.—Iowa—Minn. *Arct.—Boreal.* Jl–Au.

7. **D. cristàta** (L.) A. Gray. Fronds crowded at apex of a short stout creeping rhizome, the fertile erect, 2–10 dm. long, 7–15 cm. broad, much overtopping the sterile; scales of stipe and rhizome pale-brown, one-colored; blades linear or lanceolate, bipinnatifid; segments rather broad, oblong or triangular-oblong, obtuse, finely, usually sharply serrate or obscurely or more deeply cut into serrate lobes; sori about medial; indusia smooth. *A. cristatum* Sw. [G]. Swampy places and roadsides: Mack.—Ida.—Neb.—Va.—Newf.; Eur. *Canad.—Allegh.—Submont.—Subalp.*

8. **D. spinulòsa** (O. F. Mueller) Kuntze. Scales of stipe and rhizome light-brown, one-colored; fronds 2–9.6 dm. long; blades ovate-lanceolate to oblong; pinnae oblique to the rachis, elongate-triangular, the lower broadly and unequally triangular; pinnules oblique to the midribs, connected by a very narrow wing, acute, incisely serrate or obliquely pinnatifid; segments incised; teeth mucronate, falcate, appressed; sori submarginal, terminal on the veinlets. *A. spinulosum* Sw. [G]. Damp woods: Newf.—Va.—Minn.—Ida.—B.C.; Eur. *Boreal—Submont.—Mont.*

9. **D. intermèdia** (Muhl.) A. Gray. Fronds in a crown, 2–9.5 dm. long; scales of the stipe light-brown, mostly with darker centers; blades usually dark, often bluish-green, ovate-lanceolate to oblong; pinnae usually nearly or quite at right angles to the rachis, the lower unequally lanceolate to ovate-lanceolate, the lowest inferior segment in the basal pair commonly shorter than the next; upper pinnae lanceolate to oblong; pinnules pinnately divided, the largest not decurrent; segments dentate. *A. spinulosum intermedium* D.C. Eat. [G]. Wooded places: Newf.—N.C.—Tenn.—Wis.—Minn.—Tex. *Canad.—Allegh.*

10. **D. dilatàta** (Hoffm.) Underw. Fronds about 1–11 cm. or more long; blades triangular to ovate or broadly oblong, usually tripinnate; lower pinnae broadly and unequally ovate or triangular; the upper lanceolate to oblong or elliptic-lanceolate; pinnules lance-oblong; sori mostly subterminal on the veinlets; in the typical form scales of the rhizome are mostly heavily dark-striped, and indusia often somewhat glandular. *A. spinulosum dilatatum* Hook. [G]. Greenl.—N.C.—Calif.—Alaska; Eur. *Temp.*

11. **D. Boòttii** (Tuckerm.) Underw. Rootstock stout; fronds 4–8 dm. high; stipe with pale-brown scales; blade lanceolate, bipinnate, acuminate, slightly narrowed below, the sterile ones smaller and less divided; pinnae lanceolate, acuminate; pinnules oblong-ovate, 4–6 cm. long, the lower often pinnatifid, the upper slightly decurrent, serrate; sori medial or nearer the midrib, distinct; indusia reniform. *A. Boottii* Tuckerm. [G]. Low woods: N.S.—Va.—W. Va.—Minn. *Canad.—Allegh.* Jl–S.

8. **CAMPTOSÒRUS** Link. WALKING FERN, WALKING LEAF.

Slender ferns, with simple entire or sinuate blades, tapering at the apex. Sori oblong or linear, irregularly scattered on either side of the reticulate veins, parallel or oblique to the midrib. The outer ones often approximate in pairs. Indusia membranous.

1. **C. rhizophýllus** (L.) Link. *Fig. 10.* Rootstock short, creeping; fronds clustered, 1–4 dm. high; stipe short, light green; blade evergreen, lanceolate, entire, the base cordate and auricled, the apex long-attenuate, rooting at the tip and producing a new plantlet; sori irregular. Shaded calcareous rocks: Que.—Ga.—Ala.—Kans.—Minn. Au–O.

9. ASPLÈNIUM L. SPLEENWORT.

Ferns varying in size, with simple or compound commonly pinnate fronds. Sori oblong or linear, single or rarely a few double, borne on the veins. Indusia superior, attached laterally to the vein, opening toward the midrib or midvein. Veins free or rarely a few uniting. Scales of rhizome with dark-walled cells.

Blade not forking.
  Fronds 0.5–2.5 dm. long; pinnae ovate, elliptic, or oblong, less than 1.5 cm. long.
    Stipe dark only at the base, otherwise green as the rachis. — 1. *A. viride.*
    Stipe and rachis blackish, or dark reddish or purplish brown.
      Pinnae not auricled. — 2. *A. Trichomanes.*
      Pinnae more or less auricled at the base. — 3. *A. resiliens.*
  Fronds 2–12 dm. long; pinnae lanceolate or lance-linear, 2–12 cm. long.
    Pinnae auricled at the base, serrate; stipe and rachis blackish. — 4. *A. platyneuron.*
    Pinnae entire or slightly crenulate, not auricled; stipe and rachis green. — 5. *A. pycnocarpon.*
Blade irregularly forking. — 6. *A. septentrionale.*

1. **A. víride** Huds. Fronds tufted; blades 2.7–20 cm. long, 0.8–1.5 cm. broad, linear-lanceolate; pinnae roundish ovate or rhombic, obtuse, broadly cuneate at base, the lower side obliquely truncate; margins, excepting the basal parts, deeply crenate; sori near the margins; indusia entire or denticulate. On rocks: Newf.—Vt.—Sask.—S.D.—Wyo.—Wash.—Alaska. *Boreal.—Mont.*

2. **A. Trichómanes** L. Fronds tufted; blades 4–18 cm. long, linear; rachis faintly alate, not fibrillose; pinnae mostly oval or oval-oblong, inequilateral; margins, except the basal sides, slightly crenate; veins on both sides of the midveins usually forked; indusia commonly slightly crenulate. On rocks: Man.—Ala.—Ariz.—Alaska; Eur., the Azores, and Canary Is. *Temp.*

3. **A. resíliens** Kunze. Rootstock short, with black scales; fronds 1–3 dm. long; blade linear-oblong, pinnate; pinnae mostly opposite, 4–12 mm. long, oblong, obtuse, crenulate or entire, auricled on the upper or both margins; sori oblong, borne nearer the margin than the midrib. *A. parvulum* Mart. & Gal. [G]; not Hook. Limestone rocks: Va.—Fla.—N.Mex.—Kans.; Mex.; Jamaica. *Austral.—Trop.* Je–O.

4. **A. platyneùron** (L.) Oakes. *Fig. 11.* Fronds tufted, the fertile erect, 20–50 cm. long, 2.5–6.3 cm. broad, the sterile rosulate, much shorter; blades linear-oblanceolate, gradually reduced toward the base; pinnae auriculate-lanceolate, subfalcate, crenate, serrate, or incised, the lower oblong or deltoid; sori oblique, near the mid-veins. Among rocks or stones: s Ont.—Me.—Fla.—N.M.—Colo. *Submont.—Mont.*

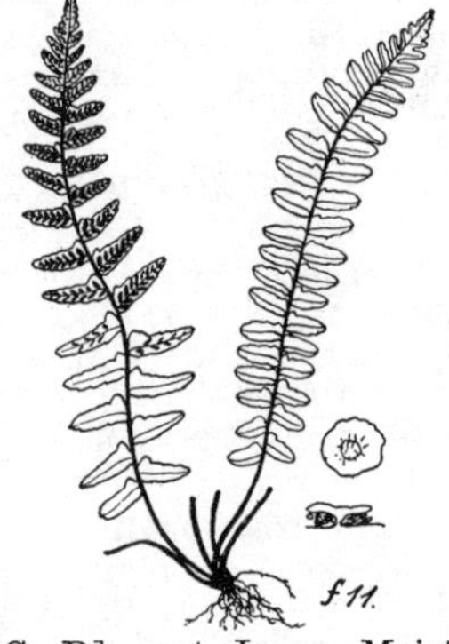

5. **A. pycnocárpon** Spreng. Rootstock stout, creeping; fronds 3–8 dm. high; stipe dark brown at the base only; blades lanceolate, pinnate, glabrous, membranous; pinnae 5–12 cm. long, linear-lanceolate, obtuse or cuneate or truncate at the base, those of the fertile fronds narrower and falcate; sori linear, slightly curved, oblique; indusium firm, concealed at maturity. *A. angustifolium* Michx. [G, B]; not Jacq. Moist woods: Que.—Ga.—Ala.—Kans.—Minn. Au–S.

6. **A. septentrionàle** (L.) Hoffm. Fronds tufted, 3.5–20 cm. long; stipes brown below; segments 2–5, oblique, linear, tapering both ways, entire or with a few oblique long narrow teeth; sori elongate, mostly 2–3 to each segment,

usually opposed in pairs; indusia entire or sparingly short-ciliate. *Belvisia septentrionalis* Mirb. On rocks: S.D.—N.M.—L. Calif.—Wyo.; Eurasia. *Submont.—Mont.*

10. ATHÝRIUM Roth. LADY FERN.

Ferns with more or less compound fronds. Scales of the rhizome with thin-walled cells. Veins free. Sori borne on the veins, indusiate, mostly oblong or linear-oblong and curved at one end over the vein, or bent back upon themselves along the other side of the vein, often horseshoe-shaped, occasionally roundish. Indusia following the shape of the sori, attached along their length at the side next the vein, opening outwardly, rarely vestigial and hidden.

Blades pinnate, the pinnae pinnatifid with crenate-serrate segments. 1. *A. thelypteroides.*
Blades bipinnate, the pinnules incised or deeply serrate. 2. *A. angustum.*

1. **A. thelypteroìdes** (Michx.) Desv. Rootstock slender, creeping; fronds 3–10 dm. high; stipe straw-colored, chaffy below; blades lanceolate, acute or acuminate at the apex, narrowed at the base; pinnae linear-lanceolate, acuminate, pinnatifid; segments oblong-lanceolate, obtuse; sori oblong; indusium light-colored, slightly curved. *Asplenium acrostichoides* Sw. [G]; not *Athyrium acrostichoideum* Bory. Rich woods: N.S.—Ga.—Kans.—Iowa. Au–O.

2. **A. angústum** (Willd.) Presl. Rootstock short and thick, densely covered with masses of old fronds; fronds 4–10 dm. high; stipe straw-colored, chaffy below; chaffs 1 cm. long, 1.5 mm. wide, linear-lanceolate, dark-brown; blade ovate-oblong to lanceolate, twice pinnate, widest near the middle; pinnae lanceolate, the pinnules confluent, doubly serrate, those of the fertile fronds usually narrower than the sterile ones, narrowly lanceolate, acute; indusium toothed, ciliate but not glandular; margin of the indusium toothed or short-ciliate, *Asplenium Filix-foemina* Am. auth. [B, G]. FEMALE FERN. Woods: Newf.—Pa.—Mo.—S.D.

11. ADIÁNTUM (Tourn.) L. MAIDEN-HAIR FERN, VENUS'-HAIR FERN.

Ferns with compound fronds having segments in the form of small leaflets, and slender, usually dark-colored shining stipes. Lobes of the leaflets reflexed, indusiform, bearing sori on the under side at the ends of the veins. Veins free.

Stipes forked into two rachises, which bear on upper side pinnae interspersed with single leaflets. 1. *A. pedatum.*
Stipes not forked at apex; blades alternately bipinnate or tripinnate; leaflets very short-stalked. 2. *A. modestum.*

1. **A. pedàtum** L. *Fig. 12.* Stipes and rachises dark chestnut-brown; fronds 2–5 dm. high; principal leaflets dimidiate-triangular-oblong, lobed on the upper side with unilaterel midveins; branches of midveins several times forked. Damp woods and in shade near water: Newf.—Ga.—Kans.—S.D.; Utah; Calif.—Mont.—Alaska; China, Japan, India. *Boreal.*

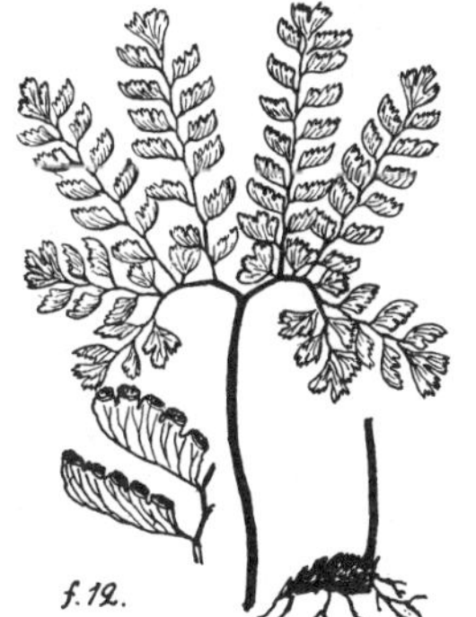
f. 12.

2. **A. modéstum** Underw. Stipes and rachises purplish brown to ebeneous; fronds 8–29 cm. long, 5–15 cm. broad, deltoid-lanceolate to deltoid-ovate, commonly tripinnate; leaflets 6–15 mm. wide or the terminal wider, nearly as long as wide, truncate, roundish or broadly cuneate at base, rounded above, 2–5-lobed, mostly 3-lobed, the incisions shallow, the margin in sterile leaflets, evenly and finely serrate-denticulate, in fertile leaflets similarly toothed between the sori; indusia largely oblong, entire to crenate. Wet places among rocks: Ariz.—s Utah—s Colo.—Tex.; S.D. *Submont.*

12. PTERÍDIUM Scop. BRAKE, BRACKEN.

Large ferns, with coarse compound fronds and creeping rhizomes. Margins of the pinnules reflexed, forming an indusium. Sporangia borne on a continuous vein-like receptacle connecting the ends of the veins. An obscure inner inferior indusium attached to the receptacle. Veins pinnate.

1. **P. aquilìnum** (L.) Kuhn. Stipes up to 9 dm. long; fronds to 12 dm. long, deltoid to ovate-deltoid, subternately decompound, the larger tripinnate below; pinnules entire, lobed, auricled or pinnatifid, the entire ones mostly oblong or linear, commonly spaced, rarely abruptly dilated at base; costae, midribs, and reflexed margins slightly hairy; surfaces otherwise glabrous or with a few scattered hairs; hairs mostly pluricellular. *Pteris aquilina* L. [G]. Woods: Newf.—Colo.—S.D.; Eur. *Submont.—Mont.*

13. CRYPTOGRÁMMA R. Br. ROCK-BRAKE.

Small ferns, with slightly dimorphous compound fronds. Margins of the pinnules reflexed over the sporangia, finally opening out flat. Sporangia without other indusia, on the upper part of the veins and extending down them. Veins free.

Fronds scattered; leaf-texture very delicate; stipes brown or brownish below or throughout. 1. *C. Stelleri.*
Fronds tufted; leaf-texture firm.
  Stipes straw-colored. 2. *C. acrostichoides.*
  Stipes chestnut-colored. 3. *C. densa.*

1. **C. Stélleri** (Gmel.) Prantl. Rhizome slender, creeping; fertile fronds 5.5–22 cm. long, the sterile usually shorter; blades ovate, bi-tripinnatifid; ultimate segments cuneate and decurrent at base, the fertile linear-oblong to lance-linear, the sterile ovate to obovate-flabelliform, crenulate. *Pellaea gracilis* Hook. On shaded, damp, mostly limestone, rock: Que.—Pa.—Minn.—Sask.—Mont.—B.C.; Colo. *Submont.—Mont.*

2. **C. acrostichoìdes** R. Br. *Fig. 13.* Fertile fronds 8–36 cm. long, long-stalked, commonly overtopping the sterile; blades ovate, the sterile tri-quadripinnatifid; sterile ultimate segments ovate-oblong or suboval, obtuse, serrulate, the fertile narrowly elliptical or linear, their reflexed margins scarcely altered. On rocks: Baffin Bay.—Ont.—Neb.—Colo.—Calif.—Alaska. *Submont.—Subalp.*

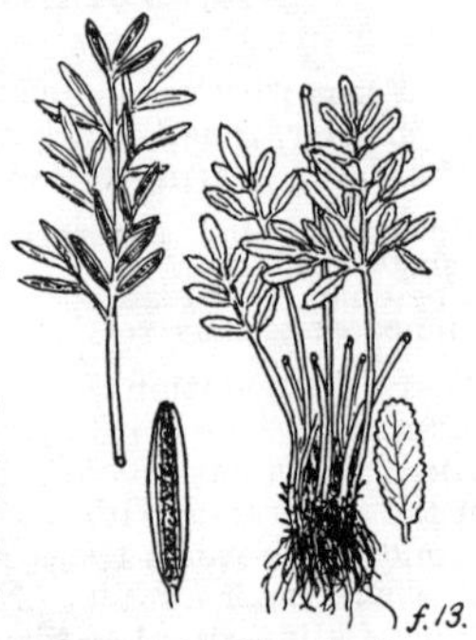
f.13.

3. **C. dénsa** (Brack.) Diels. Primary rachis, except in the upper part of the blade, and scales of rhizome chestnut-colored, like the stipe; fronds 6–29 cm. long; blades ovate or oblong-deltoid, densely tripinnate; sterile ultimate segments lance-linear, incisely serrate, the fertile linear, nearly sessile, their reflexed margins delicately indusiform, erose-toothed. *Pellaea densa* Hook. [B]. On rocks: Que.—N.M.—Calif.—B.C.

14. PELLAÈA Link. CLIFF-BRAKE.

Ferns, with compound fronds and usually dark-colored stipes, the fertile divisions usually narrower than the sterile. Sori borne on the upper part of the free veins, usually confluent in a submarginal line. Margins of the fertile segments reflexed over the sporangia, indusiform.

Segments of the blade obtuse or acute, rarely a few mucronate.
  Stipes and rachises entirely glabrous or with only few occasional long flaccid jointed hairs.
    Stipe bright chestnut-brown; leaflets scarcely bluish green, not punctate beneath. 1. *P. pumila.*
    Stipe dark red-purple, turning black; leaflets distinctly bluish green, punctate beneath. 2. *P. glabella.*

Stipe dark red-purple, turning black; leaflets distinctly bluish green, punctate beneath.
Stipes and rachises purplish black, rather thickly clothed with slender flaccid jointed hairs. 3. *P. atropurpurea.*
Segments sharp-pointed or mucronate; blades once pinnate above, bipinnate below. 4. *P. mucronata.*

1. **P. pùmila** Rydb. *Fig. 14.* Scales of the rhizomes rust-colored, short, dense; fronds tufted, 3–8 cm., seldom 10 cm. long, stipe 1–3. cm. long; blade oblong in outline, mostly simply pinnate with 2–5 pairs of pinnae; both sterile and fertile pinnae oblong, about 1 cm. long, obtuse, entire or the lower with one or two lobes at the base; membranous margin of the indusium about 0.2 mm. wide, crenulate. Dry rocks: S.D.—Alta.—Wyo.—Neb.—Kans. (?) *Submont.—Mont.*

2. **P. glabélla** Mett. Scales of the rhizomes rusty; fronds 12–25 cm. high; blade oblong, usually twice pinnate, the lowest pinnae usually 3–5-foliolate with the terminal pinnule stalked, the basal lobes often acute; membranous margin of the indusium 0.2–0.3 mm. wide, mostly entire. Dry rocks: Vt.—N.Y.—Kans.—Minn. *Canad.—Allegh.*

3. **P. atropurpùrea** Link. Scales of rhizome fulvous rust-colored; fronds tufted, 2.5–41.7 cm. long; stipes wiry; blades lanceolate or ovate-lanceolate, pinnate or usually bipinnate below; sterile segments commonly suboval; fertile segments mostly linear or sublinear, sometimes auricled, obtuse or acute; branches of midveins mostly about twice forked; indusia entire or slightly erose. On dry rocks, particularly limestone: N.H.—Ga.—Me.—Ariz.—S.D.

4. **P. mucronàta** D. C. Eat. Scales of rhizome black-striped at center, with light tawny margins; fronds tufted, 8–60 cm. long; stipes and rachises dark mahogany-colored or purplish black, smooth; segments sessile or subsessile, the sterile roundish oval, sometimes subcordate at base, the fertile linear-oblong, with strongly revolute margins. *P. Wrightiana* Hook. *P. longimucronata* Hook. *P. truncata* Goodding. On rocks: Tex.—Kans.—Calif.—L. Calif. *Son.*

15. **CHEILÁNTHES** Sw. Lip Fern.

Small ferns with compound, usually chaffy or hairy fronds. Indusium formed of the reflexed margins of pinnules, roundish and distinct or continuous. Sori roundish and distinct or more or less confluent, borne on the thickened apices of free veins.

Fronds tomentose; indusia continuous. 1. *C. Féei.*
Fronds hirsute; indusia discontinuous. 2. *C. lanosa.*

1. **C. Féei** Moore. *Fig. 15.* Scales of the rhizome cinnamon-brown, mostly black-striped at center; fronds tufted, 5–16 cm. long; stipes and rachises dark brown or blackish; blades ovate-lanceolate, tripinnate or tripinnatifid, rarely bipinnate; lower pinnae deltoid, the upper ones oblong-ovate; ultimate segments minute, the apical one slightly the largest, their lower surfaces densely tomentose; hairs whitish brown, soft, matted, flattened, jointed; indusium formed of the scarcely altered margin. *C. lanuginosa* Nutt. On or among rocks: B.C.—Minn.—Ill.—Ark.—Ariz.; Mex. *Plain—Submont.*

2. **C. lanòsa** (Michx.) Wats. Rootstock creeping, with rusty scales; fronds tufted, 1–3 dm. high;

stipe chestnut-brown, rusty-hirsute; blades oblong-lanceolate, gradually acute, bipinnate; pinnae ovate, 1–3 cm. long, densely hirsute; pinnules oblong, pinnatifid; indusia herbaceous. Rocks: Conn.—Ga.—Tex.—Kans. *E. Temp.* Jl–S.

16. **NOTHOLAÈNA** R. Br.

Small ferns, with pinnately compound fronds lacking proper indusia and almost always farinose, hairy or chaffy on the under surface. Margins of the blade sometimes inflexed at first over the sporangia. Sori roundish or oblong, marginal, borne near the ends of the free veins.

1. **N. dealbàta** (Pursh) Kunze. Rootstock short, chaffy; fronds 0.5–2 dm. high; stipe wiry, as long as the blade; blade triangular-ovate, 3–4-pinnate, the rachis brown or black; pinnae ovate-deltoid, long-stalked; ultimate segments ovate-oblong or elliptic, entire, white and powdery beneath. Calcareous rocks: Mo.—Tex.—Ariz.—Neb. *Plains—Prairies.* Je–S.

17. **POLYPÒDIUM** [Tourn.] L. Polypody.

Ferns varying in size and habit. Fronds entire to pinnately compound. Sori borne on the backs of the fronds, round to elliptical, non-indusiate, dorsal or terminal on the free or anastomosing veins, the veins free in our species.

Blade glabrous beneath; plant green.
  Rootstock sweet, with a liquorice taste; scales uniformly colored throughout; sori medial. 1. *P. hesperium.*
  Rootstock neither sweet nor with a liquorice taste; scales darkened on the back; sori near the margins. 2. *P. virginianum.*
Blade densely scaly beneath; plant grayish. 3. *P. polypodioides.*

1. **P. hespèrium** Maxon. *Fig. 16.* Rhizomes hard, liquorice-like, chaffy; stipes stramineous; blades 7.6–20 cm. long, 2.5–4.4 cm. broad, linear-oblong, deeply pinnatifid or below pinnate; lowest pinnae usually shorter than the middle ones; segments elliptical or somewhat spatulate, narrowest just above the often dilated base, broadly rounded at apex, obscurely or evidently crenate; under surface slightly glandular; veins 1–3 times forked; sori large, medial, on apices of the veinlets. B.C.—S.D.—N.M.—Ariz.—Wash. *W. Temp.*

2. **P. virginiànum** L. Rootstock soft and spongy, not sweet; fronds 1–2.5 dm. high; blade lanceolate, 3–8 cm. wide; pinnae oblong, the lowest as long as or longer than the median ones, entire or slightly toothed, obtuse; sori smaller, nearly marginal. *P. vulgare* Am. auth. [G, B]; not L. Rocky banks: Newf.—Ga.—Ala.—Man. *Canad.—Allegh.* Jl.

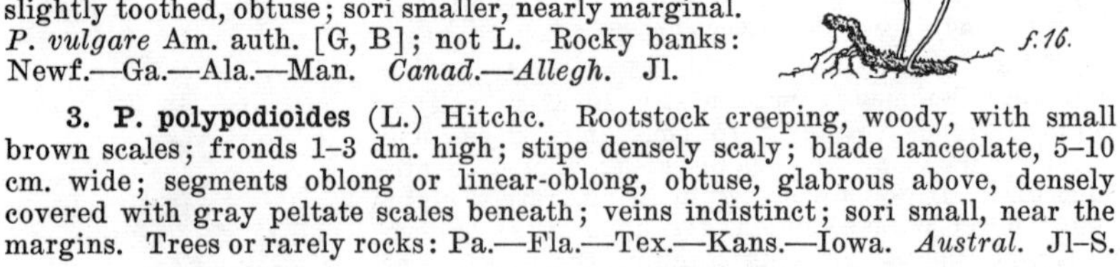
f. 16.

3. **P. polypodioìdes** (L.) Hitchc. Rootstock creeping, woody, with small brown scales; fronds 1–3 dm. high; stipe densely scaly; blade lanceolate, 5–10 cm. wide; segments oblong or linear-oblong, obtuse, glabrous above, densely covered with gray peltate scales beneath; veins indistinct; sori small, near the margins. Trees or rarely rocks: Pa.—Fla.—Tex.—Kans.—Iowa. *Austral.* Jl–S.

Family 4. **MARSILEACEAE.** Marsilea Family.

Perennial herbaceous plants rooting in mud, in ours with 2–4-foliolate stalked leaves borne on a stalked creeping rhizome. Sporocarp containing both macrospores and microspores, borne on peduncles attached to the lower parts of the petioles or to the rhizome close to them.

1. MARSÍLEA L.

Marsh or aquatic plants commonly growing in shallow water, with their leaves floating on its surface. Leaves 4-foliolate. Sporocarps crustaceous, ovoid or bean-shaped, usually with 2 teeth near the base, divided vertically into two cells, which are subdivided into transverse compartments (sori).

1. **M. vestìta** Hook. & Grev. *Fig. 17.* Petioles 1–14 cm. long; leaflets broadly cuneate, usually hairy, entire; peduncles erect or ascending, distinct from the petioles, scarcely as long as the sporocarps; sporocarps solitary, hairy, with 2 acute teeth and about 7–9 sori in each cell. *M. mucronata* A. Br. B.C.—Ia.—Ark.—Calif.

Family 5. SALVINIACEAE. Salvinia Family.

Aquatic, floating plants, bearing minute, apparently 2-ranked leaves on a more or less elongate, sometimes branching, axis. Sporocarps borne 2 or more on a common stalk, soft, one-celled, thin-walled, each containing a central often branched receptacle which bears either macrosporangia containing solitary macrospores or microsporangia containing numerous microspores.

1. AZÓLLA Lam.

Small moss-like plants, with pinnately branched stems bearing rootlets on the under side. Leaves imbricate, 2-lobed. Sporocarps borne in pairs beneath the stem, some small, ovoid or acorn-shaped, containing at base a single macrospore with a few attached bodies of uncertain function above it, the others larger, globose, containing on a basal placenta, numerous stalked microsporangia enclosing masses of microspores, which are usually furnished with anchor-shaped processes.

1. **A. caroliniàna** Willd. Plants much branched, 6–25 mm. broad; lobes of leaves ovate, the lower reddish, the upper green, reddish margined; macrospores minutely granular, accompanied by three corpuscles; processes of the microspores rigid, septate. In still waters: Ont.—Fla.—Calif.—B.C.; trop. Am.

Family 6. EQUISETACEAE. Horsetail Family.

Rush-like, verticillately branched or unbranched plants, with rootstocks and mostly hollow, jointed stems, bearing sheaths at the nodes. Spores in a terminal cone formed of verticils of stalked peltate bracts bearing on the under side 6 or 7 sporanges, which open on the inner side. Prothallium in damp places, green, variously lobed.

1. EQUISÈTUM (Tourn.) L. Horsetail.

Characters of the family.

Stem annual; spike rounded at the top; stomata scattered in the grooves of the stem.

Stems of two kinds, the fertile ones succulent, appearing earlier than the sterile ones.
- Fertile stems simple, soon withering, branchless. — 1. *E. arvense.*
- Fertile stems, when older, producing branches, only the naked apex withering.
  - Branches compound; ridges with 2 rows of hooked spinules. — 2. *E. sylvaticum.*
  - Branches simple; ridges with 3 rows of broad spinules. — 3. *E. pratense.*

Stems of one kind; branches simple or none.
- Teeth of the sheaths persistent; plant usually branched, at least in age.
  - Sheaths 7–8 mm. long; stem stout, sparingly branched.
    - Central cavity about one-sixth the diameter of the stem; sheath loose; teeth grooved, black with white margins. — 4. *E. palustre.*
    - Central cavity one-half to four-fifths the diameter of the stem; teeth dark brown.
      - Sheaths loose; centrum one-half to two-thirds the diameter of the stem; vallecular holes present. — 5. *E. litorale.*
      - Sheaths tight; centrum two-thirds to four-fifths the diameter of the stem; vallecular holes usually absent. — 6. *E. fluviatile.*
  - Sheaths less than 5 mm. long; stem slender, much branched. (Autumnal forms of) — 1. *E. arvense.*
- Teeth of the sheaths deciduous; plant unbranched or nearly so. — 7. *E. kansanum.*

Stem perennial, evergreen (except sometimes in *E. laevigatum*), mostly simple; spike apiculate, with a rigid tip; stomata in regular rows.
- Central cavity present; stem more than 6-angled, except sometimes in *E. variegatum.*
  - Teeth of the sheaths deciduous; stem tall and stout, many-grooved.
    - Stem tubercled, rough.
      - Sheath close, usually with a black or dark ring near the base, then grayish and with a narrow dark border at the base of the teeth.
        - Stem 1–2 m. high, stout; sheaths as broad as long. — 8. *E. robustum.*
        - Stem 0.3–1 m. high, slender; sheath longer than broad. — 9. *E. affine.*
      - Sheaths more ampleate above, green, with or without a narrow border, but no ring below. — 10. *E. intermedium.*
    - Stem not tubercled, smooth; sheaths enlarged upwards, with or without a dark margin, but rarely with a ring below. — 11. *E. laevigatum.*
  - Teeth of the sheaths persistent, white-margined: stem low, slender, 5–10 grooved. — 12. *E. variegatum.*
- Central cavity wanting; stem 6-angled. — 13. *E. scirpoides.*

1. E. arvénse L. *Fig. 18.* Spore-bearing stems 1–2.5 dm. high, 3–5 mm. in diameter, with loose 8–12-toothed sheaths, rarely with a few branches; cones peduncled, 2–4 cm. long, 5–10 mm. thick; sterile stems decumbent to erect, 1–3.5 dm. high, 2–3 mm. thick, 10–14-furrowed, branched; branches 3–4-angled, solid; teeth of the sheaths lanceolate, acuminate. In the fall the sterile stems sometimes bear small cones 4–10 mm. long, mostly with sterile spores (var. *serotinum*). Wet banks: Greenl.—S.C.—Calif.—Alaska; Eurasia. *Alp.—Arct.—Temp.*

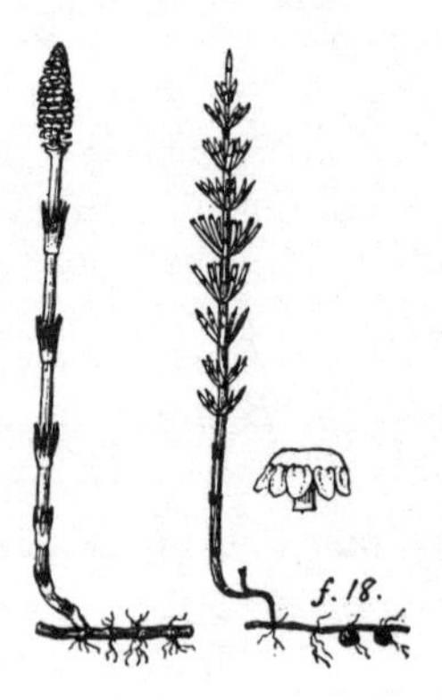

2. E. sylváticum L. Stems 1–4.5 dm. high, 8–14-ridged, 3–4 mm. thick; both the fertile and the sterile ones branched; central cavity constituting half the diameter; sheaths brown, loose, the teeth more or less coherent; primary branches, 4–5-angled, the secondary ones 3-angled; cones peduncled, 1–3 cm. long, 5–8 mm. thick. Wet shady places: Newf.—Va.—Ia.—B.C.—Alaska; Eurasia. *Submont.—Mont.—Boreal.* My–Je.

3. **E. praténse** Ehrh. Stems 2–4 dm. high, 3–5 mm. thick, 8–20-ridged, the fertile ones at first nearly unbranched, some developing short spreading branches, the sterile ones with long branches; central cavity one-third the diameter; sheaths green, loose; teeth lanceolate with dark middle; branches 3-ridged; teeth of the sheaths deltoid; cones peduncled, 1–2 cm. long, about 5 mm. thick. Alluvial soil: N.S.—N.J.—Colo.—Alaska; Eurasia. *Boreal—Submont.—Subalp.* Ap–My.

4. **E. palústre** L. Stems 2–9 dm. high, 3–5 mm. thick, 5–10-angled; branches long, ascending, hollow, 4–7-angled; sheaths widened upwards; teeth lance-subulate; cones 1.5–2.5 cm. long, 5–6 mm. thick. Wet places: Newf.—Conn.—Wyo.—Wash.—Alaska; Eurasia. *Boreal.—Submont.—Mont.* Je–Au.

5. **E. litoràle** Kühlewein. Stems diffuse to erect, 2–9 dm. high, 6–18-grooved; branches 2.5–15 cm. long, 3–5-angled; teeth dark-brown, acute, coherent in groups. Wet sandy shores: N.B.—Pa.—Minn. My–Je.

6. **E. fluviátile** L. Stems 3–15 dm. high, 4–8 mm. thick, 10–30-angled; central cavity four-fifths of the diameter; branches spreading or upcurved, long, 4–6-angled, hollow; sheaths appressed; teeth dark brown, distinct, narrow; cones short-peduncled, 1.5–2.5 cm. long, 5–7 mm. thick. *E. limosum* L. Shallow water: Newf.—N.Y.—Wyo.—Wash.—B.C. *Temp.* Je–Jl.

7. **E. kansànum** J. H. Schaffner. Stems 3–5 dm. high, 15–30-grooved, light green; sheaths long, dilated above, green, with a narrow black band at the top, rarely with a faint one below; cones short-peduncled, ovate or ovate-oblong, 1.5–2.5 cm. long, 7–8 mm. thick. Clay banks: Mo.—Utah—Mont. *Plain—Submont.*

8. **E. robústum** A. Br. Stems 1–2 m. high, 8–12 mm. thick, 16–48-angled, dark green; ridges rounded, rough, with 3 rows of tubercles; sheath tight, about as broad as long; teeth mostly persistent, lance-linear, sharply 3-angled, black, with white margins; spike sessile, 1.5–2.5 cm. long, 6–7 mm. thick, strongly apiculate. *E. hyemale robustum* A. A. Eat. [G]. Wet places: N.Y.—Md.—La.—Tex.—Calif.—B.C. *Temp.—Plain—Submont.* Je–Au.

9. **E. affìne** Engelm. Stems 3–10 dm. high, 4–8 mm. thick, 16–40-angled, dark green; ridges rounded, rough with 2 rows of tubercles; sheaths longer than broad; teeth lance-linear, sharply 3-angled, persistent or coherent by their tips and torn off; cone 1–1.5 cm. long, about 5 mm. thick, sharply apiculate. *E. robustum affine* Engelm. *E. hyemale* Am. auth.; not L. [B]. N.S.—N.Y.—Ariz.—Sask. *Boreal—Plain—Submont.* Je–Au.

10. **E. intermèdium** (A. A. Eaton) Rydb. Stems 3–12 dm. high, 5–8 mm. thick, 20–30-angled; ridges rough, with 2 rows of tubercles; sheath longer than broad, green, except a narrow black and white border, sometimes with a dark band beneath; teeth thin, brown, hyaline-margined, deciduous or persistent; cones sessile, about 1 cm. long and 5 mm. thick, apiculate. *E. laevigatum* Milde, not A. Br. Moist sandy soil: Mich.—Mo.—Calif.—Wash. *Prairie—Submont.*

11. **E. laevigàtum** A. Br. Stems 1–10 dm. high, 4–6 mm. thick, simple or with erect branches, 20–30-grooved; sheaths widened upwards, green, with a black limb; teeth mostly deciduous, with black triangular bases; cones sessile or nearly so, 1–1.5 cm. long, about 5 mm. thick, rather bluntly apiculate. Alluvial soil: Ohio—Tex.—Calif.—B.C. *Prairie—Mont.* Je–Au.

12. **E. variegàtum** Schleich. Stems 1.5–3 dm. high 2–4 mm. thick, 5–10-grooved, tufted; central cavity one-third the diameter; sheaths loose, green

below, with a dark ring above; teeth black, with broad white border, persistent, with a filiform deciduous tip; cones short-peduncled or sessile, 8–10 mm. long, about 3 mm. thick. Wet grounds: Lab.—N.Y.—Colo.—Alaska. *Boreal—Mont.*

13. **E. scirpoìdes** Michx. Stems tufted, filiform, less than 1.5 dm. high, 1–2 mm. thick, flexuose-curving; sheaths short, with 3 subulate, persistent teeth; cones 3–5 mm. long, 2 mm. thick. Damp places: Lab.—Pa.—Mont.—Wash.—B.C.; Eurasia. *Boreal—Mont.*

Family 7. **ISOËTACEAE.** QUILLWORT FAMILY.

Small water or bog plants, with a corm-like short stem and numerous crowded subulate or nearly filiform leaves. Spores of two kinds, smaller microspores (male) and larger macrospores (female), in axillary sporangia covered by the enlarged bases of the leaves, the macrospores spherical, with an equatorial ring-like ridge and three other ridges meeting at the apex, the microspores obliquely oblong, triangular in cross-section.

1. **ISÒËTES** L. QUILLWORT.

Characters of the family.

Submerged species; leaves without peripheral bast-bundles.
  Stomata absent; macrospores marked with confluent crests, more or less honeycombed. — 1. *I. macrospora.*
  Stomata present; macrospores spinulose or tubercled. — 2. *I. Braunii.*
Amphibious or terrestrial species; leaves with peripheral bast-bundles and stomata; macrospores more or less tuberculate.
  Paludose; leaves 12–45 cm. long, 1.5–3.5 mm. thick; macrospores 0.25–0.4 mm. — 3. *I. melanopoda.*
  Terrestrial; leaves 7–20 cm. long, 0.5 mm. thick; macrospores 0.4–0.6 mm. — 4. *I. Butleri.*

1. **I. macrόspora** Durieu. Leaves erect, 2 mm. thick, round, dark green, 5–15 cm. long; velum covering one-third of the unspotted sporange; macrospore 0.6–0.8 mm.; upper faces with parallel ridges, the lower ones honeycombed; microspores smooth. *I. lacustris* Am. auth. [B]; not L. Water: Lab.—N.J.—Ont.—Minn. *Huds.—Canad.*

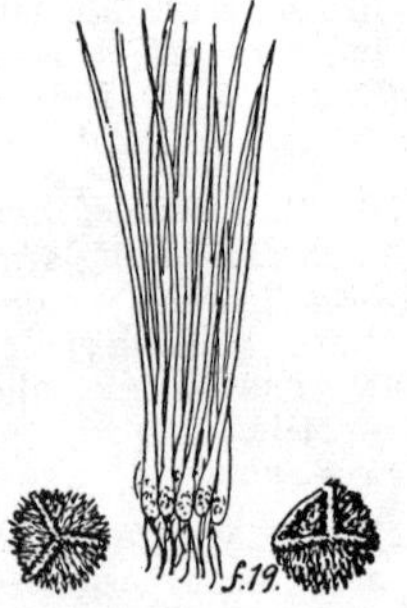

2. **I. Braùnii** Durieu. *Fig. 19.* Leaves erect or spreading, green or reddish-green, tapering, 7–20 cm. long; velum covering one-half to three-fourths of the spotted sporange; macrospores 0.4–0.5 mm., with broad, retuse spinules; microspores smooth. *I. echinospora Braunii* Engelm. [G, B]. In water: Greenl.—N.J.—Calif.—B.C. *Subarct.—Boreal.*

3. **I. melanόpoda** J. Gay. Polygamous; leaves chestnut or black at the base; velum covering less than one-fifth of the sporange, the latter spotted; macrospores 0.25–0.4 mm. broad, smooth or with low confluent tubercles; microspores spinulose. Shallow ponds and pools: Ill.—Okla.—Iowa. *Prairie.*

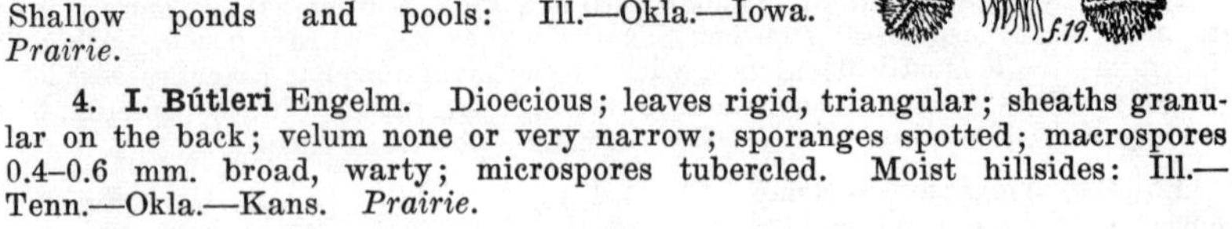

4. **I. Bútleri** Engelm. Dioecious; leaves rigid, triangular; sheaths granular on the back; velum none or very narrow; sporanges spotted; macrospores 0.4–0.6 mm. broad, warty; microspores tubercled. Moist hillsides: Ill.—Tenn.—Okla.—Kans. *Prairie.*

Family 8. **LYCOPODIACEAE.** Club-moss Family.

Low, erect or trailing plants, somewhat resembling mosses, with leafy, usually branched, often elongated stems. Leaves small, lanceolate or subulate or sometimes oblong or roundish, some bearing on their axils or on their upper surfaces solitary 1–3-celled sporanges. Spores minute, of one kind.

1. **LYCOPÒDIUM** L. Club-moss, Ground Pine.

Perennial plants, with evergreen, usually stiff, commonly imbricated, one-nerved, 4–16-ranked leaves. Main stems usually creeping, with aerial branches. Sporangia flattened, coriaceous, usually reniform, 1-celled, opening by a transverse slit around the margin, borne in the axils of ordinary or modified and bract-like leaves. Spores sulphur-colored.

- Sporanges borne in the axils of ordinary leaves, which are not arranged in distinct spikes.
  - Leaves hollow at base, all or mostly all ascending, of nearly one length. — 1. *L. Selago.*
  - Leaves flattened, not hollow at base, spreading or deflexed, longer and shorter in alternating zones.
    - Leaves broadest above the middle, erose-denticulate towards the apex. — 2. *L. lucidulum.*
    - Leaves lance-linear, broadest at the base, entire. — 3. *L. porophilum.*
- Sporanges borne in the axils of leaves arranged in spikes on the aerial branches.
  - Spikes sessile, or with stalks not more than 1 cm. long.
    - Aerial branches all simple. — 4. *L. inundatum.*
    - Aerial branches, at least in part, branched.
      - Leaves of the ultimate divisions of the aerial branches in 4 rows. — 5. *L. alpinum.*
      - Leaves of the ultimate divisions of the aerial branches in more than 4 rows.
        - Aerial branches tree-like. — 6. *L. obscurum.*
        - Aerial branches not tree-like, mostly loosely forking or forming compact tufts.
          - Leaves of the ultimate divisions of the aerial branches in 5 rows. — 7. *L. sitchense.*
          - Leaves of the ultimate divisions of the aerial branches in 8 rows. — 8. *L. annotinum.*
  - Spikes borne on bracteate peduncles more than 2 cm. long.
    - Leaves of the divisions of the aerial branches in many rows. — 9. *L. clavatum.*
    - Leaves of the divisions of the aerial branches in 4 rows.
      - Ultimate branches conspicuously flattened; leaves on the under row minute, deltoid, cuspidate. — 10. *L. complanatum.*
      - Ultimate branches narrow and slightly flattened; leaves of the under row scarcely reduced, subulate. — 11. *L. tristachyum.*

1. **L. Selàgo** L. Main stems very short, dichotomously branching into upcurved or erect forking branches forming a tuft 2.5–24 cm. high; leaves crowded, appressed or ascending or rarely a few reflexed, triangular to linear-acuminate or aciculate, acute, entire; sporophylls shorter than the leaves, triangular; plant often bearing gemmae. Rocks: Greenl.—Newf.—Me.—N.Y.—N.C.—Mont.—Wash.—Va.—Alaska.; Mex.; Eur. *Subarct.—Temp.*

2. **L. lucídulum** Michx. Stem curved or decumbent at the base, 1–3 dm. high, 1–3 times dichotomous; leaves dark green, widely spreading or deflexed, oblanceolate, alternately longer and shorter, the shorter ones more often bearing sporanges in the axils; proliferous gemmae often present, caducous. Damp woods: Newf.—S.C.—Tenn.—Iowa—Minn. Au–O. *Huds.—Canad.*

3. **L. poróphilum** Lloyd & Underw. Main stems short, upcurved, dichotomously branching into upcurved or erect forking branches forming a tuft 3–35 cm. high; leaves crowded, spreading or reflexed, linear or nearly so, entire or

rarely very minutely denticulate, those in the zones of the longer slightly broadened above the middle and similarly contracted toward the base, those in the zones of the shorter broadest at the base, but very gradually tapering; sporophylls entire or minutely denticulate above the middle, acuminate; plant often bearing gemmae. On rocks, especially sandstone: Newf.—Que.—Pa.—Ind.—Minn.—Alta.—B.C.—Wash. *Boreal.*

4. **L. inundàtum** L. Main stems very slender, simple or slightly forked, creeping above ground, oftenest arched, leafy; aerial branches 1 or 2 to each division of the main stem, erect, 1–8 cm. high, the sterile portion commonly shorter or not longer than the spikes; leaves of the main stem linear-lanceolate, entire or rarely a few slightly toothed, upcurved; leaves of the aerial branches more slender, spreading; entire; spikes solitary; sporophylls spreading, linear-deltoid, entire or toothed just above the base. Sandy bogs: Ont.—Newf.—N.J.—Pa.—Ida.—Wash.—Alaska; Old World. *Boreal.*

5. **L. alpìnum** L. Main stem creeping at or near the surface of the ground; aerial stems ascending, 2.5–11 cm. high, repeatedly branched; fertile branches the longer, terete, with subulate leaves; sterile branches dorsiventral with 4-ranked leaves, those of the upper row narrowly ovate, acute, those of the lateral rows thick, asymmetrical, falcate, acute, with decurved tips, those of the under row trowel-shaped; spikes sessile; sporophylls ovate, acute, erose. Woods: Alaska—B.C.; Que.—Lab.—Greenl.; Old World. *Subarctic.*

6. **L. obscùrum** L. Main stems creeping, buried in the ground; aerial branches 9–42 cm. long, the leaves 8-ranked on the lower division, 6-ranked on the terminal, linear-lanceolate, spreading, upcurved, twisted, acute, or mucronate; spikes sessile; sporophylls broadly ovate, papery and erose-margined, acuminate, with a subulate apex. Woods: Lab.—Newf.—N.C.—Tenn.—Minn.—Mont.—B.C.—Alaska; Asia. *Boreal.*

7. **L. sitchénse** Rupr. Main stems creeping on or near the surface of the ground; aerial branches dichotomously branching, forming tufts, 2.5–25 cm. high, their branches terete, the fertile the longer; leaves of the branchlets appressed or spreading and upcurved, linear, thick, entire, acute; spikes sessile or short-stalked, the peduncles with minute bracts; sporophylls broadly ovate, long-acuminate, subulate, erose. Cold woods: Lab.—Newf.—N.Y.—w Ont.; Alaska—B.C.—Ida.—Wash. *Subarctic.*

8. **L. annótinum** L. Main stems creeping above ground, leafy; aerial stems 2–39 cm. high, simple or 1–3 times forked; leaves spreading or reflexed, upcurved at apex, lanceolate or linear-lanceolate, serrulate, tipped with a rigid point; spikes 1–several; sporophylls broadly ovate, serrulate, erose. Wooded places: Greenl.—W.Va.—Minn.—Alta.—Colo.—Wash.—Alaska. *Arct.—Boreal.*

9. **L. clavàtum** L. *Fig. 20.* Main stems creeping above ground, leafy; aerial branches ascending, 4–33 cm. high, simply or pinnately branched; leaves many-ranked, linear, mostly bristle-tipped, entire or denticulate, on the main stems denticulate; peduncles simple or forked at apex, their bracts whorled or scattered, mostly bristle-tipped; sporophylls deltoid-ovate, acuminate or bristle-tipped, with membranous erose margins. Woods: Lab.—N.C.—Minn.—Sask.—Ida.—Ore.—Alaska; trop. Am. and Old World. *Subarct.—Boreal.*

10. **L. complanàtum** L. Main stems creeping on or slightly below the surface of the ground; aerial branches yellowish green, 4–47 cm. high, irregularly forked, their branches few, glaucous,

flattened, mostly making annual growths at their tips, with minute, decurrent leaves, the lateral leaves broad, little or not spreading at the tip, the upper narrow, incurved, the lower diminished, deltoid-cuspidate; peduncles mostly 1–2 times forked at summit; sporophylls broadly ovate, acuminate. Wooded places: Lab.—Newf.—N.Y.—Minn.—Ida.—Wash.—B.C.; Alaska. Au.

11. **L. tristàchyum** Pursh. Horizontal stems extensively creeping, 3–12 dm. long, underground, sparingly branched; branches erect or ascending, repeatedly forked, flattened; leaves 4-ranked, imbricate, appressed, decurrent, nearly all alike, those of the lateral rows thicker; peduncles 8–12 cm. long, usually twice dichotomous; spikes cylindric, 18–20 mm. long. Open woods: Me.—Ga.—Minn. *Canad.* Au.

Family 9. SELAGINELLACEAE. Little Club-moss Family.

Leafy terrestrial plants, with branching stems and 4–6-ranked leaves. Sporanges solitary in the axils of the leafy bracts, some containing macrospores, other microspores, the macrospores with a hemispheric base and a triangular-pyramidal apex.

1. SELAGINÉLLA Beauv. Little Club-moss, Selaginella.

Characters of the family.

Stem rooting only at the base; bracts of the spike broad, thin, spreading; macrospores spinulose. — 1. *S. selaginoides.*
Stem rooting mostly its whole length; bracts of the spike narrower and thicker, appressed and closely imbricate; macrospores with a network of thickened ridges.
  Plant closely branched, more or less glaucous; leaves linear-oblong; bracts ovate-triangular. — 2. *S. densa.*
  Plant more openly branched; light-green; leaves linear-lanceolate; bracts deltoid. — 3. *S. rupestris.*

1. **S. selaginoìdes** (L.) Link. Stems prostrate, creeping, slender; leaves lanceolate, acute, spreading, sparsely spinulose-ciliate; spikes thicker, ascending, 3–7 cm. high; bracts lax, ascending, lanceolate or ovate-lanceolate, strongly ciliate. Mountains: Greenl.—N.H.—Colo.—B.C. *Boreal—Mont.*

2. **S. dénsa** Rydb. *Fig. 21.* Stems densely cespitose, 4–10 cm. long, creeping, densely branched; leaves densely imbricate, pale, glaucous when young, dark-cinerescent in age, linear-oblong, 2–3.5 mm. long; apical bristle 0.8–1.5 mm. long; marginal cilia 2–12, nearly 0.1 mm. long; spike 1–3 cm. long; bracts ovate-triangular, 3 mm. long or less; apical bristle nearly 1 mm. long; cilia 10–20 on each side, less than 0.1 mm. long. *S. Engelmannii, S. Bourgeauii,* and *S. Haydeni* Hieron. Hills and mountains: Man.—N.M.—Utah—Wash.—B.C. *Plain.*

f.21.

3. **S. rupéstris** (L.) Spring. Stem cespitose, branched, decumbent or creeping, 3–15 cm. long, 2–6 cm. high; leaves imbricate, bright-green, linear-lanceolate, acute, 2–3 mm. long; apical bristles straight, about 0.6 mm. long; marginal cilia on each side; spike 1–2.5 cm. long; bracts deltoid, about 1.5 mm.; apical bristle about 0.5 mm. long. Dry rocks: Me.—Ga.—Iowa—Minn.; Eurasia. *Allegh.* Au–O.

SPERMATOPHYTA

SEED-BEARING OR FLOWERING PLANTS.

Plants with flowers containing stamens, or pistils, or both, and producing seeds containing an embryonic plant. Alternation of generations not apparent, the gametophyte being exceedingly reduced. The pollen-grains (*microspores*) contained in the anther-sacs of the stamens, after liberation and reaching the stigma of the pistil or the naked ovules, germinate and produce a tube, by means of which the egg-cell (*macrospore*) of the ovule (*macrosporange*) is fertilized.

PAGE

Ovules and seeds borne on the surface of an open bract or scale; stigmas wanting. Class I. GYMNOSPERMAE. 35
Ovules and seeds in a closed cavity (ovary), formed by one or more united modified leaves, with one or more stigmas at the end. Class II. ANGIOSPERMAE. 41
Cotyledons usually single; earlier leaves alternate; leaves mostly parallel-veined; stem endogenous. Subclass 1. MONOCOTYLEDONES. 41
Cotyledons mostly 2; earlier leaves opposite; leaves mostly netted-veined; stem exogenous. Subclass 2. DICOTYLEDONES. 246

KEY TO THE FAMILIES.

Class 1. **GYMNOSPERMAE.** NAKED-SEEDED PLANTS.

Pistillate flowers in aments; fruit a cone, either dry or berry-like.
Carpellary scales subtended by bracts, never peltate; ovules inverted; buds scaly; wings of the seeds formed from a portion of the carpellary scales. 1. PINACEAE. 35
Carpellary scales not subtended by bracts, mostly peltate or fleshy; ovules erect; buds naked; wings of the seeds, if present, a part of the seed-coat. 2. JUNIPERACEAE. 39
Pistillate flowers single or in pairs, without carpellary scales; fruit drupaceous or baccate. 3. TAXACEAE. 40

Class 2. **ANGIOSPERMAE.** SEED-VESSELED PLANTS.

Subclass 1. **MONOCOTYLEDONES.**

Small lens-shaped, ellipsoid, or flask-shaped floating aquatics without leaves. 15. LEMNACEAE. 196
Plants with true stems and leaves, the latter however, sometimes scale-like.
Perianth rudimentary or degenerate, its members often bristles or mere scales, not corolla-like, or wanting.
Flowers not in the axils of dry chaffy bracts (glumes).
Perianth of bristles or chaffy scales.
Flowers in elongate terminal spikes; fruit hidden among bristles. 4. TYPHACEAE. 41
Flowers in globose lateral spikes; fruit not hidden among bristles. 5. SPARGANIACEAE. 42
Perianth fleshy or herbaceous, or wanting.
Flowers in dense spikes subtended by an enlarged bract (spathe); fruit baccate; endosperm present; emersed water or bog plants. 14. ARACEAE. 194
Flowers, if spicate, not subtended by a spathe; fruit drupaceous; endosperm wanting; submerged water plants.
Gynoecium of distinct carpels; stigma disk-like or cup-like. 6. ZANNICHELLIACEAE. 44
Gynoecium of united carpels; stigmas 2–4, slender. 7. NAJADACEAE. 50
Flowers in the axils of dry or chaffy, usually imbricate bracts (glumes).
Leaves 2-ranked, their sheaths with their margins not united; stem mostly hollow; fruit a grain. 12. POACEAE. 58

Leaves 3-ranked, their sheaths with united margins; stems solid; fruit an achene. 13. CYPERACEAE. 140

Perianth of 2 distinct series, the inner series usually corolloid.

Gynoecium of distinct carpels.

Petals similar to the sepals; anthers long and narrow; carpels coherent. 8. SCHEUCHZERIACEAE. 51

Petals different from the sepals, in ours white; anthers short and thick; carpels not coherent. 9. ALISMACEAE. 52

Gynoecium of united carpels.

Stamens numerous; water plants with broad netted-veined floating leaf-blades. 55. NYMPHAEACEAE. 328

Stamens 3–6.

Ovary and fruit superior.

Stamens dissimilar, or only 3 with fertile anthers; endosperm mealy.

Calyx and corolla of free, very different members; stamens free. 18. COMMELINACEAE. 198

Calyx and corolla of quite similar members and partly united; stamens partly adnate to the perianth. 19. PONTEDERIACEAE. 201

Stamens alike and fertile.

Styles present, distinct or united; stigmas terminal.

Styles distinct; capsule septicidal. 20. MELANTHIACEAE. 202

Styles united, often very short or obsolete during anthesis.

Petals and sepals very unlike.

Flowers in imbricate heads; stamens 3; sepals 3, unequal, or only 2. 16. XYRIDACEAE. 197

Flowers not in heads; stamens 6; sepals 3, all alike.

Flowers spicate; corolla 2-lipped. 19. PONTEDERIACEAE. 201

Flowers cymose; corolla regular. 26. CALOCHORTACEAE. 223

Petals and sepals nearly alike; capsule loculicidal.

Sepals and petals chaffy.

Flowers in terminal, involucrate heads, monoecious or dioecious. 17. ERIOCAULACEAE. 198

Flowers not in involucrate heads, hermaphrodite. 21. JUNCACEAE. 205

Sepals and petals not chaffy.

Herbs with bulbs, corms or rootstocks.

Plants with bulbs, or corms, or short erect rootstocks.

Flowers in umbels, at first included in and later subtended by a scarious involucre. 22. ALLIACEAE. 212

Flowers solitary or racemose, or in *Leucocrinum* by shortening of the stem the inflorescence umbel-like, without involucre. 23. LILIACEAE. 215

Plants with elongate horizontal rootstocks. 24. CONVALLARIACEAE. 218

Shrubby plants with woody caudices, or trees. 25. DRACAENACEAE. 223

Styles wanting.

Flowers perfect; plants not climbing.

Leaves and bracts alternate; plants with bulbs; fruit a capsule. 26. CALOCHORTACEAE. 223

Leaves or leaf-like bracts whorled; plants with rootstocks; fruit a berry. 27. TRILLIACEAE. 224

Flowers dioecious; plants climbing or trailing. 28. SMILACACEAE. 226

Ovary and fruit wholly or partly inferior.

Flowers regular; androecium not reduced; stamens 3 or more.

Aquatic plants, dioecious or polygamous.

Ovary 1-celled, in ours with 3 parietal placentae; styles 3. 10. ELODEACEAE. 56

Ovary 6–9-celled; stigmas 6–9. 11. HYDROCHARITACEAE. 57

Land-plants.

Stamens 6; leaves not equitant.

Flowers perfect; upright herbs with parallel veins. 29. AMARYLLIDACEAE. 228

Flowers dioecious; vines with netted-veined leaves. 30. DIOSCOREACEAE. 229

Stamens 3; leaves equitant. 31. IRIDACEAE. 229

Flowers irregular, perfect; stamens 1 or 2. 32. ORCHIDACEAE. 232

Subclass 2. **DICOTYLEDONES.**

A. Corolla wanting.
I. Calyx wanting, at least in the staminate flowers.
Herbs.
Carpels distinct, except at the base; flowers spicate, perfect.
33. SAURURACEAE. 246
Carpels united; flowers axillary, or if spicate, monoecious or dioecious.
Land plants; styles distinct, cleft or foliaceous; ovary 3-celled.
95. EUPHORBIACEAE. 511
Aquatic plants; styles simple.
Leaves opposite, simple; styles 4, united in pairs; ovary 4-celled.
96. CALLITRICHACEAE. 522
Leaves distichous, filiform, or dichotomously branched; styles and cells of the ovary 2. 70. PODOSTEMONACEAE. 387
Trees or shrubs.
Fruit many-seeded; each seed with a tuft of hair.
34. SALICACEAE. 246
Fruit 1-seeded; seeds without tufts of hairs.
Fruit a nut or achene.
Fertile flowers solitary under each bract; nut naked.
35. MYRICACEAE. 256
Fertile flowers 2 under each bract; nut seated in an involucre.
38. CORYLACEAE. 262
Fruit a drupe or a samara. 133. OLEACEAE. 628
II. Calyx present, at least in the staminate or in the perfect flowers.
1. Flowers, at least the staminate, in aments or ament-like spikes.
Plant parasitic; fruit berry-like. 159. LORANTHACEAE. 752
Plant not parasitic; fruit a nut or an achene.
Staminate and pistillate flowers both in aments; fruit not with a bur or cup. 37. BETULACEAE. 258
Staminate flowers in aments; pistillate ones often solitary.
Leaves pinnate; ovules erect and orthotropous; cotyledons lobed and wrinkled. 36. JUGLANDACEAE. 256
Leaves simple; ovules pendulous and anatropous; cotyledons smooth, entire.
Pistillate flowers distinct at maturity; fruit a nut.
39. FAGACEAE. 263
Pistillate flowers ultimately united into an aggregate fruit, the individual fruits drupe-like. 41. MORACEAE. 269
2. Flowers, at least the staminate, not in aments.
a. Ovary superior.
Gynoecium of 1, or several and distinct carpels; stigma and style of each solitary.
Carpel solitary.
Ovary neither enclosed nor seated in a hypanthium or a calyx-tube.
Trees; anthers opening by 2 or 4 valves; sepals 4 or 6 in two series. 62. LAURACEAE. 351
Herbs; anthers opening by a slit.
Flowers not solitary in the axils of the leaves; land plants.
Plants with scarious stipules; flowers cymose.
51. CORRIGIOLACEAE. 314
Plants not with scarious stipules; flowers clustered.
43. URTICACEAE. 271
Flowers solitary in the axils of the leaves; aquatic plants.
54. CERATOPHYLLACEAE. 328
Ovary enclosed or seated in a hypanthium or a calyx-tube.
Hypanthium becoming fleshy in fruit, enclosing the tailless achenes; calyx 4-merous; stamens 4 or 8.
Leaves green, not scurfy; seeds pendulous.
117. THYMELAEACEAE. 562
Leaves scurfy; seeds erect. 118. ELAEAGNACEAE. 562
Hypanthium not becoming fleshy; achenes tailed; calyx 5-merous; leaves not scurfy.
(*Cercocarpus* in) 79. ROSACEAE. 403
Carpels several.
Stamens inserted below the ovary.
(Genera in) 58. RANUNCULACEAE. 331
Stamens inserted on the edge of a cup-shaped hypanthium.
(Genera in) 79. ROSACEAE. 403
Gynoecium of 2 or several united carpels; stigmas or styles 2 or several.
Ovary, by abortion, 1-celled and 1-ovuled.
Leaves with sheathing stipules. 44. POLYGONACEAE. 274

Leaves estipulate, or stipules, if present, not sheathing.
Trees or shrubs; ovary not seated in a hypanthium.
40. ULMACEAE. 267
Herbs or vines.
Stipules herbaceous; inflorescence spicate or racemose; leaf-blades palmately veined. 42. CANNABINACEAE. 271
Stipules scarious, or hyaline, or none; inflorescence cymose; leaf-blades pinnately veined.
Fruit a utricle, or achene, circumscissile or bursting irregularly or a berry.
Stipules wanting; stamens not inserted on the margin of the hypanthium.
Bracts not scarious. 45. CHENOPODIACEAE. 292
Bracts scarious. 46. AMARANTHACEAE. 304
Stipules present, scarious, or, if wanting, the stamens on the margin of a distinct hypanthium.
51. CORRIGIOLACEAE. 314
Fruit a capsule, dehiscent by apical or longitudinal valves.
52. ALSINACEAE. 316

Ovary several-celled, or with several placentae, several-ovuled.
Stamens hypogynous, inserted under the gynoecium in the perfect flowers, not on a disk in the staminate flowers.
Flowers perfect.
Stamens not tetradynamous.
Stamens 2; inflorescence spicate.
(*Besseya* in) 147. SCROPHULARIACEAE. 707
Stamens 3–10.
Fruit baccate; inflorescence racemose.
48. PHYTOLACCACEAE. 310
Fruit not baccate; inflorescence cymose or axillary.
Ovary 1-celled.
Placentae parietal; pistils 2-carpellary.
74. SAXIFRAGACEAE. 390
Placentae central or basal; pistils of 3–5 carpels.
Calyx not petaloid, of 4 or 5 distinct sepals.
52. ALSINACEAE. 316
Calyx more or less petaloid, campanulate, merely 5-lobed. (*Glaux* in) 130. PRIMULACEAE. 620
Ovary 3–5-celled.
Fruit of several carpels, laterally affixed to the receptacle, each carpel circumscissile; leaves alternate. 72. PENTHORACEAE. 389
Fruit a 3–5-seeded capsule, loculicidal or circumscissile; leaves in ours verticillate.
49. TETRAGONIACEAE. 311
Stamens tetradynamous.
(Apetalous species in) 66. BRASSICACEAE. 356
Flowers monoecious or dioecious. 95. EUPHORBIACEAE. 511
Stamens perigynous or epigynous, inserted on the margin of a hypanthium or a disk.
Fruit a samara. 103. ACERACEAE. 529
Fruit drupe-like or berry-like.
(Apetalous species in) 105. RHAMNACEAE. 532

b. Ovary inferior.
Flowers not in involucrate heads.
Fruit a berry, or a drupe, or nut-like.
Leaves reduced to minute appressed scales; plants parasitic on conifers. 159. LORANTHACEAE. 752
Leaves not scale-like and appressed.
Shrubs, with scurfy and usually silvery leaves.
118. ELAEAGNACEAE. 562
Herbs, not with scurfy leaves.
Stamens as many as the perianth-members and alternate with them, or fewer. 49. TETRAGONIACEAE. 311
Stamens as many as the perianth-members and opposite them, or twice as many.
Water plants, with whorled leaves.
122. HALORAGIDACEAE. 582
Land plants, parasitic or saprophytic, with alternate leaves.
158. SANTALACEAE. 751
Fruit a capsule.
Sepals as many as the ovary-cavities, or one half as many.
Hypanthium merely enclosing the ovary
119. LYTHRACEAE. 562
Hypanthium adnate to the ovary. 121. ONAGRACEAE. 566
Sepals not of the same number as the ovary-cavities.
Styles 2; leaves alternate. 74. SAXIFRAGACEAE. 390

Styles 6; leaves 2, basal. 160. ARISTOLOCHIACEAE. 753

Flowers, at least the staminate, in involucrate heads.

Calyx corolla-like. 47. NYCTAGINACEAE. 308

Calyx not corolla-like. 166. AMBROSIACEAE. 762

B. Corolla present.

I. Petals distinct, at least at the base.

1. Carpels solitary, or several and distinct, or united only at the base.

Stamens at the base of the receptacle, *i.e.*, hypogynous.

Trees or shrubs with monoecious flowers; flowers in dense heads. 77. PLATANACEAE. 402

Herbs, or if woody plants, the flowers not monoecious and in heads.

Plants with relatively firm stems and leaves, not succulent.

Stamens numerous; anther-sacs opening by slits.

Water plants with peltate floating leaf-blades and with or without dissected submerged leaves.

Carpels immersed in a fleshy top-shaped receptacle; sepals, petals, and stamens many. 56. NELUMBONACEAE. 330

Carpels not immersed; sepals and petals 3–5 each.

Sepals and petals 5; stamens many, spirally arranged. 58. RANUNCULACEAE. 331

Sepals and petals 3, or rarely 4; stamens 3–18, in definite series. 57. CABOMBACEAE. 330

Land plants.

Plants with perfect flowers, or rarely monoecious or dioecious, then not vines with simple leaves.

Sepals and petals 3–15, of the same number; herbs or woody vines. 58. RANUNCULACEAE. 331

Sepals 3; petals 6, in two series; shrubs or trees. 63. ANONACEAE. 352

Plants dioecious vines with simple leaves. 61. MENISPERMACEAE. 350

Stamens definite in number, in ours 6; anthers opening by hinged valves.

Herbs. 59. PODOPHYLLACEAE. 348

Shrubs. 60. BERBERIDACEAE. 349

Plants with succulent stems and leaves. 71. CRASSULACEAE. 387

Stamens on the margin of a hypanthium (the hypanthium very small in some Saxifragaceae).

Flowers regular or nearly so (actinomorphic).

Endosperm present, usually copious and fleshy.

Herbs; stipules mostly wanting.

Carpels as many as the sepals.

Carpels wholly distinct, dehiscent ventrally; succulent plants. 71. CRASSULACEAE. 387

Carpels united at the base, circumscissile. 72. PENTHORACEAE. 389

Carpels fewer than the sepals, 2 or rarely 3, distinct or only partly united; plant scarcely succulent. 74. SAXIFRAGACEAE. 390

Shrubs or trees; fruit thin-walled follicles; stipules present. (*Opulaster* in) 79. ROSACEAE. 403

Endosperm wanting or scant; stipules mostly present.

Carpels several or numerous, or, if solitary, becoming an achene. 79. ROSACEAE. 403

Carpel solitary, not becoming an achene.

Ovary 2-ovuled; fruit a drupe; leaves simple. 81. AMYGDALACEAE. 443

Ovary several-ovuled; fruit a legume; leaves pinnately compound. 82. MIMOSACEAE. 446

Flowers irregular (mostly zygomorphic).

Upper petal enclosed by the lateral ones in the bud; corolla not papilionaceous. 83. CAESALPINIACEAE. 448

Upper petal enclosing the lateral ones in bud; corolla papilionaceous. 85. FABACEAE. 451

2. Carpels several and united.

a. Ovary superior.

*** Stamens inserted at the base of the ovary or receptacle.**

† Stamens numerous.

Corolla with a filamentous crown; ovary stalked; vines with tendrils. 115. PASSIFLORACEAE. 558

Corolla without a crown; ovary sessile; herbs or shrubs without tendrils.

Sepals 2. 50. PORTULACACEAE. 312

Sepals more than 2.

Sepals imbricate.

Filaments united in 3 or more sets; leaves pellucid-punctate. 109. HYPERICACEAE. 543
Filaments distinct; leaves not punctate.
Calyx deciduous. 64. PAPAVERACEAE. 352
Calyx persistent.
Stigmas distinct or united, but not discoid; land plants; petals and sepals 4.
67. CAPPARIDACEAE. 384
Stigmas united into a disk; aquatic plants or bog-plants.
Aquatics with cordate or reniform floating leaf-blades; petals numerous.
55. NYMPHAEACEAE. 328
Bog-plants with pitcher-like leaves; petals 5 or none. 69. SARRACENIACEAE. 387
Sepals valvate.
Stamens distinct or in several groups; anthers 2-celled; fruit in ours nut-like; trees.
107. TILIACEAE. 537
Stamens monadelphous; anthers 1-celled; fruit of several carpels separating at maturity, or capsular.
108. MALVACEAE. 538

†† Stamens few, not over twice as many as the petals.
Stamens as many as the petals and opposite them.
Anther-sacs opening by hinged valves. 60. BERBERIDACEAE. 349
Anther-sacs opening by slits.
Flowers monoecious. 95. EUPHORBIACEAE. 511
Flowers perfect. 50. PORTULACACEAE. 312
Stamens as many as the petals and alternate with them, or more, sometimes twice as many.
Stamens 6 or less; petals 4; sepals 2 or 4.
Sepals 2; endosperm present; flowers irregular; stamens diadelphous. 65. FUMARIACEAE. 354
Sepals 4, rarely more; endosperm wanting; flowers regular.
Capsule 2-celled; stamens tetradynamous, rarely 2 or 4.
66. BRASSICACEAE. 356
Capsule 1-celled; stamens not tetradynamous.
67. CAPPARIDACEAE. 384
Stamens, petals, and sepals of the same number, or stamens more, usually twice as many as the sepals or petals.
Ovary 1-celled.
Ovules, or seeds, on basal or central placentae.
Sepals 2. 50. PORTULACACEAE. 312
Sepals 4–5.
Shrubs with minute alternate, scale-like leaves; flowers in terminal racemes.
111. TAMARICACEAE. 547
Herbs; leaves opposite.
Sepals distinct; petals not clawed; ovary sessile.
52. ALSINACEAE. 316
Sepals united; petals clawed; ovary more or less distinctly stipitate. 53. CARYOPHYLLACEAE. 323
Ovules, or seeds, on parietal placentae.
Stamens with united filaments and no staminodia.
108. MALVACEAE. 538
Stamens with distinct filaments.
Staminodia present. 73. PARNASSIACEAE. 389
Staminodia wanting.
Stigmas 2-cleft; plants insectivorous, with glandular-hispid leaves. 68. DROSERACEAE. 386
Stigmas entire; land plants, not insectivorous; leaves not glandular-hispid.
Styles in ours distinct.
109. HYPERICACEAE. 543
Styles wholly united.
Corolla regular or nearly so; stamens 8 or more.
Sepals and petals 4; the latter not fugaceous. 67. CAPPARIDACEAE. 384
Sepals and petals 3 or 5; the latter fugaceous. 112. CISTACEAE. 548
Corolla irregular; one petal spurred; stamens 5.
113. VIOLACEAE. 550
Ovary several-celled.

Stamens adnate to the gynoecium.
137. ASCLEPIADACEAE. 640
Stamens not adnate to the gynoecium.
Stamens with wholly or partly united filaments.
Anthers opening lengthwise; corolla regular.
Leaves punctate; filaments polyadelphous, *i.e.*, united in three or more bundles.
109. HYPERICACEAE. 543
Leaves not punctate; filaments monadelphous, *i.e.*, united in a single bundle.
Styles united around a central column from which they break at maturity.
86. GERANIACEAE. 499
Styles distinct or permanently and partly united; the tips and the stigmas free.
Leaves compound; stamens 10–15.
87. OXALIDACEAE. 501
Leaves simple; stamens 5.
88. LINACEAE. 503
Anthers opening by pores; corolla irregular.
94. POLYGALACEAE. 510
Stamens with distinct filaments.
Anthers united; flowers irregular.
89. BALSAMINACEAE. 505
Anthers distinct; flowers regular or nearly so.
Anthers opening by pores.
Gynoecium superior; fruit usually capsular.
Herbaceous saprophytes without green leaves.
127. MONOTROPACEAE. 610
Herbs with green leaves and rootstocks.
126. PYROLACEAE. 607
Shrubs. (*Ledum* in)
128. ERICACEAE. 611
Gynoecium inferior; fruit baccate or drupaceous.
129. VACCINIACEAE. 616
Anthers opening by slits.
Stigmas or styles distinct and cleft, or foliaceous.
95. EUPHORBIACEAE. 511
Stigmas or styles distinct or united, neither cleft nor foliaceous.
Style wanting; dwarf water plants.
110. ELATINACEAE. 546
Styles present; not water herbs.
Stamens 2, rarely 3.
133. OLEACEAE. 628
Stamens more than 3.
Leaves punctate with translucent dots.
92. RUTACEAE. 508
Leaves without translucent dots.
Calyx irregular, one of the sepals spurred or saccate; anthers united around the stigma.
89. BALSAMINACEAE. 505
Calyx regular; none of the sepals saccate; anthers distinct.
Style basal, arising between the nearly distinct lobes of the ovary; leaves odd-pinnate.
90. LIMNANTHACEAE. 506
Style not basal; leaves in ours abruptly pinnate.
91. ZYGOPHYLLACEAE. 507

**** Stamens inserted on the margin of a disk or hypanthium (perigynous or hypogynous).**
Styles and upper part of the ovaries distinct.
74. SAXIFRAGACEAE. 390
Styles united.
Stamens as many as the petals and opposite them.
Sepals manifest; petals involute; fruit capsular or drupaceous; ours shrubs or trees. 105. RHAMNACEAE. 532
Sepals minute or obsolete; petals valvate; fruit a berry; ours vines with tendrils. 106. VITACEAE. 534
Stamens as many as the petals and alternate with them, or more.
Hypanthium flat or obsolete; disk fleshy.

Styles united, arising in the center, between the nearly distinct lobes of the ovary; small hydrophytic plants. 90. LIMNANTHACEAE. 506
Styles not arising between the ovaries; trees or shrubs.
Plants with secreting glands in the bark.
Leaf-blades punctate with oil-glands. 92. RUTACEAE. 508
Leaf-blades not punctate. 93. SIMARUBACEAE. 509
Plants without secreting glands in the bark.
Plants with resiniferous tissue; fruit drupaceous; seeds without aril; leaves in ours compound. 100. ANACARDIACEAE. 525
Plants without resiniferous tissue.
Stigmas tufted or many-cleft; low shrubs. 99. EMPETRACEAE. 524
Stigmas entire.
Leaf-blades simple, pinnately veined.
Fruit a drupe; ovules solitary in each cell of the ovary. 98. AQUIFOLIACEAE. 523
Fruit a capsule; ovules 2 or more in each cell; seeds in ours with a conspicuous aril. 97. CELASTRACEAE. 522
Leaf-blades simple and palmately veined, or compound.
Leaves opposite.
Fruit capsular.
Flowers regular; fruit a bladdery, 3-lobed capsule; leaves pinnate. 101. STAPHYLEACEAE. 528
Flowers irregular; fruit a leathery capsule; leaves digitate. 102. HIPPOCASTANACEAE. 528
Fruit a samara. 103. ACERACEAE. 529
Leaves alternate. 104. SAPINDACEAE. 531
Hypanthium cup-shaped or campanulate; disk obsolete or inconspicuous.
Anthers opening by pores; leaf-blades 3–5-ribbed. 120. MELASTOMATACEAE. 565
Anthers opening by slits; leaf-blades 1-ribbed. 119. LYTHRACEAE. 563

b. Ovary at least partly inferior.
1. Stamens numerous.
Ovary partly inferior.
Fruit a capsule. 75. HYDRANGEACEAE. 397
Fruit a pome. 80. MALACEAE. 436
Ovary wholly inferior.
Sepals and petals 4 or 5, very unlike each other; leaves ample.
Trees or shrubs; fruit fleshy. 80. MALACEAE. 436
Herbs with rigid hairs; fruit capsular. 114. LOASACEAE. 556
Sepals and petals nearly alike, at least the latter numerous; leaves typically and in all ours mere scales or wanting; succulent plants armed with spines. 116. CACTACEAE. 558
2. Stamens not more than twice as many as the petals.
Styles wanting; stigmas sessile. 122. HALORAGIDACEAE. 582
Styles present.
Plants without tendrils.
Styles distinct.
Ovules several in each cavity of the ovary; fruit a capsule or a fleshy many-seeded berry.
Fruit, if dehiscent, valvate.
Leaves opposite; fruit a leathery capsule. 75. HYDRANGEACEAE. 397
Leaves alternate.
Fruit a fleshy berry. 76. GROSSULARIACEAE. 398
Fruit a woody, 2-beaked capsule. 78. HAMAMELIDACEAE. 402
Fruit circumscissile. 50. PORTULACACEAE. 312
Ovules solitary in each cavity of the ovary; fruit a drupe or 2–5 more or less united achenes.
Fruit drupaceous or baccate; gynoecium 1–several-carpellary; if 2-carpellary the stigmas introrse.
Ovule with a ventral raphe; leaves mostly alternate; blades lobed or compound. 123. ARALIACEAE. 584

Ovule with a dorsal raphe; leaves mostly opposite; blades entire or merely toothed. 125. CORNACEAE. 604
Fruit dry, a cremocarp; gynoecium 2-carpellary; stigmas terminal. 124. AMMIACEAE. 585
Styles united or single.
Ovary enclosed in or surpassed by the hypanthium or adnate to it.
Anthers opening by pores; fruit a berry. 129. VACCINIACEAE. 616
Anthers opening by slits; fruit a capsule.
Ovary with parietal placentae. 114. LOASACEAE. 556
Ovary with central or basal placentae.
Hypanthium merely enclosing the ovary. 119. LYTHRACEAE. 563
Hypanthium adnate to the ovary. 121. ONAGRACEAE. 566
Ovary exceeding the hypanthium, the top free. 75. HYDRANGEACEAE. 397
Plants with tendrils; fruit a pepo; leaf-blades palmately veined. 161. CUCURBITACEAE. 754

II. Petals more or less united.
A. Ovary superior.
1. Stamens free from the corolla.
Gynoecium of a single carpel; corolla zygomorphic.
Corolla papilionaceous, the upper petal enclosing the rest; stamens 5–10. 85. FABACEAE. 451
Corolla not papilionaceous, the upper petal not enclosing the rest; stamens 3 or 4. 84. KRAMERIACEAE. 451
Gynoecium of several united carpels.
Filaments united.
Stamens diadelphous. 65. FUMARIACEAE. 354
Stamens monadelphous.
Anther-sacs opening by slits; calyx and corolla regular. 87. OXALIDACEAE. 501
Anther-sacs opening by pores; calyx and corolla very irregular. 94. POLYGALACEAE. 510
Filaments distinct.
Corolla regular; anthers distinct.
Herbaceous saprophytes, without green leaves. 127. MONOTROPACEAE. 610
Herbs or shrubs with green leaves.
Corolla of essentially distinct petals, *i.e.,* united only at the base; herbs with rootstocks. 126. PYROLACEAE. 607
Corolla of distinctly united petals; shrubs. 128. ERICACEAE. 611
Corolla irregular; one of the petals free; anthers united. 89. BALSAMINACEAE. 505

2. Stamens partially adnate to the corolla.
a. Stamens as many as the lobes of the corolla and opposite them, or twice as many or more.
Ovary 1-celled, with central placentae; herbs. 130. PRIMULACEAE. 620
Ovary several-celled; shrubs or trees.
Styles and stigmas distinct; flowers monoecious or dioecious. 131. EBENACEAE. 627
Styles and stigmas united. 132. SAPOTACEAE. 628
b. Stamens as many as the lobes of the corolla and alternate with them, or fewer.
*** Corolla not scarious, veiny; fruit various, but not a pyxis.**
† Carpels distinct, except sometimes at the apex.
Styles united; stamens distinct or gynandrous. 136. APOCYNACEAE. 638
Styles distinct; stamens monadelphous. 137. ASCLEPIADACEAE. 640
†† Carpels united.
Ovary 1-celled, with central placentae.
Corolla-lobes convolute or imbricated in the bud; leaves typically opposite and simple. 134. GENTIANACEAE. 630
Corolla-lobes induplicate-valvate in the bud; leaves alternate, in ours mostly basal. 135. MENYANTHACEAE. 637

Ovary 2–3-celled, or falsely 4-celled, or if 1-celled with parietal placentae.
Stamens 2, opposite each other; corolla regular. 133. OLEACEAE. 628
Stamens usually more than 2, if only 2 not opposite each other and the corolla irregular.
Stamens 5, if only 3 or 4, not didynamous.
Fruit a capsule or berry; ovary not 4-lobed.
Styles or stigmas usually distinct.
Parasitic twining plants, with scale-like leaves. 139. CUSCUTACEAE. 651
Plants not parasitic; leaves normal.
Inflorescence not scorpioid; flowers cymose or solitary; ovary 2–3-celled.
Corolla plaited and the plaits convolute in the bud; flowers axillary, solitary or cymose-conglomerate; plants usually twining. 138. CONVOLVULACEAE. 647
Corolla merely convolute in the bud, not plaited; flowers cymose; plants never twining. 140. POLEMONIACEAE. 653
Inflorescence more or less distinctly scorpioid; ovary in ours 1-celled or imperfectly 2-celled. 141. HYDROPHYLLACEAE. 660
Styles and stigmas wholly united.
Ovules few. 138. CONVOLVULACEAE. 647
Ovules numerous.
Median axis of the gynoecium in the same plane as the axis of the stem; seeds mostly pitted. 146. SOLANACEAE. 698
Median axis of the gynoecium not in the same plane as the axis of the stem; seeds tuberculate. (*Verbascum* in) 147. SCROPHULARIACEAE. 707
Fruit of 1–4 nutlets; ovary more or less distinctly 4-lobed.
Style or stigma furnished with a glandular ring. 142. HELIOTROPIACEAE. 663
Style or stigma not furnished with a glandular ring. 143. BORAGINACEAE. 665
Stamens 4 and didynamous, or 2 or 1.
Stamens 2 and opposite each other; corolla regular. 133. OLEACEAE. 628
Stamens, if 2, not opposite each other; corolla not regular.
Carpels ripening into 2 or 4 nutlets, an achene, or a drupe.
Style apical on the lobeless ovary.
Ovary 2-celled. 144. VERBENACEAE. 676
Ovary 1-celled. 153. PHRYMACEAE. 735
Style arising between the 4 lobes of the ovary. 145. LAMIACEAE. 679
Carpels ripening into a capsule.
Placentae of the ovary axile.
Ovary 2-celled, rarely 3–5-celled; land plants.
Corolla-lobes imbricate; capsule not elastically dehiscent. 147. SCROPHULARIACEAE. 707
Corolla-lobes valvate; capsule elastically dehiscent. 152. ACANTHACEAE. 734
Ovary 1-celled; ours submerged water plants or bog plants. 148. LENTIBULARIACEAE. 729
Placentae of the ovary parietal.
Plants parasitic on the roots of other plants; leaves scale-like, not green. 149. OROBANCHACEAE. 731
Plants with green leaves, not parasitic.
Ovary and capsule 2-celled; seeds winged; plants woody. 151. BIGNONIACEAE. 733
Ovary and capsule 1-celled; seeds wingless; plants herbaceous. 150. MARTYNIACEAE. 732

**** Corolla scarious, veinless; fruit a pyxis.** 154. PLANTAGINACEAE. 736

B. Ovary inferior.

Stamens with the filaments free from the corolla.
Stamens 10; anther-sacs opening by terminal pores or chinks. 129. Vacciniaceae. 616
Stamens 5 or fewer; anther-sacs opening by longitudinal slits.
Corolla regular; anthers distinct. 162. Campanulaceae. 756
Corolla irregular; anthers united. 163. Lobeliaceae. 758
Stamens adnate to the corolla.
Ovary with 2–many fertile cavities and 2–many ovules; calyx unmodified, at least not a pappus.
Plants tendril-bearing. 161. Cucurbitaceae. 754
Plants not tendril-bearing.
Stamens as many as the corolla-lobes.
Leaves with stipules (often leaf-like and usually regarded as leaves) adnate to the stem between the leaf-bases. 155. Rubiaceae. 739
Leaves without stipules or if present these adnate to the petioles. 156. Caprifoliaceae. 744
Stamens twice as many as the corolla-lobes; low herbs with ternately dissected leaves. 157. Adoxaceae. 751
Ovary with one fertile cavity; calyx often modified into a pappus.
Flowers not in heads, often in head-like spikes or racemes. 164. Valerianaceae. 760
Flowers in involucrate heads.
Flowers all with tubular corollas or none, or only the ray-flowers with ligulate corollas.
Stamens distinct.
Flowers hermaphrodite, surrounded by a cup-like involucel; anthers versatile. 165. Dipsaceae. 762
Flowers unisexual, not involucellate; anthers basifixed. 166. Ambrosiaceae. 762
Stamens united by the anthers, or if distinct (in *Kuhnia*) the flowers hermaphrodite and anthers basifixed. 167. Carduaceae. 768
Flowers all with ligulate corollas. 168. Cichoriaceae. 886

Class 1. **GYMNOSPERMAE.** Naked-seeded Plants.

Ovules naked, inserted on the upper side of an open, mostly flat, more or less developed scale, not enclosed in an ovary. Pollen-grains divided into two or more cells, of which one produces a pollen-tube; this fertilizes the ovule directly. All trees or shrubs, most of them evergreen.

Family 1. **PINACEAE.** Pine Family.

Resinous trees or shrubs, mostly with evergreen, needle-shaped or linear leaves. Stamens several together, subtended by a scale, forming elongate aments; filaments more or less united; anthers usually 2-celled; pollen-grains globose, ellipsoid, or lobed. Pistillate aments consisting of usually numerous, spirally disposed scales subtended by bracts. Ovules inverted, usually 2 at the base of each scale. Fruit a dry cone. Seeds usually 2 at the base of each scale, often samara-like; wing formed by a part of the scale.

Leaves several together, surrounded by a sheath at the base; cones maturing the second year.
- Cone-scales with dorsal, usually spine-armed appendages; seeds with elongate wings attached to the seeds when they fall. — 1. Pinus.
- Cone-scales with inconspicuous terminal unarmed appendages.
  - Seeds with well-developed wings, which remain attached to the seeds; cones distinctly stalked, pendulous. — 2. Strobus.
  - Seeds with rudimentary wings attached to the scales; cones subsessile, spreading. — 3. Apinus.

Leaves not surrounded by sheaths; cones maturing the first year.
- Leaves in fascicles at the ends of short branches, deciduous. — 4. Larix.
- Leaves scattered along the branches, persistent.
  - Branches smooth, not roughened by persistent leaf-bases; cones erect, their scales and bracts deciduous from the persistent axis. — 5. Abies.
  - Branches roughened by persistent leaf-bases (sterigmata); cone-scales and bracts persistent.
    - Leaf-blades petioled with a single dorsal duct, flat, 2-ranked; anthers opening transversely; seeds with resin-vesicles. — 6. Tsuga.
    - Leaf-blades sessile, with two lateral ducts, in ours 4-angled and spreading in all directions; anthers opening longitudinally; seeds without resin-vesicles. — 7. Picea.

1. **PÌNUS** (Tourn.) L. Hard Pine, Pitch Pine.

Monoecious evergreen trees or rarely shrubs, with two kinds of leaves, the primary leaves chaff-like, deciduous, the secondary ones green, needle-shaped, usually with two fibro-vascular bundles (in all ours), in fascicles of 2–5, surrounded by a sheath, which is usually persistent. Staminate aments elongate, at the ends of branches of the preceding year; anthers 2-celled, opening longitudinally; pollen-grains 3-celled, the two lateral cells empty. Pistillate aments globose or oblong, sessile or nearly so, below the terminal bud or on the young twigs. Cones in ours subsessile, maturing the second autumn; scales thick, spreading at maturity, with a dorsal appendage or thickening, usually armed with a spine or at least a tubercle. Seeds samara-like, with the wing remaining attached to the seed.

Leaves 2–3 cm. long; cones 3–5 cm. long.
- Cones erect, more or less incurved; scales unarmed or nearly so, at least at muturity. — 1. *P. Banksiana.*
- Cones spreading, or somewhat reflexed; scales with evident dorsal spines. — 5. *P. Murrayana.*

Leaves 8–16 cm. long.
- Cone-scales unarmed; cones 3–5 cm. long, subterminal. — 2. *P. resinosa.*
- Cone-scales armed with a short spine-tip.
  - Cones subterminal, 6–9 cm. long. — 3. *P. scopulorum.*
  - Cones lateral, 5–6.5 cm. long. — 4. *P. echinata.*

1. **P. Banksiàna** Lamb. A tree 10–20 m., rarely 30 m. high; bark dark brown, tinged with red, irregularly furrowed; leaves in rather remote clusters of 2, dark green, 2–3 cm. long; staminate flowers yellow; pistillate ones dark purple; cones conic-ovoid, erect and incurved, 3–5 cm. long, dull purple or green, turning yellow and shining. *P. divaricata* (Ait.) Gordon [B]. GRAY or NORTHERN SCRUB PINE, BANKSIAN PINE. Sandy soil: N.S.—n N.Y.—Minn.—Alta.—Mack. *Canad.*

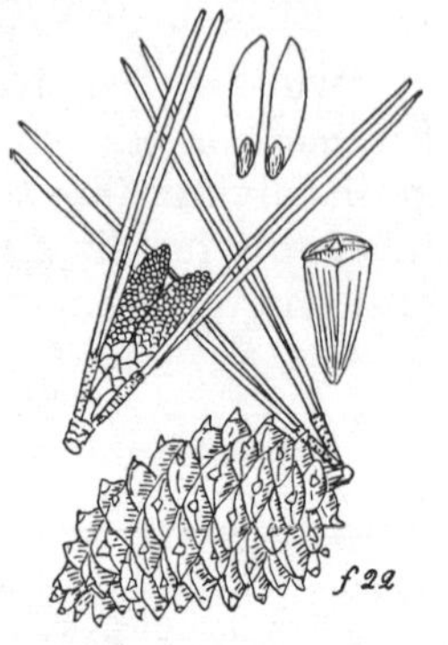

2. **P. resinòsa** Ait. A tall tree, up to 45 m. high, with smooth red bark; leaves in 2's, dark-green, 10–15 cm. long, with 2 fibro-vascular bundles; staminate aments 12–18 mm. long, early deciduous; cones ovoid, conical, usually 5 cm. long; their scales slightly thickened, without spines. RED or NORWAY PINE. Dry woods: Newf.—Pa.—Minn.—Man. *Canad.* My–Je.

3. **P. scopulòrum** (Engelm.) Lemmon. *Fig. 22.* A tree 25–30 m. high; bark thick, deeply divided into plates; leaves in 2's or 3's, yellowish green, 8–15 cm. long; staminate flowers yellow; pistillate ones purple; cones conic-ovoid, horizontal. *P. ponderosa scopulorum* Engelm. ROCK PINE, BULL PINE. Hills and mountains: S.D.—Neb.—N.M.—Ariz.—Mont. *Submont.—Mont.*

4. **P. echinàta** Mill. A tree, up to 40 m. high, with thick and furrowed and rough bark; leaves in 2's or 3's, deep green, 8–15 cm. long; staminate aments 1–1.5 cm. long; cones ovoid-conic, with thick scales and short deciduous spines. *P. mitis* Michx. YELLOW PINE. Sandy or rocky soil: N.Y.—Kans.—Tex.—Fla. *E. Temp.* Mr–Ap.

5. **P. Murrayàna** Balf. A tree 20–30 m., sometimes 45 m. high; bark of the trunk about 5 mm. thick, close and firm, covered with appressed scales, orange-brown; branches light orange; leaves yellowish green, 3–7 cm., usually about 5 cm. long; flowers orange-red; cones spreading, 3–4 cm. long, short-ovoid. *P. contorta Murrayana* Engelm. LODGE-POLE PINE. Hills and mountains: Sask. (Cypress Hills)—S.D.—Colo.—Calif.—Alaska. *Mont.*

2. **STRÒBUS** Opiz. WHITE PINE.

Monoecious evergreen trees, with two kinds of leaves as in *Pinus;* secondary leaves mostly with a single fibro-vascular bundle, in fascicles of 5, surrounded at the base by deciduous sheathing bud-scales. Staminate aments as in *Pinus.* Pistillate cones stalked, borne behind the terminal bud; scales in many series. Cones maturing the second season, drooping; scales becoming leathery, without dorsal thickening or spine. Seeds with a well-developed wing remaining attached to the seed.

1. **S. Weymouthiàna** Opiz. *Fig. 23.* A tree 20–50 m. high, with nearly smooth bark; leaves in 5's, slender, glaucous, 7–12 cm. long; cones stalked, cylindric, often somewhat curved, 10–15 cm. long, about 2.5 cm. thick, the scales slightly thickened at the apex, rounded or obtuse. *Pinus Strobus* L. [G, B]. *S. Strobus* Small [R]. Dense woods: Newf.—Man.—Iowa—Pa.—Ga. *Canad.—Allegh.* Je.

3. APÌNUS Necker. Cembra Pine.

Monoecious trees or shrubs, with two kinds of leaves as in *Pinus;* secondary leaves with a single fibro-vascular bundle, in fascicles of 5. Staminate aments as in *Pinus.* Pistillate cones subsessile below the terminal bud; scales in several series. Cones maturing the second season, spreading; scales more or less thickened, but without dorsal thickening or spine. Seed large, edible, with a very short wing or mere margin remaining attached to the scale, when the seed falls.

1. **A. fléxilis** (James) Rydb. A tree, 10–15 m. high; bark of old trunks blackish or dark-brown, deeply furrowed, that of the branches light-gray, shining, smooth; leaves in 5's, 3.5–7 cm. long; cones 8–25 cm. long; scales somewhat thickened, spreading at maturity, the exposed portion green or yellowish, the unexposed pale red. *Pinus flexilis* James. Limber Pine. Mountains: Alta.—S.D.—w Neb.—w Tex.—s Calif. *Mont.—Submont.*

4. LÀRIX (Tourn.) Adans. Larch, Tamarack.

Slender deciduous monoecious trees. Leaves needle-shaped, soft, very many in each fascicle, developed in early spring from lateral scaly buds. Staminate aments terminating short lateral branches of the preceding year, from naked buds. Anthers 2-celled, opening transversely; pollen-grains 1-celled, globular. Pistillate aments crimson, usually subtended by leaves; cone-scales persistent. Seeds samara-like; wings attached to the seeds.

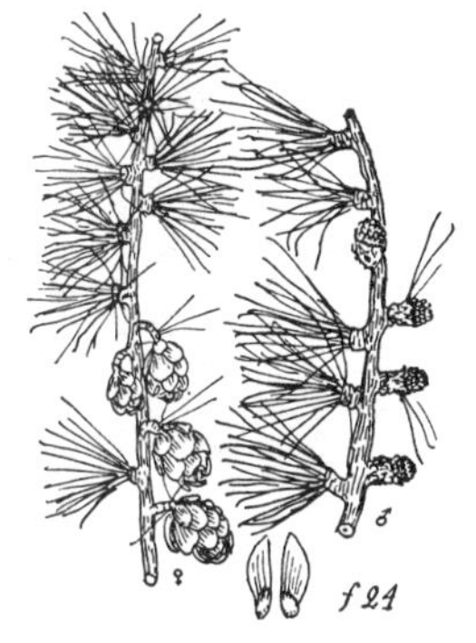
f 24

1. **L. laricìna** (Du Roi) Koch. *Fig. 24.* A tree 15–25 m. high, with a trunk 3–6 dm. in diameter; bark of the trunk separating into reddish-brown or ashy scales; twigs smooth, at first whitish, turning yellowish-brown; leaves in clusters of 12–20, somewhat 3-angled, ridged beneath, 2–3 cm. long, bright green; cones russet-brown, 1.5–2 cm. long; scales about 20, obovate, erose, twice as long as the bracts. *L. americana* Michx. Tamarack, American Larch. Swamps: Lab.—Mass.—Pa.—Ill.—Sask.—Alaska. *Boreal—Subarct.*

5. ÀBIES (Tourn.) Hill. Fir, Balsam.

Monoecious evergreen trees with spreading branches. Leaves flattened, leaving rounded scars on the smooth branches (without sterigmata), with 2 longitudinal resin-ducts and solitary fibro-vascular bundles, those of the lateral branches appearing 2-ranked by the twisting of the base, those of the cone-bearing branches usually curved upward and shorter. Staminate aments in the axils of the leaves of the preceding year, subtended by imbricate bracts; pollen-sacs opening transversely or nearly so. Pistillate aments lateral, erect; cones maturing the first year, erect; scales and bracts deciduous at maturity. Seeds samara-like; wings remaining attached to the seeds.

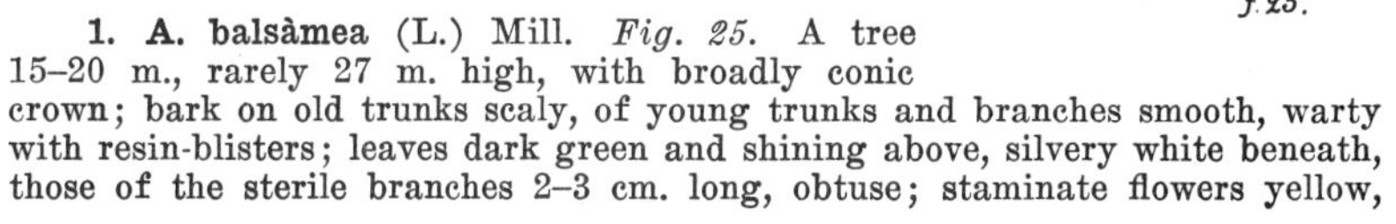
f.25.

1. **A. balsàmea** (L.) Mill. *Fig. 25.* A tree 15–20 m., rarely 27 m. high, with broadly conic crown; bark on old trunks scaly, of young trunks and branches smooth, warty with resin-blisters; leaves dark green and shining above, silvery white beneath, those of the sterile branches 2–3 cm. long, obtuse; staminate flowers yellow,

tinged with purplish; pistillate ones deeply purple; fruit oblong, puberulent, 5–10 cm. long; scales obovate, longer than broad, usually twice as long as the bracts, serrulate. BALSAM FIR. Low ground: Lab.—Mass.—Va.—Iowa—Man.—Alta.—Mackenzie. *Boreal—Subarct.*

6. **TSÙGA** (Endl.) Carr. HEMLOCK.

Monoecious evergreens with spreading or somewhat reflexed branches. Leaves flattened, with stomata only on the lower side, on the lateral branches appearing 2-ranked by the twisting of the petioles, articulate to short persistent bases (sterigmata). Staminate aments axillary to leaves of the preceding year; pollen-sacs confluent, opening by a transverse slit. Pistillate aments solitary, terminal on branches of the preceding year; bracts membranous, inconspicuous; scales in few series. Cones drooping, maturing the first year; scales at last loose and spreading, persistent. Seeds samara-like; wings attached to the seeds.

1. **T. canadénsis** (L.) Carr. A tree up to 35 m. high, with a trunk 1.5 m. thick, the bark flaky when old, the branches spreading or drooping; leaves short-stalked, 12–18 mm. long, scarcely 2 mm. wide, dark-green above, glaucous beneath; cones ovoid, 1.5–2.5 cm. long; cone-scales suborbicular. Hilly or rocky woods: N.S.—Minn.—Wis.—Ala.—Ga. *Canad.—Allegh.* Ap–My.

7. **PÌCEA** Link. SPRUCE.

Evergreen monoecious trees. Leaves needle-shaped, in ours quadrangular or nearly terete, leaving persistent bases (sterigmata) when falling, scattered, pointing in all directions. Staminate aments from the axils of the leaves of the preceding year, or rarely terminal; anther-sacs opening longitudinally. Pistillate aments terminal. Cones maturing the first season, drooping; scales closely imbricate, leathery, thin, without dorsal thickening, persistent. Seeds samara-like, with hyaline wings remaining atttached to the seeds.

Branchlets glabrous; cones 3–5 cm. long, deciduous in the first winter; scales rather thin, entire or minutely denticulate on the margins; cones cylindric, oblong, 3.5–5 cm. long. 1. *P. glauca.*
Branchlets pubescent; cones 1.5–3 cm. long, persistent for several years; cone-scales rigid, erose or dentate. 2. *P. Mariana.*

1. **P. glaùca** (Moench) Voss. *Fig. 26.* A tree 10–20 m., rarely 30 m. high, with a trunk 3–9 dm. in diameter; bark ashy-brown, scaly; branches and sterigmata glabrous; leaves incurved, acute or acuminate, blue-green; staminate flowers pale red, becoming yellow after shedding the pollen; pistillate flowers pale red or yellowish-green; cones cylindric or oblong, 3–6 cm. long; scales orbicular, slightly emarginate. *P. alba* Link. *P. canadensis* (Mill.) B.S.P. [B, G, R]. WHITE SPRUCE. River banks and hillsides: Lab.—Newf.—Me.—n N.Y.—Wisc. —S.D.—Alta.—Alaska. *Boreal—Subarctic.* The Black Hills spruce has shorter and broader cones than the typical *P. glauca.* It was erroneously referred by Sargent to *P. albertina* S. Br.

2. **P. Marìana** (Mill.) B.S.P. A tree usually 6–10 m. high, occasionally 25 m. high, with a trunk 1–3 dm. in diameter; bark grayish-brown, scaly; twigs russet-brown, short-pilose; leaves pale blue-green and glaucous, 1–2 cm. long, callous-tipped at the apex; staminate flowers dark red; pistillate ones purple; cones short, ovoid, 1–3 cm. long, ashy-brown, persistent; scales rounded, erose-dentate. *P. nigra* Link. BLACK SPRUCE. Swamps: Lab.—Newf.—N.C.—Wisc.—Sask.—Alta.—Mack. *Canad.—Boreal—Subarctic.*

Family 2. **JUNIPERACEAE.** JUNIPER FAMILY.

Evergreen dioecious or monoecious trees or shrubs. Buds naked. Leaves opposite or whorled, mostly reduced and scale-like, appressed or sometimes subulate and spreading. Perianth wanting. Aments solitary, the pistillate ones with few carpellary scales. Ovules erect, 1–several under each scale. Cones often with peltate scales, in some genera fleshy. Seeds wingless or, if winged, the wings formed by a portion of the seed-coat.

Plants mostly dioecious; cones berry-like or drupe-like with coalescent fleshy scales.
  Aments axillary; cones with smaller scales at the top; leaves all subulate and spreading. — 1. JUNIPERUS.
  Aments terminal; cones with larger scales at the top; leaves at least of the mature plants scale-like and appressed. — 2. SABINA.
Plants monoecious; cones dry; scales merely imbricate. — 3. THUJA.

1. **JUNÍPERUS** (Tourn.) L. JUNIPER.

Dioecious or monoecious shrubs or trees. Leaves in whorls of 3, subulate, ascending or spreading, without glands on their back. Staminate aments axillary, solitary; pollen-sacs several under each scale. Pistillate ament of 2–3 series of fleshy scales; ovules solitary. Cone berry-like. Seeds wingless.

Low shrub with depressed branches; leaves abruptly bent at the base, deeply channeled, abruptly acute. — 1. *J. sibirica.*
Tree or erect shrub; leaves straight or nearly so, shallowly channeled, gradually acuminate. — 2. *J. communis.*

1. **J. sibírica** Burgsd. *Fig. 28.* Shrub seldom 5 dm. high, usually with decumbent branches; bark dark red, scaly; leaves 5–12 cm. long, keeled, dark green below, white above, ascending; fruit dark blue, with a bloom, 7–9 mm. in diameter; seeds 1–3, ovate, acute, angled, about 3 mm. long. *J. nana* Willd. [B]. *J. communis montana* Ait. [G]. High mountains or dry open rocky places: Lab.—N.Y.—Mich.—N.M.—Calif.—Alaska; Eurasia. *Subalp.—Mont.*

f.28.

2. **J. commùnis** L. An erect shrub or low tree sometimes 7–8 m. high; bark dark red, scaly; leaves spreading, mostly straight, prickly-pointed, keeled, 1–2 cm. long, dark green on the lower side, white on the upper; fruit 6–7 mm. in diameter, dark blue, 1–3-seeded; seeds ovate, acute, about 3 mm. long. In the range mostly var. *depressa* Pursh. [G], a low bush, with spreading or decumbent branches. Dry hills: N.S.—N.J.—Pa.—w Neb.—N.M.—B.C; Eurasia. *Boreal.*

2. **SABÌNA** Haller. RED CEDAR.

Evergreen monoecious or dioecious shrubs or trees. Leaves alternately in 2's or 3's, scale-like and appressed, or in young plants subulate and more spreading, often with a gland on the back. Staminate aments small, solitary or 3–6 together, terminal on the branchlets; pollen-sacs 3–6 under each ovate or shield-like scale. Pistillate aments subglobose, of 2–3 series of fleshy scales. Ovules erect, solitary or sometimes 2 under each scale. Cones berry-like. Seeds 1–4, wingless.

Trees or erect shrubs, with straight peduncles.
  Tree with a conic crown; fruit maturing the first season. — 1. *S. virginiana.*
  Tree with a round crown; fruit maturing the second season. — 2. *S. scopulorum.*
Prostrate shrubs, with recurved peduncles. — 3. *S. horizontalis.*

1. **S. virginiàna** (L.) Antoine. A tree up to 30 m. high, with shreddy bark, red wood, and a pyramidal crown; leaves except in the juvenile state appressed, usually 4-ranked, acute, 1–4 mm. long, with a gland on the back; berry ovoid, 5–6 mm. thick, blue, with a bloom; seeds 3–4 mm. long, smooth, shining. *Juniperus virginiana* L. [G, BB]. Hillsides: N.B.—Man.—Neb.—Tex.—Fla. *E. Temp.* Ap–My.

2. **S. scopulòrum** (Sargent) Rydb. *Fig. 29.* A tree sometimes 10–12 m. high, with rounded crown; bark dark reddish-brown or grayish-red, fissured and scaly; leaves opposite, appressed, acute or acuminate, with obscure glands on the back, dark green, 1–1.5 mm. long; berry globose, dark blue, with a bloom; seeds 4 mm. long, acute, angled, grooved. *Juniperus scopulorum* Sargent. [BB]. Foothills and river bluffs: Alta.—Tex.—Ariz.—B.C. *Submont.*

f.29.

3. **S. horizontàlis** (Moench) Rydb. A prostrate shrub, spreading on the ground; horizontal branches sometimes 5 m. long; leaves of the mature branches ovate, opposite, acute, distinctly glandular on the back, 1–1.5 cm. long; berry-like cones on recurved peduncles, globose, 5–7 mm. long, dark-blue, with a bloom, 1–3-seeded. *J. Sabina procumbens* Pursh. *J. horizontalis* Moench [G, BB]. On banks and hillsides: N.S.—Me.—n N.Y.—ne Iowa—Wyo. —B.C. *Boreal.—Mont.—Submont.*

3. **THÙJA** L. Arbor Vitae, White Cedar.

Evergreen monoecious shrubs or trees. Leaves scale-like, 4-ranked, alternately opposite, usually with a gland on the back. Staminate aments terminal, solitary, nearly sessile between the leaves; anthers in 2 or 3 series, stalked; anther-sacs 4; pistillate aments solitary, terminal, ovoid or oblong; scales 2-ranked in several series. Cones oblong or ovoid, persistent; scales dry and flat. Seeds flat, winged on both sides.

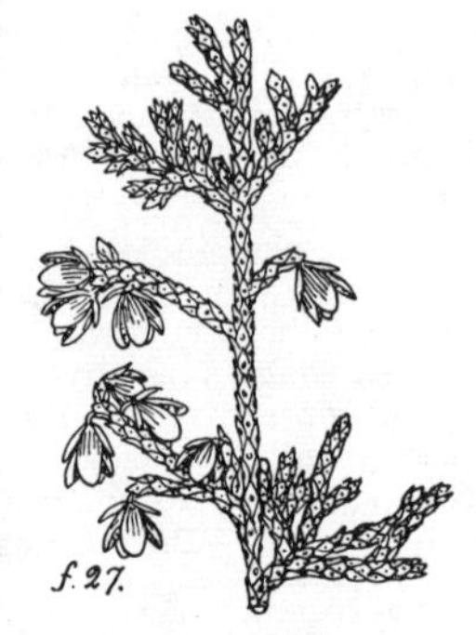
f. 27.

1. **T. occidentàlis** L. *Fig. 27.* A small tree or shrub, rarely 20 m. high, with white wood; leaves 4-ranked, those of the branchlets 5–7 mm. long, sharp-pointed, of two kinds, those of the upper and lower rows flat, acute, those of the lateral rows boat-shaped; cones oblong, 1–1.3 cm. long, brown, their scales 2 or 3 pairs, oblong to ovate, obtuse. Swamps and banks: N.B.—Man.—Minn.—N.C.—N.J. *Canad.—Allegh.* My–Je.

Family 3. **TAXACEAE.** Yew Family.

Evergreen monoecious or dioecious trees or shrubs. Buds scaly. Leaves spirally arranged, but usually 2–ranked, spreading, in ours simple and linear. Staminate flowers usually in crowded aments, in ours axillary; pollen-sacs opening longitudinally. Pistillate flowers solitary. Ovules solitary, orthotropous, sessile, without carpellary scale. Fruit berry-like. Seed nearly enclosed by the pulpy aril or naked; seed-coats woody or bony. Endosperm fleshy or mealy. Cotyledons 2.

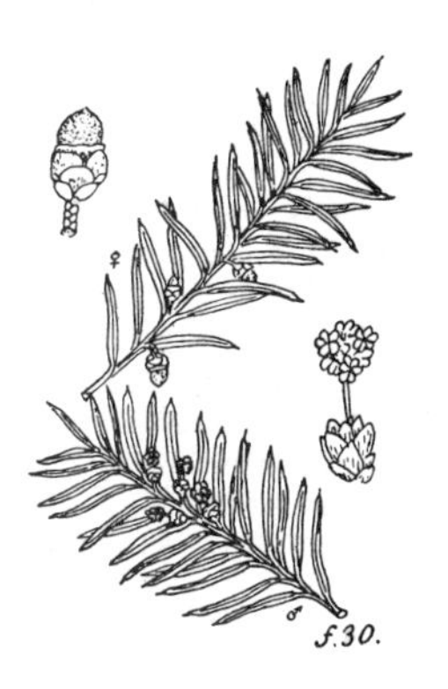

1. **TÁXUS** [Tourn.] L. Yew.

Usually dioecious evergreen trees or shrubs. Leaves linear, 2-ranked and spreading. Staminate aments short-stalked, subtended by several imbricate bracts, axillary. Pollen-sacs 6–8 under each shield-like scale. Pistillate aments consisting of a single sessile ovule subtended by imbricate bracts. Aril accrescent into a fleshy cup. Seeds nut-like.

1. **T. canadénsis** Marsh. *Fig. 30.* A low shrub, rarely more than 1.5 m. high; leaves dark green on both sides, linear, 1–2 cm. long, nearly 2 mm. wide, narrowed at the base; fruit red and pulpy, oblong, 6 mm. long, with the tip of the seed exposed. Dark woods: Newf.—Man.—ne Iowa—Va. *Canad.* Ap–My.

Class 2. ANGIOSPERMAE.

Ovules enclosed in a cavity (*ovary*), formed either by one modified infolded leaf (*carpel*) with united margins, or by several united leaves. The apex of the carpel (*stigma*) is formed of and kept moist by secretive cells; when a pollen grain falls on the stigma, it germinates and sends out a tube which penetrates the tissues of the pistil till it reaches an ovule, which it fertilizes.

Subclass I. MONOCOTYLEDONES.

Embryo with a single seed-leaf; the first leaves alternate. Stem endogenous, *i.e.*, the fibro-vascular bundles irregularly arranged in the soft tissues, without differentiation of pith, wood, and bark. Leaves usually parallel-veined, or the secondary veins running from the midrib to the margins without ramifications. Parts of the flowers mostly in 3's or multiples of 3's.

Family 4. TYPHACEAE. Cat-tail Family.

Tall water or marsh plants, with simple glabrous terete stems and creeping rootstocks. Leaves alternate, long, linear, striate, sheathing at the base. Flowers monoecious, crowded in dense terminal spike-like racemes, which are subtended by spathaceous, usually deciduous bracts; staminate spike uppermost. Perianth consisting of bristles. Stamens 2–7; filaments connate or free. Ovary 1, stipitate, 1–2-celled; styles 1 or 2. Fruit nut-like. Endosperm copious, mealy.

1. **TŸPHA** (Tourn.) L. Cat-tail, Cat-tail Flag.

Characters of the family.

Racemes with the staminate and pistillate portions usually contiguous; pollen-grains in 4's; fruiting peduncles bristle-like, 2–3 mm. long. — 1. *T. latifolia.*

Racemes with the staminate and pistillate portions usually separate; pollen of simple grains; fruiting pedicels short, 1 mm. long or less. — 2. *T. angustifolia.*

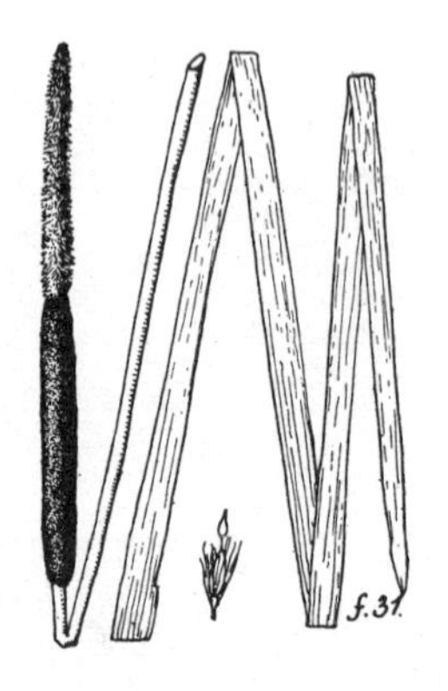

1. **T. latifòlia** L. *Fig. 31.* A stout perennial, 1–2.5 m. high; leaves flat, 5–25 mm. wide; staminate racemes light brown, with intermixed bractlets, the pistillate ones dark brown or black, without bractlets, each 1–2 dm. long; stigmas rhomboid or spatulate; fruit furrowed, bursting in water. Marshes and shallow lakes: Newf.—Fla.—Mex.—Calif.—B.C. —Mack.; Eurasia. *Temp.* Je–Au.

2. **T. angustifòlia** L. A slender perennial; stem 1–3 m. high; leaves narrowly linear, 3–15 mm. wide, striate, usually plano-convex; racemes light brown; pistillate portion 5–15 mm. in diameter, with bractlets; stigmas linear or linear-oblong; nutlets terete, not bursting in water. Marshes, mostly along the coast: N.S.—Fla.—Mex.—Calif.—Ida.—S.D.—Kans.; W. Ind., Cent. and S. Am.; Eurasia. *Temp.—Subtrop.*

Family 5. SPARGANIACEAE. BUR-REED FAMILY.

Marsh or water plants, with creeping rootstocks, fibrous roots, and linear alternate leaves sheathing at the base. Flowers monoecious, in dense globular heads, the staminate heads uppermost, generally sessile, the pistillate ones below, sessile or the lowest peduncled, often subtended by leafy bracts. Perianth reduced to a few (3–6) irregular chaffy scales. Stamens usually 5, distinct. Ovary 1- (seldom 2-) celled; style 1; stigma 1, seldom 2. Fruit nut-like, 1- or 2-celled, 1- or 2-seeded. Ovules anatropous. Endosperm copious.

1. SPARGÀNIUM (Tourn.) L. BUR-REED.

Characters of the family.

Achenes broadly obpyramidal, truncate at the summit, sessile; stigmas usually 2. — 1. *S. eurycarpum.*
Achenes fusiform or rarely slightly obovoid, stipitate; stigma 1.
  Stipe and beak each about 2 mm. long or more; fruiting heads 1.5 cm. broad or more.
    Beak straight or slightly curved; stigma linear.
      Heads all axillary; beak shorter than the body of the achene; leaves more or less keeled.
        Inflorescence branched; branches geniculate, bearing 3–7 staminate heads. — 2. *S. androcladum.*
        Inflorescence simple, or if branched, the branches strict and bearing 0–2 staminate heads. — 3. *S. americanum.*
      Heads, at least some of them, supra-axillary.
        Leaves, at least the middle ones, strongly triangular-keeled; stem usually erect, strict.
          Fruiting heads more than 2 cm. broad, distant; bracts ascending. — 4. *S. chlorocarpum.*
          Fruiting heads about 1.5 cm. broad, approximate; bracts nearly erect. — 5. *S. acaule.*
        Leaves not keeled, narrow and slender; stem weak and often floating.
          Leaves usually 5–10 mm. wide; leaves and bracts scarious-margined below; fruiting heads 1.5–2 cm. broad; achenes gradually beaked. — 6. *S. multipedunculatum.*
          Leaves 3–4 mm. wide; leaves and bracts not conspicuously scarious-margined; achenes abruptly beaked. — 7. *S. angustifolium.*

Beak gladiately curved; stigma short, oblong. — 8. *S. fluctuans.*
Stipe and beak short, less than 1 mm. long; fruiting heads 1 cm. broad; stigma oblong.
Heads all sessile, axillary, the staminate flowers distant from the pistillate ones; achenes short-beaked. — 9. *S. minimum.*
Lower pistillate heads pedicelled and supra-axillary; achenes beakless. — 10. *S. hyperboreum.*

1. **S. eurycárpum** Engelm. A stout glabrous perennial marsh-plant, 5–25 dm. high; leaves linear, 5–10 dm. long, 7–10 mm. wide; inflorescence more or less compound, the branches usually with one or two pistillate heads and several staminate ones, the former compact, in fruit 2–2.5 cm. in diameter; achenes bluntly 4- or 5-angled, the top rounded, flat or even a little depressed, abruptly contracted into the style. In swamps and along streams: Newf.—Fla.—Utah—Calif.—B.C. *Temp.* Je–Au.

2. **S. andrócladum** (Engelm.) Morong. A stout marsh-plant, 3–10 dm. high; leaves dark-green, 5–10 dm. long, 5–12 mm. wide; bracts barely concave at the base, narrowly if at all scarious-margined; inflorescence branched, the lower branches usually with 1 pistillate and 3–7 staminate heads; fruiting heads 2–2.5 mm. in diameter; body of the achenes fusiform, often constricted at the middle; beak about 3 mm. long; stigma linear, 2 mm. long. *S. americanum androcladum* Fern. & Eames [G]. Shallow water: Newf.—Fla.—Ala.—Man.; B.C.—Calif. *E. Temp.* My–S.

3. **S. americànum** Nutt. A stout marsh-plant, 3–7 dm. high; leaves 3–10 dm. long, 6–12 mm. wide; bracts shorter, conspicuously dilated and scarious-margined below; inflorescence simple; heads sessile or the lowest pistillate one peduncled, or rarely also with 1 or 2 staminate ones on the branch; fruiting heads 2–2.5 cm. in diameter; body of the achenes fusiform, 5–6 mm. long; beak 2–3 mm. long; stigma oblong, 1 mm. long. *S. Nuttallii* Engelm. *S. simplex* Am. auth. [B]. Bogs and muddy shores: N.S.—S.C.—Okla.—Minn. *E. Temp.* Jl–S.

4. **S. chlorocárpum** Rydb. *Fig. 32.* A slender marsh-plant, 3–6 dm. high, sometimes floating; leaves 3–6 dm. long, 3–7 mm. wide, at least the middle ones keeled; bracts similar, scarcely scarious-margined below; pistillate heads 2–4, sessile or the lowest one peduncled, some of them supra-axillary; staminate heads 3–7; fruiting heads 2–2.5 cm. in diameter; body of the achenes fusiform, 4–5 mm. long; beak 4 mm. long; stigma 1.5 mm. long. *S. diversifolium* Graebner, in part [G]. Marshes and rivers: N.Y.—Ind.—Iowa—Minn. *Allegh.* Jl–S.

5. **S. acaùle** (Beeby) Rydb. A low marsh-plant, 1–3 dm. high; leaves erect, 2–4 dm. long, 2–4 mm. wide, cellular-reticulate, overtopping the stem; bracts broadly scarious-margined; pistillate heads 1–3, crowded, supra-axillary, usually all sessile; staminate heads 2–4; fruiting heads 1.5 cm. in diameter; body of the achenes fusiform, 3–4 mm. long; beak 1.5–2 mm. long; stigmas linear, 1 mm. long. *S. diversifolium acaule* Fernald & Eames [G]. Muddy shores: Newf.—Va.—Iowa—S.D. *Canad.—Allegh.* Jl–S.

6. **S. multipedunculàtum** (Morong) Rydb. A rather slender marsh-plant, 3–5 dm. high; leaves linear, 2–5 dm. long, slightly keeled; inflorescence simple or a little branched; staminate heads 3–5, often close together, but distant from the pistillate ones; these 2–6, the upper ones sessile, the lower peduncled, and most often axillary; achenes fusiform, gradually acuminate above. *S.*

simplex multipedunculatum Morong. *S. subvaginatum* Meinsh., in part. In shallow water: Mack.—w Ont.—Mont.—Colo.—Calif.—B.C. *Plain—Subm.* Je–Au.

7. **S. angustifòlium** Michx. A slender, more or less floating water-plant; leaves narrowly linear, 3–6 dm. long, flat, not keeled, often dilated at the base, floating; inflorescence mostly simple; staminate heads 2–5; pistillate ones 2–4, sessile, or the lowest one peduncled and supra-axillary; achenes fusiform, abruptly acuminate at the apex. *S. simplex angustifolium* Engelm. [B]. In deep water: Newf.—Conn.—Pa.—Colo.—Calif.—B.C. *Boreal.* Je–Au.

8. **S. flúctuans** (Morong) B. L. Robinson. A slender water-plant; stem floating, 3–5 dm. long; leaves 3–6 dm. long, 3–11 mm. wide, cellular-reticulate; bracts somewhat scarious-margined below; inflorescence branched, the branches usually with 1 pistillate and 2 staminate heads, the main axis with 2–4 staminate heads; fruiting heads 2 cm. in diameter; body of the achenes fusiform, sometimes constricted at the middle; beak 2 mm. long; stigmas short, oblong. *S. androcladum fluctuans* Morong. [B]. Lakes and ponds: Me.—Conn.—Minn. *Canad.* Jl–Au.

9. **S. mínimum** Fries. A slender and floating water-plant; stem 1–3 dm. long, or in shallow water shorter and erect; leaves 1–4 dm. long, 1–7 mm. wide, thin and flat, usually floating; inflorescence simple; staminate heads 1 or 2, close together; pistillate heads 1–3, sessile or the lowest peduncled and axillary; body of the achenes obovoid, abruptly contracted into a short beak. Ponds and streams: Labr.—N.J.—Mich.—Utah—Ore.—Alaska; also Eurasia. *Boreal—Subarct.*

10. **S. hyperbòreum** Laest. A slender floating water-plant, stems 2–5 dm. long, in shallow water decumbent, 1–2 dm. long; leaves light green, flat, 1–4 mm. wide, up to 6 dm. long; inflorescence flexuose; staminate heads solitary; pistillate heads 2–4, in fruit 8–10 mm.; body of the achenes ellipsoid, beakless. Shallow water: Greenl.—Newf.—Man.—Alaska; Eurasia. *Arct.—Subarct.*

Family 6. ZANNICHELLIACEAE. Pondweed Family.

Immersed water plants, with slender, jointed, often branching stems, flat leaves, and perfect or monoecious flowers in axillary spikes or clusters. Perianth none, but flowers sometimes in hyaline envelopes. Stamens 1–4, seldom more, distinct and hypogynous in the perfect flowers, or solitary in the staminate ones. Ovaries 1–4, distinct, 1-celled and 1-ovuled. Fruit mostly drupelets or achenes.

Stamens 4, the connectives with dilated appendages; drupelets sessile. 1. Potamogeton.
Stamens 1 or 2, the connectives without appendages; drupelets manifestly stipitate.
  Stigmas sessile; anthers 2; flowers perfect, on long peduncles 2. Ruppia.
  Stigmas terminating a long style; anther 1; flowers monoecious, the two kinds together in the same axils. 3. Zannichellia.

1. POTAMOGÈTON (Tourn.) L. Pondweed, Fishweed.

Immersed water plants, with flat, 2-ranked leaves, alternate or the upper opposite, often of two kinds, floating and submerged, the former more or less coriaceous and broad, the latter pellucid, thin, and narrower. Stipules present, free or adnate to the lower part of the petiole or blade, enclosing the young flower-buds. Inflorescence spicate, axillary, mostly emersed. Stamens 4; appendages short-clawed, valvate in the bud. Ovaries 4, sessile, distinct, with a short style or sessile stigma. Fruit of 4 drupelets. Seeds crustaceous.

Species with both floating and submerged leaves.
  Submerged leaves bladeless. 1. *P. natans.*
  Submerged leaves with proper blades.

Submerged leaves of two kinds, oval or oblong ones and lanceolate and strongly curved ones; floating leaves with 30 or more nerves. — 2. *P. amplifolius.*
Submerged leaves of only one kind; floating leaves with less numerous nerves.
  Stipules free from the petioles and blades.
    Submerged leaves lanceolate.
      Submerged leaves all petioled, more than 7-nerved.
        Blades of the floating leaves rounded at the base.
          Floating leaves elliptic, not mucronate. — 3. *P. americanus.*
          Floating leaves ovate, mucronate. — 4. *P. illinoensis.*
        Blades of the floating leaves cordate or subcordate at the base. — 5. *P. pulcher.*
      Submerged leaves all sessile or the uppermost short-petioled.
        Peduncles of the same thickness as the stem; leaves not serrulate at the apex.
          Plant green; submerged leaves narrower than the floating ones. — 11. *P. heterophyllus.*
          Plant red; submerged leaves as wide as the floating ones or wider. — 6. *P. alpinus.*
        Peduncles thicker than the stem; leaves serrulate at the apex. — 7. *P. angustifolius.*
    Submerged leaves linear.
      Floating leaves less than 2.5 cm. long.
        Spikes all alike, floating, cylindric. — 8. *P. Vaseyi.*
        Spikes of two kinds, the floating ones cylindric, the submerged ones globose, on very short peduncles. — 9. *P. dimorphus.*
      Floating leaves more than 3 cm. long.
        Submerged leaves of nearly the same width throughout, coarsely reticulate in the middle. — 10. *P. epihydrus.*
        Submerged leaves broader below the middle, without reticulation. — 11. *P. heterophyllus.*
  Stipules adnate to the base of the linear-setaceous submerged leaves. — 12. *P. diversifolius.*
Species with submerged leaves only.
  Leaves with broad blades, lanceolate or oval, many-nerved.
    Leaves with entire margins.
      Leaves short-petioled or sessile, not amplexicaul. — 13. *P. lucens.*
      Leaves more or less amplexicaul.
        Leaves elongate-lanceolate, semi-amplexicaul, cucullate; apex of the embryo straight, pointing to the base of the fruit. — 14. *P. praelongus.*
        Leaves rounded-ovate to short-lanceolate, amplexicaul, not cucullate; embryo curved, pointing inside the base of the fruit. — 15. *P. Richardsonii.*
    Leaves with finely and sharply serrulate margins. — 30. *P. crispus.*
  Leaves narrowly linear to capillary.
    Stipules free from the petioles and the leaf-blades.
      Leaves 2–4 mm. wide.
        Species without glands at the base of the leaves; leaves with 3 principal nerves and several fine ones. — 16. *P. compressus.*
        Species with glands at the base of the leaves.
          Glands large and translucent; nerves mostly 3; end of the embryo curved, pointing inside the base of the fruit.
            Leaves acute; stipules 0.5–1 cm. long; spike subglobose. — 17. *P. Hillii.*
            Leaves obtuse; stipules 1.2–2 cm. long; spike oblong. — 18. *P. obtusifolius.*
          Glands small, dull; nerves of the leaves 5–7; the straight end of the embryo pointing to the base of the fruit. — 19. *P. Friesii.*
      Leaves seldom more than 2 mm. wide, often less.
        Glands absent; nutlets keeled. — 20. *P. foliosus.*
        Glands present; nutlets not keeled, or obscurely so.
          Stipules 1–2 cm. long, persistent. — 21. *P. rutilus.*

Stipules less than 1 cm. long, deciduous.
Leaves 1–1.5 mm. wide, flaccid. — 22. *P. pusillus.*
Leaves 1.5–2 mm. wide, stiff, the margins usually involute towards the apex. — 23. *P. strictifolius.*
Stipules adnate to the base of the leaves.
Leaves 1.5 mm. wide or less, with entire margins.
Nutlets prominently 3-keeled; leaves often 3–5-nerved. — 24. *P. interuptus.*
Nutlets 1–2-keeled or keel-less.
Stigma broad, sessile; nutlets indistinctly 1-keeled or keel-less.
Stipules narrow, not much broader than the leaves.
Leaves filiform, less than 0.5 mm. wide; stipular sheaths 3–8 mm. long. — 25. *P. filiformis.*
Leaves about 1 mm. wide; stipular sheaths 1–2 cm. long. — 26. *P. interior.*
Stipules, especially the lower, several times as broad as the leaves. — 27. *P. vaginatus.*
Stigma capitate, on an evident style; nutlets with 2 lateral keels but no median one. — 28. *P. pectinatus.*
Leaves several-nerved, 3–6 mm. wide, fine serrulation seen under a lens. — 29. *P. Robbinsii.*

1. **P. nàtans** L. Stem 6–15 dm. long, mostly simple; floating leaf-blades oval or ovate, abruptly short-pointed, rounded or cordate at the base, thick, 21–29-nerved, 5–10 cm. long, 2.5–5.5 cm. wide; submerged leaves bladeless and early perishing; peduncles 5–10 cm. long; spike cylindric, 3–5 cm. long, very dense; fruit 4–4.5 mm. long, 2–5 mm. thick; stone 2-grooved on the back. Ponds and streams: N.S.—N.J.—Calif.—Alaska; Eurasia. *Boreal.* Jl–S.

2. **P. amplifòlius** Tuckerm. Stem occasionally branched; floating leaf-blades (occasionally wanting) ovate or oval, acute at the apex, rounded or subcordate at the base; submerged leaves mostly short-petioled; the blades of the upper ovate or elliptic, 7–15 cm. long, 3–6 cm. wide, thin and shining; those of the lower lanceolate, often 1 dm. long, about 25-nerved, generally conduplicate and falcate; peduncles thickened upwards, 5–20 cm. long; spike cylindric, 2.5 cm. long; fruit 4–5 mm. long, 2.5 mm. thick, 3-keeled. Lakes: N.B.—Ga.—Neb.—B.C. *Temp.* Jl–S.

3. **P. americànus** Cham. & Schlecht. Stem much branched, 1–2 m. long; floating leaf-blades rather thick, elliptic, pointed at each end, 5–15 cm. long, 1.5–3 cm. wide, 11–23-nerved; submerged leaf-blades pellucid, thin, 1–3 dm. long, 0.5–2.5 cm. wide, 7–15-nerved, lanceolate to linear-lanceolate, with distinct petioles; peduncles thickened upwards, 5–7 cm. long; spike cylindric, 3–5 cm. long; fruit about 4 mm. long, obliquely obovoid, 3-keeled on the back. *P. lonchites* Tuckerm. *P. fluitans* Am. auth. Ponds and slow streams: N.B.—Fla. Calif.—B.C. *Temp.* Je–O.

4. **P. illinoénsis** Morong. Stem stout, branching; floating leaf-blades oval or elliptic, 5–15 cm. long, 4–9 cm. wide, 19–27-nerved, with a short point; submerged leaves oblong-elliptic, acute at each end, 1–2 dm. long; peduncles thickened upwards; spike 4–5 cm. long; fruit rounded-obovate, 3.5–4.5 mm. long, 3-keeled on the back. Streams: Ill.—Minn.—Neb.—Ark. *Prairies.* Je–Au.

5. **P. púlcher** Tuckerm. Stem rather stout, simple; floating leaf-blades coriaceous, ovate or rounded-ovate, acute or obtuse at the apex, cordate at the base, 4–10 cm. long, 3–7 cm. wide; submerged leaf-blades pellucid and cellular along the mid-rib, the upper lanceolate, acute or acuminate at each end, 10–15-nerved, the lower thicker, oval or oblong, or spatulate-oblong; peduncles thicker than the stem; spikes 2–4 cm. long; fruit 3–4 mm. long, 3-keeled. Ponds: Me.—Fla.—Ark.—Minn. *E. Temp.*

6. **P. alpìnus** Balbis. *Fig. 33.* Stem slender, simple or branched; floating leaf-blades rather thin, oblanceolate or spatulate, or sometimes oblong, 11–17-nerved, 5–12 cm. long, sometimes wanting; submerged leaves thin, semi-pellucid, oblong to linear-lanceolate, 7–30 cm. long, 5–20 mm. wide, 7–17-nerved, sessile or the upper short-petioled; spike cylindric, 2–4 cm. long; fruit obovoid-lenticular, reddish, 2–5 mm. long, 3-keeled. *P. rufescens* Schrader. Ponds and streams: N.S.—N.J.—Colo.—Calif.—Alaska; Eur. *Boreal—Subarct.* Jl–Au.

f.33.

7. **P. angustifòlius** Berch. & Presl. Stem slender, branching; floating leaf-blades somewhat coriaceous, elliptic, 4–10 cm. long, 12–35 mm. wide, many-nerved; submerged leaves lanceolate or oblanceolate, thin, acute or cuspidate, 5–15 cm. long, 5–30 mm. wide, 7–17-nerved; peduncles 6–15 cm. long; spike cylindric, 2.5–5 cm. long; fruit obliquely ovoid, 2.5–4 mm. long, 2 mm. thick, 3-ribbed; style short, blunt. *P. Zizii* Roth [B]. Lakes and streams: Me.—Fla.—Tex.—Calif.—B.C.; Mex., C. Am., and Eur. *Temp.—Plain.* Jl–Au.

8. **P. Vàseyi** Robbins. Stem slender, much branched, floating leaf-blades coriaceous, spatulate-ovate, 7–12 mm. long, 4–6 mm. wide, rounded at the apex; submerged leaves slender, filiform, 0.5–1 mm. wide, 2–3.5 cm. long; spike interrupted, 5–10 mm. long; peduncles 5–15 mm. long, as thick as the stem; achene 3-keeled, smooth. Lakes: Que.—Minn.—Iowa.—N.Y. *Canad.*

9. **P. dimórphus** Raf. Stem slender, simple or branched; floating leaf-blades coriaceous, lanceolate or oblong-lanceolate, acute or obtuse, 1–2.5 cm. long, 3–8 mm. wide, many-nerved; submerged leaves capillary, 1–1.5 mm. wide; emersed spikes cylindric, up to 1 cm. long, peduncled, the submerged ones almost globose, sessile or nearly so; fruit pitted, 3-keeled. *P. spirillus* Tuckerm. Ponds and streams: N.S.—Mo.—Minn. *Canad.* My–O.

10. **P. epihỳdrus** Raf. Stem slender, simple, compressed, 3–18 dm. long; floating leaves opposite; blades elliptic or obovate, 4–9 cm. long, 6–17 mm. wide, many-nerved, obtuse, short-petioled; submerged leaf-blades linear or linear-lanceolate, 5–15 cm. long, 2–4 mm. wide, coarsely reticulate along the midrib, 5-nerved; spike cylindric, 1–6 cm. long; fruit rounded-obovoid, 2.5–4 mm. long, 2–3 mm. thick, 3-keeled; style short, terminal. *P. Nuttallii* Cham. & Schlecht. Ponds and streams: Newf.—S.C.—Neb.—B.C. *Boreal.* Je–Au.

11. **P. heterophỳllus** Schreb. Stem slender, much branched, sometimes 3 m. long; floating leaf-blades oval, pointed at the apex, narrowed, rounded, or subcordate at the base, 1.5–6 cm. long, 8–28 mm. wide, 9–19-nerved; submerged leaves linear or linear-lanceolate, pellucid, 1–5 cm. long, 2–18 mm. wide, 3–9-(mostly 7-) nerved, acuminate or cuspidate; peduncles often thickened upwards; spike cylindric, 1–2 cm. long; fruit rounded or obliquely obovoid, 2–3 mm. long, 1–2 mm. thick, indistinctly 3-keeled; style short, apical. A form with the submerged leaves linear, flaccid, 5–12 cm. long, 2–6 mm. wide, is *P. heterophyllus graminifolius.* Ponds or lakes and slow streams: Lab.—Fla.—Utah—Calif.—B.C.; Eur. *Temp.* Jl–S.

12. **P. diversifòlius** Raf. Stem very slender, much branched; floating leaf-blades coriaceous, oval or elliptic and obtuse, or lanceolate-oblong and acute, 1–2.5 cm. long, 4.5–12 mm. wide; submerged leaves linear-setaceous, 2–4 cm. long, 1 mm. wide or less; stipules of the floating leaves free, those of the submerged leaves mostly adnate; emersed peduncles 6–15 mm. long; submerged ones 4–6 mm. long, clavate, as long as the spikes; emersed spikes 3.5–7 mm. long, often interrupted; fruit cochleate, about 1 mm. long, 3-keeled, the middle

keel narrowly winged. *P. hybridus* Michx. [G]. Still water: Me.—Fla.—Tex.—Calif.—Mont.; Mex. *Temp.* Je–S.

13. **P. lùcens** L. Stem thick, branched, leafy; leaves submerged, elliptic or lanceolate, or the uppermost oval, thin, shining, acute or acuminate, or rarely rounded at both ends and merely mucronate, 6–20 cm. long, 15–40 mm. wide, the ends often serrulate; peduncles 7–15 cm. long; spike cylindric, 3–6 cm. long, thick; fruit roundish, 3 mm. long, 2.5 mm. thick. Ponds: N.S.—Fla.—Mex.—Calif.; Eur., C. Am., and W. Ind. *Temp.—Trop.* Au–O.

14. **P. praelóngus** Wulfen. Stem flexuose, white, often 25 dm. long, flattened, much branched; leaves submerged, oblong-lanceolate, thin, bright green, semi-amplexicaul, 0.5–3 dm. long, 1–4 cm. wide, with 3–5 principal nerves; peduncles 7–50 cm. long, straight, as thick as the stem; spike cylindric or globose, 1–3 cm. long; fruit obliquely obovoid, 4–5 mm. long, 3–4 mm. thick; middle keel sharp; style short. Ponds and streams: N.S.—N.J.—Minn.—Calif.—B.C.; Eur. *Boreal.* Je–Au.

15. **P. Richardsònii** (Benn.) Rydb. Stem very leafy and much branched; leaves submerged, thin, lanceolate, 5–10 cm. long, 8–15 mm. wide at the broadened amplexicaul base, 13–23-nerved, acute or acuminate and incurved at the apex; peduncles 3–4 cm. long, thickened upwards; spike cylindric, 2–2.5 cm. long; fruit about 4 mm. long, 2.5 mm. thick, obscurely 3-keeled. *P. perfoliatus lanceolatus* Robbins. *P. perfoliatus Richardsonii* A. Bennett. Ponds and lakes: N.Y.—Del.—Wyo.—Calif.—Alaska. *Boreal.* Je–S.

16. **P. compréssus** L. Stem branching, much flattened, often winged; leaves submerged, linear, obtuse or mucronate, 5–30 cm. long, 2–4 mm. wide, with 3 principal veins and many fine ones; peduncles 4–10 cm. long; spike cylindric, 12–15-flowered, about 1 cm. long; fruit obovoid, with a broad base, about 4 mm. long, 2.5–3 mm. thick, 3-keeled on the back; style short, recurved. The plant often propagates by means of terminal leaf-buds. *P. zosteraefolius* Schum. [G, B]. Ponds and slow streams: N.B.—N.J.—Sask.—Ore.—B.C.; Eur. *Boreal.* Jl–S.

17. **P. Híllii** Morong. Stem slender, widely branching, flattish; leaves linear, acute, 2.5–6.5 cm. long, 1–2.2 mm. wide, 3-nerved, the lateral nerves delicate and near the margins; stipules whitish, striate, obtuse; spike capitate, on short spreading or recurved peduncles, with 3–6 fruits; fruit obliquely ovoid, 3-carinate on the back; embryo coiled. Ponds: Conn.—Mich.—S.D.—Pa. *Canad.—Allegh.* Jl–S.

18. **P. obtusifòlius** Mert. & Koch. Stem slender, branched above, more or less compressed; leaves linear, obtuse or mucronate, 5–8 cm. long, 1–4 mm. wide, usually 3-nerved, with a broad midrib, and 2 translucent glands at the base; peduncles 2–3 cm. long, slender, erect; spike ovoid, 5–8-flowered, 6–8 mm. long; fruit obliquely obovoid, 3 mm. long, 2 mm. thick, 3-keeled; style short. Still water: Que.—Minn.—Wyo.—Kans.—Pa.; Eur. *Canad.—Plains.* Jl–O.

19. **P. Frièsii** Rupr. Stem compressed, branching, 5–12 dm. long; leaves linear, 4–7 cm. long, about 2 mm. wide, obtuse or mucronate at the apex, 5- (rarely 7-) nerved; peduncles 2–4 cm. long, often thickened upwards and thicker than the stem; spike interrupted; fruit obliquely ellipsoid, about 2 mm. long, 2-grooved on the back, usually with a shallow pit on the sides; style recurved. *P. compressus* Oeder, not L. *P. mucronatus* Am. auth. Still water: N.B.—N.Y.—Iowa—B.C.; Eur. *Boreal.* Jl–Au.

20. **P. foliòsus** Raf. Stem very slender, branched, flattened, 3–10 dm. long; leaves very narrowly linear, almost filiform, 3–5 cm. long, 0.5–1 mm. wide, 3-nerved; peduncles clavate, about 1 cm. long; spike short, about 4-flowered; fruit lenticular, almost orbicular, 2 mm. long, 3-keeled on the back; middle keel winged, sinuate-dentate. *P. pauciflorus* Pursh. Ponds and streams: N.B.—Fla.—N.M.—Calif.—B.C.; W. Ind. *Temp.* Jl–Au.

21. **P. rùtilus** Wolfgang. Stem slender, usually simple; winter-buds usually wanting; leaves narrowly linear, 2–3.5 cm. long, 0.5–1 mm. wide, acuminate, 3–5-nerved, with 2 glands at the base; stipules strongly nerved; peduncles 1.5–3.5 cm. long; spike elongate, 6–8-flowered; fruit narrowly obliquely ovoid, 2 mm. long; embryo curved. Ponds: Anticosti—Man.—Minn. Me.; Eur. *Canad.*

22. **P. pusíllus** L. Stem very slender, filiform, much branched, 1.5–6 dm. long; leaves filiform, 3–12 cm. long, 0.5–1.5 mm. wide, 1–3-nerved; peduncles 5–30 mm. long, rarely 3–6 cm. long; spike 3–10-flowered; fruit obliquely ellipsoid, about 1.5 mm. long, 1 mm. thick, 2-grooved on the back, not keeled. Ponds and slow streams: N.S.—Va.—Tex.—Mex.—Calif.—B.C.—Yukon; Eur. *Temp.* Je–Au.

23. **P. strictifòlius** Benn. Stem slender, 3–5 dm. high, branched above; leaves 1–1.2 cm. long, 1.5–2 mm. wide, rather stiff, 3-nerved, the lateral nerves faint, the margin revolute; stipules deciduous, linear-lanceolate, veined and reticulate; peduncles rigid, 2.5–3.5 cm. long; spike interrupted, 6 mm. long, 3–8-fruited; fruit 2 mm. long, 1 mm. broad, obliquely elliptic, faintly carinate. Streams and lakes: Que.—Mich.—Iowa—Utah—Sask. *Boreal—Submont.* Je–Au.

24. **P. interúptus** Kit. Stem 5–13 dm. long, with spreading branches; leaves linear, 7–15 cm. long, 2–2.5 mm. wide, 3–5-nerved, transversely veined; free portion of the stipules rather short, obtuse, scarious; spike interrupted; fruit obliquely obovoid, 3.5–4 mm. long, prominently keeled, and with lateral ridges on the back. Ponds and streams: N.B.—Mich.—Neb.—Ind.; Eur. *Canad.—Allegh.* Au.

25. **P. filifórmis** Pers. Stem slender, filiform above, branching, leafy, 1–4 dm. long; leaves linear-filiform, 5–30 cm. long; free portion of the stipules 3–5 cm. long; peduncles 4–7 cm. long; spike interrupted, with 2–12 flowers in each whorl; fruit ovoid, 2–3 mm. long, nearly 2 mm. thick, not keeled; stigma sessile. Ponds and lakes: Que.—N.Y.—Mich.—Wyo.—Alta. *Canad.—Plains.* Je–Au.

26. **P. intèrior** Rydb. Stem slender, branched, 3–6 dm. long; leaves linear, 3–15 cm. long, about 1 mm. wide, obtuse, with a strong midrib and raised or revolute margins; peduncles 3–7 cm. long; spike interrupted; fruit obliquely ovoid, 2-grooved on the back; stigma subsessile; free portion of the stipules 2–4 mm. long. *P. marinus occidentalis* Robbins. Ponds and lakes, especially in alkali water: Ont.—N.M.—Utah—Nev. *Plains—Basin.* Jl–Au.

27. **P. vaginàtus** Turcz. Stem freely branching, with 2–4 branches at the node; stipules adnate to the leaf-blades, the lower broad, 2–5 times as wide as the stem, 2.5–5 cm. long, 2–5 mm. wide, obtuse, prominently veined; leaf-blades 1–4.5 cm. long, 1–2 mm. wide; spike 1–6 cm. long, interrupted; fruit obovate, 3 mm. long, 2 mm. wide; stigma subsessile, lateral, *P. moniliforme* St. John. Deep water: Lab.—N.Y.—Wis.—N.D.—Alta; Eurasia. *Boreal.* Jl–Au.

28. **P. pectinàtus** L. Stem slender, much branched, very leafy, 3–10 dm. long; leaves setaceous, attenuate at the apex, 3–15 cm. long, 0.1–0.5 mm. wide, sometimes nerveless; stipular sheath 1–2 cm. long; free portion 3–8 mm. long; peduncles filiform, 5–20 cm. long; spike interrupted, with several whorls of flowers; fruit obliquely ovoid, 3–4 mm. long. Fresh, salt, or alkali waters: N.B.—Fla.—L. Calif.—Alaska; Eur. *Temp.* Jl–S.

29. **P. Robbínsii** Oakes. Stems stout, widely branching, sometimes rooting at the nodes, 5–10 dm. long; leaves linear, acute, crowded in 2 ranks, 7–12 cm. long, 2–6 mm. wide, auricled at the point of union with the stipules; stipular sheath about 1 cm. long; free portion of the stipules 1–2 cm. long, mostly lacerate; peduncles 3–10 cm. long; spike interrupted, flowering under water;

fruit obovoid, about 4 mm. long, 3 mm. thick, 3-keeled, the middle keel sharp. Ponds and lakes: N.B.—N.J.—Minn.—Ida.—Ore.—B.C. *Boreal.* Jl–S.

30. **P. críspus** L. Stem branching, compressed; leaves 2-ranked, linear-oblong, sessile, obtuse, serrulate and crisped, 2–10 cm. long, 3–7-nerved; stipules small, scarious, obtuse; spike about 1 cm. long; fruit ovoid, 3 mm. long, 3-keeled, long-beaked. Water: Mass.—Va.—Minn.; nat. from Eur. Au.

2. RÚPPIA L. Ditch Grass.

Slender water plants, widely branched and with capillary stems, filiform, alternate, 1-nerved leaves, with membranous sheaths. Peduncles spadix-like, filiform, at first very short, at last generally much elongated and spirally coiled. Flowers consisting of 2 sessile anthers, and 4 pistils, sessile at first, in fruit long-stipitate. Fruit small, more or less obliquely ovoid drupes.

Sheaths 6–10 mm. long; drupe about 2 mm. long or less. 1. *R. maritima.*
Sheaths 20–40 mm. long; drupe 3–4 mm. long. 2. *R. occidentalis.*

1. **R. marítima** L. Stem slender, filiform, whitish; leaves 2–10 cm. long, 0.5 mm. or less wide; sheaths 6–10 mm. long, with a short free tip; peduncles in fruit sometimes 3 dm. long; drupes about 2 mm. long; beak almost straight or curved. *R. curvicarpa* A. Nels. In brackish or salt water, along the coasts: Newf.—Tex.—Alaska—L. Calif.; occasionally in the interior, Minn.—Sask.—Mex.; Eurasia and S. Am. *Temp.* Je–Au.

2. **R. occidentàlis** S. Wats. *Fig. 34.* Stem comparatively stout, 3–6 dm. long, with very short internodes and fan-like clustered branches; leaves 7–20 cm. long, and 0.3 mm. wide or less; sheaths with distinct free tips, often overlapping each other; peduncles sometimes 5 dm. long; stipes about 25 mm. long; drupes ovoid or pyriform, scarcely oblique; beak short, straight. In saline ponds: Sask.—Minn.—Neb.—B.C.—Alaska. *W. Boreal—Plain—Submont.* Jl–Au.

f. 34

3. ZANNICHÉLLIA (Mich.) L. Horned Pondweed.

Slender branching aquatics, with opposite filiform leaves and sheathing membranous stipules. Flowers monoecious, sessile, naked, usually both kinds in the same axil; the staminate ones consisting of a single 2-celled anther, borne on a pedicel-like filament, the pistillate ones of 2–6 sessile pistils in a cup-shaped involucre. Fruit nut-like, obliquely oblong, flattened, with a short slender beak, ribbed or toothed on the back. Seed orthotropous.

1. **Z. palústris** L. Stem capillary from a creeping rhizome; leaves 3–7 cm. long, 0.5 mm. or less wide, acute, thin, 1-nerved; fruits 2–6 together, 2–4 mm. long, sessile or short-pedicelled; style persistent, 1–2 mm. long. In fresh and brackish ponds and ditches: Ont.—Fla.—Tex.—Calif.—B.C.; also in the Old World. *Temp.* Jl–O.

Family 7. NAJADACEAE. Najas Family.

Slender submerged aquatic plants, with linear spinulose-toothed whorled or opposite leaves, sheathing at the base. Flowers monoecious or dioecious, solitary and axillary. Staminate flowers with a double perianth, the inner hyaline; stamen 1. Pistillate flowers of a single pistil, with 2–4 subulate stigmas. Fruit a small drupe.

1. NÀJAS L.

Characters of the family.

Leaves 1–3 mm. wide, coarsely toothed; back of the leaves and internodes spiny; plant dioecious. 1. *N. marina.*
Leaves 0.5–1 mm. wide, finely, almost microscopically serrulate; back of the leaves and internodes unarmed; plant monoecious.
Drupe shining, with 30–50 rows of indistinct reticulations. 2. *N. flexilis.*
Drupe dull, with 16–20 rows of strongly marked reticulations. 3. *N. guadalupensis.*

1. **N. marìna** L. Stem stout, compressed, generally armed with teeth twice as long as broad; leaves broadly linear, 12–45 mm. long, with 6–10 spine-pointed teeth on each margin; sheaths broadly rounded, their margins entire or with a few teeth; fruit 4–5 mm. long; epicarp as well as the dull nutlet rugose-reticulate. In lakes and ponds with fresh or brackish water: N.Y.—Fla.—L. Calif.—Calif.; also Eurasia and Austr. *Temp.—Trop.* Jl–Au.

2. **N. fléxilis** (Willd.) Rostk. & Schmidt. Stem slender, forking, unarmed; leaves narrowly linear, 1–2 cm. long, numerous and crowded, pellucid, with 20–30 minute teeth on each margin; sheaths rounded, with 5–10 teeth on each side; fruit ellipsoid, 2–3 mm. long. In ponds and slow streams: Que.—Fla.—La.—Calif.—Ore.; also Eur. *Temp.* My–Au.

f. 35

3. **N. guadalupénsis** (Spreng.) Morong. *Fig. 35.* Stem very slender, filiform, 3–6 dm. long, branched and leafy; leaves numerous, crowded, 12–25 mm. long, 0.5–1 mm. wide, with 40–50 teeth on the margins; sheaths as in *N. flexilis;* fruit about 2 mm. long. *N. microdon* R. Br. In ponds and lakes: Minn.—Neb.—Fla.—La.—Tex.—Ore.; also Mex. and Trop. Am. *Austral—Trop.—Plain.* Jl–S.

Family 8. SCHEUCHZERIACEAE. Arrow-grass Family.

Marsh plants, with terete rush-like leaves and small, perfect, spicate or racemose flowers. Perianth-segments 4 or 6, in two series. Stamens 3–6; anthers 2-celled. Carpels 3–6, 1–2-ovuled, more or less united, separating at maturity, either dehiscent or indehiscent. Seeds anatropous. Embryo straight; endosperm none.

Stem scapose; leaves all basal; flowers spicate or racemose. 1. Triglochin.
Stem leafy; flowers few, in loose racemes. 2. Scheuchzeria.

1. TRIGLÒCHIN L. Arrow-grass.

Marsh herbs, with half-round elongated linear leaves, sheathing at the base, and terminal racemes or spikes on long naked scapes. Perianth-segments in ours 6, the inner 3 inserted higher up. Stamens 6; anthers 2-celled, sessile or nearly so. Ovaries 3–6, 1-celled and 1-ovuled; style short or none; stigmas 3–6, plumose. Fruit of 3–6 cylindraceous, oblong, obovoid, or clavate carpels, united at first, at maturity separating from the base upwards. Seeds cylindraceous or ovoid-oblong, compressed or angular.

Carpels 6; fruit oblong or ovoid, obtuse at the base. 1. *T. maritima.*
Carpels 3; fruit linear-clavate, tapering at the base. 2. *T. palustris.*

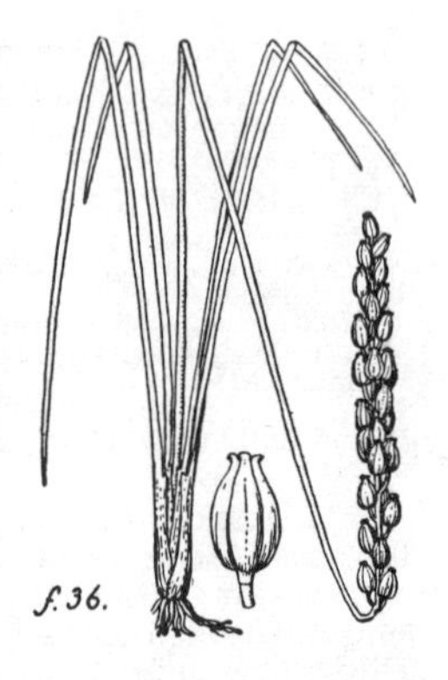

1. T. marítima L. *Fig. 36.* A stout plant, with subligneous rootstock, without stolons; leaves half-cylindric, 2 mm. wide; racemes often 4 dm. long or more; pedicels decurrent, 2–3 mm. long, in fruit ascending; fruit 5–6 mm. long, 3–4 mm. in diameter; carpels triangular, grooved on the back. In salt marshes: Lab.—N.J.—Calif.—Alaska; Eurasia. *Boreal—Subarct.—Plain—Submont.* Je–S.

2. T. palústris L. A slender plant, with short rootstock and slender stolons, 2–4 dm. high; leaves shorter than the scape, 1–3 dm. long, sharp-pointed; racemes 1–3 dm. long; pedicels slender, capillary, in fruit erect, 5–7 mm. long; stigmas sessile; fruit 6–7 mm. long. In bogs: Greenl.—N.Y.—Ind.—N.M.—Alaska; Eurasia and S. Am. *Boreal—Arct.—Mont.* Jl–S.

2. SCHEUCHZÈRIA L.

Rush-like bog plants, with perennial, creeping rootstocks. Leaves half-round below, flat above, striate, and with membranous sheaths at the base. Perianth with 6 segments in two series. Stamens 6; filaments elongate; anthers linear, basifixed. Ovaries 3, rarely 4–6, distinct or connate at the base, 1-celled, each cell with 1 or 2 ovules. Fruit of 3–6 divergent follicles, opening on the inside.

1. S. palústris L. A leafy bog plant, 1–2.5 dm. high; leaves 1–4 dm. long, the upper reduced to bracts; basal leaves with sheaths often 1 dm. long and ligules 1 cm. long; pedicels 6–20 mm. long, in fruit spreading; flowers white; segments 1-nerved, 3 mm. long; follicles 5–6 mm. long. In bogs: Lab.—N.J.—Iowa—Calif.—Alaska; Eurasia. *Boreal—Subarct.* Je–Jl.

Family 9. ALISMACEAE. Water-plantain Family.

Marsh or aquatic plants, with scapose stems, basal long-petioled sheathing, mostly cross-veined, leaves, and fibrous roots. Flowers mostly verticillate, in elongate racemes or panicles, regular, pedicelled. Sepals 3, persistent. Petals 3, deciduous. Stamens 6 or more, included. Pistils many, distinct; ovaries 1-celled and generally 1-ovuled, becoming achenes in fruit.

Carpels in a ring on a small flat receptacle. — 1. Alisma.
Carpels in several series on a convex receptacle.
  Flowers perfect.
    Style not apical; fruiting heads not bur-like; achenes turgid, beakless or nearly so. — 2. Helianthium.
    Style apical; fruiting heads bur-like; achenes flat, prominently beaked. — 3. Echinodorus.
  Flowers polygamous or unisexual, the upper ones staminate.
    Lower flowers of the inflorescence perfect. — 4. Lophotocarpus.
    Lower flowers of the inflorescence pistillate. — 5. Sagittaria.

1. ALÍSMA L. Water-plantain.

Annual or mostly perennial scapose herbs. Leaves erect or floating; blades several-veined, petioled. Flowers perfect in compound panicles. Sepals 3, usually ribbed, persistent. Petals 3, white or pinkish, spreading, deciduous. Stamens 6, two opposite each petal. Carpels few or many, in one whorl. Achenes ribbed or grooved on the back.

Achenes longer than wide, grooved on the back, their inner edges not meeting; pedicels straight, ascending.

Petals slightly longer than the sepals; corolla 3–4.5 mm. wide; achenes 1.5–2 mm. long. 1. *A. subcordatum.*
Petals much longer than the sepals; corolla 10–13 mm. wide; achenes 2.5–3 mm. long. 2. *A. brevipes.*
Achenes as wide as long, ribbed on the back, the inner edges meeting; pedicels recurved. 3. *A. Geyeri.*

1. **A. subcordàtum** Raf. Perennial; leaf-blades oblong to oval or ovate, 5 cm. long, from cuneate to cordate at the base, pointed at the apex; scape 1–10 dm. high; sepals broadly ovate to orbicular, obtuse; petals white or pinkish, 1–2 mm. long; fruiting heads 3.5–4.5 mm. broad; achenes obliquely obovate; beak ascending. *A. Plantago-aquatica* Bigel. [G,B]; not L. Water and wet places; Mass.—Fla.—Tex.—Minn. *E. Temp.* Jl–S.

f. 37.

2. **A. brévipes** Greene. *Fig. 37.* Perennial; leaf-blades oblong to ovate, 5–19 cm. long, acute at the apex, rounded, truncate or subcordate at the base; scape 1 m. high or less; sepals orbicular or rounded-ovate, 3 mm. long or more; petals 5–6 mm. long; fruiting heads 5–6.5 mm. broad; achenes broadly obovate, 2.5–3 mm. long; beak ascending. *A. superbum* Lunell. In water and wet places: N.S.—N.D.—N.M.—Calif.—B.C. *Plain—Mont.* Jl–S.

3. **A. Géyeri** Torr. Perennial; leaf-blades oblong to ovate-lanceolate, rarely lance-linear, 5–9 cm. long, acute at both ends; scapes mostly 1–5 dm. long, diffusely spreading; sepals rounded-ovate, about 2–5 mm. long; petals 2–4 mm. long; fruiting heads 4.5–5.5 mm. broad; achenes suborbicular; beaks erect. Wet places: w N.Y.—Neb.—Nev.—Ore.—Wash. *Plain—Submont.* Je–S.

2. HELIÁNTHIUM Engelm. BUR-HEAD.

Annual or perennial, scapose water-plants. Leaves erect, with linear or oblong blades, or in a Cuban species with floating leaf-blades, cordate at the base. Flowers in a few whorls, or paniculate with spreading or recurved pedicels. Flowers perfect. Sepals and petals 3, the latter white or pink. Stamens 6 or 9, with elongate filaments and short anthers. Carpels rather few, in several series on an elevated receptacle; style not apical, minute. Achenes turgid, in globose or depressed heads, ribbed or crested, beakless or nearly so.

1. **H. párvulum** (Engelm.) Small. Plant 0.5–1.5 dm. high; leaf-blades linear to elliptic, 1–3 cm. long, acute, 3-ribbed; pedicels 2–8, recurved in fruit; sepals rounded-ovate or deltoid-ovate, 1.5–2 mm. long; petals suborbicular, of the same length, emarginate; fruiting heads globose, 3–4 mm. in diameter; achenes 1–1.5 mm. long, merely ribbed. *H. tenellum* (Mart.) Britton. Shallow water: Mass.—Fla.—Tex.—Mich.—Minn?; Mex. and W. Ind. *Allegh.—Trop.*

3. ECHINÓDORUS Rich.

Annual or perennial, usually acaulescent marsh-plants. Leaves erect or ascending, with oblong to cordate, many-ribbed blades. Flowers in several whorls on the scape, with rather short pedicels. Flowers perfect. Petals and sepals 3. Stamens 10–30; anthers oblong. Carpels numerous, in several series, on a convex or globose receptacle. Style terminal. Achenes forming a bristly head, the bodies more or less flattened, ribbed, beaked by a persistent style.

Pedicels stiff, spreading; body of the achenes cuneate; beak slender, half as long as the body. 1. *E. cordifolius.*
Pedicels pliable, curved; body of the achenes falcate; beak one fourth as long as the body, incurved. 2. *E. radicans.*

1. **E. cordifòlius** (L.) Griseb. Scape 1–5 dm. high, surpassing the leaves; leaf-blades ovate to lanceolate, truncate or cordate at the base, 4–20 cm. long, 3–13-ribbed; pedicels not very unequal in length; bracts lanceolate or linear-lanceolate; sepals ovate, acute, 4–5 mm. long; petals white, twice as long; fruiting heads 4–10 mm. *E. rostratus* Engelm. Borders of ponds: Ills.—Neb.—Calif.—Fla.; also Mex., C. Am., and W. Ind. *Temp.—Trop.* F–S.

2. **E. radìcans** (Nutt.) Engelm. Scape spreading or decumbent, 3–12 dm. long; leaf-blades ovate or oblong, truncate or cordate at the base, 5–20 cm. long; pedicels very unequal, 1.5–6 cm. long; bracts subulate from a broad base; sepals ovate or rounded, obtuse, 5–6.5 mm. long; petals white, slightly longer; fruiting heads 7–8 mm. Ponds: Ill.—Tex.—Iowa—N.C.—Fla. *Temp.* Je–S.

4. LOPHOTOCÁRPUS T. Durand.

Annual or perennial acaulescent water-plant or marsh-plants. Flowers polygamous, in several whorls on the upper part of the scape, the upper ones staminate, the lower perfect. Stamens 9–15; filaments flattened; anthers oblong. Carpels numerous, in several series on a convex receptable; style slender, oblique. Achenes crested or winged.

1. **L. calycìnus** (Engelm.) J. G. Smith. Emersed water-plants; leaves 1.5–6 dm. long; blades sagittate or hastate, 6–30 cm. long, the basal lobes longer than the terminal one; scape shorter than the leaves, with 2–7 whorls of flowers; sepals suborbicular, becoming 9–15 mm. long; fruiting heads 11–16 mm.; achenes cuneate, 2–2.5 mm. long, with a stout beak and a thin dorsal wing. *Sagittaria calycina* Engelm. Shallow water: Del.—S.D.—N.M.—Ala. *Allegh.—Austral—Son.* Jl–S.

5. SAGITTÀRIA L. Arrow-head, Swan or Swamp Potato.

Perennial water or bog plants, with tuber-bearing or nodose rootstocks. Leaves long-petioled with a sagittate or lanceolate blade or, especially the earlier ones, reduced to bladeless phylloids. Flowers monoecious or dioecious, borne in verticils of 3's near the top of the scapes, pedicelled, the staminate uppermost. Sepals and petals 3, the latter large, white. Stamens numerous, inserted on the convex receptacle; anthers 2-celled, dehiscent by lateral slits. Pistillate flowers with numerous pistils. Achenes densely aggregated in globular heads, flat and often wing-margined.

Pedicels in fruit reflexed or recurved; leaf-blades tapering or subcordate at the base. — 1. *S. platyphylla.*
Pedicels in fruit ascending.
  Leaf-blades without basal lobes.
    Pistillate flowers with pedicels as long as those of the staminate ones.
      Leaf-blades triangular in cross-section. — 2. *S. cristata.*
      Leaf-blades flat.
        Filaments dilated, relatively short; achenes cuneate, the beak minute, below the top of the body. — 3. *S. graminea.*
        Filaments not dilated, slender; achenes curved-oblong, the beak horizontal, at the top of the body. — 4. *S. ambigua.*
    Pistillate flowers subsessile or with very short pedicels. — 5. *S. rigida.*
  Leaf-blades with basal lobes.
    Beak of the achene horizontal.
      Beak minute or inconspicuous.
        Beak borne below the apex of the achene. — 3. *S. graminea.*
        Beak borne at the top of the achene. — 6. *S. longiloba.*
      Beak long and flat.
        Beak of the achene much shorter than the width of the body, the upper wing-margin scarcely decurrent on the beak; pedicels of the pistillate flowers less than twice as long as the fruit and shorter than those of the staminate heads; middle lobe of the leaves ovate-lanceolate to linear acute or acuminate. — 7. *S. latifolia.*

Beak of the achene as long as the width of the body, the upper wing-margin distinctly decurrent on the beak, pedicels of the pistillate flowers at least twice as long as the fruit, as long as those of the staminate heads; terminal lobe of the leaves broadly ovate, often obtuse. — 8. *S. esculenta.*
Beak of the achene erect.
Achenes with thin and unequal wings; beak sharp, inserted over the outer edge of the ventral wing. — 10. *S. brevirostra.*
Achenes with thick and equal wings; beak blunt, inserted over the inner edge of the ventral wing. — 9. *S. cuneata.*

1. **S. platyphýlla** (Engelm.) J. G. Smith. Emersed water-plant; leaf-blades thick, lanceolate, elliptic, or ovate, 5–15 cm. long, 5–7-nerved, acute or acuminate at the apex, narrowed or subcordate at the base; scape 2–5 cm. high; whorls of the flowers 3–8; bracts ovate, 3–8 mm. long; corolla 2.5–3 cm. wide; fruiting heads 1 cm. broad; achenes obliquely obovate, 2 mm. long, winged on both margins; beak long, horizontal. Shallow water: Ala.—Kans.—Tex. *Austral.* Au–O.

2. **S. cristàta** Engelm. Plant usually submerged, 3–7.5 dm. high; leaf-blades linear-lanceolate to elliptic, 6–11 cm. long, shorter than the petioles; scape elongate; whorls of flowers 4–6, the lowest one pistillate; bracts lanceolate, 4–7 mm. long, distinct; corolla 2 cm. broad; fruiting heads 15–20 mm. in diameter; achenes cuneate-obovate, 3 mm. long, with the lower margin undulate; lateral ribs crested; beak nearly apical, horizontal. Water: Minn.—Iowa—N.Y. *Prairies.*

3. **S. gramínea** Michx. Plant emersed or submerged, 1–6 dm. high; phylloids, if present, flattened, linear-lanceolate, acute, 8–30 mm. long, 1–2 cm. wide; leaf-blades lanceolate to ovate-elliptical, acute at both ends, or very rarely truncate, with short divaricate lobes at the base, 5–15 cm. long, 3–5-ribbed; bracts ovate, acute, 3–5 mm. long, connate to the middle; fruiting heads 5–10 mm. in diameter; achenes 1.5 mm. long, dorsally crested and obliquely one-ribbed on the sides. In shallow ponds and marshes: Newf.—Sask.—Tex.—Fla. *Temp.* Ap–O.

4. **S. ambígua** J. G. Smith. Plant usually emerged, 3–8 dm. high; leaves erect, the blades lanceolate, 1–5 dm. long, 5-nerved; scape as long as the leaves or longer; whorls of flowers 8–15; bracts lanceolate, 1–1.5 cm. long, united at the base; fruiting heads 12–15 mm. broad; achenes 2 mm. long, narrowly winged, the wings procurrent into the beak. Water: Mo.—Kans.—Okla. *Ozark.* Je–Au.

5. **S. rígida** Pursh. Plant emersed or submerged and then much elongate; leaf-blades 5–20 cm. long, linear, lanceolate, or elliptic to ovate, 7–9-nerved, acute to cordate at the base, sometimes with spreading basal lobes; scape 1–8 dm. high; bracts ovate, obtuse, 4–8 mm. long, united at the base; corolla 2–3 cm. broad; fruiting heads 8–15 mm. broad; achenes cuneate, 3–4 mm. long, winged; beak stout, ascending. *S. heterophylla* Pursh [G]. Water: Que.—Tenn.—Minn.—N.J.—Kans. *Canad.—Allegh.* Je–Au.

6. **S. longíloba** Engelm. A monoecious, slender, erect perennial, 3–5 dm. high; leaf-blades sagittate; basal lobes linear-lanceolate, acuminate; fruiting heads 10–12 mm. in diameter; achenes 2 mm. long, narrowly winged on both margins; beak lateral. In shallow ponds: Neb.—Colo.—Sonora.—Tex. *Plain—Son.* Au–O.

7. **S. latifòlia** Willd. A rather stout plant, monoecious, 2–6 dm. high, with the lower verticils fertile, or rarely dioecious; leaf-blades sagittate, 5–20 cm. long, very variable, glabrous; lobes from ovate-lanceolate to linear (var. *gracilis*), acute or acuminate; bracts ovate, acute, 1–3 cm. long; petals elliptic 4–8 mm. long; fruiting heads 10–15 mm. in diameter; achenes 2.5–3 mm. long, broadly obovate, winged on both margins; beak triangular-lanceolate, acumi-

nate, making almost a right angle to the achenes. Shallow water: N.B.—Fla. —Calif.—B.C.; also Mex. and C. Am. Je–S.

8. **S. esculénta** Howell. A stout plant, 3–10 dm. high; leaf-blades broadly sagittate, 1.5–4 dm. long, mostly monoecious, lateral lobes mostly ovate, acuminate; bracts ovate, 2–3 cm. long; petals 8–15 mm. long; fruiting heads 15–20 mm. in diameter; achenes 3–3.5 mm. long, obovate-cuneate, much longer than broad; beak lanceolate. *S. variabilis obtusa* Engelm., not *S. obtusa* Thunb., nor *S. obtusa* Willd. Shallow water: N.Y.—Va.—La.—Kans.—Minn.—Ariz.—Calif.—Wash; Mex. *Temp.* Jl–N.

9. **S. cuneàta** Sheldon. *Fig. 38.* A rather weak plant growing in mud or water, glabrous, 2–4 dm. high; leaves when emersed sagittate; petiole rather stout, usually curving outwards; blade 6–18 cm. long; basal lobes narrow, lanceolate, somewhat divergent; blades in deep water less developed, sometimes not lobed, floating; bracts lanceolate or linear-lanceolate, acute, 8–20 mm. long, scarious-margined; fruiting heads 10–15 mm. in diameter; achenes 2 mm. long, winged on both margins. *S. arifolia* Nutt. [G,B]. In mud and shallow water: Me. — N.D. — Conn. — Kans. — N.M.—Calif.—B.C. *Canad.—Plain—Submont.*

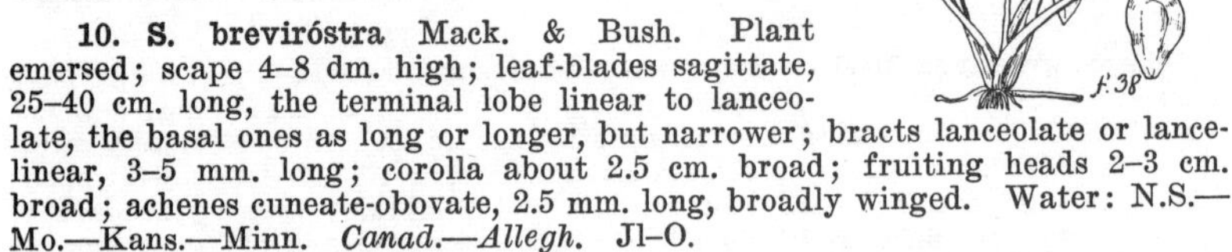

10. **S. brevirόstra** Mack. & Bush. Plant emersed; scape 4–8 dm. high; leaf-blades sagittate, 25–40 cm. long, the terminal lobe linear to lanceolate, the basal ones as long or longer, but narrower; bracts lanceolate or lance-linear, 3–5 mm. long; corolla about 2.5 cm. broad; fruiting heads 2–3 cm. broad; achenes cuneate-obovate, 2.5 mm. long, broadly winged. Water: N.S.—Mo.—Kans.—Minn. *Canad.—Allegh.* Jl–O.

Family 10. ELODEACEAE. Water-weed Family.

Submersed or floating water-plants. Leaves alternate, opposite, or whorled. Plants monoecious, dioecious or polygamous. Flowers enclosed in a spathe of 1–3, usually united bracts. Hypanthium in the pistillate flowers well developed, tubular, in the staminate flowers often shorter or obsolete. Perianth regular or nearly so. Sepals 3. Petals 3 or wanting. Stamens 3–9; filaments short, often monadelphous. Pistil single, compound. Ovary 1-celled with 2–6, usually 3, parietal placentae. Ovules numerous. Fruit indehiscent, maturing under water.

Staminate flowers with 9 stamens; plants with floating stems and whorled leaves. 1. Anacharis.
Staminate flowers with 1–3 stamens, usually 2; plants with root-stocks and basal, spirally clustered, band-like leaves. 2. Vallisneria.

1. ANACHARIS* Rich. Water-weed.

Submerged water plants with elongated branched stems, often rooting at the nodes, dioecious or polygamo-dioecious. Leaves opposite or whorled, sessile, pellucid, 1-nerved. Spathe 2-cleft at the apex, that of the staminate plant oval or obovate, sessile, stipitate, in the pistillate plant lanceolate and sessile. Sepals and petals 3 or the latter lacking. Stamens in the staminate flowers usually 9, in two series, in the inner series 3; in the hermaphrodite usually only 3. Fruit linear or lance-linear. *Elodea* Michx.; not Juss. *Philotria* Raf.

* In our treatment of this genus, we have adopted essentially the classification and nomenclature of Frère Marie-Victorin in his paper on "*L' Anacharis canadensis:* Histoire et solution d'un imbroglio taxonomique" (Contributions du Laboratoire de Botanique de l'Université de Montréal, No. 18, 1931.)

Leaves oblong or ovate-oblong, obtuse. 1. *A. canadensis.*
Leaves linear, acute.
Staminate flowers sessile, breaking off within the spathe; leaves 1 mm. wide or less. 2. *A. occidentalis.*
Staminate flowers on elongating pedicels, carrying them to the surface of the water; leaves 1–4 mm. wide. 3. *A. Planchonii.*

1. **A. canadénsis** (Michx.) Polygamo-dioecious plants; leaves oblong or ovate-oblong, obtuse, 5–10 mm. long, 2–4 mm. wide; hypanthium of the pistillate or hermaphrodite flowers 5–10 cm. long; sepals and petals elliptic, obtuse, about 1.5 mm. long; stamens usually 3, fertile, sterile, or wanting. *Elodea canadensis* Michx. *Philotria canadensis* Britton. [G]. Slow streams and ponds: Que.—Va.—Minn.—Neb. *Canad.—Allegh.* My–Au.

2. **A. occidentàlis** (Pursh) Victorin. Dioecious water plant; leaves linear, 5–8 mm. long, 1 mm. wide or less, acute; staminate spathe 5 mm. long, ovoid, sessile; sepals and petals oval, 1–1.5 mm. long; stamens 9; pistillate spathe 1 cm. long; hypanthium 2–5 cm. long; sepals and petals elliptic, 1–1.5 mm. long. Ponds and slow streams. *Philotria minor* (Engelm.) Small: Ohio—Ky.—Ark.—Kans. —N.D. *Prairie.*

f. 40.

3. **A. Planchónii** (Casp.) Rydb. *Fig. 40.* Dioecious water plant; leaves lanceolate to oblong-linear, 8–14 mm. long, 1–3.5 mm. wide; spathe of the staminate flowers obovate, contracted to a narrow base, sepals oval, 4 mm. long; petals linear-lanceolate, acuminate; spathe of the pistillate flowers linear-cylindric, 1–1.5 cm. long; hypanthium 3–15 cm. long; sepals oval, 2 mm. long; petals obovate; staminodia 3. *Elodea ioensis* Wylie. *Anacharis canadensis Planchonii* (Casp.) Victorin. Lakes: Iowa—Colo.—N.Y.—Minn. *Canad.—Allegh.* Jl–S.

2. VALLISNÈRIA (Michx.) L. Tape Grass, Eel Grass.

Aquatic, dioecious, submersed plants. Leaves basal, ribbon-like, floating. Staminate flowers numerous, crowded on a spadix, in a 2–3-parted spathe, at maturity detached and floating on the water surface. Pistillate flowers solitary, at the ends of slender peduncles, subtended by a tubular spathe. Hypanthium linear-cylindric. Petals 3. Stigmas 3, 2-cleft. Fruit cylindric, indehiscent.

1. **V. americàna** Michx. *Fig. 39.* Leaves 1–5 dm. long, 3–8 mm. wide; peduncles of the pistillate flowers 3–6 dm. long, in fruit spirally twisted; hypanthium in flower 1–2 cm. long, in fruit 5–12 cm. long; sepals elliptic, 2–3 mm. long; petals minute; stigmas sessile, cleft less than half their length. *V. spiralis* Am. auth. [G,B]. Water: N.S.—Va.—Ind.—S.D. *E. Temp.* Jl–S.

f. 39.

Family 11. HYDROCHARITACEAE. Frog's-bit Family.

Water- or mud-plants. Leaves clustered. Flowers monoecious or dioecious, subtended by spathes of distinct or united bracts. Perianth regular, of 3 sepals and 3 petals, superior. Stamens 6–12. Gynoecium of 6–15 united carpels; ovary inferior; styles as many as the cavities of the ovary; ovules numerous. Fruit indehiscent.

1. **LIMNÒBIUM** Rich. FROG'S-BIT.

Floating aquatics, mostly stoloniferous. Leaves clustered, with long petioles and broad several-nerved blades. Flowers monoecious; spathe of 2 membranous bracts. Perianth white; petals narrower than the sepals. Staminate flowers 2–4 in a spathe; stamens 6–12; filaments monadelphous. Pistillate flowers solitary, with 3–6 abortive stamens; ovary 6–9-celled, with central placentae; stigmas 6–9, 2-parted. Fruit a several-celled berry.

1. **L. Spóngia** (Bosc) Rich. *Fig. 41.* Leaf-blades ovate to orbicular or reniform, 2–5 cm. broad, obtuse at the apex, truncate or cordate at the base, with petioles 5–25 cm. long; staminate flowers on long scapes, 8–10 cm. long; pistillate scape shorter, recurved in fruit; berry 1–1.5 cm. long. Shallow water and mud: Ont.—Fla.—La.—Kans. *E. Temp.* My–Au.

Family 12. **POACEAE.** GRASS FAMILY.

Annual or perennial herbs, or in warmer climates sometimes trees or vines. Stems (culms) usually hollow except at the nodes. Leaves sheathing at the base; the sheaths usually split on the side opposite the blades. Inflorescence spicate, racemose or paniculate, consisting of spikelets, composed of usually 2-ranked bracts, called glumes, the lower 1–4, usually 2, empty, the remaining (lemma) enclosing a bract-like organ (the palet), and inside this a flower, usually consisting of 3 stamens and 1 pistil. Ovary 1-celled, 1-seeded. Styles 1–3, usually 2. Fruit a seed-like grain (caryopsis), in some exotic species nut-like or berry-like.

Spikelets falling from the pedicels entire, naked or enclosed in bristles or bur-like involucres, 1-flowered, or if 2-flowered the lower flower staminate; no upper empty glumes; rachilla not extending above the upper glume.
  Inflorescence monoecious, the staminate and pistillate flowers in different spikes or on different parts of the same spike, the staminate ones above the pistillate. Tribe 1. MAYDEAE.
  Inflorescence not monoecious; flowers all perfect, or perfect and staminate or neutral mixed.
    Spikelets round or somewhat compressed dorsally; empty glumes manifest; hilum punctiform.
      Lemma and palet hyaline, thin, much more delicate in texture than the empty glumes.
        Spikelets in pairs (or in *Sorghastrum* often in 3's), one sessile and the other pedicellate. Tribe 2. ANDROPOGONEAE.
        Spikelets not in pairs (*Alopecurus, Polypogon, Cinna,* etc.) Tribe 6. AGROSTIDEAE.
      Lemma, at least that of the perfect flower, similar in texture to the empty glumes, or thicker and firmer, never hyaline and thin. Tribe 3. PANICEAE.
    Spikelets much compressed laterally; empty glumes none or rudimentary; hilum linear. Tribe 4. ORYZEAE.
Spikelets with the empty glumes persistent, the rachilla articulated above them, 1–many-flowered; upper lemmas frequently empty; rachilla often produced beyond the upper lemma.
  Spikelets borne in an open or spike-like panicle or raceme, usually upon distinct pedicels.
    Spikelets 1-flowered.
      Empty glumes 4; palet 1-nerved. Tribe 5. PHALARIDEAE.

Empty glumes 2, rarely 1; palet 2-nerved (except in *Cinna.*)
Tribe 6. AGROSTIDEAE.
Spikelets 2–many-flowered.
Lemma usually shorter than the empty glumes; the awn dorsal and usually bent. Tribe 7. AVENEAE.
Lemma usually longer than the empty glumes; the awn terminal and straight (rarely dorsal in *Bromus*) or none.
Tribe 9. FESTUCEAE.
Spikelets in two rows, sessile or nearly so.
Spikelets on one side of the continuous axis, forming one-sided spikes.
Tribe 8. CHLORIDEAE.
Spikelets alternately on opposite sides of the axis, which is often articulated.
Tribe 10. HORDEAE.

TRIBE 1. MAYDEAE.

Staminate and pistillate flowers in different parts on the same spike, the staminate above, the pistillate portion of the spike breaking up into several 1-fruited joints. 1. TRIPSACUM.
Staminate flowers in terminal paniculate spikes, the pistillate ones in thick axillary solitary spikes. 2. ZEA.

TRIBE 2. ANDROPOGONEAE.

Rachis-internodes and pedicels sulcate, the median portion translucent, the margins thickened. 3. BOTHRIOCHLOA.
Rachis-internodes not sulcate.
Some or all of the racemes sessile. 4. ANDROPOGON.
All of the racemes more or less peduncled.
Pedicellate spikelets wanting. 5. SORGHASTRUM.
Pedicellate spikelets present and usually staminate. 6. SORGHUM.

TRIBE 3. PANICEAE.

Spikelets naked, not involucrate.
Glumes 2.
Rachis produced beyond the upper spikelet; spikelets narrow.
43. SPARTINA.
Rachis not so produced; spikelets globose or obovoid.
First glume and rachilla joint forming a swollen ring-like callus below the spikelet. 12. ERIOCHLOA.
First glume wanting or present but not forming a callus below the spikelet.
Spikelets obovoid, turgid. 44. BECKMANNIA.
Spikelets plano-convex. 7. PASPALUM.
Glumes 3.
Glumes not awned.
Spikelets in very slender 1-sided racemes, which are usually whorled or approximate. 8. SYNTHERISMA.
Spikelets in panicles or panicled racemes.
Fruit not rigid, papillose; lemma with a more or less hyaline margin, not inrolled. 9. LEPTOLOMA.
Fruit rigid; lemma not hyaline-margined. 10. PANICUM.
Glumes awned or awn-pointed. 11. ECHINOCHLOA.
Spikelets involucrate.
Involucre of bristles. 13. SETARIA.
Involucre of two spine-bearing valves. 14. CENCHRUS.

TRIBE 4. ORYZEAE.

Spikelets perfect; glumes wanting or rarely rudimentary. 15. LEERSIA.
Spikelets unisexual; pistillate spikelets above the staminate ones; glumes present.
16. ZIZANIA.

TRIBE 5. PHALARIDEAE.

Third and fourth glume enclosing staminate flowers. 17. HIEROCHLOË.
Third and fourth glumes empty.
Lateral florets reduced to small awnless scales; spikelets compressed.
18. PHALARIS.
Lateral florets consisting of an awned hairy lemma; spikelets terete.
19. ANTHOXANTHUM.

TRIBE 6. AGROSTIDEAE.

Lemma indurate when mature and very closely embracing the grain, or at least firmer than the glumes.
Spikelets with no basal callus; lemmas awnless, with inrolled margins.
24. MILIUM.

Spikelets with a basal callus; lemmas usually awned, the margins flat.
Lemma 3-awned. 20. ARISTIDA.
Lemma 1-awned or awnless.
Awn twisted and bent. 21. STIPA.
Awn not twisted.
Lemma broad; awn deciduous.
Inflorescence paniculate or racemiform. 22. ORYZOPSIS.
Inflorescence dichotomous. 23. ERIOCOMA.
Lemma narrow, awned from the tip or mucronate, closely enfolding the grain.
Rachilla not prolonged behind the palet. 25. MUHLENBERGIA.
Rachilla prolonged into a bristle behind the palet. 26. BRACHYELYTRUM.

Lemma usually hyaline or membranaceous at maturity, at least more delicate than the glumes.
Glumes compressed-carinate; stigma subplumose (*i.e.*, with short hairs all around), projecting from the apex of the nearly closed glumes; inflorescence dense, spikelike.
Rachilla of the spikelets articulated above the empty glumes, which are therefore persistent. 28. PHLEUM.
Rachilla of the spikelets articulated below the empty glumes, hence the spikelets falling off entire. 29. ALOPECURUS.
Glumes not compressed-carinate; stigma plumose, projecting from the sides of the spikelets; inflorescence an open or spike-like panicle.
Grain not permanently enclosed in the lemma and palet; pericarp opening readily at maturity. 27. SPOROBOLUS.
Grain permanently enclosed in the lemma and the palet; pericarp adherent.
Spikelets readily falling off when mature. 30. POLYPOGON.
Spikelets with the empty glumes at least persistent.
Palet 1-nerved and 1-keeled; stamen 1. 31. CINNA.
Palet 2-nerved and 2-keeled or sometimes wanting; stamens 3.
Lemma naked at the base.
Lemma awned at the tip or mucronate, closely enfolding the grain. 25. MUHLENBERGIA.
Lemma awnless or with a dorsal awn, loosely enclosing the grain.
Glumes shorter than the lemmas; spikelets larger. 32. ARCTAGROSTIS.
Glumes longer than the lemmas; spikelets small. 33. AGROSTIS.
Lemma with long hairs at the base.
Flowering glume and palet thin-membranous; rachilla prolonged behind the palet, bristle-like. 34. CALAMAGROSTIS.
Flowering glume and palet chartaceous; rachilla not prolonged behind the palet. 35. CALAMOVILFA.

TRIBE 7. AVENEAE.

Spikelets deciduous; lower flower perfect, the upper staminate, awned. 36. HOLCUS.
Spikelets as a whole not deciduous; the glumes persistent.
Awn of the lemma inserted dorsally below the teeth.
Flowers all perfect or the upper staminate.
Grain free from the palea.
Lemma with a bearded callus; spikelets more than 1 cm. long. 42. SCHIZACHNE.
Lemma without callus; spikelets less than 1 cm. long.
Lemma erose-toothed or shortly 2-lobed at the apex. 37. DESCHAMPSIA.
Lemma 2-cleft or deeply 2-toothed at the apex; teeth awn-pointed. 38. TRISETUM.
Grain furrowed, adherent to the glumes; spikelets exceeding 1 cm. in length.
Ovary not crowned by a villous appendage. 39. AVENA.
Ovary crowned by a villous appendage (awned species of) 80. BROMUS.
Upper flowers perfect, the lower staminate. 40. ARRHENATHERUM.
Awn of the lemma inserted between the teeth. 41. DANTHONIA.

TRIBE 8. CHLORIDEAE.

Plants with perfect flowers.
Spikelets with 1 (rarely 2) perfect flowers.
Spikelets deciduous as a whole; rachis articulated below the empty glumes.
Rachis produced above the upper spikelet; spikelets narrow. 43. SPARTINA.

Rachis not produced above the upper spikelet; spikelets globose. 44. BECKMANNIA.
Spikelets with at least the empty glumes persistent.
Glumes above the perfect flower none; spikes very slender.
Spikes paniculate, very slender; plant cespitose. 45. SCHEDONNARDUS.
Spikes digitate; plants extensively creeping. 47. CYNODON.
Glumes above the fertile flower one or several.
Spikelets remote. 48. GYMNOPOGON.
Spikelets imbricate.
Spikes subverticillate or digitate; fertile lemma 1-awned or awnless. 46. CHLORIS.
Spikes racemose; fertile lemma 3-awned. 49. BOUTELOUA.
Spikelets with 2–3 perfect flowers.
Spikes few, digitate. 50. ELEUSINE.
Spikes very slender, racemose, alternate.
Spikelets small, numerous and crowded, glumes thin. 51. LEPTOCHLOA.
Spikelets long, few and distant; glumes thick. 52. ACAMPTOCLADOS.
Plants dioecious; spikelets of the two sexes very unlike. 53. BUCHLOË.

TRIBE 9. FESTUCEAE

Lemmas many cleft into awn-like divisions. 54. PAPPOPHORUM.
Lemmas entire or at most 3-lobed.
Hairs on the rachilla or the lemma very long and enclosing the latter. 55. PHRAGMITES.
Hairs, if any, on the rachilla and the lemma shorter than the latter.
Stigmas barbellate on elongated styles; spikelets in threes in the axils of spinescent leaves. 56. MUNROA.
Stigmas plumose, sessile or on short styles.
Lemma 1–3-nerved.
Lemmas coriaceous in fruit; seeds beaked and exserted. 72. DIARRHENA.
Lemmas not coriaceous in fruit; seeds beakless.
Lateral nerves of the lemmas hairy.
Lemma deeply 3-lobed; internodes of the rachilla long. 57. TRIPLASIS.
Lemma entire or slightly 2-lobed; internodes of the rachilla short.
Inflorescence a short congested raceme; leaf-blades with thick cartilaginous margins. 58. ERIONEURON.
Inflorescence a panicle; leaf-blades not with cartilaginous margins.
Panicle simple or compound, the spikelets on pedicels of varying length. 59. TRIDENS.
Panicle composed of long branches, along which the appressed spikelets are arranged on short pedicels. 60. DIPLACHNE.
Lateral nerves of the lemmas glabrous.
Callus of the lemma copiously pubescent with long hairs; panicle open.
Panicle contracted; lemma broadly oval, rounded at the apex. 61. RHOMBOLYTRUM.
Panicle diffuse; lemma lanceolate and acute. 62. REDFIELDIA.
Callus of the lemma glabrous.
Second glume similar to the first one or nearly so.
Panicle narrow, dense, and spike-like, shining; its branches erect. 63. KOELERIA.
Panicle open; its branches spreading.
Rachilla continuous (except in *E. cilianensis*); lemma deciduous; palet persistent; plants of dry soil. 64. ERAGROSTIS.
Rachilla articulated; lemma and palet both deciduous with the rachilla-internodes; water plants with 2-flowered spikelets. 65. CATABROSA.
Second glume very unlike the first one, broad at the summit. 66. SPHENOPHOLIS.
Lemma 5–many-nerved.
Spikelets with two or more of the upper glumes empty, broad and enfolding each other. 67. MELICA.
Spikelets with upper glumes flower-bearing or narrow and abortive.
Stigmas arising at or near the apex of the ovary.

Spikelets borne in one-sided fascicles which are arranged in a glomerate or interrupted panicle; lemma herbaceous. 68. DACTYLIS.
Spikelets borne in panicles or racemes.
Glumes more or less compressed and keeled.
Spikelets much flattened, the lowest 1–4 flowers sterile. 71. UNIOLA.
Spikelets not much flattened with no sterile flowers at the base.
Spikelets cordate, large. 69. BRIZA.
Spikelets not cordate.
Plants dioecious; lemma of the pistillate spikelets coriaceous; palet strongly 2-keeled and serrate on the margin. 70. DISTICHLIS.
Plants with perfect flowers or in some species of *Poa* dioecious; spikelets all alike; lemma thin; palet ciliate or smooth on the margin. 73. POA.
Glumes rounded on the back, at least below.
Lemma with a basal ring of hairs, prominently 7-nerved, toothed at the apex. 74. FLUMINIA.
Lemma naked at the base.
Lemma obtuse or acutish and scarious at apex, usually toothed.
Lemma distinctly 5–7-nerved; style present. 75. GLYCERIA.
Lemma obscurely 5-nerved; style none. 76. PUCCINELLIA.
Lemma acute, pointed or 2-toothed at the apex, often awned.
Lemmas notched or 2-toothed at the apex, bearing an awn just below the notch.
Bearded callus at the base of the lemma absent. 77. BROMELICA.
Bearded callus at the base of the lemma present. 42. SCHIZACHNE.
Lemmas entire at the apex, if awned the awn terminal.
Stigmas bilaterally plumose; flowers hermaphrodite. 78. FESTUCA.
Stigmas subplumose, the branches arising on all sides; plant dioecious. 79. HESPEROCHLOA.
Stigmas plainly arising below the apex of the ovary which is tipped by a hairy cushion. 80. BROMUS.

TRIBE 10. HORDEAE.

Spikelets usually single at the nodes of the rachis.
Glumes broad, with their sides turned to the rachis.
Glumes broad, several-toothed or several-awned.
Spikelets terete, not sunken in the rachis. 81. TRITICUM.
Spikelets plano-convex, sunken in the flexuose rachis. 82. AEGILOPS.
Glumes not toothed, 1-awned or awnless.
Perennials; spikelets several-flowered. 85. AGROPYRON.
Annuals or biennials; spikelets 2-flowered. 83. SECALE.
Glumes with their back turned to the rachis. 84. LOLIUM.
Spikelets 2–6 at each node of the rachis, or if solitary the empty glumes arranged obliquely to the rachis.
Spikelets 1-flowered or with 1 or 2 rudimentary secondary flowers. 86. HORDEUM.
Spikelets 2–many-flowered.
Rachis of spikes articulated, readily breaking up into joints. 87. SITANION.
Rachis of spikes continuous, not breaking up into joints.
Empty glumes well developed. 88. ELYMUS.
Empty glumes wanting or reduced to short bristles. 89. HYSTRIX.

1. TRÍPSACUM L. GAMA GRASS. SESAME GRASS.

Tall perennial grasses, with rootstock and spikate inflorescence. Leaves with broad blades. Spike articulate, elongate, the upper portion bearing staminate, the lower pistillate flowers. Staminate spikelets 2-flowered, in pairs at each node of the spike; empty glumes 2, coriaceous and shining; lemma thinner, enclosing the hyaline palet. Pistillate flowers solitary at the nodes, 1-flowered, sunk into an excavation in the internode of the rachis; glumes 3, the outer very hard, thick, and shining, concealing the rest; stigmas 2, exserted. Grain enclosed in the rachis and the appressed outer glume.

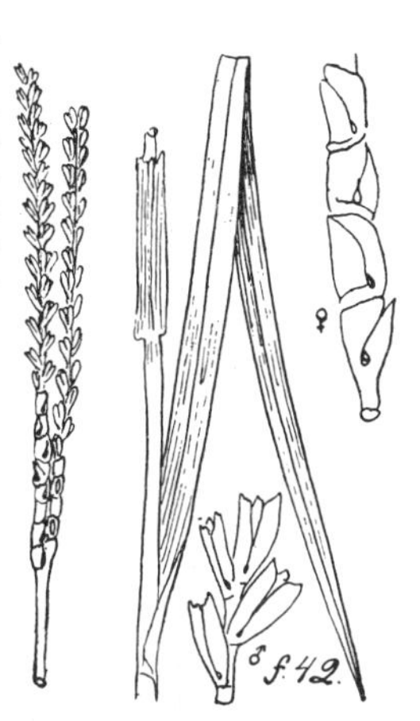

1. **T. dactyloìdes** L. *Fig. 42.* Glabrous or pubescent at the summit of the sheaths and on the upper surface of the leaf-blades; stem 1–2.5 m. high; leaf-blades 4–6 dm. long, 1–4 cm. wide; spikes 1–4 at the end of the stem, 1–3 dm. long, the margins of the excavations in the rachis enclosing the pistillate flowers ciliate, long-hairy below the spikelet; spikelets 7–10 mm. long, the staminate ones oblong, the pistillate ones rounded-ovate. Swamps and along streams: Conn.—Neb.—Fla.—Tex.—Mex. *Allegh.—Ozark.* My–O.

2. **ZÈA** L. INDIAN CORN, MAIZE.

Tall annual grasses. Leaf-blades broad. Staminate spikelets in a terminal panicle (tassel), composed of several spike-like branches, the spikes being arranged in pairs, one sessile, the other short-stalked, each 2-flowered; glumes 2, firm; lemma and palet shorter, thin. Pistillate spikelets in thick spikes (ears), arranged in several rows on the thick rachis (cob); each being 2-flowered; glumes 2; first flower sterile, with a minute palet; second flower pistillate, with developed lemma and palet; styles exceedingly long (silk).

1. **Z. Màys** L. Stems simple or branched, stout, 1–5 m. high; leaves broad, channeled, 2–5 cm. wide; ligule short, hyaline; panicle 5–10 cm. long; ears 1–3 dm. long, 4–8 cm. thick. Rarely escaped from cultivation, throughout the United States. Jl–S.

3. **BOTHRIÓCHLOA** Kuntze. BEARD-GRASS.

Perennial grasses, resembling *Andropogon*, with showy, often silvery panicles. Racemes numerous, the internodes with manifestly thickened margins; the median portion thin and translucent. Sessile spikelet of 3 glumes, the outer 2 indurate, the first 2-keeled, the second 1-keeled, the third and the lemma hyaline, the latter very narrow, gradually merging into the awn; the pedicellate spikelet staminate and similar, or sterile and reduced. Stigmas plumose. *Amphilophis* Nash.

1. **B. saccharoìdes** (Sw.) Rydb. Tufted perennial, smooth and glaucous; stems 5–10 dm. high, simple or somewhat branched; leaf-blades 1–5 dm. long, 3–7 mm. wide, rough above; racemes 1–4 cm. long, the terminal hairs 1½–2 times as long as the internodes; sessile spikelet 4 mm. long, half longer than the internodes; awn geniculate, more or less twisted, 1–1.5 cm. long; pedicellate spikelet consisting of a single glume 2–3 mm. long. *Andropogon glaucus* Torr. *A. Torreyanus* Steud. Dry soil: Tex.—Mo.—Colo.—Ariz.; Mex. *Texan—Son.* Je–Au.

4. **ANDROPÒGON** (Royen) L. BLUE-STEM, BEARD-GRASS, BROOM-GRASS.

Perennial or annual grasses, tufted or from elongated rootstocks, with spike-like racemes, disposed singly in pairs, or sometimes in 3's or more, terminating the stem or its branches. Spikelets sometimes with a ring of short hairs at the base, in pairs at each node of the articulated and frequently hairy rachis, one sessile, the other pedicellate. Sessile spikelet of 3 glumes, the outer 2 indurate, often pubescent, the third usually hyaline, the lemma entire or 2-toothed at the apex, bearing a straight, contorted, or spiral awn, or sometimes awnless; palet small, hyaline. Pedicellate spikelet usually sterile, of 1 or 2 glumes, sometimes of four glumes and enclosing a staminate or more rarely a perfect flower, or frequently entirely wanting, the first glume rarely short-awned. Stamens 1–3. Styles distinct; stigmas plumose.

Racemes usually singly; apex of the rachis internodes with a translucent cup-shaped appendage. — 1. *A. scoparius.*
Racemes usually in pairs or more; apex of the rachis not appendaged.
  Pedicellate spikelet sterile, or reduced to a mere pedicel. — 2. *A. virginicus.*
  Pedicellate spikelet staminate.
    Lemma of the sessile spikelets with a long geniculate awn more or less spiral at the base.
      Outer two glumes of the sessile spikelets more or less hispidulous all over; hairs of the rachis-internodes usually 2 mm. long or less, mostly white. — 3. *A. provincialis.*
      Outer two glumes of the sessile spikelets smooth or nearly so, except on the nerves; hairs of the rachis-internodes 3–4 mm. long, usually yellow. — 4. *A. chrysocomus.*
    Lemma of the sessile spikelets awnless or with a short straight untwisted awn.
      Marginal hairs of the pedicels and rachis-internodes copious, stiff. — 5. *A. Hallii.*
      Marginal hairs of the pedicels and rachis-internodes scant, lax, crisp, or almost wanting. — 6. *A. paucipilus.*

1. **A. scopàrius** Michx. Perennial with a strong rootstock; stems tufted, 4–15 dm. high, glabrous, scabrous, and in the western form usually glaucous; leaf-blades 4–6 cm. long, 8 mm. wide or less, scabrous and sometimes hirsute at the base; racemes 3–6 cm. long, with white-hairy internodes and pedicels; sessile spikelet 5–7 mm. long; awn of the lemma geniculate, 8–15 mm. long, twisted at the base; pedicellate spikelet 2–4.5 mm. long, tipped with an awn 1 mm. long or less. *Schizachyrium scoparius* (Michx.) Nash. [BB, R]. Dry sandy soil: N.B.—Fla.—Tex.—N.M.—Alta. *E. Temp.—Plain—Submont.* Jl–S.

2. **A. virgínicus** L. Stem tufted, 5–10 dm. high; leaf-sheaths, at least the lower ones, hirsute on the margins; blades 3–4 dm. long, 2–5 mm. wide, hirsute at the base above; upper leaf-sheath or spathe usually exceeding the racemes; racemes 2–4, usually 2 together; sessile spikelet 3–4 mm. long; awn 10–15 mm. long; the sterile pedicel with long silky hairs. Open ground; Mass. —Fla.—Tex.—Kans. (?)—Mo. *Temp.* S–O.

3. **A. provinciàlis** Lam. *Fig. 43.* Stem 1–2 m. high; sheaths glabrous, or sometimes slightly hairy; leaf-blades 1.5–6 dm. long, 5–12 mm. wide, usually glabrous and smooth beneath, slightly scabrous above and somewhat hairy at the base; racemes in 2's–6's, 5–10 cm. long; sessile spikelet 7–10 mm. long; awn 7–15 mm. long; pedicellate spikelet nearly as large, staminate, awnless. *A. furcatus* Muhl. [G, B]. Meadows: Me.—Fla.—Tex.—N.M. —Mont.—Sask. *E. Temp.—Submont.* Jl–S.

4. **A. chrysócomus** Nash. Stem 7–15 dm. high; sheaths smooth and glabrous; leaf-blades 2–3 dm. long, 7 mm. wide or less, smooth beneath, slightly scabrous above; racemes in 2's–4's, 5–9 cm. long, long-exserted; sessile spikelet about 1 cm. long; outer two glumes hispid on the nerves; awn 10–12 mm. long; pedicellate spikelet awnless. Prairies and plains: Neb.—Tex.—Colo. *Plain—Submont.* Jl–S.

5. **A. Hállii** Hack. Stem robust, 1–2 m. high, more or less glaucous, glabrous; sheaths glabrous and glaucous; leaf-blades 2–3 dm. long, 5–8 mm. wide; spikes in 2's–5's, 5–10 cm. long; hairs of the internodes about 2 mm.

long, yellowish or whitish; sessile spikelet about 8 mm. long; outer glumes glabrous at the base, pubescent towards the apex; awn straight, 4–10 mm. long or sometimes wanting; pedicellate spikelet awnless, usually larger than the sessile one, staminate. Sandy soil: N.D.—Miss.—Mex.—Mont. *Plain—St. Plains.* Jl–S.

6. **A. paucípilus** Nash. Stem 1–1.5 m. tall, stout; sheaths smooth and glabrous; leaf-blades 3 dm. long or less; racemes in 2's or 3's, 5–7 cm. long, exserted; sessile spikelets 9–10 mm. long; outer 2 glumes hispid on the nerves toward the apex, pubescent towards the summit; awn almost none; pedicellate spikelet similar to the sessile one, staminate. Dry soil: Mont.—Neb. *Plain.* Jl.

5. **SORGHÁSTRUM** Nash. INDIAN GRASS.

Perennials with rootstocks. Panicles large, with the principal branches solitary but branching near the base and hence appearing verticillate. Sessile spikelet dorsally compressed; empty glumes 3, the outer two indurate, often hairy, the third hyaline. Lemma hyaline, with a long awn, which is spiral at the base. Pedicellate spikelet wanting or rudimentary, usually represented merely by a hairy pedicel at one side of the sessile spikelet, or at the ends of the branches by two pedicels, one on each side. Styles distinct; stigmas plumose.

1. **S. nùtans** (L.) Nash. *Fig. 44.* Perennial, with a scaly rootstock; stems 1–2.5 m. high; leaf-sheaths usually smooth and glabrous; blades 3–6 dm. long, 5–13 mm. wide, very rough; panicles 2–5 dm. long, loose, the apex usually nodding; spikelets 6–8 mm. long, lanceolate, the 2 outer glumes golden-brown, the first one densely pubescent with long erect hairs; awn geniculate, 1–1.5 cm. long, closely spiral up to the bend, then loosely twisted. *Andropogon nutans* L. *Chrysopogon nutans* A. Gray. Meadows: Ont.—Fla.—Tex.—Ariz.—Sask.; n Mex. *E. Temp.—Plain—Submont—Son.* Jl–S.

6. **SÓRGHUM** Moench. JOHNSON GRASS, BROOM CORN, SUGAR CORN.

Perennials with rootstocks, or annuals. Branches of the large panicle verticillate. Spikelets in pairs or at the ends of the branches in 3's, 1 sessile and pistillate, 1 or 2 pedicelled and staminate or neutral. Empty glumes 3, the outer two indurate, the third hyaline. Lemma hyaline, awned, or awnless. Styles distinct; stigmas plumose. *Holcus* L.

Rachis of the racemes at last readily disarticulating; pedicellate spikelet usually staminate, equaling the sessile one; perennial with a rootstock. 1. *S. halapense.*
Rachis of the racemes continuous and tardily disarticulating; pedicellate spikelet usually neutral, shorter than the sessile one; annual. 2. *S. saccharatum.*

1. **S. halepénse** (L.) Pers. Perennial; stem 5–15 dm. tall; leaf-blades 2–5 dm. long, 0.5–3 cm. wide; panicle 1.5–5 dm. long, oblong to oval; sessile spikelet 4.5–5.5 mm. long, ovoid, the two outer glumes densely appressed-pubescent; awn readily deciduous, 1–1.5 cm. long; pedicellate spikelets 5–7 mm. long, lanceolate, the 2 outer glumes sparingly pubescent. *Holcus halepensis* L. [R]. JOHNSON GRASS. In fields and waste places: Pa.—Fla.—Tex.—Colo.; Calif.; Mex., C. Am. and S. Am.; escaped from cutivation. Je–D.

2. **S. saccharàtum** Moench. Annual; stem 1–1.5 m. high; sheaths glabrous or pubescent; leaf-blades 4–6 dm. long, 1–7 cm. wide, long-acuminate; panicle 1–3 dm. long, the branches erect or ascending; sessile spikelet ovate-lanceolate

to obovate, 4–5 mm. long; first glume more or less pubescent; awn 4–10 mm. long; pedicelled spikelet 2–3 mm. long. *H. Sorghum* L. SUGAR CORN. Waste places; rarely escaped from cultivation.

7. PÁSPALUM L. PASPALUM.

Usually perennials, with rootstocks. Inflorescence of one or more unilateral racemes; spikelets 1-flowered, arranged singly or in pairs alternately in two rows on one side of a flattened and more or less winged rachis. Glumes 2, rarely 3, membranous. Lemma glabrous and shining, convex, with its back turned towards the rachis, at last indurate. Styles distinct, long; stigmas plumose.

Rachis of the spikes narrow, not membranous, less than 1 mm. wide.
  Plant with one to several axillary peduncles from the upper sheath; leaves ciliate on the margins.
    Spikelets glabrous, 2 mm. long.
      Leaf-blades glabrous or puberulent, long-hairy only on the margins and sometimes on the midrib below.
        Spikelet orbicular, as broad as long, straw-colored. — 1. *P. stramineum.*
        Spikelet obovate or oval, longer than broad, green. — 2. *P. ciliatifolium.*
      Leaf-blades densely long-pubescent on both surfaces. — 3. *P. Muhlenbergii.*
    Spikelets pubescent.
      Spikelets 2 mm. long; leaves short-pubescent. — 4. *P. Bushii.*
      Spikelets 1.5 mm. long; leaf-blades densely long-pubescent. — 5. *P. setaceum.*
  Plant simple, without axillary peduncles from the upper sheath.
    Spikelets 2.5–3.5 mm. long. — 6. *P. laeve.*
    Spikelets about 4 mm. long.
      Sheaths usually glabrous. — 7. *P. glabratum.*
      Sheaths densely pubescent. — 8. *P. floridanum.*
Rachis of the spikes dilated, with membranous margins, 2 mm. wide or more. — 9. *P. mucronatum.*

1. P. stramíneum Nash. *Fig. 45.* Stem 2–8 dm. high; sheaths all but the lowest ones glabrous, except the ciliate margins; blades yellowish green, long-ciliate on the margins, 5–25 cm. long, 5–10 mm. wide; racemes 4–10 cm. long, usually in pairs; spikelets in pairs, orbicular, 2 mm. broad; first glume 3-nerved, pubescent; second glume 2-nerved, glabrous or nearly so. *P. setaceum* A. Nels., not Michx. Sandy places: Wis. — Minn. — Colo. — Tex. — Mo. *Plain—Prairie.* Je–O.

2. P. ciliatifòlium Michx. Stem tufted, erect, 4–8 dm. tall, smooth and glabrous; sheaths and leaf-blades glabrous, except the ciliate margins; blades 5–25 cm. long, 6–15 mm. wide; racemes usually single, 5–10 cm. long; spikelets 2 mm. long; first glume 3-nerved, the second 2-nerved. Sandy or rocky soil: D.C.—Kans.—Ala.—Miss.; Bermudas. *Carol.—Ozark.* Je–S.

3. P. Muhlenbérgii Nash. Stem 4–8 dm. high, erect or reclining, glabrous; sheaths pubescent to nearly glabrous; blades 0.5–2 dm. long, 7–10 mm. wide; racemes single or two together, 5–10 cm. long; spikelets 2 mm. long, in pairs, glabrous, on puberulent pedicels, obovate or oval; glumes both 3-nerved. *P. pubescens* Muhl. Fields and sandy places: Mass.—Ga.—Miss.—Kans. *Allegh.—Carol.* Je–O.

4. P. Búshii Nash. Stem erect, 8–10 dm. high; lower sheaths pubescent, the upper ciliate on the margins only; leaf-blades 5–10 cm. long, 5–15 mm. wide; racemes 2 or 3, 10–12 cm. long; spikelets in pairs; glumes both 3-nerved or the second 2-nerved. Dry soil: Mo.—Minn.—Neb.—Tex. *Prairie—Texan.* Au–O.

5. **P. setàceum** Michx. Stem tufted, 2–7 dm. high; basal leaf-sheaths densely pubescent, the upper sheaths ciliate on the margins; leaf-blades 3–15 cm. long, 2.5–5 mm. wide, erect or ascending; racemes single, 4–8 cm. long; spikelets either single or in pairs, on short pubescent pedicels, broadly obovate; first glume 3-nerved, the second usually 2-nerved. Sandy soil: N.Y.—Fla.—Miss.—Kans. *E. Temp.* Jl–O.

6. **P. laève** Michx. Stem tufted, prostrate or spreading, 3–12 dm. long; sheath compressed, glabrous; leaf-blades 5–8 mm. wide, glabrous or pubescent above near the base; racemes spreading; spikelets single, glabrous, oval, 2.5–3.5 mm. long; glumes both 3-nerved. *P. angustifolium* Le Conte. Fields and sandy soil: Md.—Fla.—Tex.—se Kans. *Austral.* Jl–O.

7. **P. glabràtum** (Engelm.) C. Mohr. Stem usually 1–2 m. long; sheaths glabrous or the basal ones sometimes hirsute; leaf-blades glabrous below, glabrous or sparingly hirsute above, 3–7 dm. long, 5–15 mm. wide; racemes usually 3–6, commonly 1–1.5 cm. long; spikelets single or in pairs, 3.5–4 mm. long, glabrous; glumes 3-nerved. *P. arundinaceum* Poir. *P. floridanum glabratum* Engelm. Low ground: Md.—Ga.—Tex.—Kans. *Austral.* My–O.

8. **P. floridànum** Michx. Stem 9–15 dm. high, stout; sheaths densely hirsute; leaf-blades densely hirsute, at least above, 3–6 dm. long, 6–10 mm. wide; racemes 2–4, erect, 8–12 cm. long; spikelets usually single, 3.5–4 mm. long, glabrous, glumes 3-nerved. Low ground: Va.—Fla.—Tex.—se Kans. *Austral.* My–O.

9. **P. mucronàtum** Muhl. Stem 1.5–8 dm. long, compressed; sheaths loose and inflated, papillose-hirsute or glabrous; blades 5–20 cm. long, 5–25 mm. wide, scabrous; racemes 10–50, 1–8 cm. long; rachis extending beyond the spikelets; spikelets singly in 2 rows, 1.2–1.5 mm. long, pubescent. *P. repens* Bergius. *P. fluitans* Kunth. In water: Va.—Fla.—Tex.—se Kans. *Carol.—Texan.* Au–O.

8. SYNTHERÍSMA Walt. CRAB-GRASS.

Annual grasses. Inflorescence composed of spike-like racemes, which are disposed in whorls, or approximate at the summit of the stems. Spikelets narrow, acute, in 2's or 3's on one side of the flat and winged or triangular rachis, one of the spikelets generally longer-pedicelled than the rest. Glumes membranous, the first small or wanting, the lemma chartaceous, glabrous and shining, at length indurated, enclosing a palet. Styles distinct; stigmas plumose. *Digitaria* Scop., not Heist.

Rachis of the racemes with wingless angles; first glume obsolete or wanting. — 1. *S. filiforme.*
Rachis of the racemes with the lateral angles winged, hence appearing flattened.
  Pedicels terete or nearly so, sparingly if at all hispidulous; lower sheaths glabrous; lemma brown in fruit. — 2. *S. Ischaemum.*
  Pedicels sharply 3-angled, the angles strongly hispidulous, as also the sheaths; lemma white in fruit.
    Spikelets about 2.5 mm. long; third glume with the nerves strongly hispid above the middle. — 3. *S. sanguinale.*
    Spikelets 3 mm. long or more; third glume with the nerves smooth or nearly so. — 4. *S. marginatum.*

1. **S. filifórme** (L.) Nash. Stem 1.5–7 dm. high; sheaths papillose or the uppermost glabrous; leaf-blades 3–20 cm. long, 1–4 mm. wide, glabrous beneath, papillose-hirsute above near the base; racemes 2–5, 2–10 cm. long; spikelets about 1.8 mm. long elliptic, acute; glumes appressed-pubescent. *Panicum filiforme* L. *Digitaria filiformis* (L.) Koeler. [G]. Sandy soil: N.H.—N.C.—Tex.—Kans.

2. **S. Ischaèmum** (Schreb.) Nash. Stems 2–5 dm. long, at last prostrate and rooting; leaf-blades 1.5–13 cm. long, 2.5–6 mm. wide, smooth and glabrous

on both sides; racemes 2–5, 2–8 cm. long; spikelets fully 2 mm. long, elliptic, acute; first glume rudimentary or wanting. *Panicum lineare* Krock., not L. *P. glabrum* Gaud. *Syntherisma humifusum* (Pers.) Rydb. *S. lineare* Nash [B], *D. humifusa* Pers. [G]. Waste places, fields and roadsides: N.S.—Fla.—Colo.—Mont; introduced from Eur. *E. Temp.—Plain—Mont.* Jl–S.

3. **S. sanguinàle** (L.) Dulac. Stem 3–10 dm. high, prostrate at the base and rooting; leaf-blades 4–20 cm. long, 4–10 mm. wide, more or less papillose-hirsute on both sides; racemes 3–10, linear, 5–18 cm. long; spikelets elliptic-lanceolate, acute; first glume minute, triangular; lemma apiculate, elliptic-lanceolate. *P. sanguinale* L. *D. sanguinalis* (L). Scop. [G]. CRAB-GRASS. Cultivated ground and waste places: Mass.—Fla.—Calif.—Wash; Mex.; naturalized from the Old World. Je–S.

4. **S. marginàtum** (Link) Nash. *Fig. 46.* Stem 5–8 dm. long, prostrate at the base and rooting at the nodes; sheaths densely papillose-hirsute; blades 5–8 cm. long, 3–10 mm. wide, flat, glabrous or pubescent; racemes 2–9, variously disposed, 2–12 cm. long; spikelets 3–4 mm. long, lance-oblong; first glume minute, triangular, glabrous; second and third glumes long-pubescent; lemma elliptic. Waste places: Md.—Kans.—Utah—Fla.; Mex.; W. Ind., C. Am., and S. Am.

9. LEPTOLÒMA Chase. FALL WITCH-GRASS.

Tufted perennials, with very diffuse panicles, which fall off and become tumble-weeds. Spikelet 1-flowered, solitary, on a long, 3-angled pedicel; first glume obsolete; second glume 3-nerved, nearly as long as the 5–7-nerved third glume; lemma cartilaginous, indurate, papillose, with a delicate hyaline margin. Grain free.

1. **L. cognàtum** (Schultes) Chase. Stem geniculate below, very brittle, 3–7 dm. high, much branched below; leaf-blades 5–8 cm. long, 4–6 mm. wide, scabrous on the margins; spikelets 2.7–3 mm. long, acuminate; second and third glumes with silky pubescence between the veins, spreading-ciliate on the margins. *Panicum cognatum* Schultes. *P. autumnale* Bosc. Dry soil and sand hills: N.H.—Fla.—Ariz.—Minn. *E. Temp.* Jl–S.

10. PÁNICUM L. PANIC-GRASS, WITCH-GRASS, TUMBLE-GRASS, TICKLE-GRASS.

Annuals or perennials. Spikelets in open or contracted panicles, 1–2-flowered, lanceolate, oblong, ovate to obovate or globose, obtuse to acute or acuminate. Glumes 3, membranous, the first and second empty, the first usually much shorter than the spikelet, sometimes minute, the third one empty or enclosing a shorter hyaline palet and often also a staminate flower, lemma shorter and usually more obtuse than the glumes, chartaceous, glabrous and shining, at length indurated, enclosing a palet. Styles distinct; stigmas plumose.

Inflorescence truly paniculate.
- Annuals.
  - Plant more or less hairy. — 1. CAPILLARIA.
  - Plant glabrous. — 2. DICHOTOMIFLORA.
- Perennials.
  - Basal leaves similar to the stem-leaves.
    - Basal leaf-sheaths terete, neither keeled nor flattened; spikelets long-pedicelled; rootstock long, scaly. — 3. VIRGATA.
    - Basal leaf-sheaths compressed, keeled; spikelets short-pedicelled along the main branches of the panicle. — 4. AGROSTOIDIA.
  - Basal leaves broader than the stem-leaves, forming a winter rosette.

Leaves linear, elongate, narrowed at both ends, not more than 5 mm. wide; secondary panicles at the base only. 5. DEPAUPERATA.
Leaves oblong-lanceolate, not very elongate.
Spikelets less than 3 mm. long. 6. LANUGINOSA.
Spikelets more than 3 mm. long.
Leaf-blades not more than 1.5 cm. wide. 7. OLIGOSANTHA.
Leaf-blades more than 1.5 cm. wide. 8. LATIFOLIA.
Inflorescence with racemiform branches. 9. OBTUSA.

1. CAPILLARIA

Spikelets 3 mm. long or less, lanceolate or elliptic.
Stem stout; panicle usually included at the base, many-branched, with numerous spikelets.
Branches of the panicle widely spreading, the well-developed pulvinus in the axil long-hairy; spikelets lanceolate, acuminate. 1. *P. barbipulvinatum.*
Branches of the panicle ascending, rarely spreading, the pulvinus glabrous or slightly pubescent; spikelets ovate or elliptic, acute. 2. *P. capillare.*
Stem slender; panicle exserted, few-branched, with comparatively few spikelets.
Spikelets lanceolate, about 0.6 mm. wide; panicle not more than one third the length of the whole plant. 3. *P. Gattingeri.*
Spikelets elliptic, about 0.8 mm. wide; panicle narrow, more than half the length of the whole plant. 4. *P. flexile.*
Spikelets 5 mm. long, ovate. 5. *P. miliaceum.*

2. DICHOTOMIFLORA

One species. 6. *P. dichotomiflorum.*

3. VIRGATA

One species. 7. *P. virgatum.*

4. AGROSTOIDIA

Rootstock present; innovations extravaginal. 8. *P. anceps.*
Rootstock absent; innovations intravaginal; plant tufted. 9. *P. agrostoides.*

5. DEPAUPERATA

Spikelets more than 3 mm. long, pointed. 10. *P. depauperatum.*
Spikelets 3 mm. long or less.
Sheaths pilose.
Spikelets blunt; branches of the panicle ascending; stems few in the tuft. 11. *P. perlongum.*
Spikelets subacute; branches of the panicle spreading; stems many in the tuft. 12. *P. linearifolium.*
Sheaths glabrous. 13. *P. Werneri.*

6. LANUGINOSA

Sheaths all or all but the lowest one glabrous.
Ligules 2–5 mm. long; spikelets 1.5 mm. long. 14. *P. Lindheimeri.*
Ligules less than 1 mm. long.
Spikelets 1.5–1.8 mm. long, rounded-oval. 15. *P. sphaerocarpum.*
Spikelets 2.2–2.8 mm. long, obovoid. 16. *P. boreale.*
Sheaths pubescent.
Pubescence spreading.
Spikelets 2.7–3 mm. long; leaf-blades papillose-hirsute on both sides. 17. *P. Wilcoxianum.*
Spikelets less than 2.6 mm. long.
Spikelets less than 2 mm. long.
Ligules 2–5 mm. long, densely hairy.
Leaf-blades stiff, glabrous above or with a few hairs. 18. *P. tennesseense.*
Leaf-blades pubescent above.
Autumnal form widely decumbent, forming mats; plant olivaceous. 19. *P. albemarlense.*
Autumnal form erect or ascending, not forming mats, plant yellowish-green.
Upper surface of the leaves long-hairy.
Stem branching very early; spikelets 1.8–1.9 mm. long. 20. *P. praecocius.*
Stem branching after the maturing of the first panicle. 21. *P. implicatum.*

Upper surface of the leaves short-hairy.	22. *P. Huachucae.*
Ligules 1–1.5 mm. long.	23. *P. tsugetorum.*
Spikelets 2–2.6 mm. long.	
Plant very villous or velutinous.	
Plant villous; stem 2.5–4.5 dm. high; ligules 4–5 mm. long.	24. *P. villosissimum.*
Plant velutinous, stem 8–13 dm. high; ligules 1 mm. long.	33. *P. scoparium.*
Plant papillose-hispid on the sheaths, sparingly hispid on the blades; autumnal state erect.	25. *P. scoparioides.*
Pubescence not spreading.	26. *P. subvillosum.*

7. Oligosantha

Panicle narrow; leaf-blades erect.	
Plant glabrous or nearly so.	27. *P. xanthophysum.*
Plant papillose-hispid.	
Spikelets slightly if at all more than 3 mm. long.	17. *P. Wilcoxianum.*
Spikelets about 4 mm. long.	28. *P. Leibergii.*
Panicle spreading, about as long as broad.	29. *P. Scribnerianum.*

8. Latifolia

Spikelets 3 mm. long; sheaths, at least the lower ones, papillose-hispid.	30. *P. clandestinum.*
Spikelets 4 mm. long; sheaths not papillose-hispid.	
Nodes bearded; plant decidedly pubescent.	31. *P. Boscii.*
Nodes not bearded; plant glabrous or nearly so.	
Panicle spreading; blades 2.5 cm. wide or more.	32. *P. latifolium.*
Panicle narrow; blades less than 2 cm. wide.	27. *P. xanthophysum.*

9. Obtusa

One species.	34. *P. obtusum.*

1. **P. barbipulvinàtum** Nash. Annual; stem at length branched and rooting at the nodes; sheaths densely pubescent with spreading hairs; papillate at the base; blades hairy on both sides, 4–13 cm. long, 5–10 mm. wide; panicles ovate, 1–2 dm. long; spikelets 3–3.5 mm. long; first glume less than one half as long as the spikelet, broadly ovate, 3-nerved; second and third glumes pubescent at the apex. Western Witch-grass. Wet sandy soil: Man.—Wisc.—Tex.—Calif.—B.C. *Plain—Submont.* Jl–S.

2. **P. capillàre** L. Erect or decumbent annual; stem 2–8 dm. high, simple or sparingly branched; sheaths papillose-hirsute; leaf-blades 1.5–3 dm. long, 5–15 mm. wide, pubescent; terminal panicle 2–3.5 dm. long; lateral ones when present smaller; spikelets 2–2.5 mm. long, somewhat acuminate or acute; first glume one fourth to one half as long as the spikelet, 5–7-nerved; second and third glabrous. Witch-grass. Dry or sandy soil and waste places: N.S.—Fla.—Tex.—Nev.—Wash.—B.C. *Temp.* Je–S.

3. **P. Gattíngeri** Nash. Stem widely spreading or decumbent, sometimes 1 m. long, branched at all nodes; panicles numerous, oval, less than 15 cm. long; spikelets about 2 mm. long, elliptic, acute, glabrous, more turgid than in the preceeding. Open grounds: Me.—Minn.—Kans.—Ky.—N.C. *Canad.—Allegh.* Au–O.

4. **P. fléxile** (Gatt.) Scribn. Stem slender, erect, 3–6 dm. high, with a few branches at the base; leaf-blades 1–2.5 dm. long, 2–6 mm. wide, erect, sometimes nearly glabrous; spikelets 3–3.5 mm. long; second and third glumes acuminate, one third longer than the fruit. Sandy soil: Pa.—Ont.—S.D.—Tex.—Fla. *E. Temp.* Au–O.

5. **P. miliàceum** L. Stem erect, 2–5 dm. high, erect or decumbent; sheaths papillose-hispid; leaf-blades 1–2.5 dm. long, 2.5 mm. wide or less; panicle dense, drooping at maturity; spikelets ovoid, turgid. Waste places: Me.—Del.—Neb.—S.D.—Ont.; adventive from the Old World or escaped from cultivation. Jl–S.

6. **P. dichotomiflòrum** Michx. Stem spreading or ascending from a decumbent base, compressed, somewhat fleshy, 3–18 dm. long; leaf-blades 2–4 dm. long, 8–15 mm. wide, scabrous above; panicles 1–4 dm. long, diffuse; spikelets secund near the ends of the branches, 3 mm. long, acute; second and third glumes pointed, longer than the fruit. *P. proliferum* Am. auth. Wet soil, fields, and waste places: Me.—Fla.—N.M.—Neb.; Bahamas, Bermudas, Cuba. *Allegh.—Trop.* Jl–S.

7. **P. virgàtum** L. Perennial with a creeping, scaly rootstock; stem 1–2 m. high, glabrous; sheaths smooth; leaf-blades elongated, 3 dm. long or more, 6–12 mm. wide, flat, rough on the margins; panicle 1.5–5 dm. long, spikelets ovate, acuminate, 4–4.5 mm. long; first glume acuminate, about half as long as the spikelet, 3–5-nerved; second glume usually longer than the rest. Meadows and plains: Me.—Fla.—Tex.—Ariz.—Nev.—Sask.; W. Ind. and C. Am. *Plain—E. Temp.* Au–S.

8. **P. ánceps** Michx. Stem erect or ascending, 6–12 dm. high; leaf-blades 1.5–5 dm. long, 6–10 mm. wide; panicle 2–5 dm. long, loose, open; branches remote, spreading; spikelets more or less secund, 3.5 mm. long; second and third glumes acuminate and curved at the apex, longer than the fruit, which is hair-tufted at the apex. *P. rostratum* Muhl. Moist soil: R.I.—Fla.—Tex.—se Kans.—Ills. *Ozark.* Jl–S.

9. **P. agrostoìdes** Spreng. Stem stout, 4–10 dm. high; leaf-blades 2–3.5 dm. long, 6–10 mm. wide; panicle 1.5–3 dm. long, ovoid, often purplish; branches stiff, ascending, naked at the base, secund; spikelets 2 mm. long, acute; pedicels short, with a few long hairs. Wet meadows: Me.—Fla.—Tex. Kans.—Minn. *E. Temp.* Jl–S.

10. **P. depauperàtum** Muhl. Stem erect or ascending, 2–4 dm. high; nodes pubescent with ascending hairs; sheaths mostly shorter than the internodes; leaf-blades 6–15 cm. long, 2–5 mm. wide, often involute; panicles not much exceeding the leaves, few-flowered, with ascending branches; spikelets about 3.5 mm. long, glabrous or sparsely pubescent, strongly nerved; second and third glumes acuminate, longer than the fruit. Dry places and open woods: N.S.—Fla.—Tex.—Kans.—Man. *E. Temp.* Je–S.

11. **P. perlóngum** Nash. Tufted pubescent perennial; stems 2–4 dm. high, simple; sheaths hirsute with long ascending hairs; leaf-blades papillose-hispid beneath, 2–3 mm. wide, the upper usually 8–14 cm. long; panicle much exserted, 4–6 cm. long, its branches nearly erect; spikelets about 3.25 mm. long and 1.5–1.75 mm. wide; first glume ovate, one third as long as the spikelet. Prairies and hills: Ind.—Tex.—Colo.—S.D.—N.D.—Man. *Plain—Prairie—Submont.* My–Jl.

12. **P. linearifòlium** Scribn. Densely tufted perennial; stem 2–4.5 dm. high, erect; sheaths more or less papillose-pilose; blades 1–3.5 dm. long, 2–4 mm. wide, usually exceeding the panicle; panicle exserted, 5–10 cm. long, few-flowered; branches flexuose, spreading; spikelets about 2.5 mm. long, sparsely pilose; first empty glume triangular, one fourth to one third as long as the spikelet. *P. Enslini* Nash, not Trin. Woods: Me.—Ala.—Tex.—Kans.—Minn. *E. Temp.* My–Jl.

13. **P. Wérneri** Scribn. Stem strict, 2.5–4.5 dm. high, in small tufts, glabrous, except a few hairs at the nodes; leaves firm, 1.5 dm. long, 3–6 mm. wide; spikelets 2.2–2.3 mm. long, glabrate or minutely puberulent; first glume orbicular, one fourth as long as the spikelet. Woods and knolls: Me.—Del.—n Tex.—se Kans.—Minn. *Canad.—Allegh.* Je–Jl.

14. **P. Lindheìmeri** Nash. Stem ascending or spreading, 5–10 dm. long, glabrous or pubescent below, elongate and spreading in autumn, with the early branches long, the later ones in short tufts; nodes swollen; sheaths much

shorter than the internodes, ciliate on the margins; leaf-blades 5–8 cm. long, 6–8 mm. wide; panicle 4–7 cm. long, nearly as wide, loosely flowered; spikelets obovate, obtuse. Sandy woods and open places: Me.—Fla.—Tex.—N.M.—Minn; also Calif. *E. Temp.* Je.–S.

15. **P. sphaerocárpum** Ell. Stem 2–5.5 dm. high, usually widely spreading, appressed-pubescent at the nodes, sparingly branched from the lower nodes late in the season; sheaths nearly as long as the internodes, loose above, ciliate on the margins; leaf-blades 6–10 cm. long, 4–14 mm. wide; panicle long-exserted, 5–10 cm. long, nearly as wide; autumnal leaves and panicles not much reduced; spikelets usually purplish; fruit china-white. Sandy ground: Vt.—Mass.—Fla.—Mex.—se Kans. *E. Temp.* Jl.–S.

16. **P. boreàle** Nash. Stem 3–5 dm. high, slender, erect, or geniculate at the base; sheaths often overlapping, ciliate on the margins; leaf-blades 6–12 cm. long, 7–12 mm. wide, erect, ciliate towards the base; panicle 5–10 cm. long, scarcely as broad, loosely flowered; spikelets pubescent, obtuse, sparingly branched from the nodes in late summer; autumnal leaf-blades and panicles not much reduced. Moist ground and open woods: Newf.—Del.—Ind.—Minn. *Canad.—Allegh.* Je–Au.

17. **P. Wilcoxiànum** Vasey. Cespitose perennial, tufted; stems 1–2.5 dm. high, sparingly pilose with long white hairs; sheaths papillose-pilose; leaf-blades 3.5–7.5 cm. long, less than 4 mm. wide, pubescent with long hairs; panicle 3–4 cm. long, ovoid, flexuose; spikelets about 2.5 mm. long, ellipsoid; first glume ovate, about one fourth as long as the spikelet; second and third glumes pubescent. Prairies: Man.—Iowa—Okla.—N.D.—Ore. *Plains—Prairie.* Je–Jl.

18. **P. tennesseénse** Ashe. Perennial, at first spreading; stem 2.5–6 dm. high; sheaths spreading-pubescent; leaf-blades 6–9 cm. long, often ciliate at the base, glabrous or nearly so above, appressed-pubescent beneath; panicle 4–7 cm. long, rather dense; spikelets 1.6–1.7 mm. long, obovate, obtuse, pubescent; first glume about one fourth as long as the spikelet; second glume shorter than the third. Open moist ground: Me.—Ga.—Miss.—Ariz.—Utah—Minn. *E. Temp.* Je–S.

19. **P. albemarlénse** Ashe. Plant olivaceous, vernal stems 2.5–4.5 dm. high, more or less geniculate, grayish-villous; blades 4.5–7 cm. long, grayish-villous beneath, puberulent as well as villous above; panicle 3–5 cm. long; spikelets 1.4 mm. long, pilose; blades of the autumnal state reduced. Meadows: Conn.—N.C.—Minn. *Allegh.*

20. **P. praecòcius** Hitchc. & Chase. Stem very slender, wiry, early branching, 1.5–4 dm. high, soon geniculate and spreading, copiously long-pilose; sheaths shorter than the internodes; ligules 3–4 mm. long; leaf-blades 5–8 cm. long, 4–6 mm. wide; primary panicle 4–6 cm. long, nearly as wide, loosely flowered with spreading or ascending branches; secondary panicles numerous, appearing before the primary one; spikelets less than 2 mm. long, obovate, turgid, long-pilose. Dry open places: Mich.—Tex.—Neb.—Minn. *Prairie—Allegh.* Je–Au.

21. **P. implicàtum** Scribn. Stem slender, 2–5.5 dm. high; sheaths papillose-pilose; ligules 4–5 mm. long; blades 3–6 cm. long, 3–6 mm. wide, pilose with long spreading hairs above, appressed-pubescent beneath; panicles 3–5 cm. long, nearly as wide, the axils long-pilose; spikelets 1.5 mm. long, obovate, obtuse, papillose-pilose; autumnal stage ascending or spreading, with fascicled branches from the nodes. Wet meadows: N.S.—Minn.—Neb.—N.J. *Canad.—Allegh.*

22. **P. Huachùcae** Ashe. Erect or ascending perennial, at length much branched; stems 2–6 dm. high; leaf-blades lanceolate or lance-linear, acuminate, pubescent beneath with short hairs, 4–10 cm. long, 5–12 mm. wide; panicle 5–10 cm. long, usually purplish; branches spreading, few-flowered; spikelets obovate, obtuse, 1.5–1.8 mm. long; first glume small, about one third as long as the spikelet; second and third glumes equal, pubescent with spreading hairs.

P. pubescens A. Gray; not Lam. Meadows: Me.—N.C.—Calif.—Mont. *Temp.* —*Mont.* Je–S.

23. **P. tsugetòrum** Nash. Spreading or ascending perennial; stem often geniculate below, densely appressed-pubescent, with intermixed longer hairs on the lower internodes; sheath less densely pubescent; blades thickish, with cartilaginous margins, 4–7 cm. long, 4–7 mm. wide, acuminate, glabrous or nearly so above, puberulent beneath; panicle 3–7 cm. long; branches flexuose, puberulent; spikelets 1.8–1.9 mm. long; first glume one third as long as the spikelet; second and third glumes equal. Sandy soil: Me.—Va.—Tenn.—Minn. *Canad.* —*Alleg.* Jl.–Au.

24. **P. villosìssimum** Nash. Stems 2.5–4.5 dm. high, erect or ascending, villous with long spreading hairs, in autumn widely spreading, geniculate at the nodes, at last prostrate, with fasciculate branches; sheaths long-hairy; ligules 4–5 mm. long; leaf-blades firm, ascending, 5–10 cm. long, 5–10 mm. wide, pilose on both sides; primary panicle 4–8 cm. long, about as wide, loosely flowered; spikelets oblong-elliptic, obtuse, papillose-pubescent. *P. atlanticum* Nash. Sandy or poor soil and open woods: Mass.—Fla.—Tex.—Minn. Je–Au. *E. Temp.*

25. **P. scoparioìdes** Ashe. Stem erect, papillose-hispid; ligules 2–3 mm. long; leaf-blades firm, 7-10 cm. long, 6–7 mm. wide; panicle densely flowered, often included, 4–7 cm. long; branches ascending or spreading, spikelets obovate, papillose-pubescent, strongly nerved; the autumnal state bears short branches at the middle and upper nodes, with involute-pointed reduced leaf-blades. Dry gravelly soil: Vt.—Conn.—Del.—Kans.—Minn. *Allegh.* Je.

26. **P. subvillòsum** Ashe. Stem slender, 1–3.5 dm. high, leafy at the base, widely spreading; stem and sheaths sparingly pilose with ascending hairs; nodes bearded; ligules 1 mm. long, with a ring of hairs; blades 4–6 cm. long, 4–6 mm. wide, pilose on both surfaces; panicle rather narrow, 3–5 cm. long; spikelets 1.9 mm. long, obtuse; autumnal state spreading and branched from the base, the leaves and panicles much reduced. *P. unciphyllum pilosum* Scribn. & Merr. Dry woods and sandy places: N.S.—Minn.—Ind.—Conn. *Canad.*

27. **P. xanthophỳsum** A. Gray. Stem ascending, 2–6 dm. high, scabrous, in late summer with erect branches from the second and third node; sheaths loose, the lower overlapping, bearded at the summit; ligules minute; leaf-blades erect, strongly nerved, 1–1.5 dm. long, 1–1.8 cm. wide; panicle long-exserted, 5–12 cm. long, narrow, few-flowered, branches erect; spikelets 4 mm. long, usually pubescent; first glume nearly half as long as the spikelet. Dry soil: Que.—Me.—N.J.—Minn.—Man. *Canad.* Je–Au.

28. **P. Leibérgii** (Vasey) Scribn. Stem 3–8 dm. high, scabrous at least below the nodes, in late summer with erect branches from the lower nodes; ligules minute; leaf-blades ascending, 8–10 cm. long, 8–12 mm. wide; panicle, 8–10 cm. long, narrow, with ascending branches; spikelets papillose-hispid; first glume half as long as the spikelet, acuminate. Prairies: N.Y.—Mo.—Kans.—N.D.—Man. *Canad.*—*Allegh.* Je–Jl.

29. **P. Scribneriànum** Nash. *Fig. 47.* Somewhat cespitose perennial; stem erect, 1.5–6 dm. high, simple or later dichotomously branched, sparingly hairy; sheath strongly papillose-hirsute; leaf-blades 5–10 cm. long, 6–12 mm. wide, glabrous and smooth above, scabrous and sparingly hairy beneath; primary panicle ovoid, 3.5–7.5 cm. long, the secondary ones much smaller and more or less included; spikelets obovoid, 3 mm. long; first glume ovate, one

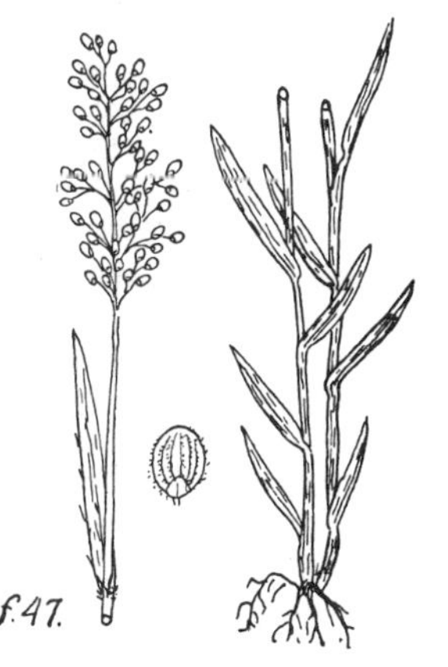
f. 47.

fourth as long as the spikelet; second and third glumes glabrous or finely pubescent. *Panicum scoparium* S. Wats.; not Lam. Meadows: Me.—Va.—Ariz.—Calif.—B.C. *Temp.* My–Jl.

30. **P. clandestìnum** L. Stems several, in very large clumps, 5–12 dm. high; in the autumn with appressed branches having shortened internodes, overlapping sheaths and crowded leaves at the summit; leaf-blades 1–2 dm. long, 1.8–2.5 cm. wide, scabrous towards the ends; panicle exserted, 1–1.5 dm. long, about as wide, densely flowered, with fascicles ascending branches, the secondary panicles of the autumnal state wholly included in the sheaths; spikelets elliptic. Moist ground: Me.—Fla.—Tex.—Minn. *E. Temp.* Je–Au.

31. **P. Bòscii** Poir. Stem erect, 3–6 dm. high, simple or dichotomously branched above; sheaths soft-pubescent; leaf-blades 5–10 cm. long, 15–35 mm. wide, cordate-clasping; panicle often included; branches spreading or ascending; spikelets few, elliptic, appressed; first glume one third to one half as long as the spikelet. *P. Porterianum* Nash. Woods: Me.—Fla.—Tex.—Minn. *E. Temp.* Je–Au.

32. **P. latifòlium** L. Stem 3–9 dm. high, glabrous in autumn, spreading and branched from the middle nodes; leaf-blades often 15 cm. long, 3 cm. wide, ciliate towards the base; panicle 8–15 cm. long, with long, few-flowered, ascending branches; spikelets obovate-elliptic, pubescent, except at the tip. Woods and sand-dunes: Me.—N.C.—se Kans.—Minn. *E. Temp.* Jl–Au.

33. **P. scopàrium** Lam. Stem 8–13 dm. high, erect, or the autumnal form spreading and branching, geniculate at the base; nodes villous with a glabrous viscid ring beneath; sheaths velvety-pubescent; leaf-blades 1.2–2 dm. long, 10–18 mm. wide; panicle long-exserted, 8–15 cm. long, the branches with viscid blotches; spikelets 2.4–2.6 mm. long, abruptly pointed, papillose-pubescent. Wet places: Mass.—Fla.—Texas.—Kans.; Cuba. *E. Temp.*

34. **P. obtùsum** H.B.K. Perennial, with a more or less tufted rootstock, producing creeping stolons; stem compressed, 2–8 dm. high, decumbent at the base, glabrous; leaf-blades 3–20 cm. long, 2–7 mm. wide, involute towards the tip, glabrous or nearly so; panicle short-exserted, 3–12 cm. long; spikelets short-pedicelled, 3–4 mm. long, obovoid, glabrous; first glume nearly as long as the spikelet; second and third glumes subequal; third glume often enclosing a staminate flower. *Brachiaria obtusa* (H.B.K.) Nash. Sandy and gravelly soil: Mo.—Kans.—Tex.—Ariz.—Colo.; also Mex. *Son.*

11. **ECHINÓCHLOA** Beauv. BARNYARD GRASS, JUNGLE RICE.

Coarse and often tall annuals. Inflorescence of several unilateral spikes or racemes. Spikelets 1- or 2-flowered, crowded in small clusters or racemes, in two rows on one side of the flat rachis. Glumes 3, membranous, hispid, the first much shorter than the spikelet, all awned or awn-pointed. Lemma chartaceous, glabrous and shining, indurate, acute or acuminate, enclosing a shorter palet. Styles distinct; stigmas plumose.

Second glume usually not awned; sheaths glabrous.
  Lemma obtuse or acutish, merely ending in a short mucro.
    Spikelets 2.5–3 mm. long; lower racemes not exceeding the internodes; bristles on the glumes weak and short. 1. *E. colona.*
    Spikelets 3–4 mm. long; lower racemes much exceeding the internodes of the rachis; bristles on the veins of the glumes longer.
      Panicle dense, purple or brown, with incurved branches; leaves 15–25 mm. broad; spikelets subglabrous or with few short slender bristles, obtuse and awnless. 2. *E. frumentacea.*

Panicle green with straight branches; leaves 15 mm. broad or less.
Panicle narrow, with short erect branches; bristles on the veins of the third glume uniform; bristles at the nodes of the inflorescence poorly developed; lower palet usually absent. 3. *E. zelayensis.*
Panicle with ascending branches; bristles on the lateral veins of the third glume stronger; bristles at the nodes of the inflorescence long and conspicuous. 4. *E. Crus-galli.*
Lemma distinctly acuminate at the apex.
Bristles on the glumes without pustulate bases; those on the midrib weak or none; glumes green or merely purplish. 5. *E. occidentalis.*
Bristles on the glumes with pustulate bases; those on the midrib well developed; glumes usually purple.
Third glume usually awnless; lemma 2–3 mm. long; branches of the inflorescence ascending. 6. *E. microstachya.*
Third glume usually distinctly awned; lemma 3–4 mm. long; branches of the inflorescence more or less spreading. 7. *E. pungens.*
Second glume usually awned or awn-pointed; third glume long-awned; lower sheath pubescent. 8. *E. Walteri.*

1. **E. colòna** (L.) Link. Stem tufted, 1–8 dm. high, often decumbent and rooting at the lower nodes; sheaths glabrous, compressed; leaf-blades flat, 2–17 cm. long, 2–8 mm. wide; racemes 3–18, one-sided, spreading, 6–30 mm. long; spikelets in 2 rows, with 1–3 spikelets at each node, hispid on the nerves, 2.5–3 mm. long. Fields and roadsides: N.J.—Fla.—Tex.—s Kans. (?). Common in the tropics. *Ozark—Trop.*

2. **E. frumentàcea** (Roxb.) Link. Annual; like *E. Crus-galli,* but stouter; stem 1 m. high or more; leaf-blades 2–3 dm. long, 15–25 mm. wide; panicle ovoid, 1–1.5 dm. long; racemes 2–5 cm. long; setae at the base of the racemes many and conspicuous; glumes brown or purple; first glume deltoid, one third as long as the second. Waste places: N.H.—N.C.—Tex.—Iowa; cultivated as BILLION-DOLLAR-GRASS or JAPANESE MILLET and rarely escaped; nat. of se Asia.

3. **E. zelayénsis** (H.B.K.) Schultes. Annual; stem 3–5 dm. high, glabrous, with erect branches; sheaths compressed; blades 0.5–2 dm. long, 5–10 mm. wide; panicle 5–10 cm. long, usually simple; racemes 2–4 cm. long, the lower longer than the internodes, the upper confluent; spikelets muticous or with a short awn; first glume broadly deltoid, strigose; second and third glumes 3.5–4 mm. long, strigillose on the face, weakly hispidulous on the lateral veins. Wet sandy soil: s Kans.—Utah—Calif.—Costa Rica; S. Am. *Son.—St. Plains.* My–O.

4. **E. Crùs-gàlli** (L.) Beauv. Annual; stem 3–10 dm. high, simple or branched below; leaf-blades 1–3 dm. long, 6–15 mm. wide; panicle ovoid, rather dense, sometimes compound; racemes 2–6 cm. long, ascending or spreading; first glume broadly deltoid, one third as long as the second, strigillose; second and third glumes strigillose on the internerves, weakly hispid on the nerves, 3–4 mm. long, acuminate; lemma with an awn 2–10 mm. long, or in var. *aristata* S. F. Gray, 1–3 cm. long. Waste places and fields: N.S.—Wash.—Calif.—Tex.—Va.; nat. from Eurasia.

5. **E. occidentàlis** (Wieg.) Rydb. Often tufted annual; stems 3–10 dm. high, glabrous; sheaths decidedly compressed; leaf-blades 1–3 dm. long, 5–15 mm. wide; panicle ovoid, 8–20 cm. long; racemes ascending, 3–7 cm. long; first glume one third as long as the second, strigose; second and third glumes ovate, acuminate, mostly awnless, 3–4 mm. long, the bristles on the marginal veins rather long, slender; lemma acuminate, 3 mm. long. *E. muricata occidentalis* Wieg. Wet sandy or alkaline soil: Me.—Wash.—Calif.—N.M.—Va.; common in the upper Mississippi basin. *Plains—Prairie.* Jl–S.

6. **E. microstàchya** (Wieg.) Rydb. Tufted annual; stems 3–8 dm. high, glabrous; leaf-blades 1–3 dm. long, 5–16 mm. wide; panicle ovoid, 5–15 cm. long; racemes 2–6 cm. long, strongly ascending; glumes dark chocolate-colored or purple; first glume deltoid, one third as long as the upper; second and third glumes strongly hispid on the veins, short-acuminate, rarely short-awned, about 3 mm. long. *E. muricata microstachya* Wieg. Rich ground and river banks: Me.—Que.—S.D.—Ariz.—Okla.—Pa. *Canad.—Allegh.* Jl–S.

7. **E. púngens** (Poir.) Rydb. *Fig. 48.* Annual; stems 4–15 dm. high, with erect branches; leaf-blades 1.5–4 dm. long, 5–20 mm. wide; panicle broadly ovoid; racemes 2–8 cm. long, more or less spreading; bristles at the nodes numerous but short; first glume more than one third as long as the second; second glume ovate, acuminate, 4–4.5 mm. long, usually dark purple or chocolate-colored; third glume similar, usually with an awn 5–15 mm. long, the nerves strongly bristly, the bristles with pustulate bases; lemma acuminate, 3–4 mm. long. *Panicum muricatum* Michx., not Retz. *P. pungens* Poir. *E. muricata* (Michx.) Fernald. Low ground and sandy shores: Me.—Minn.—Tex.—Fla.; Calif. *E. Temp.*

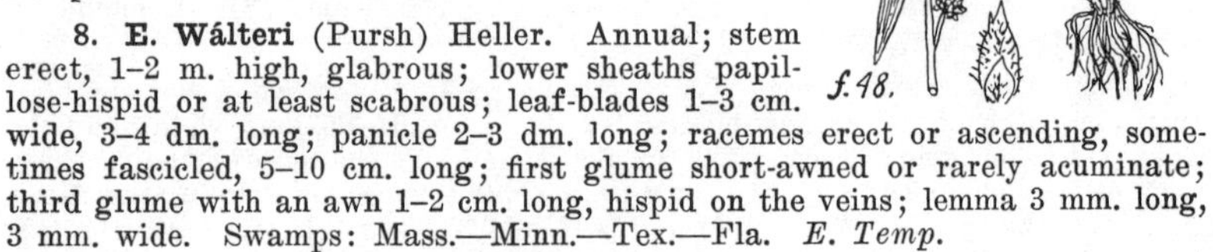

8. **E. Wálteri** (Pursh) Heller. Annual; stem erect, 1–2 m. high, glabrous; lower sheaths papillose-hispid or at least scabrous; leaf-blades 1–3 cm. wide, 3–4 dm. long; panicle 2–3 dm. long; racemes erect or ascending, sometimes fascicled, 5–10 cm. long; first glume short-awned or rarely acuminate; third glume with an awn 1–2 cm. long, hispid on the veins; lemma 3 mm. long, 3 mm. wide. Swamps: Mass.—Minn.—Tex.—Fla. *E. Temp.*

12. **ERIÓCHLOA** H.B.K.

Perennial grasses, with flat leaf-blades. Inflorescence composed of spike-like 1-sided racemes, racemosely arranged. Spikelets with an annular callus at the base, articulate below the callus, 1-flowered, in 2 rows on a flattened rachis, acute or acuminate. Empty glumes 2, membranous, acute or acuminate. Lemma shorter, glabrous and shining, indurate, awl-pointed, enclosing the palet of the same structure. Styles distinct; stigmas plumose.

1. **E. punctàta** (L.) Hamilt. Stem tufted, 2–8 dm. high, at last much branched, puberulent at the nodes; sheaths and blades glabrous or pubescent, the latter 2–30 cm. long, 3–10 mm. wide; racemes 4–20, erect, 1.5–4 cm. long; spikelets 10–30, lanceolate, 4.5–5.5 mm. long, acuminate; glumes long-pubescent; lemma 1/3–1/2 as long as long as the glumes, transversely rugose, 2.5 mm. long; awn hispid, half as long or more. Plains and prairies: Mo.—Kans.—Tex.; Mex., W. Ind. *Texan—Trop.* Jl–S.

13. **SETÀRIA** Beauv. Foxtail Grass.

Annuals or perennials. Inflorescence of dense terminal cylindric spike-like or narrowly thyrsoid panicles. Spikelets in a cluster of 1–several sterile barbed bristle-like persistent branches. Empty glumes 3, membranous, the first often very short and together with the larger second one empty, the third glume frequently longer than the second, empty, or rarely enclosing a palet and also sometimes a staminate flower, the lemma usually shorter than the spikelet, chartaceous, glabrous, shining, often transversely rugose, finally indurated, obtuse, enclosing a shorter palet. Styles distinct, elongated; stigmas plumose. *Chaetochloa* Scribn.

Perennials with a rootstock. — 1. *S. geniculata.*
Annuals.
  Bristles 5–15 at the base of each spikelet, involucrate. — 2. *S. lutescens.*
  Bristles 1–3 at the base of each spikelet, not involucrate.
    Bristles downwardly barbed. — 3. *S. verticillata.*
    Bristles upwardly barbed.
      Panicle usually 1 cm. thick or less; bristles commonly green; spikelets about 2 mm. long. — 4. *S. viridis.*
      Panicle usually 1–3 cm. thick; bristles usually purple; spikelets 2.5–3 mm. long. — 5. *S. italica.*

1. **S. geniculàta** (Lam.) Beauv. Stem 3–8 dm. high; sheaths compressed, keeled, glabrous; leaf-blades erect, 1.5 dm. long or less, 5–7 mm. wide, glabrous; spike 3–5 cm. long, 1.5–2 cm. thick; bristles green; spikelets scarcely 3 mm. long; first glume half as long as the spikelet, 3-nerved; second glume a little longer, 3–5-nerved; lemma strongly transversely rugose, elliptic. *C. geniculata* (Lam.) Millsp. & Chase. *C. occidentalis* Nash [B]. *C. perennis* (Hall & Henry) Smyth. Meadows: Conn.—Fla.—Tex.—Calif.; Mex.—Trop. Am. *E. Temp.* Jl–S.

2. **S. lutéscens** (Weigel) Stuntz. *Fig. 49.* Erect or ascending, glaucous annual; stem 3–12 dm. high, branching at the base, glabrous, compressed; leaf-blades 5–15 cm. long, 4–8 mm. wide, glabrous or with scattered long hairs at the base; spikelet broadly ovate, 3 mm. long; second glume one half to two thirds as long as the spikelet, ovate, acute; third glume equaling the lemma, which is broadly ovate, striate, transversely rugose. *S. glauca* Beauv. [G]. *C. glauca* Scribn. [B,R]. Waste places and cultivated ground: N.S.—Fla.—Tex.—Colo.—Sask.; naturalized from Eurasia. *Temp.—Plain—Submont.* Je–S.

3. **S. verticillàta** (L.) Beauv. A tufted annual; stems 3–6 dm. high; leaf-blades scabrous, 5–20 cm. long; panicle 5–10 cm. long, interrupted at the base; bristles 3–6 cm. long; spikelets 2–2.5 mm. long; first glume one third as long as the second; lemma obscurely transversely rugose. *C. verticillata* Scribn. [B]. Waste places and around dwellings: N.S.—D.C.—Mo.—Neb.—S.D.; Utah; Mex.—W. Ind.; nat. from Eur. Jl–S.

4. **S. víridis** (L.) Beauv. Erect glabrous annual; stem 2–9 dm. high, branched at the base, compressed; leaf-blades 5–25 cm. long, 4–10 mm. wide, long-acuminate, slightly scabrous on both sides; bristles slender, scabrous, 1–1.5 cm. long, green or rarely purplish; lemma elliptic, finely and faintly wrinkled below or only striate. *C. viridis* Scribn. [B,R]. Waste places and cultivated ground: Newf.—Fla.—Calif.—B.C.; Mex; naturalized from Eur. *Temp.—Submont.* Jl–S.

5. **S. itálica** (L.) Beauv. Stout, erect, somewhat glaucous annual; stem simple or branched at the base, 5–20 dm. high, glabrous; nodes bearded; leaf-blades lanceolate, 2–4 dm. long, 1.5–3 cm. wide, scabrous on both sides; bristles green or purplish, 3–10 cm. long, scabrous; lemma smooth or faintly transversely rugose, striate. ITALIAN MILLET. *C. italica* Scribn. [B,R]. Waste places and fields; escaped from cultivation: Que.—Fla.—Tex.—Colo.—S.D.; native of Eurasia. Jl–S.

14. CÉNCHRUS L. BUR-GRASS, SANDBUR, SANDSPUR.

Annuals or perennials. Spikes terminal. Spikelets 2–6, in an ovate or globose involucre, consisting of two thick hard valves, which are exteriorly armed with stout spines at the base, the involucres articulated to the rachis and readily deciduous, carrying the persistent spikelets with them. The first and second glumes empty, the first small or minute, the third equaling or longer than the second, enclosing a palet and also sometimes a staminate flower, the lemma chartaceous, firmer, enclosing a palet of similar texture and a perfect flower. Stamens 3. Styles often connate at the very base; stigmas plumose.

1. **C. paucifiòrus** Benth. *Fig. 50.* Erect or decumbent annual; stem 2–9 dm. long, branching; leaf-sheaths usually loose, compressed, smooth; blades 6–12 cm. long, 4–8 mm. wide, smooth or rough; spikes 3–6 cm. long, sometimes partly included; involucres 6–20, each enclosing 2 spikelets, 3–5 mm. broad, pubescent; spines 3–4 mm. long; spikelets 6–7 mm. long, usually exserted beyond the involucre. *C. tribuloides* Auth., not L. *C. carolinianus* Hitchc. scarcely Walt. Sandy banks and waste places, sometimes becoming a noxious weed: Me.—Fla.—Tex.—Calif.—N.D.; Mex., C. Am., and W. Ind. *Temp.—Plain—Son.* Mr–N.

15. LEÈRSIA Sw. RICE CUT-GRASS.

Perennials. Panicles usually open, rarely contracted, terminal, the branches slender. Spikelets often with a cartilaginous ring at the base, articulated below the scales, compressed, 1-flowered. Glumes 2, compressed-keeled, somewhat rigid, awnless, the first one usually ciliate on the keel, broader than the second, which encloses a perfect flower but no palet. Stamens 1–6. Styles short or slender, distinct; stigmas plumose with branched hairs. *Homalocenchrus* Mieg.

Spikelets narrowly oblong, loosely overlapping.
  Spikelets 3 mm. long or less; branches of the panicle rigid. 1. *L. virginica.*
  Spikelets 4–5 mm. long, branches of the panicle lax. 2. *L. oryzoides.*
Spikelets broadly oval, densely imbricate. 3. *L. lenticularis.*

1. **L. virgínica** Willd. Stem slender, 3–10 dm. high, much branched; leaf-blades 5–15 cm. long, 2–15 mm. wide, scabrous; panicle simple, with stiff spreading branches, 5–20 cm. long; lower glume hispid on the keel and margins; stamens 1 or 2. *H. virginicus* Britton. [B,R]. Swamps: Me.—Fla.—Tex.—Minn. *E. Temp.* Au–S.

2. **L. oryzoìdes** (L.) Sw. *Fig. 51.* Stem 3–15 dm. high, often rooting at the nodes; leaf-blades 6–25 cm. long, 4–13 mm. wide, very rough; terminal panicle 1–2 dm. long, at last exserted; lateral ones often included at the base; spikelets 4–5 mm. long, about 1.5 mm. wide, elliptic; outer glumes hispid on the keel and margins, hispidulous on the sides; stamens 3. *H. oryzoides* Mieg. [B,R]. In swamps: N. S.—Fla.—Tex.—Colo.—Sask.; also Eur. *Temp.* Au–O.

3. **L. lenticulàris** Michx. Stem simple, 6–12 dm. high, erect or decumbent at the base; leaf-blades 1–3 dm. long, 8–20 mm. wide, nearly smooth; panicle 1–1.3 dm. long, its branches lax, at last spreading; spikelets flat, 5 mm.

long, strongly bristly-ciliate; stamens 2. *H. lenticularis* Scribn. [B]. Wet ground: Va.—Fla.—Tex.—Minn. *Allegh.—Ozark.* Jl–S.

16. **ZIZÀNIA** L. WILD RICE, INDIAN RICE, WATER OATS.

Tall aquatic monoecious annuals, with long flat leaf-blades and large panicle. Spikelets 1-flowered, the pistillate ones borne on the upper branches, the staminate ones on the lower. Glumes in the pistillate spikelets wanting; lemma long-awned, with two strong lateral nerves, closely clasping the palet. First glume of the staminate spikelets 5-nerved, the second 3-nerved. Stamens 6. Grain cylindric, enclosed in the membranous lemma and palet.

Leaf-blades usually less than 1 cm. wide; staminate branches ascending, with 5–15 spikelets; body of the pistillate lemma 10–17 mm. long. 1. *Z. palustris.*
Leaf-blades 1–3 cm. wide; staminate branches spreading, with 30–60 spikelets; body of the pistillate lemma 2–3 cm. long. 2. *Z. interior.*

1. **Z. palústris** L. *Fig. 52.* Stem 7–15 dm. high, 4–5 mm. thick, striate, glabrous, except the finely velutinous nodes; leaf-blades 3–5 cm. long, 4–10 mm. (rarely 15 mm.) wide; ligules lanceolate, 3–7 (rarely 10 mm.) long; panicle 2–5 dm. long; staminate branches ascending, with 5–15 spikelets; lemmas 9–10 mm. long, often purplish, glabrous or nearly so, awn-pointed; pistillate branches erect, the lower ones with 2–6 spikelets; body of the lemma 10–17 mm. long, firm, shining, glabrous, except at the apex, and sometimes along the margins and veins; awn 2–5.5 cm. long, short-hairy, more densely so and with longer hairs at the base. *Z. aquatica angustifolia* Hitchc. *Z. aquatica* Hitchc. [G], not L. In water: N.B.—N.J.—Iowa—Man. *Canad.* Jl–S.

f. 52.

2. **Z. intèrior** (Fassett) Rydb. Stem 1–3 m. high, 5–12 mm. thick, striate, glabrous, except the nodes; leaf-blades 3–8 cm. long, 1–3 cm. wide, with a strong midrib; ligules ovate, 10–15 mm. long, usually lacerate; panicle 3–7 dm. long; staminate branches more spreading, with 30–60 spikelets; lemma 7–9 mm. long, often pale, mucronate, or awn-pointed, glabrous; pistillate branches numerous, strongly ascending, the lower with 8–30 spikelets; body of the lemma 2–3 cm. long, 2 mm. broad, glabrous except at the rather long-hairy apex; awn as in the preceeding. *Z. aquatica interior* Fassett. In water: Wis.—Ind.—Tex.—N.D. *Prairies.* Jl–S.

17. **HIERÓCHLOË** Gmel. SWEET GRASS, HOLY GRASS.

Sweet-scented perennials with rootstocks. Panicles open or contracted. Spikelets 3-flowered, the terminal flower perfect, the others staminate. Empty glumes nearly equal, acute, glabrous. Lemma 2-toothed or 2-lobed, with or without an awn. Stamens in the staminate flowers 3, in the perfect flowers 2. Styles distinct; stigmas plumose. *Torresia* R. & P. *Savastana* Schrank.

Lemmas of the staminate flowers awnless. 1. *H. odorata.*
Lemmas of the staminate flowers awned. 2. *H. alpina.*

1. **H. odoràta** (L.) Wahl. *Fig. 53.* Perennial with a creeping rootstock; stem 3–6 dm. high, smooth; lower leaf-blades elongated, 1–2 dm. long, 2–6 mm. wide, the upper ones 1–5 cm. long; panicle 5–10 cm. long, its branches spreading or reflexed; spikelets yellowish brown or purplish, 4–6 mm. long; first and second glumes acute; third and fourth glumes villous and strongly ciliate, awn-pointed. *T. odorata* Hitchc. [R]. *H. borealis* R. & S. *Savastana odorata*

Scribn. [B]. Moist places and among bushes: Lab.—N.J.—Iowa—Colo.—Ariz.—Alaska; Eurasia. *Boreal.—Alp.* Je–Jl.

2. **H. alpìna** (Sw.) R. & S. Perennial with a creeping rootstock; stem 1.5–4.5 dm. high; lower leaf-blades 7–15 cm. long, 2 mm. wide, the upper shorter, 2–4 mm. wide; panicle 1.5–3.5 cm. long, contracted; spikelets 5–7 mm. long; first staminate lemma with an awn 2 mm. long, the awn of the second about 6 mm. long; fertile lemma awn-pointed. *S. alpina* Scribn. [B]. Mountain tops; Greenl.—N.H.—N.Y.—Man.—Alaska. *Arct.—Subarct.* Jl–Au.

18. PHÁLARIS L. Canary Grass

Annuals or perennials. Panicles terminal, cylindric and spike-like, capitate, or densely thyrsoidal and somewhat interrupted. Spikelets articulated above the empty glumes, compressed, 1-flowered, crowded. Empty glumes 4, the lower 2 persistent, larger than the rest, thin-paleaceous, compressed-keeled, the keel usually more or less winged, sometimes wingless, awnless; third and fourth glumes shorter, very thin and narrowly lanceolate, sometimes reduced to bristles, or rarely one of them wanting; lemma chartaceous, at length indurated, awnless, sometimes pointed, obscurely 3–5-nerved, the midnerve sometimes obsolete, enclosing a faintly 2-nerved palet. Styles distinct; stigmas plumose.

Outer glumes not winged; inflorescence a narrow panicle. 1. *P. arundinacea.*
Outer glumes winged; inflorescence a spike or spike-like panicle.
  Spikelets narrow; third and fourth glumes much reduced; blades subulate-linear, hairy. 2. *P. caroliniana.*
  Spikelets broad; third and fourth glumes thin, membranous; blades lanceolate, glabrous, rarely sparingly hairy. 3. *P. canariensis.*

1. **P. arundinàcea** L. *Fig. 54.* Glabrous perennial, with a horizontal rootstock; stem erect, 6–15 dm. high; leaf-blades 1–2.5 dm. long, 6–16 mm. wide; panicle 7–20 cm. long, dense, 1–2 cm. thick; spikelets 5–6 mm. long; outer glumes 3-nerved; third and fourth glumes less than half as long as the lemma, which is pubescent with long appressed hairs. Wet places; N.S.—N.J.—Nev.—B.C.; Eurasia. *Boreal—Plain—Mont.* Je–Au.

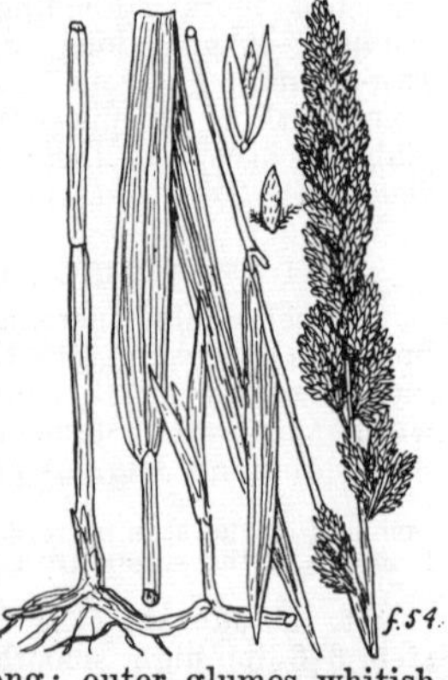

2. **P. carolinìana** Walt. Glabrous annual; stem 3–10 dm. high; leaf-blades 5–15 cm. long, 4–10 mm. wide; panicle oblong, 2.5–10 cm. long, 1–1.5 cm. thick; spikelets 5 mm. long; outer glumes 3-nerved; third and fourth glumes less than half as long as the lemma, which is acuminate and with long appressed hairs. Wet ground: Fla.—S.C.—Mo.—Colo.—Calif.; Mex. *Son.—Austral.* My–Au.

3. **P. canariénsis** L. Glabrous annual; stem 3–9 dm. high, branched at the base; leaf-blades 5–30 cm. long, 4–12 mm. wide, strongly scabrous; panicles oblong or ovoid, 1–4 cm. long; spikelets 6–8 mm. long; outer glumes whitish with green nerves; third and fourth glumes broadly lanceolate, about half as long as the pubescent lemma. Waste places: N.S.—Va.—Colo.—S.D.—Sask.; also Calif.—Ore.; naturalized from Europe and Africa. *Plain—Submont.* Je–Jl.

19. **ANTHOXÁNTHUM** L. SWEET VERNAL GRASS.

Sweet-scented annual or perennial grasses, with flat leaf-blades and narrow spike-like dense panicle. Spikelets 3-flowered, but the two lower flowers reduced to the sterile lemma, falling off with the fertile floret; glumes very unequal; sterile lemmas 2-lobed, dorsally awned, longer than the floret, fertile lemma truncate, awnless; palet faintly 1-nerved. Anthers 2. Stigmas elongate, plumose. Grain free.

1. **A. odoràtum** L. Perennial; stem 3–6 dm. high; leaf-blades 2–15 cm. long, 2–6 mm. wide, glabrous or nearly so; panicle 3–8 cm. long; spikelets brownish-green, 8–10 mm. long, spreading in anthesis; first sterile lemma short-awned below the apex; second sterile lemma with a strong bent awn near the base. Meadows and pastures: Newf.—N.C.—Miss.—Minn.; Calif.—Alaska; nat. from Europe. My–Jl.

20. **ARÍSTIDA** L. POVERTY GRASS, WIRE GRASS.

Tufted perennials, with narrow leaves. Inflorescence paniculate or racemose. Spikelets 1-flowered, narrow. Empty glumes 2, membranous, persistent, keeled, awnless, usually longer than the lemma. The latter firm, narrow, rigid, strongly convolute, with a prominent callus at the base, 3-awned at the apex; the central awn often bent and twisted, the lateral ones shorter and spreading or ascending. Styles distinct; stigmas plumose.

Awns articulate to the lemma, united at the base into a spiral neck.
- Neck or column of the awns 6 mm. long or more. — 1. *A. tuberculosa.*
- Neck or column of the awns 2 mm. long or less. — 2. *A. desmantha.*

Awns distinct, not articulate to the lemma.
- Central awn coiled at the base.
  - First glume equaling or slightly shorter than the second, which is 7–9 mm. long. — 3. *A. dichotoma.*
  - First glume much shorter than the second, which is 10–15 mm. long.
    - Lateral glumes straight and erect, 1–2 mm. long; straight portion of the central awn 5–8 mm. long. — 4. *A. Curtissii.*
    - Lateral awns more or less spreading, one half to two thirds as long as the central awn, whose straight portion is 10–15 mm. long. — 5. *A. basiramea.*
- Central awn not coiled at the base.
  - Panicle narrow, the branches short, erect or ascending.
    - Central awn strongly reflexed. — 6. *A. ramosissima.*
    - Central awn erect to spreading, not recurved.
      - First glume much shorter than the second, usually about half as long.
        - Spikelets crowded, 4–6 on short branches spikelet-bearing to near the base; awn less than 2 cm. long; annuals. — 7. *A. fasciculata.*
        - Spikelets not crowded, usually 1–3, on branches naked at the base; awns more than 2 cm. long; perennials.
          - Branches of the loose and nodding panicle slender, flexuose or curved. — 8. *A. purpurea.*
          - Branches of the erect panicle stiff and appressed.
            - Second glume of the spikelets 2 cm. long or more, 1.5–2 times as long as the lemma. — 9. *A. longiseta.*
            - Second glume of the spikelets 1.5 cm. long or less, scarcely exceeding the lemma.
              - Panicle simple, or nearly so; branches with a single spikelet, rarely with two; stem naked above. — 10. *A. Fendleriana.*
              - Panicle compound; branches bearing 2 spikelets or more; stem leafy, stout. — 11. *A. Wrightii.*
      - First glume from a little shorter to longer than the second.

Spikelets more than 2 cm. long; first glume 5–7-nerved. 12. *A. oligantha.*
Spikelets less than 1.5 cm. long; first glume 1–3-nerved.
First glume shorter than the second or equaling it.
Lemma 5–6 mm. long; central awn 6–10 mm. long. 13. *A. gracilis.*
Lemma 7–9 mm. long; central awn 15–22 mm. long. 14. *A. intermedia.*
First glume longer than the second. 15. *A. purpurascens.*
Panicle open; branches divergent.
Central awn 1.2–2 cm. long. 16. *A. divaricata.*
Central awn 2.5–3 cm. long. 11. *A. Wrightii.*

1. **A. tuberculòsa** Nutt. Stem 1.5–5 dm. high, dichotomously branched; leaf-blades 1–2 dm. long, involute; panicle 1–2 dm. long, rigid, the branches in pairs, one short, with about two spikelets, the other longer and with several spikelets; glumes 2.5 cm. long, with slender awn-tips; lemma 12–15 mm. long; awns of equal length, 3.5–5 cm. long. Sandy soil: N.H.—Ga.—Miss.—Neb.—Minn. *E. Temp.* Jl.–S.

2. **A. desmántha** Trin. & Rupr. Stem 3–6 dm. high, branched; leaf-blades 1.5–3 dm. long, less than 2 mm. wide, scabrous above, smooth below; panicle about 15 cm. long, with slender ascending branches; glumes nearly equal, 12–14 mm. long without the awn; lemma without the awn 8–10 mm. long; central awn 2.5–3 cm. long, the lateral ones a little shorter. Dry soil: Neb.—Tex. *Plain—St. Plains.* Au–S.

3. **A. dichótoma** Michx. Stem tufted, wiry, much branched at the base, 1–6 dm. high; sheaths loose; leaf-blades 3–7 cm. long, involute, scabrous; panicle 5–12 cm. long, narrow; spikelets few, the lateral panicles often sessile, inclosed in the sheaths; glumes 7–8 mm. long; lemma without the awn 6 mm. long; central awn 3–6 mm. long, the lateral ones minute. Sandy soil: Me.—Ga.—Tex.—Neb. *E. Temp.* Au–S.

4. **A. Curtíssii** (A. Gray) Nash. Tufted; stem 2–5 dm. high, branching; leaf-blades 4–16 cm. long, 1–2 mm. wide; panicle spike-like, 5–9 cm. long, branches short, erect; first glume shorter than the second, which usually equals the lemma, 7–11 mm. long; middle awn 10–15 mm. long; lateral awns 1–2 mm. long, straight. *A. dichotoma Curtissii* A. Gray. [G]. Dry soil: Va.—Wy.—Okla.—Minn. *Allegh.—Plains.* S–O.

5. **A. basiràmea** Engelm. Stems freely branching at the base, sparingly so above, 2–6 dm. high; leaf-blades 5–15 cm. long, 1.5 mm. wide; panicle loose, 3–8 cm. long, with erect branches, often partly included in the sheath; glumes acuminate, the first 8–10 mm. long, the second 12–14 mm. long; lemma without the awn about 1 cm. long; central awn 1–2 cm. long, the lateral ones 2–7 mm. long. Dry prairies: Ills.—Mo.—Okla.—Man. *Prairie.* Au–S.

6. **A. ramosíssima** Engelm. Stem tufted, 1.5–6 dm. high; leaf-blades 3–7 cm. long, setaceous, scabrous above; panicle loose, 5-10 cm. long, with few spikelets; glumes 1.5–2 cm. long, awned from the bifid apex, unequal; lemma about 2 cm. long; central awn 2–3 cm. long; lateral awns minute. Dry prairies: Ind.—Tenn.—Okla.—Kans. *Prairie.* Jl–S.

7. **A. fasciculàta** Torr. A tufted annual; stems 3–6 dm. high, branched; leaf-blades 5–15 cm. long, 2 mm. wide; panicle 5–17 cm. long, loose; branches at first strict, later more or less spreading; first glume 1-nerved, shorter than the second; lemma equaling or longer than the second; awns ascending, the lateral ones shorter than the middle one. (Mistaken for *A. oligantha* Michx.) Dry soil: Tex.—Kans.—s Utah—Ariz.—Mex. *St. Plains.—Son.* My–S.

8. **A. purpùrea** Nutt. Tufted perennial; stem 3–5 dm. high; sheaths villous at the throat; blades involute, scabrous above, 1 dm. long or less, 1–1.5 mm. wide; panicle nodding, narrow, purplish, 1–2 dm. long; branches naked

at the base; lower glume 6–8 mm. long, the second about twice as long; lemma about 1 cm. long, scabrous above. Dry hills: Kans.—Utah—s Calif.—Mex. *Son.* Mr–Au.

9. **A. longisèta** Steud. Densely tufted glaucous glabrous perennial; stem 1–4 dm. high; sheaths shorter than the internodes; leaf-blades strongly involute, 2–11 cm. long; spikelets purplish; first empty glume shorter than the lemma; lemma 12–16 mm. long; awns 6–11 cm. long. Sandy soil: Minn.—Ill.—Tex.—Ariz.—Wash.; Mex. *Prairie—Plain—Submont.* My–Au.

10. **A. Fendleriàna** Steud. *Fig. 55.* Densely tufted glabrous perennial; stem 1.5–3 dm. high, erect; sheaths smooth, confined to the base of the stem; leaf-blades involute, often curved; panicle 7–10 cm. long, strict; spikelets 12–15 mm. long; lemma 9–12 mm. long; central awn 2–3.5 cm. long; lateral awns a little shorter. *A. purpurea* Coult., not Nutt. *A. longiseta Fendleriana* Merrill. Dry soil: S.D.—Tex.—Mont.—Calif. *Son.—Plain—Submount.* Je–S.

11. **A. Wrìghtii** Nash. Stem tufted, 3.5–5 dm. high, simple; leaf-blades 3–20 cm. long, involute, often curved; panicle 1–2 dm. long, its branches more or less spreading; first glume about half as long as the second, which is 12–14 mm. long; lemma 10–12 mm. long; central awn 2.5–3 cm. long, the lateral ones a little shorter. Dry soil: Kans.—Tex.—N.M. *Son.—St. Plains.* My–O.

12. **A. oligántha** Michx. Stems tufted, wiry, 3–6 dm. high; sheaths loose; leaf-blades 3–15 cm. long, 1–2 mm. wide, smooth, involute; spikelets few, in a raceme-like inflorescence, which is often flexuose; glumes long-awned from the 2-cleft apex; lemma 17–20 mm. long, scabrous above; awns nearly equal, 4–7 cm. long. Dry sterile soil: N.J.—La.—Tex.—Neb.; also Calif. *E. Temp.* Au–O.

13. **A. grácilis** Ell. Stem 1–6 dm. high, slender; leaf-blades 3–10 cm. long, 2 mm. wide or less, involute when dry; panicle 7–17 cm. long, spike-like, slender; spikelet about 6 mm. long; lemma about equaling the second glume; central awn 6–10 mm. long, horizontal; lateral awns 2–6 mm. long. *A. longispica* Poir. Dry soil: N.H.—Fla.—Tex.—Neb. *E. Temp.* Au–S.

14. **A. intermèdia** Scribn. & Ball. Stems in small tufts, 3–7 dm. high, freely branched, geniculate at the base; leaf-blades 5–15 cm. long, rigid, involute; panicle 2–4 dm. long, slender, with short appressed branches; glumes 7–9 mm. long, scabrous; lateral awns 14–16 mm. long. Dry soil: Ind.—Mo.—Tex.—Kans. *Ozark.—Texan.* Au–S.

15. **A. purpuráscens** Poir. Stems in small tufts, glabrous, 3–6 dm. high, simple or sparingly branched; leaf-blades 1–2 dm. long, 1–4 mm. wide, usually involute towards the ends; panicle slender, 1–3 dm. long, purplish; glumes 10–12 mm. long, aristate; lemma 6–7 mm. long; awns divergent, 1.5–3 cm. long, the central one somewhat longer than the lateral ones. Sandy soil: Mass.—Fla.—Tex.—Kans.—Minn. *E. Temp.* S–O.

16. **A. divaricàta** Humb. & Bonpl. Tufted perennial; stem 3–9 dm. high, branched; sheaths rough; blades of the stem-leaves 1.5–3 dm. long, 2–4 mm. wide; those of the sterile shoots longer and narrower; panicles 3–5 dm. long, open; empty glumes acuminate, awn-pointed, 9–13 mm. long; central awn 1.2–2 cm. long, erect; lateral ones somewhat shorter, ascending. *A. Humboldtiana* Trin. & Rup. [R]. Dry soil: Kans.—Colo.—Ariz.; Mex. *St. Plains.—Son.* Jl–S.

21. STÌPA L. SPEAR GRASS, PORCUPINE GRASS, DEVIL'S DARNING-NEEDLES, FEATHER GRASS.

Usually tufted perennials. Inflorescence paniculate. Spikelets narrow, 1-flowered; flowers perfect. Glumes 2, narrow, persistent, keeled, acute, rarely awned. Lemma narrow, strongly convolute, rigid, with a strong, usually acute callus at the base and ending in a bent awn, which is spirally twisted below the knee. Styles distinct, short; stigmas plumose.

Outer glumes of the spikelets 2 cm. long or more.
  Base of the panicle exserted; lemma 20–25 mm. long; awn straight above the bend. 1. *S. spartea.*
  Base of the panicle usually included in the upper sheath; lemma 8–12 mm. long; awn slender and curled above the bend. 2. *S. comata.*
Outer glumes of the spikelets 1.5 cm. long or less.
  Panicle narrow; branches ascending.
    Lemma 5–6 mm. long; glumes hyaline; stem slender, 3–6 dm. high. 3. *S. viridula.*
    Lemma about 8 mm. long; glumes firmer; stem stout, 5–20 dm. high. 4. *S. Vaseyi.*
  Panicle loose and open; branches spreading or reflexed.
    Lemma 7–8 mm. long, its callus at the base acute. 5. *S. Richardsonii.*
    Lemma about 4 mm. long, its callus at the base short and blunt. 6. *S. canadensis.*

1. **S. spártea** Trin. *Fig. 56.* Tufted perennial; stem 6–15 dm. high, erect, simple; basal leaf-blades 2–5 dm. long, usually involute, scabrous above; stem-leaves 1.5–3 dm. long, usually flat; panicle 1–2.5 dm. long; outer glumes of spikelets 2.5–3.7 cm. long, acuminate, awn-pointed, glabrous; callus long, sharp; awn 12–20 cm. long. Prairies: Man.—Ind.—Kans.—N.M.—B.C. *Prairie—Plain—Submont.* Je–Au.

2. **S. comàta** Trin. & Rupr. Tufted perennial; stem 3–6 dm. high, glabrous; sheaths usually longer than the internodes, the uppermost inflated, enclosing the base of the panicle; leaf-blades somewhat scabrous, involute; those of the basal innovations filiform, 1–3 dm. long, those of the stem 6–15 cm. long, broader; panicles 15–20 cm. long, loose; outer glumes 18–25 mm. long, glabrous; awn 1–2 dm. long, twice bent below. Prairies and sandy places: Man.—Wis.—N.M.—Calif.—Alaska. *Prairie—Plain—Submont.* Je–Au.

3. **S. viridula** Trin. Tufted perennial; stem smooth, 3–6 dm. high; leaf-blades involute, smooth or scabrous above, the lower 1–2 dm. long; stem-leaves 6–15 cm. long; outer glumes 7–9 mm. long, prominently 3-nerved, awn-pointed; lemma 5–6 mm. long, strigose; awn 2–2.5 cm. long, twice bent, glabrous or somewhat scabrous. Plains and prairies: Man.—ne Kans.—Utah—Mont. *Prairie—Plain—Submont.* Jl–Au.

4. **S. Vàseyi** Scribn. Densely tufted perennial; stem 1–2 m. high, stout, often 1 cm. thick below; sheaths usually broad and loose, hairy at the junction of the blades; blades of stem-leaves usually flat, 3–6 dm. long, scabrous, those of the innovations narrower, involute; panicle 2.5–4 dm. long, dense; empty glumes nearly equal, lanceolate, about 1 cm. long, acuminate; lemma about 8 mm. long, silky-strigose; awn 2.5–3 cm. long, minutely scabrous. Hills and mountain sides: Tex.—Colo.—Ida.—S.D.; Mex. *Son.—Mont.* My–Au.

5. **S. Richardsònii** Link. Tufted perennial; stem slender, 5–10 dm. high; leaf-blades involute, filiform, 5–15 cm. long, smooth; panicle open, 7–12 cm. long; branches slender, with 1–3 spikelets; empty glumes purplish, 8–9 mm.

long, broadly lanceolate, acute; lemma thinly pubescent; awn 12–20 mm. long, strigillose. Hillsides and open woods: Sask.—S.D.—Colo.—Alta. *Mont.* Jl–S.

6. **S. canadénsis** Poir. Tufted perennial; stem 3–6 dm. high, smooth or somewhat scabrous; blades 5–12 cm. long, 1–2 mm. wide, scabrous; panicle 5–12 cm. long; outer glumes 4–5 mm. long, obtuse or acutish, glabrous, membranous; lemma silky-strigose; awn 8–10 mm. long, twisted but only slightly bent. *S. Macounii* Scribn. [B]. Sand hills and open woods: N.B.—Me.—Mich.—Sask. *Boreal.* Jl.

22. ORYZÓPSIS Michx. Mountain Rice.

Tufted perennials. Inflorescene paniculate with racemose branches. Spikelets 1-flowered, broad; flowers perfect. Empty glumes subequal, acute. Lemma broad, indurate, convolute, with a short and obtuse callus at the base, ending in a terminal, early deciduous, mostly straight awn. Styles distinct: stigmas plumose. Grain oblong, free.

Spikelets, exclusive of the awn, 2.5–5 mm. long; leaves slender and involute.
  Awn less than 2 mm. long, much shorter than the glume; outer glumes 3–4 mm. long. — 1. *O. pungens.*
  Awn 4–8 mm. long, much longer than the glume; glumes about 2.5 mm. long. — 2. *O. micrantha.*
Spikelets, exclusive of the awn, 6–8 mm. long; leaves broad and usually flat.
  Stem nearly naked, the leaves crowded near the base; panicle 5–8 cm. long, its branches short and erect. — 3. *O. asperifolia.*
  Stem leafy to the top; panicle 15–30 cm. long, its branches long, more or less spreading. — 4. *O. racemosa.*

1. **O. púngens** (Torr.) Hitchc. Tufted perennial; stem glabrous, 1.5–3 dm. tall, simple; sheaths shorter than the internodes, smooth; leaf-blades smooth or scabrous, the lower 1–2 dm. long, the upper 3–10 cm. long; panicle 3–6 cm. long; branches erect or ascending; spikelets 3–4 mm. long; empty glumes glabrous, whitish, faintly nerved, acute; lemma strigose, ellipsoid. *O. canadensis* Torr. *O. juncea* B.S.P. [B]. Dry rocky places: Que.—Pa.—S.D.—B.C. *Boreal—Mont.* My–Je.

2. **O. micrántha** (Trin. & Rupr.) Thurb. Somewhat tufted perennial; stem glabrous, 3–7 dm. tall, slender; leaf-blades erect, scabrous, less than 1 mm. wide, usually involute, the lower 2–3 dm. long; panicle 8–15 cm. long; empty glumes 2–2.5 mm. long, glabrous, acute; lemma a little shorter, glabrous, shining; awn 6–8 mm. long. Hillsides and among bushes: Sask.—Neb.—N.M.—Ariz.—Mont. *Plain—Mont.* Je–Au.

3. **O. asperifòlia** Michx. Tufted perennial; stem 3–5 dm. high, simple, usually scabrous; leaf-blades erect, scabrous, 1–4 dm. long, 4–10 mm. wide; panicle 5–8 cm. long, narrow, spike-like; empty glumes glabrous, 6–8 mm. long, many-nerved, apiculate; lemma whitish, sparingly hairy, except a ring of dense hairs at the base; awn 7–10 mm. long. Woods: N.S.—Pa.—S.D.—N.M.—B.C. *Boreal—Mont.* My–Je.

f. 57.

4. **O. racemòsa** (Smith.) Ricker. *Fig. 57.* Stem erect, 3–12 dm. high, tufted; leaf-blades 1–3.5 dm. long, 5–15 mm. wide, scabrous below; glumes 6–8 mm. long, acute; lemma slightly shorter, pubescent, becoming black in fruit; awn 1.5 cm. long. *O. melanocarpa* Muhl. [B]. Rocky woods: Me.—Del.—Mo.—Neb.—e S.D. *Canad.* Je–O.

23. **ERIOCÒMA** Nutt. Indian Millet, Wild Rice, Sand Rice.

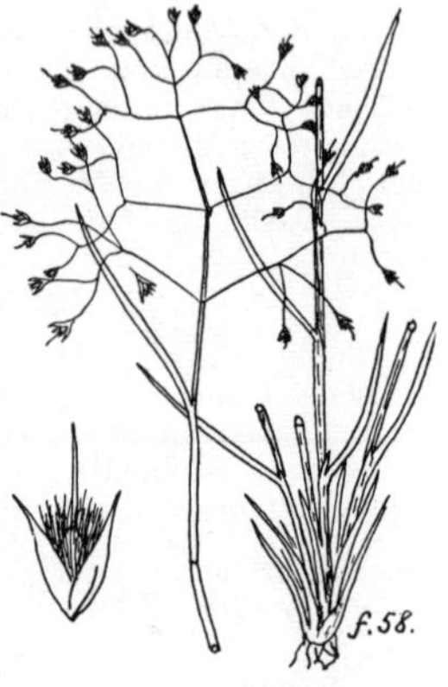

Densely tufted perennials with involute leaves. Inflorescence cymosely and divaricately branched. Spikelets 1-flowered; flower perfect. Empty glumes 2, membranous, somewhat scarious, acuminate. Lemma rather broad, indurate and convolute, densely pubescent with silky hairs, with an obtuse callus at the base, and a deciduous straight awn at the apex. Styles distinct; stigmas plumose. Grain oblong, free, enclosed in the lemma.

1. **E. hymenoìdes** (R. & S.) Rydb. *Fig. 58.* Stem 3–6 dm. high, erect, rigid, smooth; leaf-blades 1.5–3 dm. long, involute, less than 2 mm. wide; panicle 1.5–3 dm. long; branches spreading, flexuose; lower glumes 6–8 mm. long, long-acuminate, with somewhat spreading tips; lemma 4 mm. long, broadly ellipsoid, densely pubescent with white hairs. *Eriocoma cuspidata* Nutt. [B]. Bad lands and sandy places: Man.—Kans.—Tex.—Calif.—Wash; Mex. *Plain—Mont.* My–Jl.

24. **MÍLIUM** (Tourn.) L. Millet Grass.

Perennials and annuals. Inflorescence paniculate. Spikelets 1-flowered. Flowers perfect, the rachilla articulate below the floret. Glumes 2, equal, obtuse. Lemma shining, awnless, shorter than the glumes, indurate, the margins rolled in over the palet, which is similar. Grain enclosed, free. Stamens 3. Stigma plumose.

1. **M. effùsum** L. Smooth perennial; stem 1–1.5 m. high; sheaths shorter than the internodes; blades 1–3 dm. long, 8–15 mm. wide, flat; panicle 1–2 dm. long, the branches 2–6 at the nodes, spreading or reflexed; spikelets borne above the middle, 3–5 mm. long; glumes minutely scabrous, elliptic. Woods: N.S.—Minn.—Ill.—Pa.; nat. from Eurasia. Je–Jl.

25. **MUHLENBÉRGIA** Schreb.

Perennials or annuals. Inflorescence a narrow, contracted or open, diffuse panicle. Spikelets 1-flowered; flower perfect. Empty glumes 2, membranous or hyaline, persistent, keeled, acute to short-awned, the second sometimes 3-toothed. Lemma somewhat rigid, enclosing the palet, entire or 2-toothed at the apex, obtuse, acute, or awned. Stamens usually 3. Styles distinct; stigmas plumose.

Panicle contracted, narrow, spike-like, the short branches rarely spreading.
- Empty glumes awl-shaped; leafy and branched plants, with a long rootstock covered by imbricate scales.
  - Lemma not awned, sometimes merely awn-pointed.
    - Empty glumes acute, about half as long as the lemma. — 1. *M. sobolifera.*
    - Empty glumes acuminate, awn-pointed or awned.
      - Empty glumes about 3 mm. long, about equaling the lemma, sharp-pointed.
        - Panicle short-exserted or partly included; lemma 1.5 mm. long or more. — 2. *M. mexicana.*
        - Panicle long-exserted.
          - Lemma 1 mm. long or less. — 3. *M. polystachya.*
          - Lemma more than 1.5 mm. long. — 4. *M. foliosa.*
      - Empty glumes about 5 mm. long, exceeding the lemma, awned. — 5. *M. racemosa.*
  - Lemma distinctly awned.
    - Glumes about equaling the lemma.

Basal hairs not more than half as long as the lemma.
Panicle linear or filiform; spikelets without an empty glume above the perfect floret. — 6. *M. Torreyi.*
Panicle oblong or cylindric; spikelets crowded.
Spikelet with an empty lemma above the perfect flower; panicle long-exserted. — 7. *M. ambigua.*
Spikelets without an empty lemma; panicle short-exserted. — 8. *M. commutata.*
Basal hairs as long as the lemma. — 9. *M. comata.*
Glumes shorter than the lemma.
Glumes less than half as long as the lemma, the first minute or wanting. — 10. *M. Schreberi.*
Glumes one half to two thirds as long as the lemma. — 11. *M. tenuiflora.*
Empty glumes lanceolate or ovate.
Lemma awnless or nearly so.
Glumes more than half as long as the lemma, acuminate. — 12. *M. cuspidata.*
Glumes less than half as long as the lemma.
Plants with strong perennial scaly rootstocks.
Glumes less than one fourth as long as the lemma; stem diffuse, decumbent and creeping. — 10. *M. Schreberi.*
Glumes one third as long as the lemma or longer; stem erect or decumbent, tufted. — 13. *M. Richardsonis.*
Plant annual; rootstocks if any slender. — 14. *M. simplex.*
Lemma long-awned. — 11. *M. tenuiflora.*
Panicle open, its branches at last long and spreading.
Leaf-blades 2–7 cm. long; awns 1–4 mm. long; stem 1–3 dm. high.
Secondary branches of the panicle single; basal leaves short, strongly recurved. — 15. *M. gracillima.*
Secondary branches of the panicle fascicled; basal leaves not recurved. — 16. *M. pungens.*
Leaf-blades 1–3 dm. long; awns 5–20 mm. long; stem 6–10 dm. high. — 17. *M. capillaris.*

1. **M. sobolífera** (Muhl.) Trin. Stem erect or ascending, sparingly branched, 4–8 dm. high, scabrous below the glabrous nodes, leafy towards the top; leaf-blades 8–12 cm. long, 4–6 mm. wide, scabrous; panicle slender, loosely flowered, 1–1.5 dm. long; spikelets 1.5–2 mm. long; glumes abruptly cuspidate or acute, two thirds to three fourths as long as the scabrous lemma. Rocky woods: N.H.—Minn.—Okla.—Va. *Canad.—Allegh.* S–O.

2. **M. mexicàna** (L.) Trin. *Fig. 59.* Perennial; stem decumbent, prostrate, or erect, smooth, branched and very leafy; leaf-blades scabrous, 4–15 cm. long, 2–6 mm. wide; panicle 5–15 cm. long; spikelets 2.5–3 mm. long; empty glumes somewhat unequal, scabrous on the keel. Wet meadows and swamps: N.B.—N.C.—Tex.—Colo.—Wyo.—N.D. *E. Temp.—Submont.* Je–S.

3. **M. polystàchya** Mack. & Bush. Stem about 6 dm. high, much branched, erect; leaves 3–6 cm. long, 2–4 mm. wide; panicle long-exserted, 5–10 cm. long, densely flowered; spikelets nearly sessile, 2 mm. long; hairs at the base of the lemma copious, one third to half as long as the glume. Open rocky woods: Mo.—Iowa. Au–O.

4. **M. foliòsa** Trin. Stem 3–10 dm. high, erect or ascending, leafy; leaf-blades 5–15 cm. long, 3–6 mm. wide; panicle long-exserted, oblong or cylindric, 8–15 cm. long; spikelets usually purple, on erect branches; glumes about as long as the lemma. Swamps: Me.—N.D.—Okla.—Colo. *Canad.—Allegh.* Au–S.

5. **M. racemòsa** (Michx.) B.S.P. Perennial; stem erect, 3–10 dm. high, branched, smooth; leaf-blades 5–12 cm. long, 2–6 mm. wide, scabrous; panicle 5–10 cm. long, dense, usually interrupted; branches 1–2.5 cm. long, erect; spikelets much crowded; lemma one half to two thirds as long as the empty glumes, acuminate to awn-pointed. *M. glomerata* Torr. Wet places: Newf.—N.J.—Mo.—N.M.—Utah—B.C. *Boreal—Plain—Mont.* Jl–S.

6. **M. Tórreyi** (Kunth) Hitchc. Stem erect or ascending, freely branched, leafy; panicle short-exserted, 1–2 dm. long, slender; spikelets on long erect branches, green or straw-colored, 2.5–3 mm. long; glumes acuminate or sometimes aristate, shorter than the lemma; lemma scabrous, mucronate or awn-tipped. *M. sylvatica* Torr. [B]. *M. umbrosa* Scribn. Moist woods or banks; N.B.—Man.—Neb.—Okla.—N.C. *Canad.—Allegh.* Au–O.

7. **M. ambígua** Torr. Stems erect, about 3 dm. high; leaves 2.5–7.5 cm. long, 2–4 mm. wide, scabrous; panicle 2.5–7.5 cm. long, the branches appressed; glumes awn-pointed, the second 4 mm. long; lemma scabrous, villous; awn 2–3 times as long. Lake shores: Minn. Au–S.

8. **M. commutàta** (Scribner) Bush. Perennial; stem branched from the base, very leafy; leaf-blades 4–15 cm. long, 2–6 mm. wide; panicle 4–15 cm. long, 2–6 mm. wide; empty glumes unequal, glabrous on the keel; awn of the lemma 4–10 mm. long; *M. mexicana commutata* Scribner. Meadows: N. Y.—S.D.—Okla.—Va. *Canad.—Allegh.*

9. **M. comàta** (Thurb.) Benth. Perennial; stem 3–7 dm. high, erect, smooth; leaf-blades 5–12 cm. long, erect, flat, scabrous; panicle often tinged with purple, 5–10 cm. long, dense, 1–15 cm. thick; empty glumes equal or the second a little longer, 2.5–3 mm. long; lemma a little shorter; awn 6–10 mm. long. Prairies and river valleys: S.D.(?)—Mont.—Colo.—Calif.—Wash. *Submont.—Mont.* Jl–S.

10. **M. Schrèberi** Gmel. Perennial, with a creeping, scaly rootstock; stem 3–9 dm. long; leaf-blades 3–9 cm. long, scabrous; panicle 5–20 cm. long, 2–5 mm. thick, lax; empty glumes small, the first often wanting; lemma without the awn about 2 mm. long, strongly scabrous; awn 1–4 mm. long. *M. diffusa* Willd. [B]. Dry hills and woods: Me.—Fla.—Tex.—w Neb.—Minn. *E. Temp.—Plain.* Au–S.

11. **M. tenuiflòra** (Willd.) B.S.P. Stem erect or ascending, 4–8 dm. high, retrorsely pubescent below, the nodes pubescent; panicle 1.5–3 dm. long, loosely flowered; spikelets 3–4 mm. long; glumes ovate, abruptly acuminate, scabrous, one half to two thirds as long as the lemma, the awn of the latter 5–10 mm. long. Rocky woods: Mass.—Minn.—Tex.—Ala. *Canad.—Allegh.* Au–S.

12. **M. cuspidàta** (Torr.) Rydb. Somewhat tufted perennial, occasionally with elongated scaly rootstocks; stem slender, 3–6 dm. high, glabrous; leaf-blades 2.5–10 cm. long, less than 2 mm. wide, involute-setaceous at least when dry; panicle 4–12 cm. long; empty glumes about 2 mm. long, scabrous on the keel; flowering glume long-acuminate, about 3 mm. long. *Sporobolus brevifolius* (Nutt.) Scribn. [G]. Dry soil: Man.—Mo.—Colo.—Alta. *Plain—Submont.* Jl–S.

13. **M. Richardsònis** (Trin.) Rydb. More or less tufted perennial, with a strong rootstock; stem 1–4.5 dm. high, erect or decumbent at the base, slender, smooth; leaves 1–5 cm. long, usually involute; panicle 1–7 cm. long, 2–4 mm. thick; empty glumes ovate, obtuse, or abruptly acute, about 1 mm. long; lemma 2.5–3 mm. long, long-acuminate. *Vilfa Richardsonis* Trin. *S. Richardsonis* Merr. [G]. *S. brevifolius* Nash [B]. Prairies and meadows: Que.—Me.—N.M.—Calif.—B.C. *Boreal—Plain—Mont.* Je–S.

14. M. símplex (Scribn.) Rydb. Cespitose leafy annual; stem 5–15 cm. high, smooth, with short lower internodes; leaf-blades 1–4 cm. long, 1–2 mm. broad, scabrous on margins and nerves above; panicle 2–5 cm. long, 2–4 mm. thick; empty glumes less than 1 mm. long, ovate, obtuse or truncate; lemma 3-nerved, mucronate, scabrous above, 2–2.3 mm. long. *Sporobolus simplex* Scribn. Wet places: Neb.—w Kans.—N.M.—Mont. *Plain—Mont.* Au–S.

15. M. gracíllima Torr. Cespitose perennial; stem 1–4 dm. high, erect or decumbent at the base; leaves mostly basal; blades involute-setaceous, somewhat scabrous; panicle 5–20 cm. long; glumes unequal, the first about 1.5 mm. long, the second 2 mm. long, awn-pointed; lemma 2.5–3 mm. long, scabrous; awn 2–4 mm. long. RING-GRASS. Plains and foothills: Tex.—Kans. —Colo.—N.M. *Plains—Submont.* Jl–O.

16. M. púngens Thurb. Perennial, with a cespitose creeping rootstock; stems decumbent at the base, 1.5–4 dm. high; leaf-blades 2.5–5 cm. long, involute, rigid, scabrous; panicle 7–15 cm. long, open, the branches 5–7 cm. long: glumes 2–2.5 mm. long; lemma 3–4 mm. long, scabrous; awn short. BLOW-OUT GRASS. PURPLE HAIR-GRASS. Sand hills and bad-lands: Tex.—Neb.—Utah—Ariz. *Plain—Son.* Jl–Au.

17. M. capillàris (Lam.) Trin. Cespitose perennial; stem erect, 6–10 dm. high; sheaths overlapping; blades 1–3 dm. long, involute; panicle 2–4 dm. long, the branches capillary, in age spreading; spikelets purple, 4 mm. long; glumes equal, acute or the second awn-pointed, half as long as the lemma; awn of the lemma capillary. Dry sandy soil: Mass.—Mo.—se Kans.—Tex.—Fla. *E. Temp.—Ozark.* S–N.

26. BRACHYÉLYTRUM Beauv.

Perennials, with short knotty rootstocks. Spikelets few, in a narrow panicle, 1-flowered. Empty glumes small, unequal, or the first one wanting. Lemma firm, narrow, 5-nerved, ending in a long straight awn. Palet firm, nearly as long as the lemma, 2-nerved, sulcate on the back. Grain oblong, free.

1. B. eréctum (Schreber) Beauv. Stem erect, 5–10 dm. high; sheath retrorsely hispid; leaf-blades 8–15 cm. long, 1–2 mm. wide, scabrous; panicle 1–2 dm. long; spikelet, excluding the awn, 1 cm. long; *B. aristatum* Beauv. Rocky woods: Newf.—N.C.—Kans.—Minn. *Canad.—Allegh.* Jl–Au.

27. SPORÓBOLUS R. Br. DROP-SEED, RUSH-GRASS.

Perennials or rarely annuals. Inflorescence a panicle, either open or narrow and spike-like. Spikelets usually 1-flowered; flowers perfect. Empty glumes 2, the first shorter than the second. Lemma usually longer than the second glume. Palet 2-nerved, often deeply 2-cleft, about equaling the lemma. Stamens 2 or 3. Styles short, distinct; stigmas plumose. Grain free from the lemma and readily dropping off.

Panicle contracted, spike-like.
  Perennials; panicles terminal only; upper sheaths 7 cm. long or more.
    Leaves glabrous or nearly so.
      Lemma and palet pubescent below.
        Lemma two thirds as long as the palet. — 1. *S. clandestinus.*
        Lemma equaling the palet. — 2. *S. canovirens.*
      Lemma glabrous, equaling the palet. — 3. *S. asper.*
    Leaves, at least the lower ones, papillose-hirsute. — 4. *S. pilosus.*
  Annuals; panicles both terminal and lateral; upper sheaths 4 cm. long or less.
    Spikelets 4 mm. long; leaves pubescent. — 5. *S. vaginiflorus.*
    Spikelets 2.5–3 mm. long; leaves glabrous. — 6. *S. neglectus.*
Panicle open, its branches more or less spreading, at least in age.
  Perennials.
    First glume half as long as the second or less; plants not with long scaly rootstock.
      Branches of the panicle verticillate.

Spikelets 1.5 mm. long, green. 7. *S. argutus.*
Spikelets 2.5–3 mm. long, purple. 8. *S. ejuncidus.*
Branches of the panicle scattered.
Spikelets about 2 mm. long; first glume lanceolate.
Pedicels not exceeding the spikelets; first glume acute.
Sheaths naked or sparingly ciliate at the throat; empty glumes glabrous. 9. *S. airoides.*
Sheaths with a conspicuous tuft of hairs at the throat; empty glumes scabrous on the keel.
Panicle always exserted, oblong, comparatively narrow, its lower branches but little, if any, exceeding the upper ones. 10. *S. flexuosus.*
Panicle usually more or less included in the sheath, its lower branches much exceeding the upper ones. 11. *S. cryptandrus.*
Pedicels much exceeding the spikelets; first glume narrow, acuminate. 12. *S. texanus.*
Spikelets 4.5–5 mm. long; first glume subulate, usually awned. 13. *S. heterolepis.*
First glume almost equaling the second; plant with a long scaly rootstock. 14. *S. asperifolius.*
Annuals; empty glumes almost equal, ovate. 15. *S. confusus.*

1. **S. clandestìnus** (Spreng.) Hitchc. Stems tufted, 4–12 dm. high; lower leaf-blades 7–30 cm. long, 2–4 mm. wide, with scabrous margins and filiform involute tip; panicle 5–15 cm. long, often included in the uppermost sheath; first glume half as long as the lemma, the second half as long as the palet. *S. asper* Am. auth. [B], not Kunth. Sandy fields and hills: Conn.—Fla.—Tex.—Kans. *E. Temp.* Au–S.

2. **S. canóvirens** Nash. Stem 3–10 dm. high, slender; leaf-blades 2.5 dm. long or less, 1–3 mm. wide, hirsute near the base, attenuate; panicle 5–7 cm. long, slender; spikelets 5–6 mm. long; glumes acuminate, unequal; lemma and palet acute. Sandy soil: Tenn.—Kans.—Tex.—Miss. *Ozark.*

3. **S. ásper** (Michx.) Kunth. Stem stout, 4–10 dm. high; leaf-blades 1–5 dm. long, 2–4 mm. wide, pilose at the base, with a long scabrous involute tip; panicle 8–25 cm. long, partly included; spikelets 5–6 mm. long; glumes unequal, obtuse or acutish, the first half as long as the lemma. *S. longifolius* Wood [B]. Dry soil: Me.—Fla.—Tex.—N.D. *E. Temp.* Au–S.

4. **S. pilòsus** Vasey. Stem 3–4.5 dm. high, stout; leaf-blades 7–15 cm. long, 2–4 mm. wide, rigid, attenuate at the apex; panicle 5–7.5 cm. long, included at the base; spikelets 5 mm. long; glumes unequal, glabrous, obtuse; lemma glabrous, equaling the second glume, a little longer than the obtuse lemma. Dry soil: Mo.—Kans.—Tex. *Ozark.* Au–S.

5. **S. vaginiflòrus** (Torr.) Wood. Stem tufted, 2–6 dm. high, erect or spreading; leaf-blades 3–7.5 cm. long, 2 mm. wide, involute toward the end; panicles several, 2–5 cm. long, included in the sheaths; spikelets 4 mm. long; glumes subequal, acuminate, nearly as long as the acuminate scabrous-puberulent lemma. Sterile fields and waste places: Me.—Ga.—Tex.—Neb.—Minn. *E. Temp.* Au–S.

6. **S. neglèctus** Nash. Stem 1.5–3 dm. high, erect or decumbent at the base; sheaths inflated; leaf-blades 3–7.5 cm. long, 2 mm. wide, scabrous and hairy near the base above, smooth beneath, attenuate at the apex; terminal panicle 2–6 cm. long, usually more or less included in the upper sheath; lateral spikelets wholly included; glumes acute; lemma acute, glabrous, a little longer than the second glume, and equaling the lemma. Dry soil: Mass.—Ky.—Kans.—N.D. *Allegh.* Au–S.

7. **S. argùtus** (Nees) Kunth. Tufted perennial; stem 1–4 dm. high, erect or decumbent at the base; leaf-blades 2.5–5 cm. long, 2–4 mm. wide, often sparingly hairy at the base; panicle-branches 1–2.5 cm. long; spikelets 1.5 mm. long. Plains: Tex.—Kans.—Colo.—Mex.; W. Ind., C. Am. and S. Am. *Son.* Ap–O.

8. **S. ejúncidus** Nash. Stem tufted, 4–7 dm. high, leafy at the base, naked above; leaf-blades 1–2.5 dm. long, setaceous, involute; spikelets 3 mm. long; first glume about one-third as long as the second; second glume, lemma, and palet equal in length; lemma glabrous. *S. junceus* (Michx.) Kunth [G]. *S. gracilis* Merr. Dry sandy soil: Va.—Fla.—Tex.—Kans. *Ozark.* Au–S.

9. **S. airoìdes** Torr. *Fig. 60.* Densely tufted perennial; stem 5–10 dm. high, erect; leaf-blades sometimes sparingly hairy at the base, 1–3 mm. wide, involute, 5–35 cm. long; panicle 1–4 dm. long, its branches at length widely spreading; spikelets 1.5–2 mm. long; glumes acute, glabrous. Dry plains and river valleys: Tex.—Mo.—Minn.—Mont.—Wash.—Calif. *Plain—Submont.* Au–S.

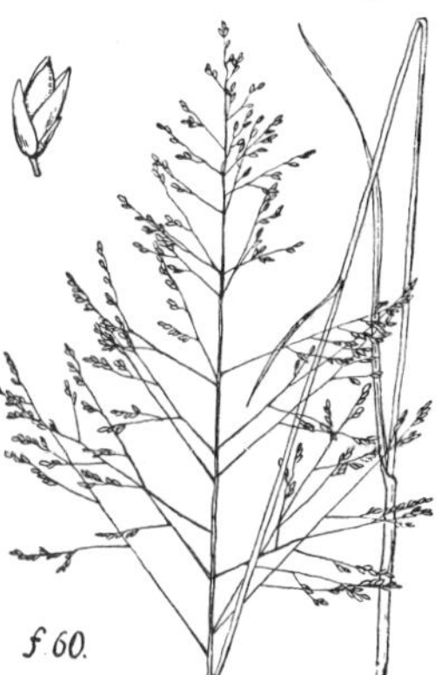
f 60.

10. **S. flexuòsus** (Thurb.) Rydb. Tufted perennial; stem 3–8 dm. high; leaf-blades 5–20 cm. long, 2–4 mm. wide, usually flat, somewhat scabrous above; inflorescence 2–4 dm. long; branches 3–6 cm. long, spreading or reflexed; spikelets about 2 mm. long, usually lead-colored; empty glumes lanceolate, somewhat scabrous. Sandy soil: Kans.—Tex.—Nev.—Ariz.; Mex. *Son.—St. Plains.* Jl–S.

11. **S. cryptándrus** (Torr.) A. Gray. More or less tufted perennial; stem 4–10 dm. long, erect, glabrous; leaf-blades 7–15 cm. long, 2–4 mm. wide, flat, glabrous beneath, scabrous above; panicle 1.5–2.5 dm. long; spikelets lead-colored, 2–2.5 mm. long; glumes scabrous on the keel. Sandy soil: Mass.—Pa.—Ariz.—Wash.; Mex. *Temp.—Submont.* Au–O.

12. **S. texànus** Vasey. Stem 3–6 dm. high, branched below; lower sheaths papillose-hirsute; leaf-blades erect, 4–15 cm. long, 3–5 mm. wide, smooth beneath, scabrous above; panicle 5–10 cm. long, included at the base; spikelets 2–2.2 mm. long; glumes glabrous, the first one narrow, acuminate, less than half as long as the second. Plains: c Kans.—Tex.—N.M. *St. Plains—Son.* Jl–A.

13. **S. heterólepis** A. Gray. Tufted perennial; stem 3–10 dm. high, erect; leaf-blades involute, glabrous, with rough margins and midribs, 2–5 dm. long; panicle 7–25 cm. long, open, exserted; branches ascending; glumes smooth and glabrous, the first one subulate, 2–3 mm. long, the second lanceolate, 4–5 mm. long, awn-pointed; lemma obtuse or acute. Dry soil: Que.—Pa.—Tex.—S.D.—Sask. *E. Temp. Plain—Submont.* Au–S.

14. **S. asperifòlius** (Nees & Meyen) Thurb. Perennial with a cespitose creeping rootstock; stem 1–5 dm. high, erect from a decumbent base; leaf-blades 2.5–10 cm. long, 2–3 mm. wide, flat, very scabrous above; panicles 7–20 cm. long, usually included at the base; branches usually spreading in age; spikelets 1.5 mm. long; glumes acute, glabrous, somewhat scabrous; lemma slightly exceeding them. Sandy or dry soil: Sask.—Mo.—Tex.—Calif.—B.C.; Mex. *Prairie—Plain—Submont.* Je–S.

15. **S. confùsus** (Fourn.) Vasey. Tufted annual; stem 1–3 dm. high, slender, branched below; sheaths short, loose, glabrous; leaf-blades 1–4 cm. long, 0.5–1.5 mm. wide; panicle open, 3–20 cm. long; branches spreading; spikelets 1–1.5 mm. long; empty glumes shorter than the lemma, glabrous or pubes-

cent; lemma pubescent. Meadows, especially in sandy soil: Neb.—Tex.—Ariz. —Ida.; Mex. *Plains—Son—Submont.* Jl–S.

28. PHLÈUM L. Timothy.

Annuals or perennials. Inflorescence a dense spike-like panicle. Spikelets 1-flowered, compressed; flower perfect. Empty glumes 2, persistent, compressed, keeled, oblique at the summit, awned. Lemma shorter, membranous, truncate, denticulate. Styles distinct, long; stigmas subplumose. Grain free.

Spikes usually elongate-cylindric; awns less than one half the length of the outer glumes. 1. *P. pratense.*
Spikes short, ovoid or oblong; awns about one half the length of the outer glumes. 2. *P. alpinum.*

1. **P. praténse** L. *Fig. 61.* Short-lived perennial, with a bulbiform thickened base; stem erect, glabrous and smooth, 3–15 dm. high; sheaths often exceeding the internodes, the upper long and close; leaf-blades 7–20 cm. long, 4–6 mm. wide, usually somewhat scabrous; spike 3–20 cm. long, 5–8 mm. thick; outer glumes 3–4 mm. long, ciliate on the keel; awns about 1 mm. long. Meadows; native but also often escaped from cultivation: N.S.—Fla.—Calif.—B.C.; Eurasia. *Temp.—Subalp.* Je–O.

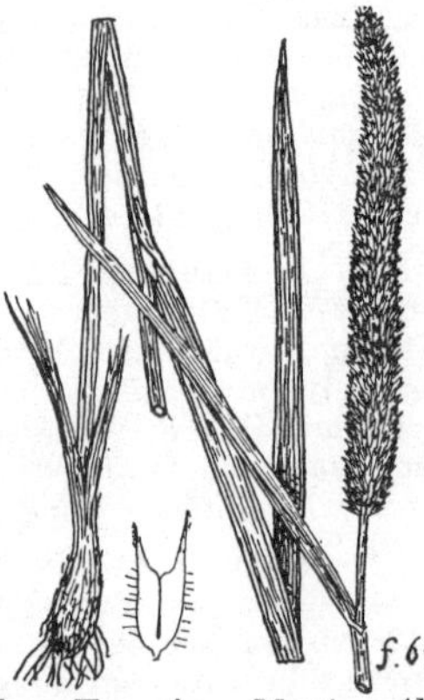

f. 61.

2. **P. alpìnum** L. Short-lived perennial; stem 1–5 dm. high, erect or sometimes decumbent at the base, smooth; sheaths often shorter than the internodes, the upper usually inflated; leaf-blades scabrous above, 2–7 cm. long, 3–8 mm. wide; spike 1–5 cm. long, 6–12 mm. thick; outer glumes 3 mm. long, strongly ciliate on the keel; awns 1.5–2 mm. long. Wet places: Lab.—N.H.—S.D.—N.M.—Calif.—Alaska; Eurasia. *Mont.—Alp.* Je–S.

29. ALOPECÙRUS L. Foxtail.

Annuals or perennials. Inflorescence a cylindric dense spike-like panicle. Spikelets articulate under the empty glumes, 1-flowered, compressed; flower perfect. Empty glumes 2, acute, sometimes short-awned, more or less united at the base, compressed-keeled. Lemma hyaline, obtuse, with a dorsal awn or point, their margins usually more or less united at the base. Styles distinct or nearly so; stigmas long, subplumose.

Glumes united one fourth their length, long-ciliate on the keel.
  Spike about 5 mm. thick; glumes 2–2.5 mm. long.
    Awn bent, twice as long as the lemma. 1. *A. geniculatus.*
    Awn scarcely exceeding the glumes. 2. *A. aristulatus.*
  Spike 8–12 mm. thick; glumes 4–6 mm. long. 3. *A. pratensis.*
Glumes united half their length, glabrous or hispidulous on the keel. 4. *A. myosuroides.*

1. **A. geniculàtus** L. Stem glabrous or nearly so, decumbent and branched at the base, 1.5–5 dm. high; leaves slightly scabrous; spike 2.5–7 cm. long; spikelets about 3 mm. long. Wet meadows, banks and streams: Newf.—S.D.—Okla.—Miss.—Fla.; Eur. *E. Temp.* My–Jl.

2. **A. aristulàtus** Michx. Somewhat tufted perennial; leaf-blades 5–15 cm. long, 3–4 mm. wide, scabrous, especially above; spike 3–8 cm. long, 4–6 mm. thick; outer glumes 2–2.5 mm. long, acutish; lemma somewhat shorter, obtuse, glabrous; awn inserted at or near the middle. *A. geniculatus* Am. auth. [B], not L. *A. fulvus* Am. auth., not Smith. Wet meadows: Me.—Pa.—Calif.—Alaska. *Plain—Subalp.* Je–S.

3. **A. praténsis** L. Perennial; stem 3–10 dm. high; sheaths loose; the upper ones inflated; leaf-blades 4–9 cm. long, 2–4 mm. wide, scabrous; panicle spike-like, 5–10 cm. long; glumes acute, long-ciliate, equaling the lemma; awn inserted one-fourth from the base of the lemma, exceeding it by about 5 mm. Meadows and pastures: N.S.—N.Y.—Kans.—Man.; nat. from Eur. Je–Jl.

4. **A. myosuroìdes** Huds. Perennial; stem erect or decumbent at the base, 3–6 dm. high, glabrous; leaf-blades 5–15 cm. long, scabrous; panicle slender, 4–10 cm. long, 4–8 mm. thick; spikelets 6–7 mm. long; awn twice as long as the glume, the latter slightly shorter than the lemma. *A. agrestis* L. [B, G]. Waste places and railroad banks: Mass.—Mo.—e Kans.—Ark.—D.C.; Calif.—Wash.; adv. from Eur.

30. POLYPÒGON Desf. BEARD GRASS.

Annuals or rarely perennials. Inflorescence a dense, contracted, spike-like panicle. Spikelets articulate below the empty glumes, crowded, 1-flowered. Flowers perfect. Empty glumes about equal, each terminating in a slender awn. Lemma thin, emarginate or 2-toothed at the apex, awned or awn-pointed on the back. Stamens 1–3. Styles short, distinct; stigmas plumose.

1. **P. monspeliénsis** (L.) Desf. Simple annual; stem 1–6 dm. high, erect or decumbent at the base; leaf-blades 4–15 cm. long, 3–6 mm. wide, scabrous panicle 2–10 cm. long; empty glumes about 2 mm. long, bearing a more or less bent awn 4–6 mm. long; lemma shorter, erose-truncate, hyaline; awn 0.5 mm. long. Waste places: S.C.—N.H.—B.C.—Calif.; Mex.; naturalized from Eur. *Plain—Submont.* Ap–Au.

31. CÍNNA L. REED GRASS.

Tall perennials with broad flat leaves. Inflorescence a large often nodding panicle. Spikelets numerous, 1-flowered, articulate under the empty glumes. Flowers perfect. Empty glumes 2, persistent, keeled, acute. Lemma similar, short-awned or awn-pointed just below the apex. Palet 1-nerved, keeled, the keel minutely ciliate. Stamen 1. Styles short, distinct; stigmas plumose.

Spikelets 4–5 mm. long; first glume much shorter than the second; panicle narrow. 1. *C. arundinacea.*
Spikelets 3 mm. long; first glume about equaling the second; panicle open. 2. *C. latifolia.*

1. **C. arundinàcea** L. Stem 5–15 dm. high, erect; leaf-blades 2–3 dm. long, 8–14 mm. wide, slightly scabrous; panicle 1.5–3 dm. long, branches ascending or in fruit erect; glumes scabrous, the second as long as the lemma; awn of the latter minute or wanting. Moist woods: Newf.—N.C.—Tex. *Canad.* Au–S.

2. **C. latifòlia** (Trev.) Griseb. Stem 6–15 dm. high, erect; leaf-blades 1–2.5 dm. long, 4–12 mm. wide, scabrous; panicle 1–2.5 dm. long; spikelets 3 mm. long; empty glumes equal, acute, hispidulous on the keel; lemma somewhat shorter, 2-toothed at the apex; awn 1–2 mm. long. *Cinna pendula* Trin. Damp woods: Newf.—N.C.—Minn.—Utah—B.C. *Submont.—Mont.* Jl–S.

32. ARCTAGRÓSTIS Griseb.

Tufted tall perennials. Leaf-blades flat. Panicle slightly branched, narrow. Spikelets 1-flowered, the rachilla articulate above the glumes, smooth, not produced above the floret. Glumes slightly unequal, acute, membranous. Lemma obtuse, or 3-toothed at the apex, 5-nerved, unarmed. Palet as long as the lemma or shorter, 2-nerved. Stamens 2 or 3. Style short. Seeds oblong, included but not adherent.

1. **A. poaeoìdes** Nash. Rootstock running, branched; stem 6–8 dm. high, scaly at the base; sheaths striate, shorter than the internodes, rough; blades erect, 1.5–2.5 dm. long 5–6 mm. wide; panicle about 1.5 dm. long, the branches fascicled, hispidulous; spikelets many, 2.5–3 mm. long; first glume 1-nerved,

two thirds as long as the 3-nerved second; lemma 5-nerved, strongly hispidulous, oblong-lanceolate; palet hispidulous, about equaling the lemma. It has been confused with the Arctic *A. latifolia.* Marshes and tundras: Yukon—Sask.—Man. *Subarctic.* Jl–Au.

33. AGRÓSTIS L. Red-top, Bent-grass, Tickle-grass.

Annuals or perennials, mostly tufted. Inflorescence paniculate. Spikelets numerous, small, 1-flowered, articulate above the empty glumes. Empty glumes 2, persistent, membranous, keeled, acute, awnless. Lemma shorter, or barely equaling the empty glumes, thin-hyaline, awnless or awned. Palet small, thin-hyaline, or often wanting. Stamens usually 3. Styles short, distinct; stigmas plumose.

Palet evident, 2-nerved, at least one fourth as long as the lemma. — 1. *A. palustris.*
Palet lacking or minute.
  Lemma awnless or with a minute awn-point.
    Panicle dense and narrow; branches ascending or erect, flower-bearing to near the base.
      Panicle lobed or interrupted; branches densely verticillate and flower-bearing to the base; empty glumes narrowly lanceolate, gradually attenuate; ligules 4–5 mm. long, lacerate; leaf-blades 6–10 mm. wide. — 2. *A. grandis.*
      Panicle usually contiguous; branches few and some naked at the base; empty glumes lanceolate, abruptly pointed; ligules 2–4 mm. long, entire or toothed; leaf-blades 2–5 mm. wide. — 3. *A. asperifolia.*
    Panicle open, branches more or less spreading in age, or sometimes reflexed.
      Stem and usually also the leaf-blades erect; panicle triangular-ovoid or pyramidal.
        Leaf-blades filiform, 1 mm. wide or less, usually involute; the basal ones numerous; upper sheaths close. — 4. *A. oreophila.*
        Leaf-blades usually flat, 1–3 mm. wide; upper sheaths loose. — 5. *A. hyemalis.*
      Stem weak, usually decumbent or prostrate at the base; blades lax; panicle oblong. — 6. *A. perennans.*
  Lemma with an awn.
    Plant annual; awn slender, barbellate, twice as long as the spikelet. — 7. *A. Elliottiana.*
    Plant perennial; awn smooth, slightly exceeding the spikelet.
      Spikelets 2 mm. long or less; branches of the inflorescence ascending. — 8. *A. canina.*
      Spikelets 2.5–3 mm. long; lower branches of the inflorescence spreading. — 9. *A. borealis.*

1. A. palústris Huds. *Fig. 62.* Tufted perennial; stems 2–15 dm. high, erect or decumbent at the base; leaf-blades 4–15 cm. long, 3–8 mm. wide; panicle upright, 5–30 cm. long, with spreading branches, the lower usually verticillate; empty glumes 2–3 mm. long, pale or purplish, nearly equal, acute; lemma two-thirds to three-fourths as long, obtuse, seldom awned. *A. alba* L. (?) [G, B, R]. Wet meadows: Newf.—Fla.—Calif.—B.C.; Eurasia. *E. Temp. Plain—Submont.* Je–O.

2. A. grándis Trin. Tufted perennial; stem 5–15 dm. high; leaf-blades 1–2 dm. long, 1–10 mm. wide, very scabrous, abruptly contracted at the ligules; panicle 1–2 dm. long, dense, contracted; empty glumes about 3 mm. long, gradually tapering at the apex, light green, scarious-margined; lemma one-half to two-thirds as long, obtuse. Wet meadows: B.C.—Mont.—Neb.—N.M.—Calif. *W. Boreal—Plain—Mont.* Je–Jl.

3. **A. asperifòlia** Trin. Tufted perennial; stem 3–6 dm. high; blades 5–15 cm. long, 2–5 mm. wide, scabrous; panicle 5–10 cm. long, narrow and dense; empty glumes broadly lanceolate, 2.5–3 mm. long, usually abruptly pointed, often tinged with purple; lemma two thirds to three fourths as long, awnless; *A. exarata minor* Hook. *A. Drummondii* Torr. Wet meadows: Man.—Kans.—N.M.—Calif.—B.C. *W. Boreal—Plain—Mont.* Je–Au.

4. **A. oreóphila** Trin. Densely cespitose perennial; stems slender, 2–4 dm. high; leaf-blades filiform-setaceous, involute; those of the stem slightly broader, about 1 mm. wide, scabrous; panicle 1–2 dm. long; lower branches 5–10 cm. long; empty glumes slightly unequal, about 2 mm. long, pale or purple, acute. *A. hyemalis geminata* Hitchc., in part. Mountains: Que.—Me.—N.Y.—Pa.—S.D—Colo.—Utah—B.C. *Boreal—Mont.—Subalp.* Jl–S.

5. **A. hyemàlis** (Walt.) B.S.P. Tufted perennial; stem 2–8 dm. high; leaf-blades 2–3 mm. wide, flat, 5–10 cm. long; panicle large and diffuse, 2–6 dm. long; spikelets crowded at the ends of the branches; empty glumes nearly equal, purplish or pale, about 2 mm. long, acute or acuminate; lemma obtuse, two-thirds to three-fourths as long. *A. scabra* Willd. Open places: Lab.—Fla.—Calif.—Alaska; Mex. *Temp.—Subalp.*

6. **A. perénnans** (Walt.) Tuckerm. Stem glabrous, smooth or nearly so, 3–8 dm. high; leaf-blades 5–20 cm. long, 2–4 mm. wide; panicle 1–2 dm. long, its branches spikelet-bearing from the middle, the ultimate ones diverging; spikelets 1.5–2 mm. long; glumes acuminate, hispidulous on the keel, unequal, longer than the lemma. Woods: Me.—S.C.—Kans.—Minn. *Canad.—Allegh.* Jl–S.

7. **A. Elliottiàna** Schultes. Stem erect, slender, 1–4 dm. high, tufted; leaf-blades 1–5 cm. long, 2 mm. wide or less, scabrous; panicle open and drooping, narrow, 5–12 cm. long; spikelets 1.5 mm. long; awn 3–4 times as long as the lemma, inserted below the middle. Dry soil; S.C—Fla.—Tex.—Kans. *Ozark—Austral—Subtrop.* My–Jl.

8. **A. canìna** L. Tufted perennial; stem 2–6 dm. high, erect, slender; basal leaf-blades involute, 3–7 cm. long, 1 mm. wide, those of the stem broader, flat; panicle open, the branches whorled; spikelets 2.5–3 mm. long; glumes subequal, acute, scabrous on the keel; lemma about 2 mm. long; awn inserted just above the middle, bent and usually exserted. RHODE ISLAND BENT. Meadows: Newf.—Alaska—Colo.—Tenn.—N.W. *Boreal—Mont.;* also Eurasia.

9. **A. boreàlis** Hartm. Stem 1–4 dm. high; leaves mostly tufted at the base; blades narrowed, 1–2 mm. wide, 5–10 cm. long; panicle pyramidal, 5–15 cm. long, the lower branches whorled; glumes purple, equal, 2.5–3 mm. long, minutely scabrous on the keel above; lemma shorter than the glumes; awn bent, 1–3 mm. longer than the spikelet, attached at the middle. Mountains and exposed places. Labr.—N.C.—Man.—Alaska. *Subarct.—Boreal.*

34. CALAMAGRÓSTIS Adans. REED-GRASS.

Erect perennials. Inflorescence paniculate. Spikelets numerous, 1-flowered; flowers perfect; the rachilla usually produced beyond the flower; the prolongation with a hair-pencil. Empty glumes 2, persistent, narrow, keeled, acute or acuminate. Lemma much shorter, thickly hyaline, with a basal ring of long hairs, and a dorsal awn. Palet small, narrow, thin, hyaline. Styles short, distinct; stigmas plumose.

Awn of the lemma geniculate, exserted; callus-hairs usually much shorter than the glume.
  Awns of the lemma greatly exceeding the empty glumes; plant tufted; leaf-blades involute. 1. *C. purpurascens.*
  Awns of the lemma about equaling the empty glumes.
    Panicle spike-like; glumes 4–6 mm. long.

Glumes sharply keeled; spikelets strongly compressed; plant stoloniferous. 2. *C. montanensis.*
Glumes not strongly keeled; spikelets not strongly compressed; plant tufted. 3. *C. rubescens.*
Panicle not spike-like; spikelets strongly compressed; plant stoloniferous. 4. *C. Pickeringii.*
Awn of the lemma straight or nearly so, included; callus-hairs usually equaling the lemma.
Panicle open, the lower branches spreading; leaf-blades usually flat; callus-hairs nearly or quite equaling the lemma.
Empty glumes 4–6 mm. long, narrow, sharp-acuminate; awn of the lemma attached below the middle, exceeding the lemma. 5. *C. Langsdorfii.*
Empty glumes 2–4 mm. long; awn of the lemma attached at or above the middle, shorter than the lemma.
Spikelets 3–4 mm. long; panicle loosely flowered. 6. *C. canadensis.*
Spikelets 2–2.8 mm. long; panicle densely flowered. 7. *C. Macouniana.*
Panicle more or less contracted; callus-hairs shorter than the lemma.
Leaves and stem harsh and scabrous; ligules 2.5–8 mm. long, erose or lacerate.
Spikelets 4–5.5 mm. long; lemma 3.5–4 mm. long; glumes long-acuminate. 8. *C. inexpansa.*
Spikelets 3–4.5 mm. long; lemma 3–3.5 mm. long; glumes acute or short-acuminate.
Leaf-blades flat or nearly so; lemma short-acuminate. 9. *C. elongata.*
Leaf-blades strongly involute; lemma acute. 10. *C. americana.*
Leaves smooth, or scabrous at the apex and on the margins, soft.
Spikelets 3–4 mm. long; glumes acuminate or sharply acute. 11. *C. neglecta.*
Spikelets 2–2.5 mm. long; glumes obtuse or acutish. 12. *C. micrantha.*

1. C. purpuráscens R. Br. Tufted perennial; stem 3–6 dm. high; leaves numerous and crowded at the base; blades of the basal leaves 2–4 dm. long, 4–5 mm. wide, scabrous, rigid, more or less involute; those of the stem 7–15 cm. long; panicle spike-like, strict, 7–10 cm. long, purplish; spikelets 6–7 mm. long; lemma slightly shorter than the glumes, acute or acuminate; awn near the base, twisted below and bent at the middle. Mountains: Greenl.—S.D.—Colo.—Calif.—Alaska. *Mont.—Alp.* Jl–S.

2. C. montanénsis Scribn. Stoloniferous perennial; stem 2–4 dm. high, stiff, erect; leaf-blades 5–15 cm. long, strongly involute, scabrous; panicle spike-like, 5–8 cm. long, dense; empty glumes narrowly lanceolate, acute, 4–6 mm. long, scabrous, pale or purple; lemma one fourth shorter, thin, finely scabrous, awn borne one third from the base. Dry plains: Sask.—Minn.—S.D.—Ida.—Alta. *Plain—Submont.* Jl–Au.

3. C. rubéscens Buckl. Tufted perennial; stem 4–10 dm. high; leaf-blades 5–15 cm. long, scabrous, involute, stiff; panicle strict, spike-like, purplish, 5–15 cm. long; glumes 4–5 mm. long, acute, minutely scabrous; lemma about as long, awned below the middle. Grassy banks: Man.—B.C.—Calif.—Wyo. Jl–Au.

4. C. Pickeríngii A. Stem 3–10 dm. high, rigid, scabrous below the panicle; sheaths smooth; leaf-blades flat, 4–10 cm. long, 4–5 mm. wide, erect; panicle 7–12 cm. long; glumes acute, longer than the obtuse lemma; awn from below the middle; callus-hairs in the typical form very short. *C. breviseta* (A. Gray) Scribn. [B]. Represented in the range by var. *lacustris,* which is stouter, usually with involute leaves, and callus-hairs one half to two thirds as long as the lemma. Wet places: Newf.—N.Y.—Minn. *Boreal.* Au–S.

5. C. Langsdórfii (Link) Trin. Perennial, with a rootstock; stem 3–6 dm. high, erect, simple; leaf-blades 1–3 dm. long, 4–8 mm. wide, scabrous, usually flat and spreading; panicle 5–15 cm. long; empty glumes 4–6 mm. long,

long-acuminate, strongly scabrous, lead-colored, brown or purplish; lemma one fourth shorter, scabrous, awned below the middle. Alpine-arctic situations: Greenl.—N.C.—n Minn.—Sask.—N.M.—Calif.—Alaska; Eur. *Arct.—Mont.—Alp.* Jl–S.

6. **C. canadénsis** (Michx.) Beauv. *Fig. 63.* Perennial, with a rootstock; stems 5–15 dm. high, erect, simple; blades 1.5–3 dm. long, 2–8 mm. wide, scabrous, usually flat; panicle 1–2 dm. long, usually purple; branches spreading; empty glumes 3–3.5 mm. long, acute, or (in var. *acuminata*) 3.5–4 mm. long and acuminate, scabrous; lemma slightly shorter, scabrous, awn-bearing near the middle; awn slender, nearly equaling the glumes. Banks and swamps: Newf. — N.C. — N.M. — Calif. — Alaska. *Boreal—Plain—Subalp.* Jl–S.

7. **C. Macouniàna** Vasey. Perennial; stem 6–10 dm. high, erect, somewhat branched below, leafy; leaf-blades flat, 1.5–2.5 dm. long, panicle 7–10 cm. long, lanceolate or conic, open; spikelets crowded on the upper part of the branchlets; empty glumes about 4 mm. long, purplish, acute, finely scabrous on the back; lemma equaling the empty glumes. Wet places: Man.—Mo.—S.D.—Wash.; also reported from Mass.—N.J. *Boreal—Plain—Submont.*

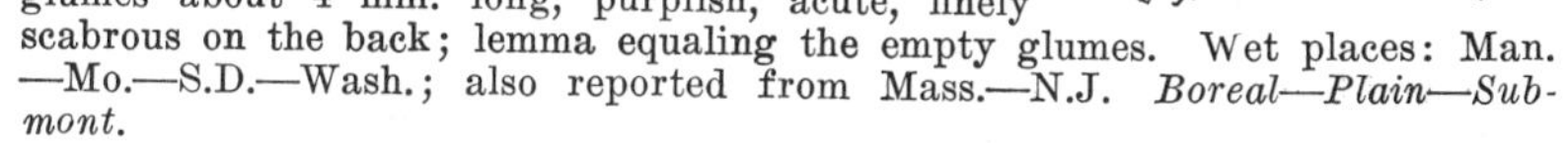

8. **C. inexpánsa** A. Gray. Perennial, with a rootstock; stem usually simple, 8–12 dm. high; leaf-blades 2–3 dm. long, 4 mm. wide or less, scabrous; panicle 1.5–2 dm. long, usually nodding at the summit; empty glumes 4–4.5 mm. long, very scabrous throughout, gradually acute; lemma scabrous, about as long, acutish; awn inserted below the middle, about equaling the lemma. Wet places: *Canad.—Allegh.* N.Y.—N.J.—S.D.—Kans.—Alta. *Plain—Submont.* Jl–Au.

9. **C. elongàta** (Kearney) Rydb. Tufted perennial; stem 7–12 dm. high; leaf-blades 1–3 dm. long, usually flat, 3–8 mm. wide, scabrous; panicle 1–2 dm. long, narrow but scarcely spike-like; glumes 3–4 mm. long, acute, very scabrous; lemma somewhat shorter, scabrous; awn attached below the middle. *C. hyperborea elongata* Kearney [B]. Wet meadows: Ont.—Pa.—Colo.—Calif.—B.C. *Boreal—Plain—Mont.* Jl–Au.

10. **C. americàna** Scribn. Cespitose perennial, with numerous marcescent sheaths at the base; stem 3–6 dm. high, scabrous above; leaf-blades 1–3 dm. long, 4–5 mm. wide, usually more or less involute, scabrous, stiff; panicle contracted, 7–15 cm. long, spike-like; empty glumes 3–4 mm. long, very scabrous; lemma nearly as long, acute, scabrous; awn attached below the middle. *C. hyperborea* Kearney, not Lange. *C. wyomingensis* Gand. Wet meadows: Lab.—Vt.—Neb.—Colo.—Calif.—Alaska. *Boreal—Plain—Submont.* Jl–Au.

11. **C. neglécta** (Ehrh.) Gaertn. Perennial, with a rootstock; stem 4–6 dm. high, slender; leaves narrow, involute, soft, 5–15 cm. long; panicle contracted, 6–10 cm. long; branches mostly erect; empty glumes about 4 mm. long, gradually acute; lemma about one fourth shorter, obtuse; awn attached near the middle. *C. stricta* Trin. Wet places: Greenl.—Me.—Minn.—Colo.—Ore.—Alaska; Eur. *Arct.—Boreal—Plain—Subalp.* Jl–S.

12. **C. micrántha** Kearney. Perennial, with a slender, creeping rootstock; stem 4–6 dm. high, slender; leaf-blades 5–20 cm. long, 1–1.5 mm. wide, more or less involute, filiform; panicle 5–8 dm. long, densely flowered, purple or pale; empty glumes 2 mm. long, scabrous; lemma slightly shorter; awn attached at or below the middle. Wet places: Sask.—Colo. *Plain—Mont.* Jl–Au.

35. **CALAMOVÍLFA** Hack. REED-GRASS, SAND-GRASS.

Tall perennials with horizontal rootstocks and elongated narrow leaf-blades. Inflorescence a panicle. Spikelets flattened, 1-flowered, crowded, the rachilla articulated above the empty glumes and not prolonged beyond the flower. Empty glumes 2, unequal, persistent, rigid, chartaceous, awnless, keeled, 1-nerved; lemma with a ring of long hairs at the base, enclosing a 2-keeled palet and a perfect flower. Stamens 3. Styles distinct; stigmas plumose.

Lemma and palet glabrous; spikelets 5–6 mm. long. 1. *C. longifolia.*
Lemma and palet pubescent; spikelets 7–8 mm. long. 2. *C. gigantea.*

1. **C. longifòlia** (Hook.) Hack. *Fig. 64.* Perennial, with a long, scaly rootstock; stem 6–20 dm. high, stout, smooth, and glabrous; leaf-blades 2–3 dm. long, stiff, more or less involute; panicle 1–4 dm. long, pale, narrow; empty glumes 6–8 mm. long, the first shorter than the second; lemma about as long as the second glume, glabrous; callus-hairs copious, half as long as the lemma. *Calamagrostis longifolia* Hook. Sandy soil: Ont.—Ind.—Colo.—Ida.—Sask. *Prairie—Plain—Submont.* Je–S.

2. **C. gigantèa** (Nutt.) Scribn. & Merr. Perennial, with a creeping rootstock; stem 1–2 mm. high; leaf-blades about 3 dm. long, 5–8 mm. wide, involute; panicle 3–7 dm. long, with ascending or spreading branches, naked at the base; second glume longer than the first and the lemma, the latter long-hairy on the back and the keel. Sandy places: Kans.—Tex.—Ariz. *Texan—Son.*

36. **HÓLCUS** L. VELVET GRASS.

Annual or perennial grasses, with rootstocks. Inflorescence paniculate, often narrow and interrupted below. Spikelets 2-flowered, the lower flower perfect, the upper staminate, the rachilla articulate below the glumes. Glumes keeled, the first 1-nerved, the second 3-nerved, acute or short-awned. Lemmas membranous, at length rigid, enclosing the palets, the first awnless, the second with a slender hooked dorsal awn. Palet 2-keeled. Style distinct; stigmas plumose. *Nothoholcus* Nash.

1. **H. lanàtus** L. Densely and softly pubescent; stem 5–10 dm. high, erect; leaf-blades 2.5–15 cm. long, 4–12 mm. wide; spikelet 4 mm. long; glumes villous; lemmas 2 mm. long, glabrous, the upper one 2-toothed, the awn inserted just below the apex. *N. lanatus* Nash [BB]. Meadows and waste places: N.S.—Minn.—e Kans.—Tenn.—N.C.; Calif.—Wash.; nat. from Eur.

37. **DESCHÁMPSIA** Beauv. HAIR-GRASS.

Perennials or annuals. Inflorescence a terminal contracted or open panicle. Spikelets 2-flowered, the rachilla articulated above the empty glumes and prolonged beyond the flowers. Empty glumes 2, persistent, keeled, acute, membranous, usually somewhat shining. Lemma thin-membranous, almost hyaline, entire or toothed, acute, obtuse, or truncate and denticulate at the apex, each bearing a slender dorsal straight or geniculate awn, which is twisted at the base. Palet narrow, 2-nerved. Stamens 3. Styles distinct; stigmas plumose.

Empty glumes not extending beyond the apex of the upper lemma; cespitose tufted perennials. 1. *D. caespitosa.*
Empty glumes extending beyond the apex of the upper lemma; perennials with rootstocks, but not tufted. 2. *D. atropurpurea.*

1. **D. caespitòsa** (L.) Beauv. *Fig. 65.* Stem 3–10 dm. high; basal leaf-blades 2–5 cm. long, 1.5–3 mm. wide, flat or in drier situations involute, rather firm; stem-leaves 3–10 cm. long; panicles 1–3 dm. long, 5–15 cm. wide; empty glumes 3.5–5 mm. long, lanceolate, acute or acuminate; lemma 3–4 mm. long, purple or lead-colored. Wet meadows and swamps: Newf.—N.J.—N.M.—Calif.—Alaska; Eurasia. *Boreal—Submont.—Subalp.* Je–S.

2. **D. atropurpùrea** (Wahl.) Scheele. Perennial, with a cespitose rootstock; stem glabrous, 1.5–5 dm. high; leaf-blades 2–5 mm. wide, 5–12 cm. long, flat and flaccid; panicle narrow, 2–10 cm. long, with short erect branches; lemma about 3 mm. long, erose-truncate at the apex; awn attached about the middle, bent. *D. latifolia* (Hook.) Vasey. *D. Hookeriana* Scribn. Wet places: Greenl.—N.H.—Colo.—Calif.—Alaska; n Eur. *Boreal—Subalp.—Alp.* Jl–S.

38. TRISÈTUM Pers. FALSE-OAT.

Annual or perennial tufted grasses, with flat leaf-blades. Inflorescence a spike-like contracted or open panicle. Spikelets usually 2-flowered, rarely 3–6-flowered, the flowers perfect, or the upper one staminate. Empty glumes membranous, unequal, acute, entire at the apex, awnless, persistent. Lemma 2-toothed at the apex, the teeth acuminate and often terminating in a bristle or slender awn; awn often twisted, inserted below the apex and arising between the teeth. Palet hyaline, 2-toothed. Styles distinct; stigmas plumose.

Glumes dull, strongly veined, the first 3-nerved, the second 5-nerved. 1. *T. interruptum.*
Glumes shining, the first 1-nerved, the second 1–3-nerved.
  Panicle spike-like, dense, its branches short and appressed. 2. *T. subspicatum*
  Panicle loose and open, its branches ascending. 3. *T. flavescens.*

1. **T. interrúptum** Buckl. Stem 5 dm. high, tufted leaf-blades 5–8 cm. long, rough, 1–2 mm. wide; panicle 4–12 cm. long, interrupted below; spikelets 4–5 mm. long; lemma with subulate teeth, the awn inserted about the middle. Dry soil: Tex.—Kans. *Texan.* Ap.–Je.

2. **T. subspicàtum** (L.) Beauv. Cespitose perennial; stems 2–4 dm. high; leaf-blades 3–15 cm. long, 2–3 mm. wide, flat or in age involute; panicle oblong, often obtuse, usually more or less purple, 2–8 cm. long; empty glumes lanceolate, acuminate or acute, slightly shorter than the lemma; awn bent and twisted, 7–8 mm. long. *T. subspicatum molle* A. Gray. *T. spicatum* (L.) Ricker [G]. Meadows and hillsides: Greenl.—N.H.—Minn.—N.M.—Calif.—Alaska; Eurasia. *Arctic—Subarct.—Subalp.—Alp.* Jl–Au.

3. **T. flavéscens** (L.) R. & S. Stem 4–8 dm. high, erect, glabrous; sheaths pubescent; leaf-blades 4–12 cm. long, 2–6 mm. wide, rough; panicle 5–12 cm. long, somewhat flexuose; spikelets 3- or 4-flowered; first glume about half as long as the second, which is about 5 mm. long; lemma 5–6 mm. long, hispidulous; awn long, bent and twisted. Waste places: N.M.—Mo.—Kans.—Miss.; nat. from Eurasia. Jl–Au.

39. AVÈNA (Tourn.) L. OATS.

Annuals or perennials. Inflorescence a contracted or open panicle. Spikelets usually large, erect or pendulous, usually 2–several-flowered, rarely 1-flowered, the rachilla articulated between the flowers, the lower flowers perfect, the upper ones often staminate or wanting. Empty glumes 2, membranous, exceeding or shorter than the lemma, persistent. Lemma 5–9-nerved, rounded on the back, the apex frequently shortly 2-toothed, bearing a dorsal twisted and geniculate awn, the upper empty ones or those enclosing staminate flowers awnless. Palet 2-cleft or 2-toothed, narrow. Styles distinct. Grain deeply furrowed, usually pubescent.

Perennials, with rootstocks; empty glumes 5–12 mm. long; lemma hairy at the base. 1. *A. Hookeri.*
Annuals; panicle open; empty glumes over 2 cm. long; spikelets 2–4-flowered.
Lemma hairy, at least at the base: awn strongly twisted. 2. *A. fatua.*
Lemma glabrous; awn scarcely twisted. 3. *A. sativa.*

1. A. Hoòkeri Scribn. *Fig. 66.* Stem 3–6 dm. high; leaf-blades flat, firm, 5–15 cm. long, 1.5–3 mm. wide, glabrous, scabrous on the margins; panicle 8–12 cm. long; spikelets 12–17 mm. long, 3–6-flowered; empty glumes about 1 cm. long, thin, greenish; lemma about 8 mm. long, brownish; awn over 1 cm. long, bent and twisted. *A. americana* Scribn. Ridges and hillsides: Man.—Minn.—S.D.—Colo.—Alta. *Boreal—Submont.—Mont.* Je–Jl.

f. 66

2. A. fàtua L. Annual; stem 3–12 dm. high, stout, glabrous; leaf-blades flat, 1–3 dm. long, 5–15 mm. wide; panicle open, 1–3 dm. long; empty glumes 2–2.5 cm. long, smooth; lemma 12–18 mm. long, in the typical form covered with long brown hairs; awn 2–2.5 cm. long. WILD OATS. A variety with lemma glabrous except at the base is *A. fatua glabrata* Peterm. Fields and waste places: Sask.—Wis.—Mo.—N.M.—Calif.—B.C.; native of Eur. and established as a weed in grain fields. *Plain—Submont.*

3. A. satìva L. Erect annual, closely resembling the preceding; stem glabrous, 6–12 dm. high; panicle open, 1–3 dm. long, usually with drooping spikelets; empty glumes 2–3 cm. long, glabrous; lemma 15–20 mm. long, glabrous; awn 2–3 cm. long. OATS. Occasionally escaped and established: Me.—Fla.—Tex.—Yukon. My–S.

40. ARRHENÁTHERUM Beauv. OAT-GRASS.

Tall perennials. Inflorescence paniculate. Spikelets 2-flowered, the lower flower staminate, the upper one perfect or pistillate, the rachilla articulated above the empty glumes and prolonged beyond the flowers. Empty glumes 2, persistent, thin-membranous, keeled. Lemma more rigid, somewhat toothed at the apex, that of the lower flower bearing near the base a long dorsal twisted and geniculate awn, that of the upper one awnless, or short-awned below the apex. Palet 2-keeled, hyaline, narrow. Stamens 3. Styles short, distinct, stigmas plumose. Grain hardly sulcate.

1. A. elàtius (L.) Beauv. Perennial, with a cespitose rootstock; stem glabrous, 6–12 dm. high, erect; leaf-blades flat, 5–30 cm. long, 2–8 mm. wide, scabrous; panicle 1–3 dm. long, narrow, with erect branches; lemmas about 8 mm. long, that of the lower flower with a bent and twisted awn about 1 cm. long. Fields and waste places: Me.—Ga.—Calif.—B.C.; nat. from Eur. Je–Jl.

41. DANTHÒNIA DC. WILD OAT-GRASS.

Usually perennial, all ours cespitose. Inflorescence a contracted or open diffuse panicle. Ligules usually represented by a hairy ring. Spikelets 3–

many-flowered, the flowers perfect or the upper ones staminate, the rachilla pilose, articulated between the glumes and prolonged beyond them. Empty glumes 2, persistent, usually extending beyond the lemma, keeled, acute or acuminate.

Lemma hairy on the back.
 Glumes 11–13 mm. long, conspicuously nerved, broad; teeth of the lemma subulate, 1–2 mm. wide. 1. *D. thermalis.*
 Glumes 8–10 mm. long, narrow; teeth of the lemma ovate, only 0.5 mm. long. 2. *D. spicata.*
Lemma glabrous on the back; teeth of the lemma aristate, 2 mm. long. 3. *D. intermedia.*

1. **D. thermàlis** Scribn. Stem 3–5 dm. high; sheaths papillate-pilose; lower leaf-blades hairy, the upper glabrous, 5–15 cm. long, involute; spikelets 5–10, 5–7-flowered; lemma 5 mm. long, thinly pilose on the back; awn about 7 mm. long. *D. spicata pinetorum* Piper. Near hot springs and in woods: B.C.—Wash.—Wyo.—S.D. *Submont.—Mont.* Jl.–Au.

2. **D. spicàta** (L.) Beauv. *Fig. 67.* Stems 3–7 dm. high, glabrous; leaf-blades scabrous, 2–15 cm. long, 2 mm. wide, usually involute, often ciliate; panicle 3–5 cm. long; spikelets 5–10, 5–8-flowered; lemma sparingly pubescent with silky hairs; teeth usually about 0.5 mm. long; awn 5–8 mm. long. Woods and hillsides: Newf.—N.C.—Kans.—N.M.—B.C. *Boreal—Submont.—Mont.* My–Jl.

3. **D. intermèdia** Vasey. Stem 3–5 dm. high; sheaths more or less pubescent at least at the mouth; panicle 3–6 cm. long; spikelets 5–10, about 5-flowered, 12–15 mm. long; empty glumes acute; lemma 6–8 mm. long, hairy on the margin; teeth about 2 mm. long; awn stout, 6 mm. long. *D. Cusickii* (Williams) Hitchc. Hillsides and mountains: Que.—N.M.—Calif.—B.C.—Sask. *Boreal—Mont.—Alp.* Jl–Au.

42. SCHIZÁCHNE Hack.

Erect perennials, with a creeping rootstock. Inflorescence paniculate. Spikelets several-flowered, the rachilla glabrous. Glumes 2, the first 3-nerved, the second 5-nerved. Lemma lanceolate, strongly 7-nerved, long-pilose on the callus, deeply bifid at the apex, awned from just below the cleft. Palet soft-pubescent on the submarginal ridges. Ovary glabrous, the style terminal, stigmas plumose. Grain dark reddish brown, smooth and shining.

1. **S. purpuráscens** (Torr.) Swallen. Stem 3–6 dm. high, glabrous; leaf-blades 2–15 cm. long, 3–6 mm. wide, scabrous above; panicle 7–12 cm. long; spikelets 3–6-flowered; empty glumes glabrous, the first 5–7 mm. long, 1-nerved, the second 6–8 mm. long, 3-nerved; lemma 7–9 mm. long; awn 8–10 mm. long. *Avena striata* Michx. [B, R], not Lam. *Trisetum purpurascens* Torr.; *Melica striata* (Michx.) Hitchc. [G]. *Bromelica striata* Farwell. In woods: N.B.—Pa.—N.M.—B.C.; Asia. *Boreal.—Mont.—Subalp* Je–Au.

43. SPARTÌNA Schreb. MARSH-GRASS.

Tall perennials, with creeping scaly rootstocks. Spikelets 1-flowered, crowded and imbricate in two rows, in one-sided spikes, the rachis extending beyond the spikelets. Empty glumes 2, firm-membranous, narrow, unequal, keeled. Lemma a little longer and broader than the second empty glume. Palet thin and almost hyaline, obscurely 2-nerved, often exceeding the lemma. Styles elongate; stigmas thread-like, papillose or short-plumose.

First glume awn-pointed, equaling the lemma; second glume long-awned. 1. *S. pectinata.*
First glume acute, shorter than the lemma; second glume acute. 2. *S. gracilis.*

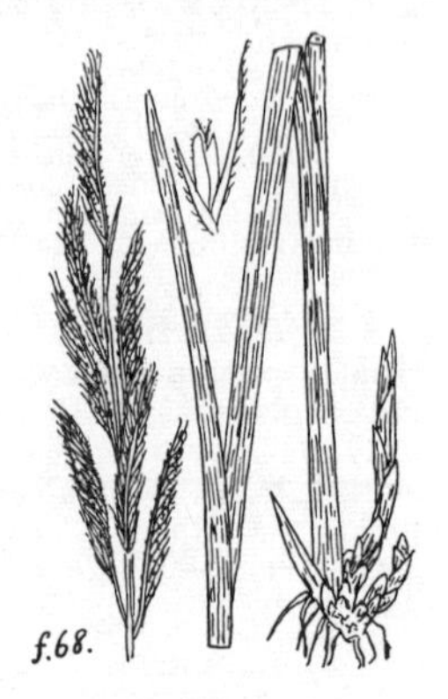

1. **S. pectinàta** Bosc. *Fig. 68.* Stem 1–2 m. high, glabrous, stout; leaf-blades 3–6 dm. long, 6–15 mm. wide, scabrous on the margin, usually flat, becoming involute; spikes 5–30, often short-peduncled, 5–12 cm. long; rachis rough on the margins; spikelets 12–14 mm. long; first empty glume equaling the lemma. *S. cynosuroides* A. Gray [B]; not Willd. *S. Michauxiana* Hitchc. [G]. In swamps and streams: N.S. — N.J. — Tex. — Colo. — Ore. — Mack. *Temp.* Au–O.

2. **S. grácilis** Trin. Stem 3–10 dm. high, glabrous; leaf-blades 2–4 dm. long, 2–6 mm. wide, flat or involute; spikes 4–8, appressed, usually short-stalked, 3–5 cm. long; spikelets 6–8 mm. long; first glume half as long as the second and the lemma. Saline soil: B.C.—Calif.—Ariz.—Kans.—Sask. *W. Temp.—Plain—Submont.* Au–S.

44. BECKMÁNNIA Host. Slough Grass.

Tall grasses with flat leaf-blades. Spikelets 1–2-flowered, orbicular, compressed, in two rows on several erect spikes. Empty glumes 2, membranous, saccate, obtuse or abruptly acute. Lemma narrow, thin-membranous. Palet hyaline, 2-keeled. Styles distinct; stigmas plumose. Grain oblong, free.

1. **B. Syzigáchne** (Steud.) Fern. *Fig. 69.* Stem 4–10 dm. high, glabrous, simple; leaf-blades 7–20 cm. long, 4–8 mm. wide, scabrous; panicle simple or compound; spikes 1–2 cm. long; spikelets 2–3 mm. long, 1–2-flowered; empty glumes saccate, abruptly acute; lemma acute or awn-pointed. *B. erucaeformis* Am. auth., not Host. Wet ground: Yukon—Calif.—N.M.—Ia.—Ont. *Prairie—Plain—Mont. W. Temp.* Jl–S.

45. SCHEDONNÁRDUS Steud.

Annuals, with involute, subulate leaf-blades. spikelets 1-flowered, acuminate, sessile, alternate on opposite sides of the long rachis, forming several long, slender spikes. Empty glumes 2, persistent, narrow, acuminate, somewhat unequal. Lemma longer, rigid, enclosing the narrow shorter palet. Styles distinct; stigmas plumose.

1. **S. paniculàtus** (Nutt.) Trelease. Annual, branched at the base; stems 2–5 dm. high, scabrous; leaf-blades 2–5 cm. long, 1–2 mm. wide, flat, stiff; spikelets nearly 3 mm. long, sessile and appressed; empty glumes hispid on the keel, the second much longer than the first and nearly equaling the lemma. *S. texanus* Steud. Sandy soil, especially river banks: Man.—Ill.—Tex.—N.M. —Sask. *Prairie—Plain—St. Plains—Submont.* Jl–S.

46. CHLÒRIS Sw.

Usually perennials. Spikelets 1-flowered, often sessile, with 1 or more empty usually awned glumes above the perfect flower, crowded in 2 rows, in verticillate or approximate spikes, the rachilla prolonged beyond the flower. Empty glumes 2, persistent, unequal, keeled, narrow, acute or acuminate, awnless or awn-pointed. Lemma acute, usually long-awned, rarely nearly awnless. Palet folded, 2-keeled. Styles distinct; stigmas plumose.

Spikes slender, naked at the base; nerves of the lemma all short-strigose. 1. *C. verticillata.*
Spikes stout; spikelets crowded at the base.
 Lateral nerves of the lemma short-hairy. 2. *C. brevispica.*
 Lateral nerves of the lemma with a tuft of long hairs extending beyond the summit of the lemma. 3. *C. elegans.*

1. **C. verticillàta** Nutt. Stem 1–4 dm. high, erect or decumbent at the base, rooting at the nodes; leaf-blades 3–7 cm. long, 2–4 mm. wide, obtuse, scabrous; spikes usually spreading, 5–11 cm. long, in 1 or 2 whorls; spikelets (without the awns) 3 mm. long; lemma 2 mm. long, its awn near the apex, scabrous, 5 mm. long. Prairies: Mo.—Neb.—Tex. *Ozark.* My–Jl.

2. **C. brevispìca** Nash. *Fig 71.* Perennial; stem 1–3 dm. high; leaf-blades 1–6 cm. long, 1–3 mm. wide, scabrous above and on the margins; spikes 6–10, finally spreading, 2.5–4.5 cm. long; spikelets (exclusive of the awns) about 2.8 mm. long; empty glumes lanceolate, acute; lemma 2.5 mm. long; awn about 2 mm. long; the fourth glume (second lemma) empty, about 1.5 mm. long, with an awn of about the same length. Sandy soil: Kans. (?)—Tex.—e Colo.—N.M. *St. Plains—Son.* Jl–Au.

3. **C. élegans** H.B.K. Stem 1.5–9 dm. high; sheaths compressed-keeled; leaf-blades less than 1 dm. long, 2–6 mm. wide, rough above; spikes 3–12, erect, 2–8 cm. long; spikelets (without the awns) 4 mm. long; midrib of the lemma abruptly contracted below the awns; awn 5–10 mm. long. Sandy soil: Mo.—Kans.—Ariz.—Mex.; W. Ind. *Ozark.—Texan—Son.—Trop.*

47. CÝNODON Rich. BERMUDA GRASS, SCUTCH GRASS.

Low diffuse and extensively creeping perennials. Spikelets numerous, in slender digitate spikes, sessile in two rows on the rachis, 1-flowered, secund. Glumes unequal, narrow, keeled. Lemma broad, boat-shaped, obtuse, ciliate on the keel. Palet as long as the lemma, ciliate on the prominent keels. Grain free. *Capriola* Adans.

1. **C. Dáctylon** (L.) Pers. *Fig. 70.* Stem 1–3 dm. high, erect, smooth and glabrous; leaf-blades 2.5–5 cm. long, 2–4 mm. wide, flat, smooth beneath, scabrous above; spikes 4 or 5, 1.5–5 cm. long; rachis flat; spikelets 2 mm. long; glumes hispid on the keel, the first shorter than the second, two thirds as long as the lemma. *Capriola Dactylon* Kuntze [B]. Fields and waste places; N.Y.—Fla.—Tex.—Kans.—W.Ind.—Mex.; cult. and nat. from Eur. *Temp.—Trop.* Jl.–S.

48. GYMNOPÒGON Beauv.

Perennials, with short, broad, rigid leaves. Spikes numerous, slender, alternate, one-sided. Spikelets remote, with 1 perfect and 0–2 neutral or staminate flowers above, the rachilla prolonged beyond the floret as a slender, often awned rudiment. Glumes narrow, subequal, scabrous on the keel. Lemma thin, bearing a slender awn just below the apex. Palet 2-keeled. Stigmas plumose. Grain linear, free.

1. **G. ambíguus** (Michx.) B.S.P. Stem 3–4.5 dm. high, decumbent at the base; sheaths short, with a villous ring at the mouth; blades 2.5–10 cm. long, 4–12 mm. wide, spreading, a little scabrous above; spikes slender, spikelet-bearing to the base, at first erect, the lower soon spreading; spikelet (without the awn) 4–5 mm. long; lemma with a hairy callus at the base; awn 4–6 mm. long. *G. racemosus* Beauv. Sandy soil; N.J.—Fla.—Tex.—Kans. *Austral.* Au–O.

49. BOUTELOÙA Lag. GRAMA, GRAMA-GRASS, MESQUITE GRASS, BUFFALO GRASS.

Perennials or annuals, mostly tufted. Spikelets few or numerous, 1- or 2-flowered, crowded in 2 rows and forming few to many one-sided, more or less curved sessile spikes; rachis usually conspicuously prolonged beyond the spikelets. Lower flowers perfect; the upper when present staminate. Empty glumes 2, narrow, acute, unequal, keeled. Lemma usually thinner and broader, 3-nerved, the nerves excurrent. Lemma of the upper 1–3 imperfect flowers borne at the end of a rachilla, 3-awned. Styles distinct; stigmas plumose.

Spikes 1–4, rarely more; spikelets 25 or more.
  Rachilla bearing the rudimentary glumes and awns with a tuft of long hairs at the apex; second glume scabrous and sparingly long-ciliate on the keel. — 1. *B. gracilis.*
  Rachilla bearing the rudimentary glumes and awns glabrous; second glume strongly papillose-hispid on the keel. — 2. *B. hirsuta.*
Spikes 12 or more; spikelets in each few, less than 12. — 3. *B. curtipendula.*

1. **B. grácilis** (H.B.K.) Lag. *Fig. 72.* Stem 1.5–4 dm. high, smooth and glabrous; leaf-blades 3–10 cm. long, 2 mm. wide or less, flat or when dry involute, usually glabrous; spikes 1–3, often strongly curved, 2–5 cm. long; spikelets numerous, about 6 mm. long; first glume about half as long as the second. *B. oligostachya* (Nutt.) Torr. [G, B]. BLUE GRAMA. Plains and prairies: Man.—Wis.—Miss.—Ariz.—Alta.; Mex. *Prairie—Plain—Submont.*

f. 72.

2. **B. hirsùta** Lag. Stem 1–5 dm. high, glabrous; leaf-blades 2–12 cm. long, 2 mm. wide or less, flat, rough, sparingly papillose-hirsute near the base; spikes 1–4, 1–5 cm. long, the rachis conspicuously prolonged beyond the spikelets; spikelets numerous, 5–6 mm. long; lemma 3-cleft to near the middle. HAIRY GRAMA. Dry or sandy soil: Minn. Ill.—Tex.—Ariz.—S.D.; Fla.; Mex. *Prairie—Plain—Submont.* Je.–Au.

3. **B. curtipéndula** (Michx.) Torr. Perennial, with more or less cespitose rootstocks; stem 3–10 dm. high, smooth and glabrous; leaves 5–15 cm. long, 2–4 mm. wide, scabrous above; spikes many, 6–15 mm. long, spreading or reflexed; spikelets 4–12, divergent, 7–10 mm. long; empty glumes scabrous especially on the keel; lemma 3-toothed at the apex. *B. racemosa* Lag. *Atheropogon curtipendulum* Fourn. [B]. TALL GRAMA. Dry soil: Ont.—N.J.—Tex.—Ariz.—Sask.; Mex. *Canad.—Plain—Submont.* Jl–S.

50. ELEUSÌNE Gaertn. YARD GRASS, WIRE GRASS, CRAB GRASS.

Annual or perennial creeping grasses. Spikes normally several, digitately arranged, or one of them inserted lower down. Spikelets numerous, crowded, imbricate. Glumes flattened, keeled, thin, shorter than the lemmas, acute. Lemmas obtuse, enclosing the compressed, 2-keeled palet; the terminal lemma empty. Stamens 3. Style distinct; stigmas plumose.

1. **E. índica** (L.) Gaertn. Stem 1.5–6 dm. high, erect or decumbent at the base; leaf-sheaths loose, overlapping; blades 7–30 cm. long, 2–6 mm. wide; spikes 2–10 at the summit of the stem, 2–8 cm. long; spikelets 3–4 mm. long, 3–6-flowered. Waste places: Mass.—Ill.—Neb.—Tex.—Fla., nat. from Old World. *Temp.—Trop.* Je–S.

51. LEPTÓCHLOA Beauv.

Annuals or perennials. Spikelets small, usually close, 2–several-flowered, rarely 1-flowered, flattened, sessile, in two rows, forming many long slender spikes. Empty glumes 2, keeled, a little unequal, usually shorter than the spikelets. Styles distinct; stigmas plumose.

1. **L. filifòrmis** (Lam.) Beauv. Annual, branched at the base; stems 3–10 dm. high, erect, glabrous; leaf-blades 5–20 cm. long, 2–6 mm. wide, scabrous; panicle 1–4 dm. long; spikes numerous, slender, ascending or spreading, the lower 5–15 cm. long; spikelets usually 3-flowered, 2 mm. long; lemma 2-toothed at the apex, ciliate on the nerves. *L. mucronata* (Michx.) Kunth. Dry soil: Fla.—Va.—Ill.—N.M.—Calif.; Mex., W.Ind. *S. Temp.—Trop.—Son.* Jl–S.

52. ACAMPTÓCLADOS Nash.

Tufted perennials, with stiff stems. Leaves flat or involute. Inflorescence a panicle with distant, scattered, widely spreading branches. Spikelets scattered, singly or in two rows, sessile, 4–6-flowered. Glumes acuminate, equal, the first 1-nerved, the second 3-nerved. Lemma acute, hard in fruit, 3-nerved. Palet compressed, 2-nerved, ciliolate, gibbous at the base. Stamens 3. Styles distinct; stigmas plumose.

1. **A. sessilispìcus** (Buckl.) Nash. Stem 2–15 dm. high; leaves mostly basal, crowded; blade 5–8 cm. long, 1.5–3 mm. wide, rough above; panicle 1–4 dm. long; lemma 4 mm. long. Prairies: Kans.—Tex. *Texan.* Jl–S.

53. BÙCHLOË Engelm. BUFFALO GRASS.

Creeping, stoloniferous, dioecious perennials. Staminate spikelets 2- or 3-flowered, crowded in 2 rows on one side of the short, flattened rachis, in small spikes. Empty glumes 2, membranous, lanceolate. Lemma similar. Pistillate spikelet 1-flowered. First empty glume membranous, usually small, the second larger, firm, concave at the base, 3-lobed at the apex. Lemma narrow, nearly hyaline, enclosing a broad, 2-nerved, convolute palet. Styles distinct, long; stigmas elongate, short-plumose. *Bulbilis* Raf.

f. 73.

1. **B. dactyloìdes** (Nutt.) Engelm. *Fig. 73.* Stoloniferous perennial; stems bearing the staminate flowers 1–3 dm. high, erect, exceeding the leaves, glabrous; those bearing the pistillate flowers 1–10 cm. long, shorter than the leaves; leaves 2–10 cm. long, 1–2 mm. wide, papillose-hirsute; staminate spikelets 2 or 3, approximate, 0.5–1.5 cm. long; spikelets 4–5 mm. long, flattened, 2–3-flowered; pistillate spikelets in the axils of the leaves, ovoid, the empty glumes indurate. *Bulbilis dactyloides* Raf. [B, R].

Plains and prairies: Minn.—Ark.—Tex.—N.Mex.—N.D.; Mex. *Prairie—Plain —Submont.* Je.–Jl.

54. **PAPPÓPHORUM** Schreber.

Perennial tufted grasses. Panicle dense, spike-like. Leaves narrow. Spikelets 1–2-flowered. Glumes acute, carinate. Lemma firm, divided into several awn-like lobes. Stamens 3, styles short; stigmas plumose.

1. **P. vaginàtum** Buckley. Stem 5–10 dm. high, erect; sheaths pubescent at the summit; blades 3 dm. long, involute; panicle 1–3 dm. long; the branches erect; spikelets 7–8 mm. long; lemma pubescent on the margins and keel. *P. apertum* Munro. Dry soil: Kans.—Tex.—Mex.—Ariz.; Mex. *St. Plains—Son.* Jl–O.

55. **PHRAGMÌTES** Trin. Reed, Cane-grass.

Tall perennials with long creeping rootstocks. Leaves broad and flat. Spikelets numerous in large terminal panicles, 3–several-flowered, the lower flower staminate, the rest perfect; rachilla articulated between the flowers and covered with long hairs. Empty glumes keeled, narrow, acute, the first much shorter than the second. Lemma similar, long-acuminate, 3-nerved. Palet hyaline, much shorter, 2-keeled. Styles short; stigmas plumose.

1. **P. commùnis** Trin. Stem glabrous, 1.5–5 m. high, stout; leaf-blades 1.5–4 dm. long, 8–50 mm. wide, flat, glabrous; panicle 1.5–3 dm. long, ample; spikelets numerous; lemma 10–12 mm. long, long-acuminate. *P. Phragmites* (L.) Karst. [B, R]. Swamps: Newf.—Fla.—Calif.—B.C.; Mex., W.Ind., Eurasia. *Temp.—Submont.* Au–O.

56. **MUNRÒA** Torr. False Buffalo Grass.

Low prostrate herbs, dichotomously branched and fasciculate at the nodes. Leaves short, rigid, crowded at the nodes and at the ends of the branches. Spikelets few, almost sessile in the axils of the leaves and almost concealed in the leaf-rosettes. Empty glumes hyaline, nerveless, acute. Lemmas larger, 3-nerved, retuse or 3-toothed at the apex, the upper one or two often sterile. Palet hyaline. Stamens 3. Styles elongated; stigmas barbellate or short-plumose. Grain free.

f. 74.

1. **M. squarròsa** (Nutt.) Torr. *Fig. 74.* Stems 1–2 dm. long; sheath short and inflated, long-hairy at the throat; leaf-blades 1–2.5 cm. long, 1–2 mm. wide, rigid, spreading, scabrous, pungent; spikelets 2–5-flowered; empty glumes shorter than the lemma, which is about 5 mm. long, 3-toothed and awn-pointed. Dry plains: Sask.—S.D.—Tex.—Ariz.—Alta. *Plain—Submont.* Au–O.

57. **TRÍPLASIS** Beauv. Sand-grass.

Tufted perennials. Panicles in our species narrow. Spikelet short-pedicellate, 2–6-flowered, the flowers perfect or the uppermost staminate, the rachis glabrous, articulate between the flowers. Empty glumes keeled, 3-nerved, shorter than the lemma. Lemma dorsally rounded at the base, 3-nerved, deeply 2-lobed at the apex, with an awn arising between the lobes. Palet shorter, with 2 ciliate keels, compressed. Styles short; stigmas plumose.

1. **T. purpùrea** (Walt.) Chapm. Tufted perennial (?); stem 3–10 dm. high, erect, prostrate or decumbent, branched below; leaf-blades 1–6 cm. long, 2 mm. wide, rigid, scabrous, sparsely ciliate; panicle 2–7 cm. long; spikelets 2–5-flowered, 5–8 mm. long; lemma oblong, 2-lobed at the apex, with erose-truncate lobes; midvein excurrent in a short point. Sandy places: Me.—Fla.—Tex.—N.M.—Colo.—Neb. *E. Temp.—Plain.* Au–S.

58. ERIONEÙRON Nash.

Tufted, low perennials. Panicles small, dense, subcapitate. Leaves thick, with thickened white margins. Spikelets several-flowered. Empty glumes narrow, acuminate. Lemma broad, 3-nerved, pubescent on the nerves below with long silky white hairs, acuminate at the apex, entire or slightly 2-toothed, the terminal awn rising between the minute teeth. Style short.

1. **E. pilòsum** (Buckley) Nash. Stem 0.5–3 dm. high, leafy mostly at the base; leaf-sheath pilose-ciliate at the summit; blades erect, thick, folded, papillose-ciliate, 2–8 cm. long; spikelets 3–8, crowded, 1–1.5 cm. long; lemma 5.5–6 mm. long, acuminate, entire or slightly toothed at the apex; awn 1–1.5 mm. long. *Tricuspis acuminata* Munro. *T. pilosa* Nash [B]. *Triodia acuminata* Vasey. Dry gravelly or sandy soil: Kans.—Nev.—Ariz.—Tex. *Son.—Submont.* Ap–O.

59. TRÌDENS R. & S.

Perennials with rootstocks, ours tufted. Panicles open, or in ours narrow, sometimes spike-like. Spikelets 3–many-flowered, the flowers perfect, or the upper one staminate. Empty glumes keeled, usually shorter than the lemma. Lemma 3-nerved, pilose on the nerves and the margins, entire or 2-toothed at the apex. Palet shorter, 2-keeled. Styles short; stigmas plumose.

Second glume 3–5-nerved. — 1. *T. elongatus.*
Second glume 1-nerved.
  Panicle open, its branches more or less spreading. — 2. *T. flavus.*
  Panicle narrow, its branches erect or strongly ascending. — 3. *T. strictus.*

1. **T. elongàtus** (Buckley) Nash. *Fig. 75.* Stem 3–10 dm. tall, scabrous; leaf-blades erect or ascending, 4–25 cm. long, 2–3 mm. wide, long-acuminate, scabrous; panicle narrow, 12–25 cm. long; spikelets 8–10-flowered, 10–14 mm. long; lemma 5–6 mm. long, obtuse or minutely 2-toothed, mucronate. *Tricuspis elongata* Nash [B]. Plains: Mo.—Colo.—Ariz.—Tex. *Ozark—Tex.—Son.* Je–N.

2. **T. flàvus** (L.) Hitchc. Stem erect, 1–2 m. high, viscid in the axils of the panicle; sheaths bearded at the summit; leaf-blades 2–6 dm. long, 5–15 mm. wide; panicle 2–4.5 dm. long, nearly as wide; spikelets purple, 7–8 mm. long, 5–8-flowered; glumes shorter than the lowest floret; the three nerves of the lemma excurrent. *Poa flava* L. *Tricuspis seslerioides* Torr. [B]. *Triodia cuprea* Jacq. Dry soil: N.Y.—Iowa—Kans.—Tex.—Fla. *E. Temp.* Jl–S.

3. **T. stríctus** (Nutt.) Nash. Stems stout, 12–15 dm. high, erect, tufted, leaf-blades 2–4 dm. long, 2–7 mm. wide, long-acuminate, smooth except on the margins; panicle 1–3 dm. long; spikelets pale or purplish, 5 mm. long, 5–8-flowered; glumes exceeding the lowest floret; only the midrib of the lemma excurrent. *Tricuspis stricta* (Nutt.) Gray [B]. Moist ground: Miss.—se Kans.—Tex. *Ozark.* Jl–S.

60. **DIPLÁCHNE** Beauv.

Tall tufted perennials. Panicles composed of several long spike-like or raceme-like branches. Spikelets linear, sessile or nearly so, the flowers perfect or the uppermost staminate. Empty glumes peristent, keeled, acute, unequal. Lemma larger, 2-toothed and mucronate or short-awned between the teeth. Palet hyaline, 2-nerved and 2-keeled. Styles distinct; stigmas plumose.

1. **D. acuminàta** Nash. Tufted perennial; stem 3–6 dm. high; leaf-blades erect, 1–3 dm. long, 3–4.5 mm. wide, scabrous; racemes numerous, erect or ascending, the longer 7–15 cm. long; spikelets 10–12 mm. long; empty glumes 8–11 mm. long; lemma 6–7 mm. long, acuminate; awn 0.75–1.33 mm. long. *Leptochloa fasciculata* Auth. [G], not Gray. Wet or moist ground: Ark.—Iowa—Neb.—Colo. *Plain.* Je–Au.

61. **RHOMBÓLYTRUM** Link. White Prairie-grass.

Perennial grasses. Panicle narrow, spike-like; spikes numerous, many-flowered. Glumes 1-nerved. Lemma broad, rounded at the summit, 3-nerved, the lateral nerves vanishing below the summit, the midrib sometimes excurrent. Stamens 3. Style short, distinct. Stigmas plumose.

1. **R. albéscens** (Vasey) Nash. Stem 2–6 dm. high; sheaths shorter than the nodes; ligule a ring of short hairs; leaf-blades 5–20 cm. long, 2–6 mm. wide, smooth; panicles 6–15 cm. long, with appressed branches; spikelets 7–11-flowered, 4.5–5 mm. long, oval; glumes broad, 1-nerved; lemma 3 mm. long, almost orbicular, erose at the summit. Gravelly soil. Kans.—Ariz.—Tex. *Son.—Texan.* Au–S.

62. **REDFIÈLDIA** Vasey. Blow-out Grass, Sand-grass.

Tall perennials, with long creeping scaly rootstocks. Panicles diffuse, with long capillary branches. Spikelets numerous, 1–3-flowered, the flowers perfect. Empty glumes about equal, 1-nerved. Lemma large, compressed-keeled, with a basal ring of hairs, 3-nerved, awn-pointed or acute. Palet shorter, 2-nerved. Styles long; stigmas short, plumose.

1. **R. flexuòsa** (Thurber) Vasey. *Fig. 76.* Stem 4–12 dm. high; leaf-blades 3–6 dm. long, 2–4 mm. wide, usually involute; panicle 2–5 dm. long; spikelets about 6 mm. long, 1–3-flowered; lemma scabrous, twice as long as the acute glabrous empty glumes. Sand hills: S.D.—Colo.—Wyo. *Plain.* Au–S.

f.76.

63. **KOELÈRIA** Pers. June Grass.

Tufted annuals or perennials. Panicles narrow, contracted, spike-like. Spikelets numerous, crowded, 2–5-flowered, the flowers perfect or the uppermost one or two staminate, shining. Empty glumes narrow, unequal. Lemma similar to the second, the upper ones gradually smaller. Palet hyaline, 2-keeled, 2-toothed. Styles very short; stigmas plumose.

Leaf-blades and sheaths glabrous, merely scabrous or the lower ones short-pubescent with mostly reflexed hairs; leaves mostly stiff.
  Stem and panicle stout; lower leaf-sheaths and sometimes the blades beneath densely retrorse-pubescent; upper leaf-blades broad, usually flat. 1. *K. latifrons.*

Stem and panicle slender; leaf-sheaths and blades merely scabrous or the lower sheaths moderately pubescent. — 2. *K. nitida.*
Leaf-blades on both sides and sheaths more or less pubescent with long spreading hairs; leaves more flaccid. — 3. *K. gracilis.*

1. K. látifrons (Domin) Rydb. *Fig. 77.* Cespitose perennial; stem stout, 3–5 dm. high, scabrous, more or less puberulent above; sheaths loose, densely retrorsely short-pubescent or in age glabrate; blades 5–10 cm. long, those of the innovations narrow and involute; those of the stem usually flat, 2–4 mm. wide, scabrous on both sides or the lower densely retrorse-pubescent beneath; panicle stout, lobed, often 1 cm. wide; spikelets mostly 2-flowered; glumes and lemma lanceolate, acute, the first glume 3–4 mm. long, the second and the lemmas 3–4 mm. long. *K. nitida latifrons* Domin. Meadows: Iowa—Okla.—Colo.—Mont.—Sask. *Prairie—Plain.* My–Jl.

2. K. nítida Nutt. Densely cespitose perennial; stem slender, minutely scabrous, 2–4 dm. high; leaves mostly all involute, slender; sheaths scabrous or the lower ones sometimes finely retrorse-pubescent; blades scabrous, involute, 5–10 cm. long, 1–2 m. wide; panicle 3–10 dm. long, rarely more than 5 mm. broad in fruit; spikelets mostly 2-flowered; glumes and lemma lanceolate, scabrous; the first glume 3–4 mm. long, the second and lemma 4–5 mm. long, acute, rarely awn-pointed (var. *munita*). *K. cristata* [G, B] and K. *gracilis*, western auth., in part. *K. elegantula* Domin. *K. gracilis glabrata* Domin, in part. Dry prairies and plains: Wisc.—Mo.—N.M.—B.C. *Prairie—Plain—Submont.* Je–Au.

3. K. grácilis Pers. Densely cespitose perennial; stem slender, 2–5 dm. high, glabrous, minutely scabrous, or in or below the inflorescence puberulent; lower sheaths more or less hirsute or pilose with long spreading hairs, or in age glabrate; blades of the innovations mostly involute, those of the upper stem often flat, 5–10 cm. long, 2–3 mm. wide, more or less pilose on both sides; panicle 5–12 cm. long, in fruit 5–8 mm. thick; spikelets 2–3-flowered; glumes lanceolate, acute, the first 2–3 mm. long, the second 3–4 mm. long; lemmas about equaling the second glume; glumes and lemmas scabrous, especially on the keel. *Aira cristata* L. *Poa cristata* L., but not *K. cristata* Pers. Meadows and banks: Ont.—Ill.—Kans.—Wyo.—Sask.; also Eurasia. *Canad.—Allegh.* Je–Au.

64. ERAGRÓSTIS Beauv. STINK GRASS, SKUNK GRASS.

Tufted annuals or perennials, sometimes prostrate or creeping; some species dioecious. Spikelets numerous, singly or in fascicles, 2–many-flowered, the flowers perfect or unisexual. Empty glumes unequal. Lemma large, 3-nerved, usually broad. Palet shorter, prominently 2-nerved and 2-keeled, often incurved and persistent on the rachis. Stamens 2 or 3. Styles short; stigmas plumose.

Plant extensively creeping, rooting at the nodes.
  Lemma less than 2 mm. long, glabrous; flowers perfect; anthers 0.2 mm. long. — 1. *E. hypnoides.*
  Lemma 3–4 mm. long, pubescent; flowers dioecious; anthers 2 mm. long. — 2. *E. reptans.*
Plant neither creeping nor rooting at the nodes; flowers perfect.
  Spikelets 2–5-flowered.
    Branches of the panicle short; stem branched above. — 3. *E. Frankii.*
    Branches of the panicle long and capillary. — 4. *E. capillaris.*

Spikelets 5–35-flowered.
Annuals, much branched, ascending or decumbent and geniculate at the base.
Spikelets 2–3 mm. broad; first glume only slightly shorter than the second.
Lower lemmas 1.5 mm. long; spikelets 2 mm. wide. — 5. *E. poaeoides.*
Lower lemmas 2 mm. long; spikelets 2.5–3 mm. wide. — 6. *E. cilianensis.*
Spikelets 1–1.5 mm. broad; first empty glume only two thirds as long as the second.
Lemma firm, dull purple or green; lateral veins prominent; spikelets about 1.5 mm. wide. — 7. *E. Purshii.*
Lemma thin, bright purple; lateral veins faint; spikelets about 1 mm. wide. — 8. *E. pilosa.*
Perennials, rigid, erect, often tufted.
Spikelets scattered on long branches.
Branches of the panicle in age widely spreading.
Pedicels long, usually exceeding the spikelets. — 9. *E. pectinacea.*
Pedicels shorter than the spikelets. — 10. *E. curtipedicellata*
Branches of the panicle erect or strongly ascending. — 11. *E. trichodes.*
Spikelets clustered on very short branches. — 12. *E. secundiflora.*

1. **E. hypnoìdes** (Lam.) B.S.P. Stem 5–45 cm. long, branched, glabrous; floral branches erect, 3–15 cm. high; leaves about 5 cm. long or less, 1–2 mm. wide, flat; spikelets 10–35-flowered, 4–15 mm. long; empty glumes unequal, the first one half to two thirds as long as the second; lemma about 2.5 mm. long; lateral veins prominent; glumes of the pistillate spikelets more acute than those of the staminate ones. *E. reptans* Am. auth.; not Nees. Sandy or gravelly shores: Vt.—Ont.—S.D.—N.M.—Fla.; B.C.—Ida.—Calif.; Mex., W.Ind. *Temp.—Plain—Submont.* Au–S.

2. **E. réptans** (Michx.) Nees. Stem branching, creeping, spreading, with erect short branches; sheaths mostly pubescent; blades spreading, 1–3 cm. long, 3 mm. wide, flat, pubescent; panicle 2–4 cm. long, oval; spikelets crowded, 12–30-flowered, 7–15 mm. long; lemma 3–4 mm. long. *E. capitata* (Nutt.) Nash. *Poa reptans* Michx. *E. Weigeltiana* (Reichenb.) Bush. Sandy soil: S.D.—La.—Tex.; Mex. *Son.* Jl–O.

3. **E. Fránkii** Steud. Stem 1.5–4 dm. high, tufted, erect or decumbent at the base; leaf-blades 5–13 cm. long, 2–4 mm. wide, rough above; panicle 5–15 cm. long, open, the branches short, ascending; spikelets ovate, 2–3 mm. long, 3–5-flowered; lemma acute, 1.5 mm. long, its lateral veins obscure. Moist places: Conn.—Minn.—Kans.—La.—Miss. Au.

4. **E. capillàris** (L.) Nees. Stem 2–5 dm. high, erect, slender; leaf-blades 7–25 cm. long, 2–4 mm. wide; panicle diffuse, 1–4 dm. long, with spreading branches; spikelets ovate, 2–3 mm. long, 2–4-flowered; lemma acute, 1.5 mm. long, its lateral veins obscure. Dry places: R.I.—Iowa—Kans.—Tex.—Ga. *E. Temp.* Au–S.

5. **E. poaeoìdes** Beauv. Stem 1–3 dm. high, decumbent below, much branched; sheaths loose, shorter than the internodes, sparingly pilose at the throat; blades 2.5–6 cm. long, 1–4 mm. wide, smooth beneath, scabrous above; panicle 5–12 cm. long, the branches more or less spreading; spikelets 5–18-flowered, less densely imbricate, the rachilla joints visible; glumes acute; lemma obtuse, with prominent marginal veins. *E. minor* Host [G]. *E. Eragrostis* (L.) Karst. [B]. Waste places and cultivated ground: Mass.—S.D.—Pa.—N.Y.; nat. from Eur. *Canad.*

6. **E. cilianénsis** (All.) Link. Stem 1–6 dm. tall; leaf-blades 7–18 cm. long, 2–6 mm. wide, flat, smooth beneath, scabrous above; panicles 5–15 cm. long; spikelets 8–35-flowered, 5–16 mm. long, very flat; empty glumes acute; lemma obtuse, 2–2.5 mm. long. *E. major* Host [B,G]. *E. megastachya* Link.

[R]. STINK GRASS. Waste places and cultivated grounds: Ont.—Fla.—Calif.—Mont.; Mex.; nat. from Eur. *Plain—Submont.* My–S.

7. **E. Púrshii** Schrad. *Fig. 78.* Stem 1–4 dm. tall, smooth; leaf-blades 4–9 cm. long, 1–2 mm. wide; panicle open, 7–20 cm. long; spikelets 5–15-flowered, 3–8 mm. long; lemma acute, firm, about 1.5 mm. long. Dry or sandy places: Ont.—Fla.—Ariz.; Mex. *E. Temp.—Plain—Submont.—Trop.*—Je–O.

8. **E. pilòsa** (L.) Beauv. Stem tufted, 1.5 dm. high, erect, slender, branched; leaf-blades 3–15 cm. long, 2 mm. wide or less; panicle 5–15 cm. long, its branches spreading, the axils usually hairy; spikelets 5–12-flowered, 3–6 mm. long, 1 mm. wide; lemma acute, the lower ones 1.5 mm. long. Cultivated ground and waste places: Me.—Minn.—Kans.—Tex.—Fla.; Mex.; Eur. *E. Temp.* Jl–Au.

f. 78.

9. **E. pectinàcea** (Michx.) Steud. Stem 3–8 dm. high, erect, rigid; leaf-blades 1–3 dm. long, 4–8 mm. wide, flat, sparingly villous at the base; panicle 1.5–6 dm. long, purple or purplish, strongly bearded in the axils; spikelets 5–15-flowered, 3–8 mm. long; empty glumes acute, subequal; lemma 1.75–2 mm. long. Dry or sandy soil: N.H.—Fla.—Tex.—N.M.—S.D. *E. Temp.—Son.—Plain.* Au–S.

10. **E. curtipedicellàta** Buckl. Stem 1.5–10 dm. high, rigid; sheaths overlapping, pilose at the summit, striate; blades 5–20 cm. long, 2–4 mm. wide, scabrous above; panicle 1–3 dm. long, 5–10 cm. wide, pyramidal, the branches widely spreading, ciliate at the axils; spikelets 5–12-flowered, 3–6 mm. long; lemmas 1.75 mm. long, with prominent lateral veins. Prairies: Kans.—Tex. *Texan.*

11. **E. trichòdes** (Nutt.) Nash. Stem 6–12 dm. high; sheaths pilose at the throat; leaf-blades 1.5–7 dm. long, 2–4 mm. wide; panicle 2–7 dm. long, narrow; lower axils sometimes bearded; spikelets usually pale, 3–9-flowered, 5–9 mm. long; empty glumes subequal; lemma acute, 2–3 mm. long. *E. capillacea* Jedwabnick. Sandy soil: Ohio—Ark.—Tex.—N.M.—Neb. *Allegh.—Plain—Son.*

12. **E. secundiflòra** Presl. Stems 1.5–10 dm. high, erect, simple; leaf-blades 5–30 cm. long, 2–4 mm. wide, flat; panicle 4–15 cm. long, the branches erect or ascending; spikelets crowded, sessile or nearly so, strongly flattened, 8–40-flowered, 6–20 mm. long, 3–5 mm. wide; empty glumes acute, equal; flowering glumes 3–3.5 mm. long, acute, usually purple-margined. *E. oxylepis* Torr. Dry or sandy soil: Mo.—La.—Tex.—N.M.—Colo.; Mex. *Osark—Texan—Son.* Au–S.

65. **CATABRÒSA** Beauv. BROOK-GRASS.

Creeping or floating aquatic perennial. Panicles open, with slender spreading or reflexed branches. Spikelets 2- (rarely 3- or 4-) flowered, with the rachilla articulate between the flowers. Empty glumes unequal, broad, thin, very obtuse. Lemma strongly 3-nerved, longer than the empty glumes. Palet 2-keeled, nearly as long. Styles distinct; stigmas plumose.

1. **C. aquática** (L.) Beauv. Stem 1–3 dm. high, bright green, flaccid; leaf-blades 3–12 cm. long, 2–6 mm. wide, flat, obtuse; panicle 3–20 cm. long, open; branches whorled; spikelets 2.5–3.5 mm. long; first glume about 1 mm., the second nearly 2 mm. long; lemma 2–2.5 mm. long, 3-nerved, erose-truncate at the apex. In water: Lab.—Que.—Colo.—Utah—Alaska; Eurasia. *Subarct.—Boreal—Subalp.* Je–Au.

66. SPHENÓPHOLIS Scribn.

Tall tufted perennials. Panicles usually narrow. Spikelets numerous, 2–3-flowered, shining. First empty glume narrow, 1-nerved, acuminate, the second much broader, obovate when spread, obtuse or truncate, 3-nerved. Lemma narrower than the second glume, obtuse or acute. Palet narrow, 2-nerved. Styles short; stigmas plumose. *Eatonia* Raf.

Second empty glume much wider than the lemma, rounded or truncate and somewhat cucullate at the apex.
  Intermediate nerves of the second glume almost as prominent as the lateral ones; leaf-blades firm, much broader than the sheaths and therefore with prominent auricles. 1. *S. robusta.*
  Intermediate nerves of the second glume faint, the lateral ones strong; leaf-blades soft, not much wider than the sheaths; auricles not prominent. 2. *S. obtusata.*
Second empty glume oblanceolate, not much wider than the lemma, obtuse or acute.
  Second empty glume rather firm, as well as the lemma obtusish. 3. *S. intermedia.*
  Second empty glume thin and with a broad, scarious margin, acutish; lemma acute. 4. *S. pallens.*

1. S. robústa (Vasey) Heller. Stem 4–10 dm. high, erect, glabrous; leaf-blades firm, dark green, 1–3 dm. long, 4–8 mm. wide, scabrous; panicle dense, usually decidedly lobed; spikelets 2.5–3 mm. long; first glume linear-subulate, about 1.5 mm. long, the second cuneate, about 2 mm. long, firm, very scabrous; lemma about 2 mm. long, obtuse. *Eatonia robusta* Rydb. River banks: Neb.—N.M.—Ariz.—Wash. *W. Temp.—Son.—Plain.* My–Jl.

f. 79.

2. S. obtusàta (Michx.) Scribn. *Fig. 79.* Stem 3–7 dm. high, erect, glabrous; leaf-blades 3–20 cm. long, 2–5 mm. wide, scabrous; panicle 5–15 cm. long, dense and spike-like, strict; spikelets crowded, 2.5–3 mm. long; first glume linear-subulate, the second cuneate, 1.5 mm. long and about as broad; lemma 1.5–2 mm. long, obtuse. *E. obtusata* A. Gray [B]. Prairies, meadows and valleys: Mass.—Fla.—Ariz.—Mont.; Mex. *E. Temp.—Son.—Plain—Submont.* Ap–Au.

3. S. intermèdia Rydb. Stem 6–8 dm. high, striate, shining; leaf-blades 8–15 cm. long, 3–5 mm. wide, usually flat; inflorescence rather narrow and dense; first empty glume about 2 mm. long, subulate, scabrous on the back. *E. intermedia* Rydb. Meadows: Man.—N.M.—Wash. *W. Temp.—Plain—Submont.* Jl–S.

4. S. pállens (Spreng.) Scribn. Stem glabrous, 3–10 dm. high; leaf-blades 5–15 cm. long, 2–6 mm. wide, scabrous; panicle 7–20 cm. long, usually nodding, lax; spikelets 3–3.5 mm. long; first empty glume subulate, slightly shorter than the second; lemma lanceolate, acute, about 3 mm. long. *E. pennsylvanica* A. Gray [B]. Open woods and among bushes: N.B.—Ga.—N.M.—B.C. *Temp.—Plain—Submont.* Je–Au.

67. MÉLICA L. MELIC-GRASS.

Perennials with rootstocks, the stem often bulbous at the base. Panicles in our species narrow, often raceme-like. Spikelets rather few, erect or nodding, 1–several-flowered, the lower flowers perfect, sometimes 1 or 2 staminate, and the upper neutral. Empty glumes membranous or hyaline, unequal in length. Lemma larger, membranous, the lateral nerves vanishing in the broad hyaline margins; upper empty lemmas gradually smaller, convolute and enclosing each other, forming an obovate or clavate mass. Palet shorter than the lemma, 2-keeled. Styles distinct; stigmas plumose.

Glumes unequal, shorter than the 3–5-flowered spikelet.
Branches of the panicle spreading; sheaths glabrous. 1. *M. nitens.*
Branches of the panicle erect; sheaths scabrous. 2. *M. Porteri.*
Glumes subequal, nearly as long as the 2-flowered spikelet. 3. *M. mutica.*

1. **M. nìtens** Nutt. Stem 8–12 dm. high, with a horizontal rootstock; leaf-blades 1–2 dm. long, 4–8 mm. wide; panicle 1.5–2.5 dm. long; spikelets 10–12 mm. long, usually 3-flowered; lemmas 7–9 mm. long, scabrous, acute; empty lemmas broad at the summit, exceeded by the fertile ones. *M. diffusa* Auth. [B], not Pursh. Rocky woods: Pa.—Minn.—Neb.—Tex.—Ky. *Allegh.—Ozark.* My–Je.

2. **M. Pórteri** Scribn. *Fig. 80.* Perennial, with a rootstock; stem 4–7 dm. high, smooth; leaf-blades 1–2 dm. long, 2–5 mm. wide, scabrous; panicle narrow, 12–15 cm. long; spikelets 4–5-flowered, nodding, 10–13 mm. long; empty glumes obtuse or acutish, the first about 3 mm., the second 5 mm. long; lemma 3–5 mm. long, scabrous. *M. parviflora* (Porter) Scribn. [B]. Plains and hills: Tex.—Neb.—Colo.—Ariz. *Plain—Submont.—Mont.* Je–S.

3. **M. mùtica** Walt. Stem from a knotted rootstock, 6–9 dm. high; sheaths scabrous; lower leaf-blades short, the upper 1–2 dm. long, 2–10 mm. wide; panicle 8–25 cm. long, with few filiform ascending branches, or reduced to a raceme; spikelets 6–8 mm. long; lemmas scabrous, obtuse; empty lemmas cucullate above, exceeded by the fertile ones. *M. diffusa* Pursh, a pubescent form. Open woods: Pa.—Minn.—Tex.—Fla. *E. Temp.* Ap–My.

68. DÁCTYLIS L. ORCHARD-GRASS.

Tall perennials, with creeping rootstocks. Panicles contracted, with the spikelets crowded at the ends of the branches in unilateral head-like clusters. Spikelets 3–5-flowered, the flowers perfect or the uppermost staminate. Empty glumes unequal, 1–3-nerved, mucronate, the second the larger. Lemma more rigid, ciliate on the keel, mucronate or short-awned. Palet nearly as long, 2-keeled. Styles distinct; stigmas plumose.

1. **D. glomeràta** L. Stem 6–12 dm. high, smooth; leaf-blades 7–25 cm. long, 2–6 mm. wide, flat, scabrous; panicle 7–20 cm. long; branches ascending or spreading in flower, erect in fruit; lemma 5–6 mm. long, pointed or short-awned. Fields and waste places: N.B.—Fla.—Calif.—B.C.; nat. from Eur. *Temp.—Mont.* Je–Jl.

69. BRÌZA L. QUAKING GRASS, QUAKE-GRASS.

Annuals or perennials, with open panicles. Spikelets few, nodding, flattened, broad, many-flowered, the flowers perfect. Glumes strongly concave, thin-membranous, 3–5 nerved, somewhat unequal. Lemmas imbricate, broader than the glumes, 5–many-nerved. Palet much shorter, hyaline, 2-keeled and 2-nerved. Styles distinct; stigmas plumose.

1. **B. máxima** L. Stem 3–5 dm. high, glabrous; leaf-blades 1–2 dm. long, 3–6 mm. wide, scabrous on the veins; spikelets 2–10, nodding, 1–2 cm. long, 8–12 mm. wide; empty glumes nearly orbicular, glabrous; lemma similar, but more or less pilose. Fields and waste places: Mass.—W. Ind.—Colo.—Calif.; adv. from Eurasia. *Plain—Submont.*

70. **DISTÍCHLIS** Raf. ALKALI GRASS, SALT-GRASS, SPIKE-GRASS.

Tufted dioecious perennials, with creeping scaly rootstocks. Inflorescence paniculate. Staminate spikelets many-flowered, very flat. Rachilla continuous. Empty glumes narrow, acute, keeled, membranous. Lemma broader, membranous. Pistillate spikelets few-flowered, less flattened. Lemma nearly coriaceous, broad. Palet compressed, the keels narrowly winged. Styles thickened at the base, moderately long; stigmas plumose.

1. **D. strícta** (Torr.) Rydb. *Fig. 82.* Stem 1–4 dm. high, erect or decumbent at the base; leaf-blades erect, 5–15 cm. long, more or less involute, long-attenuate; panicle of the pistillate plant 3–6 cm. long, dense and spike-like; spikelets 8–15 mm. long; empty glumes lanceolate, acuminate, about 5 mm. long; the second a little broader than the first, scarious-margined; lemma 5–6 mm. long, acute, straw-colored with greenish nerves; palet 4–5 mm. long; panicle of the staminate plant looser, 3–10 cm. long; spikelets more flattened, 1–2.5 cm. long, straw-colored; lemma narrower, 6–7 mm. long. *D. spicata* Coult. & Nels., not Greene. Alkaline soil: Sask.—Tex.—Ariz.—B.C.—Wash. *Columb.—Son.—Plain—Submont.* Je–S.

71. **UNÌOLA** L. SPIKE-GRASS.

Tall erect grasses. Inflorescence paniculate. Spikelets flat, 3–many-flowered, 2-edged, the flowers perfect, or the upper staminate. Glumes unequal. Lemma many-nerved, coriaceous, the lower 1–4 empty. Palet rigid, 2-keeled and winged. Stamens 1–3. Style distinct; stigma plumose. Grain compressed, free.

1. **U. latifòlia** Michx. *Fig. 81.* Stem 5–15 dm. high; leaf-blades 1–2.5 dm. long, 5–25 mm. wide, smooth except the margins; panicle 1–2.5 dm. long, with pendulous branches; spikelets many-flowered, ovate or oblong, 2–3 cm. long; lemma 9–12 mm. long, ciliate on the winged keel; stamen 1. Moist places: Pa.—Ill.—Kans.—Tex.—Fla. *Austral.* Au–S.

72. **DIARRHÈNA** Beauv.

Perennials with running rootstocks. Panicles simple, with few spikelets. Spikelets several-flowered, the upper 1 or 2 flowers sterile. Glumes unequal, the first much smaller than the second, ovate, coriaceous. Lemmas ovate, coriaceous, convex on the back, their three nerves ending in cuspidate tips. Palet ciliate. Stamens 2. Style distinct, but short; stigma plumose. Grain free, beaked, longer than the glume. *Korycarpus* Zea.

1. **D. aurundinàcea** (Zea) Rydb. Stem 5–12 dm. high, very rough below the panicle; leaf-blades 2–6 dm. long, 1–2 cm. wide, long-acuminate, panicle raceme-like, 5–10 cm. long; spikelets 3–5-flowered, 12–16 mm. long; lemma acuminate. *Korycarpus diandrus* (Michx.) Kuntze [B]. *Diarrhena americana* Beauv. [G]. Rich woods: Ohio—Neb.—Okla.—Ga. *Allegh.* Au–S.

73. PÒA L. BLUE-GRASS, MEADOW-GRASS.

Annuals or perennials, rarely dioecious, with paniculate inflorescence. Spikelets 2–6-flowered, flat, the flowers perfect or in some species unisexual, the rachis articulate between the flowers. Empty glumes persistent, strongly keeled, acute, the first usually 1-nerved, the second 3-nerved. Lemma usually longer, more or less keeled, acute or obtuse, awnless, 5-nerved, often pubescent on the keel and marginal nerve, as well as the rachis, the hairs near the base sometimes long and curled (cobweb). Palet shorter than the lemma, 2-keeled and 2-nerved. Styles short; stigmas plumose.

- Annuals, but somewhat tufted.
  - Lemma 5-nerved, not cobwebby at the base. — 1. *P. annua.*
  - Lemma 3-nerved or obscurely 5-nerved, cobwebby at the base. — 2. *P. Chapmaniana.*
- Perennials.
  - Cobweb at the base of the lemma present, sometimes rather scant; lemma usually strongly keeled.
    - Stems bulbous at the base. — 3. *P. bulbosa.*
    - Stems not bulbous at the base.
      - Intermediate nerves of the lemma prominent.
        - Lemma acute or (in *P. sylvestris*) acutish or obtuse, pubescent at least on the midrib.
          - Plants with creeping rootstock; branches of the inflorescence ascending in fruit; ligules short and truncate.
            - Lateral veins of the lemma pubescent.
              - Inflorescence open. — 4. *P. pratensis.*
              - Inflorescence narrow and dense. — 5. *P. arachnifera.*
            - Lateral veins of the lemma glabrous. — 6. *P. trivialis.*
          - Plants with short ascending rootstock; branches of the inflorescence spreading or reflexed.
            - Branches of the inflorescence spikelet-bearing from the middle and up; spikelets 3–4 mm. long; lemma usually obtuse. — 7. *P. sylvestris.*
            - Branches of the inflorescence spikelet-bearing only near the end.
              - Spikelets 5–6 mm. long; flowers pale. — 8. *P. Wolfii.*
              - Spikelets 3–4 mm. long; flowers usually purple.
                - Internerves of the lemma more or less pubescent below; plant 1–3 dm. high. — 9. *P. cenisia.*
                - Internerves of the lemma glabrous; plant 3–5 dm. high. — 10. *P. reflexa.*
        - Lemma obtuse.
          - Veins of the lemma glabrous. — 11. *P. languida.*
          - Veins of the lemma pubescent. — 7. *P. sylvestris.*
      - Intermediate nerves of the lemma faint.
        - Stem compressed; lemma obtuse, rootstock creeping. — 12. *P. compressa.*
        - Stem terete.
          - Branches of the inflorescence spreading or reflexed; lemma obtuse; rootstock creeping. — 13. *P. plattensis.*
          - Branches of the inflorescence ascending or rarely spreading; lemma acute; rootstock not creeping.
            - Marginal nerves of the lemma glabrous; branches of the inflorescence spikelet-bearing above the middle only. — 14. *P. alsodes.*
            - Marginal nerves of the lemma pubescent; branches of the inflorescence spikelet-bearing from below the middle.

Stem stout; ligule triangular; branches of the panicle at last spreading; second glume narrower than the lemma; leaves 2–5 mm. wide. — 15. *P. palustris.*
Stem slender; ligule seldom over 1 mm. long, truncate; branches of the inflorescence ascending or erect.
Spikelets green; second glumes with broad scarious margins and strong lateral nerves.
Second glume narrower than, but fully as long as the lemma, acuminate. — 16. *P. nemoralis.*
Second glume fully as wide, but shorter than the lemma, acute. — 17. *P. interior.*
Spikelets usually purple-tinged; scarious margins of the glumes scarcely evident and lateral nerves faint. — 18. *P. crocata.*
Cobweb wanting.
Spikelets rounded at the base; glumes very broad and the keel strongly arched. — 19. *P. alpina.*
Spikelets acute at the base; glumes narrow, not strongly arched on the keel.
Plants with horizontal rootstocks, not bunch-grasses.
Internerves of the lemma pubescent; stem stout; inflorescence dense; ligules acute.
Intermediate nerves of the lemma strong; plant tall; glumes 5 mm. long. — 20. *P. pratensiformis.*
Intermediate nerves of the lemma weak; glumes 3–4 mm. long.
Glumes yellowish green or straw-colored, faintly nerved. — 21. *P. arida.*
Glumes bright green or strongly 5-nerved. — 22. *P. Overi.*
Internerves of the lemma glabrous; inflorescence open.
Lemma obtuse, straw-colored, with brownish margins. — 23. *P. glaucifolia.*
Lemma acute, purplish or lead-colored. — 24. *P. glauca.*
Plant without extravaginal rootstocks, densely tufted bunch-grasses.
Spikelets strongly flattened; lemma acute and keeled on the back; plant dioecious.
Ligules 5–7 cm. long, acute or acuminate. — 25. *P. longiligula.*
Ligules short, rounded or truncate, those of the sterile shoots obsolete. — 26. *P. brevipaniculata.*
Spikelets little flattened; lemma rounded on the backs towards the apex, almost straight, obtuse.
Ligules lanceolate, acuminate or attenuate.
Plant yellowish green; spikelets yellowish or straw-colored.
Lemma merely scabrous. — 27. *P. laevigata.*
Lemma more or less strigose on the lower portion. — 28. *P. lucida.*
Plant dark green; spikelets dark green or purplish. — 29. *P. Buckleyana.*
Ligules 1–2 mm. long, truncate, rounded or abruptly acute.
Plant 2–4 dm. high; leaves mostly basal and stiff, short, seldom 8 cm. long; ligules rounded. — 30. *P. juncifolia.*
Plant taller, 4–10 dm. high, leafy; leaves longer.

Internerves of the lemma glabrous; nerves silky. — 23. *P. glaucifolia.*
Internerves of the lemma as well as the nerves scabrous.
Ligules ovate or rounded, acute or obtuse; leaves soft. — 31. *P. confusa.*
Ligules truncate; leaves stiff. — 32. *P. truncata.*

1. **P. ánnua** L. Stems usually decumbent and branched at the base or erect; ligules rounded at the apex, 2 mm. long; leaf-blades flat, flaccid, 1–10 cm. long, 1.5–3 mm. wide; panicle 1–10 cm. long; spikelets 3–5 mm. long; empty glumes smooth, the first lanceolate, acute, 1.5 mm. long, the second obtuse, nearly 2 mm. long; lemma 2.5–3 mm. long, the nerves pilose below. Waste places and cultivated ground: Lab.—Ga.—Calif.—B.C.; Mex.; nat. from Eurasia. *Plain—Mont.* My–O.

2. **P. Chapmaniàna** Scribner. Tufted annual; stems terete, mostly erect, 1–2 dm. high, rarely 3 dm. high; sheaths close; ligules 1 mm. long, truncate; blades 1–4 cm. long, 1 mm. wide; panicle oblong, 2–5 cm. long, the branches ascending; spikelets 3–7-flowered; glumes 1.5 mm. long, 3-nerved; lemma as long, the intermediate nerves obscure, the medians and the laterals pubescent below. Dry soil: Va.—Iowa—Miss.—Fla. *Carol.—Ozark.* Ap–My.

3. **P. bulbòsa** L. Perennial; stems bulbous at the base, 1–4 dm. high, glabrous; leaves short, flat or keeled; ligules oblong-lanceolate; panicles oval, dense, often viviparous, with short solitary and paired branches; spikelets 4–6-flowered; glumes 3-nerved; lemma densely cobwebby at the base. Lawns and waste places: N.D.—Ida.; adv. from Eurasia.

4. **P. praténsis** L. Stem 3–12 dm. high, smooth; leaf-blades 1–6 mm. wide, those of the stem 5–15 cm. long, the basal ones longer; panicle 5–20 cm. long, ovate or conical; spikelets 3–6-flowered, 4–5 mm. long; midvein and marginal veins of the lemma silky below, the intermediate ones glabrous. Meadows, fields and woods: Greenl.—Fla.—Calif.—Alaska; Eurasia. *Arct.—Temp.—Plain—Subalp.* Je–Au.

5. **P. arachnífera** Torr. Perennial, with a long creeping rootstock; stems somewhat tufted, 3–7 dm. high, glabrous; ligule a short membranous ring; blade smooth beneath, scabrous above, 0.5–2 dm. long, 3–10 mm. wide; panicle dense and narrow, 7–16 cm. long, 1–4 cm. wide, with ascending branches; spikelets numerous, 4–7-flowered; glumes acuminate, hispidulous on the midrib; lemmas 4–5 mm. long, densely pubescent at the base, midrib and lateral nerves pilose below the middle. Grassy valleys: S.C.—Kans.—N.M.—Tex.—Fla. *Austral.*

6. **P. triviàlis** L. Perennial, with a creeping rootstock; stems decumbent at the base, 3–10 dm. high, scabrous below the panicle; sheaths retrorsely scabrous; ligules 4–6 mm. long, acute; blades 5–15 cm. long, 2–4 mm. wide; panicle 6–15 cm. long, the branches ascending; spikelets 2–3-flowered; glumes shorter than the lemmas, scabrous on the keel; lemma 2–3 mm. long, silky only on the keel. Meadows: Que.—Iowa—Kans.—La.—S.C.; nat. from Eur.

7. **P. sylvéstris** A. Gray. Rootstock short; stem compressed, 3–10 dm. high; ligules 1 mm. long or less; blades 4–15 cm. long, 2–6 mm. wide; panicle 1–2 dm. long, the short branches spreading or reflexed; spikelets 2–4-flowered, 2.5–4 mm. long; glumes narrower than the lemmas; lemma obtuse or acutish, 2.5 mm. long; midrib pubescent throughout, the lateral veins below the middle. Woods and thickets: N.Y.—Minn.—Tex.—Fla. *E. Temp.* Je–Jl.

8. **P. Wólfii** Scribner. Rootstock short; stem 4–10 dm. high; ligules 1 mm. long, deltoid; blades 5–15 cm. long, 2 mm. wide; panicle 8–15 cm. long, the branches ascending; glumes unequal, scabrous on the keel; lemmas 4 mm.

long, the midrib and marginals silky-villous. Meadows: Wis.—Minn.—Neb.—Tenn. *Prairie.*

9. **P. cenísia** All. Stem 1–3 dm. high, slender; ligules 1–2 mm. long, truncate; blades 2–10 cm. long, 1–2 mm. wide, involute; panicle 2–8 cm. long, open; spikelets 2–4-flowered, 5–6 mm. long, purplish; glumes glabrous; lemma about 4 mm. long. Arctic-alpine situations: Greenl.—Lab.—Man.—Colo.—Alaska; Eur. *Arct.* Jl.–Au.

10. **P. refléxa** Vasey & Scribn. Stem 3–5 dm. high, slender, erect, smooth; ligules 2 mm. long, obtuse; leaf-blades 5–15 cm. long, 2–4 mm. wide; panicle 5–10 cm. long; spikelets 2–3-flowered, 3–4 mm. long, usually purple; empty glumes acuminate, smooth; lemma very acute. Wet meadows: Alta.—Man.—N.M.—Utah—Ore.—B.C. *W. Boreal.—Submont.—Subalp.*

11. **P. lánguida** Hitchc. Rootstock short; stem terete, 3–10 dm. high; sheaths compressed, shorter than the internodes; ligules 1–2 mm. long, truncate, toothed; blades 2.5–10 cm. long, 2 mm. wide; panicle nodding, 5–12 cm. long, the few long capillary branches ascending; spikelets 2–4-flowered, 3–4 mm. long; glumes unequal, glabrous; lemma 3 mm. long, obtuse, glabrous, except the short cobweb at the base. *P. debilis* Torr. [G, B]. Woods: Que.—Minn.—Kans.—Pa. *Canad.—Allegh.* Je–Au.

12. **P. compréssa** L. Perennial, with a creeping rootstock; stem 1.5–4 dm. high, decumbent at the base, much flattened, smooth; ligules truncate; leaf-blades bluish green, stiff, erect, 2–10 cm. long, about 2 mm. wide, often convolute, scabrous above; panicle narrow, with ascending short branches, spikelet-bearing to near the base; spikelets 3–5-flowered, 3–9 mm. long; empty glumes acute, nearly equal; lemma 2–2.8 mm. long, obtuse; cobweb scant. Waste places, cultivated grounds, and woodlands: N.H.—N.C.—Calif.—B.C.—Yukon; Eurasia. *Boreal—Plain—Mont.* Je–Jl.

13. **P. platténsis** Rydb. Stem 3–5 dm. high, glabrous, erect, striate; leaf-sheath 1–2 dm. long, striate, glabrous, minutely scabrous; ligule lanceolate-deltoid, acute, 2–3 mm. long; blades 1–3 dm. long, or the uppermost shorter, about 3 mm. long, strongly 7–9-nerved; panicle small, 5–10 cm. long, 5–4 cm. wide; branches spreading, flexuous, or reflexed, 1–2 cm. long, with 3–6 spikelets; spikelets mostly 3-flowered, about 5 mm. long; glumes lanceolate, nearly equal, 2–2.5 mm. long, glabrous, faintly veined; lemma 2.5–3 mm. long, obtuse, with a scant cobweb, ciliate along the keel for half its length; palet also ciliate half its length. Meadows: w Neb.

14. **P. alsòdes** A. Gray. Rootstock short; stem 2–6 dm. high, smooth; sheaths elongate, the upper sheathing the base of the inflorescence; ligules truncate; leaf-blades 1–3 dm. long, 2–5 mm. wide; panicle 1–2 dm. long, the branches in 3's or 4's, at last spreading, filiform; spikelets 3–4-flowered; glumes unequal, glabrous; lemma 4 mm. long, acute, glabrous except the midrib, which is silky and cobwebby at the base, the nerves faint. Woods and thickets: Que.—Minn.—Neb.—Tenn.—N.C. *Canad.—Allegh.* My–Je.

15. **P. palústris** L. Perennial, with a creeping rootstock; stem 4-12 dm. high, erect, smooth; leaf-blades 5–15 cm. long, 2–5 mm. wide; panicle 1.5–3 dm. long, open; spikelets 3–5-flowered, 3–5 mm. long; empty glumes glabrous, scabrous on the keel; lemma obtuse, 2–3 mm. long; midnerve and lateral nerves pubescent below. *P. triflora* Gilib. [G]. *P. serotina* Ehrh. *P. flava* Auth. [B], not L. Meadows and swampy places: Newf.—N.J.—Calif.—B.C.; Eurasia. *Boreal—Plain—Mont.* Jl–Au.

16. **P. nemoràlis** L. Rootstock more or less creeping; stem 3–8 dm. long, slender; ligules very short, truncate; blades 5–15 cm. long, about 2 mm. wide; panicles narrow, 1–2 dm. long, the branches usually ascending; spikelets 2–4-flowered; glumes subequal, much narrower than the lemmas, gradually acuminate, narrowly scarious-margined, 2–3 mm. long; lemma mostly green; midrib

and laterals sparingly pubescent, the intermediates faint; cobweb scant. Shady places: Minn.—Pa.—N. J.; adv. from Europe, native in Greenl. and Lab. Je–Jl.

17. **P. intèrior** Rydb. Tufted perennial; stem slender, 3–6 dm. high; leaf-blades 2–8 cm. long, about 2 mm. wide, glabrous; panicle 4–10 cm. long; spikelets 2–5-flowered, 3–5 mm. long; lemma 2–2.5 mm. long, scabrous on the midnerve and lateral nerves. *P. nemoralis* Scribn., not L. Woods and copses: (Isle Royale) Mich.—Minn.—Neb.—N.M.—Utah—Wash.—Alaska. *Prairie—Plain—Mont.*

18. **P. crocàta** Michx. Tufted perennial; stems strict, erect, 2–4 dm. high; leaf-blades narrow, erect, 1–2 mm. wide, 4–10 cm. long, usually involute; panicle rather narrow, 5–15 cm. long; spikelets 2–4-flowered, about 4 mm. long; lemma about as long as the glumes, pubescent on the midnerve and lateral nerves. *P. nemoralis strictior* A. Gray. Dry hills and meadows: Lab.—Vt.—Minn.—Colo.—Ariz.—Alaska. *Boreal—Submont.—Subalp.* Je–Au.

19. **P. alpìna** L. Stem 0.5–4 cm. high, erect or decumbent at the base; ligules truncate; leaf-blades 2–8 cm. long, 2–4 mm. wide, abruptly acute, flat; panicle 2–8 cm. long; spikelets 3–5-flowered, 5–6 mm. long; empty glumes broad, glabrous, acute, scabrous on the keel; lemma about 4 mm. long, obtuse, its lower half pubescent. Alpine-arctic regions in wet places: Greenl.—Que.—ne Minn.—Colo.—Utah—Alaska; Eurasia. *Arct.—Boreal.—Mont.—Alp.* Je–Au.

20. **P. pratensifórmis** Rydb. *Fig. 83.* Stem 3–6 dm. high; leaf-blades flat, 2–6 mm. wide, those of the stem 2–7 cm. long, those of the sterile shoots 1.5–3 dm. long, scabrous on the margins; panicle 5–12 cm. long, spreading in anthesis only; spikelets 3–5-flowered, 6–8 mm. long; empty glumes nearly equal, acute; lemma obtuse, silky-hairy on the nerves to the middle and pubescent all over near the base. *P. pseudopratensis* Scribn. & Rydb., not Beyer. Wet places: Minn.—Kans.—Utah. *Plain—Submont.* Je–Jl.

21. **P. árida** Vasey. Stem 2–6 dm. high; leaf-blades 2–15 cm. long, erect, flat, or becoming somewhat involute, striate; panicle narrow, dense, 7–15 cm. long; branches erect; spikelets 6–7 mm. long, 5–9-flowered; lemma obtuse, scarcely compressed. *P. andina* Nutt. *P. pratericola* Rydb. Prairies and meadows: S.D.—Okla.—N.M.—Utah—Wyo. *Prairie—Plain—Subalp.*

22. **P. Òveri** Rydb. Perennial, with horizontal rootstocks; stems about 5 dm. high, decidedly scabrous above, especially in the inflorescence; sheaths striate, scabrous; ligules lanceolate, acute, 2–2.5 mm. long; leaf-blades 6–10 cm. long, 2–4 mm. wide, scabrous on both sides; panicle narrow, 10–15 cm. long, the branches short, erect or nearly so, 2–3 cm. long, spikelet-bearing from near the base; spikelets 4–5 mm. long, 3–5-flowered; glumes green, strongly 5-nerved, about 3 mm. long; lemma 3–3.5 mm. long, acutish, faintly veined, finely pubescent, especially below, but without cobweb. Open woods: S.D.

23. **P. glaucifòlia** Scribn. & Williams. Stem about 6 dm. high, smooth, except at the nodes; leaf-blades flat, glaucous, 1.5–2 dm. long, 2–4 mm. wide; panicle open; spikelets compressed, 3–4-flowered; empty glumes ovate, obtuse or acutish, 3-nerved, scabrous on the back, about 4 mm. long; lemma obtuse, pubescent on the keel and nerves, 3–4 mm. long. Moist banks: S.D.—Colo.—Mont. *Submont.—Mont.* Jl–Au.

24. **P. glaùca** Vahl. Perennial, with a rootstock; stem strict, rigid, 1–3 dm. high, rarely higher; sheaths crowded at the base; ligules 1–2 mm. long; blades 3–5 cm. long, 2 mm. wide, flat, glaucous; panicles 3–7 cm. long, usually narrow, the branches erect or ascending, scabrous; spikelets purple or lead-colored, 2–5-flowered, 5–6 mm. long; glumes acute or acuminate, 3-nerved, scabrous on the keel towards the apex; lemma 3–3.5 mm. long, acute, villous on the keel and margins, villous below. Rocky shores and cliffs: Greenl.—Me.—Minn.—N.D.—Alaska; Eur. *Arct.—Subarct.* Jl–Au.

25. **P. longilígula** Scribn. & Williams. Stem 3–5 dm. high, with numerous persistent basal sheaths; blades of sterile shoots 1.5–2.5 dm. long, 1–2 mm. wide, flat or conduplicate; those of the stem-leaves shorter; panicle 6–12 cm. long; spikelets 4–6-flowered, 6–10 mm. long; lemma 4–5.5 mm. long, scabrous on the back, villous on the keel and marginal nerves below. Hillsides and plains: N.D.—N.M.—Calif.—Ore. *Son.—Subalp.* My–Jl.

26. **P. brevipaniculàta** Scribn. & Williams. Stem 1–3 dm. high, smooth; leaf-blades flat or conduplicate; those of the sterile shoots 1 dm. long or more; those of the stem much shorter; panicle 3–6 cm. long, 1–2 cm. wide; spikelets green or tinged with purple, 4–6-flowered, 4–6 mm. long; second glume 3.5 mm. long; lemma obtuse or acutish, 3.5–4 mm. long. Perhaps not distinct from *P. Fendleriana.* Dry meadows and mountain sides: S.D.—Colo.—Utah—N.M.—Ariz. *Son.—Mont.*

27. **P. laevigàta** Scribn. Stem erect, smooth, 5–10 dm. high; blades of the basal leaves 1–2 dm. long, about 1 mm. wide, involute; those of the stem-leaves 2–5 cm. long; panicles narrow, 1–2 dm. long; spikelets appressed, 6–19 mm. long, about 5-flowered; empty glumes nearly equal, acute, thin, scarious-margined; lemma 4–5 mm. long, linear-oblong, the apex scarious and yellowish. *P. laevis* Vasey. Dry meadows and hillsides: Mont.—N.M.—B.C. *Submont.—Mont.*

28. **P. lùcida** Vasey. Stem 3–6 dm. high, smooth; blades of basal leaves 12–15 cm. long, 1–2 mm. wide, soft; those of the stem-leaves 5–7 cm. long; panicle 1–1.5 dm. long, narrow; spikelets 6–8 mm. long, 3–4-flowered, shining, pale; empty glumes abruptly acute, unequal, 3–4 mm. long; lemma 2 mm. long, obtuse. Dry hills: Alta.—S.D.—N.M. *Submont.* Je–Jl.

29. **P. Buckleyàna** Nash. Stem 2–6 dm. high, rigid, smooth; leaf-blades 2–10 cm. long, about 2 mm. wide, flat or becoming involute in drying; panicle 3–10 cm. long, narrow; spikelets 2–5-flowered, 4–6 mm. long, dark green and tinged with purple; empty glumes acute, nearly equal, scabrous on the keel; lemma 4–5 mm. long, obtuse, scabrous above, usually more or less pubescent below. *P. tenuifolia* Buckley, in part. (?) *P. wyomingensis* Scribn. Dry plains and hills: N.D.—Colo.—Utah—Wash. *Columb.—Plain—Alp.* Je–Au.

30. **P. junciòlia** Scribn. Stem glabrous, 1.5–3 dm. high; blades of the sterile shoots about 5 cm. long, involute; those of the stem 2–5 cm. long, 1–2 mm. wide; panicle narrow; spikelets 3–5-flowered, 5–6 mm. long; empty glumes ovate, usually obtuse, subequal, 3–4 mm. long; lemma minutely scabrous on the back, 3–4 mm. long, obtuse. Plains and meadows: Wyo.—Colo.—Utah. *Plain—Submont.* Jl–Au.

31. **P. confùsa** Rydb. Stem 6–9 dm. high; basal leaf-blades 1–2 dm. long, 2–3 mm. wide, flat or involute, puberulent; stem-leaves several; blades about 1 dm. long; panicle narrow, 1–1.5 dm. long, dense; spikelets 7–8 mm. long, usually 4-flowered; empty glumes lanceolate, shining, minutely strigillose above; lemma narrow, about 3.5 mm. long, yellowish green, with brownish scarious margins. Meadows and benchlands: Alta.—Neb.—Colo. *Plain—Mont.*

32. **P. truncàta** Rydb. Stem about 9 dm. high, stiff; leaf-blades 1–2 dm. long, 2–3.5 mm. wide, scabrous on the back; panicle about 1.5 dm. long, narrow;

spikelets 3–5-flowered, 7–9 mm. long; empty glumes 5–6 mm. long, tinged with purple, scabrous on the nerves; lemma narrow, about 5 mm. long, straw-colored or tinged with purple; strigillose throughout. Hills and gulches: Colo.—S.D. —Alta. *Submont.* Jl–Au.

74. **FLUMÍNIA** Fries. SPANGLE-TOP.

Tall perennial aquatic grasses, with rootstocks. Panicle ample. Spikelets 3–4-flowered, perfect, the callus hairy. Glumes acute, thin, membranous, 3–5-nerved. Lemma firm, convex below, 5–7-nerved, the nerves unequal and excurrent as short points. Palet as long as the lemma, 2-toothed. Ovary hairy at the summit, style short, stigmas plumose. *Scolochloa* Link, not Mert. & Koch.

1. **F. festucàcea** (Willd.) Hitchc. Stem stout, 1–2 m. high; leaf-blades 2–3 dm. long; panicle 1.5–3.5 dm. long, the branches fascicled; spikelets 6–12 mm. long; glumes acute, the first shorter than the second; lemma 7-nerved, scabrous. *Scolochloa festucacea* (Willd.) Link [G, B]. Water: Man.—Iowa —Neb.—Sask.; Eur. *Prairie.* Jl–Au.

75. **GLYCÈRIA** R. Br. MANNA-GRASS.

Usually perennials with rootstocks (all ours), rarely annuals. Inflorescence paniculate. Spikelets usually numerous, ovate to linear, few–many-flowered, the glabrous rachilla articulate between the flowers. Empty glumes 2, obtuse or acute, unequal. Lemma obtuse or truncate, often denticulate, convex on the back, but not keeled, prominently nerved, hyaline at the apex. Palet 2-keeled. Styles short, distinct; stigmas plumose. *Panicularia* Fabr.

Spikelets ovate or oblong, 8 mm. long or less.	
Lemma very broadly ovate, obscurely veined.	1. *G. canadensis.*
Lemma elliptic or lance-oblong.	
Panicle elongate, narrow, with erect branches.	
Spikelets 4 mm. long, 3–4-flowered.	2. *G. Torreyana.*
Spikelets 3 mm. long, 5–6-flowered.	4. *G. rigida.*
Panicle not elongate, open.	
Lemma rounded at the apex.	
Spikelets 3 mm. long or less.	
Leaf-blades flat and lax; lemma slightly scarious-margined; branches of the inflorescence long; branches drooping.	3. *G. nervata.*
Leaf-blades conduplicate, stiff, ascending; lemma distinctly scarious-margined; branches of the inflorescence short, ascending.	4. *G. rigida.*
Spikelets 4–6 mm. long; branches spreading in fruit.	5. *G. grandis.*
Lemma truncate and toothed at the apex; panicle few-flowered, with ascending branches.	6. *G. pallida.*
Spikelets linear-cylindric, 1–2 cm. long.	
Spikelets 1–1.5 cm. long, distinctly pedicelled.	7. *G. borealis.*
Spikelets 1.5–2 cm. long, subsessile.	8. *G. septentrionalis.*

1. **G. canadénsis** (Michx.) Trin. Stem erect, 6–10 dm. high; sheaths compressed, overlapping below; leaf-blades 1.5–3.5 dm. long, 4–8 mm. wide, scabrous; panicle 1.5–3 dm. long, loose and open, with drooping branches; spikelet 5–10-flowered, ovate, resembling that of *Briza*; glumes 1-nerved; lemmas obtuse or acutish, obscurely 7-nerved. *Panicularia canadensis* Kuntze [B]. Wet places: Newf.—Minn.—Kans.—N.J. *Canad.—Allegh.* Jl–Au.

2. **G. Torreyàna** (Spreng.) Hitchc. Stem mostly simple, erect, from a running rootstock, 6–9 dm. high; sheaths close, smooth; blades 3–4 dm. long, 3–6 mm. wide, scabrous; panicle linear, 1.5–3 dm. long; spikelets 3–4-flowered, about 4 mm. long; glumes unequal, 1-nerved, the second twice as long as the first; lemmas 2 mm. long, obtuse, 7-nerved. *G. elongata* Torr. *P. Torreyana* Merrill [B]. Wet woods: Me.—Minn.—Iowa—Ky.—N.C. *Canada.*

3. **G. nervàta** (Willd.) Trin. *Fig. 84.* Stem slender, 3–10 dm. high; leaf-blades 1.5–3 dm. long, 4–10 mm. wide, smooth beneath, scabrous above; panicle 7–20 cm. long, open; spikelets 3–7-flowered, 2–3 mm. long; lemma 1.5 mm. long, obtuse or rounded at the apex, striate. *P. nervata* Kuntze [B, R]. Wet meadows and swamps: Lab.—Fla.—Calif.—Alaska; Mex. *Temp.—Submont.* Je–Au.

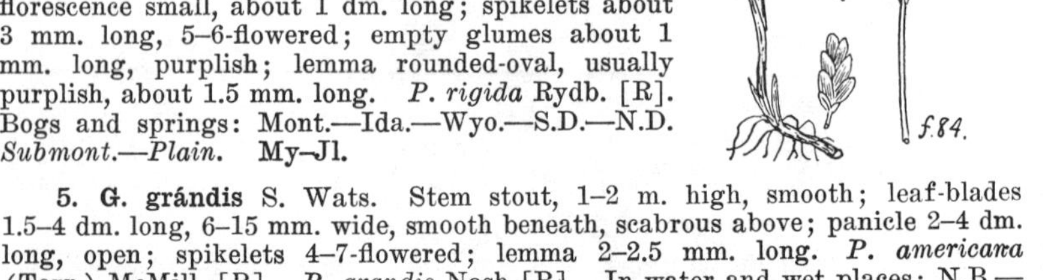

4. **G. rígida** (Nash) Rydb. Stem 3–4 dm. high; leaf-blades 5–15 cm. long, 3–4 mm. wide; inflorescence small, about 1 dm. long; spikelets about 3 mm. long, 5–6-flowered; empty glumes about 1 mm. long, purplish; lemma rounded-oval, usually purplish, about 1.5 mm. long. *P. rigida* Rydb. [R]. Bogs and springs: Mont.—Ida.—Wyo.—S.D.—N.D. *Submont.—Plain.* My–Jl.

5. **G. grándis** S. Wats. Stem stout, 1–2 m. high, smooth; leaf-blades 1.5–4 dm. long, 6–15 mm. wide, smooth beneath, scabrous above; panicle 2–4 dm. long, open; spikelets 4–7-flowered; lemma 2–2.5 mm. long. *P. americana* (Torr.) McMill. [B]. *P. grandis* Nash [R]. In water and wet places: N.B.—N.Y.—N.M.—Calif.—Alaska. *Temp.—Plain—Submont.* Je–Au.

6. **G. pállida** (Torr.) Trin. Stem slender, 3–10 dm. high, ascending at the base; sheaths loose; blades 5–15 cm. long, 2–8 mm. wide; panicle lax, 7–15 cm. long, with few branches; spikelets pale green, 4–9-flowered, 6–7 mm. long; glumes obtuse; lemma scabrous. *P. pallida* Kuntze [B]. Shallow water: N.S.—Man.—Tenn.—N.C. *Canad.—Allegh.* Jl–Au.

7. **G. boreàlis** (Nash) Batchelder. Stem glabrous, 6–15 dm. high; leaf-blades 1–2 dm. long, 2–10 mm. wide, erect, scabrous on both sides; panicle narrow, 2–5 dm. long; branches erect, smooth; spikelets 7–13-flowered; empty glumes scarious-margined, 1-nerved, the first acute, the second obtuse, erose; lemma 3.5–4 mm. long. *P. borealis* Nash [B,R]. Water and wet places: Me.—N.Y.—N.M.—Calif.—Alaska. *Boreal—Submont.—Mont.* Jl–Au.

8. **G. septentrionàlis** Hitchc. Stem 1–1.5 m. high; blades 1–2.5 dm. long, 6–8 mm. wide; panicle 2–2.5 dm. long, subflexuose; spikelets 8–12-flowered; empty glumes obtuse; lemma 4–4.5 mm. long, hispidulous. *Glyceria fluitans* Am. auth. [B]; not R. Br. Shallow water: Newf.—Va.—Calif.—B.C. *Boreal—Plain—Submont.*

76. PUCCINÉLLIA Parl. MEADOW-GRASS.

Perennials with rootstocks, ours tufted. Inflorescence paniculate. Spikelets 3–several-flowered; flowers perfect. Empty glumes 2, obtuse or acute, unequal. Lemma rounded on the back, obscurely 5-nerved. Palet nearly equaling the lemma. Styles wanting; stigmas sessile, plumose. Grain usually adherent to the palet.

Lemma 2–2.5 mm. long, distinctly veined. 1. *P. Nuttalliana.*
Lemma 2 mm. long or less, obscurely veined. 2. *P. tenuiflora.*

1. **P. Nuttalliàna** (Schultes) Hitchc. Stem 3–12 dm. high, erect, strict; leaf-blades 5–10 cm. long, 3 mm. wide or less, usually involute or the upper flat; panicle open, with spreading or ascending branches; spikelets scattered, 1–7-flowered, 3–6 mm. long; empty glumes unequal, the first acute, 1-nerved, the second obtuse or acute, 3-nerved. *P. airoides* (Nutt.) Wats. & Coult. [G, B]. *Glyceria distans airoides* Nutt. Saline soil: Man.—Kans.—N.M.—Nev.—B.C.—Mack. *Prairie—Plain—Subalp.* Je–Au.

2. **P. tenuiflòra** (Griseb.) Scrib. & Merr. Stems 4–6 dm. high; leaf-blades 6–9 cm. long, 1–2 mm. wide, involute; panicle 1–1.5 dm. long; branches capillary, ascending or spreading, scabrous, often 1 dm. long; spikelets 3–4 mm. long, 3- or 4-flowered. *Atropis tenuiflora* Griseb. Alkaline flats: Alaska—Sask.—Man. *W. Boreal—Plain—Submont.* Je–Au.

77. BROMÉLICA (Thurber) Farwell.

Broom-like perennial grasses with rootstocks. Panicles with erect or spreading branches with few spikelets. Leaves flat or convolute above. Sheath scabrous or hairy. Ligules short. Spikelets of 3–8 perfect flowers and a single sterile one at the top. Glumes shorter than the lower flowers. Lemmas prominently 7-nerved, notched or bifid at the apex, usually bearing an awn below the notch. Grain free. Stigmas plumose.

1. **B. Smíthii** (Porter) Farwell. Perennial with a rootstock; stem 7–12 dm. high, scabrous; leaf-blades 1–2 dm. long, 6–12 mm. wide, scabrous; panicle 1–3 dm. long, its branches at last spreading or reflexed; spikelets 3–6-flowered; first empty glume 4–6 mm. long, obscurely 3-nerved; second glume 6–8 mm. long, 5-nerved; lemma about 10 mm. long, strongly nerved; awn 3–5 mm. long. *Avena Smithii* Porter [B]. *Melica Smithii* Vasey [G]. *M. retrofracta* Suksd. Damp shady places: Mich.—Wyo.—Ore.—Wash. *Boreal—Mont.* Jl–Au.

78. FESTÙCA L. Fescue-grass.

Perennials, or annuals, usually tufted. Inflorescence paniculate. Spikelets 2–several-flowered, the flowers perfect or the upper ones staminate, the rachilla articulate between the flowers. Empty glumes 2, membranous, unequal, keeled, acute, the first mostly 1-nerved, the second 3-nerved. Lemma rounded on the back, at least below, acute, more or less awned, 3–7-nerved. Palet a little shorter than the lemma, 2-keeled. Stamens 1–3. Styles very short, distinct, terminal or nearly so; stigmas plumose, bilateral.

Perennials; stamens 3.
- Leaf-blades of the innovations narrow, 3 mm. wide or less, involute.
  - Innovations extravaginal, i.e., plants with creeping rootstocks and stolons.
    - Spikelets pubescent. — 1. *F. Kitaibeliana.*
    - Spikelets glabrous or scabrous. — 2. *F. rubra.*
  - Innovations infravaginal; plants bunch-grasses.
    - Body of the lemma 3–8 mm. long; leaf-blades long, persistent on the sheath.
      - Lemma (without the awn) 3–4 mm. long; plant 1–2 dm., rarely 3 dm. high; inflorescence spike-like. — 3. *F. brachyphylla.*
      - Lemma (without the awn) 5–8 mm. long; plant usually more than 3 dm. high.
        - Inflorescence dense and narrow, spike-like; leaves filiform, involute, 0.5 mm. wide or less. — 4. *F. saximontana.*
        - Inflorescence open.
          - Stem-leaves flat, firm, usually 1 mm. wide or more. — 5. *F. duriuscula.*
          - Stem-leaves filiform, less than 1 mm. wide. — 6. *F. ovina.*
    - Body of the lemma 8–12 mm. long; leaf-blades soon breaking off from the sheaths; palet notched at the apex.
      - Plant 5–15 dm. high; basal sheaths long, loose, in age more or less flattened; empty glumes narrowly lanceolate, acuminate; second glume shorter than the spikelet. — 7. *F. campestris.*
      - Plant 3–5 dm. high; basal sheaths usually short and close; glumes acute, lanceolate; second glume about equaling the spikelet. — 8. *F. scabrella.*
- Leaf-blades all flat, 4 mm. wide or wider; perennials with rootstocks.

Lemma 5–7 mm. long; spikelets 5–10-flowered; inflorescence narrow, with short erect branches. 9. *F. elatior.*
Lemma 4 mm. long or less; spikelets 3–6-flowered.
Lemma acute; panicle loose, the branches spikelet-bearing at the ends. 10. *F. obtusa.*
Lemma obtuse; panicle more compact, the branches spikelet-bearing from the middle. 11. *F. Shortii.*
Annuals; stamens usually solitary. 12. *F. octoflora.*

1. **F. Kitaibeliàna** Schultes. Stem 2–5 dm. high; blades of stem-leaves flat, 3–12 cm. long; inflorescence narrow, often secund; spikelets 8–12 mm. long, 5–9-flowered; empty glumes unequal, 2–4 mm. long, lanceolate, glabrous; lemma narrowly lanceolate, 5–6 mm. long; awn scabrous, 2–4 mm. long. *F. rubra subvillosa* Mert. & Koch [G]. Hills and mountain sides: Greenl.—N.B.—Wyo.—Ore.—Alaska; Eurasia. *Arctic—Boreal—Submont.* Je–Au.

2. **F. rùbra** L. Stem 3–10 dm. high; blades soft, green or (in var. *glaucoidea* Piper) more or less glaucous; panicle 5–20 cm. long, often narrow; spikelets 4–6-flowered, 7–8 mm. long, green or more or less glaucous, often purple-tinged; lemma 5–7 mm. long, lanceolate; awn scabrous, 1–4 mm. long. Hills: Greenl.—Va.—Colo.—Alaska; Eurasia. *Arctic—Boreal—Submont.—Subalp.* Je–Au.

3. **F. brachyphýlla** Schultes. Stem 1–2 dm. (rarely 3 dm.) high; leaf-blades narrow, involute, 1–5 cm. long; panicle 2–5 cm. long; spikelets 2–5-flowered; lemma green or purplish, about 3 mm. long, acuminate; awn scabrous, 2–3 mm. long. *F. brevifolia* R. Br. Alpine-arctic regions: Greenl.—Vt.—S.D.—N.M.—Ariz.—Calif.—Alaska. *Subalp.—Alp.—Arctic.* Jl–S.

4. **F. saximontàna** Rydb. *Fig. 85.* Densely tufted perennial; stem 2–4 dm. high, slender, glabrous; leaf-blades very slender, 3–10 cm. long, less than half a millimeter wide, strongly involute; panicle 4–10 cm. long; spikelets 4–6-flowered; lemma lanceolate, 3–5 mm. long. *Festuca pseudovina* (Beal) Rydb., not Hack. Dry ridges: Mich.—Neb.—Colo.—B.C. *W. Boreal—Mont.—Alp.* Je–Au.

5. **F. duriúscula** L. A tufted perennial; stem 3–5 dm. high; leaves flattened, 0.7–1 mm. wide, 5–12 cm. long; panicle open, 1–1.5 dm. long; spikelets 4–8-flowered; lemma about 6 mm. long; awn 3–5 mm. long. *F. ovina duriuscula* (L.) Koch [G, B]. Meadows: Newf.—Va.—Ida.—Man.; nat. from Eur. Je–Jl.

6. **F. ovìna** L. Densely tufted perennial; stem 2–6 dm. high; leaf-blades pale-green, capillary, involute, 5–12 cm. long; panicle short, open, but often contracted after blooming, 5–10 cm. long; spikelets 5–7 mm. long, 3–6-flowered; lemma 3–3.5 mm. long, attenuate; awn 1–3 mm. long. Mountains, fields and waste places: Greenl.—N.H.—ne Minn.; n Eur.; cultivated in lawns and escaped southward to N.Y.—Ky.—Iowa—S.D. Je–Jl.

7. **F. campéstris** Rydb. Stem 4–10 dm. high; leaf-blades 3–5 dm. long, scabrous; panicle 12–25 cm. long; spikelets 10–12 mm. long, 2–4-flowered; empty glumes membranous, narrowly lanceolate, acuminate; lemma very scabrous, attenuate, thicker than the empty glumes, 9–10 mm. long. *F. scabrella major* Vasey. Plains and hills: Man.—Alta.—N.D.—Wash.—Yukon. *Plain—Submont.* Je–Au.

8. **F. scabrélla** Torr. Stem 3–5 dm. high; blades of the basal leaves 1–2 dm. long, scabrous, striate; those of the upper stem-leaves short, 3–5 cm. long, often flat; panicle small, 5–15 cm. long; branches ascending; spikelets often

purple-tinged, 3-4-flowered, about 1 cm. long; lemma scabrous, 6–8 mm. long, acute. *F. Hallii* (Vasey) Piper. Mountains: Man.—Alta.—Mont.—Wash.—B.C.; Colo. *Mont.* Je–Jl.

9. **F. elàtior** L. Stem smooth, 5–12 dm. high; leaf-blades 1–6 dm. long, 4–8 mm. wide, smooth beneath, scabrous above; panicle 1–2 dm. long, rather open in anthesis, narrowed in fruit; spikelets 3–13-flowered, usually 6–8-flowered, 9–11 mm. long; lemma 5–7 mm. long, scabrid toward the apex. Meadows and copses: N.S.—N.C.—N.M.—Calif.—B.C.; cult. and nat. from Eur. Je–Au.

10. **F. obtùsa** Spreng. Stem 6–10 dm. high; sheaths short; leaf-blades 1–3 dm. long, 4–6 mm. wide, rough above; panicle 1–2.5 dm. long, lax; spikelets 5–7 mm. long, 3–5-flowered; glumes scabrous on the keel; lemma smooth, faintly nerved. *F. nutans* Spreng. [G, B]; not Moench. Woods: N.S.—Fla.—Tex.—Neb.—N.D. *E. Temp.* Je–Au.

11. **F. Shórtii** Kunth. Stem 6–12 dm. high; leaf-blades 1–2.5 dm. long, 2–6 mm. wide, rough above; panicle 7–18 cm. long; spikelets 3–6-flowered, broadly obovate in age, 5–6 mm. long; lemma smooth. Wet prairies and open woods: Pa.—Ga.—Tex.—Kans.—Minn.—Iowa. Jl–Au.

12. **F. octoflòra** Walt. Usually tufted; stems 5–40 cm. high, glabrous or retrorsely puberulent; leaf-blades involute or rarely flat, 2–10 cm. long; panicle narrow, erect, racemiform, 3–12 cm. long; spikelets 5–10 mm. long, rarely longer; lemma lanceolate, scabrous to glabrous, 4–5 mm. long, attenuate; awn straight, 1–7 mm. long. *F. tenella* Willd. In dry sandy soil: Que.—Fla.—Calif.—B.C. *Temp.—Mont.* Ap–Jl.

79. **HESPERÓCHLOA** (Piper) Rydb.

Dioecious perennials, densely tufted, producing both intravaginal innovations and stout extravaginal stolons. Panicle narrow. Spikelets 3–5-flowered, those of the staminate plant flattened, those of the pistillate one turgid. Empty glumes 2, lanceolate, subscarious, the lower 1-nerved, the upper 3-nerved. Lemma ovate, 5-nerved, acuminate. Palet shorter than the lemma. Ovary hispidulous at the apex, deeply sulcate; stigma elongate, the numerous short branches arising on all sides. Grain 2-dentate at the apex.

1. **H. Kíngii** (S. Wats.) Rydb. Stems stout, striate, glabrous, 4–10 dm. high; leaf-blades firm, flat or loosely involute, coarsely striate, 1–3 cm. long, 3–6 mm. wide; panicle narrow, erect, 8–20 cm. long; spikelets 6–10 mm. long; empty glumes nearly smooth, shining, the first 3–4.5 mm. long, the second 5–7 mm. long; lemma ovate, abruptly acuminate, scabrous, 5–8 mm. long. *Poa Kingii* S. Wats. *Festuca confinis* Vasey. Cañons and hillsides: Ore.—Calif.—Colo.—Neb.—Mont. *Submont.—Subalp.* Je–Au.

80. **BRÒMUS** L. Brome-grass, Chess, Cheat.

Annuals or perennials, with paniculate inflorescence. Spikelets usually large, often drooping, several–many-flowered; flowers perfect or the upper ones imperfect. Empty glumes 2, membranous, persistent, narrow, unequal, acute or the second one sometimes short-awned. Lemma longer than the glumes, rounded or keeled on the back, usually awned, the awn dorsal and inserted just below the 2-toothed apex. Palet shorter than the lemma, 2-keeled. Ovary crowned with a villous appendage, at the base of which arise the lateral styles; stigmas plumose.

Second empty glume 5–7-nerved; first empty glume 3-nerved.
  Lemma compressed-keeled.
    Palet less than three fourths as long as the lemma, which is scarcely toothed. 1. *B. breviaristatus.*

Palet more than three fourths as long as the lemma, which is distinctly toothed at the apex. 2. *B. unioloides.*

Lemma rounded on the back, broadly elliptic.

Annuals or biennials; introduced or naturalized species.

Lemma nearly as broad as long, awnless or with very short awn. 3. *B. brizaeformis.*

Lemma much longer than broad, conspicuously awned.

Lemma and glumes glabrous; spikelets distinctly pedicelled.

Awn much shorter than the lemma, nearly erect.

Sheaths glabrous. 4. *B. secalinus.*

Sheaths densely pubescent.

Lemma 9–10 mm. long; panicle drooping. 5. *B. commutatus.*

Lemma 7 mm. long or less; panicle not drooping.

Panicle 2–3 dm. long; spikelets less than 6 mm. wide, several on the lower branches. 6. *B. arvensis.*

Panicle 1 dm. long or less; spikelets less than 6 mm broad, 1 or 2 on the lower branches. 7. *B. racemosus.*

Awn as long as the lemma or longer, strongly divergent at maturity.

Lemma oblanceolate, 7–9 mm. long, deeply notched. 8. *B. patulus.*

Lemma obovate, about 10 mm. long, obtuse or slightly emarginate. 9. *B. squarrosus.*

Lemma and glumes more or less pubescent; spikelets nearly sessile. 10. *B. hordeaceus.*

Perennials, with rootstocks; native species. 11. *B. Kalmii.*

Second empty glume 3-nerved; first empty glume 1-nerved, except in *B. Porteri.*

Awn shorter than the lemma; plants perennial with rootstocks, all native except *B. inermis.*

Inflorescence more or less drooping.

Empty glumes decidedly pubescent.

First empty glume 3-nerved; inflorescence narrow. 12. *B. Porteri.*

First empty glume 1-nerved; inflorescence open.

Sheaths glabrous or pubescent only at the mouth.

Sheaths with a ring of dense hairs at the base of the blades. 13. *B. altissimus.*

Sheaths without a ring of hairs. 14. *B. purgans.*

Sheaths densely soft-pilose. 15. *B. incanus.*

Empty glumes glabrous or merely scabrous on the nerves; lemma ciliate on the margins, glabrous or sparingly hairy on the back; sheaths glabrous or the lower sparingly hirsute.

Panicle more or less nodding; spikelet usually green, loose at maturity, displaying the rachillae between the flowers; lemmas lanceolate-attenuate, delicately nerved; marginal bands appressed-pilose. 17. *B. ciliatus.*

Panicle scarcely nodding; spikelets bronze-tinged, densely flowered; rachillae hidden; lemmas obtuse or acutish, strongly nerved; marginal bands villous-hirsute. 16. *B. Dudleyi.*

Inflorescence not drooping.

Inflorescence narrow; its branches erect; lemma usually with awn 2–3 mm. long. 18. *B. Pumpellianus.*

Inflorescence broad; its branches spreading; lemma usually awnless. 19. *B. inermis.*

Awn longer than the lemma; introduced tufted annuals.

Spikelets numerous on slender, recurved pedicels; lemma 8–12 mm. long. 20. *B. tectorum.*

Spikelets few; pedicels not recurved; lemma 12–30 mm. long. 21. *B. sterilis.*

1. **B. breviaristàtus** (Hook.) Buckl. Tufted short-lived perennial; stem 6–12 dm. high; leaf-blades sparsely pilose throughout or glabrate, 1.5–2.5 dm. long, 6–12 mm. wide; panicle erect, narrow, 1–2 dm. long; spikelets 2.5–4 cm. long, 5–7 mm. wide, 7–9-flowered; empty glumes scabrous or scabrous-puberulent; lemma 11–14 mm. long; awn 4–5 mm. long. *B. marginatus* Nees. Meadows: B.C.—Calif.—Colo.—Neb.—N.D.—Alta.; introduced in Iowa. *Submont.—W. Temp.*

2. **B. unioloìdes** (Willd.) H.B.K. Stout, more or less tufted annual; stems 3–10 dm. high, glabrous; sheath pilose-pubescent, rarely glabrous; panicle 1.5–3.5 dm. long; branches ascending or spreading; spikelets 2–3.5 cm. long, 5–9 mm. broad, 7–11-flowered; empty glumes smooth or slightly scabrous; lemma broadly lanceolate, acute, nearly glabrous to strongly scabrous, 13–16 mm. long; awn usually present, 2 mm. long or less. Fields, prairies and waste places: Fla.—Mo.—Colo.—Calif. *S. Temp.—Son.* Ap–Au.

3. **B. brizaefórmis** Fisch. & Mey. Stem 3–6 dm. high, glabrous or slightly pubescent at the nodes; leaf-blades pubescent on both sides, 5–10 cm. long, 2–5 mm. wide; panicle 5–25 cm. long, lax, secund, nodding; spikelets 15–25 mm. long, ovate; empty glumes obtuse, smooth or minutely scabrous; lemma about 1 cm. long, very broad, smooth or minutely scabrous, broadly scarious-margined. Waste places and fields: Mass.—Del.—Calif.—B.C.; adv. from Eurasia. *Temp. —Submont.* Je–Jl.

4. **B. secalìnus** L. Stem 3–7 dm. high, smooth or pubescent at the nodes; leaf-blades 1–2 dm. long, sparsely pubescent above, glabrous beneath; panicle 8–18 cm. long, at first erect, drooping in fruit; spikelets 10–18 mm. long, 6–8 mm. broad; lemma 6–8 mm. long, obtuse, smooth or scabrous, its margins strongly incurved in fruit; awn 3–5 mm. long. CHEAT. Fields and waste places: Me.—Fla.—Calif.—Wash.; nat. from Eur. *Temp.* Ap–Au.

5. **B. commutàtus** Schrader. Stem 3–7 dm. high; sheaths pilose; blades 5–10 cm. long, pilose; panicle 1–1.5 dm. long, drooping, the lower branches with 1 or 2 spikelets; spikelets 2–2.5 cm. long, 6–7 mm. wide; glumes unequal, the second one third longer than the first; lemma about 8 mm. long; awn 8–10 mm. long, straight. Fields: Mass.—Mich.—Iowa—Mo.—Md.; Ore.—Wash.; nat. from Eur. My–Jl.

6. **B. arvénsis** L. Stem 3–9 dm. high, erect; sheaths pilose; ligules 2 mm. long; blades pilose on both sides, 7–15 cm. long, 4–6 mm. wide; panicle diffuse, the lower branches with 4–8 spikelets; spikelets 7–11-flowered, 1.5–2.5 cm. long, 3–4 mm. wide, scabrous throughout; lower glume 4–5 mm. long, the upper 7–8 mm. long; lemma 7–8 mm. long, obtuse; awn 7–10 mm. long, straight or somewhat twisted in age. Fields: N.J.—Mich.—Iowa—Mo.—Fla.; nat. from Eur. Je–Au.

7. **B. racemòsus** L. Stem 3–7 dm. high, scabrous-puberulent under the inflorescence and pubescent at the nodes; leaf-blades 7–14 cm. long, pilose-pubescent; panicle simple, somewhat nodding; spikelets 15–20 mm. long, 5–9-flowered, acute; lemma smooth or scabrous, 6–8 mm. long, very shortly bidentate at the apex; awn 5–8 mm. long, straight. Waste places: N.S.—Ga.—N.M.—Ore.—Wash.—Yukon; adv. from Eur. Je–Au.

8. **B. pátulus** Mert. & Koch. Stem somewhat geniculate at the base, 4–6 dm. high; leaf-blades pubescent, 4–10 cm. long, 2–4 mm. wide; panicle 12–20 cm. long, diffuse, somewhat drooping; spikelets 2–2.5 cm. long, 5–6 mm. wide; lemma 7–9 mm. long, hyaline-margined, emarginate, glabrous. *B. japonicus* Thunb. (?) [G]. Waste places: Mass.—Colo.—S.D.; adv. from Eur. Je–Jl.

9. **B. squarròsus** L. Stems tufted, 2–4 dm. high, smooth; sheaths retrorsely pilose; ligules 19 mm. long; blades 8–15 cm. long, 3–5 mm. wide, soft-pubes-

cent on both sides; panicle 6–12 cm. long, open, the branches drooping; spikelets 6–12-flowered, 15–20 mm. long; glumes glabrous, obtuse or acutish, the first 5-nerved, the second 7–9-nerved; lemma minutely scabrous; awn spreading, bent, 10 mm. long, scabrous. Fields: N.Y.—N.J.—Kans.—S.D.; adv. from Eur. Jl–Au.

10. B. hordeàceus L. Stem 2–8 dm. high, usually somewhat pubescent; leaf-blades 5–15 cm. long, 3–5 mm. broad, pilose or glabrate; panicle contracted, 5–10 cm. long; spikelets 5–13-flowered, 12–15 mm. long, 4–6 mm. wide; lemma obtuse, coarsely pilose, 8–9 mm. long; awn stout, 6–9 mm. long. *B. mollis* L. HAIRY CHEAT. Waste places, roadsides, fields: Me.—N.C.—Calif.—B.C.; nat. from Eur. *Temp.—Submont.* My–Jl.

11. B. Kálmii A. Gray. Stem slender, 5–10 dm. high; sheaths and leaf-blades conspicuously or sparingly villous; blades 7–17 cm. long, 5–10 mm. wide, sparingly pubescent; panicle narrow, 7–10 cm. long; spikelets drooping, 7–12-flowered, densely silky; lemma 8–10 mm. long, 7-nerved, obtuse; awn 2–3 mm. long. Dry ground: Que.—Mass.—Pa.—Mo.—S.D.—Man. *Canad.—Allegh.* Je–Jl.

12. B. Pòrteri (Coult.) Nash. Stem 5–9 dm. high, pubescent at the nodes; sheaths usually sparingly short-pilose; leaf-blades 7–30 cm. long, 3–5 mm. wide, scabrous; panicle 1–2 dm. long, drooping; spikelets 2–2.5 cm. long, 7–9-flowered; lemma 11–13 mm. long, coarsely pubescent, the apex hyaline, slightly emarginate; awn 2–4 mm. long. Hillsides and meadows: Man.—N.M.—Ariz.—Alta. *Plain—Subalp.* Je–Au.

13. B. altíssimus Pursh. Stem 6–10 dm. high, very leafy, slightly pubescent at the nodes or glabrous; leaf-blades 1–3 dm. long, 5–8 mm. wide, scabrous and sometimes sparingly hairy above, with conspicuous auricles at the base; panicle 1–3 dm. long, open, somewhat nodding; spikelets 2–3 cm. long, loosely 6–10-flowered; lemma about 1 cm. long, broadly lanceolate, pilose, densely so toward the base; awn 4–5 mm. long. *B. purgans latiglumis* Shear. *B. latiglumis* (Shear) Hitchc. [R]. Meadows: Conn.—Pa.—Kans.—Mont. *Canad.—Allegh.—Plain.* Je–Au.

14. B. púrgans L. Stem 7–14 dm. high, glabrous or pubescent at the nodes; leaf-blades 1.5–3 dm. long, 5–15 mm. wide, somewhat auricled at the base, short-pilose on the veins above, scabrous or smooth beneath; panicle lax, nodding, 1.5–2.5 dm. long; spikelets 7–14-flowered, 2–2.5 cm. long; lemma acute, 10–12 mm. long, sparsely pubescent across the back; awn 4–6 mm. long, straight. Woods and shady banks: Que.—Ga.—Okla.—Wyo.—N.D.—Man. *E. Temp.—Plain—Submont.* My–Jl.

15. B. incànus (Shear) Hitchc. Stem 5–12 dm. high; sheaths longer than the internodes; blades 1.5–2.5 dm. long, 5–15 mm. wide; panicle lax; spikelets 7–11-flowered, 2–2.5 cm. long; lemmas 10–12 mm. long, sparingly pubescent on the back; awn about 5 mm. long. Wooded hills: Pa.—Va.—Tex.—N.D. *Allegh.—Ozark.*

16. B. Dúdleyi Fernald. Stem 5–12 dm. high, erect, glabrous or sparsely pilose at the nodes; leaves glabrous on both sides or villous above, 1–2.5 dm. long, 4–12 mm. wide; panicle 5–20 cm. long, with ascending branches; spikelets 2–2.5 cm. long, 4–7-flowered; glumes scarious-margined, purplish; outer glume lanceolate, 1-nerved, the inner oblong-lanceolate, 3-nerved, lemma 1–1.2 cm. long, obtuse, the awn 2–4 mm. long. Damp places: Newf.—N.Y.—Minn.—Mont.—B.C. Jl–Au.

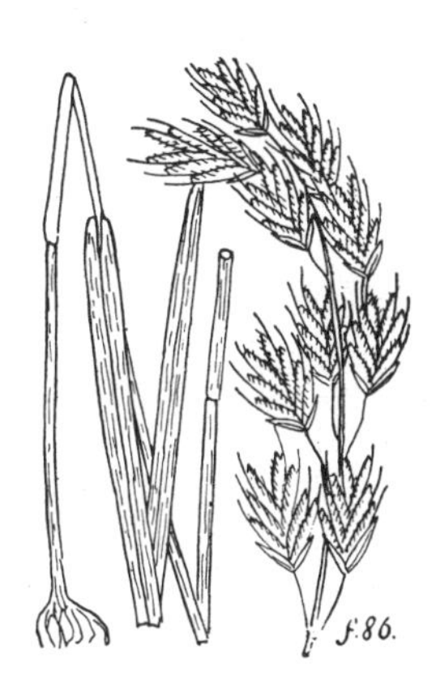

17. B. ciliàtus L. *Fig. 86.* Stem 6–13 dm. high, glabrous; leaf-blades 1.5–2.5 dm. long, 5–12 mm. wide, mostly scabrous above, glabrous beneath; panicle large, drooping, 1.5–2.5 dm. long; spikelets 2–3 cm. long, 6–11-flowered; lemma obtuse, emarginate, 12–15 mm. long; awn 3–5 mm. long, straight. *B. Richardsoni* Link. Meadows and hillsides: Newf.—Fla.—N.M.—Ariz.—Ore.—B.C. *Temp.—Plain—Subalp.* Je–Au.

18. B. Pumpelliànus Scribn. Stem glabrous, or pubescent merely at the nodes; leaf-blades 1–2 dm. long, 5–10 mm. wide, smooth beneath, scabrous or slightly pubescent above, auricled at the base; spikelets erect, 7–11-flowered, 2–3 cm. long, 5–7 mm. wide; lemma 10–12 mm. long, coarsely ciliate along the margins and across the back at the base. Meadows and hillsides: Man.—S.D.—Colo.—Calif.—Alaska. *Boreal—Submont.—Mont.*

19. B. inérmis Leyss. Stem 5–10 dm. high; leaf-blades smooth and glabrous, 1.5–2.5 dm. long, 5–10 mm. wide; panicle 1–2 dm. long; spikelets 2–2.5 cm. long, 4–5 mm. wide; lemma 9–12 mm. long, obtuse, emarginate, typically glabrous. Fields and meadows: Que.—Ohio—N.M.—Wash.—Mont.; escaped from cultivation. Je–Au.

20. B. tectòrum L. Stems 3–6 dm. high, glabrous; leaf-blades mostly pubescent throughout; panicle 6–15 cm. long; spikelets 13–20 mm. long; lemma lanceolate, scabrous-puberulent to pilose-pubescent, 11–13 mm. long, narrowly 2-toothed at the apex; awn straight, 13–15 mm. long. Waste places and sandy soil: Mass.—Va.—Miss.—Colo.—Calif.—B.C.; adv. or nat. from Eur. *Temp.—Plain—Submont.* My–Jl.

21. B. stérilis L. Stem erect or ascending, 5–10 dm. high; leaf-blades pubescent throughout; panicle 1–2 dm. long, lax, open; spikelets 2.5–3.5 cm. long, 6–10-flowered; empty glumes lance-subulate, the first 7–9 mm. long, the second 11–13 mm. long; lemma linear-lanceolate, 12–16 mm. long, scabrous; awn stout, very rough. Waste places: Mass.—D.C.—Colo.—Wash.—B.C.; adv. from Eur. *Temp.—Plain—Submont.* My–Jl.

81. TRÍTICUM L. Wheat.

Annuals or biennials with spicate inflorescence. Spikelets solitary at each node, alternate, 2–5-flowered; flowers mostly perfect. Glumes with one edge towards the rachis, broad, short, often toothed, but seldom awned at the apex. Lemma ventricose on the back, several-nerved, often awned or at least toothed at the apex. Palet shorter than the lemma, 2-keeled. Grain deeply furrowed, free, pubescent at the apex.

Spikelets crowded, overlapping, mostly with 3 fertile flowers; grain free.
  Glumes distinctly keeled only at the apex; grain dull, neither glossy nor semi-translucent. — 1. *T. aestivum.*
  Glumes distinctly keeled, almost winged, to the base; grain glossy, often somewhat translucent. — 2. *T. durum.*
Spikelets neither crowded nor overlapping, mostly with 2 fertile flowers; grain attached to the lemma and palet. — 3. *T. Spelta.*

1. T. aestìvum L. Annual or biennial; stem about 1 m. high, hollow, smooth; leaves at first rather light green, soft; spike 4–12 cm. long, 4-angled; spikelets mostly 4-flowered; lemma with or without awn. *T. sativum* Lam. Wheat. Old fields and waste places: cult. and rarely escaped.

2. **T. dùrum** Desf. Annual or rarely biennial; stem 1 m. high or more, usually with a pith; leaves smooth, whitish green, with hard cuticle; spike almost terete, thicker than in the ordinary wheat, 5–10 cm. long; lemma long-awned. DURUM or MACARONI WHEAT. Old fields and waste places: cult. and rarely escaped.

3. **T. Spélta** L. Annual, stem strict; leaves flat, elongate; spike elongate, lax, with a brittle rachis; spikelets oval; glumes equal, oblong, convex on the back, shorter than the flowers; lemmas equal, muticous or rarely aristate. Waste places: N.D.; escaped from cultivation.

82. AÈGILOPS L.

Annuals, with a cylindric inflorescence. Spikelets single and alternate, more or less sunken into the flexuose rachis, plano-convex, 2–4-flowered, the terminal 1 or 2 flowers rudimentary. Glumes equal, firm, coriaceous, convex, not keeled, many-nerved, truncate, lobed, or awned at the apex. Lemma somewhat coriaceous, not keeled, 2–3-dentate or awned. Palet 2-keeled, ciliate. Stigmas sessile, lateral. Grain pubescent at the apex, free.

1. **A. cylíndrica** Host. Stem 3–6 dm. high pubescent; leaves flat, rough; spike long-cylindric, brittle; spikelets 5–10, oblong-cylindric, partly surrounded by the rachis; glumes and lemma 2-lobed at the apex and awned between the lobes; awns of the upper spikelets 3–4 cm. long, scabrous. Fields and waste places: Kansas, adv. from central Eur. and n Africa. My–Jl.

83. SECÀLE L. RYE.

Annuals or biennials, with spicate inflorescence. Spikelets solitary at each node, alternate, 2-flowered. Flowers all hermaphrodite and fertile. Glumes subulate, awn-pointed, with one edge toward the rachis. Lemma long-awned at the apex, keeled to the base. Grain slightly compressed laterally, deeply furrowed, free, pubescent at the apex.

1. **S. cereàle** L. Annual or biennial; stem 1.5–3 m. high, glabrous; leaves scabrous, flat; spike 10–15 cm. long, 4-angled; spikelets 2-flowered; glumes subulate; lemma lanceolate, hispidulous-ciliate, 5-nerved, awned. Waste places and old fields: escaped from cultivation.

84. LÒLIUM L. DARNEL, RYE-GRASS.

Perennials with rootstocks, or annuals, with terminal 2-sided spikes. Spikelets with the edge towards the rachis, compressed, several–many-flowered; flowers perfect or the upper ones imperfect; rachilla articulate between the flowers. Glumes in the terminal spikelet 2, in the lateral spikelets 1, facing the rachis, the inner one lacking. Lemma shorter and broader than the empty glume, rounded on the back. Palet shorter than the lemma, 2-keeled. Styles distinct, short; stigmas plumose.

Empty glume shorter than the spikelet.
  Lemma lanceolate: flowers appressed to the rachis.
    Lemma unarmed; spikelets 3–10-flowered, usually half longer than the glume. — 1. *L. perenne.*
    Lemma awned.
      Perennial; spikelet 5–10-flowered, about twice as long as the glume. — 2. *L. italicum.*
      Annual.
        Spikelet 3–9-flowered, about half longer than the glume. — 3. *L. rigidum.*
        Spikelet 10–25-flowered, about thrice as long as the glume. — 4. *L. multiflorum.*
  Lemma elliptic; spikelet 3–8-flowered, not more than half longer than the glume; flowers in age spreading; annual. — 5. *L. remotum.*
Empty glume as long as or longer than the glume; lemma oblong, armed; annual. — 6. *L. temulentum.*

1. **L. perénne** L. *Fig. 87.* Stem 2–8 dm. high, smooth; leaf-blades 5–12 cm. long, 2–4 mm. wide, smooth; spike 7–20 cm. long; spikelets 5–10-flowered, 8–12 mm. long; empty glume strongly nerved; lemma 4–6 mm. long, acuminate, awnless. Waste places and cultivated ground: N.S.—Va.—N.M.—Calif.—B.C.; nat. from Eurasia. Je–Au.

2. **L. itálicum** R. Br. Stem 6–10 dm. high, the upper part and the axis of the inflorescence rough; leaves convolute in bud; spikelets 5–12-flowered, 1–2 cm. long; lemma 7–8 mm. long. Fields and roadsides: N.Y.—Iowa—N.J.; nat. from Eur. Je–Au.

3. **L. rígidum** Gaud. Annual; stem 2–6 dm. high; leaves 2–6 mm. wide, flat, 1 dm. long or more; spikelets appressed to the rachis, 3–9-flowered; lemma 6–10 mm. long, lanceolate, awnless, or in the var. *Duethiei* Hook. (the only form represented) with an awn 7–12 mm. long. Wheat-fields: N.D., introduced from the Old World. My–Au.

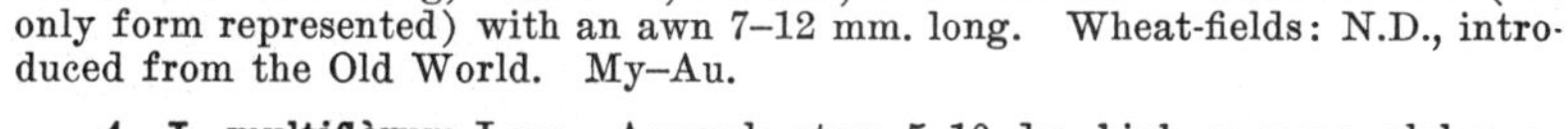

4. **L. multiflòrum** Lam. Annual; stem 5–10 dm. high or more, glabrous; lower leaves involute, the upper flat, slightly rough; spikelets large, compressed; glume blunt; lemma lanceolate, the awn about as long as the lemma. Grain-fields: N.Y.—N.J.—Mo.—N.D. Je–Au.

5. **L. remòtum** Schrank. Stem 3–8 dm. long; leaves flat, 2–3 mm. wide, rough above; spikelets somewhat spreading from the axis, 3–8-flowered, 4–5 mm. long; glume subacute, lanceolate; lemma elliptic, usually awnless, rarely awn-tipped. Flax-fields: N. D., introduced; nat. of s Eur. and e Asia. My–Jl.

6. **L. temuléntum** L. Stem 6–12 dm. high, smooth, leaf-blades 1–2.5 dm. long, 2–6 mm. wide, smooth beneath, rough above; spike 1–3 dm. long; spikelets 4–8-flowered, 10–18 mm. long; empty glume strongly veined; lemma awned or awnless, the awn sometimes as long as the lemma (var. *macrochaeton*). Poisonous. Waste places and cultivated ground: N.B.—Ga.—Calif.—B.C.; nat. from Eur. Je–Au.

85. AGROPÝRON Gaertner. Wheat-grass, Quitch-grass, Quick-grass, Quack-grass.

Perennials with terminal, 2-sided spikes; spikelets compressed, with the side of the spikelet toward the rachis, usually single at each node; flowers perfect, or the upper ones imperfect; rachilla articulate between the flowers. Glumes 2, standing at right angles to the median plane of the spike, *i.e.*, with one edge toward the rachis, usually firm. Lemma broader, rigid, rounded on the back, from obtuse to acuminate, awnless or awned. Palet shorter than the lemma, 2-keeled, ciliate on the keels. Styles very short; stigmas plumose. Fruit enclosed in the lemma, adherent to the palet.

Spikelets pectinately spreading; glumes carinate on the back, lateral veins obsolete. — 1. *A. cristatum.*
Spikelets ascending or appressed; glumes not carinate, distinctly veined.
  Tufted plants with intravaginal innovations; no horizontal stolons.
    Lemma long-awned, *i.e.*, the awn usually longer than the body of the lemma.
      Awn divergent.
        Spikelets erect, empty glumes nearly equaling the spikelet. — 2. *A. Vaseyi.*

Spikelets spreading or ascending; empty glumes half as long as the spikelet. 3. *A. spicatum.*

Awns erect.

Stem stout; spike 7–10 mm. thick, erect, but usually unilateral; spikelets (exclusive of the awns) 15–25 mm. long. 4. *A. Richardsonii.*

Stem slender; spike about 5 mm. thick, seldom unilateral; spikelets (exclusive of the awns) about 1 cm. long. 5. *A. caninoides.*

Lemma short-awned or awnless.

Spikes stout and dense, 3–8 cm. long; glumes broadest above the middle. 6. *A. biflorum.*

Spike slender and lax, 7–20 cm. long; glumes broadest below the middle.

Spikelets terete, appressed; glumes nearly as long as the spikelets. 7. *A. tenerum.*

Spikelets flattened; glumes much shorter than the spikelets. 8. *A. inerme.*

Stoloniferous plants with horizontal rootstocks, sometimes slightly tufted; innovations extravaginal.

Lemma with a long, more or less divergent awn.

Lemma pubescent. 9. *A. albicans.*

Lemma glabrous or scabrous.

Leaves mostly involute, glabrous on both sides; spikelets pale. 10. *A. Griffithsii.*

Leaves mostly flat, pilose above; spikelets usually purplish. 14. *A. Leersianum.*

Lemma awnless or with a very short erect awn; glumes usually narrowly lanceolate, acuminate.

Sheaths conspicuously pilose. 11. *A. Palmeri.*

Sheaths glabrous or nearly so.

Lemma glabrous or merely scabrous.

Spikelets erect, nearly cylindric or slightly compressed.

Spike elongate.

Glumes nearly equaling the spikelets; leaf-blades scabrous throughout. 12. *A. pseudorepens.*

Glumes much shorter than the spikelets; leaves usually pubescent above and glabrous beneath.

Lemmas acute to awn-pointed; spike usually green. 13. *A. repens.*

Lemmas with an awn as long as the body, spike usually more or less purplish. 14. *A. Leersianum.*

Spike short; glumes about half as long as the spikelets. 15. *A. riparium.*

Spikelets much flattened, spreading. 16. *A. Smithii.*

Lemma from villous to hispidulous.

Lemma short-pubescent.

Spikelets compressed, more or less spreading; lemma acuminate or strongly acute. 17. *A. molle.*

Spikelets terete or nearly so, erect; lemma obtuse or acutish. 18. *A. subvillosum.*

Lemma long-villous. 19. *A. dasystachyum.*

1. A. cristàtum (Schreb.) Gaertn. Perennial with a fibrous rootstock; stem 2–5 dm. high; leaf-blades 5–15 cm. long, 2–5 mm. wide; spike oblong, the rachis flexuose, scabrous; spikelets 3-flowered, lanceolate, short-aristate; glumes lanceolate, carinate, aristate. *Triticum cristatum* Schreb. Waste places: N. D.; adv. from e Eur. and Siberia.

2. A. Vàseyi Scribn. & Smith. Stem erect, 3–5 dm. high, wiry, glabrous; leaf-blades 3–15 cm. long, 1–2 mm. wide, usually involute; spike slender, 5–10 cm. long, often somewhat flexuose; spikelets 3–5-flowered; glumes oblanceolate or oblong, acute or acuminate, 6–8 mm. long; lemma 8 mm. long, lanceolate, acute; awn 8–10 mm. long. Hills and mountain sides: Wash.—Ore.—Neb.—Colo.—Mont. *W. Boreal—Submont.* My–Au.

3. A. **spicàtum** (Pursh) Scribn. & Smith. Stem 3–10 dm. high, glabrous; leaf-blades 5–20 cm. long, 1–2 mm. wide, involute or rarely flat; spikes 8–20 cm. long; spikelets 3–6-flowered, flattened; lemma 8–10 mm. long, scabrous above; awn stout, 12–25 mm. long. *A. divergens* Nees. BUNCH GRASS. Dry rocky hills and benchlands: B.C.—Calif.—Ariz.—Colo.—Mont.—N.D. (?) *W. Temp.—Plain—Submont.* Je–Au.

4. A. **Richardsònii** (Trin.) Schrad. Stems smooth, 6–8 dm. high; leaf-blades 8–20 cm. long, 2–6 mm. wide, scabrous, flat, often with involute tips; spikes 7–10 cm. long, one-sided; spikelets 3–4-flowered; glume 12–16 mm. long, short-awned or awn-pointed; lemma 10–13 mm. long, scabrous on the margins. *A. unilaterale* Cassidy, not Beauv. Scarcely distinct from *A. caninum* L. of Eur. Meadows and among bushes: Sask.—Minn.—Iowa—N.M.—B.C. *W. Boreal—Prairie—Plain—Mont.* Je–S.

5. A. **caninoìdes** (Ramaley) Beal. Stem 3–10 dm. high; leaf-blades 1.5–3 dm. long, 3 mm. wide, erect, involute or nearly flat; spike 1–2 dm. long, dense; spikelets 1–2 cm. long, pale or purple-tinged; glumes 7–9 mm. long, short-awned or awn-pointed; lemma 7 mm. long, scabrous. *A. caninum* Am. auth., mostly [G, B]. Mountain meadows: N.S.—Mich.—N.M.—Calif.—Mont. *Boreal —Mont.—Subalp.* Jl–Au.

6. A. **biflòrum** (Brign.) R. & S. Tufted perennial; stem 1.5–6 dm. high, often somewhat decumbent at the base; leaf-blades 5–15 cm. long, 2–6 mm. wide, flat or somewhat involute; spike 2.5–10 cm. long, 4–6 mm. wide; spikelets 3–6-flowered; lemma 5–7-nerved, 8–12 mm. long, acuminate or short-awned. *A. violaceum* (Hornem.) Lange [B]. Mountains: Greenl.—N.Y.—Neb.—N.M. —Alaska; Eurasia. *Arct.—Subarct.—Subalp.—Alp.* Jl–S.

7. A. **ténerum** Vasey. Stem 3–10 dm. high; leaf-blades 2–6 mm. wide, 7–15 cm. long, scabrous, flat or involute; spike slender, 1–1.5 dm. long, lax; spikelets 3–5-flowered; glumes 10–12 mm. long, acute to awn-pointed, 5-nerved; lemma 8–10 mm. long, lanceolate; awn 1–4 mm. long or more. RYE GRASS. Dry soil: B.C.—Calif.—N.M.—Minn. *W. Temp.—Plain—Submont.* Je–Au.

8. A. **inérme** (Scribn. & Smith) Rydb. Stems 3–10 dm. high, glabrous; leaf-blades erect, 1–2 dm. long, 2–4 mm. wide, more or less pubescent above, somewhat glaucous, flat or involute; spike 1–3 dm. long, lax; spikelets distant, 1.5–2 cm. long, 6–10-flowered; glumes 8–10 mm. long; lemma 10–12 mm. long, smooth, often mucronate or with a short awn-tip. Closely related to *A. spicatum* and *A. Vaseyi.* Plains and meadows: B.C.—Wash.—Utah—Wyo. *Submont.* Je–Au.

9. A. **álbicans** Scribn. & Smith. Stem erect, 3–6 dm. high, glaucous; leaf-blades rigid, ascending, involute, scabrous, 7–20 cm. long, 2–3 mm. wide; spike slightly nodding, 7–10 cm. long; spikelets 8–10, 5–7-flowered, 15–18 mm. long; glumes pubescent, oblanceolate, acuminate, tipped with an awn 4–6 mm. long; lemma about 9 mm. long, ovate-lanceolate; awn 12–15 mm. long. Hills and benchlands. Mont.—Wyo.—w Neb.—S.D. *Submont.* Je–Au.

10. A. **Griffíthsii** Scribn. & Smith. Stem glabrous, 3–8 dm. high, striate; leaf-blades rather rigid, mostly involute, 5–12 cm. long; spike erect, 8–15 cm. long; spikelets pale, closely 5–7-flowered; glumes 8 mm. long, with awns 3 mm. long; lemma oblong, 8–10 mm. long; awns 8–10 mm. long. Plains: Wyo.—N.D.—S.D. *Plain.* Au.

11. A. **Pálmeri** (Scribn. & Smith) Rydb. Stem 3–6 dm. high; leaf-blades glaucous, 1–2 dm. long, 3–4 mm. wide, at last involute, scabrous and often sparingly pubescent on both sides; spikes rather dense, 1–1.5 dm. long; spikelets compressed, 1.5–2 cm. long; lemma about 1 cm. long, acuminate, pilose to hispidulous-scabrous. Bottom-lands and hillsides: N.M.—Colo.—Neb.—Utah —Ariz. *Son.—Submont.* My–Jl.

12. A. **pseudorèpens** Scribn. & Smith. Stem 3–10 dm. high, glabrous; leaf-blades scabrous throughout, 12–20 cm. long, 2–6 mm. wide, flat, or involute when dry; spike 1–2.5 dm. long; spikelets rather distant, 10–16 mm. long, 3–7-flowered; lemma acuminate or awn-pointed, scabrous, about 1 cm. long. Prairies and river valleys: Neb.—Tex.—Ariz.—B.C. *Prairie—Plain—Mont.* Je–S.

13. A. **rèpens** (L.) Beauv. Stem 3–10 dm. high, glabrous, or the lower sheath somewhat hairy; leaves 7–30 cm. long, bright-green, smooth beneath, somewhat pubescent above; spikelets 3–7-flowered, 1–1.5 cm. long; lemma 1 cm. long, glabrous or scabrous; glumes strongly nerved, awn-pointed. QUACK GRASS. Fields and waste places: Newf.—Minn.—Iowa—Tex.—Fla.—Calif.—B.C.; nat. from Eur. Jl–S.

14. A. **Leersiànum** (Wulfen) Rydb. Perennial with a long running rootstock; stem 3–10 dm. high, glabrous, or scabrous in the inflorescence; lower sheath often pubescent; leaves dark green, glabrous beneath, pilose above; spike 1–2 dm. long, rather lax; spikelets 5–7-flowered, 1–1.5 cm. long, mostly purplish; glumes lance-oblong, 8–9 mm. long, mostly 7-nerved, awn-pointed; lemma about the same length, glabrous, scabrous on the margins, with an awn about as long as the body. *Triticum sepium* Thuill.; not Lam. *T. repens Leersianum* Wulf. *T. Leersianum* R. & S. Roadsides, waste places and lawns: N.S.—N.Y.—Minn.—N.D.; nat. from Eur.

15. A. **ripàrium** Scribn. & Smith. Stem glabrous, erect. 4–6 dm. high; leaf-blades flat, becoming involute, 5–15 cm. long, 2–4 mm. wide, scabrous throughout; spike 5–10 cm. long; spikelets 8–15, compressed, 5–7-flowered, spreading, 5–6 mm. long; lemma oblong-lanceolate, acute, scabrous toward the apex, acute or acuminate. River banks: Mont.—N.D.—Colo. *Submont.*

16. A. **Smíthii** Rydb. *Fig. 88.* Stem 3–10 dm. high; leaf-blades spreading, rigid, bluish green, glaucous, smooth or minutely scabrous beneath, becoming involute, 1–3 dm. long, 4–6 mm. wide; spikes 7–18 cm. long; spikelets 7–13-flowered, sometimes in pairs; lemma 8–12 mm. long, lanceolate, acute or acuminate, seldom awn-pointed, smooth or nearly so. *A. spicatum* Scribn. & Smith, in part [B]. *A. occidentale* Scribn. BLUE-JOINT; BLUE-STEM. Prairies and plains: Man.—Mo.—Tex.—Ariz.—Ore.—B.C. *Prairie—Plain—Mont.* Je–S.

f. 88.

17. A. **mólle** (Scribn. & Smith) Rydb. Stems 4–8 dm. high; leaf-blades more or less glaucous, 1–3 dm. long, 2–4 mm. wide, at last involute, scabrous; spike 1–2 dm. long, rather dense; spikelets about 2 cm. long, 6–8-flowered; lemma 8–10 mm. long, densely hispidulous. River bottoms, especially in alkaline soil: Sask.—N.M.—Utah—Wash. *Plain—Mont.* Jl–S.

18. A. **subvillòsum** (Hook.) E. Nelson. Stems 4–6 dm. high, slender, glabrous; leaf-blades 1–2 dm. long, 2–4 mm. wide, flat or at last involute, scabrous; spike rather crowded; spikelets compressed, 12–15 mm. long; lemma 6–10 mm. long, obtuse or acute. River banks and sandy soil: Sask.—Neb.—Colo.—Nev.—B.C. *Plain—Mont.* Je–Au.

19. A. **dasystàchyum** (Hook.) Scribn. Stem 3–10 dm. high; leaf-blades 5–20 cm. long, 2–6 mm. wide, flat, becoming involute in drying; spike 6–18 cm. long; spikelets slightly compressed, 4–8-flowered; lemma broadly lanceolate, 10–12 mm. long, acute or rarely short-awned. Sand-dunes: Hudson Bay—Mich.—Wis.—Ida.—Sask. *Plain—E. Boreal.* Jl–Au.

86. HÓRDEUM L. Barley, Squirrel-tail, Foxtail.

Annuals or rarely perennials, with spicate inflorescence. Spikelets alternately in 3's, rarely in 2's, at each node of the articulate rachis, sessile or short-pedicelled, 1-flowered; flower perfect or in the lateral spikelets often imperfect or rudimentary; rachilla extending beyond the flower as a long slender bristle. Glumes 2, awn-like, subulate, or lanceolate, rigid, falling together with the rachilla-joint. Lemma lanceolate, rounded on the back, awned, lobed, or in the lateral spikelets awnless. Palet shorter, 2-keeled. Styles very short, distinct; stigmas plumose.

Lateral spikelets sessile.
  Annuals.
    Lemma not awned, the awns represented by chartaceous lobes. 1. *H. aegiceras.*
    Lemma awned. 2. *H. vulgare.*
  Perennials. 3. *H. Pammeli.*
Lateral spikelets stalked; lemma awned.
  Glumes of the central spikelets lanceolate. 4. *H. pusillum.*
  Glumes of central spikelet not lanceolate.
    Lateral florets not awned, neutral or staminate. 5. *H. nodosum.*
    Lateral florets long-awned.
      Awn 2–3 cm. long. 6. *H. caespitosum.*
      Awn 4–6 cm. long. 7. *H. jubatum.*

1. **H. aegíceras** (E. Meyer) Royle. Stem 5–7 dm. high; leaf-blades glabrous, 1–1.5 cm. broad; spikes about 1 dm. long; spikelets all fertile; glumes lance-subulate, strigose; lemma 3-lobed at the apex; lateral lobes lanceolate, acute, divergent; middle lobe ovate, obtuse, slightly hooded. Pearl Barley. Moist waste grounds: cult. and often escaped.

2. **H. vulgàre** L. Annual or rarely biennial; stem 5–7 dm. high; leaf-blades broad, flat; ligules obsolete; spikes about 1 dm. long; empty glumes small, subulate, awned; spikelets all hermaphrodite; lemma lanceolate, long-awned, smooth. Barley. Waste places and old fields: sometimes escaped from cultivation.

3. **H. Pámmeli** Scribn. & Ball. Perennial; stems erect or geniculate at the base, 6–10 dm. high; leaf-blades 1.2–2 dm. long, 5–8 mm. wide; lateral spikelets nearly sessile, perfect; middle spikelet 2-flowered; glumes 2.3–3.5 cm. long, subulate-attenuate into slender awns. Prairies: Ill.—S.D.—Wyo. *Prairie—Plain.* Je–Au.

4. **H. pusíllum** Nutt. *Fig. 89.* Tufted annual; stem 1–4 dm. high, usually decumbent at the base; leaf-blades 1–7 cm. long, 1–4 mm. wide; spike 3–7 cm. long; central spikelets perfect, the lateral ones imperfect; inner glumes of the lateral spikelets lanceolate, the outer ones subulate, all awned; lemmas smooth, that of the central spikelet 6–8 mm. long, short-awned, those of the lateral spikelets smaller and stalked. Dry soil: Ont.—Ga.—Calif.—B.C. *Temp.—Plain—Submont.* Ap–Je.

5. **H. nodòsum** L. Tufted annual or perennial; stems 2–6 dm. high, erect or sometimes decumbent at the base; leaf-blades 4–12 cm. long, 2–6 mm. wide, scabrous; spike 3–8 cm. long; lemma of the central spikelet 6–8 mm. long; awn 6–12 mm. long; lemma of the lateral spikelets smaller. Meadows and waste places: Alaska—Calif.—Tex.—Yukon; Eur. *W. Temp.—Submont.—Mont.* Ap–Au.

6. **H. caespitòsum** Scribn. Cespitose perennial; stem 3–6 dm. high, smooth; leaf-blades 5–15 cm. long, 3–5 mm. wide; spike 3–5 cm. long; lemma of the central spikelet lanceolate, 6 mm. long; awn about 12 mm. long; lemma

of the lateral spikelets 2–4 mm. long, with an awn of about the same length. Meadows and springy places: Sask.—S.D.—Kans.—N.M.—Ida.—Wash. *Plain—Submont.* Je–Au.

7. **H. jubàtum** L. Tufted perennial; stem 2–8 dm. high, erect; leaf-blades 3–12 cm. long, 2–4 mm. wide, scabrous; spikes 5–10 cm. long; lemma of the central spikelet 6–8 mm. long, scabrous at the apex; awn 4–6 cm. long; lemma of the lateral spikelets 4–6 mm. long, short-awned. Dry sandy soil and prairies: Lab.—N.J.—Tex.—Calif.—Alaska. *Temp.—Plain—Subalp.* Je–S.

87. SITÀNION Raf.

Tufted perennials, with cylindric, spicate inflorescence. Spikelets sessile, several at each node of the articulate rachis, 1–5-flowered. Glumes 2, rarely entire, awn-like, usually 2–5-cleft with subulate awned divisions. Lemma broader, entire, long-awned, or 2- or 3-cleft and short-awned. Palet shorter than the glume, 2-keeled. Styles very short, distinct; stigmas plumose.

Glumes lanceolate, scarious-margined. — 1. *S. lanceolatum.*
Glumes setaceous, or cleft into setaceous divisions, not scarious-margined.
  Glumes entire; lowest flower perfect; lemma glabrous, leaves glabrous or scabrous. — 2. *S. elymoides.*
  Glumes or some of them cleft into setaceous divisions; lowest flower rudimentary.
    Lemma pubescent; sheaths and blades short-pubescent. — 3. *S. Hystrix.*
    Lemma scabrous; sheaths and blades long-pubescent. — 4. *S. strigosum.*

1. **S. lanceolàtum** J. G. Smith. Stem 3–5 dm. high, erect, smooth; leaf-blades 8–15 cm. long, 2–3 mm. wide, rigid; spikes erect, 6–10 cm. long, barely exserted; glumes 5–6 mm. long, usually 2-nerved, entire or unequally 2-cleft, the longer awn scabrous, divergent, 1–1.5 cm. long; lemma 8–9 mm. long, glaucous, entire or minutely 3-cleft; awns spreading, 2–5 cm. long. *Agropyron sitanioides* J. G. Smith. Mountains: Mont.—S.D. *Submont.* Au.

2. **S. elymoìdes** Raf. Stem 3–6 dm. high, glabrous and glaucous; leaf-blades smooth and glaucous or puberulent, 1–2 dm. long, 2–4 mm. wide, flat or involute; spike somewhat nodding, exserted or enclosed at the base in the uppermost sheath; glumes 6–8 cm. long; lemma 8–11 mm. long; awn scabrous, spreading, 5–6.5 cm. long. *S. longifolium* [G] and *S. brevifolium* J. G. Smith. Hillsides and plains: Neb.—Kans.—N.M.—Ariz.—Nev.—Mont. *Plain—Mont.* Je–S.

3. **S. Hýstrix** (Nutt.) J. G. Smith. Stems 1–3 dm. high, erect, scabrous above; leaf-blades flat or at length involute, finely pubescent throughout, 2–12 cm. long, 1–2 mm. wide; spike 5–7 cm. long, erect, flexuose; empty glumes scabrous, 3–4 cm. long; lemma 7–8 mm. long, minutely pubescent, 3-awned, the middle awn about 3 cm. long. Dry hills and "sage plains": S.D.—Colo.—Wyo.—Wash.—Ore. *Son.—Submont.* My–Jl.

4. **S. strigòsum** J. G. Smith. Stem 3–6 dm. high, glabrous or minutely pubescent above; leaf-blades 1–2.5 dm. long, 3–6 mm. wide, flat, or the lower involute, pilose; spike stout, exserted, 8–12 cm. long; glumes entire or bifid, subulate-setaceous, 5–6 cm. long; lemma 8–10 mm. long, 3-cleft above, the middle awn 5–7 cm. long, the lateral ones 1–2 mm. long. Valleys: Wash.—Ore.—N.D.—S.D. *Submont.* Je–Au.

88. ÉLYMUS L. Lyme-grass, Wild Rye, Rye-grass, Buffalo Rye.

Perennials, with rootstocks and cylindric spikes. Spikelets sessile, usually 2 or 3 at each node, but in some species mostly single, 2–several-flowered. Glumes 2, inserted more or less obliquely, rigid, awn-like, subulate or lance-linear, persistent, awn-pointed or awned. Lemma oblong or lanceolate, rounded on the back, usually awned. Palet shorter than the glume, 2-keeled. Styles very short, distinct; stigmas plumose.

Lemma long-awned.
Spike broad; spikelets spreading.
Glumes lanceolate or lance-subulate; spike dense.
Glumes lanceolate, 5–7-nerved, thick and curved at the base; awns straight; spike erect.
Lemma (including the awn) 2–3 cm. long; glumes (including the awn) 2–3.7 cm. long.
Glumes and lemma glabrous or scabrous.
Plant robust; spike scarcely exserted; lemma glabrous or nearly so. — 1. *E. virginicus.*
Plant slender; spike long-exserted; lemma scabrous-hispidulous. — 2. *E. jejunus.*
Glumes and lemma hirsute. — 3. *E. hirsutiglumis.*
Lemma (including the awn) 3.3–4.5 cm. long; glumes (including the awns) 3–4 cm. long.
Glumes and lemma villous-hirsute. — 4. *E. australis.*
Glumes and lemma scabrous or glabrous. — 5. *E. glabriflorus.*
Glumes narrowly linear-lanceolate, neither conspicuously thickened nor curved at the base; spike often nodding; awns usually arched outwards in age.
Glumes short, less than 15 mm. long. — 6. *E. diversiglumis.*
Glumes 15–30 mm. long.
Lemma hirsute or villous.
Leaves thin, flat, villous above; glumes 15–20 mm. long, long-hirsute. — 7. *E. canadensis.*
Leaves firm, usually glabrous, 5–10 mm. long; glumes 20–25 mm. long, short-hispid. — 8. *E. philadelphicus.*
Lemma hispidulous-scabrous or glabrous.
Robust; spike usually included at the base; blade 8–15 mm. wide. — 9. *E. robustus.*
Slender; spike long-exserted; blades seldom more than 5 mm. wide. — 10. *E. brachystachys.*
Glumes setaceous; spike laxer.
Lemma hirsute. — 11. *E. striatus.*
Lemma glabrous or scabrous. — 12. *E. arkansanus.*
Spike narrow; spikelets erect, appressed.
Lemma pubescent. — 13. *E. vulpinus.*
Lemma glabrous or scabrous.
Glumes lanceolate, acuminate or short-awned, 2–5-nerved. — 14. *E. glaucus.*
Glumes linear-subulate. — 15. *E. Macounii.*
Lemma awnless or short-awned, the awn less than one third as long as the body.
Glumes strongly 3–5-nerved. — 16. *E. curvatus.*
Glumes 1-nerved or indistinctly 3-nerved.
Lemma glabrous or hispidulous. — 17. *E. condensatus.*
Lemma pubescent.
Spike long; glumes lanceolate, villous. — 18. *E. arenarius.*
Spike short; glumes subulate, scabrous. — 19. *E. innovatus.*

1. **E. virgínicus** L. Stem 6–10 dm. high; sheath usually glabrous; leaf-blades 1–3 dm. long, 4–16 mm. wide, scabrous; spike 5–15 cm. long, thick, erect; lemma 6–8 mm. long, glabrous; awn scabrous, 5–18 mm. long. Along streams: N.S.—Fla.—Tex.—Sask. *E. Temp.—Plain.* Je–Au.

2. **E. jejùnus** (Ramaley) Rydb. Stem slender, 3–6 dm. high; leaf-blades 1–2 dm. long, 2–5 mm. wide, scabrous; spike 2–6 cm. long, strict; spikelets usually 2 at each node; lemma lanceolate, 7–8 mm. long, scabrous, hispidulous; awn 3–10 mm. long. *E. virginicus minor* Vasey. *E. virginicus jejunus* Ramaley. Sand hills and river banks: Minn.—Mo.—Wyo.—Mont. *Prairie—Plain—Submont.* Je–Au.

3. **E. hirsutiglùmis** Scribn. & Smith. Stem 3–10 dm. high; upper sheath often enclosing the base of the spike; leaf-blades 2–3 dm. long, 8–18 mm. wide, scabrous; spike 5–15 cm. long, stout, the rachis pubescent; spikelets usually in pairs; glumes 1–1.2 cm. long, the awn scabrous, as long or shorter; lemma

lanceolate, 8–10 mm. long, appressed-hirsute; awn scabrous, 12–16 mm. long. *E. virginicus hirsutiglumis* Hitchc. [G]. River banks: Me.—S.D.—Kans.—Va. *Canad.—Allegh.* Jl–Au.

4. **E. austràlis** Scribn. & Ball. Stem 1–1.5 m. high; upper sheath scarcely inflated; leaf-blades 2–3 dm. long, 8–12 mm. wide, sparingly villous above; spikes exserted, straight, 8–14 cm. long; spikelets slightly spreading, 2–4-flowered; glumes elongate, about 1 mm. wide, villous-hirsute; lemma villous-hirsute; awn long and straight; palet 5 mm. long. Swampy woods and banks: Mass.—Ga.—Mo.—Neb. *Allegh.*

5. **E. glabriflòrus** (Vasey) Scribn. & Ball. Stem 6–10 dm. high, smooth; sheaths loose, glabrous or hirsute; blades 1.5–3 dm. long, 7–10 mm. wide, flat, scabrous or slightly hirsute above; spike robust, erect, 1–1.5 dm. long; spikelets 3–5-flowered; glumes scabrous on the nerves, 1–1.5 cm. long; lemma minutely scabrous or glabrate, 9–12 mm. long; awn scabrous, 2–3 cm. long. Woods and thickets: Mass.—Fla.—Tex.—Neb. *E. Temp.*

6. **E. diversiglùmis** Scribn. & Ball. Stem 9–12 dm. high, glabrous; leaf-blades spreading, 1.5–2.5 dm. long, 6–12 mm. wide, scabrous on both sides; spikelets in pairs, 2-flowered; lemma linear-lanceolate, acute, 8–10 mm. long, indistinctly 3–5-nerved, hirsute or scabrous-hispidulous; awn 2–3 cm. long. Rich open places: Wis.—Wyo.—N.D. *Prairie—Plain—Submont.* Jl.

7. **E. canadénsis** L. *Fig. 90.* Stem 6–15 dm. high, glabrous; leaf-blades 1–3 dm. long, 4–20 mm. wide, scabrous; spike 1–3 dm. long, usually long-exserted, nodding; spikelets spreading, 3–5-flowered; lemma 8–14 mm. long; awn 2–5 cm. long, rough. *E. philadelphicus hirsutus* Farwell. River banks and among bushes: N.S.—Ga.—N.M.—B.C. *Temp.—Plain—Submont.* Jl–Au.

f. 90.

8. **E. philadélphicus** L. Stem 6–15 dm. high, stout; sheaths glabrous or the lower ones pubescent; blades 1–4 dm. long, 3–15 mm. wide, firm, striate, flat or convolute; spike 1–1.5 dm. long, 2–3 cm. thick, spikelets 3–4 at the nodes, 3–5-flowered; lemma 12–14 mm. long, short-hirsute; awns 3–4 cm. long. *E. glaucifolius* Muhl. *E. crescendus* (Ramaley) Wheeler. Banks: N.Y.(?)—Mich.—Sask.—Neb.—Mo. *Allegh.* Je–Au.

9. **E. robústus** Scribn. & Smith. Stem 1–2 m. high, smooth; leaf-blades 2–4 dm. long, 8–20 mm. wide, scabrous on both sides; spike 12–18 cm. long, 2.5–4 cm. thick; spikelets 3–4 at the nodes, 3–4-flowered; lemma 12–16 mm. long, attenuate above; awn 3–4 cm. long. River valleys: Minn.—Ill.—Mo.—N.M.—Mont.—Ida. *Prairie—Plain—Submont.* Je–Au.

10. **E. brachystàchys** Scribn. & Ball. Stem 3–10 dm. high, smooth; leaf-blades 1–2 dm. long, 6–10 mm. wide, semi-involute, smooth or somewhat scabrous beneath, finely scabrous above; spike dense, nodding, 8–15 cm. long; spikelets usually in 2's, 3–5-flowered; lemma 11–13 mm. long; awn scabrous, 2–4 cm. long. Prairies and plains: Mich.—Tex.—N.M.—(Black Hills) S.D.; Mex. *Plain—Submont.* Je–Au.

11. **E. striàtus** Willd. Stem 3–9 dm. high; leaf-blades 1–2 dm. long, 4–10 mm. wide, smooth or slightly scabrous beneath, pubescent above; spike exserted, nodding, 6–12 cm. long; spikelets divergent, 1–3-flowered; lemma about 6 mm. long, hispid; awn 1.5–3 cm. long. Woods and river banks: Me.—N.C.—Tex.—N.D. *Plain—Submont.* Je–Au.

12. **E. arkansànus** Scribn. & Ball. Stem slender, 6–8 dm. high, smooth; sheaths ciliate on the margins, or the lower hirsute; blades 1–2 dm. long, 5–8 mm. wide, scabrous beneath, pubescent above; spike long-exserted, 6–9 cm. long; spikelets 2-flowered; glumes divergent, subulate, 2–3-nerved, 8–10 mm. long (including the awn), scabrous; lemma narrowly lanceolate, minutely scabrous, 7 mm. long; awn slender, 2–4 cm. long. Woods: N.Y.—Minn.—S.D.—Neb.—Ark. *Allegh.—Ozark.* Jl–Au.

13. **E. vulpìnus** Rydb. Stem 5–7 dm. high, striate, erect; leaves 1–1.5 dm. long, 2–6 mm. wide, scabrous on the margins; spike exserted, 1–1.5 dm. long, 6–7 mm. thick, slightly nodding; spikelets 1–2 at each node, 4–6-flowered; lemma linear-lanceolate, 8 mm. long; awn 8–10 mm. long. Meadows: Alta.—Neb. *Plain.* Jl–Au.

14. **E. glaùcus** Buckley. Stem 6–10 dm. high, smooth; leaf-blades flat, scabrous on both sides, 6–15 mm. wide, 5–20 cm. long; spike erect, 6–15 cm. long, 5–8 mm. thick; spikelets in pairs, rarely in threes or single, 3–4-flowered; lemma scabrous towards the apex, 5-nerved, 7–10 mm. long, with a white margin, or in var. *tenuis* Vasey [*E. americanus* Vasey & Scribn.] narrower and with purplish margin; awn scabrous, 7–15 mm. long. Among bushes and in meadows: Mich.—N.M.—Calif.—B.C. *W. Boreal—Submont.—Subalp.* Je–Au.

15. **E. Macoùnii** Vasey. Stem 3–10 dm. high, smooth; leaf-blades erect, scabrous, 7–15 cm. long, 2–4 mm. wide; spike slender, 7–12 cm. long; spikelets often solitary at the nodes, 1–3-flowered; lemma scabrous above, 8–10 mm. long, 5-nerved. Meadows: Man.—Minn.—Kans.—N.M.—Utah—Alta. *Plain—Submont.*

16. **E. curvàtus** Piper. Stem erect, 6–10 dm. high, glabrous; leaf-blades 1–2 dm. long, 5–7 mm. wide, usually flat, scabrous on both sides; spike stout, erect, short-exserted, 5–12 cm. long; lemma 3–5-nerved, sparingly scabrous-hispidulous towards the apex, 8–12 mm. long, acuminate to short-awned; awn 2 mm. long or less. *E. virginicus submuticus* Hook. [G]. River banks: Man.—Ills.—Kans.—Wash. *W. Boreal—Prairie—Plain—Submont.* Jl–Au.

17. **E. condensàtus** Presl. Stems tufted, 1–2 m. high, stout, striate, scabrous or scabro-puberulent above; leaf-blades 3–10 dm. long, 5–10 mm. wide, more or less scabrous; spike 2–4 dm. long; lemma 12–15 mm. long, acuminate or awn-pointed. Hills and alkaline ground: Alta.—S.D.—Neb.—N.M.—Calif.—B.C. *W. Temp.—Plain—Mont.* Je–Au.

18. **E. arenàrius** L. Stem stout, 6–12 dm. high, with a creeping rootstock; leaf-blades firm, involute at the tip, 1–3 dm. long, 5–10 mm. wide; spike dense, 8–25 cm. long, 1.5–2 cm. thick; spikelets 3–7-flowered; glumes 15–28 mm. long, villous; lemma 5–7-nerved. Shores: Greenl.—Me.—Alaska; Mich.—Minn; Eur. *Arct.—Subarct.* Jl.

19. **E. innovàtus** Beal. Stem smooth, or pubescent just below the nodes; leaf-blades rather rigid, 5–18 cm. long, 2–8 mm. wide, scabrous on the margins and nerves beneath; spike 4–10 cm. long; spikelets 3–6-flowered, 10–15 mm. long; lemma densely pubescent, usually villous, 8–10 mm. long; awns 2–4 mm. long. *E. Brownii* Scribn. & Smith. Hills: Sask.—S.D.—Wyo.—B.C. *Submont.* Je–Au.

89. **HÝSTRIX** Moench.

Perennials, with rootstocks and terminal spikes. Spikelets in pairs or 3's, spreading, 2–several-flowered; rachilla articulate below the lemmas. Glumes in the lowest spikelets subulate, minute, elsewhere wanting. Lemma lanceolate, rigid, convolute, rounded on the back, awned. Styles very short; stigmas plumose. Grain oblong, adherent to the palet.

1. **H. pátula** Moench. Stem 6–12 dm. tall; leaf-blades 1–2.5 dm. long, 6–12 mm. wide, smooth beneath, scabrous above; spike 7–18 cm. long; spikelets at length spreading, 8–12 mm. long, exclusive of the awns; lemma 8–12 mm. long, acuminate; awn about 2.5 cm. long. *Asprella Hystrix* Willd. [B, R]. Rocky woods: N.B.—Ga.—Neb.—Sask. *E. Temp.—Plain.* Je–Jl.

Family 13. **CYPERACEAE.** SEDGE FAMILY.

Grass-like or rush-like plants, with mostly solid stems. Leaves 3-ranked, with closed sheaths and narrow blades. Flowers perfect or unisexual, in spikelets; bractlets (glumes or scales) 2-ranked or spirally arranged. Perianth composed of bristles, a sack-like organ (perigynium), or wanting. Stamens usually 3, rarely 1 or 2, or more than 3; filaments slender; anthers 2-celled. Gynoecium of 2 or 3 united carpels, but ovary 1-celled and 1-ovuled; stigmas 2 or 3. Ovules anatropous, erect. Fruit an achene; endosperm mealy.

Flowers all perfect, or at least one in each spikelet perfect.
  Glumes of the spikelets 2-ranked.
    Perianth represented by bristles; inflorescence axillary. 1. DULICHIUM.
    Perianth wanting; spikelets in terminal, solitary or umbellate heads. 2. CYPERUS.
  Glumes of the spikelets spirally imbricate.
    Base of the style persistent as a tubercle on the achene.
      Basal empty glumes several. 3. RYNCHOSPORA.
      Basal empty glumes wanting, or 1 or 2.
        Spikelets solitary; stem leafless; bristles usually present. 4. ELEOCHARIS.
        Spikelets several or numerous; stem leafy; bristles none. 5. STENOPHYLLUS.
    Base of the style not persistent as a tubercle.
      Base of the style swollen; bristles none. 6. FIMBRISTYLIS.
      Base of the style not swollen; bristles usually present.
        Flowers without any inner scales.
          Bristles much elongating in fruit, silky.
            Bristles 6, but each 4–6-cleft to near the base, therefore appearing numerous. 7. ERIOPHORUM.
            Bristles 6, simple, crisp. 8. LEUCOCOMA.
          Bristles short, or little elongating, rarely wanting. 9. SCIRPUS.
        Flowers with small inner scales between the flower and the rachis.
          Spikelets many-flowered.
            Glumes awned below the apex; flowers with 3-stalked petal-like glumes alternating with the 3 bristles. 10. FUIRENA.
            Glumes awnless or with the midrib excurrent; flowers without petal-like glumes and bristles. 11. HEMICARPHA.
          Spikelets 1–2-flowered, with several empty glumes at the base. 12. MARISCUS.
Flowers monoecious or dioecious.
  Achenes not enclosed in a perigynium, supported by a hard disk. 13. SCLERIA.
  Achenes enclosed in a perigynium; disk none. 14. CAREX.

1. **DULÍCHIUM** Rich.

Tall perennials with rootstocks, hollow jointed stems, leafy to the top. Leaves 3-ranked. Spikelets in axillary, simple or compound spikes, flat, linear, many-flowered; glumes 2-ranked, carinate, conduplicate, decurrent on the internode below. Flowers perfect. Perianth of 6–9 retrorsely barbed bristles. Stamens 3. Style persistent as a beak on the top of the achene. Stigmas 2. Achenes linear-oblong.

1. **D. arundinàceum** (L.) Britton. Stem stout, 3–10 dm. high; leaves numerous, flat, 2–8 cm. long, 4–8 mm. wide, spreading; lower sheaths bladeless; spikelets 12–25 mm. long, about 2 mm. wide, 6–12-flowered; glumes lanceolate, acuminate, strongly nerved, brownish; bristles longer than the achenes. *D. spathaceum* Pers. Wet places: Newf.—Fla.—Tex.—Ore.—B.C.; C. Am. *Temp.—Trop.—Plain.* Jl–O.

2. CYPÈRUS L. Galingale, Nut Grass.

Subscapose perennials or annuals. Leaves basal; blades narrow, grasslike. Scapes in our species simple, triangular, with one or more bracts at the summit, subtending a simple or compound, umbellate or capitate inflorescence; rays usually very unequal in length. Spikelets flat to nearly terete. Glumes 2-ranked, concave, conduplicate or keeled. Flowers perfect; perianth none. Stamens 1–3. Styles deciduous from the summit of the achenes; stigmas 2 or 3. Achenes without a tubercle.

Glumes falling away from the persistent rachis of the flat spikelets.
  Style 2-cleft; achenes lenticular.
    Spikelets yellow; superficial cells of the achenes oblong. — 1. *C. flavescens.*
    Spikelets green or brown; superficial cells of the achenes quadrate.
      Glumes membranous, dull; style much exserted. — 2. *C. diandrus.*
      Glumes subcoriaceous, shining; style scarcely exserted. — 3. *C. rivularis.*
  Style 3-cleft; achenes 3-angled.
    Wings of the rachis none or very narrow.
      Annuals, stamen 1.
        Glumes awned or mucronate. — 4. *C. inflexus.*
        Glumes acute, neither awned nor mucronate. — 5. *C. acuminatus.*
      Perennials.
        Stamen 1; spikelets ovate, 3–8 mm. long. — 6. *C. pseudovegetus.*
        Stamens 2 or 3; spikelets linear-oblong, 8–16 mm. long.
          Heads oblong; spikelets ascending; stem rough. — 7. *C. Schweinitzii.*
          Heads short; spikelets more or less spreading; stem smooth.
            Glumes broadly ovate; achene 1.5–2 mm. long. — 8. *C. Houghtoni.*
            Glumes oblong-ovate; achene 2–2.5 mm. long. — 9. *C. Bushii.*
    Wings of the rachis prominent.
      Wings of the rachis permanently adnate; perennials.
        Glumes wholly or partly purple-brown; achenes linear.
          Glumes wholly appressed. — 10. *C. rotundus.*
          Glumes with the tips more or less spreading. — 11. *C. Hallii.*
        Glumes straw-colored; achenes ovoid. — 12. *C. esculentus.*
      Wings of the rachis separating from it as an interior pair of scales. — 13. *C. erythrorhizos.*
Spikelets falling wholly away, usually leaving the two lower glumes persistent.
  Annuals: spikelets elongate, nearly terete.
    Glumes well imbricate; achenes obovoid. — 14. *C. speciosus.*
    Glumes distant, scarcely overlapping; achenes linear-oblong. — 15. *C. Engelmanni.*
  Perennials, tuber-bearing; spikelets more or less flattened.
    Achenes linear-oblong, 3–4 times as long as thick.
      Spikelets 7–25-flowered, racemose. — 16. *C. strigosus.*
      Spikelets 3-flowered, racemose. — 17. *C. ovularis.*
    Achenes oblong, twice as long as thick; spikelets 4–11-flowered, subcapitate. — 18. *C. filiculmis.*

1. **C. flavéscens** L. Stem 0.5–4 dm. high; bracts 3, very unequal, leaves 2–3 mm. wide; umbels with 2–4 very short rays; spikelets 5–15 mm. long, 1.5–2.5 mm. wide; glumes obtuse, ovate, 1-nerved; stamens 3; achenes orbicular, shining. Low ground: Me.—Mich.—Kans.—Tex.—Fla.; Mex.; Old World. *E. Temp.* Au–O.

2. **C. diándrus** Torr. *Fig. 91.* Annual; stems tufted, 5–30 cm. high; leaves about 2 mm. wide; bracts usually 3, at least one much exceeding the spikelets; spikelets in terminal capitate clusters, linear-oblong, 8–18 mm. long, flat, many-flowered; glumes ovate, green, brown, or with brown margins, obtuse, 1-nerved; stamens 2 or 3; achenes oblong, gray. Marshy places: N.B. — S.C. — Colo. — N.D. *Canad.—Allegh.—Plain—Submont.* Au–O.

3. **C. rivulàris** Kunth. Stem slender, tufted; bracts 3; umbel simple; spikelets linear or linear-oblong, 8–20 mm. long; glumes green or dark-brown, or with brown margins, obtuse, shining; stamens 3; achenes oblong, lenticular, pointed, dull. Along streams and ponds: Me.—Minn.—Kans.—N.C. *Canad.—Allegh.* Au–O.

4. **C. infléxus** Muhl. Stems slender, tufted, 2–15 cm. high; leaves 1–2 mm. wide; bracts 2 or 3, longer than the umbel; spikelets capitate or in 3-rayed, sessile umbels, linear-oblong, 6–10-flowered, 4–6 mm. long; glumes light brown, lanceolate, several-nerved. *C. aristatus* Rottb. [G]. Wet sandy soil: N.B.—Fla.—Calif.—B.C.; Mex. *Temp.—Plain—Submont.* Jl–S.

5. **C. acuminàtus** Torr. & Hook. Stems slender, tufted, 5–35 cm. high; leaves usually less than 2 mm. wide, light green; bracts much elongated; spikelets capitate, in 2–4-rayed umbels, flat, ovate-oblong, 4–8 mm. long; glumes oblong, pale green, 3-nerved, with short, more or less recurved tips. Wet soil: Ill.—La.—Calif.—Wash. *W. Temp.—Prairie—Plain.* Jl–O.

6. **C. pseudovégetus** Steud. Stem 3–10 dm. high; bracts 4–6; leaves 3–4 mm. wide, very long, keeled; umbel compound, many-rayed; spikelets ovate, 3–8 mm. long, in small heads, greenish; glumes conduplicate, keeled, 1-nerved, curved, acute, the tip recurved; stamen 1; achenes linear, pale. *C. calcaratus* Nees. Wet places: Del.—Kans.—Tex.—Fla. *S. Temp.—Ozark.* Jl–S.

7. **C. Schweinítzii** Torr. Stems tufted, rough, 3–7 dm. high; leaves 2–5 mm. wide, rough-margined; bracts 3–7, erect; spikelets spicate, in 3–9-rayed umbels, flat, linear-oblong, 6–12-flowered, 8–16 mm. long; glumes convex, light green, ovate, acute or acuminate, 9–13-nerved. Sandy soil: Ont.—Mo.—Kans.—(? N.M.)—S.D.—Sask. *Prairie—Plain.* Au–O.

8. **C. Hoùghtoni** Torr. Stem smooth, 3–6 dm. high; leaves 1–2 mm. wide; bracts 3–5, some much exceeding the inflorescence; spikelets loosely capitate in 1–5-rayed umbels, compressed, acute, 8–15 mm. long, about 2 mm. wide, 11–15-flowered; glumes chestnut-brown, shining, oblong, obtuse, strongly 11-nerved. Sandy places: Mass.—Va.—Ark.—w S.D.—Man. *Canad.—Allegh.—Plain.* Jl–Au.

9. **C. Búshii** Britton. Stem smooth, 3–6 dm. high, longer than the leaves; leaves 3–4 mm. wide, smooth; spikelets capitate in 1–5-rayed umbels, flat, linear, acute, 8–15 mm. long; glumes oblong, mucronate, shining, 11-nerved. *C. filiculmis* Coult., not Vahl. Sandy soil: Minn.—Mo.—Tex.—Colo.—Ore.—Wash. *W. Temp.—Plain—Prairie.* Jl–S.

10. **C. rotúndus** L. Rootstock scaly, tuber-bearing; stem 1–5 dm. high; bracts 3–5; leaves 3–6 mm. wide; umbel 3–8-rayed; spikelets linear, 8–20 mm. long, 2–3 mm. wide; glumes dark brown, with green rib and margins, ovate, appressed, 3-nerved; achenes linear-oblong, 3-angled. Sandy fields: N.Y.—Fla.—Tex.—Kans.—Minn.; Trop. Am.; Old World. *E. Temp.—Trop.* Jl–S.

11. **C. Hállii** Britton. Rootstock scaly; stem 6–9 dm. high; bracts 3–6; leaves 4–6 mm. wide; umbel compound; spikelets numerous, 7–15-flowered, 1–1.5 cm. long, 2–3 mm. wide; glumes ovate, acute, strongly 7–9-nerved, reddish-brown, with light margins; achenes linear-oblong, 3-angled. Wet places: Kans.—Tex. *Texan.* Jl–S.

12. **C. esculéntus** L. Rootstock scaly, tuber-bearing; stem 3–8 dm. high; bracts 3–6; leaves 4–8 mm. wide; umbel 4–10-rayed, often compound; spikelets straw-colored, 1–2.5 cm. long, 3 mm. wide; glumes ovate-oblong, 3–5-nerved; achenes obovoid, obtuse, 3-angled. Moist ground: N.B.—Minn.—Neb.—Tex.—Fla.; Alaska—Calif.; Trop. Am.; Old World. *Temp.—Trop.* Au–O.

13. **C. erythrorhìzos** Muhl. Annual; stems tufted, 0.7–6 dm. high; leaves 3–8 mm. wide, rough-margined, the lower equaling or exceeding the stem; bracts 3–7, some 3–5 times as long as the inflorescence; spikelets spicate in mostly compound umbels, linear, subacute, 6–30 mm. long, less than 2 mm. wide, compressed, many-flowered; glumes bright chestnut-brown, oblong-lanceolate, mucronate, appressed. Wet places: Mass.—Fla.—Tex.—Calif.—Wash. *Temp.—Plain.* Au–O.

14. **C. speciòsus** Vahl. Stem tufted, 1–4 dm. high, reddish towards the base; leaves 3–5 mm. wide; umbel 3–7-rayed, simple or compound; spikelets 10–30-flowered, 8–25 mm. long, 2 mm. wide; glumes ovate, obtuse, dull brown, faintly 3–5-nerved; achenes pale, ovoid, 3-angled. *C. ferax* Britt. [G], not Rich. Marshes: Mass.—Minn.—Tex.—Fla.; Calif.; Trop. Am. *Temp.—Trop.* Au–O.

15. **C. Engelmánni** Steud. Annual; stem 1.5–7 dm. high; leaves elongate, 4–6 mm. wide, rough on the margins; bracts 4–6, the longer exceeding the inflorescence; inflorescence compound; spikelets narrowly linear, 12–25 mm. long, 5–15-flowered, rachis winged; glumes distant, scarcely overlapping, greenish brown, faintly 3–5-nerved; stamens 3; achenes 3-angled. Wet ground: Mass.—N.J.—Mo.—Minn. *Allegh.* Au–O.

16. **C. strigòsus** L. Perennial, with a corm-like base, 3–10 dm. high; leaves rough-margined, 4–6 mm. wide; some of the bracts exceeding the inflorescence; spikelets spicate or subcapitate in more or less compound umbels, flat, linear, 6–25 mm. long, 2 mm. wide or less; glumes straw-colored, oblong-lanceolate, several-nerved, appressed; achenes linear-oblong, acute. Moist meadows: Me.—Fla.—Tex.—Calif.—Wash.—w S.D. *Temp.—Plain.* Au–O.

17. **C. ovulàris** (Michx.) Torr. Stem 2–8 dm. high; leaves 4–6 mm. wide; longest bracts exceeding the inflorescence; umbel simple, few-rayed; heads of spikelets 8–14 mm. broad; spikelets 4–7 mm. long; glumes ovate, green, several-nerved; achenes linear-oblong. Hills and fields: N.Y.—S.D.—Kans.—Tex.—Fla. *E. Temp.* Jl–S.

18. **C. filicúlmis** Vahl. Stem smooth, slender, 1.5–4.5 dm. high; leaves 2–4 mm. wide, keeled, rough-margined; spikelets clustered in 1–7 globose heads, linear, acute, 5–12 mm. long, 2 mm. wide; glumes ovate, pale green, strongly 7–11-nerved, appressed; achenes oblong or obovoid, apiculate, dull gray. Dry fields and hills: Ont.—Minn.—Neb.—Tex.—Fla.; Mex. *E. Temp.—Trop.* Je–Au.

3. RYNCHÓSPORA Vahl. Beaked Rush.

Caulescent perennials, with rootstocks. Spikelets oblong or fusiform; glumes spirally imbricate, the lower empty, usually mucronate or cuspidate by the excurrent midrib. Perianth of 1–24, mostly 6, barbed or scabrous bristles, or rarely wanting. Stamens usually 3. Stigmas 2, rarely wholly united. Achenes lenticular, smooth, cancellate or transversely wrinkled. Base of the style persistent, forming a tubercle, or the whole style persistent.

Style entire or 2-toothed, persistent as a subulate beak. 1. *R. corniculata.*
Style 2-cleft, only the base persistent as a tubercle.
Glumes white or nearly so; bristles 9–15. 2. *R. alba.*
Glumes brown; bristles 6. 3. *R. capillacea.*

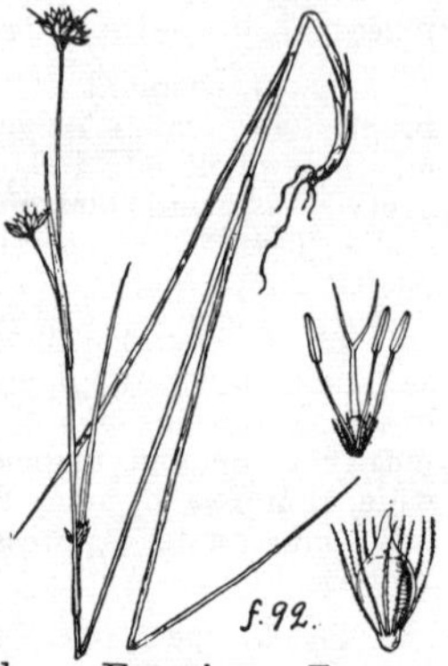

1. **R. corniculàta** (Lam.) A. Gray. Stem triangular, smooth, 1–2 m. high; leaves flat, 1.5–4.5 dm. long, 6–16 mm. wide, rough-margined; spikelets spindle-shaped, 8–12 mm. long, capitate at the ends of the rays; glumes lanceolate, thin, light-brown, acute; bristles 6, rigid, upwardly scabrous; achenes obovate, flat, 4 mm. long, dark-brown, smooth. Swamps: Del. — Minn. — Tex. — Fla. *S. Temp. — Ozark.* Jl–S.

2. **R. álba** (L.) Vahl. *Fig. 92.* Stems slender, glabrous, 1.5–2.5 dm. high; leaves bristle-like, 0.5–1 mm. wide; spikelets several, in 1–4 dense corymbose heads, narrowly oblong, 4–6 mm. long; glumes ovate or ovate-lanceolate, white, acute; bristles 9–15, equaling the achenes, which are obovate-oblong, lenticular, pale brown; tubercle triangular, flat. Bogs: Newf.—Fla.—Ky.—Ida.—Calif.—Alaska; Eurasia. *Temp—Mont.* Je–Au.

f. 92.

3. **R. capillàcea** Torr. Stem filiform, glabrous, 1.5–5 dm. high; leaves filiform, less than 0.5 mm. wide; spikelets few, in 1–3 loose clusters, acute at each end, 4–6 mm. long; glumes ovate-oblong, chestnut-brown, mucronate; bristles 6, downwardly barbed; achenes oblong, light-brown, lenticular; tubercle compressed, triangular-subulate, half as long as the achene. Bogs: Me.—S.D.—Mo.—N.J. *Canad.—Allegh.* Jl–Au.

4. ELEÓCHARIS R. Br. SPIKE-RUSH, WIRE GRASS.

Annual or perennial scapose herbs. Leaves reduced to mere sheaths, or the lower rarely blade-bearing. Scape angled or terete. Spikelets solitary, terminal, erect. Glumes spirally arranged, imbricate. Perianth of 1–12 bristles, usually retrorsely barbed. Stamens 2–3. Stigmas 2 or 3. Achene lenticular or triangular, sometimes obscurely so. Base of the styles persistent on the summit of the achenes, forming a tubercle.

Style-branches 2; achenes lenticular or biconxex.
Sheath hyaline, and scarious at the summit. 1. *E. olivacea.*
Sheath firm, not scarious at the summit.
Annuals, with fibrous roots.
Achenes black, shining. 2. *E. atropurpurea.*
Achenes pale brown, dull.
Spikelet ovoid or oblong.
Tubercle narrower than the top of the achene. 3. *E. ovata.*
Tubercle as broad as the top of the achene. 4. *E. obtusa.*
Spikelet oblong-cylindric to lance-oblong.
Spikelets narrowly oblong or subcylindric; glumes blunt, closely appressed. 5. *E. Engelmanni.*
Spikelets lance-ovoid or lance-oblong; glumes acutish, more spreading. 6. *E. monticola.*
Perennials, with rootstocks.
Basal glumes of the spikelets 2 or 3 below the thinner fertile glumes.
Tubercle elongate, much longer than broad; achenes 1.2–2.1 mm. long, narrowly obovoid or pyriform; culm subterete, rather firm. 7. *E. palustris.*
Tubercle depressed-deltoid or broadly ovate, as broad as or broader than long;

achenes 1.2–1.6 mm. long, broadly obovoid or round.
Stem firm and wiry, subterete; fertile glumes loosely ascending, narrowly ovate to lanceolate, mostly acute or attenuate. — 8. *E. Smallii.*
Stem soft, flat or compressed; fertile glumes appressed, ovate, obtuse or subacute. — 9. *E. mamillata.*
Basal glumes solitary, spathiform, usually encircling the base of the spikelet.
Stem terete or subterete, scarcely rigid; glumes usually purple, reddish or purple-tinged.
Spikelets closely many-flowered; fertile glumes often 40 or more, scarious-membranous, opaque, commonly brown or rufescent; achenes 1–1.4 mm. long. — 10. *E. calva.*
Spikelets loosely few-flowered; fertile glumes 5–30, firm-membranous or subcoriaceous, lustrous, castaneous or dark-purple. — 11. *E. uniglumis.*
Stem flattened, rigid; glumes whitish or straw-colored with brownish stripes. — 12. *E. xyridiformis.*
Style-branches 3; achenes trigonous or turgid; perennials, with rootstocks.
Achenes cancellate and longitudinally ribbed.
Spikelet compressed; stem filiform. — 13. *E. acicularis.*
Spikelet terete; stem not filiform. — 14. *E. Wolfii.*
Achenes smooth, papillose or reticulate.
Tubercle of the achenes short-conic to depressed, plainly distinguishable from the achene.
Achenes papillose.
Stem filiform; glumes obtuse. — 15. *E. tenuis.*
Stem flat; glumes acute. — 16. *E. acuminata.*
Achenes finely reticulated. — 17. *E. arenicola.*
Tubercle of the achenes long-conic, scarcely distinguishable from the body of the achene.
Perennials.
Stem flattened, 3–6 dm. high. — 18. *E. rostellata.*
Stem 3-angled, filiform, less than 1 dm. high. — 19. *E. pauciflora.*
Annuals, tufted.
Stem 1–3 dm. high, terete, grooved; spike 8–20-flowered. — 20. *E. intermedia.*
Stem less than 1 dm. high, flattened; spike 3–8-flowered. — 21. *E. pygmaea.*

1. **E. olivàcea** Torr. Perennial, with a running rootstock, tufted; stem slender, flattened, 2–10 cm. high; spikelet ovoid, 4 mm. long, 2 mm. broad; glumes ovate, thin, acute, reddish-brown, with green midrib, scarious-margined; style 3-cleft; bristles 6–8, retrorsely barbed, longer than the achene with the tubercle; achene obovoid, smooth, brown; tubercle one fourth to one third as long, conic, acute. Wet soil: Me.—N.D.(?)—Kans.—S.C. *E. Temp.* Au–S.

2. **E. atropurpùrea** (Retz.) Kunth. Stems tufted, slender, 3–10 cm. high; spikelet ovoid, subacute, 3–4 mm. long, 2 mm. thick; glumes ovate-oblong, obtuse, or the upper acute, purple-brown, with green midrib and narrow scarious margins; bristles 2–4, fragile, white, about as long as the achenes; achene 0.5 mm. long, smooth, lenticular; tubercle depressed-conic, constricted at the base. Moist soil: Iowa—Fla.—Tex.—Colo. (?); trop. Am.; Eurasia. *S. Temp.* Jl–S.

3. **E. ovàta** (Roth) R. & S. Annual; stem erect or depressed, 0.3–5 dm. high; spikelet ovoid, obtuse, 2–7 mm. long, 2–4 mm. thick; glumes oblong or narrowly ovate, obtuse, purple-brown, with pale rib and white margins; achene obovoid, about 1 mm. long; tubercle deltoid-conic, higher than broad; bristles much exceeding the achene. Wet places: N.B.—Conn.—N.D.—Iowa. *Canad.—Allegh.* Jl–S.

4. **E. obtùsa** (Willd.) Schultes. Annual; stem 0.5–7 dm. high; spikelet ovoid, obtuse, 2–13 mm. long, 2–5 mm. thick; glumes ovate-oblong to subor-

bicular, rounded at the apex, dull brown; bristles 6–8, longer than the achene; achene ovoid, pale brown; tubercle depressed-conic, one third as high as the achene. Muddy places: N.S.—Va.—Okla.—S.D.; Ore.–Ida.—B.C. *Temp.* Jl–S.

5. **E. Engelmánni** Steud. *Fig. 94.* Stems 2–4.5 dm. high, tufted; spikelet obtuse, 5–15 mm. long, 2–3 mm. thick; glumes pale-brown, with green midrib and narrow scarious margin, ovate, obtuse; bristles about 6, not longer than the achenes, or none; achene broadly obovate, smooth; tubercle broad, low, covering the top of the achene. Wet places: Mass.—N.J.—Tex.—Calif.—Wash. *Temp. Plain—Submont.* Jl–S.

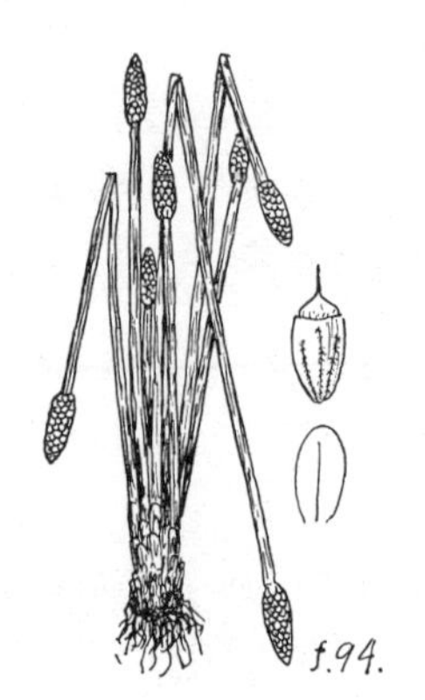

6. **E. montícola** Fernald. Stems 1–2.5 dm. high, tufted; spikelets 6–9 mm. long, 2–3.5 mm. thick, acutish; glumes chestnut-brown or purplish, with paler midribs and margins, acutish; bristles 6, equaling or slightly exceeding the achene, or (in var. *leviseta*) represented only by the unbarbed basal portion; achene as in the preceding. Wet places: Calif.—Wash.; the var. *leviseta:* Wash.—Ida.—Man. *Boreal—Submont.*

7. **E. palústris** (L.) R. & S. Perennial, with a creeping rootstock; stem terete, 1–19 dm. high, 0.5–5 mm. thick; sheaths red or brown; spike linear-cylindric to ovoid, 5–25 mm. long, 2.5–7 mm. thick; basal glumes 2 or 3, firm, ovate to oblong; fertile glumes oblong-ovate, obtuse, reddish-brown with pale margins; achenes yellowish or chestnut brown; bristles 4, usually longer than the body of the achene. Wet places: Labr.—Vt.—S.D.—Wyo.—Calif.—B.C.; Eurasia. *Subarct.—Boreal.* Au–S.

8. **E. Smállii** Britton. Stoloniferous perennial; stem firm and wiry, 0.5–2.5 mm. thick, 2.5–10 dm. high; spike slenderly attenuate to ovoid, 5–20 mm. long, 2.5–5 mm. thick, loosely flowered, basal glumes ovate or oblong, obtuse; fertile glumes ascending, with more spreading tips, acute or acuminate, thin, scarious, with 2 purple bands; achenes 0.8–1.2 mm. broad, yellowish to dark-brown; bristles fully equaling the achene or shorter or wanting. Swamps, ponds or banks: N.S.—Del.—Mo.—Neb. *Canad.—Allegh.—Prairie.*

9. **E. mamillàta** Lindb. Perennial, with a creeping rootstock; stems flat or compressed, 2–12 dm. high, sheaths pale-brown, red at the base; spike subcylindric to lanceolate, 10–30 mm. long, 2–5 mm. thick; basal glumes narrowly ovate; fertile glumes pale-brown or purplish, appressed; achenes yellowish or pale-brown, 1.2–1.6 mm. long; bristles 5–6 (rarely 8) often overtopping the tubercle; tubercle depressed-deltoid. *E. macrostachya* Britt. Marshes, ditches and shores: Ill.—La.—Tex.—Calif.—B.C.; Mex. *W. Temp.—Prairie Trop.*

10. **E. cálva** Torr. Stoloniferous perennial; stem 1–6.5 dm. high, filiform, 0.5–1.5 mm. thick, terete; sheath red or chestnut-brown; spike linear-lanceolate or narrowly ovoid; basal glume orbicular to round-ovate; fertile glumes oblong or ovate, obtuse, thin, reddish to pale brown, appressed; achenes narrowly obovoid, 1–1.4 mm. long; tubercle conical; bristles 0–4, about equaling the achene. *E. glaucescens* Am. auth. Wet shores or bogs: Que.—Fla.—Okla.—N. Mex.—Wash.; Mex. *Temp.*

11. **E. uniglùmis** (Link) Schultes. Perennial, with capillary stolons; stem 0.3–7 dm. high, 0.3–3 mm. thick, terete; sheaths reddish at the base; spike lanceolate or narrowly ovoid, 3–17 mm. long, 2–6 mm. thick, 5–30-flowered; basal glume orbicular or round-ovate, chestnut-brown, with a pale margin; fertile glumes oblong-ovate, obtuse or subacute, chestnut or purplish; achenes

obovoid, yellowish to dark brown, 1.2–1.8 mm. long; tubercle depressed-deltoid to ovoid; bristles wanting or usually short. Calcareous or alkaline shores: Greenl.—R.I.—N.D.—Wyo.—Ore.—B.C.; Eurasia. *Boreal.*

12. **E. xyridifórmis** Fern. & Brack. Perennial, with a black rootstock; stem pale, 1–5.5 dm. long, compressed, often twisted, sheath reddish at the base; spike narrowly lanceolate, 1–2 cm. long, 2–3.5 mm. thick; basal glume orbicular, coriaceous, straw-colored, 1.5–2.5 mm. broad; fertile glumes elliptic or ovate, obtuse or rounded at the apex, white-margined; achenes obovoid, straw-colored to nearly chestnut. Moist places: Kans.—Tex.—Ariz.—Nev.; Mex. *Plain—Son.*

13. **E. aciculàris** (L.) R. & S. Stems tufted, filiform, obscurely 4-angled, grooved, 5–20 cm. high; spikelets narrowly ovate, acute, 3–10-flowered, 3–6 mm. long; glumes oblong, obtuse, or the upper acute, pale green, often with two brown bands; bristles 3–4, fragile, fugaceous, shorter than the achene; achene obovoid, pale, obscurely 3-angled; tubercle conic. Wet places: Newf.—N.J.—N.M.—Calif.—B.C.; Mex.; Eurasia. *Plain—Mont.* Jl–S.

14. **E. Wólfii** A. Gray. Perennial, with a short rootstock; stem flattened, 2–5 dm. high; spikelet oblong or oblong-ovate, acute, 4–6 mm. long, 2 mm. thick; glumes ovate, obtuse or the upper acute, pale green, with purple-brown bands; bristles none; achenes obovoid, obtusely 3-angled, 9-ribbed; tubercle depressed-conic, much shorter than the achene. Wet places: Ill.—Minn.—Iowa. *Prairie.* Je–Au.

15. **E. ténuis** (Willd.) Schultes. Stems tufted, filiform, 4-angled, 2–4 dm. high; spikelets narrowly oblong, acute, 6–10 mm. long, about 2 mm. thick; glumes thin, obovate or obovate-oblong, obtuse, with greenish midvein and scarious margins; bristles 2–4, shorter than the achene, fugaceous or wanting; achenes obovoid, obtusely 3-angled, yellowish brown; tubercle conic, short. Wet places: N.S.—Fla.—Tex.—Colo.—Sask. *E. Temp.—Plain—Submont.* My–Jl.

16. **E. acuminàta** (Muhl.) Nees. Stems tufted, flattened, striate, 2–6 dm. high; spikelets ovoid or oblong, obtuse, 6–12 mm. long; glumes oblong or ovate-lanceolate, acute or obtusish, purple-brown with green midrib and hyaline margins; bristles 1–5, mostly shorter than the achene; achene obovoid, very obtusely 3-angled, light yellowish brown; tubercle depressed-conic. *E. compressa* Sull. Wet places: Que.—Ga.—La.—Colo.—B.C. *Temp.* Je–Au.

17. **E. arenícola** Torr. Stems tufted, slender, 1.5–4.5 dm. high, grooved; spikelets oblong to ovoid-oblong, 4–12 mm. long, 2.5–3.5 mm. thick; glumes oblong or ovate, thin, blunt, with a brown apex and whitish margins; bristles 6, persistent, the longer ones as long as the achene; achenes 3-angled, oblong-obovoid, faintly reticulate; tubercle thick, deltoid. Perhaps not distinct from *E. montana* H.B.K. Sandy shores: S.C.—Fla.—Tex.—Calif.—Colo. *Son.—Submont.*

18. **E. rostellàta** Torr. Perennial; stems slender, flattened, the sterile ones often reclining and rooting at the summit, grooved; spikelets oblong, 6–12 mm. long, 2 mm. thick; glumes ovate, obtuse or the upper acute, green with a darker midvein; bristles 4–8, longer than the achene; achene obovoid, finely reticulate. Marshes and wet meadows: N.H.—Fla.—Tex.—Calif.—B.C.; Mex. *Temp.—Plain—Submont.* Au–S.

19. **E. pauciflòra** (Lightf.) Link. Rootstock filiform; stems 3-angled, filiform, leafless, 7–25 cm. high; spikelet oblong, compressed, 4–10-flowered, 4–6 mm. long; glumes brown with lighter margins and midrib, lanceolate, acuminate; bristles 2–6, usually longer than the achenes; achenes obovoid, abruptly beaked, finely reticulate. *Scirpus pauciflorus* Lightf. [G, B]. Wet soil: Que.—Me.—N.Y.—N.M.—Calif.—B.C.; Eurasia. *Mont.—Subalp.* Jl–O.

20. **E. intermèdia** (Muhl.) Nees. Annual; stem filiform, tufted, diffuse, grooved, 1–3 dm. high; spikelet ovoid-oblong, acute, 2–7 mm. long; glumes oblong-lanceolate, obtuse, light purplish-brown, with green ribs; bristles longer than the achene, downwardly barbed; achene 3-angled, obovoid, light brown, finely reticulate; tubercle conic-subulate, slightly constricted at the base, one fourth to one half as long as the achene. Marshes: Ont.—N.J.—Iowa—S.D. *Canad.—Allegh.* Jl–S.

21. **E. pygmaèa** Torr. Annual; stem filiform, flattened, tufted, 2–5 cm. high; spikelet ovoid-oblong, acute, 2–3 mm. long, 3–8-flowered; glumes ovate or lanceolate, pale green, the lower obtuse, the upper acutish; bristles 6, downwardly barbed, longer than the achene; achene oblong, pointed at each end, pale, 3-angled. *Scirpus nanus* Spreng. [G, B]. Muddy places in salt marshes and around salt springs: N.S.—Minn.—Tex.—Fla.; Wash.—Calf.; Eur. *Temp.* Jl–S.

5. STENOPHÝLLUS Raf.

Scapose annuals, with narrowly linear or filiform leaves. Spikelets umbellate or capitate, rarely solitary; glumes spirally arranged, imbricate, deciduous. Flowers perfect; perianth wanting. Stamens 2 or 3. Base of the style swollen, persistent, forming a tubercle; stigmas 2 or 3. Achenes 3-angled or lenticular.

1. **S. capillàris** (L.) Britton. Annual; stems tufted, filiform, smooth, 5–25 cm. high; leaves filiform, roughish; bracts 1–3, setaceous; spikelets narrowly oblong, 5–8 mm. long, less than 2 mm. thick, in terminal, simple or compound umbels; glumes oblong, obtuse or emarginate, dark brown with green keel; achenes yellowish, transversely wrinkled. River valleys: Me.—Fla.—Calif.—Ore. (but no specimens seen from the Rockies). Jl–S.

6. FIMBRÍSTYLIS Vahl.

Annual or perennial subscapose herbs, with grass-like leaves. Spikelets capitate or in ours umbellate, terete, several- or many-flowered; glumes spirally arranged, imbricate. Flowers perfect; perianth none. Stamens 1–3. Style usually enlarged at the base, but wholly deciduous at maturity; stigmas 2 or 3. Achenes lenticular or 3-angled.

Style 2-cleft; achene lenticular; spikelets ellipsoid.
  Glumes glabrous. 1. *F. interior.*
  Glumes, at least the lower ones, pubescent or puberulent. 2. *F. puberula.*
Style 3-cleft; achene 3-angled; spikelets linear. 3. *F. autumnalis.*

1. **F. intèrior** Britton. *Fig. 93.* Perennial, with short stolons; stem thickened at the base, striate, smooth, 3–6 dm. high; leaves rough-margined, involute; spikelets in somewhat compound umbels, ovoid or ovoid-oblong, acutish, 1 cm. long or less; glumes yellowish-brown, ovate, striate, mucronate, dull; achenes broadly ovate, blunt, cancellate, chestnut brown. *F. castanea* and *F. thermalis* of Fl. Colo. Meadows: Colo.—Neb.—Tex. *Plain.* Jl–Au.

2. **F. pubérula** (Michx.) Britton. Perennial, with a stout rootstock; stem 3-angled, 2–3 dm. high; leaves involute, 2 mm. wide, more or less pubescent; spikelets in simple or compound umbels, oblong, obtuse or acutish, 6–10 mm. long, 2–3 mm. thick, glumes thin, brown with lighter midrib, broadly elliptic or oval, dull, puberulent; achenes obovate, biconvex, pale brown, striate and reticulate. Meadows: N.Y.—Fla.—Tex.—Kans. *E. Temp.*

3. **F. autumnàlis** (L.) R. & S. Annual; stems slender, tufted, flat, 0.5–4 dm. high; leaves narrowly linear, 1–2 mm. wide, glabrous; bracts 2 or 3, shorter than the umbel; spikelets 6–40 mm. long, 1 mm. thick, acute; glumes ovate-lanceolate, strongly mucronate, with strong midrib; achenes obovoid, nearly white, smooth or nearly so. Moist places: Me.—Mich.—Kans. —La.—Fla.; Trop. Am. *E. Temp.—Trop.* Je–S.

7. **ERIÓPHORUM** L. COTTON-GRASS.

Perennial bog plants, with rootstocks. Stems triangular or terete. Leaves with linear blades or some of them reduced to bladeless sheaths. Spikelets terminal, solitary, or a few in heads or umbels. Glumes spirally arranged, all usually subtending perfect flowers. Perianth of 6 members, but each 4–6-cleft to the base into long, soft bristles, exserted much beyond the glumes at maturity. Stamens 1–3. Styles deciduous; stigmas 3. Achenes 3-angled, oblong, ellipsoid or obovoid.

Spikelets solitary; involucre wanting.	
Plant stoloniferous.	1. *E. Chamissonis.*
Plant tufted, not stoloniferous.	
Upper sheaths inflated; stem rough above.	2. *E. callitrix.*
Upper sheaths not inflated; stem smooth.	3. *E. opacum.*
Spikelets several, subtended by foliaceous bracts.	
Leaf-blades triangular-channeled throughout.	4. *E. gracile.*
Leaf-blades flat, at least below the middle.	
Glumes with a prominent rib; stamens 3.	
Midrib of the glumes not prominent at the tip of the glume.	5. *E. angustifolium.*
Midrib of the glumes prominent to the very tip.	6. *E. viridicarinatum.*
Glumes striate, not ribbed; stamen 1.	7. *E. virginicum.*

1. **E. Chamissònis** C. A. Meyer. Stems 1–7 dm. high, somewhat triangular; upper sheaths inflated, bladeless; leaf-blades filiform, triangular-channeled, 3–10 cm. long; bristles often reddish-brown. *E. russeolatum* Fries. Bogs: N.B.—Ont.—Wyo.—Ore.—B.C.; Eurasia. *Mont.—Subalp.* Je–Au.

2. **E. cállitrix** Cham. Stems obtusely 3-angled, 2–5 dm. high; leaf-blades filiform, triangular-channeled; glumes thin, ovate-lanceolate or the lowest lanceolate, acuminate, pale brown. *E. vaginatum* Torr., not L. Bogs: Newf. —Mass.—Pa.—Minn.—B.C.—Alaska. *Boreal—Mont.* Je–Au.

3. **E. opàcum** (Björnstr.) Fernald. Stem terete or nearly so, 3–6 dm. high; basal leaves elongate, filiform, channeled; stem-leaves reduced to 2 or 3 close sheaths; glumes thin, ovate-lanceolate or the inner ones linear-lanceolate, acuminate. Bogs: Me.—Mass.—B.C.—Yukon; Eurasia. *Boreal—Mont.* My–Jl.

4. **E. grácile** Koch. *Fig. 95.* Stem slender, smooth, terete, 3–6 dm. high; blades of the stem-leaves 2–3 cm. long, the basal ones longer; bracts about 1 cm. long; spikelets 2–6, on pubescent peduncles; glumes ovate, gray or nearly black, acutish, with prominent midrib. Bogs: Que.—N.Y. —Colo.—Calif.—B.C.; Eurasia. *Canad.—Plain—Mont.* Je–Au.

5. **E. angustifòlium** Roth. Stem smooth, obtusely triangular above, 3–6 dm. high; blades rough-margined, 3–8 mm. wide; bracts 2–4, often black at the base; spikelets 2–12, ovoid or oblong; peduncles smooth; glumes ovate-lanceolate, acute or acuminate, purple-green or brown. *E. polystachyon* L., in part. *E. ocreatum* A. Nelson. Bogs: Newf. —Me.—Ill.—N.M.—Ore.—Alaska; Eurasia. *Boreal—Submont.—Subalp.* Je–Au.

6. **E. viridicarinàtum** (Engelm.) Fernald. Stem 3–6 dm. high; leaf-blades 2–6 mm. wide, the upper 15 cm. long or less; bracts not black at the base; spikelets 5–30; peduncles fine-hairy; glumes ovate-lanceolate; achenes oblong-obovoid. *E. polystachyon* Am. auth., mainly. Bogs and wet meadows: Newf.—Ga.—Ohio—Wyo.—B.C. *Boreal—Submont.* Jl–Au.

7. **E. virgínicum** L. Stem obscurely triangular above, smooth, 4–10 dm. high; leaves flat, 2–5 mm. wide, channeled towards the apex; glumes ovate, acute, brown with green center, about 5-nerved. Bogs: Newf.—Man.—Neb.—Fla. *E. Temp.* Je–S.

8. LEUCOCÒMA (Ehrh.) Rydb. Alpine Cotton-grass.

Perennial bog plants, with rootstocks. Stems triangular. Spikelets solitary, usually subtended by a subulate bract and attached slightly obliquely. Glumes spirally arranged, all usually subtending perfect flowers. Perianth of 6 white soft bristles, at maturity exserted far beyond the glumes. Stamens 3. Styles deciduous. Achenes obovoid, somewhat 3-angled, without tubercle. *Trichophorum* Pers., in part.

1. **L. alpina** (L.) Rydb. Stems numerous, not tufted, 1.5–2.5 dm. high; leaves subulate, 6–20 mm. long, triangular, channeled; lower sheaths often bladeless; glumes oblong-lanceolate, yellowish brown with slender midvein. *Eriophorum alpinum* L. Cold bogs: Newf.—Conn.—Mich.—Minn.—B.C.; Eurasia. *Mont.—Alp.* Je–Au.

9. SCÍRPUS L. Bulrush, Club-rush, Tule.

Annual or perennial, caulescent or scapose herbs. Leaves grass-like, or in many species reduced to basal sheaths. Spikelets terete or somewhat flattened, solitary, capitate or umbellate. Glumes spirally arranged, some of the lower empty. Flowers perfect. Perianth of 1–6 barbed, pubescent, or smooth bristles, rarely wanting. Stamens 2–3. Style wholly deciduous, not swollen at the base, or rarely its base persistent as a small tip. Stigmas 2 or 3. Achenes triangular or lenticular, rarely plano-convex.

Involucre of a single bract.
  Spikelets solitary, rarely two together.
    Bract scarcely exceeding the spikelet, often shorter; bog plants.
      Stem terete; bristles of the flowers smooth. — 1. *S. caespitosus.*
      Stem 3-angular; bristles of the flowers upwardly barbed. — 2. *S. Clintoni.*
    Bract at least twice as long as the spikelet; aquatic plants. — 3. *S. subterminalis.*
  Spikelets normally more than one, usually several, sometimes numerous.
    Spikelets few, 1–12, appearing lateral.
      Spikelets in a capitate cluster.
        Stem not sharply angled.
          Annuals; stem slender.
            Achenes strongly transversely rugose. — 4. *S. Hallii.*
            Achenes smooth or slightly rough.
              Achenes biconvex or lenticular; bristles surpassing the achenes. — 5. *S. debilis.*
              Achenes plano-convex; bristles minute or wanting. — 6. *S. Smithii.*
          Perennials; stem stout. — 7. *S. nevadensis.*
        Stem sharply 3-angled; perennials.
          Achenes plano-convex; bristles not exceeding the achenes.
            Spikelets acute; bracts long; glumes awned. — 8. *S. americanus.*
            Spikelets obtuse; bracts short; glumes mucronate. — 9. *S. Olneyi.*
          Achenes 3-angled; bristles longer than the achenes. — 10. *S. Torreyi.*

Spikelets in spike-like clusters. 11. *S. rufus.*
Spikelets numerous in small clusters of 1–7, arranged in compound umbels; perennials with stout rootstocks.
Style 2-cleft; achenes obovate, plano-convex, brown.
Achenes 2 mm. long, nearly as long as the glumes; spikelets ovoid. 12. *S. validus.*
Achenes 3 mm. long, distinctly shorter than the glumes; spikelets oblong-cylindric. 13. *S. acutus.*
Style 3-cleft; achenes obcordate, 3-angled, yellowish. 14. *S. heterochaetus.*
Involucre of two or more leaves with flat blades; perennials, with rootstocks.
Spikelets 3–15, capitate or umbellate, rather large.
Spikelets in a single head. 15. *S. paludosus.*
Spikelets in an irregular umbel, with 1–3 spikelets at the end of each ray. 16. *S. fluviatilis.*
Spikelets numerous, in compound umbels or in umbellate heads, relatively small.
Bristles downwardly barbed, not much exceeding the fruit.
Style-branches 2; achenes plano-convex, mostly 4. 17. *S. microcarpus.*
Style-branches 3; achenes oblong, 3-angular; bristles 6.
Bristles twice as long as the achene; leaves 4–6 mm. wide. 18. *S. polyphyllus.*
Bristles scarcely exceeding the achene; leaves 6–16 mm. wide.
Plant dark green; glumes acute; achenes pale brown. 19. *S. atrovirens.*
Plant pale; glumes rough-awned; achenes straw-colored. 20. *S. pallidus.*
Bristles 6, smooth, bent or curled.
Bristles at maturity scarcely exceeding the glumes; glumes with a strong green midrib, produced into a sharp point. 21. *S. lineatus.*
Bristles at maturity much exceeding the glumes.
Spikelets all sessile in glomerules of 3–15; involucels reddish-brown. 22. *S. cyperinus.*
Spikelets all or at least some of the lateral ones of each group pedicelled.
Involucels and glumes brown; plant stout, 12–15 dm. high; leaves 5–10 mm. wide. 23. *S. pedicellatus.*
Involucels and glumes dark lead-colored or nearly black; plant slender, 5–12 dm. high; leaves 2–5 mm. wide. 24. *S. atrocinctus.*

1. **S. caespitòsus** L. Stems light green, filiform, 1–4 dm. high; basal sheaths numerous, the upper one bearing a short blade; spikelet ovoid-oblong, 4 mm. long; glumes yellowish-brown, ovate, obtuse; achenes oblong, 3-angled, brown, acute. Bogs and wet places: Greenl.—Ill.—Minn.—Colo.—Wash.—Alaska; Eurasia. *Arct.—Boreal—Mont.—Alp.* Je–Au.

2. **S. Clíntoni** A. Gray. Tufted perennial; stem slender, 1–4 dm. high, rough on the angles; sheaths imbricate, some of them bearing a short subulate leaf, the uppermost with a linear blade; spikelets solitary, ovoid; bract subulate, about as long as the spikelet; glumes ovate, pale brown; bristles 3 6, usually exceeding the 3-angular achene. Fields and thickets: N.B.—N.C.—Minn. *Canad.—Allegh.* Je–Au.

3. **S. subterminàlis** Torr. Stem slender, terete, nodulose, 3–10 dm. high; leaves very slender, 1.5–6 dm. long, 0.5–1.5 mm. wide; spikelet oblong-cylindric, 6–15 mm. long; bract subulate, erect, 1–2.5 cm. long; glumes ovate-lanceolate, acute, light brown, with green midrib; bristles 6, rarely equaling the achenes; these obovoid, 3-angled, dark brown, smooth. Ponds and streams: Newf.—S.C.—Mich.—Minn.—Ida.—Wash.—B.C. *Temp.—Submont.* Jl–Au.

4. **S. Hállii** A. Gray. Stems slender, smooth, obtusely 3-angular, 1–3 dm. high; upper sheath bearing a filiform blade 1–6 cm. long; spikelets 1–7, capi-

tate, oblong-cylindric, obtuse, 6–12 mm. long; bract 2–10 cm. long; glumes ovate-lanceolate, greenish brown, acuminate; bristles none. Wet places: Mass.—Fla.—Tex.—Colo.—Utah—S.D.—Mex. *E. Temp.—Plain—Submont.* Jl–S.

5. **S. débilis** Pursh. Annual; stem slender, triangular, tufted, 1.5–5 dm. high; spikelets 1–12, in capitate clusters, ovoid-oblong, subacute; bract linear, 3–10 cm. long; glumes yellowish-brown, with green midrib, ovate; bristles 4–6, downwardly barbed, as long as the achene; achene plano-convex, broadly obovate or suborbicular, dark brown, shining. Wet soil: Me.—S.D.—Neb.—Ala.—Ga. *E. Temp.* Jl–S.

6. **S. Smíthii** A. Gray. Stem terete, slender, 0.5–4 dm. high, dull green; bract erect; spikelets 1–5, ovoid, acutish, 5–10 mm. long; glumes oblong-oval; style 2-cleft; achenes cuneate-obovate; bristles 1 or 2, minute or wanting. Wet shores: Me.—Pa.—Ill.—Minn. *Canad.* Jl–S.

7. **S. nevadénsis** S. Wats. Stems 2–4 dm. high; leaves several, mostly basal, convolute; spikelets 1–5, ovoid or ovoid-oblong, 6–18 mm. long; bract flattened above, rough-margined; glumes broadly ovate, obtuse or acute, chestnut-brown, smooth and shining; achenes minutely reticulate. Wet places, especially in alkaline soil: Sask.—N.D.—Wyo.—Calif.—Wash. *Son.* Je–Jl.

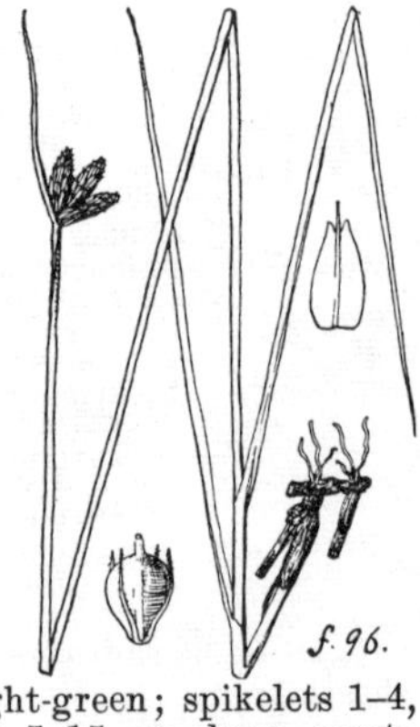

8. **S. americànus** Pers. *Fig. 96.* Stems erect, 3–12 dm. high; leaves 1–3, narrowly linear, keeled; spikelets 1–7, oblong, acute, 8–25 mm. long; bract 2–10 cm. long; glumes broadly ovate, brown, often emarginate or 2-cleft, awned; achenes smooth, dark-brown. Fresh or saline swamps: Newf.—Fla.—Tex.—Calif.—B.C. *Temp.—Plain—Submont.* Je–S.

9. **S. Ólneyi** A. Gray. Stems stout, 6–20 dm. high; leaves 1–3, narrow, 2–12 cm. long; spikelets 5–12, oblong, obtuse, 5–8 mm. long; bract short, stout, 1–4 cm. long; glumes oval or orbicular, dark brown, with green midrib, emarginate or mucronate, brown. Salt marshes: N.H.—Fla.—Tex.—Calif.—Ore.; W. Ind.; Mex.; C. Am. *Temp.—Trop.—Plain Son.* Je–S.

f. 96.

10. **S. Tórreyi** Olney. Stem triangular, nodulose, 5–10 dm. high; leaves narrowly linear, nodulose, light-green; spikelets 1–4, oblong, narrowed at each end, 10–16 mm. long; bract 5–15 cm. long, erect; glumes ovate or lanceolate, chestnut-brown, shining, mucronate; bristles longer than the achene, downwardly barbed; achenes obovoid, light brown, 3-angled, shining. Swamps: Vt.—Man.—N.D.—Iowa—Pa. *Canad—Allegh.* Jl–S.

11. **S. rùfus** (Huds.) Schrad. Stem somewhat compressed, 7–40 cm. high; leaves half-terete, channeled, 1–8 cm. long, less than 2 mm. wide; bract erect, longer than the spike; spikelets red-brown, ovoid-oblong, 6 mm. long, in a 2-ranked spike; glumes lanceolate, 1-nerved; bristles 3–6, shorter than the achene, deciduous, upwardly barbed; achenes oblong, pointed at each end, plano-convex, 3–4 mm. long. Marshes: N.B.—Man.; Eur. *Canad.* Jl–Au.

12. **S. válidus** Vahl. Stem stout, terete, smooth, spongy, 1–3 m. high, sometimes 2 cm. thick, sheathed below; spikelets 5–12 mm. long, 3–4 mm. thick; glumes ovate or suborbicular, slightly pubescent, with strong midrib; achenes plano-convex. *S. lacustris* Am. auth., not L. COMMON BULRUSH. In water: Newf.—Fla.—Calif.—B.C.; W. Ind.; Mex. *Temp.—Plain—Mont.* Je–S.

13. **S. acùtus** Muhl. Similar to *S. validus;* basal sheaths fimbrillate on the margins; spikelets 20 mm. long, 4 mm. thick; glumes ovate, short-awned, viscid above; achenes biconvex. *S. lacustris occidentalis* S. Wats. *S. occiden-*

talis Chase. In water: Newf.—N.Y.—Mo.—N.M.—Calif.—B.C. *Temp.—Plain —Submont.* Jl–S.

14. **S. heterochaètus** Chase. Similar to *S. validus;* stem slender, 1–2 m. high, sheathed below; bracts glabrous; spikelets usually solitary on the slender branches, ovoid or ellipsoid, acutish, 8–15 mm. long, about 5 mm. thick; glumes ovate, glabrous, often erose-margined. In water: Vt.—Mass.—Neb.—Ida.—Ore. *Boreal.* Jl–S.

15. **S. paludòsus** A. Nels. Stems slender, smooth, sharply triangular, 3–6 dm. high; leaves pale green, smooth, 2–4 mm. wide; bracts 2 or 3; spikelets oblong-cylindric, mostly acute, 15–25 mm. long; glumes ovate, puberulent or glabrous, pale brown, 2-toothed at the apex, awned; achenes lenticular, obovate, yellowish brown. *S. campestris* Britt. [G]; not Roth. *S. interior* Britt. *S. Brittonianus* Piper. Salt marshes: Que.—N.J.—Kans.—Tex.—Calif.—B.C. *Boreal.—W. Temp.—Plain—Submont.* My–Au.

16. **S. fluviàtilis** (Torr.) A. Gray. Stem stout, 3-angled, 9–20 dm. high; leaves 8–15 mm. wide; bracts 3–5, erect or spreading, often 2 dm. long; spikelets 1–3 at the ends of the spreading or drooping rays, oblong-cylindric, acute, 15–25 mm. long, 7 mm. thick; glumes ovate, scarious, puberulent, with a recurved awn; bristles 6, downwardly barbed, as long as the achene; achene sharply 3-angled, obovoid, dull. Shallow water: Que.—N.D.—Kans.—N.J. *Canad.—Allegh.* Je–S.

17. **S. microcàrpus** Presl. Stem 1–1.5 m. high; sheaths often tinged with red; leaves rough-margined, often 1 m. long or more; spikelets ovoid-oblong, acute, 3–4 mm. long; glumes brown, with green midvein; stamens 2; achenes oblong-obovate, nearly white. *S. rubrotinctus* Fern. [G]. Swamps: Newf.—Conn.—N.M.—Calif.—Alaska. *Boreal. Plain—Submont.* Jl–S.

18. **S. polyphýllus** Vahl. Stem sharply 3-angled, 3–12 dm. high; leaves 4–6 mm. wide, rough-margined; bracts 3–6, the longer exceeding the inflorescence; spikelets capitate at the ends of the rays, ovoid, 3 mm. long; glumes ovate, obtuse, mucronulate, light-brown; bristles 6, longer than the achene, downwardly barbed above the middle; achene obovoid, 3-angled, dull, short-pointed. Swamps and wet meadows: Mass.—Minn.—Ark.—Ala. *E. Temp.* Jl–S.

19. **S. atróvirens** Muhl. Stem triangular, leafy, 6–14 dm. high; leaf-blades elongate, more or less nodulose, rough on the margins, dark green, 6–12 mm. wide; spikelets ovoid-oblong, acute, 4–10 mm. long; glumes greenish brown, ovate-oblong, acute; bristles rarely wanting. Swamps: N.S.—Ga.—N.M.—La.—Sask. *E. Temp.—Plain—Submont.* Je–Au.

20. **S. pállidus** (Britton) Fernald. Stem triangular, 1–1.5 m. high; leaf-blades elongate, pale, 6–15 mm. wide, somewhat nodulose; spikelets oblong to oblong-cylindric; glumes pale, ovate, acute, tipped with an awn half as long as the body. Wet ground: Man.—Kans.—Tex.—N.M.—Wyo. *Plain—Submont.* Je–Au.

21. **S. lineàtus** Michx. Stem 3-angled, 3–15 dm. high; leaves 4–10 mm. wide, rough-margined; spikelets mostly solitary at the ends of the raylets, oblong, 6–10 mm. long, 2 mm. thick; glumes ovate or oblong, reddish brown, with green midrib; bristles 6, smooth, entangled, longer than the achene, equaling the glume or longer; achene oblong, narrowed at each end, 3-angled, short-beaked. Swamps: N.H.—Kan.—Tex.—Ga.; Ore. *Temp.* Je–S.

22. **S. cyperìnus** (L.) Kunth. Stem smooth, obtusely 3-angled or nearly terete, 6–18 dm. high; leaf-blades elongate, 4–6 mm. wide, rough-margined; bracts 3–6, their bases often brown or black; umbels irregularly compound; spikelets ovoid-oblong, obtuse, 3–10 mm. long; glumes ovate or lanceolate, acute or subacute. Swamps: Newf.—Fla.—La.—Sask. *E. Temp.—Plain.* Au–S.

23. S. pedicellàtus Fernald. Stem stout, 2–4 mm. thick; leaves pale green, firm; inflorescence large, 1–2.5 dm. high, the rays nodding, spikelets 3–6 mm. long; bristles in fruit whitish-brown. Swamps: Que.—N.Y.—Minn. *Canad.* Jl–Au.

24. S. atrocínctus Fernald. Stem 5–12 dm. high, smooth, terete; leaves bright-green, 2–5 mm. wide; bracts blackish at the base; spikelets 2.5–6 mm. long; glumes lance-ovate, acute; achenes 3-angled, white, ovate. Swamps: Newf.—N.J.—Ia.—B.C. *Boreal.—Submont.—Mont.* Je–Au.

10. FUIRÈNA Rottb. UMBRELLA GRASS.

Perennial herbs, with 3-angled leafy stems. Spikelets many-flowered, terete, in terminal and axillary clusters. Glumes spirally imbricate, the lower 1 or 2 empty. Flowers perfect. Perianth represented by 3 ovate, oblong, or cordate, stalked, often awned scales and usually 3 downwardly barbed bristles. Stamens 3. Style 3-cleft, deciduous. Achenes 3-angled, usually more or less stalked, smooth.

Perianth-scales awned from the apex. — 1. *F. squarrosa.*
Perianth-scales awned on the back, below the apex. — 2. *F. simplex.*

1. F. squarròsa Michx. Stems tufted, glabrous, 5–50 cm. high; leaves flat, glabrous or nearly so, or the lower sheaths pubescent; spikelets 1–10, in capitate clusters; glumes oblong, brown, pubescent; perianth-scales oblong, narrowed at each end; achene yellow-brown. Meadows, swamps, and shores: Mass.—Neb.—La.—Fla. *E. Temp.* Jl–S.

2. F. símplex Vahl. Stems tufted, 1–5 dm. high, glabrous; leaves glabrous or ciliate, the sheaths pubescent; perianth-scales ovate-oblong, obtuse to notched at the apex, obtuse to subcordate at the base; achene white. Moist soil: Mo.—Neb.—Tex.; Mex. *Prairie—Texan.* Je–S.

11. HEMICÁRPHA Nees & Arn.

Dwarf annual tufted herbs. Leaves narrow, often setaceous. Spikelets terete, solitary or in small clusters. Glumes spirally arranged. Flowers perfect, with a small translucent scale between each and the axis. Perianth wanting, *i.e.*, bristles none. Stamen 1. Style 2-cleft, not swollen at the base.

Glumes merely mucronate.
  Glumes broadly ovate. — 1. *H. Drummondi.*
  Glumes oblong or oval. — 2. *H. micrantha.*
Glumes narrowed into an awn as long as the body. — 3. *H. aristulata.*

1. H. Drummóndi Nees. Stems longer than the leaves, compressed; bract usually 1; spikelets ovoid, 2–5 mm. long; glumes ovate or rhombic, with a broad green midrib, mucronate; achenes narrowly obovoid, ashy, with a few remote papillae. Damp sand: Ont.—Ill.—Kans.—Tex.—Ark. *Prairie—Texan.* Jl–O.

2. H. micrántha (Wahl.) Britton. Dwarf tufted annual; stems 2–15 cm. high; bract solitary; spikelets 1–3, ovoid, 2–4 mm. long; glumes oblong or obovate, brown; achenes cylindric, reticulate, papillate. Sandy places: N.H.—Fla.—Wash.; Mex.—S.Am. *Temp.—Trop.* Jl–S.

3. H. aristulàta (Coville) Smyth. Stems longer than the setaceous glabrous leaves, about 2 dm. high; bracts 1–3, sometimes 2 cm. long; spikelets 4–8 mm. long; glumes rhombic-obovate, brown, abruptly contracted into a subulate awn about as long as the body; achenes narrowly obovate, black. *H. intermedia* Piper. Sandy banks: Minn.—Iowa—Kans.—Tex.—Calif.—Wash. *W. Temp.—Son.—Submont.* Jl–S.

12. **MARÍSCUS** (Hall.) Zinn. TWIG-RUSH, SAW-GRASS.

Perennial herbs, with rootstocks. Spikelets umbellate. Glumes imbricate, the lower empty, the middle ones usually subtending imperfect flowers, the upper with fertile flowers. Perianth none. Stamens 2 or 3. Style 2- or 3-cleft, deciduous. Achenes ovoid or globose, smooth or longitudinally striate. Tubercle none. *Cladium* P. Br.

1. **M. mariscoìdes** (Muhl.) Kuntze. Stem obscurely 3-angled, smooth, 3–10 dm. high; leaves 2 mm. wide, concave; umbels 2 or 3, compound; spikelets oblong, narrowed at each end, 5 mm. long, in small heads; glumes chestnut-brown, ovate; upper flowers perfect, the middle ones staminate; stamens 2; style 3-cleft; achene ovoid, striate, 2 mm. long. *Cladium mariscoides* Torr. [G]. Marshes: N.S.—Minn.—Fla. *E. Temp.* Jl–S.

13. **SCLÈRIA** Berg. NUT-RUSH.

Leafy perennial herbs. Flowers monoecious, the spikelets in small clusters. Pistillate spikelets 1-flowered, the staminate ones several- to many-flowered. Glumes imbricate, the lower 1–3 empty, and sometimes also the upper ones in the pistillate spikelets. Perianth wanting. Style 3-cleft, deciduous. Ovary often supported on a disk. Stamens 1–3. Achenes globose or ovoid, more or less bony, in our species white.

Flower-clusters terminal, and usually also 1 or 2 in the upper axils; hypogynous disk present.
  Achenes smooth. 1. *S. triglomerata.*
  Achenes papillose. 2. *S. pauciflora.*
Spikelets in an interrupted spike; hypogynous disk wanting; achenes transversely wrinkled and reticulate. 3. *S. verticillata.*

1. **S. triglomeràta** Michx. Stem 4–10 dm. high, 3-angled; leaves flat, glabrous or nearly so, 3–9 mm. wide; hypogynous disk covered with a white crust; achene shining, 2–3 mm. high, obtuse, obscurely 3-angled. Meadows and thickets: Vt.—Wis.—Kans.—Tex.—Fla. *E. Temp.* Jl–S.

2. **S. pauciflòra** Muhl. Stem slender, 2–5 dm. high, 3-angled; leaves less than 2 mm. wide, the sheaths often densely puberulent; hypogynous disk 3-angled, with 6–9 small tubercles; achene 1–2 mm. thick, subglobose. Barrens and dry meadows: N.H.—Ohio—Kans.—Tex.—Fla. *E. Temp.* Je–S.

3. **S. verticillàta** Muhl. *Fig. 97.* Stem filiform or slender, 1–5 dm. high, 3-angled, smooth; leaves 1 mm. wide or less; spikelets 4–6 together in several clusters forming an interrupted spike; achene globose, 1 mm. thick, crustaceous. Moist meadows: Mass.—Minn.—Tex.—Fla.; Mex.; W. Ind. *E. Temp.—Trop.* Jl–S.

14. **CÀREX** (Rupp.) L. SEDGE.*

Grass-like sedges, perennial by rootstocks. Culms mostly triangular, often strongly phyllopodic or aphyllopodic. Leaves 3-ranked, the upper (bracts) elongate or short, and subtending the spikes of flowers or wanting. Plants monoecious or sometimes dioecious; flowers solitary in the axils of scales (glumes). Spikes one to many, either wholly pistillate, wholly staminate,

* The treatment of the genus *Carex* is partly copied from K. K. Mackenzie's revision in Rydberg's Flora of the Rocky Mountains, and partly adapted from this treatment in the second edition of Britton and Brown's Illustrated Flora.

androgynous or gynaecandrous. Perianth none. Staminate flowers of three (or rarely two) stamens, the filaments filiform. Pistillate flowers of a single pistil, with a style and two or three stigmas. Achene completely surrounded by the perigynium, or rarely rupturing it in ripening, 3-angled, lenticular or plano-convex. Rachilla occasionally developed.

Achenes lenticular and stigmas two; lateral spikes sessile; terminal spike partly pistillate, or if staminate, the lateral spikes short, or heads dioecious.
- Spike one.
  - Spike orbicular to short-ovoid. 1. CAPITATAE.
  - Spike linear. 3. DIOICAE.
- Spikes more than one.
  - Perigynia not white-puncticulate.
    - Rootstocks long-creeping, the culms arising singly or few together.
      - Perigynia not thin or wing-margined, the beak obliquely cut.
        - Stems not branching.
          - Spikes densely aggregate into a globular-ovoid head, appearing like one spike. 2. FOETIDAE.
          - Spikes distinct. 5. DIVISAE.
        - Stems becoming decumbent and branching. 4. CHORDORRHIZAE.
      - Perigynia thin or wing-margined, the beak bidentate. 6. ARENARIAE.
    - Rootstock not long-creeping, the culms densely cespitose.
      - Spikes androgynous.
        - Perigynia abruptly contracted into the beak.
          - Spikes few (10 or less); perigynia green or tinged with reddish brown. 7. MUHLENBERGIANAE.
          - Spikes numerous; perigynia yellowish or brownish.
            - Perigynia yellowish; opaque part of leaf-sheath transversely rugulose. 8. MULTIFLORAE.
            - Perigynia brownish; opaque part of leaf-sheath not transversely rugulose. 9. PANICULATAE.
        - Perigynia tapering into the beak. 10. STIPATAE.
      - Spikes gynaecandrous or pistillate or rarely staminate.
        - Perigynia at most thin-edged.
          - Perigynia spreading at maturity. 11. STELLULATAE.
          - Perigynia appressed. 12. DEWEYANAE.
        - Perigynia narrowly to broadly wing-margined. 13. OVALES.
  - Perigynia white-puncticulate. 14. CANESCENTES.

Achenes triangular or lenticular; if lenticular, lower lateral spikes conspicuously peduncled, or terminal spike staminate and lateral spikes elongate.
- Achenes strongly constricted at base, rounded at apex. 15. PHYLLOSTACHYAE.
- Achenes not strongly constricted at base, pointed at apex.
  - Spike one; perigynia rounded and beakless at apex. 16. POLYTRICHOIDEAE.
  - Spikes one or more; when one, perigynia not both rounded and beakless at the apex.
    - Perigynia both coriaceous and shining, the beak obliquely cut.
      - Spike solitary. 17. OBTUSATAE.
      - Spikes several. 18. NITIDAE.
    - Perigynia not both coriaceous and shining.
      - Spike one; perigynia triangular, glabrous, not reflexed or flattened. 19. RUPESTRES.
      - Spikes one to many; when one, perigynia differing from above.
        - Perigynia closely enveloping the achene, strongly tapering at base pubescent or puberulent; bracts sheathless or nearly so.
          - Spike normally one.
            - Spikes androgynous; leaf-blades very narrow. 20. FILIFOLIAE.
            - Spikes dioecious. 21. SCIRPINAE.
          - Spikes two or more.
            - Perigynia obtusely triangular; plant glabrous. 22. MONTANAE.
            - Perigynia acutely triangular; plant pubescent. 23. TRIQUETRAE.
        - Perigynia not as above; or if so. bracts strongly sheathing.
          - Lowest bract long-sheathing, its blades rudimentary.
            - Leaf-blades flat; perigynia puberulent or pubescent. 24. DIGITATAE.
            - Leaf-blades filiform; perigynia glabrous. 25. ALBAE.

Lowest bract sheathless or long-sheathing; if long-sheathing, its blade well-developed.
Lowest bract strongly sheathing; perigynia never strongly bidentate with stiff teeth.
Achenes lenticular; stigmas two.
26. BICOLORES.
Achenes triangular; stigmas three.
Glumes not dark-tinged.
Pistillate spikes short-oblong to linear, erect.
Perigynia with few to many strong nerves or else nerveless.
Perigynia tapering at base, triangular, closely enveloping the achenes.
Rootstock long-creeping.
27. PANICEAE.
Rootstock not long-creeping.
28. LAXIFLORAE.
Perigynia rounded at base, suborbicular in cross-section, loosely enveloping achenes.
29. GRANULARES.
Perigynia finely many-striate.
Perigynia tapering at the base, constricted at the apex; obtusely triangular, closely enveloping the achene.
30. OLIGOCARPAE.
Perigynia rounded at both ends, terete.
31. GRISEAE.
Pistillate spikes elongate, linear to cylindric, slender-peduncled, the lower drooping.
Perigynia beakless or short-beaked; terminal spike gynaecandrous.
32. GRACILLIMAE.
Perigynia conspicuously and strongly beaked.
Stems strongly reddish-tinged at the base, aphyllopodic. 33. DEBILES.
Stems not strongly reddish-tinged at base, phyllopodic.
Spikes slender, few-flowered; perigynia 4 mm. long or less, not inflated, the beak not becoming bidentate.
34. CAPILLARES.
Spikes dense, many-flowered; perigynia longer, more or less inflated, the beak becoming bidentate.
35. LONGIROSTRES.
Glumes dark-tinged. 36. FRIGIDAE.
Lowest bract sheathless, or sheathing; if sheathing, perigynia strongly bidentate with stiff teeth.
Foliage pubescent; perigynia not bidentate.
Terminal spikelet staminate.
37. PALLESCENTES.
Terminal spikelet gynaecandrous.
38. VIRESCENTES.
Foliage glabrous, or if rarely pubescent, perigynia bidentate.
Perigynia beakless or very short-beaked; achenes triangular.
Perigynia glaucous; terminal spikelet staminate.
39. LIMOSAE.
Perigynia not glaucous; terminal spikelet gynaecandrous.
Glumes dark-colored.
40. ATRATAE.
Glumes not dark-colored.
41. SHORTIANAE.
Perigynia with strongly bidentate beak, or if not, achenes lenticular.
Achenes lenticular; perigynia dull.
Glumes obtuse to acuminate, not aristate; achenes not constricted at the middle.
42. ACUTAE.
Glumes long-aristate; achenes constricted at the middle. 43. CRYPTOCARPAE.
Achenes triangular, or if rarely lenticular, perigynia shining.

Perigynia coriaceous, little if at all inflated, often pubescent; bracts sheathless. 44. HIRTAE.
Perigynia glabrous, often inflated; if rarely coriaceous, the bracts sheathing.
  Perigynia little inflated; lower bract strongly sheathing. 45. EXTENSAE.
  Perigynia little to much inflated; lower bract not strongly sheathing.
    Spike one. 46. PAUCIFLORAE.
    Spikes more than one.
      Perigynia less than 10 mm. long.
        Perigynium body ovoid or globose; not truncate.
          Perigynia finely and closely ribbed. 47. PSEUDO-CYPEREAE.
          Perigynia coarsely ribbed or nerveless. 48. PHYSOCARPAE.
        Perigynium body obovoid or obconic, abruptly contracted at the apex. 49. SQUARROSAE.
      Perigynia 10 mm. long or more. 50. LUPULINAE.

1. CAPITATAE.

Represented by one species. 1. *C. capitata.*

2. FOETIDAE.

Represented by one species. 2. *C. incurva.*

3. DIOICAE.

Represented by one species in our range. 3. *C. gynocrates.*

4. CHORDORRHIZAE.

Represented by one species. 4. *C. chordorrhiza.*

5. DIVISAE

Rootstocks slender, light brownish; culms obtusely triangular, normally smooth; leaf-blades narrowly involute.
  Perigynia long-beaked; heads dioecious or nearly so. 5. *C. Douglasii.*
  Perigynia short-beaked; heads androgynous. 6. *C. Eleocharis.*
Rootstocks stout; culms acutely triangular, normally rough above.
  Perigynia not strongly nerved ventrally; leaf-sheaths hyaline. 7. *C. praegracilis.*
  Perigynia strongly nerved ventrally; upper leaf-sheaths green-striate opposite the blades. 8. *C. Sartwellii.*

6. ARENARIAE.

Represented by one species. 9. *C. siccata.*

7. MUHLENBERGIANAE.

Sheath tight, often thickened at the mouth, inconspicuously if at all septate-nodulose.
  Perigynia corky-thickened at the base, widely spreading or reflexed at maturity.
    Stigma slender, not twisted, long, light red; perigynia light green, tapering into the beak, slightly hyaline at the mouth. 10. *C. rosea.*
    Stigma stout, twisted, short, deep red, abruptly contracted into the beak, conspicuously white-hyaline at the mouth. 11. *C. convoluta.*
  Perigynia not corky at the base, mostly ascending, more rarely spreading.
    Beak of the perigynia obliquely cleft on the back, scarcely 2-dentate; glumes not concealing the perigynia. 12. *C. vallicola.*
    Beak of the perigynia strongly 2-dentate.
      Heads 15–35 mm. long, the lower spikelets distinct.

Perigynia covered by the brownish or reddish cuspidate glumes. — 13. *C. Hookeriana.*
Perigynia not wholly covered by the glumes, hence conspicuous.
Perigynia spreading, 3–3.5 mm. long; bracts not dilated at the base; glumes about as long as the narrow perigynia.
Perigynia strongly ribbed, convex ventrally. — 14. *C. Muhlenbergii.*
Perigynia nerveless or nearly so, flat ventrally. — 15. *C. plana.*
Perigynia ascending, 4 mm. long; bracts broadly dilated at the base; lower glumes exceeding the broad perigynia. — 16. *C. austrina.*
Heads 8–20 mm. long, spikelets densely capitate.
Body of the glumes about as long as the broadly ovate perigynium. — 17. *C. mesochorea.*
Body of the glumes exceeded by the perigynium.
Perigynia elliptic-ovate; leaves 2–4 mm. wide. — 18. *C. cephalophora.*
Perigynia orbicular-ovate; leaves 1–2 mm. wide. — 19. *C. Leavenworthii.*
Sheath loose and membranous, conspicuously septate-nodulose, easily breaking.
Stem sharply 3-angled, neither flattened nor winged.
Perigynium dull, the beak 1/4–1/3 as long as the body.
Body of the perigynium ovate, tapering gradually into the beak, faintly nerved on the back; teeth 1 mm. long. — 20. *C. gravida.*
Body of the perigynium globose, abruptly beaked, strongly nerved, teeth 0.5 mm. long. — 21. *C. Lunelliana.*
Perigynium deep green, the beak at least half as long as the ovate body or longer.
Bracts not developed; glumes half as long as the perigynium body. — 22. *C. cephaloidea.*
Bracts, at least some of them, developed; glumes equaling the perigynium body.
Leaves 3–4 mm. wide; spikelets aggregate or approximate. — 23. *C. aggregata.*
Leaves 5–10 mm. wide; lower spikelets distant. — 24. *C. sparganioides.*
Stems more or less flattened, narrowly winged. — 25. *C. alopecoidea.*

8. Multiflorae.

Leaves exceeding the stem; beak of the perigynia equaling the body. — 26. *C. vulpinoidea.*
Leaves shorter than the stem; beak of the perigynia shorter than the body. — 27. *C. brachyglossa.*

9. Paniculatae.

Stems loosely cespitose; sheaths not copper-tinged at the mouth; head little interrupted; perigynia 2–2.75 mm. long, shining, not concealed by the scales. — 28. *C. diandra.*
Stems densely cespitose; sheaths copper-tinged at the mouth; head interrupted; perigynia 2.5–4 mm. long, dull, concealed by the scales. — 29. *C. prairea.*

10. Stipatae.

Perigynia 4–5 mm. long; beak 1–2 times as long as the body; sheath not red-dotted.
Sheath cross-rugulose, easily broken, not thickened at the mouth. — 30. *C. stipata.*
Sheath not cross-rugulose, not easily broken, thickened at the mouth. — 31. *C. laevivaginata.*
Perigynia 6–7 mm. long; beak 3–4 times as long as the body. — 32. *C. Crus-corvi.*

11. Stellulatae.

Spikelet solitary or rarely with a small additional one at the base. — 33. *C. exilis.*

Spikelets more than one.
Beak of the perigynia one fourth as long as the body, the teeth short; ventral suture inconspicuous. 34. *C. interior.*
Beak of the perigynia longer, sharply bidentate; ventral suture conspicuous.
Staminate flowers terminal or basal or in separate spikelets; perigynia abruptly contracted into the short very rough beak. 35. *C. sterilis.*
Staminate flowers mostly at the base of the terminal spikelet; perigynia sharp-edged, slightly contracted into a beak, fully half as long as the body. 36. *C. angustior.*

12. DEWEYANAE.

Represented by one species. 37. *C. Deweyana.*

13. OVALES.

Bracts or lower bract conspicuous, several–many times the length of the head. 38. *C. sychnocephala.*
Bracts not conspicuous, rarely slightly exceeding the head.
Beak of the perigynia flattened and serrulate to tip, often strongly bidentate.
Glumes about the length of the perigynia and nearly of the same width above, the perigynia nearly entirely concealed.
Head stiff, the spikes approximate. 39. *C. adusta.*
Head not stiff, flexuous or moniliform. 40. *C. aenea.*
Glumes shorter than perigynia and noticeably narrower above, the upper part of perigynia largely exposed.
Perigynia subulate to lanceolate, at least 2½ times as long as wide.
Perigynia subulate, the margin at the base almost obsolete. 41. *C. Crawfordii.*
Perigynia lanceolate, the margin conspicuous at the base.
Leaves at most 3 mm. wide, those of the sterile shoots few, ascending. 42. *C. scoparia.*
Leaves broader, those of the sterile shoots numerous, widely spreading.
Tips of the perigynia appressed or ascending; spikelets 7–12 mm. long. 43. *C. tribuloides.*
Tips of the perigynia widely spreading or recurved; spikelets 4–8 mm. long.
Inflorescence dense, oblong; stem stiff, stoutish. 44. *C. cristatella.*
Inflorescence loose, elongate; stem weak, slender. 45. *C. projecta.*
Perigynia ovate-lanceolate or broader, at most twice as long as wide.
Perigynia narrowly to broadly ovate, 3–4 mm. long.
Perigynia brownish; spikes closely aggregate, rounded at base. 46. *C. Bebbii.*
Perigynia green; spikes contiguous to widely separate, usually tapering at base.
Leaf-blades 2.5 mm. wide or less; perigynia spreading-ascending; culms slender. 47. *C. tenera.*
Leaf-blades 2–6 mm. wide (averaging 4 mm.); perigynia appressed-ascending; culms stout. 48. *C. normalis.*
Perigynia ovate or broader, 3.75–6 mm. long.
Perigynia 4–5.5 mm. long, thickish, nerveless or nearly so on the inner face. 49. *C. brevior.*
Perigynia 5.5–7.5 mm. long, thin, prominently 10–16 nerved on the inner face. 53. *C. Bicknellii.*
Beak of the perigynia slender, terete, and scarcely, if at all, serrulate towards tip, obliquely cut, at times becoming obscurely bidentate.

Scales about the length of the perigynia, and of nearly the same width above, the perigynia nearly entirely concealed.
  Stem and head stiff and rigid. — 50. *C. xerantica.*
  Stem slender; head flexuous or moniliform. — 51. *C. praticola.*
Scales shorter than perigynia and noticeably narrower above, the upper part of perigynia largely exposed. — 52. *C. festivella.*

14. CANESCENTES.

Spikes androgynous; perigynia unequally biconvex. — 54. *C. disperma.*
Spikes gynaecandrous; perigynia plano-convex.
  Lowest bract bristle-like, much prolonged, many times exceeding its 1–5-flowered spike; spikes widely separate. — 55. *C. trisperma.*
  Lowest bract much shorter or none; spikes several–many-flowered, the upper approximate.
    Spikes 2–4, subglobose, closely approximate, forming an ovate or subglobose head; perigynia scarcely beaked; scales white-hyaline. — 56. *C. tenuiflora.*
    Spikes one–many, the lower more or less strongly separate; perigynia shortly to strongly beaked; scales darker.
      Perigynia broadest near middle; beak short, smooth or moderately serrulate.
        Beak of the perigynia smooth or very nearly so; glumes obtuse to acutish, strongly tinged with reddish brown or chestnut; spikes closely approximate.
          Spikelets approximate, the lower less than 4 mm. wide. — 57. *C. Heleonastes.*
          Spikelets widely separate, 4–5 mm. wide. — 58. *C. norvegica.*
        Beak of the perigynia serrulate, or if smooth, glumes acutish to cuspidate and scarcely, if at all, tinged with reddish brown; lower spikes remote.
          Plant glaucous; leaf-blades 2–4 mm. wide; spikes many-flowered; perigynia scarcely beaked, appressed-ascending, with emarginate or entire orifice. — 59. *C. canescens.*
          Plant not glaucous; leaf-blades 1–2.5 mm. wide; spikes fewer-flowered; perigynia distinctly beaked, loosely spreading, with minutely bidentate orifice. — 60. *C. brunnescens.*
      Perigynia ovate, broadest near the base; beak conspicuous, strongly serrulate. — 61. *C. arcta.*

15. PHYLLOSTACHYAE.

Glumes (except the lowest) not leaf-like, not enveloping the perigynia, green with hyaline margins.
  Body of the perigynia oblong; pistillate flowers 3–10. — 62. *C. Willdenovii.*
  Body of the perigynia globose; pistillate flowers usually 2 or 3. — 63. *C. Jamesii.*
Glumes leaf-like, half enveloping the perigynia, without hyaline margins.
  Perigynia 5 6 mm. long, with a long smooth beak 2 mm. long; leaf-blades deep green. — 64. *C. durifolia.*
  Perigynia 4 mm. long, with a serrulate beak 1 mm. long; leaf-blades light green and glaucous. — 65. *C. saximontana.*

16. POLYTRICHOIDEAE.

Represented by one species. — 66. *C. leptalea.*

17. OBTUSATAE.

Represented by one species. — 67. *C. obtusata.*

18. NITIDAE.

Represented by one species in our range. — 68. *C. supina.*

19. RUPESTRES.

Represented by one species. — 69. *C. rupestris.*

20. FILIFOLIAE.

Represented by one species. 70. *C. filifolia.*

21. SCIRPINAE.

Glumes very minutely hyaline-margined; perigynia whitish-pubescent. 71. *C. scirpoidea.*
Glumes very broadly hyaline-margined; perigynia yellowish-hirsute. 72. *C. scirpiformis.*

22. MONTANAE.

Basal spikelets absent.
  Plant long-stoloniferous.
    Perigynia 2 mm. thick or more when mature, terete. 73. *C. heliophila.*
    Perigynia 1.5 mm. thick when mature, the body round-triangular in cross-section. 74. *C. pennsylvanica.*
  Plant without long stolons.
    Glumes suborbicular, obtuse or mucronate, half as long as the hirsute perigynium-body. 75. *C. Peckii.*
    Glumes ovate, acuminate or cuspidate, nearly equaling the appressed-pubescent perigynium-body. 76. *C. varia.*
Basal spikelets present.
  Lower bract leaf-like.
    Staminate spikelets 2–6 mm. long; perigynia 3 mm. long or less. 77. *C. deflexa.*
    Staminate spikelets 6–12 mm. long; perigynia 4 mm. long or more. 78. *C. Rossii.*
  Lower bract scale-like.
    Perigynia membranous, puberulent; leaves slender, light green.
      Perigynia 2.5–3 mm. long; beak less than half as long as the body; achenes globose-ovoid. 79. *C. umbellata.*
      Perigynia 3.25–4.5 mm. long; beak more than half as long as the body; achenes oblong-obovoid. 80. *C. rugosperma.*
    Perigynia subcoriaceous, glabrous, except the beak; leaves stiff, deep green. 81. *C. tonsa.*

23. TRIQUETRAE.

Represented by one species. 82. *C. hirtiflora.*

24. DIGITATAE.

Basal spikes present; scales abruptly cuspidate. 83. *C. pedunculata.*
Basal spikes absent; scales not abruptly cuspidate.
  Staminate spike 3–6 mm. long; scales obtuse, one half as long as the perigynia. 84. *C. concinna.*
  Staminate spike 8–22 mm. long; scales acute to acuminate, from little shorter than to exceeding the perigynia. 85. *C. Richardsonii.*

25. ALBAE.

Represented by one species in our range. 86. *C. eburnea.*

26. BICOLORES.

Represented by one species. 87. *C. aurea.*

27. PANICEAE.

Beak of the perigynia none or very short.
  Sheaths of the bracts short; plant glaucous; leaf-blades involute; spikes approximate. 88. *C. livida.*
  Sheaths of the bracts long; plant not glaucous; leaf-blades flat; spikes distant.
    Stem aphyllopodic, purplish at the base; plant loosely stoloniferous. 89. *C. colorata.*
    Stem phyllopodic, not strongly purplish at the base; rootstock very slender, deep-seated.
      Fertile-culm blades 3–5, 2–3.5 mm. wide; perigynia 3.5 mm. long or less; spikes linear. 90. *C. tetanica.*
      Fertile-culm blades 6–10, 3–7 mm. wide; perigynia longer; spikes oblong or linear-oblong. 91. *C. Meadii.*
Beak of the perigynia straight, prominent. 92. *C. vaginata.*

28. LAXIFLORAE.

Sheaths and base of the stem strongly purplish; staminate glumes purplish. — 93. *C. plantaginea.*
Sheaths not purple-tinged, the base of the stem rarely so; staminate glumes never purplish.
  Perigynia acutely triangular, short-tapering at the base.
    Perigynia short-beaked; second bract and leaves overtopping the stem; leaf-blades 2.5–5 mm. wide, erect. — 94. *C. digitalis.*
    Perigynia beakless or nearly so; second bract and leaves not overtopping the stem; leaf-blades 4–8 mm. wide, spreading. — 95. *C. laxiculmis.*
  Perigynia obtusely triangular, long-tapering at the base, smooth.
    Perigynia strongly 24–45-nerved.
      Apex of the perigynia broad, curved sideways.
        Spikelets laxly flowered; leaf-blades 0.7–3 cm. wide. — 96. *C. albursina.*
        Spikelets densely flowered; leaf-blades 3–15 mm. wide.
          Basal sheaths brown; bracts usually overtopping the stem; staminate spikelet subsessile. — 97. *C. blanda.*
          Basal sheaths purple; bracts rarely overtopping the stem; staminate spikelet long-stalked. — 98. *C. laxiflora.*
      Apex of the perigynia narrow and sharp; stem wing-margined. — 99. *C. ormostachya.*
    Perigynia 15–21-nerved. — 100. *C. leptomeria.*

29. GRANULARES.

Stems tufted; bracts elongate, overtopping the spikelets; staminate spikelet short-stalked.
  Basal leaves 5–15 mm. wide; perigynia obovoid, erect or ascending, nerved, 2–2.5 mm. long. — 101. *C. Shriveri.*
  Basal leaves 3–9 mm. wide; perigynia broadly obovoid, soon squarrose, ribbed, 2.3–4 mm. long. — 102. *C. granularis.*
Stems with long-creeping rootstock; bracts short; staminate spikelet long-stalked. — 103. *C. Crawei.*

30. OLIGOCARPAE.

Represented by one species. — 104. *C. oligocarpa.*

31. GRISEAE.

Perigynia elliptic, 1.5 mm. wide; leaves 2 mm. wide or less; sheaths of the bracts and peduncles rough. — 105. *C. conoidea.*
Perigynia oblong, 1 mm. wide; leaves 3–7 mm. wide; sheaths of bracts and peduncles smooth. — 106. *C. grisea.*

32. GRACILLIMAE.

Plant glabrous; perigynia 3.5 mm. long or less. — 107. *C. gracillima.*
Plant pubescent; perigynia 4–5 mm. long. — 108. *C. Davisii.*

33. DEBILES.

Perigynia smooth or puberulent.
  Leaf-blades pubescent; spikelets dense. — 109. *C. castanea.*
  Leaf-blades glabrous; spikelets very slender. — 110. *C. arctata.*
Perigynia tuberculate-hispid. — 111. *C. assiniboinensis.*

34. CAPILLARES.

Represented by one species in our range. — 112. *C. capillaris.*

35. LONGIROSTRES.

Represented by one species in our range. — 113. *C. Sprengelii.*

36. FRIGIDAE.

Perigynia strongly narrowed at the base; beak prominent, serrulate. — 114. *C. misandra.*
Perigynia rounded at the base; beak very short. — 115. *C. atrofusca.*

37. PALLESCENTES.

Represented by one species in our range. 116. *C. abbreviata.*

38. VIRESCENTES.

Represented by one species. 117. *C. Bushii.*

39. LIMOSAE.

Perigynia ovoid-elliptic, thick; pistillate glumes broadly oval, obtuse or mucronate; stems obtusely triangular. 118. *C. rariflora.*
Perigynia broader, much-flattened; pistillate glumes ovate, tapering at the apex, to long-cuspidate; stems sharply triangular.
  Plant strongly stoloniferous; leaf-blades involute, glaucous, 3 mm. wide or less; glumes little exceeding the perigynia. 119. *C. limosa.*
  Plant tufted; leaves flat, not glaucous, wider; glumes much exceeding the perigynia. 120. *C. paupercula.*

40. ATRATAE.

Terminal spikelet in some plants pistillate and linear-cylindric or staminate only at the apex, in others staminate. 121. *C. Hallii.*
Terminal spikelets staminate or gynaecandrous, neither pistillate nor linear-cylindric.
  Glumes not exceeding the perigynia or slightly so; stem central; sheaths not filamentose.
    Perigynia neither flattened nor 2-edged. 122. *C. Halleri.*
    Perigynia flattened and 2-edged.
      Spikes sessile or short-stalked, erect, contiguous. 123. *C. Parryana.*
      Spikes, except the uppermost, decidedly stalked, often drooping and distant. 124. *C. atratiformis.*
  Glumes awned, decidedly exceeding the perigynia; stems lateral in the clump; sheaths strongly filamentose. 125. *C. Buxbaumii.*

41. SHORTIANAE.

Represented by one species. 126. *C. Shortiana.*

42. ACUTAE.

Stems aphyllopodic.
  Beak of perigynia very short, not twisted.
    Plants in clumps or tussocks; stolons short, ascending.
      Perigynia elliptic, longer than the glumes. 127. *C. stricta.*
      Perigynia obovate or orbicular, shorter than the glumes. 128. *C. Haydeni.*
    Plants with numerous long, horizontal stolons, forming sods.
      Perigynia green, granular-rough; lower sheath filamentose, not septate-nodulose; stem slender. 129. *C. strictior.*
      Perigynia rough only towards the apex; lower sheath septate-nodulose, not filamentose; stem stout. 130. *C. Emoryi.*
  Beak of perigynia prominent, twisted when dry; stolons very short or none. 131. *C. torta.*
Stems phyllopodic.
  Beak of the perigynia, if present, not bidentate.
    Stems single or in small clumps, strongly stoloniferous; glumes very dark, with slender midrib; strict and low plants. 132. *C. concolor.*
    Stems densely cespitose; glumes with broad light-colored center; tall and slender plants.
      Leaves 1–3 mm. wide; staminate spikelet single; rootstock not long-stoloniferous. 133. *C. lenticularis.*
      Leaves broader; staminate spikelets 2 or more; stolons long.
        Perigynia elliptic, broadest near the middle. less than 3 mm. long, 1–1.5 mm. wide. 134. *C. aquatilis.*
        Perigynia strongly obovate, broadest at the apex, 3 mm. long, 1.75 mm. wide. 135. *C. substricta.*
  Beak of the perigynia decidedly bidentate; perigynia strongly ribbed. 136. *C. nebraskensis.*

43. Cryptocarpae.

Awns of the pistillate glumes erect; perigynia not inflated. — 137. *C. maritima.*
Awns of the pistillate glumes spreading; perigynia somewhat inflated. — 138. *C. crinita.*

44. Hirtae.

Beak of the perigynia much shorter than the body, the teeth 1 mm. long or less.
  Perigynia densely to sparsely pubescent.
    Stem aphyllopodic, strongly purplish and filamentose at the base; perigynia strongly nerved. — 139. *C. lacustris.*
    Stem phyllopodic, neither purplish nor filamentose at the base; perigynia impressed-nerved. — 140. *C. impressa.*
  Perigynia densely or strongly pubescent.
    Nerves of the perigynia obscured by pubescence, the teeth of beak short.
      Leaf-blades flat, more than 2 mm. wide. — 141. *C. lanuginosa.*
      Leaf-blades involute, 2 mm. wide or less. — 142. *C. lasiocarpa.*
    Nerves of the perigynia prominent, the teeth of beak prominent, slender. — 143. *C. Houghtonii.*
Beak of the perigynia including teeth nearly as long as body, the teeth 1.5 mm. long or more.
  Perigynia glabrous.
    Perigynia ovoid, the teeth less than 2 mm. long, erect or spreading; scales acute to aristate; leaf-blades glabrous. — 144. *C. laeviconica.*
    Perigynia lanceolate or ovoid-lanceolate, the teeth 2–4 mm. long, widely spreading; scales long-aristate; leaf-blades pubescent beneath. — 145. *C. atherodes.*
  Perigynia hairy. — 146. *C. trichocarpa.*

45. Extensae.

Perigynia 2–3 mm. long, the beak scarcely half as long as the body. — 147. *C. viridula.*
Perigynia 4–6 mm. long, the beak about as long as the body.
  Glumes hidden, scarcely tinged with brownish-red at maturity; perigynia 4–4.5 mm. long. — 148. *C. cryptolepis.*
  Glumes conspicuous, strongly brownish-red; perigynia 5–6 mm. long. — 149. *C. flava.*

46. Pauciflorae.

Represented by one species. — 150. *C. pauciflora.*

47. Pseudo-Cypereae.

Perigynia suborbicular in cross-section, more or less inflated. — 151. *C. hystricina.*
Perigynia obtusely triangular, scarcely inflated, closely enveloping achene.
  Teeth of perigynia erect, 1 mm. long; body of the beak 1 mm. long. — 152. *C. Pseudo-Cyperus.*
  Teeth of perigynia spreading or recurved, 1.5–2 mm. long; body of the beak 1.5–2 mm. long. — 153. *C. comosa.*

48. Physocarpae.

Pistillate glumes not rough-awned (except sometimes the lowest).
  Perigynia scarcely inflated; beak entire or emarginate; stigmas normally two and achenes lenticular.
    Perigynia lanceolate; fertile culms filamentose at the base; rootstocks creeping. — 154. *C. miliaris.*
    Perigynia ovoid; fertile culms rarely if ever filamentose at the base; plant loosely stoloniferous. — 155. *C. physocarpa.*
  Perigynia from little to much inflated; beak bidentate; stigmas normally three and achenes triangular.
    Pistillate spikes globose or short-oblong, 5–15-flowered. — 156. *C. oligosperma.*
    Pistillate spikes oblong to cylindric, 15–many-flowered.
      Beak of perigynia rough. — 157. *C. Tuckermani.*
      Beak of perigynia smooth.
        Lower perigynia not reflexed; bracts moderately exceeding the spikes.

Perigynia ascending; lower sheaths more or less strongly filamentose; culms sharply triangular.
Perigynia oblong-ovoid, tapering into the beak; spikes loosely flowered at base. 158. *C. Raeana.*
Perigynia ovoid or globose-ovoid, contracted into the beak; spikes more closely flowered. 159. *C. vesicaria.*
Perigynia spreading at maturity; lower sheaths not filamentose; culms bluntly triangular below the spikes. 160. *C. rostrata.*
Lower perigynia reflexed; bracts many times exceeding the spikes. 161. *C. retrorsa.*
Pistillate glumes rough-awned. 162. *C. lurida.*

49. SQUARROSAE.

Glumes exceeding the perigynia; terminal spikelet small, normally staminate. 163. *C. Frankii.*
Glumes much shorter than the perigynia; terminal spikelet gynaecandrous.
Glumes acuminate or awned; spikelets oval; beak of perigynia spreading. 164. *C. squarrosa.*
Glumes obtuse; spikelets oblong-cylindric; beak of perigynia appressed. 165. *C. typhina.*

50. LUPULINAE.

Pistillate spikelets globose or subglobose; style straight.
Glumes mostly strongly awned; pistillate spikelets 1–12-flowered; leaves 2.5–7 mm. wide, the sheath prolonged. 166. *C. intumescens.*
Glumes usually obtuse; pistillate spikelets 6–30-flowered; leaves 5–9 mm. wide, the sheath not prolonged. 167. *C. Asa-Grayi.*
Pistillate spikes oblong or cylindric; style abruptly bent. 168. *C. lupulina.*

1. CAPITATAE: Cespitose, the rootstocks somewhat elongate. Leaf-blades filiform. Spikelet solitary, ovoid, androgynous, densely flowered, bractless. Perigynia plano-convex, sharp-edged, not inflated, the walls thinnish, essentially nerveless, sessile, the smooth beak conspicuously hyaline-tipped, in age bidentulate. Achenes lenticular. Stigmas 2.

1. C. capitàta L. Stems 1–2 dm. high; leaf-blades filiform, involute, about 0.5 mm. wide; spike with 6–25 ascending perigynia; glumes ovate-orbicular, obtuse; perigynia broadly ovoid, 2–2.5 mm. long, and about as broad, abruptly beaked. An arctic species also occurring very locally on mountain summits southward: Greenl.—N.H.—Man.—Alta.—Calif.—Alaska; Mex.; Eur. *Arct.—Alp.* Je–Au.

2. FOETIDAE: Cespitose with a creeping brown rootstock. Stems solitary or in small clumps. Leaves narrow. Spikelets several, androgynous, in globose or oblong-ovate heads. Perigynia spreading, plano-convex, loose, stipitate at the base, the beak obliquely cut. Achenes lenticular, stigmas 2.

2. C. incúrva Lightf. Stem 3–20 cm. high; leaves 2 mm. wide, usually curved; spikelets 2–5, sessile, 5–15 mm. broad; glumes brown with silvery margins, shorter than the perigynia; perigynia ovate, 3 mm. long; beak conic, rough. Barren lands: Greenl.—Newf.—Man.—Alaska; Eur. *Arctic.*

3. DIOICAE: Rootstocks elongate. Stems slender. Leaf-blades filiform. Spikelet solitary, bractless, staminate, pistillate, or androgynous, narrow. Perigynia at length widely spreading, plano- or bi-convex, not inflated, subcoriaceous, glabrous, nerved, rounded and spongy at base, sharp-edged, strongly beaked, the apex hyaline, in age bidenticulate. Achenes lenticular. Stigmas 2.

3. C. gynócrates Wormskj. Stems 1–3 cm. high, smooth; rootstocks horizontal; leaves bristle-like; spike often wholly staminate or pistillate; 5–15 mm.

long, 2–5 mm. wide; glumes sharp-pointed, reddish brown; perigynia few, 3 mm. long, biconvex, nerved, narrowed into a rough beak half as long as the body. Sphagnum swamps: Greenl.—N.Y.—Mich.—Colo.—Alaska; Eur.; Siberia. *Arctic—Alp.—Mont.* Southward local. My–Au.

4. CHORDORRHIZAE: Stems elongate, prostrate, branching, the rootstocks and roots little developed. Leaf-blades narrow. Spikelets 2–10, androgynous, few-flowered, closely aggregated into an ovoid head. Bracts absent or inconspicuous. Perigynia very thick, plano-convex, oblong-ovate, smooth, shining, coriaceous, closely many-nerved, rounded on the margins, rounded and spongy at the base, the short beak obliquely cut. Achenes lenticular, closely enveloped. Stigmas 2.

4. **C. chordorrhìza** Ehrh. Stems 1.5–3 dm. high; head 1–1.5 cm. long; leaves 2–3 mm. wide; spikelets 2–4; perigynia broadly ovoid, with a very short beak 4–5 mm. long. Sphagnum bogs: Anticosti—Me.—Ind.—Sask.—Mack.—B.C.; Eurasia. *Boreal.* My–Jl.

5. DIVISAE: Rootstocks long-creeping, dark-colored, tough, the stems arising singly or in small clumps at intervals, mostly stiff, dark-tinged at the base, aphyllopodic. Leaf-blades narrow. Spikelets few to many, more or less closely aggregated into an oblong or oblong-ovoid head, androgynous or dioecious, ovoid or oblong. Heads in some species dioecious or nearly so. Lowest one or several of the lower bracts developed, short-prolonged, the others bract-like. Perigynia appressed-ascending, plano-convex, smooth, often shining, coriaceous, more or less nerved on outer and nerved or nerveless on inner surface, sharp-edged but not wing-margined, rounded and spongy at base, the obliquely cut beak in age bidenticulate. Achenes lenticular, closely enveloped. Stigma 2.

5. **C. Douglásii** Boott. Stems 6–30 cm. high; leaf-blades 1–2.5 mm. wide; staminate spikes linear-elliptic, 8–15 mm. long, 2.5–4 mm. wide; pistillate spikes wider; glumes ovate to lanceolate, yellowish-brown, with hyaline margins and lighter center, concealing the perigynia; perigynia lanceolate, strongly nerved, 4 mm. long, 1.75 mm. wide, light brownish; styles elongate. *C. irrasa* Bailey. Dry or alkaline soil: Man.—Neb.—N.M.—Calif.—B.C. *Plain—Mont.* My–Au.

6. **C. Eleócharis** Bailey. Stems 5–20 cm. high; leaf-blades 1.5 mm. wide at the base, involute above; spikes few, densely aggregated into a head 7–15 mm. long; glumes broadly ovate, brownish with hyaline margins; perigynia few to a spike, ovate, lightly nerved on both surfaces, 3 mm. long, 1.75 mm. wide, blackish at maturity; styles short. *C. stenophylla* Am. auth.; not Wahl. Dry soil: Man.—Iowa—Kans.—N.M.—Utah—Yukon. *Plains—Mont.* Je–Jl.

7. **C. praegrácilis** W. Boott. Stems 2–5 dm. high; leaf-blades 1.5–3 mm. wide, flattened or channeled; head linear-oblong to ovate-oblong, 1–5 cm. long, 6–12 mm. wide, the 5–15 spikelets densely aggregated, androgynous, with 4–10 perigynia; glumes ovate-lanceolate, acute to cuspidate, light brownish, with hyaline margins, nearly concealing the perigynia; perigynia 3–4 mm. long, 1.5 mm. wide, dark brown at maturity, nerved on the outer, nearly nerveless on the inner surface. *C. marcida* Boott. *C. usta* Bailey. *C. alterna* (Bailey) Clarke. *C. camporum* Mackenzie. Meadows, widely distributed and variable: Man.—Iowa—Kans.—N.M.—Calif.—B.C.; Mex. *Prairie—Plains—Submont.—W. Temp.* My–Au.

8. **C. Sartwéllii** Dewey. Stems 4–7 dm. high, rough above; leaf-blades 2.5–4 mm. wide, flat; head oblong, 3–5 cm. long, about 1 cm. wide, the numer-

ous spikelets densely aggregated, the larger with 15–20 perigynia; perigynia lance-ovate, 2.5–4 mm. long, 1.5–2 mm. wide, finely nerved, the beak much shorter than the body. Marshes and bogs: Ont.—N.Y.—Ill.—Colo.—Alta. *Plain—Submont.—Boreal.* My–Au.

6. ARENARIAE: Rootstocks long-creeping, dark-colored, tough, the stems arising singly or in small clumps at intervals, mostly stiff, dark-tinged at base, aphyllopodic. Leaf-blades narrow. Spikelets several to many, more or less closely aggregated, some or all gynaecandrous or staminate, some usually pistillate or androgynous. Lower bract or bracts short-prolonged; the others bract-like. Perigynia appressed-ascending, plano-convex, sharp or wing-margined, at least above, conspicuously rough-beaked, the beak sharply bidentate. Achenes lenticular. Stigmas 2.

9. C. siccàta Dewey. Stems 2–9 dm. high, rough above; leaf-blades 2–3 mm. wide, flat or channeled; head linear-oblong, 2–3.5 cm. long, 5–10 mm. wide, the 6–12 spikelets closely aggregated, except the lower 1–3; perigynia few to several, ovate-lanceolate, 5–6 mm. long, 2 mm. wide, much flattened, several-nerved, the beak as long as the body. Dry soil and hills: Me.—N.J.—Nev.—Ariz.—Wash.—Sask. *Boreal—Mont.—Submont.* My–Jl.

7. MUHLENBERGIANAE: Densely cespitose or in a few species the rootstocks elongate, tough and dark-colored. Stems not flattened. Opaque part of sheaths neither transversely rugulose nor red-dotted. Spikelets usually ten or less, androgynous or pistillate but never gynaecandrous, rarely at all compound. Bracts from little to strongly developed. Perigynia plano-convex, green or chestnut-tinged or light brownish in age, from appressed to reflexed, often strongly spongy at base, nerveless or sparingly nerved, narrowly sharp-margined, conspicuously beaked, the beak sharply bidentate. Achenes lenticular. Style-base more or less thickened. Stigmas 2.

10. C. ròsea Schk. Stems slender, rough above, 3–7 dm. high; leaf-blades flat, soft, 1–3 mm. wide; spikelets 2–8, subglobose, 3–8 mm. thick, glumes ovate-oblong to ovate-orbicular, white-hyaline; perigynia 2–15, ovoid-lanceolate, flat, bright-green, nerveless, shining, 2–4 mm. long; beak stout, one fourth as long as the body. Woods and thickets: Newf.—Ga.—La.—N.D. *E. Temp.* My–Jl.

11. C. convolùta Mackenzie. Stems densely cespitose, 3–6 dm. high, strict, rough above; leaf-blades flat, 1.5–3 mm. wide, deep green; spikelets 4–7, the upper contiguous, the lower distant; lowest bracts setaceous; glumes broadly ovate, obtuse, white-hyaline with green midrib; perigynia 6–20, plano-convex, deep-green, nerveless, 3.25–4.5 mm. long, *C. rosea pusilla* Peck. Dry woods: Me.—S.C.—Kans.—S.D.—Man. *E. Temp.* My–Jl.

12. C. vallícola Dewey. Stems 2.5–6 dm. high, rough above; leaf-blades about 1 mm. wide, flat or channeled; head 15–20 mm. long, 7 mm. wide; perigynia 1–10, oblong-elliptic, 3.5 mm. long, 2 mm. wide, not nerved, margined above, the short beak minutely serrulate. *C. brevisquama* Mackenzie. Dry slopes: S.D.—Wyo.—Ore. *Submont.* My–Je.

13. C. Hookeriàna Dewey. Stems 2.5–5 dm. high, rough above; leaf-blades channeled, 1.5–2.5 mm. wide; head 1.5–2 cm. long, about 7 mm. wide; perigynia few, strictly appressed, lanceolate, 4 mm. long, 1.5 mm. wide, nerveless, margined, the serrulate beak half the length of the body. Dry soil: Man. —N.D.—Alta. *Boreal—Plains.* Je.

14. C. Muhlenbérgii Schk. Stems tufted, erect, sharply 3-angled, 3–10 dm. high; leaf-blades 2–5 mm. wide, involute in drying; bracts bristle-form; spikelets 4–10, ovoid or subglobose, in an oblong head, 2–3.5 cm. long; glumes hyaline with a green midrib, ovate-lanceolate, cuspidate or awn-tipped; perigynia ovate-oval, 3 mm. long, contracted into a beak half as long as the body. Field and hills: Me.—Fla.—Tex.—Minn. *E. Temp.* My–Jl.

15. **C. plàna** Mackenzie. Stems cespitose, 3–9 dm. high, sharply triangular, rough above; leaf-blades 2–3 mm. wide, 1–3 dm. long, flat or channeled at the base; spikelets 5–10, densely aggregate into an oblong head 2–4 cm. long; glumes ovate, greenish-hyaline, with a 3-nerved green center, cuspidate or aristate; perigynia 3–3.5 mm. long, 2–2.25 mm. wide, plano-convex, several-ribbed dorsally, serrulate; beak 1 mm. long. Hillsides: Me.—D.C.—Okla.—Neb. *Allegh.—Ozark.*

16. **C. austrìna** Mackenzie. Stems slender, sharply 3-angled, 3–7 dm. high; leaf-blades 2.5–4.5 mm. wide; bracts dilated at the base; spikelets many, ovoid or subglobose, aggregate, forming a head 15–30 mm. long; glumes hyaline, ovate, awned, several-nerved; perigynia 4 mm. long, body suborbicular, contracted into a beak half as long. Sunny places: Mo.—Kans.—Tex. *Ozark.* Ap–Jl.

17. **C. mesochòrea** Mackenzie. Stems slender, erect, rough above, about 2.5 dm. high; leaves 2.5–4 mm. wide; spikelets few, subglobose, densely clustered into an ovoid head, 10–17 mm. long, 10 mm. thick; glumes ovate, cuspidate, narrower than the perigynia; perigynia 3.5 mm. long, the body ovate, contracted into a beak half as long. *C. mediterranea* Mackenzie, not C. B. Clarke. Dry places: D.C.—Tenn.—Mo.—Kans. *Ozark.* My–Jl.

18. **C. cephalóphora** Muhl. Stems slender, erect, 2–6 dm. high; leaf-blades 2–4 mm. wide; spikelets few, subglobose, densely clustered into an ovoid head, 8–15 mm. long; glumes ovate, thin, green or in age slightly yellowish, acuminate or cuspidate; perigynia 2.5 mm. long, nerveless, narrowed into a beak shorter than the body. Dry fields and hills: Me.—Fla.—Tex.—S.D.—Man. *E. Temp.* My–Jl.

19. **C. Leavenwórthii** Dewey. Stems very slender, 1.5–5 dm. high; leaf-blades 1–3 mm. wide; spikelets 4–7, densely crowded in a short oblong head, 8–15 mm. long; glumes ovate, acute or short-cuspidate, shorter than the perigynia; perigynia globose-ovate, 2–2.5 mm. long, narrowed into a beak one-fourth as long as the body. Meadows: N.J.—Fla.—Tex.—Kans.—Iowa. *S. Temp.* My–Jl.

20. **C. grávida** Bailey. *Fig. 98.* Stems 4–8 dm. high, sharply triangular, rough above, cespitose from short rootstocks; leaf-blades flat, 3–14 mm. wide; head 2–3.5 cm. long, 8–14 mm. wide, ovoid or oblong; perigynia 10–20, dull, ascending, broadly ovate or suborbicular, 3–4 mm. long, 2–3 mm. wide, little nerved, the beak one third the length of the body. Prairies and plains: Ont.—Ohio—Kans.—Wyo.—N.D. *Allegh.—Plain—Submont.* My–Jl.

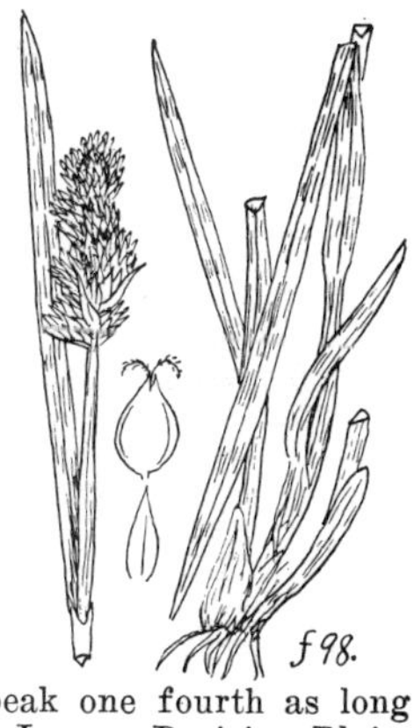

f98.

21. **C. Lunélliàna** Mackenzie. Stems 3–4 dm. high, from a short dark rootstock, sharply 3-angled; leaf-blades light-green, flat, 4–8 mm. wide, 1–2 dm. long; head ovoid, dense, 1–3 cm. long, 8–15 mm. wide; spikelets 6–12, only the lower distinguishable; glumes ovate, tawny, tinged with a green 3-nerved center, cuspidate, shorter than the perigynia; perigynia plano-convex, 4.25 mm. long, the body orbicular, greenish, contracted into a serrulate beak one fourth as long as the body. Dry prairies: Ill.—Mo.—Tex.—N.M.—Iowa. *Prairie—Plain.* My–Jl.

22. **C. cephaloìdea** Dewey. Stems slender or stoutish, erect, very rough above; leaf-blades flat, 4–8 mm. wide, thin and lax; spikelets 4–8, subglobose, aggregated into an elongate cluster, 2–4 cm. long; glumes ovate, membranous, cuspidate or awned, half as long as the body of the perigynia; perigynia ovate, deep green, 4 mm. long, sharp-edged, tapering into a beak half as long as the body. Woods and thickets: N.B.—N.J.—Ill.—Minn. *Canad.—Allegh.* My–Jl.

23. **C. aggregàta** Mackenzie. Stems slender, triangular, 6 dm. high or less; leaf-blades 3–4 mm. wide; spikelets many, globose, densely aggregated into an elongate head, 2.5–3.5 cm. long, 1 cm. thick; glumes ovate, hyaline, with green midrib, acuminate or cuspidate, narrow; perigynia 3 mm. long, the body ovate, tapering into a beak nearly as long as the body. Dry woods: N.Y.—D.C.—Okla.—Kans.—Iowa. *Allegh.—Ozark.* My–Jl.

24. **C. sparganioìdes** Muhl. Stem stout, rough above, sharply 3-angled, 2–10 dm. high; leaf-blades broad and flat, 5–10 mm. wide, shorter than the stem; spikelets 6–12, deep green, oblong or subglobose, 5–8 mm. in diameter, the upper aggregate, the lower separate; glumes ovate, hyaline, acute or cuspidate, equaling the perigynia; perigynia flat, ovate, 3 mm. long, 2 mm. wide, narrowly wing-margined; beak rough, half as long as the body. Woods and thickets: N.H.—Va.—Kans.—Iowa. *Canad.—Allegh.—Ozark.* Je–Au.

25. **C. alopecoìdea** Tuckerm. Stems stout, 4–10 dm. high; leaf-blades flat, 3–8 mm. wide; spikelets androgynous, up to 10, in a compact or interrupted head, 2.5–5 cm. long; glumes ovate, brown-tinged, acuminate or short-awned, about as long as the perigynia; perigynia ovate or ovate-lanceolate, rounded at the base, 3–4 mm. long, tapering into rough beak nearly as long as the body. Meadows: Me.—N.J.—Iowa—Minn. *Canad.—Allegh.* Je–Jl.

8. Multiflorae: Densely cespitose. Stem sharply triangular. Opaque part of sheaths usually transversely rugulose, red-dotted. Spikelets numerous, small, androgynous or pistillate, but never gynaecandrous, the lower more or less compound. Bracts frequently conspicuous. Perigynia plano-convex, not very thick, yellowish or yellowish brown, appressed-ascending or somewhat spreading, not thick-walled, somewhat spongy at base, short-stipitate, sharp-margined, more or less nerved, conspicuously rough-beaked, the beak bidentate. Achenes lenticular. Stigmas 2. Style-base more or less enlarged.

26. **C. vulpinoìdea** Michx. Stem 3–9 dm. high, very rough above, exceeding the leaves; leaf-blades long-tapering, 2–5 mm. wide; head 2–12 cm. long, with very many small spikelets; pistillate glumes strongly awned; perigynia yellowish at maturity, 2–3 mm. long, the beak about the length of the body. Swampy places: N.B.—Fla.—Tex.—Colo.—Ore.—B.C. *Temp.—Plains.* Je–Au.

27. **C. brachyglóssa** Mackenzie. Stems 3–12 dm. high, brown at the base; leaf-blades flat, bright green, 2–4 dm. long, 3–6 mm. wide, rough on the margins; head 2–7 cm. long, greenish-yellow when young, brown in age; spikelets numerous, confluent; glumes ovate, reddish brown, with hyaline margins, and green 3-nerved center, awn-tipped; perigynia ovate, 2–2.7 mm. long, abruptly contracted into the beak. *C. xanthocarpa* Bickn.; not Regel. Fields and pastures: Me.—Va.—Kans.—Iowa. *Canad.—Allegh.* Je–Au.

9. Paniculatae: Densely or loosely cespitose. Stems not flattened. Opaque part of sheaths strongly red-dotted. Spikes numerous, small, androgynous or pistillate, but never gynaecandrous, the lower compound or decompound. Bracts usually inconspicuous. Perigynia thick, strongly convex on the dorsal, and often somewhat convex on the ventral face, brownish or chest-

nut in age, ascending or spreading, coriaceous, spongy at base, stipitate, narrowly margined, more or less nerved, conspicuously rough-beaked, the beak bidentate. Achenes lenticular. Style-base more or less thickened. Stigmas 2.

28. **C. diándra** Schrank. Stems slender, sharply triangular, loosely cespitose, 3–7 dm. high; leaf-blades 1.25 mm. wide; head 2.5–5 cm. long, somewhat compound; the lower spikelets more or less separated; glumes acute, brownish, with lighter midvein and hyaline margins; perigynia somewhat biconvex, dark chestnut, shining, nerveless on inner face, spreading at maturity. Wet meadows: Lab.—Pa.—Colo.—B.C.—Alaska; Eurasia. *Mont.—Boreal.* My–Jl.

29. **C. praìrea** Dewey. Stems sharply triangular, 5–10 dm. high; leaf-blades with slightly revolute margins; head 4–8 cm. long, decompound, the lower 3–5 branches separated; glumes acute or acuminate, light brown, with broad light-colored center and hyaline margins; perigynia 1.25 mm. wide, nerveless, or lightly nerved on inner face, ascending, plano-convex or nearly so, reddish brown, dull, the beak serrulate. Wet meadows: Que.—N.J.—Ia.—Neb.—Sask. *E. Boreal.* My–Jl.

10. STIPATAE: Densely cespitose or with more or less elongated rootstocks. Stem triangular or somewhat flattened. Opaque part of sheaths usually transversely rugulose or red-dotted. Spikes few to many, androgynous to pistillate, but never gynaecandrous, the lower from simple to compound. Bracts little developed. Perigynia plano-convex, yellowish or yellowish brown, appressed-ascending to spreading, not thick-walled, but strongly spongy at base, stipitate, strongly many-nerved, the margins nearly obsolete on the lower half, conspicuously beaked, the beak bidentate. Achenes lenticular. Style-base more or less thickened. Stigmas 2.

30. **C. stipàta** Muhl. Stems 3–10 dm. high, strongly serrulate above, growing in dense clumps; leaf-blades 4–8 mm. wide, the opaque part of sheaths thin, cross-rugulose; head 2.5–10 cm. long, the lower spikelets often separate; glumes light brownish, hyaline-margined; perigynia serrulate. Swamps and wet meadows: Newf.—Fla.—Utah—Tex.—Calif.—B.C. *Temp.—Plain—Submont.* My–Au.

31. **C. laevivaginàta** (Kükenth.) Mackenzie. Stems 3–8 dm. high, slightly winged, leaf-blades flaccid, flat, 1–3 dm. long, 3–6 mm. wide; head compound, linear-oblong or oblong, 2.5–6 cm. long, 1–15 cm. wide; spikelets numerous; glumes deltoid-lanceolate, hyaline with 3-nerved green center, awned, narrower than the perigynia; perigynia 4–10, lanceolate, green or yellow at maturity, 4.5–5 mm. long, sharp-edged, abruptly stipitate. Swampy woods: Mass.—Fla.—Mo.—Minn. *E. Temp.*

32. **C. Crús-còrvi** Shuttlew. Stems in clumps, stout, 3-angled, very rough above, 6–12 dm. high; leaf-blades flat, 5–12 mm. wide, rough-margined; spikelets yellowish brown, staminate above, numerous in a large branching cluster 1–3 dm. long; glumes ovate or lanceolate, thin, shorter than the perigynia; perigynia lanceolate, strongly nerved, 3–5 mm. long, stipitate, with a disk-like base. Swamps: Ind.—Fla.—Tex.—Neb.—Minn. *E. Temp.*

11. STELLULATAE: Densely cespitose. Stems triangular. Sheaths not red-dotted nor cross-rugulose. Spikelets 2–10, or by reduction one, gynaecandrous, pistillate, or in a few species staminate, not compound. Bracts inconspicuous. Perigynia plano-convex, with orbicular, ovate, or broadly oval bodies, green, yellowish brown, or brown, spreading or reflexed at maturity, 2.5–3.5 mm. long, strongly spongy at base, sharp-edged nearly, if not entirely, to the rounded or truncate base, not puncticulate, nerved on the outer, nerved or nerveless on the inner surface, the beak bidentate or obliquely cut. Achenes lenticular. Stigmas 2.

33. **C. éxilis** Dewey. Stems densely cespitose, slender, 2.5–6 cm. high; leaves involute-filiform, shorter than the stem; spikelet usually solitary, terminal, gynaecandrous, occasionally dioecious; glumes ovate, acute, shorter than the perigynia; perigynia ovoid-ellipsoid, brownish, about 3 mm. long, strongly several-nerved on the outer face, spreading or reflexed at maturity; beak slender, rough, half as long as the body. Bogs: Labr.—Del.—Mich.—Minn. *Canad.* My–Jl.

34. **C. intèrior** Bailey. Stems 2–3.5 dm. high, slender and wiry; leaf-blades 1–2 mm. wide; spikelets 2–4, the lateral pistillate, subglobose, with 3–10 perigynia, the upper long-tapering and staminate at base; glumes half the length of the perigynia, ovate-orbicular, very obtuse, brownish, hyaline-margined all around, the center lighter-colored, the midvein not reaching the tip; perigynia ovate, plump, abruptly beaked, 2.5 mm. long, 1.5 mm. wide, nerveless or obscurely nerved on the inner face, sparingly serrulate on the upper margins. Boggy places: Me.—Fla.—Ariz.—Calif.—B.C.; n Mex. *Temp.—Plain—Subalp.* My–Au.

35. **C. stérilis** Willd. Stems densely cespitose, slender, 3–6 dm. high, sharply angled, rough; leaf-blades 1–3 dm. long, 1.25–2.5 mm. wide, canaliculate; heads 2–3 cm. long, spikelets 3–6, approximate or the lowest separate; pistillate spikelets subglobose, 4–6 mm. long; glumes ovate, obtuse, 3-nerved on the green center, brown-chestnut, perigynia ovate or ovate-lanceolate, 3 mm. long, yellowish-brown, abruptly contracted into the short beak. Swamps: Newf.—N.J.—Ill.—Minn. *Canad.* Je.

36. **C. angústior** Mackenzie. Stems very slender, 1–3 dm. high; leaf-blades 0.5–1.5 mm. wide; spikelets 2 or 3, with 3–15 perigynia, the terminal long-clavate; glumes as long as the body of perigynia, acute, acuminate or cuspidate, ovate, yellowish brown-tinged, with hyaline margins and midvein prominent to tip; perigynia divaricate, 2.5–3.5 mm. long, with raised margins, the beak serrulate. *C. stellulata* var. *angustata* Carey. Boggy places: Newf.—N.C.—Colo.—Calif.—B.C. *Temp.—Submont.* Je.

12. DEWEYANAE: Densely cespitose. Stems triangular. Sheaths not red-dotted nor cross-rugulose. Spikelets 3–8, gynaecandrous, pistillate or rarely staminate, simple. Lower one or two bracts often conspicuous. Perigynia plano-convex, with narrowly ovate or linear-oblong bodies, light or yellowish green, 3.5–5.5 mm. long, appressed, strongly spongy at base, only upper half sharp-edged, round-tapering at base, nerved on the outer face, nerved or nerveless on the inner face, the beak bidenticulate to deeply bidentate. Achenes lenticular. Stigmas 2.

37. **C. Deweyàna** Schw. Stems slender, weak, 2.5–9 dm. high, little brownish-tinged at base, strongly roughened beneath head; leaf-blades 2–5 mm. wide; spikelets 2–7, ovate-oblong or subglobose, the lower more or less separate, with 3–15 perigynia; lower bract conspicuous; glumes white-hyaline, with green midvein, mostly cuspidate; perigynia nerveless on the inner face, shallowly bidentate. Woods: Lab.—Pa.—Ia.—Colo.—B.C.—Mack. *Boreal—Plain—Submont.* My–Au.

13. OVALES: Densely cespitose or (rarely) with short-prolonged rootstocks. Stems triangular. Opaque part of sheaths not red-dotted or cross-rugulose, but sometimes green-striate. Spikelets two or three up to twenty, with several to many perigynia, the terminal gynaecandrous, the lateral pistillate or gynaecandrous, simple, the inflorescence varying from capitate to moniliform. Lower bracts from inconspicuous to very conspicuous. Perigynia varying from scale-like or flat (except when distended by the achene) to thick and plano-convex, the body subulate to reniform, narrowly to broadly wing-margined, appressed or ascending or spreading, little corky-thickened at base, prominently beaked, the beak bidentate, or obliquely cut becoming biden-

ticulate, usually serrulate on margins. Style-base scarcely thickened. Achenes lenticular. Stigmas 2.

38. **C. sychnocéphala** Carey. Stems 1–6 dm. high, very smooth; leaf-blades 1.5–3 mm. wide; spikelets 6–15, densely aggregated into an oblong-ovoid head 1.5–3 cm. long; glumes lanceolate, half the length of the perigynia; perigynia 5–6 mm. long, nerved on both faces, the beak deeply bidentate. Meadows and thickets: Ont.—N.Y.—Ia.—Mont. *Canad.—Plain.* Jl–Au.

39. **C. adústa** Boott. Stems 2.5–7 dm. high, nearly smooth; leaf-blades 2–3 mm. wide; spikelets 3–15, the lower only slightly separate; glumes ovate, straw-colored, with greenish midvein; perigynia broadly ovate, 4–5 mm. long, 2–2.5 mm. wide, rather abruptly beaked. *C. pinguis* Bailey. Dry soil: Newf.—N.Y.—Minn.—Sask.—Mack. *Plain.* Je–S.

40. **C. aènea** Fernald. *Fig. 99.* Stems 4–9 dm. high, rough beneath the head; leaf-blades 2.5–4 mm. wide; spikelets 3–12, 7–24 mm. long, 5–7 mm. thick; glumes ovate, light brownish-tinged, hyaline-margined; perigynia ascending, ovate, 4 mm. long, 2 mm. wide, rounded at base, tapering into a beak half the length of the body. *C. foenea* Auth.; not Willd. Dry places and open woods: Lab.—Conn.—Alta.—B.C.—Yukon. *Boreal—Mont.* My–Jl.

41. **C. Crawfórdii** Fernald. Stems 2–6 dm. high, rough beneath the head; leaf-blades 1–3.25 mm. wide; spikelets 3–12, 5–11 mm. long, 3–5 mm. wide, rather closely aggregated, well-defined; glumes lance-ovate, acute or acuminate, brownish; perigynia erect-ascending, brownish, 4 mm. long, about 1 mm. wide, obscurely nerved, tapering into a beak half the length of the body. Open places: Newf.—Conn.—Mich.—Ida.—B.C. *Boreal.* Je–S.

42. **C. scopària** Schkuhr. Stems 2–8 dm. high, rough beneath the head; leaf-blades 1.5–3 mm. wide; spikelets 3–10, 6–16 mm. long, 4–6 mm. wide, aggregate or scattered, clearly defined; glumes ovate, acute, brownish tinged; perigynia erect-ascending, brownish, usually 5.5–6.5 mm. (occasionally 4–5.5 mm.) long, 1.2–1.9 mm. wide, nerved, tapering into a beak half the length of the body. Moist soil: Newf.—Fla.—Colo.—B.C. *Temp.—Plain—Submont.* My–Au.

43. **C. tribuloìdes** Wahl. Stems 3–10 dm. high, stout, erect; leaf-blades flat, 3–8 mm. wide; spikelets 6–20, obovoid, blunt, densely clustered, sometimes separate, 7–12 mm. long; glumes lanceolate, straw-colored, acute, half as long as the perigynia, perigynia thin, greenish-brown, flat, ascending or erect, spreading or recurved, 3–5 mm. long, with a wing-margined beak. Meadows: Vt.—Fla.—La.—Minn. *E. Temp.* Jl–S.

44. **C. cristatélla** Britton. Stem stout, 3–10 dm. high, rough above; leaves 3–7 mm. wide; heads 6–15, globose or subglobose, 4–8 mm. broad, densely aggregated into an oblong head, 2.5 cm. long or more; glumes straw-colored, shorter than the perigynia; perigynia spreading or ascending, 3–4 mm. long, narrowly wing-margined, tapering into the serrulate beak. Meadows and thickets: Vt.—Pa.—Mo.—Neb. *Canad.—Allegh.* Jl–S.

45. **C. projécta** Mackenzie. Stems erect-triangular, slender, 4–10 dm. high, in large clumps; leaf-blades 3–7 mm. wide, shorter than the stem; spikelets 8–15, straw-colored, subglobose, 5–8 mm. long, alternately and loosely arranged, forming a flexuose spike, 2.5–5 cm. long; glumes ovate-lanceolate,

obtuse or acutish, straw-colored, narrower and shorter than the perigynia; perigynia ascending, wing-margined, 3–5 mm. long, tapering into a beak shorter than the body. Damp soil: N.S.—N.Y.—Minn.—N.D. *Canad.* My–Jl.

46. **C. Bébbii** Olney. Stems 2–8 dm. high, rough beneath the head; leaf-blades 2–4.5 mm. wide; spikelets 5–10, 4–9 mm. long, 3–6 mm. wide; glumes ovate, acute, brownish; perigynia ascending, narrowly ovate, 3–4 mm. long, 1.5–2 mm. wide, rounded at base, obscurely nerved, tapering into a beak half the length of the body. Low ground: Newf.—N.J.—Neb.—Colo.—Wash.—B.C. *Boreal—Plain—Submont.* Je–Au.

47. **C. ténera** Dewey. Stems 3–7 dm. high, slender, rough beneath head; spikelets 3–8, 4–5 mm. thick; glumes acutish, hyaline with green midvein, light brownish-tinged; perigynia ascending or spreading, ovate, thick, 3–4 mm. long, 1.5–2.5 mm. wide, rounded at base, nerved, tapering into a beak half the length of the body. *C. straminea* Am. auth., in part. Woodlands: N.B.—Mass.—Ky.—S.D.—Mont.—Canadian Rockies. *Mont.—Boreal.* Je–Au.

48. **C. normàlis** Mackenzie. Stems triangular, rough above, 6–10 dm. high, in dense clumps; leaf-blades 3–6 mm. wide, much shorter than the stem; spikelets 4–12, green or brownish, subglobose, 6–9 mm. long, clavate at the base, aggregated into a spike 2.5–5 cm. long; glumes ovate, nearly as wide as but shorter than the perigynia; perigynia ovate, wing-margined, 3–4 mm. long, rounded at the base, conspicuously nerved on the outer face, tapering into a rough beak half as long as the body. Woodlands: Que.—N.C.—Kans.—Man. *Canad.—Allegh.*

49. **C. brévior** (Dewey) Mackenzie. Stems 3–12 dm. high, rough beneath the head; leaf-blades 2–4 mm. wide; spikelets 3–10, oblong to globose, 4–9 mm. wide, 7–15 mm. long, aggregate or separate; glumes acute; perigynia ovate to orbicular, thick, 4–5.5 mm. long, 2.5–3.5 mm. wide, spreading or ascending, faintly nerved on inner face, the beak about one third the length of the body. *C. festucacea* Auth., in part. Dry soil: N.B.—Fla.—Tex.—N.M.—B.C. *Temp.—Plain—Submont.* Ap–Au.

50. **C. xerántica** Bailey. Stems 3–6 dm. high, nearly smooth; leaf-blades 2–3 mm. wide; spikelets 3–6, approximate but distinct, ellipsoid, 8–14 mm. long, 5 mm. wide; glumes silvery-hyaline, darker tinged; perigynia 4–5.5 mm. long, 2–2.5 mm. wide, nearly nerveless on inner face, round-tapering at the base, tapering into a beak one third the length of the body. Prairies and plains: Man.—Sask.—Colo.—Kans. *Plains.* Jl.

51. **C. praticola** Rydb. Stems slender, 2.5–6 dm. high, roughened beneath the head; leaf-blades 1–2 mm. wide; spikelets 2–6, elliptic, 6–16 mm. long, 5 mm. wide; glumes somewhat brownish-tinged, shining; perigynia appressed, ovate-lanceolate, 4.5–6.5 mm. long, narrowly winged, round-tapering at the base, nearly nerveless on inner face, tapering to a short hyaline-tipped beak. *C. pratensis* Drejer. Meadows and open woods: Greenl.—Me.—N.D.—Colo.—Calif.—Alaska. *Submont.—Mont.—Boreal.* Je–Au.

52. **C. festivélla** Mackenzie. Stems rather stout, 2–7 dm. high, roughened above; leaf-blades 2.5–4 mm. wide; spikelets 3–8, densely aggregate, ovoid-orbicular, 5–9 mm. long, 4–8 mm. wide; glumes ovate, dark reddish brown with narrow hyaline margins; perigynia numerous, appressed, 3.5–5 mm. long, 1.75–2.75 mm. wide, ovate, lightly nerved on the inner face, rounded at the base, narrowed into a beak one third the length of the body. *C. festiva* auth., as to Rocky Mountain plant. Meadows and mountain sides: Que.—S.D.—N.M.—Ariz.—Nev.—B.C. *Boreal—Submont.—Mont.* Je–Au.

53. **C. Bicknéllii** Britton. Stems loosely tufted, 6–12 dm. high; leaves 1–3 dm. long, 2.5–4.5 mm. wide, spikelets 3–7, ovoid or subglobose, 8–18 mm. long, approximate or the lowest separate, brownish or greenish; glumes lance-

ovate, obtuse or acute, straw-colored or brownish, with hyaline margins, shorter and narrower than the perigynia; perigynia spreading-ascending, 5.5–7.5 mm. long, 3–4 mm. wide; body broadly ovate or suborbicular, thin, the rough beak one fourth to one third as long as the body. Dry soil: Me.—N.J.—Tex.—S.D.—Man. *Canad.—Allegh.* Je–Jl.

14. CANESCENTES: Cespitose, but in some species with slender stolons. Culms triangular. Sheaths not cross-rugulose. Spikes 1–10, with few to many perigynia, the terminal gynaecandrous, the lateral pistillate or gynaecandrous, simple. Bracts inconspicuous. Perigynia plano-convex, white-puncticulate, lanceolate, ovate, oval, or obovoid, appressed to spreading, beakless to prominently beaked, more or less nerved on both sides, not winged or margined, but acute-edged above, nearly or entirely filled by the lenticular achene. Style-base not enlarged. Stigmas 2.

54. **C. dispèrma** Dewey. Stems weak, 1.5–6 dm. high, in large clumps, but with slender rootstocks; leaf-blades 1–1.5 mm. wide; spikelets distant or the upper aggregate, 1–5-flowered; perigynia ovoid-elliptic, 2 mm. long, 1.5 mm. wide, finely nerved, the minute beak smooth. *C. tenella* Schk. Bogs: Lab.—N.J.—Ind.—N.M.—Calif.—B.C.—Alaska; Eurasia. *E. Temp.—Submont.—Subalp.* Je–Au.

55. **C. trispèrma** Dewey. Stems filiform, weak, 1.5–8 dm. long, in large clumps, but with slender rootstocks; leaf-blades 1–2 mm. wide; perigynia oblong, 2.5–4 mm. long, nearly 2 mm. wide, finely nerved, narrowed at the apex, the minute beak smooth or nearly so. Swamps and wet woods: Lab.—Md.—Neb.—Sask. *Boreal.* Je–S.

56. **C. tenuiflòra** Wahl. Stems slender, weak, 2–6 dm. long, loosely cespitose and stoloniferous; leaf-blades 0.5–2 mm. wide; perigynia oblong-obovoid, 2.5–3.5 mm. long, 1.5 mm. wide, obscurely nerved. Bogs: Lab.—Me.—Minn.—Alaska; Eurasia; *Boreal.* Je–Au.

57. **C. Heleonástes** Ehrh. Stems stiff, erect, 1–4.5 dm. high; leaf-blades 1–2 mm. wide; spikelets 2–5, with 5–10 perigynia; the latter 3 mm. long, 1.25 mm. wide, several-nerved, tapering into the short beak. Bogs: Ont.—Alta.—B.C.—Mack.; Eurasia. *Subalp.—Boreal.* Jl–Au.

58. **C. norvègica** Willd. Stems slender, smooth, 1.5–4 dm. high; leaf-blades 2.5 mm. wide or less, glaucous; spikelets 3–6, brown, oblong to subglobose, the upper close together, the lower distant, 6–12 mm. long, 3–6 mm. thick, the uppermost clavate at the base; glumes broadly ovate, reddish-brown, obtuse, shorter than the perigynia; perigynia ascending, 2.5–3 mm. long, coriaceous, obovoid, blunt-edged, abruptly narrowed to the base; beak short, smooth. Salt meadows: Lab.—Mass.—Man.; Eur. *Arct.*

59. **C. canéscens** L. Stems in large clumps, slender, 2.5–8 dm. high; spikelets 4–9, 3–12 mm. long, 3–5 mm. wide; perigynia 1.8–2.8 mm. long, 1–1.8 mm. wide, faintly few-nerved. Swamps and bogs: Lab.—Va.—Calif.—Alaska; Eurasia; S. Am.; Australia. *Temp.—Submont.—Subalp.* My–Au.

60. **C. brunnéscens** (Pers.) Poir. Stems slender, 2–5 dm. high; spikelets 4–8, 4–10-flowered, subglobose, 4–13 mm. in diameter; perigynia mostly smaller than in the preceding species. Wet places, banks and open woods: Lab.—N.C.—Colo.—Wash.—Alaska; Eur. *Boreal—Submont.—Subalp.* Je–Au.

61. **C. árcta** Boott. Stems slender, erect, 1.5–8 dm. high; leaf-blades 2–4 mm. wide; spikelets 5–15, many-flowered, aggregated into a head 1.5–3 cm. long; perigynia ascending or somewhat spreading, 2–3 mm. long, many-nerved. *C. canescens* var. *oregana* Bailey. Swamps and wet woods: N.B.—N.Y.—Minn.—Mont.—Calif.—B.C. *Boreal—Submont.* Je–Au.

15. Phyllostachyae: Stems densely cespitose, weak, compressed, dilated below the spikes, mostly much exceeded by the leaves. Spikes 2–4, androgynous, the rachis zigzag, dilated; one spike terminal, the others basal on often much elongated peduncles; staminate portion with small tight glumes; pistillate portion loosely one–several-flowered, the glumes elongated and conspicuous. Perigynia glabrous, 2-keeled, but otherwise nerveless, nearly round in cross-section, more or less beaked, the beak flattened-triangular, the orifice hyaline. Achenes triangular, stipitate, the sides convex, filling the perigynia, the apex rounded. Stigmas 3, short. Style jointed at the base with achene, soon withering.

62. C. Willdenòvii Schk. Stems up to 2.5 dm. high, somewhat serrulate; leaves rather stiff, 2–3 mm. wide, 1–4 dm. long, overtopping the inflorescence; spikelets 1–5, 2.5 cm. long or less, the uppermost on stalks 7–17 cm. long, the lower nearly basal; glumes acute to awned, several-nerved, the lower 1 or 2 commonly bract-like; perigynium body oblong, smooth, 2–3 mm. long, narrowed into a flattened 2-edged rough beak as long as the body. Dry woods and thickets: Mass.—Fla.—Tex.—Man. *E. Temp.* Ap–Jl.

63. C. Jamèsii Schw. Stem slender, 5–20 cm. high; leaf-blades 1–3.5 dm. long, 2–3 mm. wide, flat, deep green; spikelets 2 or 3, androgynous, one or two of them on slender stalks; lower glumes bract-like, foliaceous, the upper shorter, sometimes not exceeding the perigynia; perigynia 2 or 3, the body subglobose, 2 mm. thick, abruptly tipped by a rough beak longer than the body. Dry woods: Ont.—W.Va.—Kans.—Iowa. *Allegh.* Ap–My.

64. C. durifòlia Bailey. Stems up to 2.5 dm. long; leaf-blades green, not glaucous, 1.5–3 dm. long, 2.5–6 mm. wide; lower pistillate glumes as in the next; perigynia 5–6 mm. long, the body oval, the beak smooth, about as long as the body. *C. Backii* Boott. Woods: Me.—N.Y.—Man.—B.C. *Boreal.*

65. C. saximontàna Mackenzie. Stems up to 1.5 dm. long; leaf-blades glaucous, 1.5–3 dm. long, 3–5 mm. wide; lower pistillate glumes leaf-like, saccate; perigynia 4 mm. long, the body oblong-globose, the beak barely 1 mm. long, the margins weakly serrulate. "*C. Backii* Boott" Fl. Colo. Woods and thickets: Colo.—Wyo.—w Neb.—Alta. *Submont.* Jl.

16. Polytrichoideae: Densely tufted. Stems slender. Leaf-blades narrow. Spikelet solitary, linear, androgynous, bractless. Rachis straight, not dilated. Perigynia appressed, membranaceous, the upper part empty, oblong-elliptic, many-nerved (not 2-ribbed), compressed-triangular, beakless. Achenes triangular. Style slender, flexuous, its base not enlarged. Stigmas 3, short.

66. C. leptàlea Wahl. Rootstocks elongated; stems 2–6 dm. high, obscurely triangular; leaf-blades 0.5–1.25 mm. wide; spikelets 4–15 mm. long, 2–3 mm. wide; perigynia 1–10, 2.5–4.25 mm. long, not concealed by glumes. *C. polytrichoides* Willd. Bogs and wet meadows: Lab.—Fla.—Tex.—Colo.—Calif.—Alaska. *Temp.—Plain—Mont.* Je–Jl.

17. Obtusatae: Rootstocks long, creeping. Stems low, slender. Leaf-blades involute. Spike solitary, linear, androgynous, bractless. Perigynia appressed-ascending, ovoid, glabrous, many-nerved, coriaceous, shining, substipitate, obscurely triangular, the beak hyaline-tipped, obliquely cut, at length bidentate. Achenes triangular. Stigmas 3.

67. C. obtusàta Liljebl. Stems 6–20 cm. high, rough above; sheaths sparingly filamentose; leaf-blades 1–1.5 mm. wide; spikelets 5–12 mm. long, 3–6 mm. wide, the upper two thirds staminate; perigynia 1–6, impressed-nerved, 3–4 mm. long, exceeding the glumes. Dry hills and ridges: Sask.—S.D.—Colo.—B.C.; Eurasia. *Mont.—Subalp.* Je–Au.

18. NITIDAE: Stems slender, leafy at the base. Leaf-blades narrow. Terminal spike linear, staminate. Lateral spikelets 1–3, small, pistillate, subglobose to oblong, few–many-flowered, sessile or short-peduncled. Lowest bract squamiform, sheathless or sub-sheathing. Glumes reddish brown with hyaline margins. Perigynia ovoid, turgid-triangular, smooth, shining, coriaceous, round-tapering at the base, tipped with the short, cylindric, hyaline-tipped, obliquely cut beak. Achenes sessile, triangular, closely enveloped, short-apiculate. Style short, jointed with the apex of achene, thickish. Stigmas 3, long.

68. **C. supìna** Wahl. Long-stoloniferous; stems 6–25 cm. high; leaf-blades 1–1.5 mm. wide; pistillate spikelets 1–2, sessile, 4–10-flowered, 4–8 mm. long, subglobose or short-oblong; perigynia nerveless (except keels), 2.5–3.5 mm. long. Dry soil: Greenl.—Minn.—Sask.—Mack.; Eurasia. *Boreal—Arctic.* My–Jl.

19. RUPESTRES: Stoloniferous. Stems low. Leaf-blades narrow, flattened. Spikelets 1 and androgynous, or 2–4, with the terminal staminate and linear, the lateral pistillate, approximate or more or less separate, few-flowered, sessile or short-peduncled. Bracts of the pistillate spike leaflet-like, sheathless or nearly so. Pistillate glumes dark chestnut to purplish brown, with hyaline margins. Perigynia appressed-ascending, obovoid, glabrous, membranous, not polished, rounded and sessile at base, 2-ridged, obscurely triangular in cross-section, abruptly contracted into a hyaline-tipped, obliquely cut, straight beak. Achenes closely enveloped, triangular, with concave sides, apiculate. Style-base thickened. Stigmas 3.

69. **C. rupéstris** All. Stems 7–10 cm. high, sharply triangular, rough; sheaths not filamentose; leaf-blades 2–3 mm. wide; spikelets 1–1.5 cm. long, 3–6 mm. wide, the upper half staminate; perigynia 6–15, finely many-nerved, 4 mm. long, concealed by the glumes. Arctic-alpine situations: Greenl.—Que.—Colo.—Alta.—Alaska; Eurasia. *Alp.—Subalp.* Je–Au.

20. FILIFOLIAE: Densely cespitose. Leaf-blades filiform or narrow. Spikelet solitary, linear or linear-oblong, androgynous, densely flowered, bractless. Perigynia more or less triangular, nerveless except for the two lateral ribs, not stipitate, puberulent or pubescent, the hyaline-tipped beak obliquely cut. Achenes triangular. Stigmas 3.

70. **C. filifòlia** Nutt. Stems 8–30 cm. high; spikelets 1–2 cm. long, with 5–10 perigynia; pistillate glumes with broad bright white-hyaline margins; perigynia 3 mm. long, puberulent, abruptly and minutely stout-beaked. Plains and ridges: Sask.—Tex.—N.M.—Wash.—Yukon. *Plains—Mont.* Ap–Jl.

21. SCIRPINAE: Rootstocks creeping. Stems leafy below. Leaf-blades narrow. Spikelet usually 1, linear, staminate or pistillate, many-flowered, occasionally with an additional spike and normally with an empty scarcely sheathing squamiform bract a short distance below the spike. Perigynia triangular or flattened-triangular, membranous, 2-keeled, pubescent or puberulent, tapering at the base, contracted at apex into the short-cylindric entire or bidenticulate beak. Achenes triangular, with flat sides, sessile. Style slender, slightly enlarged at the base, not jointed. Stigmas 3, short.

71. **C. scirpoìdea** Michx. Rootstocks stoutish; stems 2–3.5 dm. high; leaf-blades 1–2 mm. wide; pistillate spikelets 1.5–3 cm. long, 2.5–5 mm. wide; glumes ovate, chocolate-brown, ciliate and puberulent; perigynia 3 mm. long, not blackish. Arctic-alpine, along streams: Greenl.—Man.—Colo.—N.Y.—Mich.—B.C.—Alaska; Eur. *Arctic—Alp.—Subalp.* Je–Au.

72. **C. scirpifórmis** Mackenzie. Rootstocks stoutish; stem 2.5–4.5 dm. high; leaf-blades 2–3 mm. wide; pistillate spikelets 2–4 cm. long, 4–5 mm.

wide; scales ovate, brownish, strongly pubescent and ciliate at the apex; perigynia 2.5 mm. long, not blackish. Mountains: Alta.—N.D. *Mont.* Jl.

22. MONTANAE: Stems slender, leafy at the base. Leaf-blades narrow, rough above. Terminal spikelet linear, normally staminate. Lateral spikelets 1–5, small, pistillate, or sometimes androgynous, subglobose to oblong, closely few–many-flowered, approximate and sessile or short-peduncled, or in some species radical and long-peduncled. Lowest bract squamiform or leaflet-like, sheathless or subsheathing. Glumes often reddish brown-tinged, acute to cuspidate. Perigynia membranaceous, ascending, the body pubescent, at least at the base of the beak, obovoid to elliptic, triangular, 2-keeled, strongly stipitate at the base, abruptly contracted into a cylindric or terete, emarginate to deeply bidentate beak, hyaline-tipped. Achenes normally triangular, the sides concave, closely enveloped, short-apiculate. Style short, jointed with apex of achene, thickish. Stigmas normally 3, long.

73. **C. helióphila** Mackenzie. Stems 12–25 cm. high; leaf-blades 1–2 mm. wide; lowest bract squamiform; pistillate spikelets 1–2 (rarely 3), subglobose, 5–15-flowered; perigynia 3.5 mm. long, 2 mm. wide, orbicular in cross-section, the beak deeply bidentate, in age strongly hyaline-tipped. Prairies and plains: Me.—Ill.—Mo.—N.M.—Alta. *Prairies—Plains—Submont.* Je–S.

74. **C. pennsylvánica** Lam. Stems slender, 1–3.5 dm. high; leaf-blades 1–3 mm. wide; staminate spikelet sessile or short-stalked, 1–2.5 cm. long; pistillate spikelets 1–4, short-oblong, sessile; glumes ovate, purplish, acute or cuspidate, fully equaling the perigynia; perigynia 4–20, broadly obovoid, about 2 mm. long, 1 mm. broad, short-pubescent, ribbed on two sides; beak one fourth as long as the body. Dry soil: N.B.—Tenn.—N.D. *Canad.—Allegh.* My–Je.

75. **C. Péckii** Howe. Loosely cespitose, the stems 1.25–6 dm. high; leaf-blades 3 mm. wide or less; lowest bract short; staminate spikelet sessile, 3 mm. long or less; pistillate spikelets 2–4, subglobose, 2–8-flowered, closely contiguous or the lower a little separate;perigynia oblong-obovoid, 3–4 mm. long, 1 mm. wide, grayish-pubescent, the beak one fourth the length of the body; glumes (except lower) half the length of the perigynia, reddish brown, with broad white-hyaline margins. *C. albicans* Auth.; not Willd. Open woods: Que.—N.J.—S.D.—B.C.—Alaska. *Boreal.*

76. **C. vária** Muhl. Stems filiform, rough above, 1–5 dm. high; leaf-blades 1–2.5 mm. wide; staminate spikelet 4–12 mm. long, sessile; pistillate spikelets 1–4, mostly close together and sessile, 3–7 mm. long; glumes ovate, green or purplish brown, acuminate or cuspidate, about as long as the perigynia; perigynia 4–12, oblong-ovoid, short-pubescent, about 2 mm. long; beak minutely 2-toothed, half as long as the body. Dry soil: N.S.—N.C.—Tex.—Neb.—Man. *E. Temp.* My–Jl.

77. **C. defléxa** Hornem. Loosely stoloniferous; stems very slender, 2–12 cm. high; leaf-blades 1–2 mm. wide; staminate spikelet inconspicuous, 2–4 mm. long, 0.5–1 mm. wide; pistillate spikelets subglobose, 2–8-flowered; perigynia 1 mm. wide, nearly orbicular in cross-section. Dry soil: Greenl.—N.Y.—Mich.—Man.—B.C.—Alaska. *Arct.—Boreal.* My–Au.

78. **C. Róssii** Boott. Rootstock stout; stems wiry, 5–25 cm. high; leaf-blades 1–2.5 mm. wide; staminate spikelet usually conspicuous, 3–10 mm. long, 1 mm. wide; pistillate spikelets globose to short-oblong, 2–12-flowered; perigynia up to 3.4 mm. long, 1.25 mm. wide, nearly orbicular in cross-section. Dry soil: Mich.—Man.—Colo.—Calif.—B.C. *Boreal.—Submont.—Subalp.* Ap–Au.

79. **C. umbellàta** Schkuhr. Densely cespitose; stems up to 1.5 dm. high, much exceeded by leaves; leaf-blades 1.5–3 mm. wide; non-basal pistillate spikelet usually present, oblong, globose; basal spikelet oblong; perigynia 2.25–3.25 mm. long, 1.25 mm. wide, rounded-triangular in cross-section, the beak obscurely

bidentate, less than half the length of the body; achenes light brown. *C. umbellata* var. *brevirostris* Boott. Dry sunny places: Alta.—B.C. *Boreal—Plain—Submont.* Ap–Jl.

80. **C. rugospèrma** Mackenzie. Stems 1–3 dm. high, slender, leaf-blades 1.5–2 mm. wide, 2 dm. long or less; staminate spikelet solitary, short-peduncled, 8–12 mm. long; pistillate spikelets 1 or 2, subsessile, at the base of the staminate one, globose-oblong, 4–7 mm. long; basal spikelets oblong, on long filiform peduncles, 4–10 cm. long; glumes ovate-lanceolate, acuminate or short-awned; perigynia 2–4, triangular, 3.25–4.25 mm. long, the body oblong, stipitate, finely pubescent; beak nearly as long. *C. umbellata* Am. auth. Dry soil: N.S.—Md.—Minn. *Canad.* My–Jl.

81. **C. tónsa** (Fernald) Bickn. Stems closely tufted, 2.5–10 cm. high, filiform; leaf-blades 2–4 mm. wide, stiff, spreading in age; staminate spikelet solitary, 1.5 cm. long or less; pistillate spikelets 1 or 2, sessile below the staminate one, and 1–3 pistillate peduncled spikelets from the base, 4–8 mm. long; glumes ovate-lanceolate, acuminate; perigynia 4 mm. long or more, the body oblong, stipitate; beak as long as the body. Dry soil: N.S.—D.C.—Ind. —Minn. *Canad.—Allegh.* My–Jl.

23. TRIQUETRAE: Cespitose, leafy toward the base. Leaves narrow, often pubescent. Terminal spikelet staminate, linear; lateral spikelets 2–4, pistillate, approximate or separate, sometimes also some basal pistillate ones. Bracts sheathing or sheathless, pistillate glumes greenish or hyaline or tinged with brown. Perigynia ascending, obovoid, short-pubescent, sharply triangular, stipitate, abruptly contracted into the short beak; achenes 3-angular, closely enveloped.

82. **C. hirtiflòra** Mackenzie. Cespitose stoloniferous plant, pubescent all over; leaves flat, soft, 3–7 mm. wide; lower bract 2.5–7.5 cm. long; staminate spikelet subsessile; pistillate spikelets 2–4, oblong-cylindric, erect, 6–20 mm. long, the upper sessile, the lower separate and stalked; glumes obovate, scarious-margined, rough-awned; perigynia sharply triangular, obovoid, stipitate, densely pubescent; beak about 4 mm. long. Woods and thickets: N.D.—N.J.—Kans. —N.D. *Canad.—Allegh.* My–Au.

24. DIGITATAE: Stems slender, leafy at the base. Leaf-blades narrow, the sheaths usually strongly purplish. Terminal spikelet linear, staminate. Lateral spikelets 1–5, approximate, or separated or sometimes radical, oblong to linear, 5–20-flowered in few rows, the peduncles included or exserted. Bracts sheathing, more or less strongly purplish-tinged, subspathaceous, the blade absent or rudimentary. Pistillate glumes strongly tinged with purplish or reddish brown. Perigynia membranaceous, appressed, oblong-obovoid, pubescent to glabrate, triangular, long-tapering into the stipitate base, abruptly contracted into the minute beak, the orifice entire or nearly so. Achenes triangular, closely enveloped. Style-base short, thickened, jointed with the apex of the achene. Stigmas 3, early deciduous.

83. **C. pedunculàta** Muhl. Densely matted, rather bright green; stems slender, diffuse, strongly purple-tinged at the base, 0.7–3 dm. long; leaf-blades 2–3 mm. wide; terminal spikelet long-stalked, usually pistillate at the base; lateral spikelets few-flowered, filiform-stalked, spreading or drooping, scattered, some appearing basal; perigynia sharply 3-angled, puberulent or in age glabrate, stipitate, tipped with a minute entire beak; glumes obovate, purplish, with green midrib, abruptly cuspidate. Dry woods: Newf.—S.D.—Va.—Iowa. *Canad.—Allegh.* My–Jl.

84. **C. concínna** R. Br. Strongly stoloniferous; stems 5–15 cm. high; leaf-blades 2–2.5 mm. wide; pistillate spikelets 2–3, rather closely 4–10-flowered; glumes broadly ovate, hyaline-margined; perigynia densely pubescent, 3 mm. long, nerveless. Dry soil: Newf.—Que.—Sask.—Colo.—B.C.—Alaska. *Mont.—Boreal.* Je–Jl.

85. C. Richardsònii R. Br. Strongly stoloniferous; stems 30 cm. high or less; leaf-blades 2–4 mm. wide; pistillate spikelets 2 or 3, approximate or the lowest remote, rather closely 8–20 flowered; glumes strongly hyaline-margined; perigynia 2.5–3 mm. long. Dry soil: Ont.—N.Y.—Ill.—S.D.—Alta. *Mont.—Boreal.* My–Je.

25. ALBAE: With long slender stolons. Stems very slender, obtusely triangular, leafy near the base. Leaf-blades of the fertile stem very narrow, involute. Terminal spikelet linear, staminate. Lateral spikelets 2–3, oblong or linear, loosely 4–8-flowered, all long-peduncled, linear or oblong, often overtopping the staminate one. Bracts sheathing, subspathaceous, bladeless, white-hyaline above. Glumes white-hyaline, yellowish-tinged. Perigynia erect, obovoid, triangular, glabrous, nerved, tapering at the base, abruptly minutely beaked, the orifice entire, hyaline. Achenes triangular, closely enveloped, the sides concave. Style-base very short, bulbous-thickened, jointed with the achene. Stigmas 3, long.

86. C. ebúrnea Boott. Stems 1–3.5 dm. high; leaf-blades 0.5 mm. wide or less; staminate spikelets sessile, 3–8 mm. long; perigynia membranaceous, obsoletely nerved, 2 mm. long, exceeding the scales. Dry sandy or rocky soil, especially limestone rocks: N.B.—Va.—Tenn.—Neb.—B.C.—Mack. *Boreal.—Plain—Mont.* My–Au.

26. BICOLORES: Stoloniferous. Stems central, slender, leafy towards the base. Leaf-blades narrow. Basal sheaths light brown. Terminal spikelet linear, staminate or occasionally gynaecandrous. Lateral spikelets 2–5, pistillate, rather closely few–many-flowered in few ranks on erect exserted peduncles. Bracts sheathing, not colored or dark-auricled, the blades elongated, leaf-like. Glumes reddish or purplish brown-tinged. Perigynia ascending, broadly oval or ellipsoid, circular in cross-section, nerved, glabrous, golden-yellow or white-pulverulent at maturity, tapering or rounded at the base, essentially beakless. Achenes lenticular, apiculate, closely enveloped. Style-base slender, short, jointed with the achene. Stigmas 2.

87. C. aùrea Nutt. *Fig. 100.* Stems 0.3–4 dm. high; leaf-blades 2–4 mm. wide; pistillate spikelets 4–20-flowered; glumes usually reddish brown-tinged; perigynia subumbonate, 2 mm. wide; style not exserted or persistent. Wet places: Newf.—Pa.—Neb.—N.M.—Calif.—B.C. *Boreal—Plain—Subalp.* Je–Jl.

27. PANICEAE: Stoloniferous. Stems central, slender, leafy towards the base. Basal sheaths brownish- or purplish-tinged. Terminal spikelets staminate, linear or linear-oblong. Lateral spikelets 1–5, pistillate, loosely to rather closely several–many-flowered, in few or several ranks, on erect, exserted or included peduncles. Bracts sheating, not colored or dark-auricled, the blades developed. Glumes purplish or reddish brown-tinged. Perigynia ascending or spreading, ovoid or obovoid, membranaceous, obtusely triangular, slightly inflated, glabrous, puncticulate, light or olive-green, pointed or beaked, the orifice entire or nearly so. Achenes triangular. Style-base slender, continuous with achene. Stigmas 3.

88. C. lívida (Wahl.) Willd. Stems 1.5–5 dm. high; leaf-blades 2 mm. wide or less; pistillate spikes closely 5–15-flowered; perigynia 3.75 mm. long, 1.75 mm. wide, faintly nerved, narrowed above, exceeding the glumes. Sphagnum bogs: Lab.—N.J.—Mich.—Alta.—Calif.—B.C.—Alaska; Eur. *Boreal.* My–Jl.

89. **C. coloràta** Mackenzie. Stems slender, rough above, 3–6 dm. high; leaves flat, 3–4 mm. wide; staminate spikelet clavate, more or less peduncled; pistillate spikelets 2–3, erect, distant, linear, 1.5–3.5 cm. long, 4 mm. thick, peduncled; glumes ovate, obtuse to cuspidate, shorter than the perigynia; perigynia 6–15, narrowly obovoid, 3.5–4 mm. long; beak 0.5 mm. long, curved. Woodlands: N.Y.—Pa.—Mo.—Man. *Canad.—Allegh.*

90. **C. tetánica** Schk. Rootstocks elongate, deep-seated, slender; stems 1.5–6 dm. high, slender; stem-leaves usually 3–5, 2–3.5 mm. wide; staminate spikelet long-stalked; pistillate spikelets 1–3, linear, 6–20-flowered, distant, erect or the lower filiform-stalked and drooping; perigynia 3 mm. long, 2 mm. wide, obtusely triangular, prominently many-nerved, abruptly minutely beaked; glumes usually shorter than the perigynia, obtuse or mucronate. Meadows and wet woods: Mass.—D.C.—Kans.—Sask. *Canad.—Allegh.*

91. **C. Meàdii** Dewey. Resembling the last, but stouter; stems 2–5 dm. high; stem-leaves usually 6–10, 3–7 mm. wide; staminate spikelets long-stalked; pistillate spikelets 1–3, oblong, densely 8–30-flowered, distant, erect, the lower long-stalked; perigynia more than 3 mm. long, 2 mm. wide, obtusely triangular, prominently many-nerved, tipped with a minute slightly bent beak; glumes obtuse to cuspidate. Meadows and prairies: N.J.—Ga.—Tex.—Neb.—Sask. *Plain.*

92. **C. vaginàta** Tausch. Stems 1.5–8 dm. high, weak; leaf-blades 2–5 mm. wide; pistillate spikelets loosely 3–20-flowered; perigynia 4 mm. long, nearly 2 mm. wide, faintly nerved, the orifice purplish-tinged, 2-toothed; glumes purplish-tinged, exceeded by the perigynia. Boggy woods: Lab.—N.Y.—Minn. —B.C.—Alaska; Eurasia. *Mont.—Boreal.* Je–Au.

28. LAXIFLORAE: Cespitose. Fertile stems mostly lateral, the sterile shoots leafy, conspicuous. Basal sheaths brownish- or purplish-tinged. Terminal spikelet staminate, linear. Lateral spikelets 2–5, pistillate or androgynous, loosely to closely few–many-flowered, in few–several ranks, on erect to drooping, included or exserted peduncles. Bracts sheathing, the sheaths green or purplish-tinged, the blades leaf-like or sometimes reduced. Glumes green with hyaline margins or more or less colored. Perigynia ascending, membranaceous, triangular, usually nerved, closely enveloping the achene, glabrous or hispidulous, tapering at the base, short-beaked or beakless, the orifice entire. Achenes triangular. Style-base slender, continuous with the achene. Stigmas 3.

93. **C. plantagínea** Lam. Stems slender, 1.5–6 dm. high, leaf-blades of the sterile shoots 1–2.5 cm. broad, remaining over winter, those of the fertile stems rudimentary; staminate spikelets long-stalked, purple; pistillate spikelets 3 or 4, stalked; glumes broadly deltoid, white-hyaline with a green midrib; perigynia oblong-elliptic, short-beaked, 3–4 mm. long, longer than the glumes. Woods: N.B.—N.C.—Ill.—Man. *Canad.—Allegh.* Ap–Je.

94. **C. digitàlis** Willd. Stems slender, 1–4.5 dm. high, smooth; leaf-blades 2–5 mm. wide, flat; bracts leaf-like; staminate spikelet stalked, linear; pistillate spikelets 2–4, linear, 1–2.5 cm. long, the upper subsessile, the lower slender-stalked, drooping; glumes lanceolate, scarious-margined, acute to short-awned; perigynia 3–12, ovoid-oblong, sharply 3-angular, brown when ripe, 3 mm. long. Woods and thickets: Me.—Fla.—Tex.—Minn. *E. Temp.* My–Jl.

95. **C. laxicúlmis** Schw. Stems ascending or diffuse, 1.5–6 dm. long; leaves of the sterile shoots elongate, 4–12 mm. long, those of the stem shorter; staminate spikelet linear, long-stalked; pistillate spikelets 2–4, oblong, 6–12 mm. long, drooping on slender peduncles; glumes ovate, green, cuspidate; perigynia 5–10, broadly ovoid-oblong, sharply 3-angular, 3–4 mm. long, longer than the glumes. *C. copulata* (Bailey) Mackenzie. Woods and thickets: Me.—Va.—Mo. —Iowa. *Canad.—Allegh.* Me–Je.

96. **C. albursìna** Sheldon. Stems nearly smooth, strongly flattened and winged, 2–6 dm. high; basal leaves 0.7–3 cm. wide; stem-leaves narrower; bracts similar to the latter; staminate spikelet sessile, linear; pistillate spikelets 2–4, linear, distant, more or less stalked, 1–3 cm. long, laxly flowered; glumes broadly oblong, scarious-margined, shorter than the perigynia; perigynia obovoid, obtusely 3-angled, 3–4 mm. long. Woods: Que.—Va.—Mo.—Minn. *Canad.—Allegh.* Jl–Au.

97. **C. blánda** Dewey. Stems 1.5–6 dm. high, pale green, brownish at base; sterile shoots developing conspicuous culms; leaf-blades 3–14 mm. wide, the sheaths with much crisped margins; staminate spikelets short-stalked or sessile, the glumes rarely reddish brown-tinged; pistillate spikelets 2–4, oblong or linear-oblong, rather closely 8–25-flowered, the upper two contiguous, sessile or nearly so, the lower distant and stalked; perigynia obovoid, 2.5–3 mm. long, strongly nerved, contracted into a short, stout, outwardly bent beak; glumes with broad white scarious margins, the lower strongly awned. Me.—Va.—Tex. —N.D. *E. Temp.—Plain.* My–Jl.

98. **C. laxiflòra** Lam. Stems erect or reclining, slender, 1.5–6 dm. long; leaf-blades 3–7 mm. wide, soft; bracts leaf-like; staminate spikelet usually stalked, its glumes rarely red-tinged; pistillate spikelets 2–4, distant, linear-cylindric, loose, 1–3 cm. long, 3–4 mm. thick; glumes ovate, scarious-margined, usually shorter than the perigynia, at least the lowest awned; perigynia obovoid, oblique, 2.5–3 mm. long; beak short, stout, bent. Meadows and thickets: Que. —Fla.—Tex.—Minn. *E. Temp.* My–Jl.

99. **C. ormostàchya** Wieg. Stems angled, 1–6 dm. high; leaves narrow, 2.5–8 mm. wide; bracts leaf-like; staminate spikelet peduncled, the glumes often purplish; pistillate spikelets 12–25 mm. long, lax, moniliform; glumes broad, subacute; perigynia 2.5–3.5 mm. long, plump. Woods and banks: Que. —N.Y.—Minn. *Canad.* My–Jl.

100. **C. leptomèria** Fernald. Stems slender, 2–6 dm. high, retrorsely scabrous; basal leaves broad, 5–10 mm. wide; cauline leaves narrowed, 2.5–6 mm. wide; bracts leaf-like; staminate spikelet small, 1–2.5 cm. long, sessile; pistillate spikelets peduncled; glumes obovate, usually mucronate; perigynia ellipsoid, 3–4 mm. long, substipitate, fragile. Low woods: Labr.—N.C.—Tenn. —Minn.—Man. *Canad.*

29. Granulares: Stems central, slender, leafy. Leaf-blades flat. Basal sheaths brownish. Terminal spikelet staminate, linear. Lateral spikelets 2–4, pistillate, closely many-flowered in several ranks, erect, on more or less strongly exserted peduncles. Bracts sheathing, not colored or dark-auricled, the blades leaf-like. Glumes ovate, pointed, slightly ferruginous. Perigynia ascending, elliptic to ovoid, membranaceous, glabrous, with many elevated nerves, suborbicular in cross-section, loosely enveloping the achenes, rounded at the base, short-tapering and minutely beaked, the orifice entire or subemarginate. Achenes triangular, strongly apiculate, jointed with the slender style. Stigmas 3, short.

101. **C. Shrìveri** Britton. Stems 1.5–7 dm. high; leaf-blades 4–16 mm. wide, somewhat glaucous; staminate spikelet short-stalked; bracts overtopping the stems, the ligules little elongated; perigynia narrowly obovoid, 2.5 mm. long, 1.5 mm. wide, the beak minute, entire; apiculation of the achene abruptly bent. Moist meadows: Me.—Va.—Kans.—Sask. *Boreal.* Je–Jl.

102. **C. granulàris** Muhl. Stems slender, smooth, 1.5–5 dm. high; leaf-blades flat, roughish, 3–9 mm. wide, bracts leaf-like; staminate spikelet sessile or short-stalked; pistillate spikelets 2–5, distant or the upper two contiguous, narrowly oblong, 1–3 cm. long, dense, the lowest long-peduncled; glumes narrowly ovate, thin, shorter than the perigynia; perigynia 10–50, ovoid or obovoid,

terete, swollen, strongly nerved, 2 mm. long; beak short, subentire. Moist meadows: N.B.—Fla.—La.—Man. *E. Temp.* My–Jl.

103. **C. Cràwei** Dewey. Stems from long creeping rootstocks, stiff, glabrous or nearly so, 0.7–4 dm. high; leaf-blades 2–4 mm. wide, rather stiff; bracts similar, short, rarely overtopping the spikelets; staminate spikelets long-stalked; pistillate spikelets 1–4, distant, oblong, 10–45-flowered, sessile or stalked; perigynia ovoid, ascending, obscurely many-nerved, 3–3.5 mm. long, suborbicular in cross-section, rounded at the base, tapering into a very short beak; glumes obovate, obtuse to cuspidate. Meadows and banks, limestone regions: N.S.—N.J.—Kans.—Wyo.—Alta. *Temp.* My–Jl.

30. OLIGOCARPAE: Stems slender, leafy, the leaf-blades flat; terminal spikelet staminate, linear; lateral spikelets 2–4, pistillate, erect on more or less exserted peduncles. Bracts sheathing, the blade leaf-like; glumes ovate, pointed. Perigynia ascending, oblong, elliptic or obovoid, glabrous, subcoriaceous, obscurely 3-angled in cross-section, tapering at the base, the mouth entire.

104. **C. oligocárpa** Schk. Stems slender, 1.5–5 dm. high, rough; leaf-blades 2.5–3.5 mm. wide, spreading; bracts leaf-like; staminate spikelet linear, stalked or subsessile; pistillate spikelets 2–4, erect, distant, 8–25 mm. long, 4 mm. thick, the upper sessile, the lower slender-stalked; glumes ovate, cuspidate, longer than the perigynia; perigynia obovoid, many-striate, 3.5–4 mm. long; beak short, straight or oblique. Dry woods: Vt.—D.C.—Okla.—Kans. *Canad.—Allegh.* My–Jl.

31. GRISEAE: Stems slender, leafy, the basal sheaths brownish or purplish; leaf-blades flat. Terminal spikelet staminate, linear; lateral spikelets 2–4, pistillate, many-flowered, erect, peduncled. Bracts leaf-like, sheathing. Glumes ovate, pointed. Perigynia ascending, oblong, glabrous, finely striate, rounded at each end, terete, beakless or nearly so, the mouth entire or emarginate. Achenes 3-angular, apiculate, jointed to the style.

105. **C. conoìdea** Schk. Stems rough, 1.5–7.5 dm. high; leaf-blades 2–4 mm. wide, staminate spikelet long-stalked; pistillate spikelets 1–3, distant, erect, oblong, 3–25 mm. long, peduncled, glumes ovate, scarious-margined, acuminate or awned, the lower longer than the perigynia; perigynia oval, terete, 3 mm. long. Meadows: N.S.—N.C.—Iowa—Minn. *Canad.—Allegh.* My–Je.

106. **C. grísea** Wahl. Stems glabrous, stout, smooth, 3–7 dm. high; leaf-blades flat, soft, spreading, 4–7 mm. wide; bracts similar; staminate spikelet subsessile, pistillate spikelets 3–5, dense, oblong, 8–25 mm. long, the upper sessile, the lower slender-stalked and distant; glumes ovate, scarious-margined, cuspidate or awned; perigynia oblong, 3 mm. long, terete. Woods and thickets: Me.—N.C.—Tex.—Neb. *Canad.—Allegh.* My–Jl.

32. GRACILLIMAE: Stems slender, leafy, the leaves flat, often hairy. Bracts leaf-like. Terminal spikelet gynaecandrous, rarely staminate; lateral spikelets 2–4, elongate, linear, at least the lower long-peduncled and drooping. Glumes white-hyaline with green midrib. Perigynia oblong or ovoid, 2-keeled, beakless or short-beaked, the mouth truncate or emarginate. Achenes 3-angular with concave sides, apiculate.

107. **C. gracíllima** Schw. Stems 3–10 dm. high, smooth; leaf-blades 3–9 mm. wide, dark-green; spikelets 3–5, narrowly cylindric, 2.5–6 cm. long, 3 mm. thick, slender-stalked, nodding; glumes thin, ovate-oblong, obtuse or cuspidate, scarious-margined; perigynia ovoid-oblong, glabrous, 3 mm. long, rounded at the apex, beakless. Moist woods and meadows: Newf.—N.C.—Ill.—Minn.—Man. *Canad.—Allegh.* My–Jl.

108. C. Davísii Schw. & Torr. Stem 4–10 dm. high; leaves 3–6 mm. wide, flat, pubescent; spikelets 3–5, linear-oblong or the terminal one clavate, dense, 3–4 cm. long, 6 mm. thick or less, slender-stalked, spreading or drooping; glumes ovate-lanceolate, long-awned, spreading; perigynia oblong-ovoid, swollen, glabrous, 4–5 mm. long; beak very short, minutely 2-toothed. Thickets and meadows: Mass.—Ga.—Tex.—Minn. *Allegh.—Ozark.* My–Jl.

33. DEBILES: Stems aphyllopodic, strongly purplish-tinged at base, tufted, slender, leafy. Leaf-blades flat. Terminal spikelet normally staminate. Lateral spikelets 2–5, elongated, narrowly linear, slender-peduncled, the lower drooping. Bracts green-sheathing, the blades leaf-like. Perigynia appressed or ascending, lanceolate to ovoid, membranaceous, obsoletely nerved, rather closely enveloping the achene, tapering to a well-developed conic beak, obliquely cut at orifice and strongly hyaline-tipped, at length bidentate. Achenes triangular, apiculate, joined with the slender style. Stigmas 3.

109. C. castànea Wahl. Stems slender, 3–10 dm. high, rough above, purple at the base; leaf-blades 2.5–6 mm. wide, pubescent; bracts linear-filiform; staminate spikelet short-stalked; pistillate spikelets 1–4, approximate, oblong or oblong-cylindric, many-flowered, 1–2.5 cm. long, 6 mm. thick, slender-stalked, drooping or the upper spreading; glumes chestnut, thin, ovate or lanceolate, acute or cuspidate, shorter than the perigynia; perigynia glabrous, pale brown, oblong-lanceolate, slightly inflated, 3-angular, tapering into a minutely toothed beak half as long as the body. Thickets and banks: Newf.—N.Y.—Minn. *Canad.* Je–Jl.

110. C. arctàta Boott. Stems glabrous, slender, 3–7 dm. high, rough above; leaves flat, rough-margined, 5–10 mm. wide, staminate spikelet short-stalked, linear; pistillate spikelets 2–5, linear, 2.5–7 cm. long, 3–4 mm. thick, loose; glumes ovate, cuspidate or short-awned, shorter than the perigynia; perigynia lanceolate, stipitate, deep-green, 2.5–4.5 mm. long, 3-angular; beak short, hyaline-tipped, 2-toothed. Woods and thickets: Newf.—Pa.—Mich.—Minn. *Canad.* My–Je.

111. C. assiniboinénsis W. Boott. Stems 2.5–7.5 dm. high; leaf-blades 1–2 mm. wide; staminate spikelet long-stalked; pistillate spikelets loosely 1–8-flowered; perigynia narrowly lanceolate, 6 mm. long, 2 mm. wide; glumes lanceolate, scarious-margined, awned. Wet soil: Man.—Minn.—S.D.—Sask. *Plain.* Je.

34. CAPILLARES: Stems phyllopodic, light brown and leafy at base, tufted, very slender. Blades narrow. Terminal spikelet staminate. Lateral spikelets 2–4, linear, drooping on long capillary peduncles, few-flowered in few ranks. Bracts green-sheathing, the blades developed. Perigynia appressed, ovoid, membranaceous, triangular, closely enveloping the achene, tapering into the slender conic beak, truncate and entire at orifice. Achene triangular, short-apiculate, jointed with the slender style. Stigmas 3.

112. C. capillàris L. Stems 0.5–6 dm. high, obtusely triangular, smooth; leaf-blades flat, 1–2 mm. wide; pistillate spikelets 2–12-flowered; glumes broadly hyaline-margined; perigynia 2.5–3 mm. long, the beak straight. Greenl.—N.Y.—Minn.—N.M.—Nev.—B.C.—Alaska; Eurasia. *Arct.—Alp.—Subalp.* Jl–Au.

35. LONGIROSTRES: Stems phyllopodic, tufted, leafy towards the base. Blades flat. Terminal spikelet staminate. Lateral spikelets 2–5, linear-oblong or oblong-cylindric, peduncled, many-flowered in several ranks. Bracts green, sheathing, the blades developed. Perigynia appressed to spreading, membranaceous, globose-triangular, somewhat inflated, prominently beaked, the apex bidentate, the teeth weak, scarious. Achenes triangular, apiculate, the apiculation very abruptly bent, jointed with the slender style. Stigmas 3.

113. **C. Sprengélii** Dewey. *Fig. 101.* Stems 2.5–9 dm. high, roughish above, strongly fibrillose at the base; leaf-blades 2.5–4 mm. wide; pistillate spikelets 10–40-flowered; glumes acute to cuspidate; body of the perigynia 2.5–3 mm. long, short-oblong, exceeded by the slender beak. *C. longirostris* Torr. Alluvial banks: N.B.—N.J.—Colo.—Alta. *Plain—Submont.* My–Jl.

36. Frigidae: Stems phyllopodic, tufted, the leaves clustered near the base. Spikelets staminate, pistillate, androgynous or gynaecandrous. Bracts green-sheathing, the blades developed or rudimentary. Glumes dark-tinged, usually with light midvein and margins. Perigynia appressed or ascending, flat to flattened-triangular, not inflated, dark-tinged, beaked, the beak hyaline at orifice, more or less bidentate. Achenes triangular, short-apiculate, pointed with the straight slender style. Stigmas 3.

114. **C. misándra** R. Br. Stems 3–40 cm. high; leaf-blades 2–3 mm. wide, much shorter than the stems; sheaths purplish-tinged, the blades short; lateral spikelets 1–3, filiform-stalked, drooping, 6–14 mm. long, 4 mm. wide, closely 5–20-flowered; glumes blackish, with narrow hyaline margins; perigynia lanceolate, narrowed at both ends, serrulate above. High summits or arctic regions: Greenl.—Ont.—Alaska; Colo.; arctic Eurasia. *Arct.—Alp.* Jl–S.

115. **C. atrofúsca** Schk. Stem obtusely 3-angular, slender, 1–3 dm. high; leaf-blades 2–3 mm. wide, mostly clusterd at the base, 2.5–6 cm. long; bracts sheathing, dark; terminal spikelet slender-stalked, staminate or gynaecandrous; pistillate spikelets 2 or 3, approximate, slender-stalked, drooping, 8–18 mm long, 7–9 mm. thick; glumes oblong-ovate, obtuse or acute, black, with lighter midrib; perigynia 15–30, ovate-oval, blackish, 5 mm. long, very short-beaked. Arctic situations: Greenl.—Lab.—Man.—Alaska. *Arctic.*

37. Pallescentes: Stems aphyllopodic, tufted, leafy towards the base. Leaves pubescent. Terminal spikelet staminate, linear. Lateral spikelets pistillate, closely 10–30-flowered in several rows. Bracts leaf-like, sheathless or nearly so. Perigynia green, rounded-triangular in cross-section, enveloping the achene, many-nerved, glabrous, rounded at the base, beakless or short-beaked. Achenes triangular, apiculate, the style thick, very short. Stigmas 3.

116. **C. abbreviàta** Prescott. Stems 2.5–5 dm. high; leaf-blades 1.5–3 mm. wide; pistillate spikelets 1–3, short-oblong, 6–16 mm. long, 6 mm. wide, sessile or short-stalked, approximate; perigynia obovoid, 2.5–3 mm. long, strongly nerved, depressed at the apex, tipped by a short slender entire beak. *C. Torreyi* Tuckerm. Dry soil: Man.—Minn.—Colo.—Sask. *Plains—Submont.* Je–Jl.

38. Virescentes: Stems aphyllopodic, tufted. Leaves pubescent at least on the sheaths. Terminal spikelet gynaecandrous; lateral spikelets pistillate, closely many-flowered. Bracts sheathless or nearly so, leaf-like; glumes cuspidate. Perigynia green, round-triangular in cross-section, several-nerved, glabrous or hairy, the mouth entire or nearly so. Achenes 3-angular, apiculate, with a short thick style.

117. **C. Búshii** Mackenzie. Stem slender, rough above, 4–7 dm. high; leaf-blades 2–3 mm. wide, glabrous or pubescent beneath, the sheaths pubescent; bracts leaf-like; spikelets 2 or 3, oblong, dense, sessile, 6–20 mm. long, 5–8 mm. thick, erect, clustered at the summit; glumes lanceolate, the lower rough-cuspi-

date, longer than the perigynia, obovoid, terete, 2.5–3 mm. long, swollen, coarsely ribbed, the mouth entire or emarginate. Meadows: Mass.—D.C.—Tex. —Kans. *Allegh.—Ozark.*

39. LIMOSAE: Stems slender, leafy below. Leaf-blades narrow. Terminal spikelet staminate, linear. Lateral spikelets 1–4, pistillate, distant, oblong, closely several–many-flowered in several ranks, drooping, on slender peduncles. Bracts leaflet-like, essentially sheathless (in our species). Perigynia appressed, coriaceous, ovoid or elliptic, glaucous, papillose, beakless or nearly so. Achenes triangular, closely enveloped, the style short, straight, exserted, its base not enlarged. Stigmas 3.

118. **C. rariflòra** (Wahl.) Smith. Stems slender, erect, 1–3.5 dm. high, smooth, from an elongated rootstock; leaf-blades 1.5–2.5 mm. wide, flat; bracts short, purple at the base; staminate spikelet long-stalked; pistillate spikelets 1–3, narrowly oblong, 6–16 mm. long, 4–5 mm. thick, nodding or ascending, slender-stalked; glumes broadly oval, purple-brown, with green midvein, half enveloping the perigynia; perigynia 3–18, slightly inflated, 3.5 mm. long, slightly 2-edged, beakless, entire at the mouth. Wet places: Greenl.—Me. —Man.—Alaska; Eurasia. *Arctic.*

119. **C. limòsa** L. Stems 1.5–6 dm. high, sharply triangular; pistillate spikelets 1–2.5 cm. long, 5–8 mm. thick, 8–30-flowered; glumes acute or short-cuspidate; perigynia broadly ovate, flattened, 2-edged, 2.5 mm. long, 2 mm. wide, several-nerved. Sphagnum bogs: Lab.—N.J.—Iowa—Mont.—Wash.—Alaska; Eurasia. *Boreal—Mont.* My–Au.

120. **C. paupércula** Michx. Stems 1–8 dm. high, sharply triangular; pistillate spikelets 0.5–2 cm. long, 4–8 mm. thick, 6–25-flowered; glumes long-acuminate or awned; perigynia subglobose, 2.5–3 mm. long, 2.25 mm. wide, few-nerved. *C. magellanica* Am. auth. Sphagnum bogs: Labr.—Pa.—Minn.—Colo.—Utah—Alaska; Eurasia. *Boreal—Submont.—Mont.* Au.

40. ATRATAE: Stems leafy below. Terminal spikelet staminate or gynaecandrous, the lateral 1–10, normally pistillate, from sessile, erect, and closely approximate, to long-peduncled, nodding, and distant. Bracts sheathless or nearly so, dark-colored at the base, the blades short. Scales dark-tinged. Perigynia membranaceous, straw-colored, often dark-tinged, elliptic to broadly obovate, circular in cross-section to much flattened, papillose to puncticulate, glabrous, abruptly short-beaked or beakless, the orifice entire or bidentate. Achenes triangular, the style slender, straight, often exserted, its base not enlarged. Stigmas 3.

121. **C. Hállii** Olney. Stoloniferous, the stems 1–3.5 cm. high, fibrillose at the base; leaf-blades 2.5–4 mm. wide; spikelets 1–5, often all pistillate, or the terminal staminate below or throughout, the lateral oblong or linear-oblong, erect, approximate; glumes obtuse to mucronate, concealing the perigynia; perigynia 2.5 mm. long, 2 mm. wide, flattened and sharp-edged, lightly few-nerved, the beak minute, bidenticulate. *C. Elrodi* M. E. Jones. *C. Parryana* Bailey (Fl. Rocky Mts.), not Dewey. Mountain meadows: Man.—N.D.—Colo.—Alta. *Plain—Mont.* My–Jl.

122. **C. Hálleri** Gunn. Stems 1.5–6 dm. high, slender; leaf-blades 1–3 mm. wide; spikelets 2–4, clustered, erect, sessile or short-peduncled, 4–10 mm. long, closely 8–25-flowered; scales black, shorter than the perigynia; perigynia minutely bidentate. *C. alpina* Sw. Rocky places: Greenl.—Que.—Minn.—S.D. —N.M.—Alaska; Eurasia. *Alp.—Mont.* Jl–Au.

123. **C. Parryàna** Dewey. Stem phyllopodic, 1–5 dm. high, leafless above; leaf-blades 2–4 mm. wide, revolute-margined; spikelets 1–4, dense, linear-

cylindric, 6–30 mm. long, 3 mm. thick, the uppermost sessile, staminate or gynaecandrous, the lowest short-stalked; glumes ovate, obtuse or mucronate, dark brown with lighter margins; perigynia broadly obovate, pale, 2.5 mm. long, minute papillose, minutely beaked, emarginate. Valleys: Minn.—Alta. *Boreal.* Je–Au.

124. **C. atratifórmis** Britton. Stem 3–9 dm. high, slender, often nodding, roughened above, slightly fibrillose at the base; leaf-blades 3 mm. wide; spikelets 3–4, oblong, 1–2.5 cm. long, 4–6 mm. wide, closely 10–30-flowered; glumes dark reddish brown; perigynia 2.5–3 mm. long, 1.5 mm. wide, rounded at the base, round-tapering at the apex. Along streams: Lab.—Me.—Alta.—Yukon. *Mont.—Boreal.* Je–Au.

125. **C. Buxbaùmii** Wahl. Densely cespitose; stems 2–9 dm. high, slender but stiff, strongly reddish-purple at base; leaf-blades 2–4 mm. wide; spikelets 8–40 mm. long, 8 mm. wide, the perigynia numerous, 3–4 mm. long, glaucous-green, lightly many-nerved, the apex minutely bidentate; glumes awned, exceeding the perigynia. *C. Holmiana* Mackenzie. Bogs: Greenl.—Ga.—Ark.—Colo.—Calif.—Alaska; Eurasia. *Temp.—Mont.—Submont.* My–Jl.

41. SHORTIANAE: Stems leafy. Leaves glabrous, flat. Spikelets 4 or 5, the terminal staminate or gynaecandrous, the lateral ones pistillate or gynaecandrous, with few staminate flowers, cylindric, more or less peduncled, many-flowered. Bracts sheathless or nearly so, the lower leaf-like. Pistillate glumes 3-nerved, mucronate or awned. Perigynia obovoid or ovoid, compressed-triangular to subglobose, subsessile or short-stipitate, abruptly short-beaked, the orifice entire or nearly so; achenes loosely enveloped, triangular with concave sides.

126. **C. Shortiàna** Dewey. Clumps medium or larger, with short rootstocks; stems 3–9 dm. high, stout, sharply 3-angular, phyllopodic, brown at the base; leaf-blades flat, 4–8 mm. wide, rough-edged; terminal spike erect, subsessile, the upper half to three fourths pistillate, 2–3.5 cm. long; lateral spikes 4 or 5, the lower distant, slender-peduncled, 1.3–5 cm. long; glumes ovate, mucronate to acuminate, reddish-brown, with 3-nerved green center; perigynia 20–60, spreading, 2.5–3 mm. long, olive-green, transversely corrugated; beak minute, bent. Moist woods: Pa.—Va.—Okla.—Iowa. *Allegh.—Ozark.* My–Jl.

42. ACUTAE: Stems leafy below, aphyllopodic or phyllopodic; terminal 1–several spikes staminate (rarely gynaecandrous), linear, the remaining pistillate, linear to cylindric or oblong, closely many-flowered, sessile or peduncled. Bracts sheathless (rarely short-sheathing), leafy or squamiform, biauriculate and often darkened at the base. Perigynia membranaceous to coriaceous, plano- or bi-convex, or turgid, elliptic to obovate, puncticulate, margined, beakless or abruptly minutely beaked, the orifice entire to deeply bidentate. Achenes normally lenticular. Style slender, straight, sometimes exserted, its base not enlarged. Stigmas normally 2.

127. **C. strícta** Lam. Stems aphyllopodic, in dense clumps, sharply 3-angular, 3–12 dm. high; leaves rough-margined, 2–4 mm. wide; lower bracts leaf-like; staminate spikelets solitary, rarely 2, peduncled; pistillate spikelets 2–5, linear-cylindric, often staminate at the apex, 1–10 cm. long, 2–4 mm. thick, sessile, ascending; glumes dark, with green margins and midrib, oblong or lanceolate, obtuse to acuminate, perigynia ovate-elliptic, 2.5 mm. long, minutely beaked, entire at the mouth. Swamps: Newf.—N.C.—Ill.—Minn. *Canad.—Allegh.* Jl–S.

128. **C. Hàydeni** Dewey. Stems slender, up to 6 dm. high; leaves 2–3 mm. wide, rough-margined; staminate spikelet sessile; pistillate spikelets 2–4, linear, 1–3 cm. long, 4 mm. thick, subsessile, ascending, sometimes with stami-

nate flowers at the top; glumes lanceolate, purplish, spreading, acute, exceeding the perigynia; perigynia suborbicular, obtuse, 1.5 mm. broad; beak minute, entire. Swamps: N.B.—N.J.—N.C.—Minn. *Canad.—Allegh.* Je–Jl.

129. **C. stríctior** Dewey. Stems 3.5–9 dm. high, sharply 3-angular, purplish at the base; leaf-blades flat, revolute-margined, 2.5–3.5 mm. wide, papillate; terminal spikelet peduncled, 2–3 cm. long; pistillate spikelets mostly 3, the uppermost approximate, sessile, often staminate at the apex, the lowest distant and peduncled, 1.5–6 cm. long, 4.5 mm. wide; glumes lanceolate, obtuse or acute, narrower than the perigynia, purplish-black or in age reddish-brown; perigynia 50–150, appressed, 2.25–2.75 mm. long, ovate, red-dotted, 2-edged; beak minute, subentire. Swamps: N.S.—N.C.—Iowa—Minn. *Canad.—Allegh.* Je–Jl.

130. **C. Emòryi** Dewey. Cespitose, long-stoloniferous; stems 5–10 dm. high, sharply triangular, rarely reddened at the base; leaf-blades green, 3–7 mm. wide; staminate spikelets 1–3, slender; pistillate spikelets 2–4, sessile or short-peduncled, more or less strongly separate, narrowly linear, 4–10 cm. long, 3–5 mm. wide; glumes lanceolate, sharp-pointed, with broad light-colored center and hyaline margins, narrower than the perigynia; perigynia very numerous, oval or ovate, 2.5 mm. long, rounded at the base, sessile or substipitate, few-nerved, the marginal nerves not prominent, puncticulate, abruptly minutely but noticeably beaked. Swales and river banks: N.J.—D.C.—Tex.—N.M.—Colo.—N.D.—Man. *Plain—Submont.*

131. **C. tórta** Boott. Stems thick at the base, 4–10 dm. high; leaf-blades 4 mm. wide, those of the stem short; staminate spikelet usually 1, peduncled; pistillate spikelets 3–6, erect to drooping, the upper sessile, linear, 2–7 cm. long, 4 mm. thick, loosely flowered at the base; glumes ovate-oblong, obtuse or acute, dark with a green midrib; perigynia oblong, nerveless, 2–3 mm. long. Beds of streams: Lab.—Mass.—Mo.—Minn.—Mack. *E. Boreal.* Je–Jl.

132. **C. cóncolor** R. Br. Stems 1–5 dm. high, smooth; leaf-blades 3–7 mm. wide, with revolute margins; terminal spikelets staminate, linear; lateral spikelets 1–4, short-oblong to linear-cylindric, 0.5–4 cm. long, 3–5 mm. wide, sessile or short-stalked; glumes obtuse or acutish, blackish with lighter midvein; perigynia 2.5–3 mm. long, nerveless or nearly so, the orifice entire. *C. rigida* Good. *C. Bigelovii* Torr. Arctic-alpine regions: Greenl.—N.H.—Alta.—B.C. —Alaska; Eurasia. *Arct.—Subarct.—Alp.* Jl–Au.

133. **C. lenticulàris** Michx. Densely cespitose; stems 3–6 dm. high, somewhat roughened; staminate spikelets 1–2, often partly pistillate; pistillate spikelets, 2–5, sessile or short-peduncled, approximate or the lower remote, 1–5 cm. long, 3–4 mm. wide; glumes obtuse, with broad light-colored center; perigynia 2.5 mm. long, 1.5 mm. wide, minutely beaked. River and lake shores: Lab.—Mass.—Neb.—Mack. *Boreal.* Je–S.

134. **C. aquátilis** Wahl. Rootstock slender, stoloniferous; stems 2–7 dm. high, sharply triangular above, slender, reddened at the base; leaf-blades 2–4 mm. wide; staminate spikelets 1 or 2, slender; pistillate spikelets 2–4, sessile or short-peduncled, separate, linear, 1.5–6 cm. long, 4–6 mm. wide; glumes oblong-obovate to lanceolate, obtuse or acutish, blackish, 1-nerved; perigynia elliptic-obovate, 2.5 mm. long, 1.25 mm. wide, nerveless, but with a median ridge, substipitate, puncticulate, reddish-dotted all over, granular, very minutely beaked. *C. variabilis* Bailey. *C. rhomboidea* Holm. Wet meadows and swamps: Greenl. —Man.—Que.—N.M.—Calif.—Alaska; Eurasia. *Boreal.—Mont.—Subalp.* Je–Au.

135. **C. substrícta** (Kükenth.) Mackenzie. Cespitose and stoloniferous; stems 6–14 dm. high, sharply triangular, reddened at the base; leaf-blades often glaucous, 4–8 mm. wide; staminate spikelets 2 or 3, slender; pistillate spikelets 2–4, sessile or short-peduncled, separate, linear, 2–7 cm. long, 4–6

mm. wide; glumes lanceolate, sharp-pointed, reddish brown with broad light-colored center, narrower than the perigynia; perigynia 3 mm. long, obovate, tapering to the stipitate base, obscurely nerved, but the marginal nerves prominent, resinous-dotted, abruptly very minutely beaked. *C. aquatilis* Am. auth., not Wahl. Swamps: Newf.—N.Y.—Neb.—Alta. *E. Boreal. Plain.* Je–Au.

136. **C. nebraskénsis** Dewey. Rootstocks creeping and stoloniferous; stems 2.5–10 dm. high; leaf-blades 4–8 mm. wide, flat; staminate spikelets 1–2; pistillate spikelets 2–5, sessile or short-peduncled, 1.5–6 cm. long, 6–9 mm. wide, contiguous or the lower separate; spikelets lanceolate, obtusish to acuminate, blackish with light midvein; perigynia ascending, 3–3.5 mm. long, 2 mm. wide, rounded at the base. Meadows and swamps: S.D.—Kans.—N.M.—Calif.—B.C. *Plain—Submont.—W. Temp.* My–Jl.

43. Cryptocarpae: Short-stoloniferous; staminate spikelets 1 or 2, linear, the other spikelets pistillate, linear to oblong, closely-flowered, the lower or all peduncled. Bracts sheathless, the upper auriculate at the base. Glumes 3-nerved, cuspidate or aristate. Perigynia coriaceous or membranous, plano-convex or biconvex, turgid, 2-ribbed, abruptly minutely beaked or beakless, subentire at the mouth. Achenes lenticular, apiculate, constricted at the middle. Stigmas 2.

137. **C. marítima** O. F. Müller. Stems sharply 3-angular, 3–7 dm. high; leaves 3–10 mm. wide, rough-peduncled; pistillate spikelets 2–4, ovoid to oblong, often staminate at the apex, 2–7 cm. long, 8–15 mm. thick, drooping, peduncled; glumes green, lance-subulate, ciliate-scabrous, 2–8 times as long as the perigynia; perigynia oval to suborbicular, pale, 3 cm. long; beak short. Salt marshes: Labr.—Mass.—Man.; Eur. Je–Au.

138. **C. crinìta** Lam. Stems usually rough, 6–15 dm. high; leaves flat, rough-margined, 3–10 mm. wide; staminate spikelets 1 or 2, stalked, sometimes pistillate at the base; pistillate spikelets 2–6, cylindric, 2.5–10 cm. long, 6–12 mm. thick, stalked, drooping, glumes abruptly rough-awned, 1–3-ribbed, brownish, spreading, 2–6 times as long as the perigynia; perigynia suborbicular, spreading, 2–3.5 mm. long, nerveless; beak short. Wet woods: Newf.—Fla.—Tex.—Minn. *E. Temp.* Je–Au.

44. Hirtae: Stems stout, leafy. Rootstocks with long stolons. Leaves septate-nodulose. Spikelets 3–10, the upper 1–5 staminate, slender, the others pistillate, many-flowered, erect. Bracts leaf-like, equaling or exceeding the stem, often sheathing. Pistillate glumes acute or aristate, ovate or lanceolate. Perigynia mostly ascending, coriaceous, ovoid or oblong-ovoid, somewhat inflated, nearly orbicular in cross-section, many-nerved, often hairy, round-tapering at the base, tapering into a bidentate beak. Achenes triangular, often stipitate, the faces flat or deeply concave, apiculate, continuous with the straight or slightly flexuous slender style. Stigmas 3.

139. **C. lacústris** Willd. Stems stout, purplish-tinged and filamentose at the base, 6–12 dm. high; leaf-blades 5–12 mm. wide, somewhat glaucous; staminate spikelets 1–5, slender; pistillate spikelets 2–5, 3–10 cm. long, 1 cm. wide; glumes purplish-tinged, acute to aristate; perigynia ovoid, 6 mm. long, 2.5 mm. wide, strongly nerved. In swamps: Newf.—Del.—Iowa—Man. *E. Boreal.* Je–Jl.

140. **C. impréssa** (S. H. Wright) Mackenzie. Stems 4–10 dm. high, smooth; leaves nodulose, scabrous, 4–8 mm. wide, flat; lower bracts leaf-like; staminate spikelets 2–4, linear; pistillate spikelets 2–4, 2–7 cm. long, 10 mm. thick, short-peduncled, erect; glumes ovate, the lower aristate and longer than the perigynia, the upper acute and shorter, straw-colored or purple-tinged; perigynia narrowly ovoid, 6 mm. long, firm, tapering into a 2-toothed beak. Swamps: Ont.—Fla.—Tex.—Neb. *E. Temp.*

141. **C. lanuginòsa** Michx. Stems stoutish, more or less reddened and filamentose at the base, 6–9 dm. high, sharp-angled and rough above; staminate spikelets 1–3, distant; pistillate spikelets 1–3, 1–5 cm. long, 5–7 mm. wide; glumes acuminate or aristate; perigynia ovoid, the beak bidentate. *C. Watsoni* Olney. Swampy places: N.S.—D.C.—Mo.—N.M.—Calif.—B.C. *Temp—Plain—Mont.* Je–Jl.

142. **C. lasiocárpa** Ehrh. Stems slender but stiff, strongly reddened and filamentose at the base, 6–9 dm. high, obtusely angled, smooth; staminate spikelets 1–3, distant; pistillate spikelets 1–3, 1–5 cm. long, 5–7 mm. wide; glumes acute or short-awned; perigynia oval-ovoid, 2 mm. wide, the beak bidentate. *C. filiformis* Auth., not L. Swamps: Newf.—N.J.—Iowa—Colo.—Ida.—B.C.; Eur. *Boreal—Plain—Submont.* Je–Au.

143. **C. Houghtònii** Torr. Long-stoloniferous; stems stoutish, 3–8 dm. high; leaf-blades 3–7 mm. wide; staminate spikelets 1–3, distant; pistillate spikelets 2 or 3, 1–4 cm. long, 7–12 mm. wide, rather loosely 15–35-flowered; glumes lanceolate, short-awned; perigynia ovoid, 3 mm. wide. Sandy or rocky soil: N.S.—Me.—Minn.—Sask. *Boreal.* My–Au.

144. **C. laevicónica** Dewey. Stems 6–12 dm. high, very rough above; leaf-blades 3–6 mm. wide, glabrous; sheaths puberulent towards the top, breaking and becoming filamentose; staminate spikelets 2–6, distant; pistillate spikelets 2–4, densely flowered, except at the base, 2.5–10 cm. long, 10–16 mm. wide; perigynia ovoid, glabrous, 8–10 mm. long, 3–4 mm. wide. Marshes: Man.—Ill.—Mo.—Kans.—Mont.—Sask. *Prairie—Plain.* My–Jl.

145. **C. atheròdes** Spreng. Stems stout, 6–15 dm. high, often roughish above; leaf-blades 5–12 mm. wide, often pubescent beneath, as are the sheaths; staminate spikelets 2–6, distant; pistillate spikelets 3–5, densely flowered except at the base, 3–10 cm. long, 12–16 mm. wide; perigynia 8–12 mm. long. *C. aristata* R. Br. Marshes: Ont.—N.Y.—Mo.—Colo.—Utah—Ore.—Yukon. Eurasia. *Plain—Submont.* Je–S.

146. **C. trichocárpa** Muhl. Stem 6–12 dm. high, rough above; leaf-blades rough-margined; staminate spikelets 2–6, long-stalked; pistillate spikelets 2–4, cylindric, densely-flowered, 2.5–10 cm. long, 10–15 mm. thick, the upper sessile, the lower slender-stalked; glumes hyaline, acute or aristate, half as long as the perigynia; perigynia ovoid, pubescent, prominently many-nerved, 8–10 mm. long; beak stout, conspicuously 2-toothed. Wet meadows and marshes: Que.—Pa.—Kans.—Minn. *Canad.—Allegh.* Je–Au.

45. Extensae: Stems slender, but strict, obtusely triangular, leafy towards the base. Leaves septate-nodulose, the blades narrow. Spikelets 2–10, the terminal normally staminate, the others pistillate, suborbicular to oblong, densely flowered, 3 cm. or less long, the upper sessile and approximate, the lower remote, peduncled, erect. Bracts leafy, more or less sheathing. Pistillate glumes ovate, mostly reddish, copper- or chestnut-tinged. Perigynia ascending, spreading or deflexed, membranaceous, smooth, many-nerved, somewhat inflated, obscurely triangular, rounded at the base, contracted into a bidentate beak, the teeth very erect. Achenes triangular, with flat faces, continuous with the very slender erect or flexuous style. Stigmas 3.

147. **C. virídula** Michx. Densely cespitose, not yellowish green, the stems 0.7–4 dm. tall, smooth, bluntly triangular; leaf-blades 1.5–3 mm. wide, canaliculate, the sheaths not prolonged at throat; staminate spikelet sessile or short-peduncled; pistillate spikelets 2–10, aggregate or the lower separate and exsert-peduncled, 4–12 mm. long, 4–7 mm. wide; glumes ovate, much shorter than the perigynia, obtuse or acutish; perigynia whitish-tipped, tapering at the base. *C. Oederi* Am. auth., in part. Lake and river banks: Newf.—N.J.—Minn.—Colo.—Utah—B.C.—Calif. *Boreal—Plain—Mont.* Je–Au.

148. **C. cryptólepis** Mackenzie. Stems cespitose, slender 2–6 dm. high, 3-angular, phyllopodic; leaf-blades flat, 1.5–3 mm. wide, staminate spikelet subsessile or peduncled, 7–8 mm. long; pistillate spikelets 3 or 4, the upper approximate, often staminate at the apex, the lower slender-peduncled, 10–20 mm. long, 7–10 mm. wide; glumes lanceolate, acute, greenish-yellow with a greenish midvein, 4–4.5 mm. long, the body obovoid, inflated, terete; beak slender, smooth. Wet meadows: Newf.—N.J.—Ind.—Minn. *Canad.* Je–Au.

149. **C. flàva** L. Yellowish green; stems 1.5–6 dm. tall, smooth or nearly so; leaf-blades 2–5 mm. wide; staminate spikelet sessile or stalked; pistillate spikelets 1–4, aggregate or the lower separate, 6–18 mm. long, 9–12 mm. wide; glumes ovate, strongly reddish-tinged, conspicuous at maturity; perigynia 5–6 mm. long, the beak deflexed, reddish-tipped. Wet meadows: Newf.—N.J.—Ohio—Mont.—B.C. Eur. *Boreal—Mont.—Submont.* Je–S.

46. PAUCIFLORAE: Long-stoloniferous. Stems slender. Leaf-blades narrow. Spikelet solitary, androgynous, bractless, narrow. Pistillate glumes soon falling. Perigynia few, short-stipitate, soon widely spreading or reflexed, obscurely triangular, straw-colored, subulate-beaked, obliquely cut. Achenes triangular, linear-oblong. Stigmas 3.

150. **C. pauciflòra** Lightf. Stems 10–25 cm. high, rough on the angles; stem-leaves 2–3; perigynia 1–6, 6–7 mm. long, obscurely several-nerved, smooth. Sphagnum swamps: Newf.—Pa.—Minn.—Wash.—Alaska; Eurasia. *Boreal—Mont.* Je–Au.

47. PSEUDO-CYPEREAE: Stems tall, generally stout, acutely angled, leafy below. Leaf-blades flat, septate-nodulose. Spikelets 3–9, the upper 1–3 slender, staminate, the others normally pistillate, densely flowered, the upper approximate, the lower remote and strongly peduncled, often nodding. Bracts leaf-like, much exceeding the culms, mostly not sheathing. Pistillate glumes aristate. Perigynia spreading or reflexed, membranaceous or stiff, triangular or circular in cross-section, 3–8 mm. long, closely many-ribbed, greenish straw-color, smooth, stipitate, contracted into a rigid, slender beak. Achenes triangular, continuous with the slender often flexuous style. Stigmas 3, short.

151. **C. hystricìna** Muhl. Stems 3–9 dm. tall, reddened at the base, rough above; leaf-blades 3–8 mm. wide; staminate spikelet slender-stalked, the glumes rough-awned; pistillate spikelets 1–4, densely many-flowered, 1–6 cm. long, 10–14 mm. wide, the lower slender-stalked; glumes rough-awned; perigynia 5–6 mm. long, 15–20-nerved. Swampy soil: Newf.—Ga.—Tex.—Ariz.—Calif.—Wash. *Temp.—Plain.* Je–Au.

152. **C. Pseùdo-Cypèrus** L. Stems glabrous, stout, sharply 3-angular, 6–10 dm. high; leaf-blades nodulose, 4–10 mm. wide; staminate spikelet short-stalked, the glumes rough-awned; pistillate spikelets 2–5, linear-cylindric, densely many-flowered, slender-stalked, 2.5–6 cm. long, 8–12 mm. thick; glumes linear with a broad base, rough-awned; perigynia short-stipitate, lanceolate, prominently many-ribbed, at least reflexed; beak short, 2-toothed. Bogs: N.S.—N.Y.—Mich.—Minn.; Eurasia. *Canad.* Je–Au.

153. **C. comòsa** Boott. Stems stout, up to 15 dm. tall, sharply angled; leaf-blades 6–14 mm. wide; staminate spikelet as in the last; pistillate spikelets as in the last but 12–14 mm. wide; glumes very rough-awned; perigynia rigid, closely many-ribbed, reflexed when mature. Swamps: N.S.—Minn.—Fla.—La.—Minn.—Calif.—Wash.—Ida. *Temp.—Plain—Submont.* Je–Au.

48. PHYSOCARPAE: Stems mostly tall and stout, leafy below. Leaves not hairy, septate-nodulose. Spikelets 2–10, the upper 1–5 staminate, the others normally pistillate, subglobose to linear-cylindric, generally closely many-flowered, erect, short-peduncled, more or less remote. Bracts leaf-like, much exceeding the inflorescence, normally sheathless. Perigynia ascending, spreading, or even reflexed, membranaceous, smooth, from little to much inflated,

suborbicular in cross-section, coarsely many-ribbed or nerveless, contracted into a beak, the beak entire to bidentate. Achenes much shorter than the perigynia, triangular or lenticular, continuous with the tortuous style. Stigmas 3 or 2.

154. **C. miliàris** Michx. Stems 3–6 dm. tall, smooth, little filamentose at the base; leaf-blades 2 mm. wide; staminate spikelets 1 or 2; pistillate spikelets 1–3, oblong-cylindric, 8–25 mm. long, 5–7 mm. wide; perigynia at most faintly nerved, 2–3 mm. long, exceeding the scale. Borders of lakes and streams: Lab.—Me.—B.C. *Boreal.* Jl–Au.

155. **C. physocárpa** Presl. Stems 3–8 dm. tall, more or less roughened; leaf-blades 1.5–3.5 mm. wide; staminate spikelets 1 or 2; pistillate spikelets 1 or 2, separate, oblong-cylindric, 1.5–3.5 cm. long, 6–8 mm. wide, closely 25–75-flowered; glumes ovate-lanceolate, sharp-pointed, brownish with lighter midvein, hyaline at the apex; perigynia obscurely nerved, 4.5–5 mm. long, 1.75–2.25 mm. wide, exceeding the scales, minutely beaked. Arctic and subarctic places: Man.—Colo.—Mack.—Alaska; Eurasia. *Alp.—Mont.* Je–Au.

156. **C. oligospérma** Michx. Long-stoloniferous; stems slender, 4.5–6 dm. high, exceeding the leaves; leaf-blades 3 mm. wide, involute in age; staminate spikelet stalked, linear; pistillate spikelets 1–2, subglobose or short-oblong, mostly 5–15-flowered; glumes acute or slightly mucronate, much shorter than the perigynia; perigynia ovoid, inflated, strongly few-nerved, yellowish green, shining, 5 mm. long, 3 mm. in diameter, the beak minutely bidentate. In bogs: Lab.—Pa.—Man.—Mack. *E. Boreal.* Je–S.

157. **C. Tuckermáni** Dewey. Stems rough above, 4–10 dm. high; leaf-blades 3–5 mm. wide, nodulose; bracts leaf-like; staminate spikelets 2 or 3, the uppermost peduncled; pistillate spikelets 2–4, stout, 12–18 mm. thick, 2.5–5 cm. long, the upper sessile, the lower stalked and spreading; glumes lanceolate, acute to short-cuspidate, less than half as long as the perigynia; perigynia much inflated, broadly ovoid, prominently nerved, abruptly contracted into a subulate, 2-toothed beak. Bogs and meadows: N.B.—N.J.—Iowa.—Minn. *Canad.—Allegh.* Je–Au.

158. **C. Raeàna** Boott. Stems 3–6 dm. high, very slender, triangular and roughened above, reddened at the base; leaf-blades flat, 2 mm. wide, little nodulose; staminate spikelets 1 or 2; pistillate spikelets 1–3, narrowly cylindric, 1.5–5 cm. long, 5 mm. wide, short-peduncled; spikelets lanceolate, acuminate; perigynia 5–6 mm. long, closely enveloping the achene, yellowish green, strongly few-nerved. Lake and river shores: Me.—Que.—Sask. *Boreal.* Je–Au.

159. **C. vesicària** L. *Fig. 102.* Stems 3–9 dm. high, acutely angled and rough above; leaf-blades 3–6 mm. wide; staminate spikelets 2–4; pistillate spikelets 1–3, sessile or short-peduncled, erect, oblong-cylindric, 2.5–7 cm. long, 6–15 mm. wide, many-flowered; glumes ovate or lanceolate, acute, acuminate, or short-awned; perigynia 5–8 mm. long, yellowish green, 8–10-nerved. *C. monile* Tuckerm. Wet meadows and swamps: Que.—Pa.—Ohio—Calif.—B.C.; Eurasia. *Plain.* Je–Au.

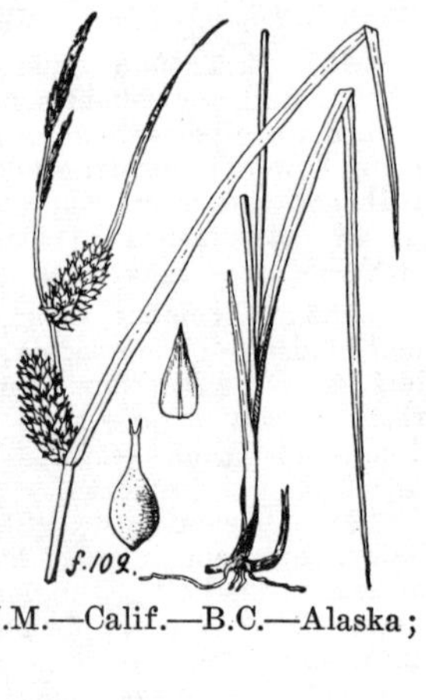

f.102.

160. **C. rostràta** Stokes. Stems stout, 3–12 dm. high; leaf-blades 2–12 mm. wide; staminate spikelets 2–4; pistillate spikelets 2–4, erect, cylindric, densely flowered, sessile or short-peduncled, 5–15 cm. long, 6–20 mm. wide; glumes lanceolate, awned or acute; perigynia ovoid, spreading at maturity, few-nerved, 4–8 mm. long, abruptly beaked. *C. utriculata* Boott. Boggy places: Lab.—Del.—N.M.—Calif.—B.C.—Alaska; Eurasia. *Plain—Mont.* Je–S.

161. **C. retrórsa** Schw. Stems stout, 3–10 dm. high, smooth or slightly roughened above; leaf-blades 5–10 mm. wide; staminate spikelets 1–3; pistillate spikelets 3–8, contiguous and sessile or nearly so, or the lower distant and long-stalked, cylindric, densely many-flowered, 2.5–7.5 mm. long, 14–20 mm. wide; glumes lanceolate, acute or acuminate; perigynia ovoid, few-nerved, 7–10 mm. long, tapering into the beak. "*C. lupulina* Muhl." of Fl. Colo. Swamps and wet places: Newf.—Pa.—Ia.—Colo.—Wash.—B.C. *Plain—Submont.* Jl–O.

162. **C. lùrida** Wahl. Stems erect, smooth, 1.5–10 dm. high; leaf-blades elongate, rough, 4–7 mm. wide; staminate spikelet elongate, sessile or peduncled; pistillate spikelets 1–4, globose to oblong-cylindric, 1–6 cm. long, 15–20 mm. thick, the upper sessile and erect, the lower peduncled, spreading or drooping; glumes rough-awned, ovate; perigynia inflated, ovoid, strongly 10-ribbed, 6–9 mm. long, yellowish green, longer than the glumes; the beak long, subulate. Swamps and wet meadows: N.S.—Fla.—Tex.—Neb.—Minn. *E. Temp.* Je–O.

49. Squarrosae: Cespitose and stoloniferous; stems leafy; blades flat, septate-nodulose. Terminal spikelet staminate or gynaecandrous. Lateral spikelets 1–5, pistillate or gynaecandrous, ovoid or oblong-cylindric, densely many-flowered. Bracts setaceous or leaf-like. Perigynia yellowish-green or green, obconic, inflated, several-nerved, membranous, abruptly truncately contracted into a bidentate beak with stiff teeth. Achenes triangular, continuous with the flexuose persistent style. Stigmas 3, short.

163. **C. Fránkii** Kunth. Stem leafy, tufted, 3–7 dm. high; leaf-blades rough, 3–8 mm. wide; staminate spikelets stalked or subsessile, sometimes pistillate at the apex; pistillate spikelets 3–6, dense, cylindric, 1–4 cm. long, 8 mm. thick, the upper subsessile, the lower slender-peduncled; glumes linear-subulate, very rough, longer than the perigynia; perigynia green, slightly inflated, 4 mm. long, obconic, the summit depressed, the beak subulate. Swamps: N.Y.—Ga.—La.—Tex.—Kans. *SE. Temp.* Je–S.

164. **C. squarròsa** L. Stems slender, rough above, 6–10 dm. high; leaf-blades 3–6 mm. wide, rough-margined; bracts leaf-like; spikelets 1–3, often 1, erect, stalked, oval, the pistillate portion 15–30 mm. long, 12–22 mm. thick, the uppermost staminate at the base; glumes scarious, lanceolate, acuminate or awn-tipped; perigynia yellowish green, squarrose or the lower reflexed, inflated, firm, obovoid, abruptly contracted into the beak. Swamps: Ont.—Ga.—Ark.—Neb. *E. Temp.* Je–S.

165. **C. typhìna** Michx. Stem 6–10 dm. high; leaf-blades 4–10 mm. wide, rough-margined; bracts leaf-like; spikelets 1–6, oblong-cylindric, dense, the pistillate portion 2.5–4 cm. long, 8–16 mm. thick, often staminate at both ends; glumes oblong-lanceolate, obtuse; perigynia dull straw-colored, obovoid, the lowest spreading or reflexed, inflated, truncately contracted into the 2-toothed beak. Swamps: Que.—Ga.—La.—Iowa. *E. Temp.* Jl–Au.

50. Lupulinae: Stems stout and leafy. Leaf-blades strongly septate-nodulose. Terminal 1–3 spikelets staminate, linear. Lateral spikelets 1–5, pistillate, or staminate at the apex, densely flowered. Bracts sheathless or lowest sheathing. Perigynia large, greenish or dull straw-colored, ovoid, strongly ribbed, strongly inflated, membranous or subcoriaceous, tapering or contracted into the deeply bidentate beak, the teeth stiff. Achenes triangular with concave sides, continuous with the persistent style. Stigmas 3, short.

166. **C. intumèscens** Rudge. Stems tufted, 4–10 dm. high; leaf-blades dark green, roughish, 3–7 mm. wide, bracts leaf-like; staminate spikelet narrow, stalked; pistillate spikelets 2, sessile or short-stalked, subglobose; glumes narrowly lanceolate, aristate, half as long as the perigynia; perigynia 1–12, spreading, 10–20 mm. long, inflated, many-nerved, contracted into the subulate beak. Boggy places: Newf.—Fla.—Tex.—Man. *E. Temp.* My–O.

167. C. Àsa-Gràyi Bailey. Stem glabrous, stout, 6–10 dm. high; leaf-blades dark-green, 5–9 mm. wide; bracts leaf-like; staminate; spikelet long-stalked, narrow; pistillate spikelets 1 or 2, globose, dense, 2.5 cm. broad; glumes ovate, obtuse to short-cuspidate, scarious, one third as long as the perigynia; perigynia 6–30, ovoid, 12–18 mm. long, glabrous or hispidulous, inflated, many-ribbed, 7 mm. long. Swamps: Vt.—Ga.—Mo.—Iowa. *Canad.—Allegh.* Je–S.

168. C. lupulìna Muhl. Stems stout, 3–12 dm. high; leaf-blades nodulose, 5–15 mm. wide; staminate spikelets solitary, rarely several, subsessile or stalked, pistillate spikelets 2–5, densely many-flowered, sessile to long-stalked, oblong, 2.5–7 cm. long; glumes lanceolate, acuminate or aristate, much shorter than the perigynia; perigynia ascending, inflated, many-nerved, 10–20 mm. long, 4–7 mm. thick, contracted into the subulate beak. Swamps: N.B.—Fla.—Tex.—Kans.—Man. *E. Temp.* Je–S.

Family 14. ARACEAE. Arum Family.

Fleshy, perennial, mostly acaulescent herbs, with rootstocks. Leaves basal, usually petioled, with broad, simple or rarely divided, reticulate-veined blades, or narrowly linear with sheathing bases. Flowers perfect, monoecious, or dioecious, borne in a dense fleshy spike (spadix); this subtended by or enclosed in a large foliaceous or colored bract (spathe). Perianth of scale-like members or wanting. Stamens 4–10, with very short filaments and thick connectives. Gynoecium of a single carpel or of several united carpels. Ovaries 1–several-celled; stigmas terminal; ovules 1 to several in each cavity. Fruit a berry or a utricle.

Spadix terminal, with an oblong sheathing spathe; leaves petioled, with broad blades.
  Spathe not fleshy; flowers without perianth, on a more or less elongate spadix.
    Flowers dioecious, covering only the lower part of the spadix; spathe more or less involute or convolute, green or purplish.
      Leaves palmately divided, in ours 3-divided; spathe involute below, hooded above; spadix terete. 1. Arisaema.
      Leaves pedately divided; spathe convolute throughout, elongate; spadix more or less flattened. 2. Muricauda.
    Flowers perfect, covering the whole spadix; spathe open, the upper part white. 3. Calla.
  Spathe fleshy; flowers with 4 hooded sepals on a globose spadix. 4. Symplocarpus.
Spadix borne somewhat laterally on the leaf-like scape; spathe leaf-like, continuous to the scape; leaves narrowly linear, equitant. 5. Acorus.

1. ARISAÈMA Martens. Jack-in-the-pulpit, Indian Turnip.

Scapose, fleshy herbs, with bitter corms. Leaves basal, petioled, the blades palmately 3–5-divided. Spadix terete, bearing the flowers on the lower portion, the upper portion naked, club-shaped. Spathe cornucopia-like, convolute below, hooded above. Flowers dioecious, without a perianth. Stamens 4, nearly sessile, the anthers opening by a continuous slit at the apex. Pistil 1; stigma peltate; ovary 1-celled. Fruits red, globose, berry-like, in a small head.

f.103.

1. A. triphýllum (L.) Schott. *Fig. 103.* Corm subglobose, 2–5 cm. thick; leaves 2–15 dm. long including the petiole, the blade 3-divided, the segments elliptic or obovate, 8–30 cm. long, acute or acuminate; lower portion of the spathe whitish, the upper green or brown, or striped with purple; fruits bright-red, 10–12 mm. thick. Swamps and moist places: N.S.—N.D.—Kans.—Tex.—Fla. *E. Temp.* Ap–Je.

2. **MURICAÙDA** Small. GREEN DRAGON, DRAGON-ROOT.

Acaulescent herbs, with corms. Leaves basal, long-petioled, with pedately divided blades. Spathe convolute throughout, long-attenuate. Spadix flattened, flower-bearing on the lower portion, the upper portion sterile, slender, long-exserted. Flowers monoecious or dioecious, without perianth. Ovary turbinate, 6–8-ovuled. Fruits berry-like, brightly colored, in dense heads.

1. **M. Dracóntium** (L.) Small. Leaves 3–10 dm. long, the segments of the blade 5–15, oblong or oblanceolate, 1–3 cm. long, entire, pointed or acuminate; spathe white or greenish, 3–10 cm. long; spadix 5–20 cm. long, with a whip-like prolongation, in the monoecious plant with the pistillate flowers at the base; berries bright red, 6–8 mm. thick. *Arisaema Dracontium* (L.) Schott. [G, B]. Moist woods: Me.—Minn.—Tex.—Fla. *Canad.—Allegh.*

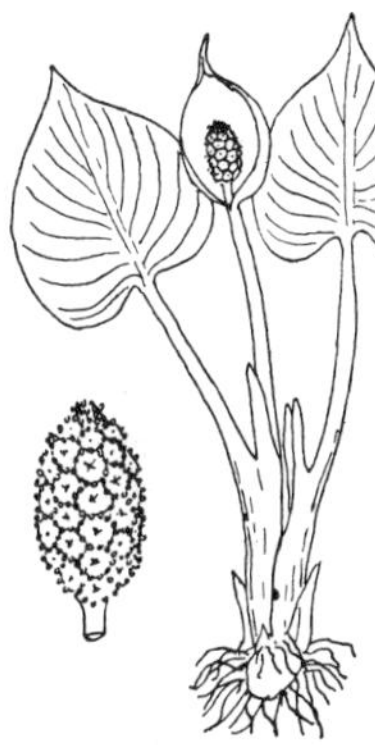

3. **CÁLLA** L. CALLA, WATER ARUM.

Acaulescent herbs, with bitter rootstocks. Leaves basal, petioled; blades ovate to rounded-cordate. Spathe ovate or lanceolate, open, concave. Spadix cylindrical, densely covered with flowers throughout. Flowers perfect or the upper ones staminate, without perianth. Stamens 6; anthers opening by slits. Ovary 1-celled; style very short; stigma circular. Fruits berrylike, obconic, depressed, in dense heads.

f.104.

1. **C. palústris** L. *Fig. 104.* Petioles 1–2 dm. long, the leaf-blades ovate or cordate, 5–12 cm. long, 3–10 cm. wide, abruptly acute; scape 1–2 dm. long; spathe 2.5–6 cm. long, green below outside, white inside; berries red. Bogs: N.S.—Man.—Iowa—Va.; Eurasia. *Canad.—Allegh.* Jl–Au.

4. **SYMPLOCÁRPUS** Salisb. SKUNK CABBAGE, MIDAS' EARS.

Acaulescent stinking herbs. Leaves basal, petioled; blades thick, leathery, oblong or ovate. Scape short and thick. Spathe thick, fleshy, leathery, shell-like, twisted at the apex. Spadix oblong or globose, thick, enclosed in the spathe. Flowers perfect. Perianth of 4 hooded sepals. Stamens 4; anthers opening lengthwise, extrorse. Ovary sunken in the spadix, 1-celled; style pyramidal; stigma minute. Fruits adnate to the fleshy perianth. *Spathyema* Raf.

1. **S. foètidus** (L.) Nutt. Leaf-blades ovate or oblong, truncate or subcordate at the base, 2–10 dm. long; petiole short, stout; spathe preceding the leaves, 8–15 cm. long, greenish or yellowish, mottled with purple; spadix globose, 2–2.5 cm. long, becoming 1–1.5 dm. in fruit; fruits green, in a globose spike. *Spathyema foetida* (L.) Raf. [B]. Swamps: N.S.—Man.—Mo.—Fla. *E. Temp.* F–Ap.

5. **ÁCORUS** L. SWEET FLAG, CALAMUS-ROOT.

Erect swamp plants, with long horizontal rootstocks. Leaves sword-shaped, equitant. Scapes 3-angled, bearing a lateral spadix and a foliaceous spathe, continuous with the scape. Flowers perfect, densely crowded on the spadix. Perianth with 6 membranous concave divisions. Stamens 6. Pistil solitary; ovary 2–4-celled, each 2–8-ovuled; stigma depressed-capitate. Fruit berry-like.

1. **A. Cálamus** L. Leaves linear, erect, 5–20 dm. long, 1–2 cm. wide, long-attenuate, 2-ranked; scape 1–2 m. high; spathe 2–7.5 dm. long; spadix 5–7 cm. long, about 1 cm. thick, lance-cylindric; plant seldom fruiting. Swamps and streams: N.S.—Fla.—Tex.—Colo.—Ida.; Eurasia. *E. Temp.* My–Jl.

Family 15. **LEMNACEAE.** Duckweed Family.

Minute floating perennial aquatics, consisting of a fleshy or membranaceous, loosely cellular, thallus-like stem, without leaves and in our genera with or without rootlets. The new stems or fronds are produced from two lateral depressions or pouches or a terminal one; the new frond being attached to the old one by a short, slender stalk, soon separating. Flowers very rare. Inflorescence consisting of 1 pistillate and 1 or 2 staminate flowers, borne on the edge on the upper surface of the frond. Staminate flower of a single stamen, with 2–4 pollen-sacks. Pistillate flowers of a single flask-like pistil, with 1–several ovules. Fruit a 1–6-seeded utricle.

Fronds with rootlets.
  Rootlets solitary, without fibrovascular bundles. — 1. Lemna.
  Rootlets several, each with a fibrovascular bundle. — 2. Spirodela.
Fronds without rootlets. — 3. Wolffia.

1. **LÉMNA** L. Duckweed.

Frond disk-like, 1–5 nerved or nerveless. Stipe attached to the basal margins of the frond. Rootlet solitary, without fibrovascular bundles. Anthers dehiscent transversely.

Fronds long-stalked, mostly submerged, forming large masses. — 1. *L. trisulca.*
Fronds short-stalked or sessile, floating on the surface.
  Fronds pale and usually strongly gibbous beneath. — 2. *L. gibba.*
  Fronds green or purplish, not gibbous beneath.
    Frond oblong-obovate, indistinctly 3-nerved.
      Frond usually purple-tinged beneath, 2–3 mm. long; root-cap pointed; spathe open. — 3. *L. perpusilla.*
      Frond rarely reddish or purplish beneath, 3–5 mm. long, rarely less; root-cap obtuse or subtruncate; spathe sack-like. — 4. *L. minor.*
    Frond oblong or elliptical, indistinctly 1-nerved or nerveless.
      Frond unsymmetrical, without papules. — 5. *L. cyclostasa.*
      Frond symmetrical, with a row of papules along the midvein. — 6. *L. minima.*

1. **L. trisúlca** L. Fronds usually submerged, seldom floating, usually several generations attached together, oblong to oblong-lanceolate, slightly unsymmetrical and falcate, dentate towards the upper end, 5–10 mm. long, 2–3 mm. wide. In springs and running water: N.S.—N.J.—Tex.—Calif.—B.C.; Old World and Australia. *Temp.—Mont.* Jl–Au.

2. **L. gíbba** L. Fronds solitary or 2–4 in a group, orbicular to obovate, 2–5 mm. long and 2–4 mm. wide, thick, convex and slightly keeled above, with large air-cavities in the gibbous portion, unsymmetrical, 3–5-nerved. In ponds: Neb.—Tex.—Ariz.—Calif.; Mex.; Old World and Australia. *Temp.—Subalp.*

3. **L. perpusílla** Torr. Fronds in colonies of 1–6, thick, usually obliquely obovate, unsymmetrical, hollow, contracted at the base, often papillose along the midrib. Ponds and streams: Mass.—N.D.—Kans.—Fla. *E. Temp.* Je–Jl.

4. **L. mìnor** L. Fronds solitary or a few together, round to oval, symmetrical, thickish, convex on both sides, sometimes slightly keeled above and with a row of papules along the the midrib. In stagnant water and slow streams: Lab.—Fla.—Calif.—Alaska; Mex.; Old World and Australia. *Temp.—Mont.*

5. **L. cyclostàsa** (Ell.) Chev. Fronds solitary or 2–8 in a group, oblong to obovate-oblong, usually somewhat falcate, 2.5–4.5 mm. long, 1–1.5 mm. wide, usually unsymmetrical at the base. *L. valdiviana.* [G]. In pools and streams: Mass.—Fla.—Calif.—Ida.; Mex.—S. Am. *Trop.—Temp.—Submont.*

6. **L. mínima** Philippi. Fronds solitary or in groups of 2–4, oblong or elliptical, symmetrical, 1.5–4 mm. long, 1–2.75 mm. wide, both surfaces convex or the lower flat, with a thin margin around the frond. In pools: Ga.—Kans.—Wyo.—Calif.; Mex.—S. Am. *Trop.—Temp.—Submont.*

2. SPIRODÈLA Schleiden. LARGER DUCKWEED.

Frond, disk-like, several-nerved. Stipe attached peltately to the frond, back of and under the basal margin. Rootlets several, each with a solitary fibro-vascular bundle. Anthers dehiscent longitudinally.

1. **S. polyrhìza** (L.) Schleiden. *Fig. 105.* Frond, solitary or in colonies of 2–5, roundish obovate, flat on both sides, green above, generally purplish beneath, 5–15-nerved, 2.5–4.5 mm. wide and 2.5–8 mm. long ; rootlets 4–16. *Lemna polyrhiza* L. In pools and ponds: N.S.—B.C.—Fla.—Mex.; Old World and S. Am. *Temp.*

3. WÓLFFIA Horkel.

Frond globose or ellipsoid, or in some exotic species flattened, without rootlets, propagating from a slit or funnelform pouch at the margin near the base. Flowers 2, bursting from the center of the frond, one staminate of one stamen, one pistillate of a single pistil, with a globular ovary, short style, and depressed stigma. Fruit a spherical utricle. Seed orthotropus.

Frond globose or ellipsoid, not at all flattened, not punctate. 1. *W. columbiana.*
Frond flattened along the margins above, with a low papilla at the center. 2. *W. papulifera.*

1. **W. columbiàna** Karst. Fronds soon detached from its parent plant, 0.5–1.5 mm. long, convex above, nearly submerged; stomata 1–10; tissue of uniform cells. Lakes and pools: Mass.—Minn.—Neb.—Kans.—La.—Fla. *E. Temp.* Je–Jl.

2. **W. papulífera** Thompson. Frond, only half submerged, slightly unsymmetrical, obliquely broadly obovate, 1 mm. broad, the under surface strongly gibbous; stomata numerous. Pools: Mo.—Kans. *Ozark.*

Family 16. XYRIDACEAE. YELLOW-EYED GRASS FAMILY.

Annual or perennial, scapose herbs. Leaves basal, sword-like, mostly 2-ranked. Flowers in a dense terminal spike, perfect, subtended by closely imbricate bracts. Sepals 3, the two lateral ones keeled, persistent, the third deciduous. Corolla white or yellow; petals 3, fugaceous. Stamens 3; usually alternating with 3 staminodia. Pistil 3-carpellary, 1-celled or imperfectly 3-celled; placentae parietal; stigmas 3. Fruit a 3-valved capsule. Seeds numerous.

1. XỲRIS L. YELLOW-EYED GRASS.

Characters of the family.

1. **X. flexuòsa** Muhl. *Fig. 106.* Perennial; leaves narrowly linear, 5–30 cm. long, often spirally twisted, 2-edged above; spike oval or subglobose, 5–10 mm. long; lateral sepals margined, 4–5 mm. long; keel ciliate, bearded at the apex; corolla yellow. Swamps: Me.—Minn.—Tex.—Ga. *E. Temp.* Jl–S.

Family 17. **ERIOCAULACEAE.** Pipewort Family.

Perennial or rarely annual, mostly caulescent herbs. Leaves narrow, clustered. Scape simple, with a sheathing bract. Flowers monoecious, the staminate above the pistillate, or rarely dioecious, in dense, involucrate, terminal heads, each flower with a minute bract. Perianth usually of 2 series; sepals and petals each 2 or 3. Stamens as many or twice as many as the sepals. Pistil 2- or 3-carpellary; ovary 2- or 3-celled; styles 2 or 3. Ovules solitary in each cavity, orthotropous. Fruit a loculicidal capsule.

1. **ERIOCAÙLON** L. Pipewort.

Perennial herbs with spongy tissue. Leaves basal, attenuate. Heads woolly. Sepals 2 or 3. Petals as many, in the staminate flowers with a gland at the middle, in the pistillate flowers much reduced and narrow. Stamens 4–6; anthers 2-celled. Pistil solitary, 2- or 3-celled; styles 2 or 3. Capsule thin-walled.

f. 107

1. E. articulàtum (Huds.) Morong. *Fig. 107.* Leaves spreading, 2–6 cm. long, longer in submerged plants, pellucid; scape 3–40 cm. long, 4–8-ridged, the sheath surpassing the leaves; heads globose, 8 mm. thick or less, gray; petals of the staminate flowers ovate or oval, ciliate all around; receptacle and filament glabrous; seeds oblong. *E. septangulare* With. Still water: Newf.—Minn.—Tex.—Fla. *E. Temp.* Jl–O.

Family 18. **COMMELINACEAE.** Spiderwort Family.

Somewhat succulent herbs, with fibrous or tuberous-thickened roots and alternate leaves sheathing at the base. Flowers perfect, subtended by leaf-like sheathing bracts. Sepals 3, herbaceous. Corolla regular or irregular, with 3, usually showy, colored petals. Stamens 6 or 5, sometimes 2 or 3 of these sterile. Gynoecium of 2 or 3 united carpels; ovary superior, 2- or 3-celled; styles united. Fruit a loculicidal capsule.

Petals all alike; perfect stamens 6; bracts leaf-like. 1. Tradescantia.
Petals unlike in shape and size; perfect stamens 3 or 2; bracts spathe-like. 2. Commelina.

1. **TRADESCÁNTIA** (Rupp.) L. Spiderwort.

Caulescent perennials, with alternate narrow leaves. Cymes umbel-like, terminal, subtended by usually 2 bracts, similar to the leaves. Sepals 3, nearly equal, distinct. Petals 3, showy, blue, rose-color or white, sessile, similar and equal, delicate. Stamens 6, all perfect; filaments filiform, in ours long-hairy. Capsule loculicidal, 3-valved. Seeds 3–6, more or less sculptured.

Stem usually less than 1 dm. high; bracts longer than the leaves. 1. *T. brevicaulis.*
Stem 1–10 dm. high; bracts shorter than the leaves.
 Stem erect, stout, the cymes merely terminal; leaves linear, or nearly so.
  Foliage bright green; sepals more or less pilose or villous.
   Pedicels glandular-villous or glabrate.
    Bracts broader than the leaves, strongly gibbous at the base; stem 2–3 dm. high, usually simple; pod densely glandular. 2. *T. bracteata.*

Bracts not broader than the leaves, not strongly gibbous at the base; stem usually over 3 dm. high, branched.
Pod densely glandular-pubescent; sepals broadly lanceolate, densely glandular; plant bright green; branches shorter than the stem proper. — 3. *T. occidentalis.*
Pod glabrate, pubescent only at the apex; pedicels glabrous or slightly glandular above. — 4. *T. universitatis.*
Pedicels and sepals villous, not glandular.
Stem usually 3–4 dm. high, glabrous or nearly so. — 5. *T. virginiana.*
Stem 1–2 dm., rarely 3 dm. high, decidedly pilose. — 6. *T. hirsutiflora.*
Foliage glaucous; sepals glabrous or with a tuft of hairs at the apex. — 7. *T. reflexa.*
Stem flexuose; cymes both terminal and axillary; leaves lanceolate. — 8. *T. pilosa.*

1. **T. brevicaùlis** Raf. Stems usually clustered, erect, 1–8 cm. high; leaf-blades linear, 1.5–3 dm. long, the sheaths villous, imbricate; bracts 2, nearly equal, longer than the leaves; pedicels 3.5–5.5 cm. long; sepals ovate, about 1 cm. long, obtuse; corolla 2 cm. broad; petals orbicular, purplish-blue, obtuse, delicately veined. Hillsides and woods: Ill.—Kans.—Tex.—Tenn. *Prairie.* Je–Au.

2. **T. bracteàta** Small. *Fig. 108.* Stem erect, 1–3 dm. high, simple, or sparingly branched, glabrous; leaves linear, 1–2 dm. long, long-acuminate; sepals ovate-lanceolate; corolla about 2.5 cm. broad. Wet meadows: Minn.—Mo.—Kans.—S.D. *Prairie—Plain—Submont.*

3. **T. occidentàlis** (Britton) Smyth. Stem stout, branched, erect, 2.5–4 dm. high; leaves narrowly linear, 2–3 dm. long, attenuate, with rather loose sheaths, glabrous; bracts 5–15 cm. long; sepals ovate-lanceolate, acute, about 1 cm. long; petals blue or rose-colored, 12–15 mm. long. *T. ramifera* Lunell. Wet meadows: Iowa—Mo.—Colo.—N.D. *Prairie—Plain—Submont.*

4. **T. universitàtis** Cockerell. Stem 3–4 dm. high, branched; leaves ascending, linear, attenuate, 2–3 dm. long; bracts narrower, 1–2 dm. long; sepals narrowly lanceolate, about 1 cm. long, acuminate; petals blue, broadly ovate, acute, 15 mm. long. Scarcely distinct from the preceding. Wet places: Mont.—S.D.—Neb.—Colo. *Plain—Submont.*

5. **T. virginiàna** L. Stem stout, erect, 2–4 dm. high, simple; leaf-blades 1–7 dm. long, acuminate, curved, the sheaths sometimes slightly ciliate; bracts 2, usually unequal; pedicels 2.5–5 cm. long; sepals ovate to elliptic, 12–18 mm. long, villous with long non-glandular hairs; corolla dark blue, purple, or white, 3–4 cm. broad; petals 1.5–2 cm. long, suborbicular; capsule 5–7 mm. long, glabrous. Hillsides and along streams: N.Y.—Minn.—Kans.—Ark.—N.C. *E. Temp.* My–Au. Je–S.

6. **T. hirsutiflòra** Bush. Stem stout, 1–2 dm., rarely 3 dm. high, densely pilose, yellowish green; leaves linear, 5–15 mm. wide, pilose with long hairs, 1.5–2 dm. long; bracts 2, nearly equal, 1 dm. long or more, somewhat arched; inflorescence 10–15-flowered; pedicels 2–3 cm. long, pilose; sepals linear-lanceolate, obtuse, 12–18 mm. long, densely pilose; corolla bright blue to rose-purple; petals 10–15 mm. long, rounded-oval. Sandy places: Kans.—Tex. *Texan.* Ap–May.

7. **T. refléxa** Raf. Stem 4–9 dm. high, usually branched, sometimes purplish; leaf-blades 2–5 dm. long, long-attenuate, the sheaths glabrous or slightly villous; bracts 2, unequal, finally reflexed; pedicels 2–2.5 cm. long, recurved; sepals oblong or elliptic, 8–10 mm. long, hooded; corolla blue, 2–3 cm. broad; petals suborbicular; capsule 5–6 mm. long, glabrous, constricted above the middle. Sandy or clayey soil: S.C.—Minn.—Kans.—Tex.—Fla. *Allegh.—Austral.* Je–Au.

8. **T. pilòsa** Lehm. Stem stout, 4–8 dm. high, often pubescent, branched; leaf-blades lanceolate, 1.3–2.5 cm. long, ciliate; pedicels 1.5–2 cm. long, villous-pilose or glabrate; sepals ovate or oblong, 7 mm. long, hooded, villous-pilose; corolla blue, 2.5–3 cm. broad; petals suborbicular; capsule globose, 5 mm. long, constricted at the middle, pilose at the apex. Shaded hillsides and thickets: Ohio—Kans.—Tenn.—Ala.—Fla. *Allegh.—Austral.* Je–Au.

2. COMMELÌNA (Plum.) L. Day-flower, Dew-flower.

Mostly perennials, with alternate leaves. Bracts usually spathe-like, folded. Flowers irregular, in small cymes. Sepals 3, unequal, the larger two more or less united. Petals 3, usually blue, two larger than the third. Fertile stamens 3, rarely 2; filaments slender, glabrous; sterile stamens 2 or 3, smaller than the fertile ones. Seeds 2 in each cell, one above the other, or solitary.

Margins of the spathe not connate at the base; anterior two cells of the capsule 2-seeded.
- The two posterior petals blue, the anterior one white, much smaller, narrower; posterior cell of the capsule empty. — 1. *C. communis.*
- All petals blue, the anterior the smaller; posterior cell of the capsule 1-seeded and indehiscent. — 2. *C. longicaulis.*

Margins of the spathe connate at the base.
- The two posterior petals blue, the anterior white, much smaller; cells of the capsule all 1-seeded.
  - Leaves lanceolate; seeds oblong; posterior petals 12–25 mm. long. — 3. *C. erecta.*
  - Leaves linear-lanceolate; seeds round; posterior petals 10–18 mm. long. — 4. *C. crispa.*
- All petals blue, the anterior one only slightly smaller; posterior cells of the capsule 1-seeded, the other two 2-seeded. — 5. *C. virginica.*

1. **C. commùnis** L. Stem branched, 3–10 dm. long, creeping; leaf-blades lanceolate, 4–12 cm. long, acute or acuminate; spathe 2–4 cm. long, ovate, twice as long as broad; sepals 4–5 mm. long; anterior petals white, rhombic-ovate or rhombic-lanceolate, the posterior ones reniform, clawed; seeds 3.5–4 mm. long. Waste ground around gardens: Mass.—Mo.—Kans.—Tex.—N.C.; native of e Asia. Jl–O.

2. **C. longicaùlis** Jacq. Stem usually branched from the base, spreading or creeping, 3–10 dm. long; leaf-blades lanceolate, 3–10 cm. long, acute; sheath sometimes ciliate or erose; spathe 2–3 cm. long, often twice as long as broad, acuminate; seeds about 2.5 mm. long, reticulate. *C. nudiflora* Am. auth.; not L. Moist and waste ground: N.J.—Kans.—Tex.—Fla.; Trop. Am. *E. Temp.—Trop.* Jl–O.

f. 109.

3. **C. eréеcta** L. Stem 1–8 dm. long, pubescent at least at the nodes; leaf-blades 6–12 cm. long, more or less pubescent; sheaths hirsute and ciliate; spathe 2–2.5 cm. long, short-acuminate; capsule 3–4 mm. long and broad. Banks: N.Y.—Kans.—Tex.—Fla. *E. Temp.* Au–O.

4. **C. críspa** Wooton. *Fig. 109.* Perennial, with tuberous roots; stem 3–8 dm. high, branched, finely

villous-pubescent above; leaves linear-lanceolate, 3–7 cm. long, 4–6 mm. wide; spathe 1.5–2 cm. long, strongly curved, pubescent and bearded at the base; sepals broadly elliptic-rotund, the upper one slightly narrower and half as long; upper two petals broadly reniform, 1–1.5 mm. long, bright blue, the third white, lanceolate, 2–3 mm. long; fertile stamens 3; staminodia 3. Sandy places: Ind.—Neb.—Colo.—Ariz.—Tex. *Prairie—Plain—Son.* Jl–Au.

5. **C. virgínica** L. Stem 2–12 dm. high, simple or nearly so; leaf-blades lanceolate or oblong, 1–2 dm. long, often scabrous; spathe usually clustered, deltoid, 2–2.5 cm. long; petals ovate; capsule 7–8 mm. long. *C. hirtella* Vahl. Wet soil, swamps, and shores: Pa.—Mo.—Kans.—Tex.—Fla. *Allegh.—Austral.* Je–S.

Family 19. **PONTEDERIACEAE.** Pickerel-weed Family.

Perennial bog or water plants, with rootstocks and alternate or basal leaves. Flowers perfect, more or less irregular. Sepals and petals each 3, both colored and partly united. Stamens 3–6; filaments partly adnate to the perianth. Pistil of 3 united carpels; ovary 3-celled, or by abortion 1-celled. Fruit a many-seeded capsule or a 1-seeded utricle.

Flowers 2-lipped; stamens 6; fruit a 1-seeded utricle. 1. Pontederia.
Flowers regular; stamens 3; fruit a many-seeded capsule.
  Stamens unequal, the 2 posterior ones with ovate anthers, the third with a sagittate anther; capsule incompletely 3-celled; leaf-blades broad, long-petioled. 2. Heteranthera.
  Stamens all alike; anthers all sagittate; capsule 1-celled, with 3 parietal placentae; leaf-blades linear, translucent, sessile. 3. Zosterella.

1. **PONTEDÈRIA** L. Pickerel-weed, Wampee.

Water herbs, with horizontal rootstocks. Leaves leathery, many-nerved, the basal ones long-petioled, the stem-leaves few. Flowers perfect, many in a dense spike subtended by a bractlike spathe. Perianth ephemeral, corolla-like, 2-lipped, blue or white, the upper lip of 3 broad, the lower of 3 narrow lobes. Stamens 6, the filaments filiform, adnate to the perianth, the anthers introrse, rarely versatile. Ovary 3-celled, but 2 of the cavities empty; style filiform. Fruit a 1-celled, 1-seeded utricle, enclosed in the accrescent base of the perianth.

f. 110.

1. **P. cordàta** L. *Fig. 110.* Stem 3–12 dm. high, mostly simple; petioles of the basal leaves long, with sheathing stipules inside their bases, those of the stem leaves shorter; blades ovate or lanceolate, 1–2 dm. long, obtuse, entire, cordate or sagittate at the base; perianth 1–1.5 cm. long; lobes of the upper lip oblong, shorter than the curved tube, the middle one with 2 yellow spots; lobes of the lower lip linear; stamens and style blue or white. Borders of streams, ponds, and swamps: N.S.—Minn.—Tex.—Fla. *Temp.* Je–S.

2. **HETERÁNTHERA** R. & P. Mud-plantain.

Creeping or floating water plants. Leaves long-petioled, with broad, ovate, cordate, rounded, or reniform blades, sometimes reduced to phyllodes. Flowers perfect, solitary or few, subtended by a spathe-like bract. Perianth white or blue, with a slender tube. Stamens 3; anthers introrse, that of the anterior stamen sagittate, those of the other two ovate. Ovary incompletely 3-celled; ovules numerous in two rows on each placenta.

Leaf-blades ovate or oval, rarely slightly cordate; spathe 1-flowered. 1. *H. limosa.*
Leaf-blades rounded-cordate, as broad as long; spathe 2–5-flowered. 2. *H. peduncularis.*

1. **H. limòsa** (Sw.) Willd. *Fig. 111.* Stem floating, 1–5 dm. long, branched; leaves petioled; petioles 3–20 cm. long; blades ovate to oblong-ovate, 1–3 cm. long, rounded or subcordate at the base; spathe 1-flowered; perianth blue or white; tube 1.5–2 cm. long; lobes linear. Shallow water: Va.—Fla.—Mex.—Colo.—S.D.; W. Ind. and Trop. Am. *E. Temp.—Trop.—Plain.* Je–Au.

2. **H. pedunculàris** Benth. Stem creeping in mud, 0.5–4 dm. long; petioles 5–20 cm. long; blades 3–6 cm. broad, deeply cordate at the base, mostly acute, raceme distinctly exserted from the sheath; perianth pale blue, the tube 7–8 mm. long, slightly curved, the lobes shorter. *H. reniformis* Auth. [G, B]; not R. & P. Shallow water: Mo.—Neb.(?)—Kans.—Tex.—La; Mex. *Prairie—Trop.*

3. ZOSTERÉLLA Small. Water Star-grass.

Submersed grass-like plants. Leaves sessile, narrowly linear, translucent. Flowers perfect, 1 or 2, subtended by a bract. Perianth yellow, with a slender tube. Stamens 3; anthers all sagittate, introrse. Ovary 1-celled, with 3 parietal placentae; ovules numerous, 2-ranked. *Schollera* Schreb.

1. **Z. dùbia** (Jacq.) Small. Stem floating or creeping, 2–10 dm. long, branched; leaves linear, flat, sheathing at the base, 1–2 dm. long; perianth-tube 1 mm. thick; lobes linear, 8–12 mm. long. *Heteranthera graminea* (Michx.) Vahl. *H. dubia* (Jacq.) MacM. [G, B]. Still water: Ont.—Fla.—Mex.—Ore.; W. Ind. *E. Temp.—Trop.—Plain.* My–S.

Family 20. MELANTHIACEAE. Bunch-flower Family.

Leafy-stemmed perennials, with rootstocks or solid bulbs. Flowers racemose or paniculate, perfect, dioecious, or polygamous. Petals and sepals each 3, distinct or nearly so. Stamens 6; filaments often partly adnate to the base of the sepals and petals; anthers versatile. Pistil of 3 united carpels; ovary 3-celled; styles 3. Fruit a septicidal capsule.

Plants with rootstocks.
  Anthers oblong or ovate, 2-celled; petals and sepals glandless.
    Anthers introrse; flowers involucrate with 3 bractlets. 1. Tofieldia.
    Anthers extrorse; flowers not involucrate. 2. Chamaelirion.
  Anthers cordate or reniform, confluently 1-celled.
    Sepals and petals each with 2 glands near the base. 3. Melanthium.
    Sepals and petals glandless; leaves broad. 4. Veratrum.
Plants with bulbs; petals and sepals with a more or less distinct gland.
  Ovary partly inferior; gland obcordate. 5. Anticlea.
  Ovary wholly superior; gland obovate or semiorbicular. 6. Toxicoscordion.

1. TOFIÉLDIA Huds. Scottish Asphodel.

Perennial herbs, with short rootstocks, fibrous roots, 2-ranked, linear, equitant leaves, and small flowers in a terminal raceme. Flowers involucrate by 3 more or less united bractlets below the calyx. Petals and sepals oblong or obovate, subequal, persistent, glandless. Stamens 6; anthers ovate, introrse. Capsule septicidal to the base, many-seeded.

Stem glabrous, scapiform; seeds unappendaged. 1. *T. palustris.*
Stem viscid-pubescent, at least above; seeds appendaged. 2. *T. glutinosa.*

1. **T. palústris** Huds. A glabrous plant, with a scape-like, leafless stem, 3–15 cm. high, and a few basal leaves, 2–10 cm. long; raceme in flower short, oblong, dense; flowers greenish; petals and sepals obovate, obtuse; capsule oblong-globose, minutely beaked. In wet places: Greenl.—Que.—Minn.—B.C.; Alaska; Eur. *Arct.—Subarct.—Alp.—Subalp.* Je–Jl.

2. **T. glutinòsa** (Michx.) Pers. Stems 1.5–5 dm. high, viscid-pubescent; leaves linear, 5–20 cm. long; panicle 2–4 cm. long, somewhat interrupted; pedicels usually 3 or 4 together, 3–10 mm. long, ascending; flowers whitish; sepals and petals oblong or oblong-spatulate, 3–4 mm. long, obtuse; capsule oblong, 6–7 mm. long. Bogs: Newf.—Minn.—N.C. *Canad.—Allegh.* Je–Jl.

2. CHAMAELÍRION Willd. Blazing Star, Devil's Bit.

Glabrous caulescent herbs, with rootstocks. Leaves broadened upwards, thickish, mostly basal. Flowers dioecious, in elongate, spike-like racemes. Petals and sepals white, narrow, 1-nerved, nearly equal. Stamens 6, rudimentary in the pistillate flowers. Ovary 3-celled. Capsule elongate, 3-sided. Seeds 6–12 in each cavity, winged.

1. **C. lùteum** (L.) A. Gray. Stem 2–12 dm. high, simple. Basal leaves spatulate or oblong-spatulate, 5–20 dm. long, tapering below into a broad petiole; stem-leaves oblanceolate to linear, few; raceme 5–20 cm. long, the staminate ones nodding at the tip; pedicles 1–5 mm. long; perianth white; capsule oblong, 7–10 mm. long. Open woods: Mass.—Mich.—Neb.—Ark.—Fla. *Allegh.—Austral.*

3. MELÁNTHIUM L. Bunch-flower.

Coarse caulescent herbs, with thick rootstocks. Leaves mostly basal, narrow. Flowers monoecious or polygamous, paniculate. Perianth white, cream-colored or greenish; sepals and petals clawed, spreading, the blades with 2 glands at the base. Stamens 6, filaments shorter than the perianth. Ovary 3-celled, sessile; styles 3. Ovules numerous. Capsule 3-celled, 3-lobed. Seeds flat, winged.

1. **M. virgínicum** L. *Fig. 112.* Stem 7–15 dm. high, leafy below the middle; leaves linear, 1–4 dm. long, attenuate, pubescent; panicle narrow, 1–5 dm. long, the branches short, ascending; perianth greenish-yellow, turning brownish, 1.5–2 cm. long; capsule broadest near the middle, 1–15 cm. long. Swamps and meadows: R.I.—Minn.—Kans.—Tex. Fla. *Allegh.—Austral.* Je–Au.

4. VERÀTRUM (Tourn.) L. White Hellebore, False Hellebore, Skunk Cabbage

Tall poisonous perennials, with broad, strongly veined and plaited leaves, and stout rootstocks. Flowers paniculate, generally polygamous, *i. e.*, the upper perfect and the lower staminate. Petals and sepals each 3, subequal, glandless and clawless. Stamens 6, free; anthers cordate, their sacs confluent. Fruit a many-seeded, slightly inferior capsule. Seeds flat, broadly winged.

Plant very leafy to the top; leaves oval; perianth yellowish green, the segments ciliate-serrulate. 1. *V. viride.*
Plant leafy at the base; basal leaves oblanceolate; perianth greenish purple, the segments entire. 2. *V. Woodii.*

1. **V. víride** Ait. Stem 6–25 dm. high, pubescent; leaves broadly oval or the upper elliptic, 1–3 dm. long, pubescent, acute or acuminate, narrowed below into the sheathing base; panicle 2–6 dm. long, pubescent, with spreading branches; pedicels shorter than the bracts; perianth yellowish green, 2–2.5 cm. broad; sepals and petals oblong, ciliate and serrulate; capsule, 2–2.5 cm. long, oblong. Swamps and deep woods: Que.—Minn.—Tenn.—Ga. *Canad.—Allegh.* My–Jl.

2. **V. Woòdii** Robbins. *Fig. 113.* Stem slender, 6–15 dm. high; leaves mostly basal, oblanceolate, sheathing, 5–8 cm. wide; stem-leaves small, linear-lanceolate; panicle 3–6 dm. long, with ascending branches; perianth greenish-purple, 1.2–1.5 cm. broad; sepals and petals oblanceolate, obtuse, entire; capsule pubescent when young, 1.2–1.5 cm. long. Woods and hills: Ind.—Iowa—Kans.—Ark. *Ozark.* Je–Jl.

5. **ANTICLÈA** Kunth. WHITE CAMAS.

Bulbiferous glabrous perennials. Leaves linear, sheathing. Flowers perfect, greenish or yellowish white, in racemes or panicles. Petals and sepals nearly alike, elliptic to obovate, withering-persistent, bearing an obcordate gland within above the narrowed base, perigynous. Stamens free; anthers cordate or reniform. Ovary partly inferior. Seeds numerous, angled.

Petals and sepals white or straw-colored, greenish only on the midrib, not clawed. 1. *A. elegans.*
Petals and sepals greenish, the former more or less contracted into a broad claw. 2. *A. chlorantha.*

1. **A. élegans** (Pursh) Rydb. *Fig. 114.* Stem glabrous, light green, 3–6 dm. high; leaves 1–3 dm. long, 5–15 mm. wide, keeled; flowers racemose, or rarely paniculate, dirty-white; bracts ovate to lanceolate, often membranous-margined; petals and sepals obovate or oval, obtuse; capsule ovoid. *Z. elegans* Pursh. *Z. glaucus* Nutt., in part. *Z. dilatatus* Greene. In meadows: Sask.—N.D.—Neb.—N.M.—Nev.—Alaska. *Plain—Subalp.* Je–Au.

2. **A. chlorántha** (Richardson) Rydb. Stem glabrous, green, 3–10 dm. high; leaves 1–3 dm. long, 5–12 mm. wide, keeled; flowers paniculate, about 1 cm. long; sepals and petals usually 7-nerved, elliptic; capsule ovoid. *Zygadenus chloranthus* Richardson [G]. *Z. glaucus* Nutt., in part. Rocky or stony places, often on limestone: N.B.—N.Y.—Ill.—S.D.—Sask. *Canad.—Allegh.* Je–Au.

6. **TOXICOSCÒRDION** Rydb. POISON CAMAS, DEATH CAMAS.

Bulbiferous, glabrous, more or less poisonous perennials, with narrow linear conduplicate leaves and yellowish, racemose or paniculate flowers. Flowers perfect. Petals and sepals bearing an obovate or semiorbicular gland at or above the base of the blade. Ovary wholly superior. Stamens more or less adnate to the base of the petals and sepals; anthers subreniform, confluently 1-celled. Seeds numerous.

Petals and sepals rounded or obtuse at the apex.
Petals without claws; glands obovate; filaments free. 1. *T. Nuttallii.*
Petals clawed, subcordate at base; glands ill-defined, semiorbicular; filaments slightly adnate at the base. 2. *T. gramineum.*

Petals and sepals acute at the apex; all leaves with sheaths; sepals cuneate at the base and short-clawed or subsessile. 3. *T. acutum.*

1. **T. Nuttállii** (A. Gray) Rydb. Stem 3–7 dm. high; inflorescence racemose or paniculate; leaf-blades 1–5 dm. long, often falcate; pedicels slender, longer than the bracts; perianth yellowish-white; sepals and petals ovate-oval, notched, 6–8 mm. long; capsule conic, 8–12 mm. long. *Zygadenus Nuttallii* S. Wats. [G]. Prairies: Tenn.—Kans.—Tex. *Ozark.* My–Je.

2. **T. gramíneum** Rydb. Stem 2–3.5 dm. high; leaves linear, 1–2 dm. long, 3–9 mm. wide; flowers light yellow; sepals broadly ovate, obtuse; petals ovate, obtuse, subcordate at the base. *Z. gramineus* and *Z. intermedius* Rydb. Hills and meadows: Man.—S.D.—Colo.—Utah—Ida. *Submont.* Je–Jl.

3. **T. acùtum** Rydb. Stem 3–5 dm. high; leaves narrowly linear, scabrous on the margin, about 2 dm. long and 4–5 mm. wide; petals and sepals 4–5 mm. long, both acute at the apex and at the base; glands obovate or cuneate. *Z. acutus* Rydb. Hillsides: S.D. *Submont.* Je–Jl.

Family 21. **JUNCACEAE.** RUSH FAMILY.

Grass-like herbs. Flowers perfect, regular, inconspicuous; sepals and petals each 3, similar, scale-like. Stamens 6 or 3, rarely 4 or 5; anthers introrse. Pistil of 3 united carpels; ovary 1- or 3-celled; stigmas 3. Fruit a loculicidal capsule. Seeds 3–many, often apiculate or tailed.

Leaf-sheaths open; capsule 1–3-celled, with axile or parietal placentae; seeds many. 1. JUNCUS.
Leaf-sheaths closed; capsule 1-celled with basal placentae: seeds 3. 2. LUZULA.

1. **JÚNCUS** (Tourn.) L. RUSH, WIRE GRASS.

Caulescent or rarely scapose swamp plants, with glabrous leaves and stems. Leaves with laterally or vertically flattened, or terete leaf-blades, or the latter sometimes wanting; sheaths with free margins. Cymes paniculate, corymbiform, or capitate. Flowers subtended by a bract and sometimes also by 2 bractlets. Stamens 6 or 3. Capsule 1-celled or by the intrusion of the placentae 3-celled. Seeds often tailed or apiculate, reticulate or ribbed, many.

Lower bracts of the inflorescence terete, erect, appearing like a continuation of the stem; inflorescence therefore apparently lateral. I. EFFUSI.
Lower bracts not appearing as a continuation of the stem, or if so, channeled on the upper side; inflorescence terminal.
  Leaves neither septate nor equitant.
    Leaves not fistulose.
      Flowers bracteolate, inserted singly on the branches of the inflorescence; leaves narrowly linear, either flat or subterete and channeled; stamens 6.
        Perennials, with short cespitose rootstocks; stems simple. II. TENUES.
        Annuals; stem branching; seed apiculate. III. BUFONII.
      Flowers not bracteolate, in true heads on the branches of the inflorescence; leaves flat, often grass-like; perennials with stoloniferous rootstocks. V. GRAMINIFOLII.
    Leaves fistulose (*i.e.*, hollow); flowers few in small heads; lower sheath bladeless; seeds caudate; stamens 6. IV. ALPINI.
  Leaves septate.
    Leaves terete, not equitant.
      Septa poorly developed; heads 1–3. IV. ALPINI.
      Septa well developed; heads usually several. VI. NODOSI.
    Leaves equitant, laterally flattened so that one edge is towards the stem; seeds apiculate; perennials with creeping rootstocks. VII. ENSIFOLII.

I. EFFUSI.

Stamens 3; stem, sepals, and petals green; cyme many-flowered. 1. *J. effusus.*
Stamens 6; cyme rather few-flowered.
  Stem light green, striate when dry on account of the free hypodermal fibro vascular bundles; sepals and petals green. 2. *J. filiformis.*

Stem dark green or at the base purplish, not striate; sepals and petals dark-brown.
Sepals short-acuminate, about equaling the acute or obtusish petals. — 3. *J. ater.*
Sepals long-acuminate, longer than the acute petals. — 4. *J. litorum.*

II. Tenues.

Leaf-sheaths covering one half of the stem or more; sepals and petals obtuse, incurved at the tip. — 5. *J. Gerardi.*
Leaf-sheaths covering one fourth of the stem or less; sepals and petals acute or acuminate.
Seeds long-caudate; leaves subterete, grooved above. — 6. *J. Vaseyi.*
Seeds apiculate or blunt.
Capsule reddish or chestnut-brown, much exceeding the sepals; leaves grooved above. — 7. *J. Greenei.*
Capsule green or straw-colored, shorter than or equaling the sepals; leaves flat.
Auricles at the summit of the sheaths membranous, whitish.
Auricles scarcely produced beyond the insertion, scarcely scarious; petals and sepals scarcely spreading. — 8. *J. interior.*
Auricles conspicuously produced beyond the point of insertion.
Capsule oblong, narrow. 3-celled, equaling the perianth or nearly so, sepals and petals erect or appressed.
Stem stout; leaves short and broad (1.5–2 mm. wide), sepals and petals 4–5 mm. long, scarious at the base only, stramineous; flowers in an open cyme. — 9. *J. brachyphyllus.*
Stem slender; leaves narrow and long; sepals and petals 3.5–4 mm. long, scarious to the apex, fuscous; flowers few, congested. — 10. *J. confusus.*
Capsule ovate or oval, 1-celled, three fourths as long as the petals or less; petals and sepals spreading. — 11. *J. tenuis.*
Auricles cartilaginous, yellowish brown; inflorescence greenish; capsule ovate; petals and sepals spreading. — 12. *J. Dudleyi.*

III. Bufonii.

One species. — 13. *J. bufonius.*

IV. Alpini.

Leaf-sheaths not auricled. — 14. *J. castaneus.*
Leaf-sheaths auricled. — 15. *J. stygius.*

V. Graminifolii.

Stamens 3; anthers brown; capsule not mucronate.
Petals lanceolate, setiform-acuminate. — 16. *J. setosus.*
Petals obovate, obtuse, hyaline-margined. — 17. *J. marginatus.*
Stamens 6; anthers yellow; capsule mucronate. — 18. *J. longistylis.*

VI. Nodosi.

Stamens 6.
Flowers solitary or paired, accompanied or replaced by leaves. — 19. *J. pelocarpus.*
Flowers in glomerules, not accompanied by leaves.
Inflorescence with short branches; flowers echinate-spreading or the lowest of the heads reflexed; capsule narrowly lanceolate.
Heads 7–10 mm. in diameter; leaf-blades erect; petals usually longer than the sepals. — 20. *J. nodosus.*
Heads 10–16 mm. in diameter; leaf-blades usually spreading; sepals longer than the petals. — 21. *J. Torreyi.*
Inflorescence with elongated branches; flowers erect-ascending.
Sepals acuminate; branches of the inflorescence divergent.
Flowers brown; capsule dark brown, 3–4 mm. long, gradually tapering at the apex. — 22. *J. articulatus.*

Flowers greenish; capsule pale-brown, 2.5–3 mm. long, abruptly mucronate. — 23. *J. amblyocarpus.*
Sepals blunt or mucronate; branches of the inflorescence erect or nearly so. — 24. *J. Richardsonianus.*
Stamens 3.
Seeds not caudate.
Capsule attenuate at the apex, much exceeding the sepals.
Heads 2–7-flowered; capsule not subulate. — 25. *J. diffusissimus.*
Heads many-flowered; capsule subulate. — 26. *J. scirpoides.*
Capsule neither attenuate nor exceeding the sepals.
Capsule 1/2–2/3 as long as the sepals, gradually acuminate; sepals subulate, much exceeding the petals. — 27. *J. brachycarpus.*
Capsule about equaling or exceeding the sepals, abruptly mucronate; sepals and petals subequal.
Heads 1–50, on ascending branches; flowers 3–3.5 mm. long. — 28. *J. acuminatus.*
Heads 200–300, on divergent branches; flowers 2–2.5 mm. long. — 29. *J. nodatus.*
Seeds caudate.
Fruit 2–3 mm. long; sepals obtuse; seeds short-caudate. — 30. *J. brachycephalus.*
Fruit 4 mm. long; sepals and petals acuminate; seeds distinctly caudate.
Capsule about equaling the sepals; short-pointed; seeds 1–2 mm. long; inflorescence short. — 31. *J. canadensis.*
Capsule much exceeding the sepals, gradually tapering; seeds 1 mm. long; inflorescence elongate. — 32. *J. brevicaudatus.*

VII. Ensifolii.

One species. — 33. *J. saximontanus.*

1. J. effùsus L. Stem soft, pliant, 3–12 dm. high; leaves mostly basal, reduced to the mere sheaths, the inner sheaths bristle-pointed; cymes many-flowered, diffuse, much branched; bractlets broadly ovate; flowers 2–2.5 mm. long; sepals and petals lanceolate, acute, as long as the capsule; capsule greenish-brown, triangular-ovoid, retuse and pointless; seeds small, 0.5 mm. long, with short pale apiculations, longitudinally reticulate. Swamps: Newf.—Man. —Tex.—Fla.; B.C.—Calif.; nat. from Eur.

2. J. filifórmis L. Stem 1–5 dm. high, slender, 1–2 mm. thick; leaves basal and reduced to brown sheaths; inflorescence 6–10-flowered, open; sepals 4–5 mm. long, lanceolate, acute, slightly exceeding the obtuse or acutish petals; capsule obovoid, green, about equaling the petals, 3-celled; seeds acute, but scarcely apiculate. In wet places: Greenl.—Pa.—Utah—Wash.—Alaska; Eurasia. *Arct.—Boreal—Plain—Submont.* Je–Au.

3. J. àter Rydb. *Fig. 115.* Stem dark green, 2–6 dm. high, about 2 mm. thick; sheaths loose, dark brown; flowers 5–20; sepals and petals 5–6 mm. long; capsule obpyramidal, long-mucronate, 3-celled. *J. balticus montanus* Engelm. Mountain valleys: Alaska — Calif. — N.M. — Kans. — N.D. — Mont. *Plain—Subalp.* Jl–Au.

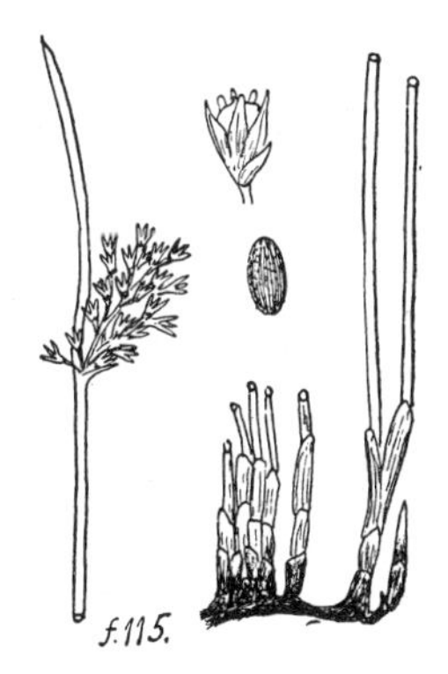

4. J. lítorum Rydb. Stem 3–10 dm. high, rigid; sheath loose, dark brown; inflorescence rather open, 4–9 cm. long; flowers numerous, chestnut-brown or dark green; sepals ovate-lanceolate, 5–6 mm. long, sharp-pointed, equaling the similar but obtuse petals; capsule ellipsoid, deep brown, obtuse or mucronate; seeds large, obtuse. *J. balticus littoralis* Engelm. [G.]; not *J. littoralis* C. A. Meyer. Sandy beaches: N.S.—Pa.; also along the Great Lakes, N. Y.—Minn.

5. **J. Gerárdi** Loisel. Stem rigid, scarcely flattened, 1.5–8 dm. high; cyme narrow, usually longer than the bract; flowers 3–4 mm. long; sepals and petals oval-oblong, about equaling the capsule; capsule ovoid, obtuse, and mucronate; seeds 0.4–0.5 mm. long, ribbed and cross-lined. Salt marshes: Me.—Minn.—N. Y.; Eur. *Canad.* Je–S.

6. **J. Vàseyi** Engelm. Stems tall and stiff, 3–8 dm. high, terete; leaf-blades three fourths as long as the stem; inflorescence 1–3.5 cm. long, with erect 2–4-flowered branches; sepals and petals 3.5–4 mm. long, green or straw-colored, lanceolate, erect, acute, and narrowly scarious-margined; capsule oblong-cylindric, obtuse. Moist shores and wet woods: Me.—Iowa—Colo.—Sask. *E. Boreal—Mont.* Je–Au.

7. **J. Greènei** Oakes & Tuckerm. Stem rigid, 2–8 dm. high; leaves nearly terete, deeply channeled; cymes 1–6 cm. long, dense, many-flowered, secund, shorter than the bract; sepals and petals 4–5 mm. long, lanceolate, light brown, acute, appressed; seeds 0.5 mm. long, ribbed and delicately cross-lined. Sandy soil, along the coast and Great Lakes: N.B.—N.J.—Minn. *Canad.* Je–S.

8. **J. intèrior** Wieg. Stems tall, stout, 5–10 dm. high, nearly terete, coarsely grooved; leaves several, with blades one third as long as the stem, 1–1.25 mm. wide; inflorescence large and open, 3–10 cm. long, many-flowered; sepals and petals 3–4 mm. long, nearly equal, lanceolate, very acute, narrowly scarious-margined, erect; capsule oblong, obtuse, imperfectly 3-celled. Dry woods and prairies: Ill.—Ark.—Tex.—N.M.—Wyo. *Prairie—Plain—Submont.* My–Au.

9. **J. brachyphýllus** Wieg. Stem stout, 4–5 dm. high, slightly compressed, grooved; leaf-blades short, one fourth to one third as long as the stem, broad and flat, 1.5–2 mm. wide; inflorescence short, many-flowered, 2–6 cm. long; sepals lanceolate, very acute, straw-colored; petals similar but scarious-margined all around; capsule narrowly oblong, obtuse or retuse. Meadows: Ark.—Okla.—Ida. *Ozark—Plain—Submont.* My–Jl.

10. **J. confùsus** Coville. Stem slender, 4–5 dm. high, light green, slightly grooved; leaf-blades filiform, two thirds as long as the stem or less, flat or involute; sepals and petals lanceolate, acute, straw-colored, with dark stripes on each side, scarious-margined; capsule oblong, a little shorter than the petals, retuse. *J. tenuis congestus* Engelm., in part. Meadows: B.C.—n N.M.—Neb.—Sask. *W. Temp. Plain—Mont.* Je–Au.

11. **J. ténuis** Willd. Stem 2–6 dm. high, spreading, slightly flattened, striate; leaf-blades nearly as long as the stem, 1–1.25 mm. wide, flat, soft; inflorescence open, 1–7 cm. long, with the flowers near the ends of the branches; petals and sepals 3–4.5 mm. long, lanceolate, very acute, green, with white scarious margins. Roadsides and meadows: Newf.—Fla.—Tex.—Mont.—Ore.—Wash. *Temp.—Submont.* My–S.

12. **J. Dúdleyi** Wieg. Stem 3–10 dm. high, stiff, erect, strongly striate; leaf-blades half as long as the stem or less, narrow, flat or involute; inflorescence small, dense, 2–5 cm. long; sepals and petals 4–5 mm. long, lanceolate, acute, yellowish green, with scarious margins; capsule ovoid, somewhat shorter than the petals, rounded and apiculate at the apex, imperfectly 3-celled. Damp places and meadows: Me.—N.Y.—N.M.—Ariz.—Wash.—Alta.; Mex. *Boreal—Submont.* My–Au.

13. **J. bufònius** L. Stem 0.5–2 dm. (seldom 3 dm.) high; leaf-blades flat, 0.25–1 mm. wide; flowers inserted singly on the branches; sepals 4–6 mm. long, narrowly lanceolate, subulate-attenuate, scarious-margined; petals somewhat shorter, less attenuate; capsule shorter than the petals, obtuse, 3-celled. Wet places: Greenl.—Fla.—Calif.—Alaska; Mex.; also Old World. *Arct.—Temp.—Trop.—Mont.* Ap–S.

14. **J. castàneus** Smith. Stems erect, terete, 1–3 dm. high, more or less leafy; blades channeled, 3–10 cm. long, 1–2 mm. thick; sepals linear-lanceolate, about 5 mm. long, acute, chestnut-brown; petals obtuse; ovate-prismatic, trigonous, acute, imperfectly 3-celled, almost black. Arctic-alpine regions: Greenl.—Newf.—Man.—N.M.—Alaska; Eurasia. *Arct.—Subarct.—Subalp.—Alp.* Je–Au.

15. **J. stýgius** L. Stem 1–3 dm. high, naked above; leaves filiform, 2–10 cm. long; heads 1–4, equaling the bract, 1–4-flowered; sepals lanceolate, acute, 3–5 mm. long, reddish; petals obtuse, three fourths as long as the capsule; capsule trigonous, ovoid, mucronate, acute or acuminate, pale; seeds oblong, with a loose coat, caudate at each end, 2–2.5 mm. long. Newf.—Minn.—N.Y.; Eur. *Subarct.—E. Boreal.* Jl–Au.

16. **J. setòsus** (Coville) Small. Stems loosely tufted, 3–7 dm. high; leaves numerous, with auriculate sheaths, 2–5 mm. wide; panicle 3–10 cm. long; heads usually 20–100; flowers 3–3.5 mm. long; sepals lanceolate, setiform-acuminate; stamens 3, shorter than the perianth; anthers reddish-brown; capsule globose-oblong, blunt, as long as the perianth. *J. marginatus setosus* Coville [G]. Wet places in woods: Ark.—Neb.—Tex. *Ozark—Texan.*

17. **J. marginàtus** Rostk. Stem erect, 2–7 dm. high, from a bulbous and stoloniferous base; leaves linear, 1–3 mm. wide; heads in simple or compound cymes, 3–12-flowered; sepals and petals oblong, the former acute or short-awned, the latter longer, 3–5 mm. long, obtuse; capsule as long as the petals, subglobose; seeds 0.3 mm. long, caudate at each end, strongly ribbed. Sandy places: Me.—Minn.—Neb.—Tex.—La. Jl–S.

18. **J. longístylis** Torr. Stem erect, slender, compressed, 3–5 cm. high, 1–1.5 mm. thick; leaf-blades 1–2 mm. wide, 5–10 cm. long; inflorescence of 1–6 heads; heads 5–12-flowered; sepals as in the preceding; petals broader, more obtuse, and with broader margins; capsule shorter than the perianth, oblong, brown, trigonous above, obtuse. Meadows: Man.—S.D.—Neb.—N.M.—Calif.—Ore.; Mex. *Prairie—Mont.* Je–Au.

19. **J. pelocárpus** E. Meyer. Stem slender, 0.5–5 dm. high, branching above, with spreading cymes; leaves few, nearly filiform, slightly septate; flowers 2.5 mm. long, greenish, tinged with red; sepals and petals obtuse, oblong, the latter longer than the former, but shorter than the capsule; capsule slender, attenuate, 1-celled; seeds 0.5 mm. long, ovoid, short-pointed. Sandy wet places: Newf.—Minn.—N.J. *Canad.* Au–S.

20. **J. nodòsus** L. Stem 1.5–6 dm. high, erect, terete or somewhat compressed, leafy; leaf-blades erect, 0.5–1 mm. thick, terete, 5–15 cm. long, septate; heads 1–30, 8–20-flowered; sepals 2.5–3.5 mm. long, lanceolate, subulate; petals similar but slightly longer; capsule lanceolate-subulate, 3-gonous, 1-celled, straw-colored, exceeding the perianth. Wet soil: N.S.—Va.—N.M.—Nev.—B.C.—Mack. *Temp.—Submont.* Jl–S.

21. **J. Tórreyi** Coville. Stem stout, 3–9 dm. high, 2–4 mm. thick, leafy; leaf-blades 1–3 mm. thick, terete, septate; heads 1–20, congested, 30–80-flowered; sepals and petals lance-subulate, about 5 mm. long; capsule subulate, 3-gonous, 1-celled. *J. nodosus megacephalus* Torr. Wet places, especially in sandy soil: w N.Y.—Miss.—Calif.—S.D. *Temp.—Mont.* Jl–S.

22. **J. articulàtus** L. Stem erect, 1.5–6 dm. high, with a creeping rootstock; leaves 1–3, slender; blades septate, slightly compressed; cymes short, 2–8 cm. long, the branches spreading; heads 3–10-flowered; flowers 2.5–3 mm. long; petals slightly longer than the sepals, shorter than the conic, 3-celled capsule. Sandy shore: Newf.—N.J.—Neb.—Man.; Europe. *Canad.* Jl–Au.

23. **J. amblyocárpus** Rydb. Stem 1.5–2.5 dm. high, leaves 5–8 cm. long, 2 mm. thick; septate, slightly compressed; branches of the inflorescence spreading; sepals greenish, broadly margined, 2.5 mm. long, acute or obtusish, equaling the obtuse petals; capsule broader than in the preceding, obtuse, mucronate, only slightly exceeding the petals. *J. articulatus obtusatus* Engelm. Brackish beaches, Me.—N.J.—Neb.—Minn. *Canad.* Jl–S.

24. **J. Richardsoniànus** Schult. Stem erect, 1.5–5 dm. high; blades terete or slightly compressed, septate, 0.5–1 mm. thick; inflorescence with 5–25 heads, which are 4–6 mm. thick, 3–12-flowered; sepals 2–2.5 mm. long, pale greenish, acutish; petals oblong, acute; capsule ovoid-oblong, slightly exceeding the perianth. *J. alpinus insignis* Fries. [G]. In wet soil: N.S.—Pa.—Neb.—Wash.—Alaska. *Boreal—Plain—Mont.* Jl–S.

25. **J. diffusíssimus** Buckley. Stem slender, 2.5–6 dm. high; leaves 1–2 dm. long, slightly compressed; cymes diffuse, loosely dichotomous; heads numerous, 2–7-flowered; flowers greenish or brown, 3 mm. long; sepals and petals subequal, linear-subulate, 4–5.5 mm. long, 1-celled; seeds oblong, reticulate, acute at the base, abruptly acuminate at the apex. Wet places: Ind.—Kans.—Tex.—Miss.—Ga. *Allegh.—Austral.* Je–Jl.

26. **J. scirpoìdes** Lam. Stem erect, 3–10 dm. high, slender; stem-leaves usually 2, terete, septate, with wide and open sheaths; cymes loose; heads few or many, densely flowered, 6–13 mm. thick, 15–40-flowered; flowers pale green, 3–4 mm. long; sepals and petals subulate, bristle-pointed, equaling the oblong-triangular 1-celled capsule; seeds ovoid, pointed at each end. Wet sandy soil: N.Y.—Fla.—Kans.—Tex. *E. Temp.* Jl–S.

27. **J. brachycárpus** Engelm. Stem erect, 4–10 dm. high; stem-leaves usually 2, terete; cymes crowded; heads 2–10, densely flowered, 7–11 mm. thick; flowers pale green, 4 mm. long; sepals subulate, exceeding the petals; capsule ½–⅔ as long as the sepals, gradually tapering. Damp soil: Mass.—N.C. —Mo.—Tex. *E. Temp.* Je–Au.

28. **J. acuminàtus** Michx. Stem tufted, slender, erect, 3–7 dm. high; leaves 1–2 dm. long; cymes loose and spreading; heads few and large, 0.5–1 cm. broad, 5–many-flowered; sepals and petals lance-subulate, greenish, in age straw-colored, as long as the capsule; capsule ovoid-prismatic, short-pointed, 1-celled; seeds ribbed and reticulate, 0.3–0.4 mm. long, acute at each end. Wet places: Me.—Minn.—Tex.—Ga.; Mex. *E. Temp.—Trop.* My–Au.

29. **J. nodàtus** Coville. Stem stout, 5–12 dm. high; leaves 1–2, terete, conspicuously septate, 2–6 dm. long; cymes 1–3 dm. long; heads very numerous, 5–8-flowered; petals and sepals lance-subulate, 2–2.5 mm. long; capsule equaling or exceeding the sepals, ovoid-prismatic, 1-celled, straw-colored; seeds fusiform. *J. robustus* (Engelm.) Coville [G, B]; not S. Wats. Swamps: Ills. —Kans.—Tex.—La. *Prairie—Texan.* Je–Jl.

30. **J. brachycéphalus** (Engelm.) Buch. Stem 3–7 dm. high, 2–4 leaved, tufted; leaf-blades up to 2 dm. long; panicle 5–15 cm. long, with spreading branches; heads 2–5-flowered; sepals shorter than the petals and the capsule; flowers greenish or light brown; capsule ovoid-oblong, 3-angled, 1-celled. Marshes and wet shores: Me.—Man.—N.D.—Ill.—Pa. *Canad.* Jl–S.

31. **J. canadénsis** J. Gay. Stem tufted, stout, 4–12 dm. high; leaves 2–4, with loose sheaths, the blades 1–2.5 dm. long; cymes compound, spreading; heads numerous, many-flowered; sepals and petals narrowly lanceolate, awl-pointed, greenish or light brown, 3–4 mm. long, shorter than the capsule; capsule lanceolate, 3-sided, 1-celled; seeds shining, striate above, long-caudate at each end. Marshy places: Newf.—N.D.—La.—Ga. *E. Temp.* Au–O.

32. **J. brevicaùdatus** (Engelm.) Fernald. Stem slender, 1.5–7 dm. high; cymes contracted, 2.5–15 cm. long; heads 3–7-flowered; sepals acute, 2.5 mm.

long; petals obtuse, shorter than the capsule, prismatic, deep brown. Muddy or wet places: Newf.—Man.—Minn.—W.Va.—N.J. *Canad.* Je–S.

33. **J. saximontànus** A. Nels. Stem winged, 2–5 dm. high; leaf-blades 3–20 cm. long, 2–4 mm. wide; inflorescence of 2–15 heads; these about 1 cm. thick; sepals and petals lanceolate, acuminate, about 3 mm. long, dark brown; capsule oblong, mucronate, a little shorter than the perianth, dark brown. *J. xiphioides montanus* Engelm. Meadows and wet places: B.C.—Calif.—N.M.—S.D.—Alta. *Submont.—Mont.* Je–S.

2. LÙZULA DC. WOOD-RUSH.

Perennial caulescent herbs, with rootstocks, glabrous or sparingly pubescent. Leaf-sheaths closed; blades grass-like. Inflorescence umbel-like, capitate, or spike-like. Flowers always subtended by usually lacerate or dentate bractlets. Stamens 6. Capsule 1-celled; ovules and seeds 3, basal. Seeds reticulate, sometimes apiculate but never tailed. *Juncoides* (Dill.) Adans.

- Flowers on slender pedicels in a corymbiform or umbelliform inflorescence.
  - Inflorescence umbelliform; flowers 3–4.5 mm. long, usually solitary on the branches. — 1. *L. saltuensis.*
  - Inflorescence a compound, corymbiform cyme; flowers 2 mm. long. — 2. *L. parviflora.*
- Flowers subsessile, crowded in head-like or spike-like clusters.
  - Spikelets peduncled, forming a corymb.
    - Stem not bulblet-bearing at the base; branches of the inflorescence strongly ascending. — 3. *L. multiflora.*
    - Stem bulblet-bearing at the base; branches of the inflorescence more divergent. — 4. *L. bulbosa.*
  - Spikelets subsessile, forming a compound spike. — 5. *L. spicata.*

1. **L. saltuénsis** Fernald. Plant loosely cespitose, stoloniferous; stem 1–4 dm. high; leaves lance-linear, hairy, the lower 0.5–1 cm. wide; umbel simple, the branches loosely ascending or spreading; sepals and petals lanceolate, pale brown or straw-colored, with hyaline margins, shorter than the conic-ovoid pointed capsule. *L. vernalis* A. Gray; not DC. *J. pilosum* Coville [G]; not Kuntze. Woods and banks: Newf.—Sask.—Minn.—Ga. *Canad.—Allegh.* Ap–My.

2. **L. parviflòra** (Ehrh.) Desv. *Fig. 116.* Stem erect, terete, 3–5 dm. high; leaf-blades broadly linear, 5–15 cm. long, 5–12 mm. wide, glabrous except the mouth of the sheath; flowers singly or rarely 2 or 3 together; sepals and petals lanceolate or ovate-lanceolate, acute; capsule ovate, 3-gonous, obtuse, exserted. *Juncoides parviflorum* Coville [R, B]. Meadows, hillsides, and thickets: Greenl.—N.Y.—N.M.—Calif.—Alaska; Eurasia. *Arct.—Boreal—Mont.—Subalp.* My–S.

3. **L. multiflòra** (Ehrh.) Lej. Stem slender, 2–5 dm. high; leaf-blades 5–10 cm. long, 1–4 mm. wide; inflorescence with erect or strongly ascending branches; heads 4–10, globose or oval, about 6 mm. thick, 8–16-flowered; sepals and petals 2.5–3 mm. long, lanceolate, mucronate, brown or reddish, with paler margins, sometimes white; capsule obovate, 3-gonous, obtuse or retuse, nearly equaling the perianth. *L. campestris multiflora* Celak. [G]. *J. intermedium* Rydb. [R]. Hills and mountains: Greenl.—N.Y.—N.M.—Calif.—B.C.; Eur. *Arct.—Boreal—Mont.* Je–Au.

4. **L. bulbòsa** (Wood) Rydb. Stem 1–4 dm. high; leaves glabrous or somewhat arachnoid on the margins of the blades and at the top of the sheaths;

inflorescence umbel-like, the peduncles very unequal; sepals and petals ovate-lanceolate or lanceolate, 2–2.5 mm. long, brownish, acuminate; capsule rounded-obovoid. *L. campestris bulbosa* Wood [G]. *J. bulbosum* Small. Woods and thickets; D.C.—Kans.—Tex.—Ga. *Austral.—Ozark.*

5. **L. spicàta** (L.) DC. Stem 1–3 dm. high, erect; leaf-blades 4–6 cm. (rarely up to 12 cm.) long, 1–3 mm. wide; inflorescence spike-like, usually nodding; sepals and petals 2–3 mm. long, lanceolate, aristate-acuminate; capsule broadly ovoid, acute, about two thirds as long as the perianth. *J. spicatum* Kuntze [B]. Hills and mountain sides: Greenl.—N.H.—N.M.—Calif.—Alaska; Eurasia. *Arct.—Boreal—Mont.—Alp.* Jl–Au.

Family 22. ALLIACEAE. Onion Family.

Perennial scapose herbs, with bulbs or corms, and narrow basal leaves. Flowers in terminal umbels subtended by or enveloped in a scarious involucre. Sepals and petals each 3, very similar, corolla-like, usually membranous. Stamens 6. Pistil of 3 united carpels; ovary superior, 3-celled; styles united. Fruit a loculicidal, 3-celled capsule.

Sepals and petals distinct or nearly so.
  Ovules and seeds 1 or 2 in each cell; plants with onion-like odor.
    Ovules 2 in each cell; leaves narrow, appearing with the scape. 1. Allium.
    Ovules and seeds solitary in each cell; leaves broad, disappearing before the the flowering time. 2. Validallium.
  Ovules and seeds several in each cell; plants destitute of onion-like odor. 3. Nothoscordium.
Sepals and petals united to about the middle; filaments united into a tube. 4. Androstephium.

1. ÁLLIUM (Tourn.) L. Onion, Garlic, Leek, Chives.

Perennial bulbous plants, with a characteristic alliaceous smell. Bracts scarious, more or less connate. Petals and sepals free or slightly united at the base. Stamens adnate to the bases of the petals and sepals; anthers introrse. Ovary sessile, 3-celled; style filiform, usually deciduous; stigmas minute; ovules 2 in each cell.

Bulb crowning a persistent rootstock; outer coat more less fibrous.
  Leaves terete and hollow; umbels dense, subcapitate, erect. 1. *A. sibiricum.*
  Leaves flat or channeled, not hollow; umbels nodding; petals and sepals obtuse or acute.
    Leaves rounded-convex on the back, not keeled, lunate in cross-section. 2. *A. recurvatum.*
    Leaves almost flat or keeled, somewhat broadly V-shaped in cross-section; leaves 3–5 mm. wide. 3. *A. cernuum.*
Bulbs without a rootstock.
  Leaves not hollow.
    Outer bulb-coat fibrous.
      Umbels bulblet-bearing; flowers few or sometimes none. 4. *A. canadense.*
      Umbels not bulblet-bearing.
        Capsule not crested; involucre usually 3-leaved.
          Scape 3–6 dm. high; pedicels 1.2–2.5 cm. long; perianth-segments thin. 5. *A. mutabile.*
          Scape 1–2 dm. high; pedicels 8–12 mm. long; perianth-segments rigid in fruit. 6. *A. Nuttallii.*
        Capsule crested.
          Plant 1–3 dm. high; pedicels 8–12 mm. long; petals and sepals about 5 mm. long. 7. *A. textile.*
          Plant 2–6 dm. high; pedicels 12–25 mm. long; petals and sepals 6–8 mm. long. 8. *A. Geyeri.*
    Outer bulb-coat not fibrous, but often more or less reticulate.
      Ovary 6-crested; umbels not bulblet-bearing. 9. *A. stellatum.*
      Ovary not crested; umbels usually bulblet-bearing. 10. *A. vineale.*
  Leaves hollow. 11. *A. Cepa.*

1. **A. sibíricum** L. Bulb small, oblong-ovoid, often oblique, about 1 cm. thick; scapes 3–6 dm. high; leaf-blades 6–20 cm. long; bracts usually 2, ovate, about 2 cm. long; petals and sepals bright rose-colored, with dark midrib, fully 1 cm. long, lanceolate, acuminate; stamens included; capsule not crested. *A. Schoenoprasum sibiricum* (L.) Hartm. [G.] Rich soil: Me.—N.Y.—Colo.—Ore.—Alaska; Asia. *Boreal—Mont.* Je–Au.

2. **A. recurvàtum** Rydb. Bulb oblong-ovoid, 1–1.6 cm. thick; scape slender, 3–5 dm. high, almost terete; leaf-blades 1–2 dm. long, 1–3 mm. wide, thick; involucre 2-leaved, almost 2 cm. long; petals and sepals elliptic-ovate, obtuse, 5 mm. long, rose-colored with darker midvein. *A. cernuum obtusum* Cockerell. Dry hills and mountain-sides: Alta.—S.D.—N.M.—B.C. *Plain—Mont.* My–Au.

3. **A. cérnuum** Roth. Bulb lance-ovoid, 1–2 cm. thick, often purplish; scape 3–6 dm. high, stout; leaf-blades 1–2 dm. long, 3–7 mm. wide, rather thin; involucre 2-leaved, rarely more than 1 cm. long; petals and sepals light pink or white, with faint midrib, otherwise as in the preceding. On banks and hill-sides: N.Y.—W.Va.—Colo.—Wash.—Sask. *Boreal—Plain—Submont.* Je–Au.

4. **A. canadénse** L. Bulbs ovoid, 1–2 cm. thick; scape 2–6 dm. high, stout; leaf-blades 1–5 dm. long, 3–8 mm. wide, rounded on the back; involucre 2- or 3-leaved; petals and sepals obtuse, pink or white, 4–6 mm. long; filaments as long as the petals. In meadows and fields: Me.—Fla.—La.—Colo.—Minn. *E. Temp.—Plain.* My–Je.

5. **A. mutábile** Michx. Bulb about 2 cm. thick, the coat conspicuously fibrous; leaves narrowly linear, channeled, 1–4 dm. long; scape longer than the leaves; bracts 2 or 3, ovate to oblong, acuminate; petals and sepals pink or white, 5–7 mm. long, lanceolate, acute or acuminate; filaments dilated at the base, shorter than the perianth. *A. lavandulare* Bates. Fields and borders of woods: Va.—Neb.—Tex.—Fla. *Austral.* Ap–Je.

6. **A. Nuttállii** S. Wats. Bulb solitary, ovoid, about 1.5 cm. thick; scape 1–3 dm. high; leaves 2–3 mm. wide, 1–1.5 dm. long; bracts 1–1.5 cm. long; petals and sepals ovate or ovate-lanceolate, acute or acuminate, rose or white, 4–6 mm. long; stamens shorter than the petals. Plains and prairies: S.D.—Kans.—Ariz.—Ida. *Plain—Prairie.* My– Je.

7. **A. téxtile** Nels. & Macbr. Bulbs usually solitary, ovoid, 1–2 cm. thick; scape slender, terete, 1–3 dm. high; leaf-blades 2–4 mm. wide, 1–1.5 dm. long; bracts fully 1 cm. long; petals and sepals ovate-lanceolate, acute or acuminate, 4–6 mm. long; capsule with 6 small rounded crests. *A. reticulatum* Fraser [G]; not Presl. Plains and dry hills: Man.—Minn.—N.M.—Ariz.—Alta. *Plain—Submont.* Ap–S.

8. **A. Géyeri** S. Wats. *Fig. 117.* Bulbs usually solitary, ovoid, 1.5–2.5 cm. thick; scape 3–6 dm. high, rather stout; leaf-blades 2–4 mm. wide, 1–2 dm. long; bracts 1.5 cm. long or more; petals and sepals pink or white, ovate-lanceolate, acuminate, 6–8 mm. long; capsule with 6 rather prominent crests. *A. reticulatum deserticola* M. E. Jones. *A. deserticola* Woot. & Standl. *A. dictyotum* Greene. Plains and valleys: Wash.—Ariz.—N.M.—N.D.—Mont. *Plain—Mont.* My–S.

9. **A. stellàtum** Ker. Bulb ovoid, 1–2 cm. thick; coats thin, mostly reddish; reticulations fine and close, elongated-rectangular; scapes 2–5 dm. high; leaves several; blades 1–3 dm. long, 1.5–2.5 mm. wide, nearly flat; bracts 2, about 1 cm. long; pedicels 1–2 cm. long; sepals rose-color, ovate, usu-

ally acute, 4–6 mm. long; petals ovate-oblong, mostly obtuse, slightly longer. On rocky banks: Man.—Ill.—Mo.—Kans.—Sask. *Prairie—Plain.* Jl–S.

10. **A. vineàle** L. Bulb ovoid, with membranous but scarcely reticulate coat; stem 3–9 dm. high, covered with sheaths below; leaves terete, fistulose, slender, channeled; umbels densely bulbiferous; filaments dilated, the alternate ones with a cusp on each side. Moist meadows: Mass.—Kans.—Va.; escaped from cult.; native of Asia. Je.

11. **A. Cèpa** L. Tall with fistulose leaves; scape 3–8 cm. tall, overtopping the leaves at the base, swollen near the middle; flowers white or bluish in large umbels; bulb membranous, rounded, spathe 2- or 3-valved, acuminate; pedicels 4–5 times as long as the flowers, sepals and petals spreading, the interior filaments 2-toothed below; escaped from cult.; nat. of the Orient. Jl.–S.

2. **VALIDÁLLIUM** Small. WILD LEEK.

Perennial, bulbous, scapose herbs, with alliaceous smell. Bulbs clustered. Leaves basal, relatively broad, appearing before the scape. Flowers perfect, in terminal umbels subtended by 2 scarious bracts. Perianth white or pale, persistent; sepals and petals each 3, distinct. Stamens 6; filaments slender; anthers introrse. Ovary 3-celled; stigma depressed. Ovules and seeds solitary in each cell. Capsule 3-lobed.

1. **V. tricóccum** (Ait.) Small. Bulbs with fleshy-membranous outer coat; leaves 2, basal, disappearing before the flowers, the blades oblong to elliptic, 1–3 dm. long, tapering at the base; scape 1–4 dm. high; sepals and petals white, oblong to oval, 6–7 mm. long, obtuse; filaments subulate, as long as the perianth; capsule 3-lobed, about 6 mm. broad; seeds smooth, black, shining. *Allium tricoccum* Ait. [G, B]. Woods and hillsides: N.B.—Minn.—Tenn.—Ga. *Canad.—Allegh.* Je–Jl.

3. **NOTHOSCÓRDIUM** Kunth. FALSE GARLIC.

Scapose bulbous herbs, without alliaceous smell. Bulbs with membranous coat. Leaves basal, with narrow blades. Umbel solitary, terminal, subtended by 2 bracts. Flowers perfect. Perianth yellow, greenish, or white. Sepals and petals each 3, narrow, 1-nerved, free. Stamens 6, adnate to the base of the perianth; filaments subulate or filiform; anthers introrse. Ovary 3-celled; style filiform; stigma capitate. Ovules several in each cavity. Capsule 3-celled, 3-lobed, loculicidal. Seeds black, angled or flattened.

1. **N. bivàlve** (L.) Britton. Leaves narrowly linear, 1–4 dm. long; scape usually overtopping the leaves; bracts lanceolate, acuminate; umbel 3–15-flowered; sepals and petals white, oblong-lanceolate, acute, 10–12 mm. long; capsule obovoid. *Allium striatum* Jacq. *N. striatum* Kunth. Sandy soil: Va.—Neb.—Tex.—Fla.; Mex. *Allegh.—Austral.—Trop.* Mr–Jl.

4. **ANDROSTÉPHIUM** Torr.

Scapose herbs, with membranous coated bulbs. Leaves basal, with elongated narrow blades. Bracts several. Flowers perfect, short-pedicelled. Petals and sepals blue or rose-colored, united to about the middle into a funnelform tube. Stamens 6, adnate to the perianth-tube; filaments united at least to the middle into a tube, which bears tooth-like lobes between the free portion of the filaments; anthers introrse. Ovary 3-celled; style filiform. Seeds several in each locule. Capsule 3-angled. Seeds few, black.

1. **A. caerùleum** (Scheele) Greene. *Fig. 118.* Bulb ovoid, with thin outer coat; leaves narrowly

linear, overtopping the scape; scape 1–3 dm. high; umbel 2–6-flowered; pedicels ascending, 1–3 cm. long; perianth lilac or violet, 2–3 cm. long; lobes oblong or oblong-lanceolate, as long as the tube; stamens included; capsule more than 1 cm. long and broad; seeds flat, 4–6 mm. broad. *A. violaceum.* Torr. Prairies: Kans.—Tex. *Texan.* Mr–Ap.

Family 23. **LILIACEAE.** Lily Family.

Perennial herbs, mostly caulescent, with bulbs, corms, or short rootstocks. Flowers in terminal racemes, corymbs, panicles, or rarely solitary. Sepals and petals each 3, similar, petaloid, sometimes partly united. Stamens 6. Pistils of 3 united carpels; ovary superior, 3-celled; styles united. Fruit a loculicidal capsule.

Plants with rootstocks and fibrous-fleshy roots; perianth-segments united below into a tube.
  Flowers white, subumbellate on subterranean pedicels; perianth-tube very long. 1. Leucocrinum.
  Flowers yellow or orange at the end of a leafless scape; perianth tube short. 8. Hemerocallis.
Plants with bulbs or corms, either leafy-stemmed or scapiferous; petals and sepals distinct or nearly so.
  Bulb scaly; plant tall, leafy.
    Anthers versatile; petals and sepals oblanceolate, clawed, with a linear nectariferous groove. 2. Lilium.
    Anthers fixed near the base, slightly if at all versatile; petals and sepals obovate-oblanceolate, not clawed; nectary a shallow pit. 3. Fritillaria.
  Bulb tunicated.
    Perianth campanulate or turbinate, the segments distinct.
      Anthers strictly basifixed; leaves 2, basal or nearly so; flowers nodding. 4. Erythronium.
      Anthers versatile; scapose plants.
        Filaments flattened. 5. Ornithogalum.
        Filaments filiform. 6. Camassia.
    Perianth globose or ovoid, the segments united. 7. Muscari.

1. **LEUCOCRÌNUM** Nutt. Star of Bethlehem, Mountain Lily.

Low acaulescent herbs, with short rootstock and fleshy-fibrous roots. Leaves basal, numerous, surrounded by scarious sheaths. Flowers in umbel-like sessile clusters, with pedicels and ovaries under ground. Petals and sepals each 3, alike, united below into a long tube. Stamens 6; filaments adnate below to the tube of the perianth; anthers linear, attached near the base, introrse; style much elongated, filiform; stigma 3-lobed. Capsule triangular, obovoid.

1. **L. montànum** Nutt. Leaves thick, numerous, 1–2 dm. long, 2–8 mm. broad; flowers 4–8; perianth white; tube 3–8 cm. long; lobes linear-oblong, about 2 cm. long; capsule truncate, 6–8 mm. long, 12–18-seeded. Hills and plains: Mont.—S.D.—n N.M.—Calif.—Ore. *Plain—Submont.* My–Je.

2. **LÍLIUM** (Tourn.) L. Lily.

Tall, leafy herbs, with thick scaly bulbs and large funnelform or campanulate flowers. Petals and sepals each 3, similar, distinct, each with a nectariferous groove at the base within. Stamens 6; filaments filiform or subulate; anthers linear, versatile, longitudinally dehiscent. Ovary 3-celled, many-ovuled; style long, somewhat clavate; stigma 3-lobed. Capsule oblong or obovoid; seeds numerous, flat, horizontal, in 2 rows in each cavity.

Flowers erect; sepals and petals narrowed below into a distinct claw.
  Leaves lanceolate, nearly all verticillate. 1. *L. philadelphicum.*
  Leaves linear, all or nearly all scattered. 2. *L. umbellatum.*
Flowers nodding or at least spreading; sepals and petals not clawed.
  Leaves all or most of them verticillate, not bulbiferous in their axils.
    Sepals and petals spreading from the middle, mostly yellow, and spotted. 3. *L. canadense.*
    Sepals and petals recurved, orange or red and spotted. 4. *L. michiganense.*
  Leaves all alternate, bulblet-bearing in their axils. 5. *L. tigrinum.*

1. L. philadélphicum L. Stem 3–10 dm. high; leaves in whorls of 3's to 8's, or the uppermost alternate, elliptic, 3–10 cm. long, rough-margined; flowers 1–4, reddish-orange, 5–6 cm. long; blades of the petals and sepals oblong to oval, spotted with purple below the middle, obtuse or abruptly pointed; capsule obovoid, 2–4 cm. long. Open woods: Me.—Man.—S.D.—Neb.—N.C. *Canad.—Allegh.* Je–Jl.

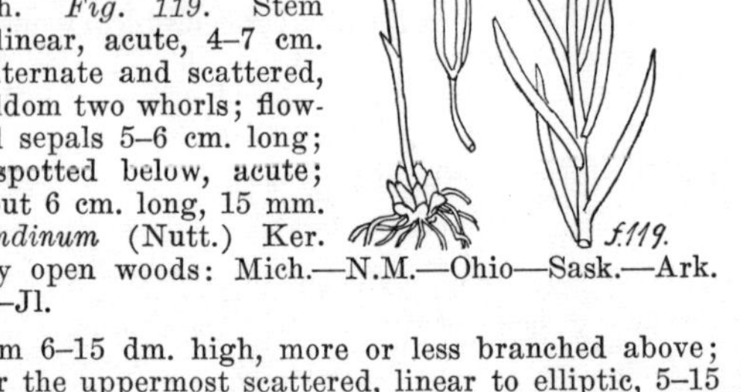

2. L. umbellàtum Pursh. *Fig. 119.* Stem leafy, 3–6 dm. high; leaves linear, acute, 4–7 cm. long, 2–7 mm. wide, mostly alternate and scattered, the uppermost forming one, seldom two whorls; flowers 1–3, umbellate; petals and sepals 5–6 cm. long; blade oval, red or orange, spotted below, acute; capsule almost cylindrical, about 6 cm. long, 15 mm. thick. *L. philadelphicum andinum* (Nutt.) Ker. *L. lanceolatum* Fitzp. In dry open woods: Mich.—N.M.—Ohio—Sask.—Ark. *Allegh.—Plain—Submont.* Je–Jl.

3. L. canadénse L. Stem 6–15 dm. high, more or less branched above; leaves in whorls of 4's–10's or the uppermost scattered, linear to elliptic, 5–15 cm. long, acute or acuminate, scabrous on the margins and the veins beneath; flowers 1–15, usually yellow, variegated and often spotted; sepals and petals 4.5–7.5 cm. long, broadest above the middle; capsule oblong-prismatic, 3–5 cm. long. Meadows and swamps: N.S.—Minn.—Neb.—Ala.—Ga. *Canad.—Allegh.* Je–Jl.

4. L. michiganénse Farwell. Stem 1–2 m. high; leaves in whorls of 3's–10's or the uppermost alternate, elliptic to lanceolate, 5–18 cm. long, 12–20 mm. wide, acuminate at each end, rough on the veins beneath; flowers 1–10, red or orange, nodding, spotted with purple or crimson; sepals and petals 5–8 cm. long, oblong-lanceolate, strongly recurved; capsule oblong, 3–5 cm. long. Confused with *L. superbum* L. [B,G]. Swamps and meadows: Ont.—Mich.—Ky.—Mo.—Minn. *Allegh.* Jl–Au.

5. L. tigrìnum Andr. Stem stout, purplish, white-pubescent above, 5–15 dm. high; leaves all scattered, lanceolate, 1–1.5 dm. long, 1–2 cm. wide, bearing black bulblets in their axils; flowers 5–25, orange-red, nodding, 7–11 cm. long; petals and sepals lanceolate, recurved, purple-spotted. Around dwellings: Me.—N.D.—Va.; escaped from cultivation; nat. of Japan and China. Jl–Au.

3. FRITILLÀRIA L. Tiger Lily, Leopard Lily.

Simple leafy herbs, with thick-scaly bulbs. Flowers open, campanulate, large, nodding; petals and sepals 3, nearly equal, oblong or ovate, deciduous, each with a nectariferous pit at the base. Stamens 6, hypogynous, free; anthers linear or oblong, attached at the base, not versatile. Ovary 3-celled; ovules numerous; style slender, 3-cleft. Capsule obovoid, globose or cylindric, 6-angled. Seeds numerous, flat, margined or winged.

1. F. atropurpùrea Nutt. *Fig. 120.* Stem slender, 1–4 dm. high, leafless below, leafy above, 1–4-flowered; leaves mostly scattered, alternate or the upper verticillate, narrowly linear, 3–8 cm. long, 3 mm. wide; flowers dark purple, mottled with yellowish green; petals and sepals elliptic to linear, 5–25 mm. long; capsule about 15 mm. long and as wide. *F. linearis* Coult. & Fish. On hillsides among bushes: Wash.—Calif.—N.M.—N.D.—Neb. *W. Temp.—Plain—Submont.* My–Je.

4. **ERYTHRÒNIUM** L. Dog-tooth Violet, Adder-tongue, Star-strikers.

Low herbs, with membranous-coated corms and simple scapiform stems bearing two leaves below. Flowers solitary or few, nodding; petals and sepals lanceolate, distinct, with a nectariferous groove. Stamens 6, hypogynous, free; anthers oblong to linear, attached at the end. Ovary 3-celled; ovules numerous in each cell; style filiform, or thickened above, often 3-cleft. Capsule obovoid or oblong, in ours bluntly 3-angled.

Not propagating by offshoots from the stem; flowers more than 2 cm. long.	
Propagating by offshoots from the base of the bulb; sepals and petals more or less recurved.	
Perianth yellow within; styles wholly united.	1. *E. americanum.*
Perianth white or pink within; styles united to above the middle.	2. *E. albidum.*
Propagating by sessile basal bulbs; sepals and petals erect or merely slightly spreading, white or tinged with lavender.	3. *E. mesachoreum.*
Propagating by offsets from the middle of the stem; flowers less than 2 cm. long, rose-colored with yellowish base.	4. *E. propullans.*

1. **E. americànum** Ker. Leaves 2, just above the ground, elliptic, 1–1.5 dm. long, usually mottled above; scape 1–3 dm. high; sepals and petals 3–3.5 cm. long, spotted within, the former linear-lanceolate, the latter linear-elliptic; capsule broadly obovoid, 1–1.5 cm. long. Thickets and moist woods: N.S.—Minn.—Ark.—Fla. *E. Temp.* Mr–My.

2. **E. álbidum** Nutt. *Fig. 121.* Leaves near the ground, elliptic, 5–15 cm. long, usually mottled with dark and light green, acute; scape 1–3 dm. high; petals and sepals 3–3.5 cm. long, linear or linear-lanceolate; capsule oblong or oblong-obovoid, 1–2 cm. long. Thickets and open woods: Ont.—Minn.—Tex.—Ga. *Allegh.* Mr–My.

3. **E. mesochòreum** Knerr. Leaves 2, near the ground, oblong, 8–18 cm. long, not spotted, usually folded; scape 1–3 dm. high; flowers nodding; sepals and petals 2.5–4 cm. long, lance-linear; capsule obovoid, 2–3 cm. long. Prairies: Iowa—Neb.—Okla. *Prairie.* Mr–My.

4. **E. propúllans** Gray. Leaves borne on a distinct decumbent stem; blades oblong, acute, 5–10 cm. long; stem 1.5–2 dm. long; flowers nodding; green or slightly mottled; perianth about 1.2 cm. long; sepals and petals oblanceolate, not recurved. Rich woods: Ont.—Minn. *Allegh.* My.

5. **ORNITHÓGALUM** L. Star-of-Bethlehem.

Scapose herbs, with coated bulbs. Leaves narrow, fleshy. Flowers perfect, in bracted racemes or corymbs. Sepals and petals subequal, white (or yellow), often greenish without, persistent. Stamens 6; filaments flattened; anthers versatile, introrse. Ovary 3-celled; ovules several or many in each cavity; stigma capitate, 3-lobed. Capsule 3-lobed, loculicidal.

1. **O. umbellàtum** L. Scape 1–3 dm. high; leaves dark green, with a light midrib, 2–5 mm. wide; flowers corymbose, erect; sepals and petals oblong-lanceolate, 1–2 cm. long, white within, green with white margins without. Fields and around dwellings: N.H.—Neb.—Kans.—Va.; nat. from Eur., or escaped from gardens. My–Je.

6. CAMÁSSIA Lindl. CAMASH, BLUE CAMAS, WILD HYACINTH, SWAMP SEGO.

Perennial herbs, with scapiform stems and edible bulbs. Leaves basal, with elongated blades. Flowers in terminal racemes. Sepals and petals each 3, alike, distinct, blue, white, or purple. Stamens 6; filaments filiform, adnate to the base of the petals and sepals; anthers versatile, introrse. Ovary 3-celled; style filiform; stigma 3-lobed; ovules numerous in each cavity. Capsule broad, 3-angled. Seeds black, shining. *Quamasia* Raf.

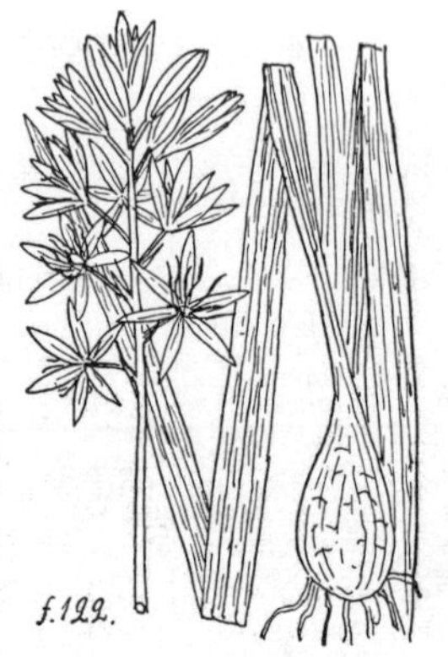

1. **C. esculénta** (Kerr) B. L. Robinson. *Fig. 122.* Stem 1.5–7 dm. high; leaves linear, keeled, 2–5 dm. long; bracts longer than the pedicels; sepals and petals pale blue, 3-nerved, 1–1.5 cm. long; capsule acutely triangular, subglobose in outline. *Camassia Fraseri* Torr. *Q. hyacinthina* (Raf.) Britton [B]. Thickets and meadows: Pa.—Minn.—Kans.—Tex.—Ga. *Allegh.—Austral.* Ap–My.

7. MUSCÀRI (Tourn.) Mill. GRAPE HYACINTH.

Acaulescent herbs with coated bulbs. Leaves arising from the bulb. Flowers perfect, regular, blue, pink, or white, borne in dense racemes on a leafless scape. Perianth globose or ovoid, with united segments. Stamens 6, included, with short introrse anthers. Style short. Capsule loculicidal, 3-celled, each cell with 2 black seeds.

1. **M. botryoìdes** (L.) Mill. Leaves linear, 6–10 mm. broad, 1–2 cm. long, channeled; raceme 2–4 cm. long; perianth globular, blue, 3–5 mm. long, with recurved teeth. Meadows and thickets: N.H.—Minn.—Kans.—Va.; escaped from cultivation; nat. of Asia. Ap–Je.

8. HEMEROCÁLLIS L. DAY LILY.

Perennials, with fibrous roots and keeled 2-ranked leaves. Flowers perfect, at the end of a leafless scape, bracted. Perianth funnel-form, the segments united below, the tube enclosing the ovary. Segments 6, reflexed-spreading, yellow or orange. Stamens 6; filaments long, inserted in the throat; anthers linear, introrse. Style slender; stigma capitate. Capsule 3-angled, 3-valved, transversely wrinkled.

1. **H. fúlva** L. Leaves channeled, 8–12 mm. wide, scape 1–2 m. high, 6–15-flowered; perianth orange, the tube 3–4 cm. long, the sepals flat and acutish, the petals wavy and obtuse. Meadows and along streams: N.B.—Va.—Kans.; escaped from cultivation and naturalized; native of Eurasia. Je–Au.

Family 24. **CONVALLARIACEAE.** LILY-OF-THE-VALLEY FAMILY.

Perennial herbs, with rootstocks and alternate or rarely basal leaves. Flowers perfect, in terminal or axillary racemes, panicles, umbels, or rarely solitary. Sepals and petals 3 or 2, similar, distinct or partly united, inferior. Stamens 6 or 4. Gynoecium of 3 or 2 united carpels; ovary 3- or 2-celled; styles united. Fruit a berry or rarely a capsule.

Leaves reduced to scales; stems intricately branched. 1. ASPARAGUS.
Leaves normal; stem or scape simple or sparingly branched.
  Stem leafy; leaves alternate.
    Sepals and petals distinct.
      Flowers white, in terminal racemes or panicles; anthers introrse; stem simple.

Petals and sepals 3; stamens 6. 2. SMILACINA.
Petals and sepals 2; stamens 4. 3. MAIANTHEMUM.
Flowers axillary or terminal, solitary or in small umbelliform clusters; anthers extrorse or opening laterally; stem branched.
Fruit a berry; filaments longer than the anthers.
Flowers axillary, greenish white or rose-colored; filaments slender; anthers acute. 4. STREPTOPUS.
Flowers terminal, yellow; filaments dilated; anthers obtuse. 5. DISPORUM.
Fruit a capsule; filaments shorter than the anthers.
Sepals and petals with a ridge on each side of the deep nectary; capsule rounded to retuse at the apex, 3-lobed. 6. UVULARIA.
Sepals and petals without ridges and nectary; capsule acute at each end, 3-winged. 7. OAKESIELLA.
Sepals and petals partially united into a tube; flowers axillary. 8. POLYGONATUM.
Stem scapiform; leaves basal; flowers in terminal umbels, or solitary and terminal. 9. CLINTONIA.

1. ASPÁRAGUS (Tourn.) L. ASPARAGUS.

Caulescent herbs, with rootstocks. Stem first simple, scaly, fleshy, edible, later intricately branched. Leaves reduced to scales, later with filiform branches clustered in the axils. Flowers perfect, axillary, solitary or in some species racemose or umbellate. Perianth nodding; sepals and petals alike, distinct or slightly united at the base. Stamens 6, inserted at the base of the perianth; filaments filiform; anthers introrse. Ovary 3-celled; stigmas 3; ovules 3 in each cavity. Fruit a pulpy berry.

1. **A. officinàlis** L. Stem 4–20 dm. high, at last with widely spreading branches; flowers mostly solitary, with recurved pedicels, in the axils of the scales; perianth greenish, 4–6 mm. long; sepals and petals linear, obtuse, erect, only the tips spreading; berry subglobose, red, 6–8 mm. broad. Waste places and salt marshes: N.B.—Man.—N.D.—Colo.—La.—Ga.; nat. from Eur. or escaped from cultivation. Je.

2. SMILACÌNA Desf. WILD SPIKENARD, WILD LILY-OF-THE-VALLEY.

Caulescent herbs, with elongated rootstocks. Leaves broad, several-nerved. Flowers in terminal racemes or panicles. Sepals and petals white or greenish white, distinct or nearly so. Stamens 6; filaments subulate; anthers introrse. Ovary 3-celled; styles short; stigma 3-lobed; ovules 2 in each cavity. Berry globose. Seeds 1 or 2, with a thin testa. *Vagnera* Adans.

Inflorescence paniculate. 1. *S. racemosa.*
Inflorescence racemose.
Leaves 6–12, sessile. 2. *S. stellata.*
Leaves 2–4, sheathing the low stem. 3. *S. trifolia.*

1. **S. racemòsa** (L.) Desf. Stem somewhat angled, finely puberulent above, 4–9 dm. high; blades elliptic or oval, 7–15 cm. long, 3–6 cm. wide, finely puberulent beneath; sepals and petals oblong, 2 mm. long or more; berry 5–6 mm. thick. *V. racemosa* Morong. [B,R]. Woods and thickets: N.S.—Ga.—Colo.—B.C. *Canad.—Allegh.—Submont.* My–Jl.

2. **S. stellàta** (L.) Desf. *Fig. 123.* Stem glabrous, 2–5 dm. high, strict; leaves sessile, minutely puberulent beneath, 5–13 cm. long; sepals and petals 3–5 mm. long; berry green, with 6 black stripes, turning black, 6–10 mm. thick. *S. stellata* (L.) Morong. [B,R]. In moist soil, meadows or copses: Newf. — Va. — Colo. — Alta. *Boreal. — Submont. — Subalp.* My–Jl.

3. **S. trifòlia** (L.) Desf. Stem slender, 0.5–4 dm. high; leaves sessile, oval, oblong, or oblong-lanceolate, with sheathing bases, acute or acuminate; raceme few-flowered, strict; sepals and petals oblong or oblong-lanceolate, obtuse, about 3 mm. long; berry dark red, 5–6 mm. in diameter. *V. trifolia* (L.) Morong. [B,R]. *V. pumila* Standl., a small form with rather short style. In bogs and wet woods: Lab.—N.J.—Minn.—Mack. *Boreal—Mont.* My–Je.

3. MAIÁNTHEMUM Wigg. Two-leaved Solomon's Seal.

Caulescent herbs, with slender rootstocks. Leaves mostly 2 or 3, with broad, several-nerved blades. Inflorescence terminal, racemose. Sepals and petals white, similar, each 2, distinct, spreading. Stamens 4, hypogynous; filaments narrowly linear; anthers versatile, introrse. Ovary 2-celled; stigmas 2; ovules 2 in each cavity. Fruit a subglobose pulpy berry; seeds 1 or 2. *Unifolium* Adans.

1. **M. canadénse** Desf. Stem slender, 1–3-leaved (usually 2-leaved), 5–18 cm. high, glabrous; leaves ovate or ovate-lanceolate, 2–8 cm. long; berry pale red, speckled, about 4 mm. thick. *U. canadense* Greene [B,R]. Moist woods and thickets: Lab.—N.C.—Minn. *Boreal.* My–Jl. *M. canadense interius* Fernald. More robust than the species, with upper part of the stem and lower surface of the leaves pubescent, stem sharply angled and the lower stem-leaves more distinctly petioled. N.Y.—Iowa—S.D.—Mack.

4. STRÉPTOPUS Michx. Twisted-stalk.

Caulescent perennial herbs, with horizontal rootstocks. Leaves many-nerved, broad, sessile or clasping. Flowers racemose on 1–3-flowered, extra-axillary peduncles. Sepals and petals much alike, greenish or purplish, each 3, distinct, with spreading or recurved tips. Petals keeled. Stamens 6, hypogynous; filaments flattened, very short; anthers sagittate, opening by lateral slits. Ovary 3-celled; stigmas 3-lobed or entire; ovules in 2 rows in each cavity. Berry oval or globose; seeds numerous.

Flowers greenish white; anthers 1-pointed. 1. *S. amplexifolius.*
Flowers rose or purple; anthers 2-pointed. 2. *S. longipes.*

1. **S. amplexifòlius** (L.) DC. *Fig. 124.* Stem 3–10 dm. high, flexuose-branched; leaves ovate or ovate-lanceolate, glabrous and glaucous, 5–13 cm. long; perianth campanulate, greenish white, 8–12 mm. long; sepals and petals lanceolate, attenuate; berry globose-ellipsoid, scarlet, rarely white, 8–12 mm. long. Damp woods: Greenl.—N.C.—N.M.—Ore.—Alaska; Eur. *Arctic—Boreal—Submont.—Subalp.* Je–Au.

f. 124

2. **S. lóngipes** Fernald. Stem 3–10 dm. high, somewhat pubescent above, from long slender rootstock; leaves oblong to ovate, 5–12 cm. long, ciliate, sessile; pedicels 1–2.5 cm. long, pubescent, usually 1-flowered; sepals and petals 8–12 mm. long, lanceolate, attenuate, papillose inside, with recurved tips; berry globose, red, 10–12 mm. long. Cold woods: Ont.—Man.—Iowa—Mich. *Canad.* My–Je.

5. DÍSPORUM Salisb.

Caulescent branched perennial herbs with rootstocks. Leaves broad, many-nerved, sessile or clasping, often oblique. Flowers terminal, solitary or in small subumbellate clusters, drooping. Sepals and petals each 3, whitish or greenish yellow, narrow, distinct. Stamens 6, hypogynous; filaments filiform, or somewhat flattened; anthers extrorse. Ovary 3-celled; stigma entire or 3-cleft; ovules 2 or more in each cavity. Fruit thick, juicy, (in ours) bright red or orange.

f.125.

1. **D. trachycárpum** S. Wats. *Fig. 125.* Stem 3–6 dm. high, more or less flexuose, more or less pubescent; leaves sessile, ovate or oval to ovate-lanceolate, 3–9 cm. long, short-acuminate; perianth narrowly campanulate, ochroleucous, 10–15 mm. long; fruit depressed-globose, deeply 3-lobed, 8–10 mm. thick. *Prosartes trachycarpa* S. Wats. *D. majus* Britt., in part. Cañons and hillsides: Man.—N.M.—Ariz.—B.C. *Plain—Subalp.* My–Je.

6. UVULÀRIA L. Bellwort.

Leafy-stemmed, slightly fleshy herbs, with rootstocks. Stems terete, forking, scaly at the base. Leaves alternate, perfoliate, broad, many-nerved, smooth on the margins. Flowers solitary, peduncled at the ends of the branches. Perianth drooping, yellow; sepals and petals each 3, with a deep nectary at the base and two callous ridges along its margins. Stamens 6; filaments free; anthers erect, elongate. Ovary 3-celled, 3-angled; styles united to the middle; stigmas 3. Capsule oblong, 3-lobed, truncate at the apex. Seeds 1–3 in each cavity, covered with a thin pale aril.

Sepals and petals papillose within; stamens shorter than the style. 1. *U. perfoliata.*
Sepals and petals smooth within; stamens longer than the style. 2. *U. grandiflora.*

f.126.

1. **U. perfoliàta** L. *Fig. 126.* Stem 1–8 dm. high; leaves glaucescent, glabrous, oblong to oval, 3–15 cm. long, acute or abruptly short-acuminate; perianth pale-yellow, 2–2.5 cm. long, the segments linear or linear-oblanceolate; connective of the anthers acute; capsule 1 cm. long, its lobes 2-beaked at the apex. Woods and thickets: Que.—Man.—S.D.—Miss.—Fla. *Canad.—Carol.* My–Je.

2. **U. grandiflòra** Smith. Stem 2–8 dm. high; leaves oblong or oblong-oblanceolate, 5–15 cm. long, pubescent beneath, at least when young, acuminate; perianth lemon-yellow, 3–4 cm. long, the segments linear-oblong, acute or acuminate; connective of the anthers obtuse; capsule 1 cm. long, with obtuse lobes. Woods and thickets: Que.—S.D.—Kans.—Tenn.—Ga. *Canad.—Allegh.* Ap–Je.

7. OAKESIÉLLA Small. Smaller Bellwort.

Leafy-stemmed, firm-fleshy herbs, with elongate rootstocks. Stems angled, branched. Leaves sessile, leathery, with scabrous margins. Flowers 1 or 2, opposite the leaves. Perianth drooping, yellow; sepals and petals each 3, narrow, without nectaries and ridges within. Stamens included; filaments flattened; anthers elongate. Ovary 3-celled; styles partly united. Ovules several in each cavity. Capsule elliptic or oval, acutish at each end. Seeds globose, with a swollen spongy raphe. *Oakesia* S. Wats.

1. **O. sessilifòlia** (L.) Small. Stem 1–4 dm. high, simple or branching above; leaves elliptic, 3–8 cm. long, acute at each end, pale beneath; perianth

greenish-yellow, 1.5–2 cm. long, the segments smooth, obtuse; capsule oval, stalked, 2.5 cm. long. *Uvularia sessilifolia* L. *Oakesia sessilifolia* (L.) S. Wats. [G]. Woods and thickets: N. B.—S.D.—Ark.—Ala.—Ga. *Canad.—Allegh.* My–Je.

8. POLYGONÀTUM (Tourn.) Hill. SOLOMON'S SEAL.

Caulescent perennial herbs, with jointed rootstocks. Leaves in our species broad, many-nerved, sessile. Flowers in axillary 1–few-flowered racemes. Petals and sepals each 3, partly united, the free portion shorter than the tube, greenish or pinkish. Stamens 6, included; filaments partly adnate to the perianth-tube; anthers sagittate, introrse. Ovary 3-celled; stigma mostly capitate; ovules 2–6 in each cavity. Berry subglobose, dark blue or black. Seeds with a horny endosperm. *Salomonia* Heist.

Leaves pubescent on the veins beneath; racemes 1–2-flowered.
  Flowers with a yellowish white tube. — 1. *P. pubescens.*
  Flowers green. — 2. *P. boreale.*
Leaves glabrous on both sides; racemes 2–10-flowered.
  Leaves lance-elliptic with 5–9 stronger nerves, tapering at the apex; filaments usually papillate. — 3. *P. commutatum.*
  Leaves ovate-oval, with 9–15 stronger nerves, abruptly short-acuminate; filaments glabrous. — 4. *P. giganteum.*

1. **P. pubéscens** (Willd.) Pursh. Stem glabrous, 3–4.5 dm. high; leaves 9–13, ovate or lance-ovate, 5–7.5 cm. long, green above, with 5–7 prominent nerves; peduncles 1–1.2 mm. long, 1- or 2-flowered; perianth 8–10 mm. long, cylindric; filaments filiform, papillate, shorter than the anthers. *P. biflorum* A. Gray [G,B]; not (Walt.) Ell. Dry woods: Me.—S.C.—Ind.—Minn. *Canad. —Allegh.* My–Jl.

2. **P. boreàle** Greene. Stem 3–4 dm. high; leaves ovate-elliptic, 7–10 cm. long, glabrous and pale green above, glaucous beneath, with 5–7 stronger nerves; peduncles slender, mostly 1-flowered; perianth 8 mm. long, deep green. Oak and maple woods: Me.—Ind.—Minn. *Canad.* My.

3. **P. commutàtum** (Schultes) Dietr. Stem 9–12 dm. high, cylindric, about 1 cm. thick; leaves 7.5–15 cm. long, 4–6 cm. wide; racemes 2–8-flowered, 3–5 cm. long; peduncles only slightly flattened, bent at the base; perianth 14–16 mm. long, whitish. Wet fields and open woods: Mass.—Va.—Tenn.—Okla.—N.D. *Allegh.* My–Jl.

f.127.

4. **P. gigantèum** Dietr. *Fig. 127.* Stem stout, 1–7 m. high, 1–2 cm. thick below; leaves 1–2 dm. long, 6–12 cm. wide; racemes 2–15 flowered; peduncles strongly flattened and usually arching throughout; perianth greenish white, 15–20 mm. long. Moist thickets: Mass.—D.C.—Ala.—Okla.—Utah.—N.D. *Allegh.* My–Jl.

9. CLINTÒNIA Raf.

Subacaulescent perennial, with creeping rootstocks. Leaves basal or nearly so, broad, many-nerved. Flowers in terminal umbels on an almost leafless scape, or in a western species the umbel reduced to 1 or 2 flowers. Petals and sepals each 3, similar, petaloid, distinct. Stamens 6; filaments filiform; anthers versatile. Ovary 2- or 3-celled; style slender; stigma 2- or 3-lobed. Berry ovoid, thin.

1. **C. boreàlis** (Ait.) Raf. Leaves oblong to oval, 1–3 dm. long, deep green, ciliate; scape erect, longer than the leaves, pubescent above; flowers greenish yellow, nodding, 1.5–2 cm. long; petals and sepals linear or lance-linear, pubescent without; berries ellipsoid or subglobose, 8–9 mm. broad, blue. Deep woods: Newf.—Lab.—Man.—Minn.—N.C. *E. Temp.* My–Je.

Family 25. **DRACAENACEAE.** YUCCA FAMILY.

Shrubby plants or trees, with woody trunks or caudices, very leafy at the apex. Leaves narrow, rigid, often with marginal filaments or finely toothed. Flowers mostly perfect, or polygamo-dioecious, racemose or paniculate. Petals and sepals 3, similar. Stamens 6. Gynoecium of 3 united carpels. Ovary superior, 3-celled; styles very short, united or obsolete. Fruit a loculicidal capsule, or fleshy and indehiscent.

1. **YÚCCA** (Rupp.) L. YUCCA, SPANISH BAYONET, SOAP-WEED, SOAP-ROOT, GRASS CACTUS.

Coarse plants, with woody trunks or caudices. Leaves firm, narrow, rigidly pointed, commonly with thread-like fibers along the edges, or serrulate or entire-margined. Flowers in terminal racemes or panicles, drooping. Sepals and petals each 3, distinct or slightly united at the base, usually white. Stamens 6, hypogynous; filaments enlarged above. Ovary 3-celled or imperfectly 6-celled, or 1-celled; style turgid; ovules numerous. Capsule either dry and dehiscent, or fleshy and indehiscent. Seeds numerous, thin, flat.

Fruit a dry capsule. 1. *Y. glauca.*
Fruit fleshy. 2. *Y. baccata.*

1. **Y. glaùca** Nutt. *Fig. 128.* Subacaulescent or branching with decumbent stems; leaves rigid, 6–12 mm. wide, 2–4 dm. long, white-margined, finely but sparingly filiferous; inflorescence 1–2 m. high, simple or somewhat branched; sepals and petals greenish white, oval to lanceolate, acute, 4–5 cm. long; capsule oblong, usually not constricted, somewhat roughened, brown. *Y. angustifolia* Pursh. Plains and hillsides: Iowa—Tex.—Ariz.—Mont. *Plains—Submont.* My–Jl.

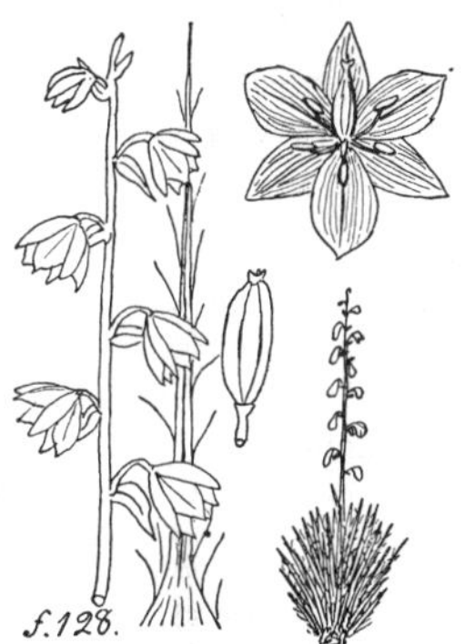
f.128.

2. **Y. baccàta** Torr. Low, usually with stout prostrate branched caudex; leaves rigid, spreading, about 6 dm. long and 5 cm. wide, concave, shagreen-roughened, with narrow brown margins, coarsely filiferous; sepals and petals lanceolate, about 7.5 cm. long; style slender, elongate; fruit large, sometimes 2 dm. long, oblong- or conical-ovoid, pendent, fleshy. Dry plains: Tex.—Kans.—Colo.—Ariz.—Nev. *St. Plains—Son.—Submont.* Ap–Je.

Family 26. **CALOCHORTACEAE.** MARIPOSA LILY FAMILY.

Perennial herbs, with coated corms and narrow leaves. Flowers perfect, regular, showy. Sepals 3, narrow, herbaceous. Petals 3, broad, gland-bearing within near the base and often bearded within, petaloid. Stamens 6, hypogynous. Gynoecium of 3 united carpels; ovary 3-celled, many-ovuled; styles united, almost none; stigmas 3, recurved. Fruit a septicidal, 3-angled capsule.

1. **CALOCHÒRTUS** Pursh. SAGO LILY, MARIPOSA LILY, BUTTERFLY LILY.

Characters of the family.

Anthers acute; gland broader than long. 1. *C. Gunnisonii.*
Anthers obtuse; gland not broader than long. 2. *C. Nuttallii.*

1. **C. Gunnisònii** S. Wats. *Fig. 129.* A strict plant, 2–5 dm. high, without bulblet; leaves several, slender, mostly involute, 5–15 cm. long; flowers 1–2; petals broadly cuneate-obovate, 2–4 cm. long, white or cream-colored, tinged and streaked with purple, yellow and purple-dotted around the gland; capsule narrowed at both ends, about 3 cm. long. In meadows: S.D.—Neb.—N.M.—Ariz.—Ida. *Submont.—Mont.* My–Jl.

2. **C. Nuttállii** Torr. & Gray. A strict plant, similar to the preceding, 2–5 dm. high, but with a bulblet borne in the axil of the lowest leaf, 1–4 cm. above the bulb; leaves and flowers resembling those of the preceding, but petals often narrower, an sometimes acutish; capsule tapering at both ends. *C. Watsoni* M. E. Jones. Dry hillsides: S.D.—Neb. —N.M.—Calif.—Wash. *Submont.—Mont.* My–Jl.

Family 27. TRILLIACEAE. Trillium Family.

Perennial caulescent or scapose herbs, with rootstocks. Leaves and bracts whorled, broad and often netted-veined. Flowers perfect, solitary, terminal, or in terminal umbels. Sepals 3 (or in exotic genera 4), green, distinct; petals of the same number, in ours white or purplish. Stamens 6–8; filaments short. Gynoecium of 3 or 4 united carpels; ovary 3- or 4-celled; stigmas sessile. Fruit a 3- or 4-celled, lobed berry.

Flowers several in a terminal umbelliform cluster; sepals and petals nearly alike; leaves and leaflike bracts forming two whorls. — 1. Medeola.
Flowers solitary; sepals and petals very different; leaves (or leaflike bracts) forming a single whorl of 3's. — 2. Trillium.

2. MEDÈOLA L. Indian Cucumber-root.

Caulescent herbs, with rootstocks. Leaves and bracts several, netted-veined, nearly similar, forming two whorls, the former near the middle, the latter at the end of the stem. Flowers umbellate, on recurved pedicels. Sepals and petals each 3, pale greenish-yellow, similar. Stamens 6; anthers shorter than the slender filaments. Ovary 3-celled; stigma recurved, thread-like, deciduous. Fruit a berry.

1. **M. virginiàna** L. Stem 2–9 dm. high; leaves 4–10, elliptic to obovate, 5–15 cm. long, sessile, acuminate; bracts similar but smaller and comparatively broader; flowers 2–15; pedicels 2–3 cm. long; sepals oblong or oval, 8–10 mm. long; berry globose, 8–15 mm. broad, dark-purple. Woods and banks: N.S.—Minn.—Ala.—Fla. *E. Temp.* Je.

2. TRÍLLIUM L. Wake-robin, Birth-root.

Fleshy herbs, with short stout rootstocks and a whorl of 3 netted-veined leaves (or leaf-like bracts) near the end of the scape. Flowers solitary, 3-merous, pedicelled or sessile. Sepals persistent. Petals early withering or deciduous, white or purple.

Flowers sessile.
  Leaves sessile; sepals not reflexed.
    Petals purple or red.
      Petals not clawed; stem smooth. — 1. *T. sessile.*
      Petals clawed; stem pubescent above. — 2. *T. viridescens.*
    Petals greenish, linear. — 3. *T. viride.*
  Leaves petioled; sepals reflexed. — 4. *T. recurvatum.*
Flowers peduncled.
  Leaves subsessile; fruit 6-angled, winged.

Petals obovate or oblanceolate, 4–6 cm. long, white, turning rose-colored. 5. *T. grandiflorum.*
Petals ovate or lanceolate, less than 4 cm. long.
Anthers in anthesis exceeding the stigma; petals often brown-purple. 6. *T. erectum.*
Anthers in anthesis surpassed by the stigma; petals white or pink.
Pedicels 8–30 mm. long, recurved; filaments nearly equaling the anthers. 7. *T. cernuum.*
Pedicels 40–60 mm. long, usually horizontal; filaments scarcely half as long as the anthers. 8. *T. declinatum.*
Leaves petioled; fruit 3-lobed, not winged. 9. *T. nivale.*

1. **T. séssile** L. Stem 1–2 dm. high; leaves oval or suborbicular, 4–8 cm. long, rounded at the base, not mottled; sepals lanceolate, 2–3 cm. long; petals narrowly elliptic, sessile, acutish, purple; stamens half as long as the petals; filaments dilated at the base, shorter than the anthers; berry ovoid, 1–1.5 cm. long. Woods: Pa.—Minn.—Kans.—Miss.—Fla. *Allegh.—Austral.* Ap–My.

2. **T. viridéscens** Nutt. Stem 4–6 dm. high, pubescent above; leaves ovate, elliptic, or suborbicular, 1–1.5 dm. long, acuminate, 5-ribbed; sepals narrowly linear-lanceolate, 4–5 cm. long, acute; petals clawed, the claw 1.5–2 cm. long, the blade narrowly lance-linear, twice as long; filaments 2.5–5 mm. long, the anthers 12–15 mm. long. Copses and hillsides: Ark.—Kans. *Ozark.* Ap–My.

3. **T. víride** Beck. Stem 1–2 dm. high, rough-pubescent near the top; leaves oblong or ovate, 5–10 cm. long, acute or obtusish, usually blotched, pubescent on the nerves beneath; sepals linear or lance-linear, 2–4 cm. long; petals 2–4.5 cm. long, linear to linear-elliptic, light green or purplish green; stamens about one third as long as the petals; filaments flattened, only 2–3 mm. long. Woods: Mo.—Kans.—Miss.—Ala.—N.C. *Austral.* Ap–My.

4. **T. recurvàtum** Beck. Stem 1–4 dm. high, smooth; leaves ovate-lanceolate to oval or suborbicular, 5–10 cm. long, rounded or subcordate at the base, often mottled, petioled; sepals lanceolate, 2–3 cm. long, acute; petals clawed, ovate to obovate, 3–4 cm. long; filaments one half or one third as long as the anthers; berry strongly winged above, 1.8 cm. long. Woods: Ohio—Minn.—Ark.—Miss.—Tenn. *Allegh.—Ozark.* Ap–Je.

5. **T. grandiflòrum** (Michx.) Salisb. Stem 2–5 dm. high, smooth; leaves oval or rhombic-oval, 5–15 cm. long, acuminate, 5-ribbed, cuneate at the base, acuminate at the apex: pedicels 5–8 cm. long, erect or ascending; sepals lanceolate, 2.5–5 cm. long, acute; petals oblanceolate to obovate, white or pink, much longer, crisp; filaments shorter than the anthers; berry subglobose, 2–2.5 cm. thick, becoming black. Woods and hillsides: Que.—Minn.—Mo.—N.C. *E. Temp.* My–Je.

6. **T. eréctum** L. *Fig. 130.* Stem 3–6 dm. high, smooth; leaves broadly rhombic, 5–15 cm. long, acuminate; pedicels 3–10 cm. long, erect or declined; sepals oblong or lanceolate, acuminate, 2–3.5 cm. long, spreading; petals lanceolate to ovate, usually brown-purple; filaments stout, 3–4 mm. long; berry about 2 cm. long, ovoid, purple or black. Woods: N.S.—Man.—Mo.—N.C. *E. Temp.* Ap–Je.

f.130.

7. **T. cérnuum** L. Stem 2–6 dm. high, smooth; leaves rhombic, often broader than long, acuminate, 5–12 dm. long; sepals lanceolate, 2–2.5 cm. long, recurved; petals white or pink, elliptic to ovate, a little longer than the sepals; berry ovoid, 1.5–2 cm. long. Woods: Newf.—Man.—N.D.—Mo.—Ga. *E. Temp.* Ap–Je.

8. **T. declinàtum** (A. Gray) Gleason. Stem 2–5 dm. high; leaves broadly rhombic, 5–10 cm. long; sepals lanceolate, 2–2.5 cm. long; petals ovate-oblong, 2–3.5 cm. long, white; fruit whitish or pinkish. Woods: Ohio—Mich.—Man.—Iowa—Mo. *Allegh.—Ozark.* Ap–Je.

9. **T. nivàle** Riddell. Stem 5–12 cm. high; leaves 2–5 cm. long, oval or ovate, obtuse, petioled; sepals oblong, obtuse; petals oblong or oval, obtuse, longer than the sepals, 1.5–3 cm. long; berry globose, 6–8 mm. long. Thickets: Pa.—S.D.—Neb.—Ky. *Allegh.* Mr–My.

Family 28. SMILACACEAE. SMILAX FAMILY.

Vines, with several-ribbed and netted-veined leaf-blades, articulate to the petioles. Flowers dioecious, in peduncled axillary umbels. Sepals and petals each 3, green, with spreading tips. Stamens 6; filaments flattened; anthers introrse. Gynoecium of 3 united carpels; stigmas 3, sessile. Fruit a berry, 1–6-seeded. Endosperm bony.

Ovules 2 in each cavity of the ovary; stem herbaceous, unarmed. 1. NEMEXIA.
Ovules solitary in each cavity of the ovary; stem woody, usually prickly. 2. SMILAX.

1. NEMÉXIA Raf. CARRION FLOWER, SMILAX.

Unarmed vines, ours with herbaceous stems. Leaves alternate, with membranous, several-ribbed blades and usually with tendrils. Flowers polygamo-dioecious, in axillary umbels. Sepals and petals 3, similar. Stamens 6, in the pistillate flowers reduced. Ovary 3-celled, in the staminate flowers abortive. Ovules 2 in each cell. Fruit a berry, with 3 bands of strengthening cells. Seeds 1–6.

Leaves glaucous beneath, short-acuminate; berry dark blue.
  Leaves without tendrils, the blades broadly elliptic. 1. *N. ecirrhata.*
  Leaves with tendrils, the blades ovate or cordate.
    Leaves glabrous. 2. *N. herbacea.*
    Leaves pubescent beneath. 3. *N. lasioneuron.*
Leaves shining-green beneath, strongly acuminate; berry black. 4. *N. pulverulenta.*

1. **N. ecirrhàta** (Engelm.) Small. Stem mostly erect, 3–6 dm. high; leaves mostly approximate at the end of the stem, usually without tendrils; blades ovate, acute or short-acuminate, pubescent beneath, strongly ribbed in age; sepals of the staminate flowers obovate or oblanceolate, 4–5 mm. long; anthers shorter than the filaments; berry globose, 9–11 mm. broad, purple-black. *Smilax ecirrhata* S. Wats. [G]. Woods: Md.—Ohio—Minn.—Okla.—S.C. *Austral—Ozark.* My–Je.

2. **N. herbàcea** (L.) Small. Stem elongate, climbling, 1–5 mm. long, glabrous; leaf-blades ovate, triangular, or lanceolate, short-acuminate, 4–8 cm. long, rounded or truncate at the base, glabrous; peduncles surpassing the leaves; sepals oblong or oblanceolate-oblong, acute, filaments twice as long as the anthers; berry subglobose, bluish black, 6–8 mm. broad. *S. herbacea* L. [G]. Woods and thickets: N.B.—Man.—Neb.—La.—Fla. *E. Temp.* Ap–Je.

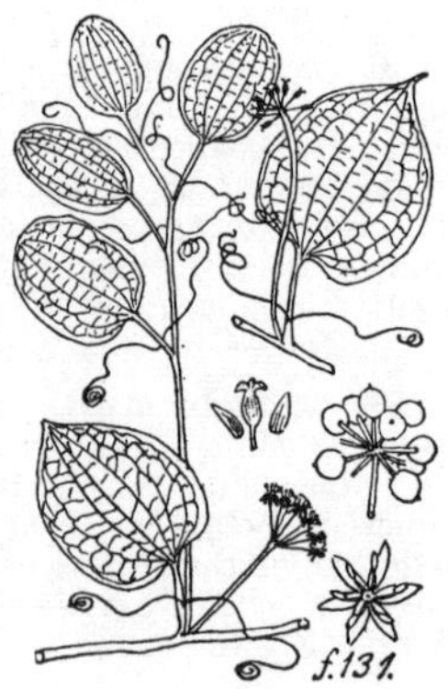
f.131.

3. **N. lasioneùron** (Hook.) Rydb. *Fig. 131.* A herbaceous vine, 1–2 m. long; petioles 2–5 cm. long; leaf-blades ovate-cordate, abruptly short-acuminate, 5–10 cm. long, 5–9-ribbed, rather thin, glabrous above, scabrous-hirsutulous on the veins beneath; peduncles 4–7 cm. long; umbel many-flowered; flowers greenish; petals and sepals oblong, 4 mm. long;

fruit 8–10 mm. thick, globose, purple with a bloom. *Smilax lasioneuron* Hook. *S. herbacea* Coult. Woods: Ont.—Ga.—Kans.—Colo.—Wyo.—Sask. *Canad.—Allegh.* My–Je.

4. **N. pulverulénta** (Michx.) Small. Stem elongate, climbing, 1–5 m. long; leaf-blades ovate to elliptic, short-acuminate, 8–15 cm. long, cordate at the base, puberulent on the veins beneath; peduncles as long as the leaves or longer; sepals of the staminate flowers oblong-ovate, acute, 5 mm. long; berry subglobose, 8–10 mm. broad. *S. herbacea pulverulenta* (Michx.) A. Gray [G]. Alluvial soil: N.Y.—Ont.—Mo.—Neb.—N.C. *Allegh.* Ap–Je.

2. **SMÌLAX** L. CATBRIER, GREENBRIER, SMILAX.

Woody vines, with prickly-armed stems and often tuberous rootstocks. Leaves alternate, petioled, leathery, several-ribbed, the petioles with coiled tendrils. Flowers dioecious, in axillary umbels; pedicels with bractlets. Sepals and petals each 3, similar, greenish. Stamens 6, in the pistillate flowers reduced. Ovary 3-celled, in the staminate flowers wanting or abortive; ovules solitary in each cavity. Fruit a berry, with 3 bands of strengthening cells. Seeds 1–3.

Peduncles of the pistillate flowers much longer than the subtending petioles.
  Leaves glaucous beneath; peduncles less than twice as long as the petioles. — 1. *S. glauca.*
  Leaves not glaucous beneath.
    Branchlets angular, often square; peduncles barely twice as long as the subtending petioles. — 2. *S. Bona-nox.*
    Branchlets subterete; peduncles 2–4 times as long as the petioles.
      Stem densely bristly, at least below; rootstock cylindric; peduncles 2–5 cm. long. — 3. *S. hispida.*
      Stem unarmed, or with a few stout prickles; rootstock tuberous; peduncles 5–7 cm. long. — 4. *S. Pseudo-china.*
Peduncles of the pistillate flowers not longer than the subtending petioles. — 5. *S. rotundifolia.*

1. **S. glaùca** Walt. Stem terete, high-climbing, armed with stout prickles, the branches angled, usually unarmed; leaf-blades broadly ovate, cuspidate, entire, glaucous beneath, 5–7-ribbed, 3–15 cm. long; peduncles 1–3 cm. long; flowers 6–12; pedicels 4–8 mm. long; sepals and petals greenish, linear-oblong, 4 mm. long; berries subglobose, 6–8 mm. broad, bluish black, with a bloom. Sandy soil: Mass.—Neb.—Tex.—Fla. *E. Temp.* My–Je.

2. **S. Bòna-nóx** L. Stem high-climbing; leaf-blades mostly leathery, often persistent, ovate to hastate or fiddle-shaped, 3–12 cm. long, acute or cuspidate at the apex, pellucid-punctate, 5–9-ribbed, smooth or erose-prickly on the margins and the ribs beneath; peduncles 1.5–3 cm. long, flattened; flowers 15–45; pedicels 4–8 mm. long; sepals and petals deep-green, those of the staminate flowers 4–5 mm. long, those of the pistillate ones much smaller; berry globose, 4–6 mm. broad, black, 1-seeded. Thickets: Mass.—Kans.—Tex.—Fla. *E. Temp.* Ap–Jl.

3. **S. híspida** Muhl. *Fig. 132.* Stem terete, 1–5 m. high; leaf-blades thin, broadly ovate to oblong, 5–15 cm. long, cuspidate, 5–9-ribbed, erose-denticulate on the margins and ribs beneath, obtuse to subcordate at the base; flowers 10–25; pedicels 4–8 mm. long; petals and sepals greenish yellow, those of the staminate flowers 4 mm. long; berries globose, 6 mm. broad, bluish black. Thickets and rocky banks: Ont.—S.D.—Neb.—Tex.—S.C. *E. Temp.* My–Jl.

4. **S. Pseùdo-china** L. Stem high-climbing; leaf-blades leathery, persistent, ovate or suborbicular, sometimes 2-lobed at the base, acute or cuspi-

date, bristly-toothed on the margins and on the ribs beneath, 7–9-ribbed, 5–13 cm. long; flowers 12–40; pedicels 6–8 mm. long; sepals and petals dark green, those of the staminate plant 5 mm. long, those of the pistillate one shorter; berries globose, 4–6 mm. broad. Thickets: Md.—Neb.—Tex.—Fla. *Carol.—Ozark.* Mr–Au.

5. **S. rotundifòlia** L. Stem 2–6 m. high, with sharp prickles; branches sharply angled, prickly; leaf-blades ovate to orbicular, 3–15 cm. long, cuspidate at the apex, rounded or cordate at the base, rough on the ribs beneath, punctate; peduncles 5–12 mm. long, flattened; flowers 5–25; pedicels 2–8 mm. long; sepals and petals greenish yellow, 3–4 mm. long or in the pistillate flowers 5 mm. long; berries subglobose, 6 mm. broad, 1–3-seeded, bluish black. Thickets: Ont.—Minn.—Kans.—Tex.—Fla. *E. Temp.* Ap–Je.

Family 29. **AMARYLLIDACEAE.** AMARYLLIS FAMILY.

Perennial fleshy plants, with bulbs, corms, rootstocks, or woody caudices. Leaves basal, usually sheathing. Flowers perfect, racemose, paniculate, umbellate or solitary. Sepals and petals each 3, epigynous, often united into a tube below, petaloid. Stamens 6. Gynoecium of 3 united carpels; ovary inferior, 3-celled, or rarely only partially so; styles united. Fruit a 3-celled capsule or berry.

Perennial herbs, with rootstocks or corms; flowers several.
  Ovary and capsule half-inferior; flowers spicate. 1. ALETRIS.
  Ovary and capsule wholly inferior; flowers solitary or subumbellate. 2. HYPOXIS.
Perennials, with bulbs; flowers solitary on the scape. 3. COOPERIA.

1. **ÁLETRIS** L. STAR-GRASS, COLIC-ROOT.

Caulescent herbs, with hard rootstocks. Leaves mainly basal, flat, spreading; stem-leaves much reduced. Flowers perfect, in elongate racemes. Perianth white or yellow, campanulate or cylindric, 6-lobed. Stamens 6; filaments partly adnate to the perianth-tube, short; anthers introrse. Ovary 3-celled, half-inferior; ovules numerous in each cavity, anatropous. Capsule 3-celled, enclosed in the persistent perianth, loculicidal. Seeds numerous, ribbed.

1. **A. farinòsa** L. Stem 3–10 dm. high; leaves narrowly oblong or broadly linear, 5–30 cm. long; racemes 1–3 dm. long; perianth white or cream-colored, cylindric, 7–9 mm. long, constricted above the middle; lobes ovate, spreading; capsule ovoid, long-beaked. Sandy soil: Me.—Minn.—La.—Fla. *Allegh.—Carol.* My–Jl.

2. **HYPÓXIS** L. STAR-GRASS.

Acaulescent small herbs, with corms or short rootstocks and narrow grass-like basal leaves. Flowers perfect, regular. Sepals and petals each 3, equal, distinct above the ovary, yellow or whitish, the sepals green without. Stamens 6; filaments adnate only to the base of the perianth, short. Capsule 3-celled, thin.

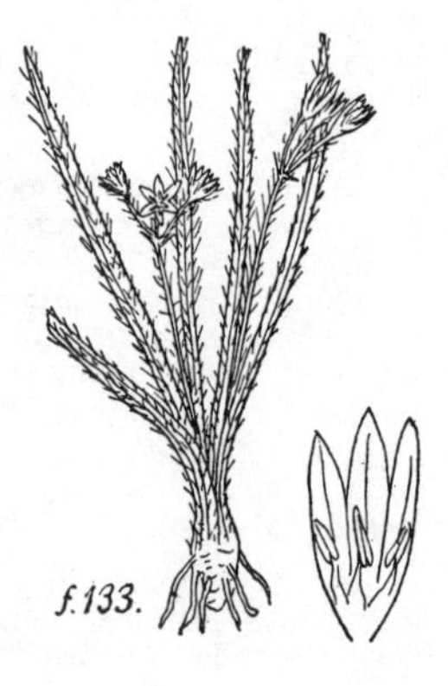
f. 133.

1. **H. hirsùta** (L.) Coville. *Fig. 133.* Leaves narrowly linear, 2–5 mm. wide, longer than the scape, more or less villous; scape 5–15 cm. high, 1–6-flowered; flowers umbellate, 6–10 mm. long, bright yellow within, villous without. *H. erecta* L. Meadows: Me.—Fla.—Tex.—Sask. *E. Temp.—Plain.* My–O.

3. COOPÉRIA Herb. PRAIRIE LILY.

Acaulescent herbs, with coated bulbs. Leaves basal, very narrow, grass-like. Scape simple, 1-flowered. Flowers perfect, subtended by a bractlike spathe. Perianth salverform, the tube elongate, cylindric, the lobes 6, spreading above. Stamens 6; filaments adnate to the throat of the perianth; anthers narrow, erect. Ovary 3-celled; style filiform; stigma slightly 3-lobed. Ovules numerous, in two rows in each cavity. Capsule 3-lobed, 3-celled, loculicidal. Seeds black.

1. **C. Drummóndii** Herb. *Fig. 134.* Bulb subglobose, 2.5 cm. thick; leaves narrowly linear, 1–3 dm. long; scape 1.5–3 dm. high, terete; perianth whitish, the tube slender, 7–13 cm. long; lobes ovate, elliptic, or oval, 1–2 cm. long, acute; ovary sessile; capsule 1 cm. thick. Prairies: Kans.—N.M.—Tex.; northern Mex.

Family 30. DIOSCOREACEAE. YAM FAMILY.

Perennial herbaceous or woody vines with tuberous, knotted rootstocks. Leaves often opposite or whorled below, alternate above, petioled, several-ribbed and netted-veined. Flowers in axillary spikes, racemes, or panicles, perfect, monoecious, or dioecious, regular. Perianth calyx-like; sepals and petals each 3, partially united. Stamens 3–6, inserted at the base of the perianth. Gynoecium of 3 united carpels; ovary inferior, 3-celled; ovules 1 or 2 in each cavity. Fruit a 3-winged, 3-valved capsule, or berry-like.

1. DIOSCORÈA (Plum.) L. YAM, WILD YAM-ROOT.

Herbaceous twining vines. Leaves alternate, at least above, petioled, cordate or broadest below the middle. Flowers dioecious or monoecious. Staminate flowers with deciduous perianth and 3–6 stamens, the pistillate ones with persistent perianth. Ovary inferior; styles 3. Capsule broadly winged, dehiscent through the wings. Seeds flat, winged.

1. **D. paniculàta** Michx. *Fig. 135.* Stem 1–5 m. long, branched; leaf-blades thin, ovate, 5–15 cm. long, acuminate, cordate at the base, pubescent beneath, or glabrous on both sides, 9–11-ribbed; staminate flowers paniculate, 3 mm. broad, the pistillate ones in simple racemes; capsule 1.5–2.5 cm. long, broadly winged. *D. villosa* Auth. [G]; not L. Thickets: R.I.—Minn.—Kans.—Tex.—Fla. *E. Temp.* Je–Jl.

Family 31. IRIDACEAE. IRIS FAMILY

Perennial herbs, with elongated or bulb-like rootstocks and narrow, equitant, 2-ranked leaves. Flowers perfect, regular or nearly so. Sepals and petals each 3, often dissimilar but both colored, distinct or united below. Stamens 3, opposite to the sepals; anthers extrorse. Gynoecium of 3 united carpels; ovary inferior, 3-celled; styles distinct, sometimes petal-like. Fruit a loculicidal capsule. Seeds numerous.

Styles alternate with the stamens; perianth-lobes subequal.
Style-branches 2-cleft; stem from a coated bulb; stamens monadelphous at the base. 1. NEMASTYLIS.
Style-branches not 2-cleft; stem from a rootstock or clump of fibrous roots.
Filaments distinct; rootstock creeping; stigmas dilated. 2. BELAMCANDA.
Filaments monadelphous; rootstock short or obsolete, with a cluster of fibrous roots; stigmas filiform. 3. SISYRINCHIUM.
Styles opposite to and arching over the stamens; lobes of the perianth unequal, the sepals spreading or recurved, the petals erect to spreading. 4. IRIS.

1. NEMÁSTYLIS Nutt.

Caulescent herbs, with coated bulbs. Leaves alternate, narrow, folded. Flowers perfect, solitary or several from 2 herbaceous bracts. Perianth-lobes 6; spreading, nearly equal. Stamens 3, the filaments more or less dilated. Ovary 3-celled; styles alternate with the stamens, each 2-cleft, the branches filiform. Ovules numerous. Capsule elongate, loculicidal, 3-valved at the apex.

1. N. acùta (Bart.) Herb. Stem 1–6 dm. high, simple or sparingly forked; leaves linear, 1–3.5 dm. long; flowers 2 or 3 from the spathe; bracts unequal, the outer shorter; perianth blue or purple, 4–5 cm. broad; lobes broadest below the middle; capsule obovoid, 1–1.5 cm. high. *N. geminiflora* Nutt. Prairies: Tenn.—Kans.—Tex.—La. *Ozark.* Ap–Je.

2. BELAMCÁNDA Adans. BLACKBERRY LILY.

Caulescent herbs, with horizontal rootstocks. Leaves alternate, sheathing the stem. Flowers perfect, in small clusters, disposed in corymbs. Perianth usually mottled, the lobes 6, united to the top of the ovary or slightly beyond it, persistent-withering. Stamens 3; filaments distinct, adnate to the base of the perianth; anthers elongate. Ovary 3-celled; styles 3, undivided, slender, alternate with the stamens. Ovules numerous. Capsule loculicidal, 3-valved to the base, the valves reflexed. The exposed seeds resembling a blackberry, persistent on the axial placenta. *Gemmingia* Fabr.

1. B. chinénsis (L.) DC. Stem 3–12 dm. high, leafy below; leaves 1–5 cm. long, acute or acuminate, their upper faces united above the middle; bracts similar, much smaller, their faces not united; flowers pink or reddish, mottled with crimson or purple, 3–5 cm. broad; lobes oblong, narrowed at the base; capsule turbinate, 2–2.5 cm. long. *G. chinensis* (L.) Kuntze. Roadsides and banks: Conn.—Neb.—Ala.—Ga.; nat. from Asia. Je–Jl.

3. SISYRÍNCHIUM L. BLUE-EYED GRASS.

Perennial grass-like herbs, ours tufted, with short rootstocks and fibrous roots. Leaves narrowly linear. Scape 2-edged or 2-winged. Flowers in terminal clusters from spathes of 2 bracts. Petals and sepals in ours blue or white, alike, spreading. Filaments monadelphous. Style-branches filiform, alternate with the stamens. Ovules few in each cavity. Capsule subglobose, readily opening at the apex.

Stem usually branched above, bearing 2 or more peduncled spathes. 1. *S. graminoides.*
Stem simple, with a single spathe.
Perianth-segments not emarginate, narrowed to the aristulate tip; stems and leaves mostly under 1 mm. wide; capsule obovoid or subglobose, 3–5 mm. long. 2. *S. septentrionale.*
Perianth-segments more or less retuse or abruptly contracted to the aristulate apex.
Stems and leaves mostly 1.5–2.5 mm. wide; capsules 4–6 mm. high. 3. *S. angustifolium.*
Stems and leaves mostly 1.5 mm. or less in width; capsules 2–4 mm. high.
Bracts of spathe smooth; valves of capsule veinless. 4. *S. mucronatum.*
Bracts of spathe minutely scaberulous; valves of capsule sparsely venose. 5. *S. campestre.*

1. **S. graminoìdes** Bickn. Stems often tufted, 1.5–6 dm. high, bright green, turning black in drying; leaves grass-like, 1–4 dm. long, 1.5–6 mm. wide, the edges minutely serrulate; scapes flattened, winged, as broad as the leaves, mostly forking into two peduncles, 5–20 cm. long; spathes green, the bracts equal or unequal, acute or acuminate, 15–20 mm. long; perianth blue, 1 cm. long; capsule subglobose, 3–6 mm. high. *S. gramineum* Curtis [G]; not Lam. Meadows: Newf.—Minn.—Tex.—Fla. *E. Temp.* My–Je.

2. **S. septentrionàle** Bickn. Stem 1–2.5 dm. high, slender, 0.5–1 mm. wide, barely margined; leaves 0.5–1 mm. broad; spathe purplish or green; outer bract 2.5–4 cm. long, the inner 1.5–2 cm. long, both hyaline-margined; perianth 4–7 mm. long, pale rose or violet. Wet meadows: Man.—N.D.—Wash. —B.C. *W. Boreal.—Plain.* Je–Jl.

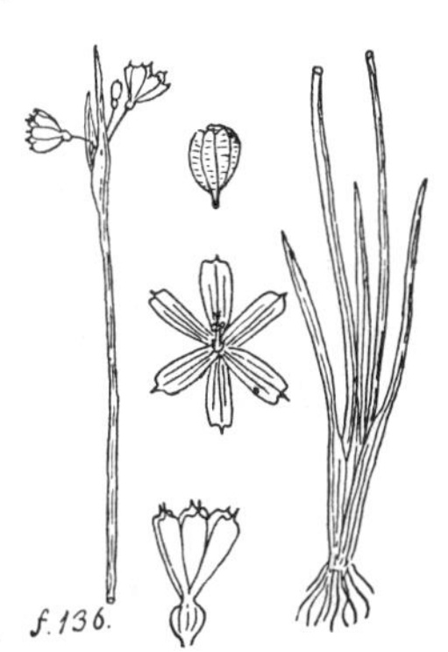
f. 136.

3. **S. angustifòlium** Mill. *Fig. 136.* Stem stiff, glaucous, 1–5.5 dm. high, simple or rarely branched, 1–3 mm. wide, winged; leaves 1–3.5 mm. wide; spathe green or slightly purplish; outer bract 2–6.5 cm. long, obscurely hyaline-margined; inner bract 1.5–3 cm. long, hyaline-margined; flowers deep violet, 10–12 mm. long; capsule ellipsoid to subglobose. *S. montanum* Greene. Meadows: Newf.—Va.—Colo.—B.C. *Boreal—Plain—Mont.* Je–Au.

4. **S. mucronàtum** Michx. Stems numerous, 1–4.5 dm. high, 0.5–1.5 mm. wide, margined or narrowly winged; leaves dull green, slender, 1–1.5 mm., rarely 2 mm. wide; spathe usually red-purple, rarely green; outer bract 1.8–5.5 cm. long; inner bract 1–1.5 cm. long, scarious-margined; perianth deep purplish blue, rarely white, 6–14 mm. long; capsules subglobose. Wet places: Ont.—D.C.—Pa.—Alta. *Canad.—Allegh.—Plain.* My–Jl.

5. **S. campéstre** Bickn. Stem usually 1.5–2 dm. high, erect, stiff, glaucous, 0.5–1.5 mm. wide, narrowly winged; leaves 1–1.5 mm., rarely 2 mm. wide; spathe pale purple or green; outer bract 2.5–4.5 cm. long, narrowly hyaline-margined below; inner bract 1.2–2 cm. long, white-hyaline on the margins; perianth 8–14 mm. long, pale blue or white; capsule trigonous-subglobose. Prairies: Man.—Wis.—Mo.—La.—N.M.—N.D. *Prairie—Plain—St. Plains.*

4. **ÌRIS** L. Blue Flag, Fleur-de-lis.

Herbs, with creeping, horizontal rootstocks. Leaves sword-shaped or linear. Flowers solitary or in terminal panicles. Sepals and petals highly colored, in ours blue; the former spreading or recurved, the latter usually smaller and erect. Stamens 3; filaments adnate below to the base of the sepals. Ovary 3-celled; styles petal-like arching over the stamens; stigmas under the usually 3-lobed tips. Ovules numerous. Capsule elongated, 3- or 6-angled. Seeds in 1 or 2 rows, vertically flattened.

Leaves 0.5–1 cm. wide; petals 2-lobed at the apex; capsule bluntly 6-angled. — 1. *I. missouriensis.*
Leaves 1–2.5 cm. wide; petals entire.
  Capsule bluntly 3-angled.
    Blade of the sepals ovate to reniform, purplish blue, spotless or the spot greenish or greenish yellow; ovary 1–2 cm. long in flower; style-branches not auriculate. — 2. *I. versicolor.*
    Blade of the sepals oblong-ovate, bright blue with a pubescent yellow spot the base; ovary 1.8–3.8 cm. long in flower; style-branches auriculate. — 3. *I. virginica.*
  Capsule sharply 6-angled. — 4. *I. foliosa.*

1. **I. missouriénsis** Nutt. *Fig. 137.* Stem 2–10 dm. high, about 5 mm. in diameter; leaves 1–5 dm. long, 5–10 mm. broad; perianth pale blue, variegated. glabrous, crestless; sepals broadly oblanceolate, about 8 cm. long; petals oblanceolate, erect, about 6 cm. long; capsule oblong-elliptic, about 4 cm. long and 1.5 cm. thick, 6-ridged. Meadows, marshes, and along streams: N.D.—N.M.—Calif.—B.C. *Plain—Mont.—W. Temp.* My–Jl.

2. **I. versícolor** L. Stem 8–10 dm. high; leaves glaucous, thick, shorter than the stem, acuminate up to 4.5 cm. long; perianth blue or lilac, variegated with white, yellow, or purplish; tube 1–1.5 cm. long; sepals broadly spatulate, much longer than the petals, 7–9 cm. long; petals flat, oblanceolate; appendages of the stigmas overlapping, erose; capsule oblong-prismatic, 3–4 cm. long. Swamps and river banks: Newf.—Man.—Min.—DC. *Canad.* My–Jl.

3. **I. virgínica** L. Stem 5–9 dm. high, branched above; leaves 1–5 cm. broad; bracts of the spathe 4–12 cm. long, sometimes foliaceous; perianth bright blue, variegated with yellow and purple; tube thick, 1–1.5 cm. long; petals oblong-spatulate, 3–7 cm. long, notched at the apex; capsule 3–4 cm. long: Swamps: D.C.—Fla.—Tex.—Neb.—Minn. *Allegh—Austral.* My–Jl.

4. **I. foliòsa** Mack. & Bush. Stem 3–5 dm. high, decumbent, zigzag above; leaves sometimes as long as the stem, attenuate, sword-shaped; outer bract 1–1.5; perianth violet-blue, variegated with white or yellow; sepals spatulate, 8 cm. long; petals erect, narrower, nearly as long; capsule prismatic, 4–5 cm. long, 2 cm. thick. *I. hexagona* Western reports. Swamps: Ky.—Ohio—Kans.—Tex. *Ozark.* Ap–My.

Family 32. **ORCHIDACEAE.** Orchis Family.

Perennial herbs, usually succulent, with corms, bulbs, or rootstocks, and tuberous or fibrous roots. Flowers perfect, irregular. Sepals 3, similar or nearly so, the lower two sometimes united. Petals 3; the two lateral ones similar; the median one (the lip) usually very dissimilar, sometimes prolonged below into a spur, usually inferior by twisting of the ovary. Stamens 3, of which 1 or 2 are abortive, adnate to the pistil and forming a column. Fertile anthers usually solitary, in a few genera 2, usually 2-celled, containing 2 or 3 waxy or powdery pollinia, these pollen-masses usually stalked and often attached at the base to a viscid gland. Gynoecium of 3 united carpels; ovary inferior, 1-celled, twisted. Capsule 3-valved. Seeds numerous. Endosperm wanting.

Fertile stamens 2; lip a large inflated sack.
  Lower two sepals distinct; lip with an almost conical obtuse prolongation below. — 1. Criosanthes.
  Lower two sepals united; lip rounded-saccate.
    Lip entire; plant leafy throughout. — 2. Cypripedium.
    Lip 2-cleft; stem scapiform; leaves 2, basal. — 3. Fissipes.
Fertile stamen 1.
  Pollinia caudate at the base, attached to a viscid disk or gland.
    Gland enclosed in a pouch-like fold.
      Sepals united above into a hood; lip entire. — 4. Galeorchis.
      Sepals free; lip 3-lobed. — 5. Orchis.
    Gland not enclosed in a pouch.
      Lip not fringed or cut-toothed.
        Valves of the anthers dilated below, enclosing the glands at the base; lip with a tubercle near the base. — 7. Perularia.

Valves not dilated at the base.
Stigma with 3 appendages; lip 3-toothed. 8. DENSLOVIA.
Stigma without appendages.
Gland surrounded by a thin membrane; lip toothed at the apex. 6. COELOGLOSSUM.
Gland naked; lip entire.
Sepals 3–5-nerved; plants with rootstock and fibro-fleshy roots.
Stem scapiform; leaves 1 or 2, basal; anther-sacs divergent; plants in ours with rootstocks.
Basal leaves 2; ovary straight. 9. PLATANTHERA.
Basal leaf 1; ovary arcuate. 10. LYSIELLA.
Stem leafy; anther-cells parallel or nearly so; plants with fleshy-fibrous roots. 11. LIMNORCHIS.
Sepals 1-nerved; plants with rounded or oblong, undivided biennial corms. 12. PIPERIA.
Lip fringed or parted or cut-toothed. 13. BLEPHARIGLOTTIS.
Pollinia not produced into caudicles.
Pollinia granulose or powdery.
Flowers comparatively large, solitary or few; anthers incumbent on a long column.
Leaves not grass-like; lip free.
Flowers terminal; lip crested. 14. POGONIA.
Flowers axillary; lip not crested. 15. TRIPHORA.
Leaves grass-like.
Flowers solitary, terminal; lip adherent to the base of the column. 16. ARETHUSA.
Flowers several, racemose; lip free. 17. LIMODORUM.
Flowers small, numerous, in spikes or racemes; anthers erect, articulate to a short column.
Anthers not operculate.
Leaves green, borne on the stem.
Leaves alternate; spike mostly twisted. 18. SPIRANTHES.
Leaves 2, opposite; spike not twisted. 19. LISTERA.
Leaves white-reticulate, basal. 20. GOODYERA.
Anthers operculate, leaves alternate. 26. EPIPACTIS.
Pollinia smooth or waxy.
Plants with corms or solid bulbs, rarely with coralloid roots; leaves not scale-like.
Leaves unfolding before or with the flowers.
Lip flat; flowers racemose.
Leaves cauline; column short; pollinia clavate. 21. MALAXIS.
Leaves basal; column elongated; pollinia globose. 22. LIPARIS.
Lip saccate; flower solitary. 23. CYTHEREA.
Leaf 1, unfolding after the flowering time. 24. APLECTRUM.
Plants with coralloid roots, bulbless; leaves reduced to scales. 25. CORALLORRHIZA.

1. CRIOSÁNTHES Raf. RAM'S-HEAD, RAM'S-HEAD LADY'S SLIPPER.

Perennial caulescent herbs, with rootstocks. Leaves sessile, alternate, several-veined, plaited. Flowers solitary, terminal. Sepals 3, distinct, spreading. Lateral petals narrow, spreading. Lip saccate, obtuse-conic, with the margins incurved, formed a rounded opening.

1. **C. arietìna** (R. Br.) House. Stem 2–3 dm. high, 1-flowered; leaves 3 or 4, elliptic or lanceolate, 5–10 cm. long; sepals lanceolate, 15–20 mm. long, brownish; petals linear, greenish brown, about as long; lip about 1.5 cm. long, white, veined and variegated with red. *Cypripedium arietinum* R. Br. [B,G]. Cold damp woods: Que.—N.Y.—Minn.—Sask. *Boreal.* My–Au.

2. CYPRIPÈDIUM L. LADY'S SLIPPER.

Caulescent herbs, with rootstocks and fibrous roots. Leaves sessile, several-nerved, plaited. Flowers solitary or several in a terminal raceme. Sepals 3, the lateral two united under the lip. Lateral petals narrow, spreading. Lip conspicuous, an inflated round sack, with rounded opening, the margins incurved. Column declined, glabrous, bearing a fertile anther on each side, and a petaloid sterile stamen above, arching over the style. Pollen pulpy or granular, without tails or glands. Stigma terminal, broadest at the base. Capsule ribbed.

Sepals obovate or oval, not longer than the lip.
  Lip 3–4 cm. long, variegated with purple and white. 1. *C. hirsutum.*
  Lip about 1.5 cm. long, white or light magenta, with purple spots. 2. *C. passerinum.*
Sepals lanceolate, attenuate, often longer than the lip.
  Lip white, sterile stamen lanceolate. 3. *C. candidum.*
  Lip yellow, sterile stamen triangular.
    Lip broader than high; sterile stamen obtuse.
      Sepals 3–5 cm. long; lip 2–3 cm. long; stigma acute. 4. *C. parviflorum.*
      Sepals more than 5 cm. long; lip 3.5–5 cm. long; stigma obtuse. 5. *C. pubescens.*
    Lip higher than broad; sterile stamen deltoid, acute. 6. *C. flavescens.*

1. **C. hirsùtum** Mill. Stem erect, 3–6 dm. high; leaves oval or elliptic, 8–20 cm. long, acute, villous-pubescent; flowers 1–3; sepals white, rounded-ovate or oval; lateral petals white, lanceolate, barely as long as the sepals; lip much inflated; stamens ovate-cordate. *C. Reginae* Walt. [B]. *C. spectabile* Salisb. Swamps and woods: Newf.—Man.—N.D.—Kans.—Ga. *Canad.—Allegh.* Je–S.

2. **C. passerìnum** Richardson. Stem 1–2.5 dm. high, often retrorsely villous; leaves oval to lanceolate, 5–10 dm. long, more or less viscid-villous; flowers 1–3; sepals 1–1.5 cm. long, the lower one slightly 2-cleft; petals oblong, obtuse, about 1.5 cm. long; lip about 1.5 cm. long, obovate, white, with purple spots inside. Pine woods: Que.—Alta.—B.C.—Alaska—Mack. *Boreal—Mont.* Je–Jl.

3. **C. cándidum** Muhl. Stem 1.5–3 dm. high; leaves oval to lanceolate, acute or acuminate, 7–12 cm. long; flowers solitary; sepals ovate-lanceolate, greenish, purple-spotted; petals lanceolate, greenish, somewhat longer, twisted; lip white, striped with purple inside, 2 cm. long. Bogs and meadows: N.Y.—Man.—S.D.—Neb.—N.J. *Canad.* My–Jl.

4. **C. parviflòrum** Salisb. *Fig. 138.* Stem 2–6 dm. high, leafy, glandular-puberulent, usually 1-flowered; leaves oval to broadly lanceolate, often acuminate, 5–12 cm. long, puberulent; upper sepal lanceolate, acuminate, about 3 cm. long and 1 cm. wide; petals narrowly linear-lanceolate, longer than the sepals. Woods: Newf.—Ga.—Colo.—B.C. *Temp.—Submont.* My–Jl.

5. **C. pubéscens** Willd. Stem 3–10 dm. high; leaves oblong or elliptic, acute or acuminate, glandular-pubescent or glabrate; flowers 1–3; sepals lanceolate, acuminate, greenish-yellow, dotted and streaked with purple; lateral petals linear, attenuate, twisted, longer than the sepals; lip yellow, veined with purple. *C. parviflorum pubescens* (Willd.) Knight [G]. Woods: N.S.—Man.—N.D.—Neb.—Ga. *E. Temp.* My–Jl.

6. **C. flavéscens** Red. Stem 3–5 dm. high, glandular-puberulent, 1–3-flowered; leaves oval, short-acuminate, 5–12 cm. long; upper sepal lanceolate, 3–4 cm. long, 1 cm. wide, twisted at the tip; lower sepals narrower; petals narrowly lance-linear, longer than the sepals; lip 2.5–3 cm. long, higher than broad. *C. bulbosum* Mill.; not L. *C. parviflorum* Sims; not Salisb. *C. pubescens* Sweet [B]; not Willd. Open woods: Me.—Man.—Iowa—Mo.—Va. *Canad.—Allegh.* My–Jl.

3. FÍSSIPES Small. MOCCASIN FLOWER, NOAH'S ARK.

Acaulescent herbs. Leaves 2, basal, plaited, spreading. Flowers usually solitary on the simple scape. Sepals 3, but the lateral 2 wholly united, forming a single body under the lip, greenish. Lateral petals similar to the sepals. Lip showy, inflated, split its whole length in front. Column declined, glandular-pubescent, bearing a sessile anther on each side and a rhomboidal sterile stamen above. Pollen granular, without gland, not caudate. Stigma broadest at the apex.

1. **F. acaùlis** (Ait.) Small. Leaves 2, oblong-elliptic or oval, 1–3 dm. long, glandular-pubescent; scape 1–5 dm. high, 1- or 2-flowered; sepals oblong or lanceolate, 3–4 cm. long; lateral petals narrower and longer; lip obovoid, 4–6 cm. long, pink or white, with darker veins, the upper part crested within with long white hairs. *Cypripedium acaule* Ait. [G,B]. Sandy woods and bogs: Newf.—Man.—Minn.—Tenn.—N.C. *Canad.—Allegh.* My–Je.

4. GALEÓRCHIS Rydb. Showy Orchid.

Acaulescent fleshy herbs, with short rootstocks and coarse fleshy roots. Flowers in a terminal spike on a scaly scape. Leaves 2, basal, fleshy. Sepals 3, converging, united above into a kind of galea. Lateral petals converging, similar to the sepals and partially adnate to them. Lip united with short column below, with a short spur. Anther 2-celled, the sacs contiguous, slightly divergent. Pollen-masses granular, one in each sac, produced into a slender tail, this attached to a small gland. Style with a knob-like projection under the anther. Stigma hollow. Gland enclosed in a pouch. Capsule elongate, beakless.

1. **G. spectàbilis** (L.) Rydb. *Fig. 139.* Leaves oblong-elliptic to oval, obtuse, narrowed below into a sheathing base; scape 1–3 dm. high; spike 3–12-flowered; perianth violet-purple, 2–5 cm. long; spur obtuse, 1.5 cm. long. *Orchis spectabilis* L. [G]. Woods: N.B.—Minn.—Kans.—Ga. *Canad.—Austral.* My–Je.

5. ÓRCHIS (Tourn.) L. Orchis.

Perennial herbs, with digitately cleft tubers, or in our species with rootstocks and fibrous roots. Flowers in terminal spikes. Sepals distinct, equal, spreading, 3-nerved. Petals ascending, in ours narrower. Lip connate with the column, produced below into a spur, usually 3-cleft. Column short. Fertile anther solitary, 2-celled; sacs contiguous and slightly diverging; pollinia granulose, one large mass in each sac, produced into a slender caudicle, attached to a small gland, which is enclosed in a pouch.

1. **O. rotundifòlia** Pursh. Perennial, with a rather slender scaly rootstock; stem 1–2.5 dm. high, scape-like; leaf solitary, near the base, orbicular or oval, 3–7 cm. long; spike 2–6-flowered; flowers 12–15 mm. long; sepals elliptic, 6–7 mm. long, rose-colored, the lateral ones spreading; upper petals similar, but slightly shorter; lip white, purple-spotted, 3-lobed, with a large middle lobe, notched at the apex; spur slender, curved, shorter than the lip. Cold damp woods: Greenl.—N.Y.—Minn.—B.C.—Yukon. *Subalp.—Boreal.—Arct.* Je–Jl.

6. COELOGLÓSSUM Hartm.

Caulescent herbs, with 2-cleft biennial tubers. Leaves alternate, sessile. Flowers perfect in elongate terminal spikes, greenish. Sepals distinct, converging, thus forming a hood. Lateral petals narrow, erect; lip obtuse, 2- or 3-toothed at the apex, produced below into a sack-like spur. Column short. Fertile anther solitary; anther-sacs 2. Pollinia solitary, with long caudicles. Gland small, surrounded by a thin membrane.

1. **C. bracteàtum** (Willd.) Parl. Stem 1.5–6 dm. high; leaves 3–5, obovate, oblanceolate, oval, or the upper narrowly lanceolate, 5–15 cm. long; bracts linear-lanceolate, longer than the flowers; flowers green or greenish; sepals lanceolate, about 6 mm. long; lip 6–8 mm. long, oblong or somewhat cuneate, 3-toothed at the apex, the middle tooth smallest; spur less than half as long as the lip. *Habenaria bracteata* (Willd.) R. Br. [B]. Wet meadows and woods: N.B.—N.C.—Iowa—Colo.—B.C. *Mont.—Boreal.* My–S.

7. PERULÀRIA Lindl.

Caulescent herbs, with a cluster of fibrous roots. Leaves alternate. Flowers small, yellowish or greenish, in an elongate spike. Sepals and lateral petals spreading, broad. Lip lanceolate, with a median tubercle and a tooth on each side near the base. Spur slender, straight. Valves of the anthers horizontal, opening upwards, dilated at the base, enclosing the gland. Pollinia granulose, produced at the base into a caudicle.

1. **P. flàva** (L.) Farwell. *Fig. 140.* Stem 3–6 dm. high; leaves elliptic, oblong, or lanceolate, 1–3 dm. long, clasping at the base; bracts surpassing the flowers; sepals ovate or rounded-ovate, 6 mm. long; lateral petals similar, but shorter, crisped; lip longer than the petals, erose, truncate; spur longer than the lip, obtuse; capsule oblong, 8–10 mm. long. *Habenaria flava* (L.) A. Gray [G]. *H. virescens* Spreng. Low ground: N.S.—Minn.—La.—Fla. *E. Temp.* Je–Jl.

8. DENSLÒVIA Rydb. Wood-orchid.

Caulescent herbs. Leaves alternate. Flowers in stout spikes, with small bracts. Sepals free, spreading. Lateral petals smaller than the sepals. Lip 3-toothed at the apex. Spur longer than the lip. Anther with parallel and approximate sacs and naked glands. Pollinia granular, short-caudate. Stigma with 3 oblong appendages, the middle one between the orbicular glands, as high as the anther.

1. **D. clavellàta** (Michx.) Rydb. *Fig. 141.* Roots coarse, clustered; stem 1–5 dm. high, naked above; leaves reduced to scales, except 1 or 2 near the middle of the stem, these oblong or oblong-oblanceolate, 5–12 cm. long; spike 2–5 cm. long, few-flowered; perianth greenish-white; sepals ovate or oval, 2–2.5 mm. long, obtuse; lateral petals similar; lip 3–4 mm. long, 3-toothed at the apex; spur clavate, longer than the ovary. *H. clavellata* (Michx.) Spreng. [G]. *H. tridentata* Hook. *Gymnadeniopsis clavellata* Rydb. [B]. Newf.—S.D.—La.—Fla. *E. Temp.* Jl–Au.

f. 141.

9. PLATÁNTHERA Rich.

Subacaulescent herbs, with fleshy rootstocks or tubers. Leaves almost basal, 2 in number, broad and many-nerved. Flowers in terminal spikes, white or greenish. Sepals distinct, large, spreading; upper sepal in ours broadly cordate, the lateral ones obliquely ovate. Lateral petals small and narrow. Lip entire, narrow, produced below into a slender spur. Fertile anther solitary; sacs diverging, with beak-like bases projecting forward. Pollinia with their caudicles laterally affixed to orbicular naked glands. Stigma without appendages. *Lysias* Salisb.

Scape bracted. 1. *P. orbiculata.*
Scape naked. 2. *P. Hookeri.*

1. **P. orbiculàta** (Pursh) Lindl. *Fig. 142.* Scape stout, 3–6 dm. high, bearing several lanceolate bracts; basal leaves 1–1.5 dm. long, 8–12 cm. wide; raceme loosely many-flowered; lateral petals obliquely lanceolate; lip linear, 12–15 mm. long; spur 2–4 cm. long. *Habenaria orbiculata* (Pursh) Torr. *Lysias orbiculata* Rydb. Rich woods: Newf.—N.C.—Minn.—Wash.—B.C. *Mont.—Boreal—Allegh.* My–Au.

f.142.

2. **P. Hoòkeri** (Torr.) Lindl. Scape 12–16 cm. high, ebracteate; leaves orbicular to elliptic, 3.5–10 cm. long; raceme strict, 8–20-flowered; flowers yellowish-green; sepals ovate-lanceolate, 8 mm. long; lateral petals linear-subulate; lip lanceolate, about 1 cm. long, spur slender, acute, 2 cm. long. *Habenaria Hookeri* Torr. [G]. *L. Hookeriana* Rydb. [B]. Woods: N.S. — Man. — Iowa — N.J. *Canad.* Je–Jl.

10. LYSIÉLLA Rydb.

Small herbs, with rootstocks. Stem scapiform, with a single obovate leaf at the base. Flowers greenish yellow. Upper sepal round-ovate, erect, surrounding the broad column; lateral sepals reflexed-spreading. Petals lanceolate, smaller; lip entire, linear-lanceolate, deflexed. Spur slightly curved, shorter than the ovary. Anther-sacs divergent, wholly adnate, arcuate; glands small, their faces incurved. Pod obovoid.

1. **L. obtusàta** (Pursh) Rydb. Stem slender, 1–2.5 dm. high, glabrous; leaf 5–12 cm. long; spike 2–6 cm. long, loosely flowered; flowers about 1 cm. long; spur longer than the lip, slender, straight. *Habenaria obtusata* Richardson [G]. Boggy places in the woods: Newf.—N.Y.—Minn.—Colo.—Alaska. *Mont.—Subalp.* Je–Au.

11. LIMNÓRCHIS Rydb. Bog Orchid.

Leafy-stemmed plants, with elongate fusiform root-like tubers and fleshy-fibrous roots. Flowers whitish or greenish or tinged with purple. Upper sepal ovate to almost orbicular, erect, 3–7-nerved; lateral sepals from linear to ovate-lanceolate, free from the lip, 3-nerved, seldom 4–5-nerved, spreading or often somewhat reflexed. Lateral petals erect, lanceolate, 3-nerved, oblique at the base. Lip entire, flat or slightly concave, reflexed, free, from linear to rhombic-lanceolate, obtuse. Column short and thick. Anther-sacs parallel, opening in front.

Flowers greenish or purplish; lip lanceolate, only slightly dilated at the base.
  Spur decidedly clavate, thickened and obtuse at the apex, shorter than the lip. 1. *L. viridiflora.*
  Spur slender, scarcely thickened towards the apex, often acutish, equaling or slightly exceeding the lip.
    Plant tall and stout; sepals 4–6 mm. long. 2. *L. media.*
    Plant slender; sepals 2.5–4 mm. long. 3. *L. huronensis.*
Flowers white or nearly so; lip rhomboid-dilated at the base.
  Spur exceeding the lip, slightly if at all clavate. 4. *L. dilatata.*
  Spur usually shorter than the lip and clavate. 5. *L. borealis.*

1. **L. viridiflòra** (Cham.) Rydb. Stem 2–5 dm. high; lower leaves 5–15 cm. long, oblanceolate, obtuse; spike short and dense; flowers 10–12 mm. long, light green; lip lanceolate, obtuse, less than 5 mm. long. *Habenaria hyperborea* Coult., not R. Br. Bogs: Alta.—S.D.—Colo.—Alaska. *Submont.—Mont.* Je–Au.

2. **L. mèdia** Rydb. Stem very stout, 4–8 dm. high; leaves lanceolate, acute, 1–2 dm. long; spike long, densely flowered; flowers divaricate, about 15 mm. long; lip about 6 mm. long. Bogs: Que.—N.Y.—S.D. *H. dilatata media* Ames [G]. *Boreal—Submont.*

3. **L. huronénsis** (Nutt.) Rydb. Stem slender, 2–3 dm. high; leaves oblanceolate, obtuse, the upper lanceolate, acute; bracts longer than the flowers; flowers light green, almost erect; sepals oblong, 1.5–2 mm. long; lip linear-lanceolate; spur slender, scarcely clavate, curved, longer than the lip; often acutish; pod 1 cm. long. *Habenaria hyperborea* Am. auth. [G]. Swamps: Newf.—N.Y.—Minn.—S.D.—N.D. *Canad.* Je–Jl.

4. **L. dilatàta** (Pursh) Rydb. *Fig. 143.* Stem slender, tall, leafy, 3–6 dm. high; leaves lanceolate, 7–20 cm. long, the lower obtuse, the upper acute; spike 5–25 cm. long, usually rather lax; flowers white, 15–18 mm. long; lip about 7 mm. long. *Habenaria dilatata* (Pursh) Hook. Bogs: Lab.—N.Y.—Neb.—Sask. *Canad.* Je–Au.

f. 143.

5. **L. boreàlis** (Cham.) Rydb. Stem tall and leafy, 4–8 dm. high; lowest leaves oblanceolate and obtuse, the rest lanceolate, acute, 5–15 cm. long; spike often rather dense, 1–2 dm. long; flowers 10–14 mm. long; lip rhombic-lanceolate, obtuse, about 5 mm. long. *Habenaria dilatata* Coult., not Hook. Bogs: Mont.—S.D.—Colo.—Wash.—Alaska. *Submont.—Mont.—W. Boreal.* Je–S.

12. PIPÉRIA Rydb. Wood Orchid.

Somewhat leafy-stemmed plants, but the leaves usually near the base and withering at or before the anthesis. Tubers spherical or rounded-ellipsoid. Flowers greenish or white; sepals and petals 1-nerved or very obscurely 3-nerved, the upper erect, the lateral ones spreading. Lateral petals free, lanceolate or linear-lanceolate, oblique. Lip linear-lanceolate to ovate, obtuse, truncate or hastate at the base, concave, united with the bases of the lower sepals, bordered with an erect margin which connects the lip with the column. Anther-sacs parallel, opening nearly laterally.

1. **P. unalascénsis.** (Spreng.) Rydb. Stem strict, slender, 3–5 dm. high, leafy only near the base; leaves oblanceolate, obtuse or acutish, 1–1.5 dm. long; stem-leaves bract-like; spike long and lax, 1–3 dm. long; flowers greenish, 8–10 mm. long; petals and sepals 2–4 mm. long; lip oblong, obtuse; spur filiform or slightly clavate. *Habenaria unalascensis* (Spreng.) S. Wats. Damp woods: Que.—Mont.—Colo.—Calif.—B.C.—Alaska. *Boreal—Mont.* Je–Au.

13. BLEPHARIGLÓTTIS Raf. Fringed Orchid.

Caulescent herbs, with fleshy or tuberous roots. Leaves alternate. Flowers numerous, in a terminal spike, the bracts sometimes leaf-like. Sepals spreading or reflexed. Lip fringed, 3-parted or laciniate. Spur longer than the lip. Anthers with widely separate and usually diverging sacs, their beak-like bases supported by the arms of the stigma, projecting forward and upward. Gland naked, pollinia granular.

Lip not 3-lobed, narrow, not dilated upward; flowers white. 1. *B. albiflora.*
Lip 3-lobed, at least the middle lobe dilated upward.
  Lobes of the lip narrow, 1 mm. broad or less, only slightly dilated; flowers greenish yellow. 2. *B. lacera.*
  Lobes of the lip broad, fan-shaped.
    Flowers lilac. 3. *B. psycodes.*
    Flowers white. 4. *B. leucophaea.*

1. **B. albiflòra** Raf. Stem 3–10 dm. high; lower leaves linear-oblong, 1–2.5 dm. long, the upper lanceolate, shorter; spike 3–10 cm. long, rather loose; sepals suborbicular, 6–7 mm. long; lateral petals smaller, toothed or fringed at the apex; lip 6–8 mm. long, fringed; spur 1–2.5 cm. long. *Habenaria Blephariglottis* (Willd.) Torr. [G]. *B. Blephariglottis* Rydb. [B]. Woods: Newf.—Minn.—Miss.—Fla. *E. Temp.* Jl–Au.

2. **B. lácera** (Michx.) Farwell. Stem 3–9 dm. high; lower leaves oblong, 8–20 cm. long, the upper lanceolate, shorter; spike 5–30 cm. long, thick; sepals ovate or suborbicular, 5–6 mm. long; petals linear, obtuse; lip 3-cleft, the lobes narrow, fringed, often with few segments; spur clavate, 14–16 mm. long. *H. lacera* (Michx.) R. Br. [G]. Swamps and open woods: Newf.—Minn.—Neb.—Ark.—Ala. *E. Temp.* Jl–Au.

f.144.

3. **B. psycòdes** (L.) Rydb. *Fig. 144.* Stem 3–10 dm. high; leaves oval to oblong-lanceolate, 5–25 cm. long or the upper shorter; spike 5–15 cm. long; sepals ovate, 8 mm. long; lateral petals oblong or oblanceolate, with a few teeth on the upper side; lip 3-parted, 8–12 mm. long, the lobes fan-shaped, deeply fringed; spur 15 mm. long, slightly clavate at the apex. *H. psycodes* (L.) Sw. [G]. Meadows and swamps: Newf.—Man.—S.D.—Neb.—N.C. *Canad.—Allegh.* Jl–Au.

4. **B. leucophaèa** (Nutt.) Farwell. Stem 3–8 dm. high, stout; leaves oblong or oblong-lanceolate, 1–2 dm. long or the upper shorter; spike 5–12 cm. long, relatively few-flowered; perianth white; sepals ovate or oval, 6–10 mm. long; petals obovate or spatulate, erose-toothed; lip 3-parted, 12–16 mm. long, the lobes cuneate, divided to the middle into a fringe; spur longer than the ovary. *H. leucophaea* (Nutt.) A. Gray [G]. Wet meadows: N.S.—S.D.—Neb.—Ala. *E. Temp.* Je–Jl.

14. POGÒNIA Juss. SNAKE MOUTH.

Glabrous caulescent herbs, with slender rootstocks. Leaves alternate, the basal ones long-petioled. Flowers solitary or few at the end of the stem; perianth rose-colored. Sepals and lateral petals distinct, erect or ascending. Lip erect, from the base of the column, fringed and crested, spurless. Column elongate, clavate. Anther-sacs parallel. Pollinia one in each sac, powdery-granular, not caudate. Stigma disk-like. Capsule erect, ribbed.

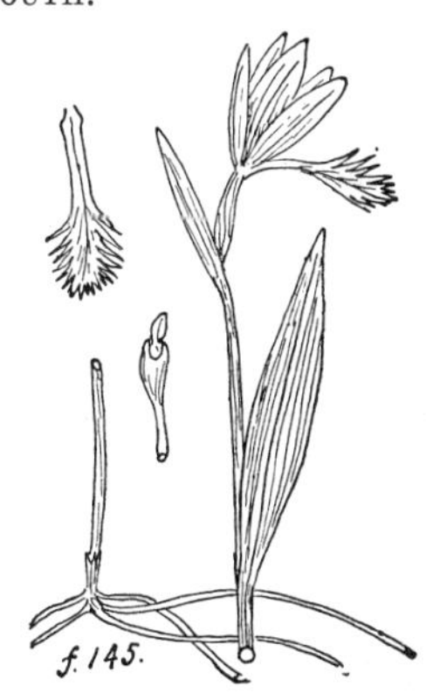
f.145.

1. **P. ophioglossoìdes** (L.) Ker. *Fig. 145.* Stem 2–5 dm. high; basal leaves long-petioled, the blades oblong or elliptic, 8–13 cm. long; stem-leaves 1 or 2, oblong-lanceolate to elliptic, clasping, 3–8 cm. long; flowers nodding; sepals linear-elliptic, 1.5–2 cm. long, acute; lateral petals elliptic or linear-cuneate, about as long; lip spatulate, enclosing the column at the base, with a yellow or white crest,

capsule 1.5–2 cm. long. Swamps and meadows: Newf.—S.D.—Kans.—Ala.—Fla. *E. Temp.* Je–Jl.

15. TRÍPHORA Nutt. NODDING-CAP.

Caulescent herbs, with fleshy tubers. Leaves alternate, clasping. Flowers axillary. Sepals and lateral petals free, ascending, nearly equal. Lip slightly clawed, erect, 3-lobed, not crested, spurless. Column elongate, clavate. Anther-sacs parallel. Pollinia one in each sac, powdery-granular, not caudate. Stigma disk-like. Capsule stalked, drooping.

1. **T. trianthóphora** (Sw.) Rydb. Stem 5–30 cm. high; leaves 2–10, ovate or suborbicular, 5–20 mm. long; flowers 1–7, pedicelled, soon drooping, pale-purple, 10–15 mm. long; sepals and lateral petals subequal, obtuse, converging, elliptic; lip 3-lobed, about as long, narrowed into a claw below; capsule oval, 10–12 mm. long. *Pogonia pendula* Lindl. *P. trianthophora* (Sw.) B.S.P. [G]. Rich woods: Me.—Wis.—Kans.—Fla. Au–S.

16. ARETHÙSA L. ROSE-LIP.

Caulescent herbs, with corms. Leaves mostly reduced to scales, the uppermost usually better developed, linear. Flowers terminal, usually solitary. Sepals cohering at the base, converging above and forming a hood. Lateral petals similar. Lip broadly dilated, recurved, crested on the inner face. Column narrow, coherent with the lip below, winged and dilated above. Anther incumbent on the column; sacs approximate. Pollinia 2 in each sac, powdery-granulate. Capsule elongate, erect, angled or ribbed.

f. 146.

1. **A. bulbòsa** L. *Fig. 146.* Corm 1–2 cm. thick; stem 1–3 dm. high; basal leaf solitary, linear, 1–1.5 dm. long; stem-leaves mostly scale-like, the uppermost linear; flowers 1 or rarely 2, subtended by 2 small bracts, rose-purple, 2.5–5 cm. long; sepals and petals oblong or linear; lip dilated to the fringed apex, often blotched, bearded with 3 lines of white hairs; capsule 2–2.5 cm. long, 6-ribbed. Bogs: Newf.—Minn.—Ind.—N.C. *Canad.—Allegh.* My–Je.

17. LIMODÒRUM L. GRASS-PINK.

Scapose herbs, with corms. Leaves basal, linear, the first year appearing alone, solitary, the second year 1 or 2, appearing with the scape. Flowers in terminal racemes or spikes, showy. Sepals and lateral petals distinct, spreading, similar or nearly so. Lip spreading, dilated, clawed, bearded on the upper surface. Column elongate, incurved, winged above. Anther sessile, lid-like, 2-celled. Pollinia one in each sac, granular. Capsule erect. *Calopogon* R. Br.

1. **L. tuberòsum** L. Leaves linear, acute, 1–4 dm. long; scape 2–9 dm. high; spike 3–20-flowered, not dense; bracts shorter than the ovary; perianth pink-purple; sepals lance-oblong, 1.5–2 cm. long; lateral sepals similar, a little longer; lip fan-shaped, erose or crenulate at the apex, crested with pink, yellow, or orange hairs. *Calopogon pulchellus* (Sw.) R. Br. [G]. Swamps: Newf.—Minn.—Mo.—Fla. *E. Temp.* Je–Jl.

18. SPIRÁNTHES Rich. LADIES' TRESSES.

Perennial herbs, with tuberous-thickened or fleshy-fibrous roots. Leaves alternate, in some species mostly basal. Flowers in terminal spiral spikes. Sepals distinct or coherent above or united with the similar lateral petals. Lip concave, sessile or slightly clawed, bearing two callosities at the base. Column oblique, arched. Fertile anther solitary, erect, without a lid. Pollinia one in each sac. Stigma with a beak. *Ibidium* Salisb. *Gyrostachys* Pers.

Flowers by twisting of the rachis of the spike forming a single spiral.
  Leaves basal or nearly so, fugacious before flowering time, with ovate or elliptic blades. — 1. *S. gracilis.*
  Leaves extending up the stem, linear, the upper persistent. — 2. *S. vernalis.*
Flowers in 2 or 3 spirals.
  Sepals, at least the lateral ones, free; lip not constricted or rarely slightly so.
    Lip truncate at the base, without callosities, or with small spreading ones at the base. — 3. *S. plantaginea.*
    Lip cuneate at the base, with 2 stout nipple-shaped callosities at the base.
      Corolla white; spike blunt; the lower bracts shorter than the corolla. — 4. *S. cernua.*
      Corolla yellowish; spike acute; the bracts longer than the corollas. — 5. *S. ochroleuca.*
  Sepals and petals more or less connivent into a hood; lip decidedly constricted at or above the middle. — 6. *S. stricta.*

1. **S. grácilis** (Bigel.) Beck. Stem scape-like, 2–6 dm. high, glabrous or pubescent above, covered with remote clasping scales; basal leaf short-petioled, 1.5–5 cm. long; spike slender, 3–15 cm. long; flowers white, 4–5 mm. long; lip shorter than the sepals, oblong, undulate or crenulate toward the apex, green in the center, with 2 nipple-shapped callosities at the truncate base; capsule slightly curved, 4–5 mm. long. *I. gracile* House. [BB]. *G. gracilis* Kuntze [B]. Open woods: N.S.—Man.—Kans.—La.—Fla. *E. Temp.* Jl–S.

2. **S. vernàlis** Engelm. & Gray. Stem 1.5–5.5 dm. high, densely pubescent above; leaves 7–15 cm. long, 8–9 mm. wide, tapering at each end; raceme slender, 8–15 cm. long; bracts longer than the ovaries, hyaline-margined; flowers yellowish, 8–10 mm. long; lip ovate or ovate-oblong, constricted about the middle, pubescent beneath. *I. vernale* Small [BB]. Sandy woods: Mass.—Kans.—Tex.—Fla. *Allegh.—Austral.* Au–S.

3. **S. plantagínea** (Raf.) Torr. Stem 1–4 dm. high; leaves linear or oblong, 5–15 cm. long, the uppermost reduced to scales; spike 2–10 cm. long, 1–1.5 mm. thick; flowers white, 6–7 mm. long; lip yellow within, oblong, undulate. *S. latifolia* Torr. *S. lucida* (A. A. Eaton) Ames. [G]. *G. plantaginea* Britton [B]. *I. plantagineum* House [BB]. Woods and shaded banks: N.B.—Minn.—N.C. *Canad.—Allegh.* My–Jl.

4. **S. cérnua** (L.) Richard. Stem 1–6 dm. high; leaves linear, 5–30 cm. long, the upper reduced to scales; spike 5–15 cm. long, 2–2.5 cm. thick; flowers fragrant, white, 8–10 mm. long; lip oblong, slightly broader at the base, 7–9 mm. long, crenulate at the apex; capsule 6–8 mm. long. *I. cernuum* (L.) House [BB]. *G. cernua* Kuntze [B]. Meadows and swamps: Newf.—Minn.—Tex.—Fla. *E. Temp.* S–O.

5. **S. ochroleùca** Rydb. Stem 3–5 dm. high; densely pubescent above; lower leaves linear, 1–2 dm. long, the upper stem-leaves reduced; spike dense, 5–15 cm. long, 1.5–2 cm. thick; flowers yellowish-green or ochroleucous, about 1 cm. long; lip rounded at the apex; callosities nipple-shaped, hairy. *S. cernua ochroleuca* (Rydb.) Ames [G]. *G. ochroleuca* Rydb. [B]. Hillsides: Me.—S.D.—Tex.—Ga. *E. Temp.* S–O.

6. **S. stricta** Rydb. *Fig. 147.* Stem 1.5–3.5 dm. high, glabrous; lower leaves 7–20 cm. long,

linear or linear-oblanceolate, the upper bract-like; spike 5–10 cm. long, 8–14 mm. thick, dense; flowers 3-ranked, white or greenish, 6–8 mm. long, fragrant; lip oblong, broad at the base, contracted below the dilated crisp terminal portion. *Gyrostachys stricta* Rydb. [B]. *S. Romanzoffiana* Coult. [G]; not Cham. *I. strictum* House [BB]. Wet places and rich hillsides: Newf.—Pa.—Colo.—Calif.—Alaska. *Boreal—Plain—Mont.* My–S.

19. **LÍSTERA** R. Br. TWAY-BLADE, TWIFOLE.

Perennials, with rootstocks and fleshy-fibrous roots. Leaves 2, opposite, near the middle of the stem. Flowers in terminal racemes, greenish or purplish. Sepals and lateral petals distinct, almost alike, reflexed. Lip notched or 2-cleft at the apex, and often with a pair of teeth or auricles near the base. Fertile anther 1, erect, joined to the back of the column, without a lid. Pollinia 2, powdery, attached to a minute gland. Column wingless. *Ophrys* (Tourn.) L.

Lip broad, slightly 2-cleft at the apex, with a mucro in the sinus. 1. *L. cordata.*
Lip narrow, 2-cleft for about half its length into linear-lanceolate lobes, without a mucro. 2. *L. convallarioides.*

1. L. cordàta (L.) R. Br. *Fig. 148.* Stem slender, 1–2 dm. high, glabrous or slightly pubescent above the leaves; leaves inserted at the middle, rounded-reniform, about 2 cm. long, 2–2.5 cm. wide, strongly veined; flowers purplish, 5–6 mm. long; sepals and upper petals oblong, 1.5–2 mm. long; lip 4–5 mm. long. Damp woods: Greenl.—Newf.—Mass.—Minn.—Mont.—N.M.—Ore.—Alaska; Eur. *Arct.—Boreal—Submont.—Subalp.* Je–Au.

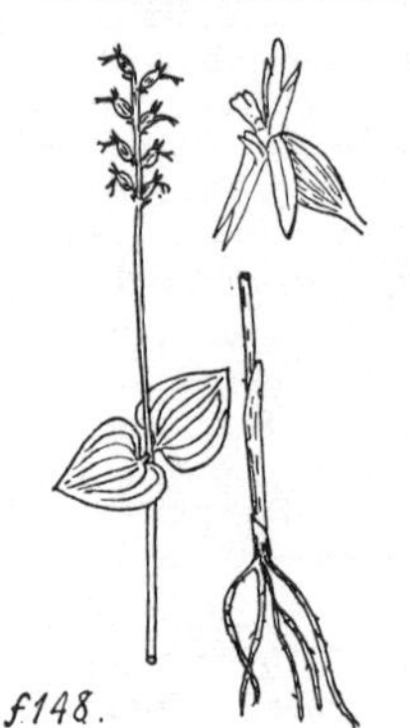
f148.

2. L. convallarioìdes (Sw.) Torr. Stem 1–2 dm. high, glabrous below, glandular above the leaves; leaves borne above the middle of the stem, broadly oval, obtuse, 3–5 cm. long, rounded at the base; lip about 9 mm. long, narrowly cuneate, retuse. Moist woods: Greenl.—Newf.—Mass.—Minn.—Man.—Ida.—Calif.—Alaska.; Eur. *Arct.—Boreal—Mont.*

20. **GOODYÈRA** R. Br. RATTLESNAKE PLANTAIN.

Caulescent perennial herbs, with rootstocks and fleshy-fibrous roots. Leaves alternate, mostly basal, strongly reticulate and often blotched. Flowers in terminal spikes, white or pink, inconspicuous. Lateral sepals distinct, spreading, the upper united with the lateral petals. Lip sessile, concave or gibbous, without callosities. Fertile anther solitary, short-stalked on the column, without a lid. Pollinia 2, one in each sac, attached to a small disk. Column straight, rather short. *Peramium* Salisb.

Lip evidently saccate, with recurved margins.
  Spike lax, 1-sided; tip of the lip recurved. 1. *G. ophioides.*
  Spike dense; tip of the lip straight. 2. *G. pubescens.*
Lip scarcely saccate, with incurved margins. 3. *G. decipiens.*

1. G. ophioìdes (Fernald) Rydb. Scape 1–2 dm. high, glandular-pubescent; leaves 1–2 cm. long, broadly ovate, dark green, usually with white blotches, most conspicuous along the cross-veins; flowers greenish white, 4–5 mm. long; upper sepal concave, with a short strongly recurved tip; anthers blunt. *Goodyera repens* Am. auth. *Epipactis repens ophioides* (Fernald) A. A. Eaton [G]. Cold mossy woods: N.S.—N.C.—S.D.—N.M.—Alaska. *Boreal—Mont.* Jl–Au.

2. **G. pubéscens** (Willd.) R. Br. Stem 1.5–4 dm. high, more or less pubescent; leaves dark green, with 5–7 white ribs and fine white cross-reticulations, oblong to ovate, 2–5 cm. long; flowers 4–5.5 mm. long; lip subglobose, with a blunt tip; anthers blunt. *P. pubescens* MacMill. [B]. *E. pubescens* A. A. Eaton [G]. Dry woods: Newf.—Minn.—Tenn.—Fla. *E. Temp.* Au–S.

3. **G. decípiens** (Hook.) Hubbard. *Fig. 149.* Scape 2–4 dm. high, glandular-pubescent, leaves ovate-lanceolate, 4–6 cm. long, acute at both ends, often without blotches, spike one-sided; flowers 7–9 mm. long; upper sepal concave, ovate-lanceolate, the tip long, often recurved; anthers ovate, long-pointed. *G. Menziesii* Lindl. *P. Menziesii* Morong. *P. decipiens* Piper [R]. *E. decipiens* (Hook.) Ames [G]. Woods: Que.—N.H.—Minn.—Ariz.—Calif.—Alaska. *Boreal—Submont.—Mont.* Jl–S.

21. MALÁXIS Soland. ADDER'S MOUTH.

Caulescent perennial herbs, with corms. Leaves 1–5, basal or on the stem. Flowers in terminal racemes, white or greenish. Sepals distinct, spreading. Lateral petals very narrow, often filiform, spreading. Lip broad, cordate or auricled at the base. Fertile anther 1, 2-celled, erect, inserted between the auricles. Pollinia 4, two in each sac, waxy, tailless and without glands. Column embracing the lip. *Microstylis* Nutt. *Achroanthes* Raf.

Leaves solitary, near the middle of the stem.
  Lip ovate, acuminate. — 1. *M. brachypoda.*
  Lip broadly obcordate, with a tooth in the sinus. — 2. *M. unifolia.*
Leaves 2–5, basal. — 3. *M. paludosa.*

1. **M. brachýpoda** (A. Gray) Fern. Corm small, ovoid; stem 1–1.5 dm. high, glabrous, striate; leaf-blade 3–5 cm. long, oval, elliptic, or ovate; pedicels 2–4 mm. long, erect; capsule ellipsoid, 6 mm. long. *Microstylis monophylla* Auth. [G], not Lindl. *A. monophylla* Greene [B]. Woods: Que.—Pa.—Colo.—Man. *Canad.—Plain—Submont.* Jl–Au.

2. **M. unifòlia** Michx. Corm globose; stem 1–2.5 dm. high; leaf-blade ovate or nearly orbicular, 2–6 cm. long; flowers greenish, about 2 mm. long; pedicels 6–10 mm. long. *Microstylis ophioglossoides* Nutt. *M. unifolia* (Michx.) B.S.P. [G]. *Achroanthes unifolia* Raf. [B]. Woods: Newf.—Fla.—Ala.—Neb.—Sask. *Plain—E. Temp.* Jl–Au.

3. **M. paludòsa** (L.) Sw. Stem scapelike, slender, 7–10 cm. high; leaves ovate, obtuse, sepals lanceolate, spreading; lateral petals much smaller; lip 3-nerved, lanceolate, apiculate, shorter than the sepals. Ottertail County, Minn.; adv. from Eur.

22. LÍPARIS Rich. TWAY-BLADE.

Low, scapose, perennial herbs, with corms. Leaves 2–6, basal. Flowers in terminal racemes, greenish. Sepals narrow, distinct. Lateral petals distinct, filiform. Lip dilated, often bearing two tubercles above the base. Column elongated, incurved, margined above. Fertile anther solitary, terminal, lid-like. Pollinia 4, two in each sac, waxy; each pair slightly united, without tails and glands. *Leptorchis* Thouars.

Racemes few-flowered; lip shorter than the petals. — 1. *L. Loeselii.*
Racemes many-flowered; lip as long as the petals. — 2. *L. liliifolia.*

1. **L. Loesélii** (L.) Rich. Corm globose; scape 5–20 cm. high, ribbed; leaves basal, elliptic or elliptic-lanceolate, 5–15 cm. long, obtuse; flowers greenish; lip obovate, pointed, shorter than the sepals; capsule about 1 cm. long, wide-angled on thick erect pedicels. *Leptorchis Loeselii* MacM. [B]. Woods Woods and thickets: N.S.—Md.—Neb.—Sask. *Plain—Canad.—Allegh.*

2. **L. liliifòlia** (L.) Rich. Corm subglobose; scape 1–2.5 dm. high, striate; leaves oval or ovate, 5–12 cm. long; flowers purplish; sepals lanceolate; lateral petals filiform; lip large, 10–12 mm. long, cuneate-obovate; capsule clavate, 12 mm. long. *Leptorchis liliifolia.* (L.) Kuntze. [B]. Woods and thickets: Me.—Minn.—Ala.—Ga. My–Jl.

23. CYTHERÈA Salisb. CALYPSO, VENUS' SLIPPER.

Acaulescent perennial herbs, with corms. Leaf solitary, basal, petioled, with a broad blade. Flower solitary, terminal, showy. Sepals and lateral petals distinct, similar. Lip large, saccate, with two short spurs near the apex, hairy within. Column dilated, petal-like, bearing a lid-like anther just below the apex. Pollinia two in each sac, united, sessile on a broad gland. *Calypso* Salisb.; not Thouars.

1. **C. bulbòsa** (L.) House. *Fig. 150.* Corm nearly globose, 1 cm. in diameter; scape 5–15 cm. high; leaf-blade rounded-ovate, 2–4 cm. long, nearly as wide, acute at the apex, rounded or subcordate at the base; flowers variegated, purple, pink, and yellow; petals and sepals 10–15 mm. long, with 3 purple lines; lip large, 15–20 mm. long. *Calypso borealis* Salisb. [G,B]. Cold woods: Lab.—Me.—Mich.—Colo.—Calif.—Alaska; Eur. *Submont.—Subalp.* My–Je.

24. APLÉCTRUM Nutt. ADAM AND EVE, PUTTY-ROOT.

Acaulescent perennial herbs, with a chain of corms, representing each a season's growth. Stem scapelike, leafless, a solitary long-petioled leaf being produced from the corm in the fall. Flowers in terminal racemes, showy. Sepals and lateral petals almost similar, narrow. Lip slightly clawed, spurless, with 3 ridges. Fertile anther solitary, a little below the summit of the column. Pollinia 4.

1. **A. spicàtum** (Walt.) B.S.P. Scape glabrous, 3–6 dm. high, with a few scales; leaf-blade ovate or elliptic or obovate, 1–1.5 cm. long; flowers dull yellowish brown, streaked with purple, fully 2 cm. long; lip shorter than the petals, obtuse, somewhat 3-lobed; capsule oblong-ovoid, 2 cm. long. *A. hyemale* Nutt. Woods: Ont.—Ga.—Mo.—Sask. *Boreal.* My–Je.

25. CORALLORRHÌZA R. Br. CORAL-ROOT.

Leafless saprophytic herbs, with coralloid roots. Stems scaly, colored. Flowers in terminal spikes. Sepals nearly equal, the lateral ones united with the foot of the column, and often forming a short spur, partly or wholly adnate to the top of the ovary. Lateral petals similar to the sepals. Lip broad, spreading, adnate to the base of the 2-edged or 2-winged column. Fertile anther solitary, terminal, lid-like. Pollinia 4, waxy, free.

Spur small, adnate to the ovary.
  Lip with 2 lobes or teeth below the middle, brownish or yellowish.
    Lip unspotted, its teeth or lobes very small; spur very small. 1. *C. trifida.*

Lip spotted, its lobes prominent, spur manifest. — 2. *C. maculata.*
Lip entire, ovate, sinuate.
Perianth 15–20 mm. long; plant light yellow; stem only lightly thickened below. — 3. *C. ochroleuca.*
Perianth 6–8 mm. long; plant light brown to madder-purple; stem bulbous at the base. — 4. *C. odontorhiza.*
Spur lacking; petals and sepals 12–14 mm. long, strongly striate. — 5. *C. striata.*

1. **C. trífida** Chatelain. Stem glabrous, 1–3 dm. high, usually yellowish; racemes 3–7 cm. long, 3–12-flowered; flowers greenish yellow or greenish brown, about 1 cm. long; lip oblong, yellowish or whitish; spur a sac, a small protuberance; capsule 8–12 mm. long. *C. innata* R. Br. *C. Corallorrhiza* (L.) Karst. [B,R]. Woods: N.S.—Ga.—Colo.—Alaska; Eur. *Temp.—Submont.—Mont.* My–Je.

2. **C. maculàta** Raf. *Fig. 151.* Stem 2–5 dm. high, purplish, with many scales; racemes 10–30-flowered; flowers 12–18 mm. long, brownish purple; lip white, spotted and lined with purple, oval in outline, deeply 3-lobed, crenulate, the middle lobe longer; spur about 2 mm. long; capsule 15–20 mm. long. *C. multiflora* Nutt. [B,R]. Woods: N.S.—Fla.—Calif.—Alaska. *Temp.—Plain—Mont.* Je–Au.

f 151.

3. **C. ochroleùca** Rydb. Stem 2–4 dm. high, not bulbous-thickened at the base; raceme 10–15-flowered; flowers 15–20 mm. long, light yellow, unspotted; petals and sepals 7–8 mm. long, oblong-lanceolate, acute, with prominent mid-vein. Woods: w Neb.—Colo.—Utah. *Submont.* Je–Jl.

4. **C. odontorhìza** Nutt. Stem about 1.5 dm. high, 6–20-flowered; sepals and petals lanceolate, about 4 mm. long, scarcely spreading; lip 2.5–3 mm. long, white, spotted with magenta-crimson, oval or broadly obvate; capsule globular-ovoid. Woods: Me.—Mich.—Iowa.—Mo.—Fla. *E. Temp.* Au–S.

5. **C. striàta** Lindl. Stem stout, 2–5 dm. high, purple; raceme 10–25-flowered; flowers purple with dark veins; lip oval or obovate, entire, undulate, about as long as the petals; column half as long as the petals; capsule ellipsoid, reflexed, 15–20 mm. long. Woods: Ont.—N.Y.—Mich.—Wyo.—Calif.—B.C. *Boreal—Submont.* My–Jl.

26. EPIPÁCTIS R. Br. HELLEBORINE.

Caulescent herbs, with rootstocks and fibrous roots. Leaves alternate, plaited, many-nerved, clasping. Flowers in bracted terminal racemes. Sepals distinct, strongly keeled. Lateral petals distinct, ovate. Lip free, not spurred, concave below, constricted in the middle, the terminal portion dilated and petal-like. Column short. Fertile anther 1, erect, operculate; its sacs contiguous. Pollinia granulose, 2-parted, becoming attached to the beaked stigma. *Serapias* L., in part.

1. **E. gigantèa** Dougl. Stem 3–10 dm. high, nearly glabrous, puberulent in the inflorescence; lower leaves ovate, sessile, the upper narrowly lanceolate, 5–20 cm. long, acute or acuminate; flowers 3–10, greenish, veined with purple; lip 15–18 mm. long; the terminal portion ovate-lanceolate, many-crested; capsule oblong, 15 mm. long. *S. gigantea* A. A. Eaton. Rich woodlands: Mont.—S.D.—w Tex.—Calif.—B.C. *W. Temp.—Son.—Submont.* Je–Au.

Subclass 2. **DICOTYLEDONES.**

Embryo mostly with 2 seed-leaves, if with only one seed-leaf, the first leaves opposite. Stem exogenous, *i.e.*, differentiated into pith, wood, and bark; fibro-vascular bundles arranged in concentric layers around the pith, the new bundles being formed between the wood and the bark. Leaves usually pinnately or palmately veined, with ramifying veins. Parts of the flowers mostly in 5's, less commonly in 4's or 2's, seldom in 3's.

Family 33. **SAURURACEAE.** LIZARD'S-TAIL FAMILY.

Perennial herbs, with rootstocks. Leaves alternate, entire. Flowers in spikes or racemes, without perianth. Stamens 3–8. Gynoecium of several distinct or nearly distinct carpels. Fruit capsular or berry-like.

1. SAURÙRUS L. LIZARD'S-TAIL.

Marsh-plants with slender rootstocks. Leaves petioled; blades cordate or ovate, with converging ribs. Racemes with drooping tips, opposite to the leaves. Stamens 6–8; filaments distinct, filiform. Carpels 3 or 4, united at the base. Fruit capsular.

1. **S. cérnuus** L. *Fig. 152.* Stem 3–12 dm. high; leafblades 5–7 cm. long, acuminate, cordate at the base; 5–9-ribbed; racemes 1–2 dm. long; flowers fragrant; filaments white; fruit depressed. Swamps: R.I.—Fla.—Tex.—Minn. *E. Temp.* Je–Au.

Family 34. **SALICACEAE.** WILLOW FAMILY.

Dioecious trees or shrubs, with soft wood and simple alternate leaves. Flowers in bracted aments, solitary in the axils of scale-like bractlets. Perianth represented by one or more glands or a cup-like disk. Stamens one or more. Pistil solitary, with 2–4 parietal placentae, becoming a dehiscent capsule; seeds numerous, bearing a tuft of hairs at the apex; endosperm wanting.

Bractlets incised; flowers with a cup-shaped disk; stamens usually more than 10; stigmas expanded; winter-buds with several scales. 1. POPULUS.
Bractlets entire or denticulate; flowers with one or more glands; stamens 2–7, in most species only 2; stigmas narrow; winter-buds with but 1 scale each. 2. SALIX.

1. PÓPULUS (Tourn.) L. POPLAR, COTTONWOOD, ASPEN.

Trees, with furrowed bark and often resinous buds of more than one scale. Leaves petioled, usually with broad, toothed or even lobed blades, and caducous stipules. Flowers in drooping aments; bractlets membranous, lobed or fimbriate at the apex, usually caducous. Disk present, oblique, entire, toothed, or lobed. Stamens adnate to the disk; filaments distinct, glabrous; anthers purple, orange, or red. Ovary sessile in the disk; style short; stigmas 2–4, dilated. Ovules anatropous. Capsule opening by 2–4 valves.

Leaves not white-tomentose beneath, toothed or rarely entire.
Petioles strongly flattened laterally.
Leaf-blades rounded, acute or short-acuminate; lobes of the styles filiform.

Leaf-blades coarsely dentate, silky-floccose when young; bud-scales tomentose. — 1. *P. grandidentata.*
Leaf-blades finely crenate, glabrous from the beginning; bud-scales glabrous or ciliate. — 2. *P. tremuloides.*
Leaf-blades broadly deltoid, cordate, or ovate, abruptly acuminate, coarsely toothed, especially about the middle.
Leaf-blades of the mature twigs with an open rounded sinus, toothless or few-toothed at the base and apex; glands at the apex of the petioles often present.
Blades few-toothed (teeth 5–15 on each margin); twigs light yellow, winter buds puberulent. — 3. *P. Sargentii.*
Blades many-toothed (teeth 15–30 on each margin); twigs grayish or reddish green; winter buds usually glabrous. — 5. *P. virginiana.*
Leaf-blades of the mature twigs rounded or cuneate at the base, toothed to near each end; glands at the apex of the petiole not present. — 4. *P. Besseyana.*
Petioles terete or nearly so.
Petioles at least half as long as the blades; blades cordate, ovate, or broadly lanceolate.
Leaves green on both sides, abruptly long-acuminate. — 6. *P. acuminata.*
Leaves paler beneath, acute or somewhat acuminate.
Twigs, petioles, and veins glabrous. — 7. *P. tacamahacca.*
Twigs, petioles, and veins of the lower surface of the leaves more or less pubescent. — 8. *P. candicans.*
Petioles about one-third as long as the blade or less; leaf-blades lanceolate, usually acute at the base. — 9. *P. angustifolia.*
Leaves white-tomentose beneath, often lobed. — 10. *P. alba.*

1. **P. grandidentàta** Michx. A tree 15–25 m. high, with smooth light green bark; leaf-blades rounded-ovate, truncate or broadly cuneate at the base, 6–10 cm. long, in age glabrate; bracts silky, irregularly 4–7-cleft; capsule papillose, conic, 2-valved, 6 mm. long. Rich woods: N.S.—N.C.—Tenn.—Iowa —Minn. *Canad.—Allegh.* Ap.

2. **P. tremuloìdes** Michx. A slender tree, with light green or whitish bark, up to 30 m. high; petioles 4–6 cm. long; leaf-blades rounded or subcordate at base, 2–6 cm. broad, green and glabrous, ciliate on the margin; bracts silky, deeply 3–5-cleft into linear lobes; stamens 6–12; stigma-lobes linear; capsule conic, warty. The western tree may be distinct, and then should be known as *P. aurea* Tidestr. QUAKING ASPEN. Cold places: Newf. N.J.—S.D.—Tenn.—N.M.—Nev.—Alaska. *Subarct.* — *Boreal* — *Submont.* — *Subalp.* Mr–My.

3. **P. Sargéntii** Dode. *Fig. 153.* A tree sometimes 30 m. high, with gray trunk and straw-colored or light yellow branches; petioles as long as the leaf-blades; blades broadly deltoid or cordate, abruptly acuminate at the apex, usually with an open concave-rounded sinus at the base, coarsely dentate, glabrous and shining; aments 5–8 cm. long, muricate, on pedicels 4–6 mm. long. *P. deltoides occidentalis* Rydb. [B]. *P. occidentalis* Rydb. WESTERN or RIVER COTTONWOOD. River bottoms: Sask.—Mo.—Ariz.—Alta. *Plain—Submont.* Mr–Ap.

f. 153.

4. **P. Besseyàna** Dode. A tree about 10 m. high, with gray trunk and yellowish terete branches; petioles about as long as the leaf-blades; blades abruptly acuminate, 5–10 cm. long, glabrous and shining, slightly paler beneath; aments 7–8 cm.

long; pedicels shorter than the fruit; capsule ovoid, about 5 mm. long, muricate, 3-valved; stigma dilated. Valleys: N.D.—Mont.—Ida. *Submont.* Ap–Je.

5. **P. virginiàna** Fourg. A tree, sometimes 35 m. high, with gray furrowed trunk and brownish green branches; leaves with long petioles, the blades broadly ovate or deltoid, short-acuminate, serrate-crenate, usually with a rounded open sinus at the base, usually with a pair of glands at the base, glabrous; staminate aments 8–12 cm. long; pistillate aments in fruit 2–3 dm. long; capsule ovoid, 8–10 mm. long, as long as the pedicel or longer. *P. monilifera* Ait. *P. deltoides* Sargent [G,B], scarcely Marsh. COTTONWOOD. NECKLACE POPLAR. Along streams: Que.—Man.—e Kans.—Tenn.—Fla. *Canad.—Allegh.* Cult. and escaped west to Man.—Colo.—Okla. Mr–Ap.

6. **P. acuminàta** Rydb. A tree up to 20 m. high, with brownish or gray trunk and straw-colored branches; petioles 3–7 cm. long; leaf-blades rhombic-lanceolate to ovate, usually cuneate at the base, finely crenate, green and glabrous on both sides; aments 3–5 cm. long, the pistillate in fruit 10–15 cm. long; capsule ovoid, blunt, 6–8 mm. long, papillose; pedicels in fruit 3–5 mm. long. RYDBERG'S or SMOOTH-BARKED COTTONWOOD. River banks: Sask.—N.D.—Tex.—Ariz.—Mont. *Plain—Submont.* Ap–My.

7. **P. tacamahàcca** Mill. A tree sometimes 25 m. high, with reddish gray trunk and light brown or gray branches; petioles 2–5 cm. long; leaf-blades ovate to ovate-lanceolate, dark green and shining above, pale beneath, acute or acuminate at the apex, rounded or cuneate at the base, crenulate, 7–12 cm. long; aments 5–10 cm. long. *P. balsamifera* Du Roi [G,B,R]; not L. BALSAM POPLAR. Along streams and lakes: Newf.—N.Y.—Colo.—Nev.—Alaska. *Boreal—Submont.—Mont.* Ap.

8. **P. cándicans** Michx. A tree sometimes 30 m. high, with gray trunk and round branches, pubescent when young, in age glabrate; petioles pubescent or ciliate; leaf-blades broadly ovate, acute or acuminate at the apex, cordate or truncate at the base, 6–15 cm. long, 3–12 cm. wide, dark green above, pale beneath, crenate; aments 6–12 cm. long; capsule oblong-ovoid, smooth, short-stalked. BALM OF GILEAD. Roadsides and river banks: Newf.—N.J.—S.D.—Alta.—Alaska. *Boreal.* Ap–My.

9. **P. angustifòlia** James. A slender tree, sometimes 20 m. high, with upright branches and greenish bark; petioles 6–12 mm. long; leaf-blades lanceolate to ovate, acute or gradually acuminate at the apex, cuneate, rounded or rarely subcordate at the base, green above, paler beneath, finely crenate, 5–12 cm. long; aments 2–6 cm. long; capsule ovoid, rugose. *P. coloradensis* Dode. *P. fortissima* Nels. & Macbr. NARROW-LEAVED or BLACK COTTONWOOD. Along streams: Sask.—S.D.—N.M.—Ida.; n Mex. *Plain—Mont.* Ap–My.

10. **P. álba** L. A large tree, sometimes 35 m. high, with light gray bark; petioles shorter than the blades, subterete; leaf-blades densely white-tomentose on both sides, becoming glabrate and dark green above, broadly ovate or orbicular in outline, 3–5-lobed and irregularly toothed, 6–10 cm. long; aments 3–6 cm. long; capsules elliptic, subsessile. WHITE or SILVER POPLAR. In yards and along roadsides: N.B.—Va.—N.M.—Utah; cult. and escaped, native of Eurasia. Mr–My.

2. SÀLIX (Tourn.) L. WILLOW.

Trees or shrubs, usually with flaky bark and slender, flexible twigs; bud of only one scale. Leaves alternate, petioled or subsessile, pinnately veined, usually finely toothed. Bractlets of aments entire or rarely denticulate, with one or more gland-like nectaries at the base of the stamens or the pistil. Stamens 1–7, usually 2; anthers usually yellow. Ovary more or less stipitate; Stigmas 2, often 2-cleft, with narrow branches. Capsule dehiscent at the apex into 2 valves.

Capsule glabrous, or slightly silky when young in nos. 10–12.
Filaments hairy, at least below; bractlets caducous, light yellow; style very short, obsolete, or none; aments in all ours on short leafy branches.
Stamens 3–7; stipe slender, 2–5 times as long as the nectaries.
Petioles without glands; leaves remotely serrulate, thin. I. AMYGDALINAE.
Petioles with glands; leaves densely glandular-serrate, firmer. II. PENTANDRAE.
Stamens 2; stipes usually very short.
Leaf-blades lanceolate, acuminate. III. ALBAE.
Leaf-blades linear or linear-lanceolate, acute, rarely acuminate. IV. LONGIFOLIAE.
Filaments glabrous; bractlets persistent, rarely light yellow. V. CORDATAE.
Capsule hairy.
Filaments hairy; leaf-blades linear to lanceolate, not very veiny; bractlets caducous. IV. LONGIFOLIAE.
Filaments glabrous, or if somewhat hairy, leaf-blades oval or orbicular, very veiny; bractlets persistent.
Capsule rostrate, distinctly stipitate; style none or short, always shorter than the stipe.
Stipe equaling or exceeding the bractlets; style 0.5 mm. long or less. VI. ROSTRATAE.
Stipe shorter than the bractlets; style usually 1 mm. long. VII. CAPREAE.
Capsule scarcely rostrate, subsessile or, if stipitate, stipe shorter than the style.
Leaves permanently silky or tomentose beneath. VIII. ARGENTEAE.
Leaves neither silky-white nor tomentose beneath. IX. ARCTICAE.

I. AMYGDALINAE.

Leaves narrowly lanceolate; petioles short.
Leaves green beneath. 1. *S. nigra.*
Leaves glaucous beneath. 2. *S. Wardii.*
Leaf-blades broadly lanceolate; petioles slender, 6–15 mm. long. 3. *S. amygdaloides.*

II. PENTANDRAE.

Leaves narrowly oblanceolate, acute, drying blackish; bractlets of the staminate aments ovate, and of pistillate lanceolate, neither toothed. 4. *S. erythrocoma.*
Leaves lanceolate to ovate; bractlets of the staminate aments obovate or spatulate, usually toothed; those of the pistillate aments oblong.
Bractlets densely white-villous; stipe of the capsules scarcely twice as long as the nectaries. 5. *S. serissima.*
Bractlets sparingly hairy; stipe of the capsules 3–4 times as long as the nectaries.
Leaves ovate-lanceolate or ovate; glands of the leaves very prominent; those on the petioles often stalked. 6. *S. lucida.*
Leaves lanceolate; glands of the leaf-blades not very prominent; those on the petioles always sessile. 7. *S. Fendleriana.*

III. ALBAE.

Stipe of the capsule in fruit 1–3 mm. long; style evident. 8. *S. fragilis.*
Stipe of the capsule in fruit less than 1 mm. long; stigma sessile. 9. *S. alba.*

IV. LONGIFOLIAE.

Capsule glabrous, or slightly silky when young.
Leaves permanently silky. 10. *S. exigua.*
Leaves glabrous in age or nearly so.
Leaves linear-lanceolate; bractlets ovate or obovate, obtuse. 11. *S. interior.*
Leaves narrowly linear; bractlets lanceolate, acute. 12. *S. linearifolia.*
Capsule permanently silky; leaves densely silky-villous when young. 13. *S. Wheeleri.*

V. CORDATAE.

Leaves dark green above; young branches not light yellow; aments on short leafy branches.
Mature leaves rather thin, dull; length of the blade less than three times its width. 14. *S. pyrifolia.*
Mature leaves firm, dark green, shining above, pale beneath; length of the blade three times the width or more.

Stipe in fruit 2–4 times as long as the bractlets; mature leaves slightly serrate. — 15. *S. Mackenziana.*
Stipe slightly if at all exceeding the bractlets.
Mature leaves lanceolate, serrate.
Twigs glabrous or nearly so; leaves not glaucous beneath; capsule 6–7 mm. long. — 16. *S. cordata.*
Twigs permanently pubescent; leaves glaucous beneath; capsule 8–10 mm. long. — 17. *S. missouriensis.*
Mature leaves oblong or oblanceolate, entire. — 18. *S. pedicellaris.*
Leaves yellowish green; young branches light yellow; aments naked or subtended by a few leaves. — 19. *S. lutea.*

VI. ROSTRATAE.

Leaves linear to lanceolate or oblanceolate.
Leaves sharply serrate, narrowly lanceolate, 5–10 cm. long. — 20. *S. petiolaris.*
Leaves entire or nearly so, linear, less than 5 cm. long.
Aments on short leafy branches.
Leaves more or less silky, oblanceolate; branches purplish, usually with a bloom. — 21. *S. Geyeriana.*
Leaves not silky, linear; branches without a bloom. — 22. *S. gracilis.*
Aments sessile, naked, appearing before the leaves; leaves tomentose beneath.
Leaf-blades 5–10 cm. long; fruiting aments 2–4 cm. long. — 23. *S. humilis.*
Leaf-blades 1–5 cm. long; fruiting aments about 1 cm. long. — 24. *S. tristis.*
Leaves ovate to obovate or broadly oblanceolate.
Bracts fuscous, obovate or cuneate.
Twigs glabrous.
Mature leaves obovate or oblong-obovate or lanceolate, indistinctly or irregularly toothed; pistillate catkins dense. — 25. *S. discolor.*
Mature leaves oblong, regularly dentate; pistillate catkins long, laxer. — 26. *S. prinoides.*
Twigs densely velutinous when young. — 27. *S. eriocephala.*
Bracts yellow, linear-oblong or lanceolate.
Mature leaves thin, glabrous. — 28. *S. perrostrata.*
Mature leaves firm, pubescent or tomentose beneath. — 29. *S. Bebbiana.*

VII. CAPREAE.

Capsule 8–10 mm. long; at least the staminate aments not leafy-bracted; leaves obovate, entire or nearly so, densely silky and fulvous, almost velutinous when young, in age glabrate above. — 30. *S. Scouleriana.*
Capsule 6 mm. long or less; aments leafy-bracted; leaves silvery-silky beneath. — 31. *S. candida.*

VIII. ARGENTEAE.

Aments more than 1 cm. broad; ovary over 5 mm. long.
Aments less than 1 cm. broad; ovary less than 5 mm. long; leaves oblong, lanceolate, or oblanceolate.
Bractlets yellow or brown; stipes 3–4 times as long as the nectaries; style less than 0.5 mm. long. — 32. *S. argyrocarpa.*
Bractlets fuscous; stipes less than twice as long as the nectaries.
Branches without a bloom.
Leaves lanceolate, dentate; capsule blunt at the apex. — 33. *S. sericea.*
Leaves linear-oblanceolate; minutely denticulate; capsule conic at the apex. — 34. *S. arbusculoides.*
Branches with a bloom. — 35. *S. pellita.*

IX. ARCTICAE.

Aments on leafy peduncles.
Twigs white-villous or silky; branches yellow or grayish. — 36. *S. brachycarpa.*
Twigs puberulent or glabrate.
Bractlets fuscous; leaves elliptic or oblong, obtuse or acute, usually reticulate, glabrous only in age. — 38. *S. glaucops.*
Bractlets yellow or brown; leaves usually short-acuminate, glabrous except when very young, not reticulate. — 37. *S. nudescens.*
Aments naked from lateral buds.
Leaves oblanceolate, strongly veined. — 39. *S. Nelsonii.*
Leaves elliptic, not strongly veined. — 40. *S. planifolia.*

1. **S. nìgra** Marsh. A tree 10–20 m., rarely 30–40 m. high; leaf-blades 6–12 cm. long, bright green on both sides, finely serrate, usually more or less falcate; aments appearing with the leaves, the staminate ones 4–6 cm. long; bractlets ovate, finely villous; pistillate aments 4–8 cm. or in fruit 10–12 cm. long; bractlets oblong, finely pubescent; capsule about 3 mm. long; stipe 1–1.5 mm. long. BLACK WILLOW. River-banks and wet places: N.B.—Fla.—Tex.—N.D.—Man. *E. Temp.—Plain.* Ap–My.

S. amygdaloides × nigra Glatf. This hybrid is intermediate between the two parents, with narrower leaves, less glaucous beneath, and with shorter petioles than in *S. amygdaloides,* but with broader blades and longer petioles than in *S. nigra.*

2. **S. Wárdii** Bebb. A tree up to 10 m. high, with reddish brown bark; leaf-blades lanceolate to oblong-lanceolate, 6–18 cm. long, somewhat pubescent and strongly veiny beneath; aments appearing with the leaves, the staminate ones 5–10 cm. long, the pistillate ones often shorter; capsule globose-ovate, 4.5–6 mm. long. Shores: Md.—Va.—Kans. *Allegh.—Ozark.* Mr–My.

3. **S. amygdaloìdes** Anders. *Fig. 154.* A tree up to 20 m. high; twigs yellowish; leaf-blades 8–12 cm. long, acuminate, finely serrate, paler and somewhat glaucous beneath; aments 3–5 cm. long; bractlets oblong or lanceolate, somewhat pubescent; capsule about 5 mm. long; stipe 1–2 mm. long. PEACH-LEAVED WILLOW. River and lake banks: Que.—N.Y.—Mo.—Tex. — Ariz. — Ore. — B.C. *W. Temp.—Plain—Submont.* Ap–My.

f.154.

4. **S. erythrocòma** Barratt. A shrub or small tree (?); bark of the twigs dark, purplish brown; leaf-blades 3–5 cm. long, finely glandular-serrate, reddish pubescent when young, in age glabrate on both sides; aments 2–4 cm. long. *S. arguta erythrocoma* Anders. Banks of rivers and lakes: Man.——B.C.—Mack. *W. Boreal.*

5. **S. seríssima** (Bailey) Fernald. A shrub 2–4 m. high; leaf-blades lanceolate, acute or acuminate, when mature dark green, shining above, pale beneath, thick and firm, 4–10 cm. long, closely and finely glandular-serrate; aments 1–2.5 cm. long; capsules 7–12 mm. long. *S. pallescens hirtisquama* Anders. AUTUMN WILLOW. Swamps: Mass.—N.J.—S.D.—Sask. *E. Boreal.* My–Jl.

6. **S. lùcida** Muhl. A tall shrub or sometimes a tree 6–8 m. high; leaf-blades mostly abruptly long-acuminate, acute or rounded at the base, sharply glandular-serrate, green and shining, slightly paler beneath, 7–12 cm. long; aments 2–5 cm., or the pistillate even 7 cm. long; bractlets sparingly pubescent; capsule 5 mm. long. SHINING WILLOW. Swamps and banks: Newf.—N.J.—Ky.—Neb.—Alta. *Canad.—Allegh.* Ap–My.

7. **S. Fendleriàna** Anders. A shrub or small tree, nearly 10 m. high; leaf-blades lanceolate, long-acuminate, finely glandular-dentate, acute at the base, 6–15 cm. long; aments 2–5 cm. long; bractlets toothed toward the apex; stamens 5–9; capsule 5 mm. long. *S. lasiandra caudata* (Nutt.) Sudw. Along mountain streams: B.C.—Alta.—S.D.—N.M.—Calif. *W. Temp.—Submont.—Mont.*

8. **S. frágilis** L. A tall tree, with rough gray bark, up to 25 m. high; twigs reddish green, brittle at the base; leaves lanceolate, long-acuminate, sharply serrate, glabrous, 8–15 cm. long, glandular at the base; staminate catkins 2–4 cm. long; stamens 2, rarely 3 or 4; pistillate catkins 7–12 cm. long,

loose; capsule long-conic. Along streams: Mass.—N.J.—Iowa—S.D.; escaped from cult., nat. of Eur. Apr.–My.

9. **S. álba** L. A large tree sometimes 30 m. high; leaf-blades serrulate, silky-pubescent on both sides when young, pale beneath, 5–12 cm. long; aments on short leafy branches, 3–6 cm. long, 4–5 mm. thick; bractlets of the staminate aments obovate, those of the pistillate ones lance-oblong; capsule almost sessile, 4–5 mm. long. WHITE WILLOW. Moist ground, along streams: N.B.—Pa.—Colo.; escaped from cultivation; native of Eur. Ap–My. *S. alba vitellina* (L.) Koch, with yellow branches and the mature leaves glabrous but white beneath and *S. alba caerulea* with olive-brown branches and bluish green leaves are cultivated and often escaped.

10. **S. exígua** Nutt. A shrub or small tree sometimes 6–7 m. high; twigs yellowish or light brown, silky-pubescent when young; leaf-blades linear to narrowly linear-lanceolate, 2–8 cm. long, yellowish, silky; staminate aments 2–6 cm., the pistillate ones 3–5 cm. long; bractlets of the former obovate, of the latter broadly lanceolate; capsule elongate-ovoid, slightly silky when young, soon glabrate, 4–5 mm. long. *S. luteosericea* Rydb. [B]. SANDBAR WILLOW. Sandbars: Sask.—Minn.—Okla.—Ariz.—B.C.—Ore. *W. Temp.—Plain—Mont.* My–Jl.

11. **S. intèrior** Rowlee. A shrub 1–4 m. high, occasionally a tree 6–9 m. high; twigs pubescent when young; leaf-blades 6–10 cm. long, less than 1 cm. wide, lance-linear, remotely dentate; aments 2–5 cm. long; bractlets glabrous or hairy toward the base, ovate or obovate, rarely oblong; capsule appressed-silky when young, soon glabrate. *S. longifolia* Muhl. [G]. *S. fluviatilis* Auth. [B]; not Nutt. SANDBAR WILLOW. River banks and shores: Que.—Va.—La.—Tex.—Sask. *E. Temp.—Plain.* Je–Jl.

12. **S. linearifòlia** Rydb. A low shrub, with reddish bark; leaf-blades narrowly linear, 4–10 cm. long, remotely dentate; aments loosely flowered, 3–4 cm. long; bractlets linear-lanceolate, acute, almost glabrous; capsule almost glabrous, 3–4 mm. long. *S. rubra* Richards. *S. tenerrima* (Howell) Heller. Sandbars: Man.—Okla.—Colo.—Wash. *Prairie—Plain.* My–Je.

13. **S. Wheèleri** (Rowlee) Rydb. A shrub with brown bark; young twigs silvery-silky; leaves linear-lanceolate to linear, 5–10 cm. long, 5–15 mm. wide, silvery-silky when young, sharply and distantly denticulate with ascending teeth; staminate catkins often clustered, 2–4 cm. long; bracts yellow, oblanceolate, acute, pilose; pistillate catkins 3–4 cm. long, lax; capsule narrow, silky, subsessile; stigma sessile; bracts deciduous. *S. interior Wheeleri* Rowlee. Gravelly shores and islands: N.B.—Pa.—Mo.—Iowa. *Cand.—Allegh.* Je–Jl.

14. **S. pyrifòlia** Anders. Usually a shrub, about 3 m. high, sometimes a small tree up to 8 m. high; leaf-blades thin, ovate, oval, or obovate, acute or obtuse at the apex, rounded or subcordate at the base, dark green above, glaucous beneath, 5–8 cm. long; aments 2–5 cm. long; bractlets persistent, brownish or yellowish, silky-villous; capsule 6–8 mm. long; stipe about 3 mm. long; style very short. *S. balsamifera* (Hook.) Barratt. [G,B]. *S. Columbiae* Nels. & Macbr. BALSAM WILLOW. Swamps: Lab.—N.Y.—Minn.—Mont. *E. Boreal.* My.

15. **S. Mackenziàna** Barratt. A small tree or shrub, sometimes 10 m. high; bark brown, glabrous; leaf-blades oblanceolate or lanceolate, 5–10 cm. long, acuminate at the apex, crenate or nearly entire, dark green above, pale beneath; aments 4–6 cm. long; bractlets obovate, hairy towards the base; capsule glabrous, 5–7 mm. long; stipe 3–4 mm. long; style about 0.5 mm. long. DIAMOND WILLOW. Along streams: Sask.—Man.—Ida.—Calif.—B.C.—Mack. *W. Boreal—Submont.* Ap–Je.

16. **S. cordàta** Muhl. A shrub 1.5–4 m. high; leaf-blades from oblong-lanceolate and subcordate at the base to narrowly lanceolate and acute at the

base [var. *augustata* (Pursh.) Anders., the form found in the Rocky Mountains], often somewhat silky-pubescent when young; staminate aments about 2 cm. long, the pistillate ones 4–6 mm. long in fruit; bractlets fuscous, white-silky; capsule glabrous, 6–7 mm. long; stipe 1–2 mm. long; style minute. HEART-LEAVED WILLOW. Wet ground: N.B.—Va.—Colo.—Calif.—B.C. *Temp.—Plain—Mont.* Ap–My.

17. **S. missouriénsis** Bebb. A tree or large shrub, 5–15 m. high, with black bark; leaf-blades lanceolate or oblanceolate, acuminate, finely glandular-serrulate, pubescent when young, 7–15 cm. long; staminate aments about 3 cm. long, the pistillate ones 7–10 cm. long; style very short. River banks: Mo.—Minn.—S.D.—Neb.—Okla. *Ozark.—Prairie.* Mr–Ap.

18. **S. pedicellàris** Pursh. A shrub 1 m. high or less; blades oblong, elliptic or sometimes oblanceolate, obtuse, entire, 2–4 cm. long, 8-16 mm. wide; aments about 2 cm. long; bractlets obtuse, slightly villous; capsule glabrous, about 5 mm. long; stipe about 2 mm. long; style minute. *S. myrtilloides* Am. auth. [B], not L. BOG WILLOW. In bogs: N.B.—N.J.—Iowa—Wash.—B.C. *Boreal—Mont.* Ap–My.

19. **S. lùtea** Nutt. A shrub or small tree 5–6 m. high, with gray bark; leaf-blades lanceolate, 3–7 cm. long, more or less acuminate, somewhat pubescent when young; aments subsessile, 2–5 cm. long; bractlets brownish, white-villous; capsule about 5 mm. long; stipe 1–2 mm. long; style about 0.5 mm. long. YELLOW WILLOW. River banks and wet places: Man.—Neb.—Utah—Mont.—Alta. *Plain—Submont.—Mont.* My–Je.

20. **S. petiolàris** Smith. A shrub 1–3 m. high; leaf-blades acuminate at both ends, slightly silky when young, in age dark green and shining above, glaucous beneath; aments appearing before the leaves, naked or nearly so, about 2 cm. long; bractlets yellow with dark tips, white-pilose; capsule 4–6 mm. long, subconic; stipe 2–3 mm. long; style obsolete. Swamps: N.B.—Tenn.—Ills.—S.D.—Sask. *E. Boreal.* My.

S. candida × petiolaris. Leaves resembling mostly *S. petiolaris* in shape but *S. candida* in pubescence; capsule long-tapering as in *S. petiolaris* but shorter and with shorter stipe. N.D.

21. **S. Geyeriàna** Anders. A shrub 2–3 m. high; leaf-blades linear-oblanceolate, 2–6 dm. long, densely silky-strigose when young, less so or sometimes glabrate in age, somewhat paler beneath; aments on very short leafy branches, 1–2 cm. long; bractlets oblong, yellowish, sparingly short-villous; capsule subconic, short-pubescent, 5–6 mm. long; stipe 1–2 mm. long; style obsolete. *S. macrocarpa* Nutt. Creek banks and mountain valleys: B.C.—Mont.—S.D.—Colo.—Ore. *Submont.—Mont.* Ap–Je.

22. **S. grácilis** Anders. A shrub 1–2 m. high; leaf-blades 3–5 cm. long, about 4 mm. wide, linear or lance-linear, at first tomentulose, soon glabrous and green above, slightly glaucous beneath, entire or denticulate; aments on short leafy branches, lax, 2–3 cm. long; bractlets oblong, yellowish with dark apex; capsule elongate-conic, 5–6 cm. long, grayish silky; stipe 3–4 mm. long; style obsolete. *S. rosmarinifolia* Hook. River banks: Sask.—Wis.—Iowa. *E. Boreal.*

23. **S. hùmilis** Marsh. A shrub, 0.5–3 m. high; leaf-blades oblanceolate, 5–10 cm. long, downy at first, but glabrescent above, soft-tomentose and very veiny beneath, undulate and revolute-margined; catkins ellipsoid; capsules narrowly conic; style obsolete. Dry plains and barrens: Newf.—Ga.—Okla.—N.D. *E. Temp.* Ap–My.

24. **S. trístis** Ait. A shrub, about 0.5 m. high; leaf-blades small. 1–5 cm. long, narrowly oblanceolate, tomentose beneath, revolute-margined; catkins ovoid; capsules ovoid-conic, with long beak; style obsolete. Sandy plains: N.S.—Fla.—Tenn.—N.D. *E. Temp.* Mr–Ap.

25. **S. díscolor** Muhl. A shrub or low tree up to 7 m. high; leaf-blades oblong oblanceolate, acute at both ends, irregularly serrate or entire, bright green above, glaucous beneath, glabrate, 4–10 cm. long, 1.5–3 cm. wide; aments appearing before the leaves, dense, 3–5 cm. long, or the pistillate ones 4–7 cm. in fruit; capsule elongate-conic, about 1 cm. long, long-silky; stipe about 2 mm. long; stigma obsolete. PUSSY WILLOW. Swamps and wet places: N.S.—Del.—Mo.—S.D.—Sask. *E. Boreal.* Mr–Ap.

S. cordata × discolor. Intermediate between the two parents, the capsule shorter, less attenuate and less pubescent than in *S. discolor,* the catkins often subtended by small leaves. Minn.

S. discolor × humilis. Catkins shorter than in *S. discolor;* capsule less attenuate and with shorter stipe; leaves more or less pubescent beneath. Minn.

26. **S. prinoìdes** Pursh. A shrub 1–3 m. high, with slender glabrous branches; leaves elliptic, 4–10 cm. long, serrate, green on both sides, acute; staminate catkins resembling those of *S. discolor;* pistillate catkins elongate, loose, 4–5 cm. long, sparingly silky; capsule thinly tomentose; style long; stigma lobed. *S. discolor prinoides* Anders. [G]. Wet meadows: N.Y.—Va.—S.D. *Canad.—Allegh.* Ap–My.

27. **S. eriocéphala** Michx. A shrub or low tree, 2–5 m. high; branches dark, densely velutinous when young; leaves elliptic, 6–10 cm. long, serrate, acute, green and shining above, glaucous and pubescent, especially on the veins beneath, often ferruginous; staminate catkins 1–2 cm. long, very dense, yellowish silky; pistillate catkins 3–4 cm. long in fruit; capsule narrowly conic, tapering into short beak, 9–10 mm. long. *S. discolor eriocephala* Robins. & Fern. [G]. Mass.—Pa.—Iowa.

28. **S. perrostràta** Rydb. A shrub 1–4 m. high; leaf-blades obovate-lanceolate or oblanceolate, when young finely silky, in age glabrate, 2–4 cm. long, 1–1.5 cm. wide, undulate or entire, light green above, pale beneath; aments usually on very short leafy branches, 1–3 cm. long; bractlets sparingly silky; capsule elongate-conic, 7–8 mm. long, finely pubescent; stipe 2–3 mm. long; style obsolete. BEAKED WILLOW. River valleys: Hudson Bay—Neb.—Colo.—Utah—Alaska. *W. Boreal—Submont.—Mont.* My–Je.

29. **S. Bebbiàna** Sarg. A shrub 2–6 m. high, sometimes a tree 8 m. high; leaf-blades elliptic, oblong, or oblong-lanceolate, acute, blunt, or short-acuminate, rounded at the base, sparingly serrate or entire; aments sessile, sometimes subtended by a few small leaves, 2–5 cm. long; bractlets sparingly long-villous; capsule elongated-conic, finely pubescent, about 8 mm. long; stipe 2–3 mm. long; style obsolete. *S. rostrata* Richardson [G]. BEAKED WILLOW. Valleys, river banks, and hillsides: Anticosti—N.J.—N.M.—Ariz.—Calif.—B.C. *Temp.—Plain—Mont.* Ap–My.

30. **S. Scouleriàna** Barratt. A shrub or tree, occasionally 9 m. high; young twigs from densely velutinous to almost glabrous; leaf-blades obovate, rounded or abruptly acute at the apex, cuneate at the base, at maturity thin, dark yellowish green and lustrous above, pale, glaucous and more or less pubescent beneath, 4–10 cm. long; aments sessile, naked, 2–4 cm. long, about 1 cm. thick; capsule subconic, about 1 cm. long, densely white-villous; stipe short; style very short. *S. flavescens* Nutt. *S. Nuttallii* Sargent. Along streams: Sask.—S.D.—N.M.—Calif.—Yukon. *Submont.—Mont.* Mr–Je.

31. **S. cándida** Flügge. A shrub 6–15 dm. high; leaf-blades oblong or oblong-lanceolate, thick, sparingly repand-denticulate or entire, acute at both ends, dark green above, 5–10 cm. long, 6–16 mm. wide; aments sessile and usually naked, appearing before the leaves, 2–5 cm. long; capsule subconic, densely white-tomentose, 6–7 mm. long; stipe less than 1 mm. long; style about 1 mm. long. HOARY WILLOW. Bogs: Lab.—N.J.—Colo.—Ida.—Alta. *Boreal—Mont.* My.

32. **S. argyrocárpa** Anders. An erect shrub 1–6 dm. high; leaf-blades oblong or oblanceolate, acute at both ends, entire or crenulate, bright green and glabrous above, silvery-silky beneath, 3–5 cm. long, 6–12 mm. wide; aments on short leafy branches, dense, 1–2.5 cm. long; bractlets villous; capsule 2–3 mm. long, white-villous; stipe 1–2 mm. long. SILVER WILLOW. Mountains and hills: Lab.—N.H.—Mack. *E. Boreal.* Je–Jl.

33. **S. serícea** Marsh. A shrub 2–4 m. high; twigs purplish, puberulent; leaves densely silky-pubescent beneath when young, lanceolate, acuminate, serrulate with gland-tipped teeth, when mature dark green above, glaucous beneath, 6–10 cm. long; catkins appearing before the leaves, sessile, sometimes leafy-bracted at the base, the staminate ones 2 cm. long, the pistillate 2–4 cm. long in fruit; capsule ovoid-oblong, obtuse, 4 mm. long; stigma sessile. Swamps and banks: Me.—Va.—Iowa. *Canad.—Allegh.* My.

S. cordata × sericea. Similar to *S. cordata* but twigs cinereous-canescent; leaves glaucescent beneath and sparsely hairy; capsules silky when young, glabrate in age. *S. cordata myricoides* (Muhl.) Cary [G]. Mass.—Va.—Iowa.

S. petiolaris × sericea. Like *S. sericea* but the leaves less silky, more sharply serrate and the capsule narrower and more acute. *S. petiolaris subsericea* Anders. N. Y.—Mich.—Iowa.

34. **S. arbusculoìdes** Anders. An erect shrub less than 1 m. high; leaf-blades elliptic-lanceolate, acute at both ends, glabrous and green above, silky beneath, minutely serrulate or entire, 2–5 cm. long; aments appearing with the leaves, 2–3 cm. long, usually sessile, naked or subtended by a few leaves; capsule 3–4 mm. long, subsessile. Swamps: Hudson Bay—Canadian Rockies—Alaska. *Subalp.—Subarct.*

35. **S. pellìta** Anders. A low shrub; leaf-blades oblanceolate, 3–7 cm. long, with entire, somewhat revolute margins, acute or obtuse at the apex, tapering at the base, glabrous or slightly pubescent above when young; aments sessile and naked, 2–3 cm. long; bractlets obovate, brown with black tip, or wholly blackish, silky; capsule 3–4 mm. long, villous, subsessile; style nearly 1 mm. long. River banks: Newf.—Me.—Que.—Sask.—Alta. *E. Boreal.* Je.

36. **S. brachycárpa** Nutt. A shrub 1 m. high or less; bark yellow; twigs densely villous; leaf-blades oblong to oval or oblanceolate, 1–3 cm. long, entire-margined; aments 1–2 cm. long; bracts yellow, obovate, villous; capsule 4–5 mm. long, villous, subsessile; style less than 0.5 mm. long. *S. stricta* (Anders.) Rydb. [B]. Wet places in the mountains: B.C.—Colo.—Alta; Man.—Que. *Boreal—Submont.—Subalp.* Jl–Au.

37. **S. nudéscens** Rydb. A shrub 1–3 m. high; leaf-blades usually short-acuminate, 3–6 cm. long, glabrous or slightly pubescent but green above, glaucous beneath; aments 2–5 cm. long; bractlets oblong, brown, sparingly white-silky; capsule elongate, subconic, 6–8 mm. long; stipe very short; style nearly 1 mm. long. *S. glaucops glabrescens* Anders. *S. Austinae* Rydb. [R]; not Bebb. Wet places on the mountains: Alta.—S.D.—Wyo.—B.C.—Yukon. *Submont.—Mont.* Je–Jl.

38. **S. glaùcops** Anders. A shrub 1–2 mm. high; twigs usually more or less villous; leaf-blades green above, paler beneath, entire-margined, 3–6 cm. long; aments 2–3 cm. long; bractlets fuscous, oblong, often acutish, short-villous; capsule grayish villous, about 5 mm. long, subsessile; style about 0.5 mm. long. Mountains: Alta.—N.M.—Utah—Calif.—Yukon. *Submont.—Subalp.* Je–Jl.

39. **S. Nelsònii** Ball. A shrub 1–3 m. high; leaf-blades acute at both ends, 8–15 mm. wide, 3–6 cm. long, entire or sparingly crenate, dark green and shining above, glaucous beneath; aments 1.5–3 cm. long; bractlets black or nearly so, ovate, acute, long-silky; capsule silky pubescent, sessile, 5–7 mm. long; style nearly 0.5 mm. long. Bogs and river banks: S.D.—Colo.—Utah—Alta. *Mont.—Subalp.* Je.

40. S. planifòlia Pursh. A shrub 1–3 m. high; leaf-blades at first silky, soon glabrate, dark green and shining above, glaucous beneath, acute at both ends, usually entire, 2–5 cm. long; aments 2–6 cm. long; bractlets black, obovate, obtuse, silky-pilose; capsule sessile, 5–6 mm. long, grayish silky; style about 0.5 mm. long. *S. chlorophylla* Anders. [B]. *S. phylicifolia* Am. Auth. [G]; not L. Mountain swamps: Lab.—N.H.—N.M.—Utah—Alaska. *Subarct.—Mont.—Subalp.* Je–Au.

FAMILY 35. **MYRICACEAE.** BAYBERRY FAMILY.

Monoecious or dioecious aromatic shrubs or trees. Leaves alternate, resinous-dotted. Flowers in scaly catkins, without perianth. Staminate catkins elongate. Stamens 2–8 on each bract; filaments short; anthers extrorse. Pistillate catkins short; gynoecium of 2 united carpels; ovary 1-celled; stigmas 2, filiform. Fruit nutlike.

Plants dioecious; pistillate catkins ovoid, the flowers subtended by 2–4 bractlets; leaves entire or serrate. 1. MYRICA.
Plants mostly monoecious; pistillate catkins globular, their flowers subtended by 8 bractlets; leaves pinnatifid, with rounded lobes. 2. COMPTONIA.

1. **MYRÌCA** L. SWEET GALE.

Shrubs with entire or dentate leaves. Staminate flowers with 4–8 stamens, the pistillate ones in ours subtended by 2 persistent short bractlets. Fruit resinous-dotted.

1. M. Gàle L. *Fig. 155.* Shrub 1–5 m. high; leaves cuneate, serrate above the middle, pale, 2–5 cm. long, thick; staminate catkins 12–20 mm. long, the pistillate ones about 8 mm. long; bracts densely imbricate. Bogs and ponds: Lab.—Va.—Minn.—Ore.—Alaska. *Boreal.* Ap–My.

2. **COMPTÒNIA** Banks. SWEET FERN.

Low, mostly monoecious shrubs, having pinnatifid leaves with rounded lobes. Staminate flowers with 3–6 stamens, the pistillate ones subtended by 8 persistent awl-shaped bracts, accrescent and in fruit becoming a burr-like involucre surrounding the smooth nut.

1. C. peregrìna (L.) Coulter. An aromatic finely pubescent shrub, 3–10 dm. high; leaves linear-oblong in outline, 4–10 cm. long, acute; lobes oblique, obtuse; staminate catkins 1–3 cm. long; bracts deltoid-reniform; nut 2 mm. long. *C. asplenifolia* Gaertn. [G]. Hillsides: N.S.—N.C.—Tenn.—Minn.—Man. Ap–My.

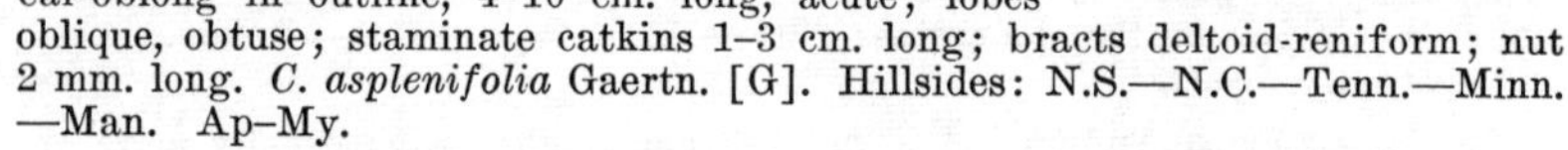

FAMILY 36. **JUGLANDACEAE.** WALNUT FAMILY.

Monoecious trees, with alternate, pinnately compound leaves and no stipules. Staminate flowers in elongate catkins; calyx 2–6-lobed, bearing several rows of stamens. Pistillate flowers solitary or a few together, terminal; gynoecium incompletely 2–4-celled, enclosed in an involucre. Fruit a nut, enclosed in a more or less fleshy husk.

Staminate catkins sessile or nearly so; nut sculptured, enclosed in an indehiscent husk. 1. JUGLANS.
Staminate catkins usually long-peduncled; nut not sculptured, enclosed in a more or less splitting and dehiscent husk. 2. CARYA.

1. JÙGLANS L. WALNUT, BUTTERNUT.

Trees with hard wood. Leaves pinnate, with many sessile leaflets. Staminate catkins elongate, sessile or nearly so; stamens 12–40; filaments free. Pistillate flowers solitary or few together; calyx 4-toothed; petals 4, minute. Styles 2, short. Stigmas club-shaped and fringed. Fruit with a fibrous-fleshy indehiscent pericarp (husk) and an irregularly furrowed endocarp or nutshell. Seeds with 2-lobed cotyledons.

Fruit elongate, acute, pubescent; nut 4-angled. — 1. *J. cineria.*
Fruit globose, obtuse, glabrous; nut not angled. — 2. *J. nigra.*

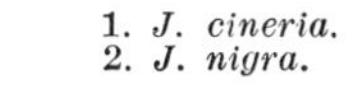

1. J. cinèrea L. A tree 15–30 m. high, with pale heart-wood and gray bark; leaflets 7–17, lanceolate to oblong-lanceolate, serrate, rounded at the base, downy, especially beneath; fruit 8–12 cm. long, acute; nut deeply sculptured, 2-celled at the base. WHITE WALNUT, BUTTERNUT. Rich woods: N.B.—Ga.—Ark.—Kans.—N.D. *Canad.—Allegh.—Ozark.* Ap–My.

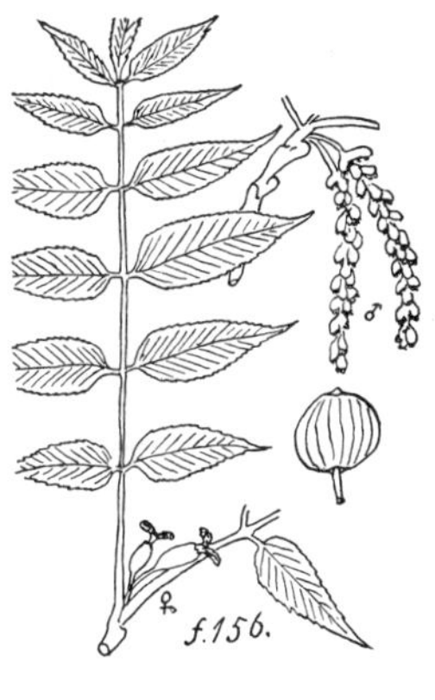

2. J. nìgra L. *Fig. 156.* A tree up to 50 m. high, with dark brown heart-wood and rough dark bark; leaflets 11–23, lanceolate, taper-pointed at the apex, rounded or subcordate at the base, serrate, 8–10 cm. long, glabrous above, minutely downy beneath and on the petiole; fruit globose, 5–8 cm. long, glabrous; nut 4-celled at the top and bottom, with thick ridges. BLACK WALNUT. Woods: Mass.—Fla.—Tex.—Minn. Ap–My.

2. CÀRYA Nutt. HICKORY.

Trees with hard wood and pinnate leaves. Leaflets serrate. Staminate flowers in clustered or branched catkins. Calyx 3-lobed. Stamens 3–10, in several rows. Pistillate flowers in spikes at the ends of leafy twigs. Calyx of a single sepal, adnate to the ovary. Stigmas spreading; ovary inferior. Fruit globose or ellipsoid, its husk becoming dry, more or less deeply 4-valved. Nut smooth, 2–4-celled at the base. *Hicoria* Raf.

Sterile catkins clustered, without common peduncle, developed from separate scaly buds; fruit elongate; nut 2-celled below. — 1. *C. Pecan.*
Sterile catkins in 3's, on a common peduncle, from the same bud as and below the season's leafy shoot; fruit subglobose; nut 4-celled at the base.
  Bud-scales valvate; leaflets falcate; nut bitter, with a thin husk. — 2. *C. cordiformis.*
  Bud-scales imbricate; leaflets not falcate; nut mostly sweet and with a thick husk, except *H. glabra.*
    Bark shaggy, separating into long plates.
      Leaflets 3–5, rarely 7; nut rounded or notched at both ends. — 3. *C. ovata.*
      Leaflets 7–9; nut acute at both ends. — 4. *C. laciniosa.*
    Bark close, not separating into plates.
      Leaves densely pubescent; husk of the fruit thick, freely splitting to the base. — 5. *C. alba.*
      Leaves glabrous or nearly so; husk of the fruit thin, not splitting to the base. — 6. *C. glabra.*

1. C. Pècan (Marsh.) C. K. Schneid. A tree 25–50 m. high; bark scaly; leaflets 5–9, oblong-lanceolate to ovate-lanceolate, 8–15 cm. long, falcate; fruit oblong, 3.5–6 cm. long; nut olive-shaped, smooth. *H. Pecan* Britt. [B]. *C. illinoensis* (Wang.) K. Koch [G]. PECAN. Along streams: Ind.—Ala.—Tex.—Kans.—Iowa; Mex. *Ozark—Texan.* Ap–My.

2. C. cordifórmis (Wang.) K. Koch. A tree 15–30 m. high; bark close, furrowed; leaflets 5–11, lanceolate or oblong-lanceolate, 8–15 cm. long, pubescent when young; fruit ovoid, slightly 6-ridged, 2.5–3 cm. long; husk thin, tardily 4-valved; nut smooth, acute, thin-walled. *H. minima* (Marsh.) Britton [B]. *C. amara* Nutt. *H. cordiformis* Britton [BB]. BITTER NUT. Damp woods: Que.—Fla.—Tex.—Minn. *E. Temp.* My–Je.

C. cordiformis × Pecan. Resembling *C. Pecan* in habit and foliage, but with fewer leaflets and the staminate aments peduncled; the nut is intermediate in form. *C. Brownii* Sarg. Ark.—Kans.

3. C. ovàta (Mill.) K. Koch. *Fig. 157.* A tree 20–40 m. high; leaflets mostly 5, lance-oval or lanceolate, 10–15 cm. long, minutely downy beneath when young, taper-pointed; fruit subglobose, 3–5 cm. long; nut slightly angular, thin-shelled. *H. ovata* Britton [B]. SHELL-BARK or SHAG-BARK HICKORY. Rich woods: Que.—Fla.—Tex.—Minn. *E. Temp.* My.

4. C. laciniòsa (Michx. f.) Loud. A tree 20–40 m. high; leaflets 7–9, downy beneath, oblong-lanceolate, 1–2 dm. long; fruit oblong-ovoid, 5–8 cm. long; nut 4-ribbed above, thick-shelled. *C. sulcata* Nutt. *H. laciniosa* Sarg. [B]. KING NUT. Rich woods: N.Y.—Tenn.—Okla.—Kans.—Iowa. *Allegh.—Ozark.* My.

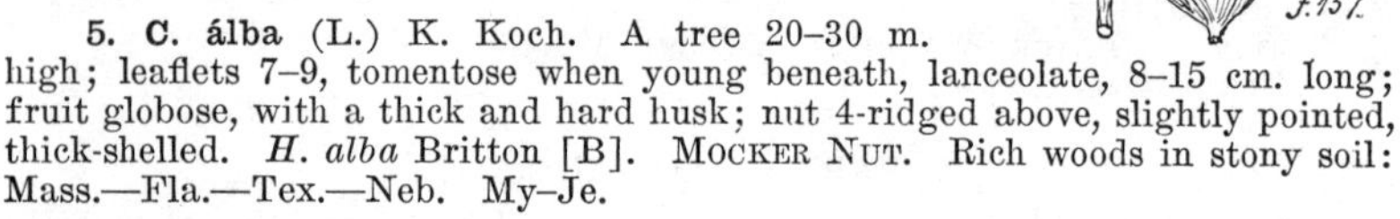

5. C. álba (L.) K. Koch. A tree 20–30 m. high; leaflets 7–9, tomentose when young beneath, lanceolate, 8–15 cm. long; fruit globose, with a thick and hard husk; nut 4-ridged above, slightly pointed, thick-shelled. *H. alba* Britton [B]. MOCKER NUT. Rich woods in stony soil: Mass.—Fla.—Tex.—Neb. My–Je.

6. C. glàbra (Mill.) Spach. A tree 20–40 m. high; leaflets 3–7, oblong-lanceolate, 7–15 cm. long, long-acuminate, glabrous; fruit subglobose or obovoid, 3.5–5 cm. long, with thin husk; nut thick-shelled, angled, bitter. *H. glabra* Britton [B]. PIGNUT HICKORY. Woods: Me.—Fla.—Tex.—Minn. *E. Temp.* My–Je.

FAMILY 37. **BETULACEAE.** BIRCH FAMILY.

Monoecious trees or shrubs, with simple alternate leaves. Staminate flowers in long drooping aments, each bract subtending 2 or 3 flowers; calyx present. Pistillate flowers also in aments, but the aments seldom drooping, without calyx; pistils 2 or 3 at the base of each bract. Fruiting aments cone-like; fruit a small nut or samara; seed solitary.

Bracts of the mature pistillate ament membranous, in ours 3-lobed, deciduous with the fruit. 1. BETULA.
Bracts of the mature pistillate aments thickened and woody, persistent. 2. ALNUS.

1. **BÉTULA** (Tourn.) L. BIRCH.

Shrubs or trees, with resinous aromatic bark; branchlets with transverse lenticels. Staminate aments slender, pendulous; calyx irregularly 2- or 4-lobed; stamens 2, inserted at the base of the calyx; filaments very short, branched at the apex; anther-sacs separate. Pistillate aments erect or drooping, oblong or cylindric; bracts elongate, in ours 3-lobed, 3-flowered, accrescent, deciduous, Fruit small, samara-like or nut-like, flat; outer seed-coat thin, produced into a wing or margin; seeds solitary, pendulous.

Fruiting aments sessile at the ends of short branches; leaves with 9–12 pairs of lateral veins; tall trees.
  Bark brown; leaves shining above; fruiting bractlets with divergent lobes. 1. *B. lenta.*
  Bark yellowish or gray; leaves dull above; fruiting bractlets with ascending lobes. 2. *B. lutea.*
Fruiting aments peduncled; leaves with 4–9 pairs of lateral veins.
  Bark of the trunk chalky or silvery white; trees.
    Leaf-blades glabrous on both sides, rhombic in outline, acuminate. 3. *B. caerulea.*
    Leaf-blades more or less pubescent beneath, ovate or subcordate, acute or short-acuminate.
      Leaf-blades broadly ovate, mostly cordate at the base, with spreading teeth; middle lobe of the bractlets narrow, rounded at the apex, about twice as long as the lateral lobes. 4. *B. cordifolia.*
      Leaf-blades narrowly ovate, rounded or cuneate at the base, with the teeth directed forward; middle lobe of the bractlets acute, slightly longer than the lateral ones. 5. *B. papyrifera.*
  Bark of the trunk reddish or brown.
    Samara-wing broader than the body; twigs and branches glandular-resiniferous, not hairy; trees. 6. *B. fontinalis.*
    Samara-wing narrower than the body.
      Leaf-blades 3–6 cm. long, rhombic in outline, acute at both ends, tomentose or pubescent beneath; trees. 7. *B. nigra.*
      Lafe-blades 1–3 cm. long, elliptic, oval, or obovate in outline, rounded or obtuse at the apex; shrubs.
        Twigs and branches pubescent with scattered long hairs, slightly glandular; leaf-blades obovate, cuneate at the base. 8. *B. glandulifera.*
        Twigs and branches glabrous, densely glandular-granuliferous; leaf-blades rounded at both ends. 9. *B. glandulosa.*

1. **B. lénta** L. A tree up to 25 m. high; bark of the trunk brown, in age becoming ashy, exfoliating; leaf-blades ovate to oblong-lanceolate, 4–12 cm. long, short-acuminate, doubly serrate, rounded or cordate at the base, paler and silky on the veins beneath; pistillate aments erect, sessile, in age 1.5–3 cm. long, oblong; bractlets 4 mm. long. CHERRY BIRCH, BLACK BIRCH. Woods: Newf.—Fla.—Tenn.—(?)Iowa. *E. Temp.* Ap–My.

2. **B. lùtea** Michx. f. A tree up to 30 m. high, with yellowish or silvery bark, which separates into thin layers; leaf-blades ovate or oblong-ovate, 5–10 cm. long, acuminate, sharply and doubly serrate, rounded or cordate at the base, glabrous above, pubescent on the nerves beneath; pistillate aments about 1.5 cm., or in fruit 2–4 cm. long, sessile, oblong; nuts with an orbicular-obovate wing. YELLOW BIRCH, GRAY BIRCH. Woods: Newf.—Ga.—Tenn.—Iowa—Man. *Canad.—Allegh.* Ap–My.

3. **B. caerùlea** Blanchard. A tree up to 20 m. high; bark separating into sheets, white; leaf-blades thin, long-pointed, cuneate or truncate at the base, glabrous on both sides, or slightly hairy on the veins, blue-green, 5–8 cm. long, sharply serrate; fruiting catkins drooping, cylindric, 2–3 cm. long, blunt; lateral lobes of the bracts divergent, rounded-ovate, the terminal one short; seeds broadly winged. *B. pendula* Fern. [G]; not Roth. BLUE BIRCH. Woods: Me.—S.D.—Man. *Canad.* Je.

4. **B. cordifòlia** Regel. Small tree, 5–15 m. high, with white flaky bark, the twigs green, becoming brown, somewhat glandular; leaf-blades 5–10 cm. long, broadly ovate, usually cordate at the base, sharply toothed, dark green and glabrous above, paler beneath and hairy in the axils of the leaves; staminate-catkins 4–5 cm. long, drooping; pistillate catkins peduncled, 4–6 cm. long, about 12 mm. thick; bracts 7–8 mm. long, the terminal lobe oblong, rounded at the apex, twice as long as the oval lateral ones. *B. alba cordifolia* Fern. [G]. Mountains and hills: Newf.—Me.—Vt.—ne Minn.—(Turtle Mts.) N.D.—Man. *Canad.—Subarct.* Mr–Je; fr. Au–S.

5. **B. papyrífera** Marsh. *Fig 158.* A tree up to 25 m. high; leaf-blades 3–11 cm. long, irregularly or doubly serrate, glabrous and dark green above, pubescent beneath; staminate aments 5–10 cm. long, 2 or 3 at the ends of the branches; pistillate aments peduncled, solitary on small lateral leafy branches, 2–4 cm. long. *B. Andrewsii* A. Nels. *B. alba* Fern.; not L. PAPER BIRCH, CANOE BIRCH. Cold woods: Lab.—N.J.—ne Iowa—Colo.—Alaska. *Submont.—Subalp.* Ap–My.

B. glandulifera × **papyrifera.** A shrub or shrub-like tree; leaf-blades up to 6 cm. long, thick, firm, dull bronze-green above, paler and sparsely hairy beneath, rhombic-ovate or oval, acute at both ends; fruiting aments 2–4 cm. long, slender-stalked; middle lobe of the bractlets longer than the ascending rounded lateral ones. *B. Sandbergii* Britton. Swampy places: Sask.—Mont. *Plain.*

6. **B. fontinàlis** Sarg. A tree occasionally 10–12 m. high, often growing in clumps and shrub-like; blades sharply double-serrate, entire towards the base, soon glabrous, 2–5 cm. long; staminate aments 5–7 cm. long, several; pistillate aments 2–3 cm. long; terminal lobe of the bractlets lanceolate, decidedly longer than the lateral ones. *B. occidentalis* Nutt.; not Hook. MOUNTAIN BIRCH, WATER BIRCH, SWAMP BIRCH, BLACK BIRCH. River banks: Man.—Neb.—N.M.—Utah—Calif.—B.C.—Yukon. *W. Temp.—Plain—Mont.* My–Je.

7. **B. nìgra** L. A tree up to 30 m. high, with bark varying from brown to yellowish; leaf-blades rhombic-ovate, 3–6 cm. long, acute, finely serrate or incised, entire at the cuneate base, glabrous above, tomentose or pubescent beneath, especially on the veins; pistillate aments about 2 cm. long, in fruit 3–4 cm. long, cylindric; bractlets pubescent, the lateral lobes nearly as long as the terminal one; nut 3.5 mm. long, with a narrow ciliate wing. RIVER BIRCH, RED BIRCH. Banks of streams and swamps: Mass.—Fla.—Tex.—Minn. *E. Temp.* Ap–My.

B. nigra × **papyrifera.** Tree resembling *B. papyrifera* with grayish white, somewhat flaky bark; leaves large, some more simple-toothed as in that species, some decidedly double-toothed as in *B. nigra,* as well as the twigs decidedly pubescent; bracts of the pistillate aments with the short middle lobe of *B. nigra* but less pubescent. New Albion, Iowa.

8. **B. glandulífera** (Regel) Butler. A shrub 1–5 m. high; leaf-blades dentate-crenate, with rounded-ovate or on the vigorous shoots triangular teeth, 1.5–3 cm. long, sparingly hirsute when young, soon glabrous, dark green above, yellowish or reddish green beneath; pistillate aments 1–2 cm. long, 6–7 mm. thick; bractlets glabrous; lobes all oblong and obtuse at the apex, of nearly the same length; wing one fourth to one half as wide as the oval nut. *B. pumila glandulifera* Regel [G]. Boggy places: Ont.—Iowa—Sask. *Prairie—Plain—Submont.* My.

B. glandulifera × **lutea.** Shrub or small tree 3–6 m. high, with grayish brown bark, with wintergreen flavor; leaves ovate to obovate, acuminate, 2.5–6 cm. long, glabrous above; lighter beneath, pubescent on the veins, serrate-crenate; fruiting catkins oblong, 1.5–2.5 cm. long, 10–12 cm. thick; middle lobe of the bracts deltoid, the lateral ones broad; wing of the seeds half as broad as the body. *B. lutea* × *pumila glandulifera* Rosend. Tamarack swamps: Minn. to Mich.

9. **B. glandulòsa** Michx. A shrub 0.3–2 m. high; twigs brown and densely glandular-resiniferous; leaf-blades 1–2.5 cm. long, rounded at both ends or sometimes acutish at the base, bright green above, paler beneath; staminate aments usually solitary, 1–1.5 cm. long; pistillate aments 1–2.5 cm. long, 4–5 mm. thick; bractlets glabrous; lobes oblong, obtuse, the lateral ones shorter,

spreading, ascending and curved upward; wing of the samara very narrow. Bog Birch, Scrub Birch. Wet places: Greenl.—Me.—Minn.—Colo.—Ore.—Alaska. *Boreal.—Submont.—Subalp.* Je–Jl.

2. **ÁLNUS** (Tourn.) Hill. Alder.

Shrubs or trees, with astringent smooth bark. Staminate aments drooping; calyx irregularly lobed; stamens as many as the lobes and opposite to them; anthers introrse. Pistillate aments ovoid or oblong, cone-like; bracts subtending 2 flowers, becoming woody, truncate or lobed at the apex, persistent. Fruit minute, nut-like, winged or wingless, with thin outer coat. Seeds solitary.

Nut broadly winged; flowers developed with the leaves.
  Twigs, peduncles and lower surface of the leaves permanently soft-pubescent. 1. *A. mollis.*
  Twigs and peduncles glabrous or nearly so; leaves glabrous beneath or slightly pubescent on the veins. 2. *A. crispa.*
Nut wingless, merely acute-margined; flowers developed before the leaves.
  Mature leaves pubescent, and usually glaucous beneath. 3. *A. incana.*
  Mature leaves glabrous beneath, except on the veins and in their axils, green on both sides.
    Leaves more or less lobed, doubly serrate. 4. *A. tenuifolia.*
    Leaves finely and sharply serrate, not lobed. 5. *A. rugosa.*

1. **A. móllis** Fernald. Shrub or small tree; young branches and peduncles pubescent; mature leaves 4.5–10 cm. long, closely serrate with unequal teeth, permanently covered beneath with soft short plush-like pubescence; mature fertile aments 1.2–2 cm. long, 9–12 mm. thick. Downy Alder. *A. Alnobetula* Am. auth. [B]. Cold bogs: Newf.—Man.—Mass. *Canad.*

2. **A. críspa** (Ait.) Pursh. Shrub, glabrous or nearly so; leaves round-oval or ovate, sometimes subcordate at the base, 3–6 cm. long, glutinous beneath, irregularly and sharply serrate; fertile catkins 1–1.5 cm. long in fruit. Green Alder. *A. viridis* and *A. Alnobetula* Am. auth. [B] in part. Shores and bogs: N.B.—N.C.—Minn. *Canad.—Allegh.*

3. **A. incàna** (L.) Moench. Shrub or small tree, rarely 6 m. high; leaf-blades broadly elliptic to ovate, rounded at the base, sharply and doubly serrate, dark green above, downy and ferruginous or glaucous, prominently veined beneath, 5–13 cm. long; staminate aments 3–8 cm. long, the pistillate in fruit 1–1.5 cm. long; nut orbicular, thick-margined. Speckled Alder. Swamps and along streams: Newf.—Pa.—Neb.—Sask.; Eurasia. *E. Boreal.* Ap–My.

4. **A. tenuifòlia** Nutt. *Fig. 159.* A tall shrub or small tree, sometimes 8–10 m. high; twigs pubescent when young; leaf-blades ovate or oval, obtuse or acute at the apex, rounded or subcordate at the base, 5–10 cm. long; staminate aments 2.5–4.5 cm. long, filaments equaling the calyx; pistillate aments when ripe 1–2 cm. long, ellipsoid; nut 2.5 mm. long. *A. incana virescens* S. Wats. River Alder. Along streams: Alaska—Calif.—N.M.—N.D.—Yukon. *W. Temp.—Submont.—Mont.* Ap–My.

5. **A. rugòsa** (Du Roi) Spreng. Shrub or small tree, with smooth bark; leaves obovate, acute at the base, finely serrulate, with impressed nerves, sparingly pubescent on the veins beneath, 7–13 cm. long; staminate aments 5–10 cm. long; pistillate aments in fruit ovoid, 1–2 cm. long. Smooth Alder. Wet places: Me.—Minn.—Tex.—Fla. *E. Temp.* Mr–Ap.

FAMILY 38. **CORYLACEAE.** HAZELNUT FAMILY.

Monoecious trees or shrubs, with simple alternate leaves. Staminate aments elongate, drooping, each bract subtending a single flower; calyx wanting; filaments distinct, branched at the apex. Pistillate aments short; bracts foliaceous at maturity, each subtending 2 or 3 flowers; calyx present, sometimes represented by a cup. Fruit 1–3 nuts, enclosed in the bracts. Seeds solitary by suppression.

Pistillate flowers many in a cylindric ament; nuts small.
 Fruiting bracts open, 3-lobed. 1. CARPINUS.
 Fruiting bracts bladder-like, enclosing the nut. 2. OSTRYA.
Pistillate flowers few in a head-like ament; nut large, in a leaf-like involucre. 3. CORYLUS.

1. **CARPÌNUS** (Tourn.) L. HORNBEAM, IRONWOOD, BLUE BEECH.

Trees or shrubs, with close-grained wood and smooth pale bark. Staminate aments short, drooping; stamens 3–20, inserted on a broad bract. Pistillate aments loose, terminal; bracts enlarging in fruit, foliaceous, 3-lobed; pistils 2, at the base of each bract. Fruit a small flattened nut, ribbed; outer coat thin, the inner bony.

1. C. caroliniàna Walt. A small tree or shrub, 5–15 m. high; bark bluish gray; leaf-blades ovate-oblong, acuminate, double-serrate, glabrous above, pubescent on the veins beneath and in their axils, 3–15 cm. long; staminate aments 2–5 cm. long; bracts 3 mm. long, ovate; pistillate aments 1–1.5 cm. long, in fruit 2–3 cm. long; nut 5–6 mm. long. Along streams: N.S.—Fla.—Tex.—Minn. *E. Temp.*

2. **ÓSTRYA** (Mich.) Scop. IRONWOOD, HOP-HORNBEAM.

Trees with close-grained, hard wood and scaly bark. Staminate aments clustered, drooping; bracts subtending 3–14 stamens; filaments very short, branched at the tips; anther-sacs separate, pilose above. Pistillate aments terminal, solitary; bracts subtending two flowers, developing into bladdery sacks. Calyx denticulate. Nut ovoid, flattened, obscurely ribbed.

f.160.

1. O. virginiàna (Mill.) K. Koch. *Fig. 160.* A tree 6–18 m. high; twigs light green and pubescent at first; leaf-blades ovate or oblong-lanceolate, acuminate at the apex, sharply serrate, 7–15 cm. long; staminate aments 3–7 cm. long; bractlets triangular-ovate, acuminate; pistillate aments about 8 mm., in fruit 4–6 cm. long; nut 5–8 mm. long, shining. Dry woods: N.S.—Fla.—Tex.—N.D.—Man. *E. Temp.—Submont.* Ap–My.

3. **CÓRYLUS** (Tourn.) L. HAZELNUT.

Shrubs or trees, with branched stem and smooth bark. Staminate aments pendulous, very long, solitary or in clusters; bracts enclosing 4–8 stamens; filaments short, forked at the apex; anther-sacs separate, pilose at the apex. Pistillate aments inconspicuous, clustered at the tips of the branches; each bract enclosing 2 bractlets and an incompletely 2-celled ovary. Nut large, enclosed in a leafy involucre formed by the more or less united bracts.

Bractlets nearly distinct, leaf-like, not produced into a beak. 1. *C. americana.*
Bractlets united, in fruit prolonged into a beak. 2. *C. rostrata.*

1. **C. americàna** Walt. A shrub 1–3 m. high, with pubescent twigs; leaf-blades oval or suborbicular-cordate, 5–15 cm. long, short-acuminate or acute, glabrous above, finely pubescent beneath; involucral bracts nearly distinct, incised, foliaceous; nut broadly obovoid, 1.5 cm. long. AMERICAN HAZELNUT. Woods and thickets: Me.—Ga.—Kans.—N.D.—Man. Ap–My.

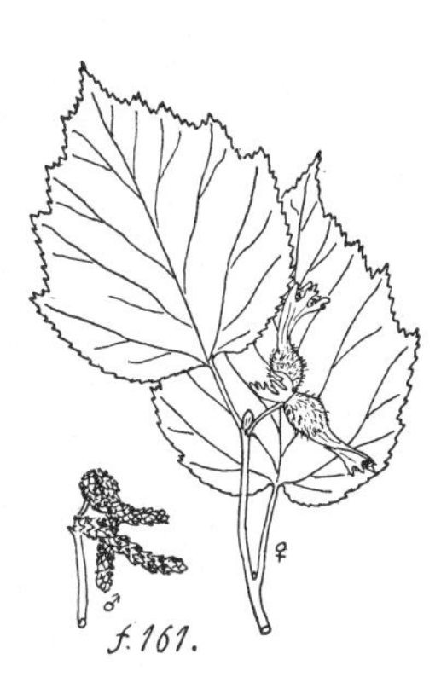

2. **C. rostràta** Ait. *Fig. 161.* A shrub 1–2 m. high, with brown branches; leaf-blades ovate or oval, sharply serrate, glabrous or with scattered hairs above, sparingly pubescent especially on the veins beneath, 5–12 cm. long; involucral bracts bristly-hairy, united and prolonged into a tubular beak, laciniate at the summit; nut ovoid, scarcely compressed, striate. BEAKED HAZELNUT. Thickets: N.S.—Ga.—Colo.—N.D. *E. Temp.—Submont.* Ap–My.

FAMILY 39. **FAGACEAE.** BEECH FAMILY.

Monoecious trees or shrubs, with simple leaves. Staminate flowers in elongate or head-like aments; calyx of 4–7 partially united sepals; stamens 4–20. Pistillate flowers in longer or shorter spikes; calyx of partly united sepals. Gynoecium of 3–7 united carpels; styles as many, but usually only one ovule maturing. Fruit a one-seeded nut, enclosed or seated in a scaly or spiny, in ours cup-like involucre.

Staminate flowers in slender catkins; cup 1-flowered; nut terete. 1. QUERCUS.
Staminate flowers capitate; cup usually 2-flowered; nuts 3-angular. 2. FAGUS.

QUÉRCUS (Tourn.) L. OAK.

Trees or shubs, with hard coarsely grained wood. Leaf-blades entire, toothed, or lobed, firm-membranous or leathery, sometimes evergreen. Staminate aments elongate, drooping, clustered; calyx campanulate, 4–7-lobed; stamens 6–12; filaments filiform. Pistillate flowers solitary or in lax spikes; calyx urn-shaped or cup-shaped. Ovary usually 3-celled; styles 3; ovules 2 in each cell, but seldom more than one maturing in each pistil. Fruit a leathery, 1-seeded nut (*acorn*), partly enclosed in an accrescent scaly involucre (*cup*).

Leaves and their lobes not bristle-tipped; acorn-shell not pubescent within; fruit maturing in the autumn of the first year; stigmas sessile.
  Leaves pinnatified or pinnately lobed (WHITE OAKS).
    Mature leaves glabrous and glaucous beneath; cup very shallow. 1. *Q. alba.*
    Mature leaves finely pubescent beneath; cup at least one third as long as the acorn.
      Scales of the cup not caudate-acuminate. 2. *Q. stellata.*
      Scales of the cup, at least the inner ones, caudate-acuminate, forming a fringe around the acorn.
        Cup 20 mm. wide, conspicuously fringed; acorns subglobose or round-ellipsoid 15–20 mm. in diameter; tall tree 10–50 m. high. 3. *Q. macrocarpa.*
        Cup 10 mm. wide, slightly fringed, only the innermost scales caudate-attenuate; acorns ovoid, about 10 mm. in diameter; shrub 1–6 mm. high, often with corky branches. 4. *Q. mandanensis.*
  Leaves sinuate to dentate, but not lobed (CHESTNUT OAKS).
    Fruit long-peduncled; peduncles often longer than the petioles; leaf-blades sinuate. 5. *Q. bicolor.*
    Fruit sessile or nearly so; leaf-blades dentate.
      Leaf-blades mostly lanceolate, with 5–13 teeth on each margin; trees. 6. *Q. Muhlenbergii.*

Leaf-blades mostly obovate, with 3–7 teeth on each margin; shrubs. — 7. *Q. prinoides.*
Leaves and their lobes or teeth bristle-tipped; acorn-shell pubescent within; fruit maturing the second season; style elongate (RED AND BLACK OAKS).
Leaves more or less lobed.
Leaves pinnately 5–9-lobed or 5–9-cleft.
Cup saucer-shaped, much broader than high.
Cup 15–25 mm. broad, leaves dull, their lobes not twice as long as the width of the middle portion of the blade. — 8. *Q. maxima.*
Cup 8–15 mm. broad; leaves shining; their lobes 2–4 times as long as the width of the middle portion of the blade.
Acorn subglobose or short-ovoid, little if at all longer than thick. — 9. *Q. palustris.*
Acorn ovoid, decidedly longer than thick. — 10. *Q. Schneckii.*
Cup turbinate or hemispheric, not much if at all broader than high.
Scales of the cup closely appressed; inner bark gray or reddish.
Cup brown, 15–20 mm. wide; scales finally glabrous and shining. — 11. *Q. coccinea.*
Cup 12–18 mm. wide, ashy with persistent pubescence.
Leaves 3–5 lobed; the lobes 1–2 toothed. — 12. *Q. texana.*
Leaves 5–7-lobed; the lobes again lobed or toothed. — 13. *Q. ellipsoidalis.*
Inner scales of the cup loosely imbricate; inner bark orange. — 14. *Q. velutina.*
Leaves 3–5-lobed above the middle; obovate in outline. — 15. *Q. marilandica.*
Leaves entire, oblanceolate or lance-oblong. — 16. *Q. imbricaria.*

1. Q. álba L. A tree up to 45 m. high, with gray or ashy, flaky bark; leaf-blades obovate in outline, 1–2 dm. long, pale green above, glaucous beneath; pinnatifid into 5–9, usually 7, obtuse lobes and rounded sinuses, the upper lobes broad and often again round-lobed; cup saucer-shaped, 15–20 mm. broad; scales tubercled; acorn 15–25 mm. long, 3–4 times as long as the cup. WHITE OAK. Woods: Me.—Fla.—Tex.—Neb.—Minn. *E. Temp.* My–Jl.

Q. alba × macrocarpa. Leaves hairy, less deeply lobed and less obovate than in *Q. macrocarpa,* more like those of *Q. alba* in outline but puberulent beneath, the cup almost devoid of mossy rim and the acorn more like *Q. alba.* *Q. Bebbiana* C. K. Schneid. Mo.–Minn.

2. Q. stellàta Wang. A tree 20–30 m. high, with narrowly ridged bark; leaf-blades broadly obovate in outline, 1–2 dm. long, dark green and shining above, finely tomentose beneath, with usually 5 broad, divergent lobes and rounded sinuses, the upper 3 lobes again round-lobed; cup hemispheric, 15–20 mm. broad, with broad, flat, spreading scales; acorns ovoid, 15–20 mm. long, pubescent at the apex. *Q. minor* (Marsh.) Sarg. [B]. POST OAK. Woods: Mass.—Fla.—Tex.—Kans.—Iowa. *E. Temp.* My–Jl.

3. Q. macrocárpa Michx. A tree 10–50 m. high; leaf-blades obovate, irregularly round-lobed, usually pinnately divided below the middle, bright green and shining above, grayish tomentulose beneath, 1–2 dm. long; fruit short-peduncled; cup hemispheric, 1–2.5 cm. in diameter; scale floccose, thick, ovate or lanceolate, the upper subulate-tipped; acorns 1.5–2.5 cm. long, ovoid. BUR OAK, MOSSY-CUP OAK. Rich soil: N.S.—Pa.—Tex.—(? Wyo.)—S.D.—Sask. *E. Temp.—Plain—Submont.* My–Je.

Q. bicolor × macrocarpa. Resembles *Q. macrocarpa* in habit, but the leaves have more numerous and shorter lobes and turn colors as in *Q. bicolor,* and the fruit is like that of the latter. *Q. Schuettei* Trelease. Minn.

Q. macrocarpa × Muhlenbergii. Intermediate between the two parents, the leaves inclined to be less obovate and more elliptic in outline and with more numerous and less deep lobes than in *Q. macrocarpa,* more decidedly lobed with more rounded lobes than in *Q. Muhlenbergii.* *Q. Hilli* Trelease. Kans.–s Minn.

4. **Q. mandanénsis** Rydb. *Fig. 162.* Shrub or low tree, 1–5 m. high, with gray bark; the branches often corky; leaves 5–15 cm. long, 3–7 cm. wide, obovate in outline, 5–7-lobed, panduriform, the terminal lobe large, once or twice lobed with rounded divisions, dark green and somewhat stellate above, pale, often almost white, and densely and finely stellate beneath; fruit subsessile or sessile; cup hemispheric, about 1 cm. broad, scales slightly corky, the innermost with subulate-caudate tips; acorns ovoid, 10–12 mm. long. *Q. obtusiloba depressa* Nutt.; not *Q. depressa.* It has been mistaken for *Q. utahensis.* It probably hybridizes with *Q. macrocarpa.* Hills and bluffs: Minn.—Neb.—Wyo.—Mont.—Sask. *Plain—Submont.*

5. **Q. bícolor** Willd. A tree up to 35 m. high; leaf-blades obovate in outline, cuneate at the base, sinuately lobed or toothed, 1–1.5 cm. long, soft-downy and white-hoary beneath; cup saucer-shaped or hemispheric, one third to one half as high as the acorn; scales pubescent and the inner acuminate; acorn 2–3 cm. long, ovoid. *Q. platanoides* (Lam.) Sudw. [B]. WHITE SWAMP OAK. Borders of streams: Me.—Ga.—Ark.—e Kans.—Minn. *Canad.—Allegh.* My–Je.

6. **Q. Muhlenbérgii** Engelm. *Fig. 163.* A tree 20–50 m. high, with gray or white, scaly bark; leaf-blades lanceolate or oblanceolate, 5–20 cm. long, acuminuate, coarsely serrate, glabrous and glossy above, glaucous and puberulent beneath; cup hemispheric, 15 mm. broad; scales thickened; acorn ovoid, 12–20 mm. long, twice as long as the cup. *Q. acuminata* (Michx.) Houba. [B]. CHESTNUT OAK. Limestone soil: Vt.—Fla.—Tex.—Minn. *E. Temp.* My–Je.

7. **Q. prinoìdes** Willd. A shrub or small tree, 1–5 m. high, with pale bark; leaf-blades obovate, 5–15 cm. long, coarsely toothed, glabrous and shining above, grayish-tomentose beneath; acorn ovoid oblong, 15–20 mm. long, twice as long as the cups. CHINQUAPIN OAK, SCRUB CHESTNUT OAK. Sandy or rocky places: Me.—N.C.—Tex.—Minn. *Canad.—Allegh.* Ap–My.

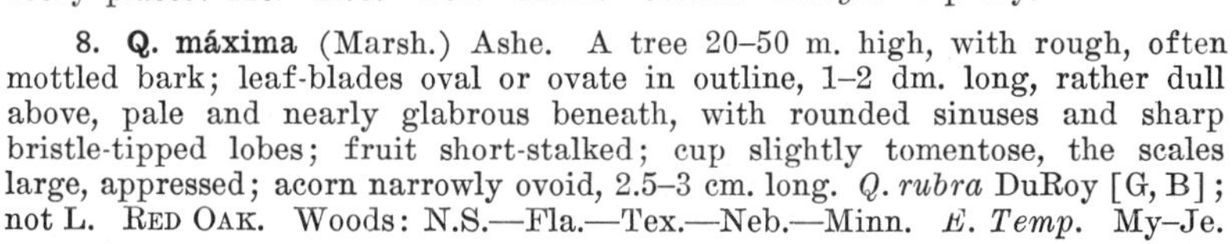

8. **Q. máxima** (Marsh.) Ashe. A tree 20–50 m. high, with rough, often mottled bark; leaf-blades oval or ovate in outline, 1–2 dm. long, rather dull above, pale and nearly glabrous beneath, with rounded sinuses and sharp bristle-tipped lobes; fruit short-stalked; cup slightly tomentose, the scales large, appressed; acorn narrowly ovoid, 2.5–3 cm. long. *Q. rubra* DuRoy [G, B]; not L. RED OAK. Woods: N.S.—Fla.—Tex.—Neb.—Minn. *E. Temp.* My–Je.

Q. ellipsoidalis × maxima. Resembling most *Q. maxima* in leaf-form and size and shape of the fruit, but the leaves are more deeply lobed, the cup deeper and acute at the base. Lever Park, Cedar Rapids, Iowa.

9. **Q. palústris** Münchh. A tree 20–60 m. high, with rough brown bark; leaf-blades broadly oval or ovate in outline, with roundish sinuses and divergent, again cleft, bristle-tipped lobes, 6–15 cm. long, glabrous on both sides, or with hair-tufts in the vein-axils beneath; cup saucer-shaped, 12–15 mm. broad, finely pubescent; acorn 10–15 mm. long. SWAMP OAK, PIN OAK. Swamps: Mass.—Fla.—Okla.—Neb.—Wis. *E. Temp.* My–Je.

10. **Q. Schnéckii** Britton. A shrub or tree 5–30 m. high, with reddish bark; leaf-blades broadly oval or obovate in outline, 6–20 cm. long, glabrous and shining above, pale and with hair-tufts in the axils beneath, pinnately 5–7-lobed, with rounded sinuses and again lobed and brittle-tipped divisions; cup saucer-shaped, 12–18 cm. broad, with appressed scales; acorn 1.5–2.5 cm. long. *Q. texana* Sarg. [G]; not Buckley. Open woods: Ind.—N.C.—Fla.—Tex.—Iowa. *Carol.—Ozark.* Ap–My.

11. **Q. coccínea** Wang. A tree up to 50 m. high, with the outer bark gray, the inner reddish; leaf-blades broadly ovate or oval in outline, 1–2 dm. long, deep green and shining above, paler beneath, 5–7-lobed, with rounded sinuses and spreading lobes and spinulose-tipped divisions; cup turbinate; scales appressed; acorn oblong-ovoid, 1.5–2 cm. long. SCARLET OAK. Dry woods: Me.—N.C.—Neb.—Minn. *Canad.—Allegh.* My–Je.

Q. coccinea × velutina. Leaves like those of *Q. coccinea* in texture and shape but usually hairy beneath in the axils of the veins, winter buds more hairy, and cup more like *Q. velutina* with larger and looser inner scales. Iowa.

12. **Q. texàna** Buckl. A small tree with spreading branches; leaf-blades 5–10 cm. long; lobes triangular or ovate, 1–2-toothed, the terminal one much larger; fruit 1.5 cm. long, sessile; cup hemispheric, 9–11 mm. broad; acorn oblong, 8–10 mm. thick. Dry, rocky soil: Tex.—Kans. *Texan.*

13. **Q. ellipsoidàlis** E. J. Hill. *Fig. 164.* A tree about 30 m. high, with smooth close gray bark; leaf-blades broadly obovate in outline, glabrous and shining in age, 5–7 oblong, with coarsely laciniate lobes; cup turbinate, one third to one half as long as the acorn; acorn small, ellipsoid, 1.2–2 cm. long. Clayey soil: Mich. — Ill. — Iowa — Minn. — Man. *Allegh.* My.

Q. ellipsoidalis × velutina. Resembling a broad-leaved *Q. ellipsoidalis* in leaves and acorns but the capsules pubescent as in *Q. velutina;* the lobes of the leaves are broader than the sinuses. *Q. paleolithiticola* Trelease. Wis.—Iowa.

14. **Q. velùtina** Lam. A tree up to 50 m. high, with thick furrowed bark; leaf-blades obovate in outline, 1–3 dm. long, glabrous and shining above, rusty-pubescent beneath, with hair-tufts in the axils of the veins, pinnatifid, with few-toothed lobes; cup turbinate, 18–23 mm. broad; upper scales pubescent, thin, squarrose when dry; acorn ovoid or hemispheric, 12–20 mm. long, pubescent. BLACK OAK, QUERCETON. Dry uplands: Me.—Fla.—Tex.—Neb.—Minn. *E. Temp.* My–Je.

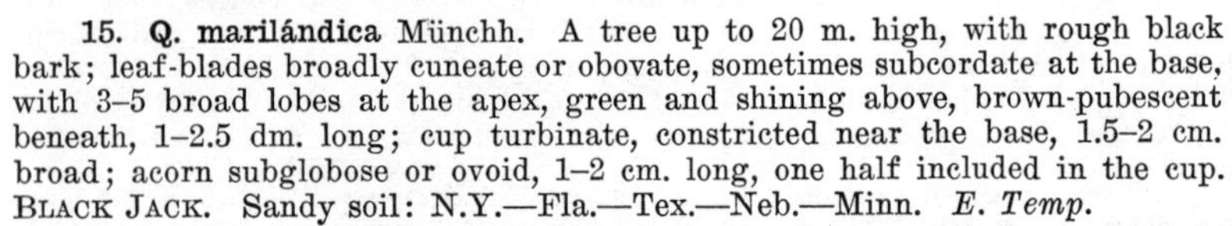

15. **Q. marilàndica** Münchh. A tree up to 20 m. high, with rough black bark; leaf-blades broadly cuneate or obovate, sometimes subcordate at the base, with 3–5 broad lobes at the apex, green and shining above, brown-pubescent beneath, 1–2.5 dm. long; cup turbinate, constricted near the base, 1.5–2 cm. broad; acorn subglobose or ovoid, 1–2 cm. long, one half included in the cup. BLACK JACK. Sandy soil: N.Y.—Fla.—Tex.—Neb.—Minn. *E. Temp.*

Q. marilandica × velutina. Intermediate between the two parents. The leaves resembling most those of *Q. velutina* but the lobes 3–5, broader, more rounded, less cuspidate, and the base of the leaves inclined to be less cuneate. *Q. quinqueloba* Engelm. *Q. Bushii* Sarg. Ill.—Mo.—Okla.

16. **Q. imbricària** Michx. A tree up to 30 m. high, with brown fissured bark; leaf-blades oblanceolate to lanceolate, usually entire or sometimes few-toothed, green and shining above, softly tomentose beneath; cup nearly

hemispheric, 1.5–2 cm. broad; scales obtuse; acorn subglobose, 1–1.5 cm. long. SHINGLE OAK. Dry woods: Pa.—Ga.—Ark.—Kans.—Neb. *Ozark.*

Q. imbricaria × palustris. It is intermediate between the two parents; the leaves are usually lobed but the lobes are short and triangular, directed forward. *Q. exacta* Trelease. Iowa.

Q. imbricaria × marilandica. Its leaves resemble *Q. marilandica* in texture, but the blades are oblong or slightly broader upwards and with 3 terminal lobes. *Q. tridentata* Engelm. Mo.—Kans.

2. FÀGUS (Tourn.) L. BEECH.

Trees or rarely shrubs, with alternate, serrate leaves. Staminate flowers capitate, with scale-like deciduous bracts; calyx campanulate, 5–7-cleft; stamens 8–16; filaments slender. Pistillate flowers usually in pairs at the end of a short peduncle; bractlets numerous, the outer subulate, the inner 4 united at the base, forming a 4-lobed involucre; calyx 6-lobed. Ovary 3-celled, styles 3, filiform. Nuts usually 2 in the prickly, 4-valved involucre.

1. **F. grandifòlia** Ehrh. *Fig. 165.* Large tree, with smooth bark; leaves ovate or ovate-oblong, coarsely toothed, cuneate at the base, acuminate at the apex, 5–10 cm. long, silky when young, in age glabrate; bur 1–2 cm. long, densely tomentose, the prickles spreading or recurved. *F. americana* Sweet [B]. Rich woods: N.B.—Va.—Tex.—Minn. *Canad. Allegh.—Ozark.* Ap–My.

FAMILY 40. ULMACEAE. ELM FAMILY.

Monoecious, polygamous, or hermaphrodite trees or shrubs, with simple leaves, often 2-ranked and oblique at the base. Flowers in cymes or racemes. Calyx of 3–8 sepals, more or less united at the base, imbricate. Stamens of the same number, opposite the sepals. Pistil of 2 united carpels; ovary 2-celled. Fruit a samara, drupe, or nut. Endosperm scant or wanting.

Fruit a samara; embryo straight. 1. ULMUS.
Fruit a drupe; embryo with conduplicate cotyledons. 2. CELTIS.

1. ÚLMUS (Tourn.) L. ELM.

Trees or shrubs, with furrowed, often corky bark. Leaves 2-ranked, oblique, straight-veined, serrate, deciduous. Flowers perfect, in axillary clusters. Calyx membranous, 4–9-lobed, usually 5-lobed, campanulate. Stamens exserted; filaments filiform or slightly flattened; anthers extrorse, emarginate at both ends. Ovary sessile or short-stalked, 1-celled or rarely 2-celled; stigmas often incurved, introrse. Samaras orbicular or oblong, winged all around. Seeds flattened.

Fruit on slender pedicels, pendulous.
  Branches without corky wings; faces of the fruit glabrous. 1. *U. americana.*
  Branches with more or less developed corky wings; faces of the fruit pubescent.
    Flowers fascicled; leaf-blades 2–7 cm. long. 2. *U. alata.*
    Flowers racemose; leaf-blades 5–12 cm. long. 3. *U. Thomasi.*
Fruit short-pediceled, not pendulous.
  Buds covered with rusty hairs; leaves ciliate, double-serrate; fruit pubescent. 4. *U. fulva.*

Buds pale, not rusty-hairy; leaves not ciliate, simply serrate; fruit glabrous. 5. *U. pumila.*

1. **U. americàna** L. *Fig. 166.* A large tree, sometimes 35 m. high, with glabrous or sparingly pubescent twigs; leaves oval or obovate, abruptly acuminate at the apex, obtuse or acutish at the oblique base, sharply, usually doubly serrate, slightly rough above, pubescent or glabrate beneath, 5–12 cm. long; samara ovate-oval, reticulate-veined, 10–12 mm. long, glabrous except the ciliate margins. AMERICAN OR WHITE ELM. Moist soil: Newf.—Fla.—Tex.—se Mont.—Sask. *E. Temp.—Plain—Submont.* Mr–Ap.

2. **U. alàta** Michx. A tree up to 20 m. high; twigs glabrous; leaf-blades narrowly elliptic to ovate-lanceolate, acute or short-acuminate, doubly serrate, obtuse or subcordate at the base, downy beneath; pedicels 4–8 mm. long; samara 5–6 mm. long. WAHOO or WINGED ELM. Along streams: Va.—Fla.—Tex.—Kans. *Allegh.—Carol.—Ozark.* Mr.

3. **U. Thómasi** Sargent. A tree up to 30 m. high; buds and twigs pubescent; leaf-blades oval or ovate, doubly serrate, short-acuminate; pedicels slender; samara oval, 12–15 mm. long, pubescent. CORK ELM, ROCK ELM. *U. racemosa* Thomas [G, B]; not Borckh. Rich soil: Que.—Ken.—Mo.—n Kans.—Neb.—S.D. *Allegh.* Ap.

4. **U. fúlva** Michx. A tree 15–25 m. high; bark rough, fragrant; inner bark mucilaginous; buds and twigs pubescent; leaf-blades ovate-oblong, soft-downy beneath; samara 15–18 mm. broad, pubescent. SLIPPERY ELM, RED ELM. Rich woods: Que.—Fla.—Tex.—N.D. *E. Temp.* Mr–Ap.

5. **U. pùmila** L. Small tree with slender pubescent branches; leaves oval-elliptic, short-petioled, firm, glabrous above, pubescent when young beneath; stamens with violet anthers; fruit obovate, deeply cleft at the apex. Roadsides and pastures: Kansas; escaped from cultivation; nat. of Asia.

2. CÉLTIS (Tourn.) L. HACKBERRY.

Shrubs or trees, with thin smooth or corky-ridged bark. Leaves oblique, serrate or entire, 2-ranked, membranous or leathery. Flowers polygamo-monoecious or monoecious, axillary, the staminate solitary or clustered, the pistillate usually solitary. Calyx 4- or 5-lobed, deciduous. Stamens 4 or 5; filaments incurved; anthers extrorse. Ovary sessile, 1-celled, with 2 recurved stigmas. Drupe globose or ellipsoid, with scant pulp and bony stone.

Leaf-blades neither coriaceous, nor pale beneath, nor strongly rugose.
  Leaf-blades lanceolate, entire or nearly so. 1. *C. laevigata.*
  Leaf-blades ovate, distinctly toothed.
    Leaves smooth above or nearly so.
      Leaf-blades acute or abruptly short-acuminate; drupe subglobose. 2. *C. occidentalis.*
      Leaf-blades long-acuminate; drupe ellipsoid. 3. *C. canina.*
    Leaves very rough above; drupe subglobose. 4. *C. crassifolia.*
Leaf-blades thick, subcoriaceous, strongly reticulate and rugose, often yellowish-green, paler beneath.
  Pedicels 15–20 mm. long, more than twice as long as the drupe; leaves more or less toothed. 5. *C. rugulosa.*
  Pedicels 4–10 mm. long, less than twice as long as the drupe; leaves subentire. 6. *C. reticulata.*

1. **C. laevigàta** Willd. A large tree, up to 30 m. high; bark of the trunk light gray, corky-ridged; leaf-blades lanceolate or oblong-lanceolate, 6–12 cm. long, long-acuminate, dark green and smooth above; pedicels slender, 1–2 cm.

long; drupe subglobose, 5–7 mm. thick, purple-black or orange. *C. mississippiensis* Bosc. [G, B]. River banks: Ind.—se Kans.—Tex.—Fla. *SE Temp.* Ap–My.

2. **C. occidentàlis** L. A small tree, sometimes 20 m. high; bark of the stem gray, corky-ridged; leaf-blades ovate to ovate-lanceolate, pubescent on the veins beneath, thin, very oblique at the base, usually sharply serrate, short-acuminate, 3–10 cm. long; pedicels in fruit 1–2 cm. long, nearly glabrous; fruit globose, 7–10 mm. in diameter, purple or blackish. Rocky places: Que.—N.C.—Okla.—S.D.—Man. *Canad.—Allegh.—Ozark.—Plain.* Ap–My.

3. **C. canìna** Raf. A small tree, up to 30 m. high; bark gray, corky-ridged; leaf-blades ovate or rarely lance-ovate, light green, 8–15 cm. long, coarsely toothed, rather gradually acuminate, smooth above or slightly pubescent when young, hairy beneath; pedicels in fruit longer than the petioles; fruit ellipsoid, about half longer than thick, purple, 1 cm. long. Hillsides: N.Y.—S.D.—Kans.—Pa. *Allegh.* Ap–My.

4. **C. crassifòlia** Lam. A tree occasionally 40 m. high; bark corky-roughened and warty; leaf-blades ovate or ovate-lanceolate, dark green, short-acuminate, usually coarsely toothed, 3–10 cm. long, hirsute beneath, especially on the veins, pedicels in fruit 1–2 cm. long; fruit globose or nearly so, 8–10 mm. in diameter, black. *C. occidentalis crassifolia* A. Gray [G]. Woods and hillsides: Mass.—S.C.—Colo.—N.D. *Canad.—Allegh.—Plain—Submont.* Ap–Je.

5. **C. rugulòsa** Rydb. *Fig. 167.* A tree 5–10 m. high, with round crown; twigs somewhat pubescent when young; leaf-blades broadly ovate, oblique, 4–7 cm. long, somewhat cordate at the base, short-acuminate, sharply serrate except at the base and apex, dark green, shining, and slightly scabrous above, brownish or yellowish green, dull beneath; fruit globose, about 8 mm. in diameter, brownish; style short but evident. *C. rugosa* Rydb., not Willd. *C. reticulata vestita* Sarg. Valleys in the foothills: Colo.—w Kans.—Okla. *Plain—Submont.* My.

6. **C. reticulàta** Torr. A shrub 1–5 m. high; bark gray, corky-ridged; branchlets densely pubescent, brownish-gray; leaf-blades broadly ovate, acute, cordate at the base, 2–4 cm. long, entire or with a few broad teeth, shining above, pale brown and densely hirsutulous beneath; fruit globose, 6–10 mm. in diameter, red. Dry rocky places: Tex.—Kans.—Colo.—N.M. *Son.—S. Plains.* Ap.

FAMILY 41. **MORACEAE.** MULBERRY FAMILY.

Shrubs or trees, with milky sap. Leaves mostly alternate, with deciduous stipules. Flowers monoecious or dioecious, in short spikes or heads, or (as in *Ficus*) inclosed in a hollow fleshy receptacle. Perianth of 4 or 5, partly united sepals. Stamens 3–5, inserted at the base of the calyx. Styles or stigmas 1 or 2; ovary 1- or 2-celled. Fruit aggregate; achenes enclosed in the fleshy calyx or receptacle.

Staminate and pistillate flowers both in spikes; leaves 3-ribbed. 1. MORUS.
Staminate flowers spicate, the pistillate ones in dense heads; leaves 1-ribbed, pinnately veined. 2. MACLURA.

1. MÒRUS (Tourn.) L. MULBERRY.

Monoecious or dioecious trees or shrubs, with thin scaly bark. Leaves alternate, 3-ribbed, toothed or 3-lobed, deciduous. Flowers spicate, staminate spikes elongate, drooping; sepals 4; stamens 4, opposite the sepals; anthers introrse. Pistillate aments short and dense; sepals 4, the outer ones larger, accrescent, enclosing the fruit. Ovary 1-celled. Drupes coalescent, forming a syncarp.

Leaves glabrous beneath or sparingly pubescent on the veins.
  Fruit white or pinkish. .... 1. *M. alba.*
  Fruit black. .... 2. *M. nigra.*
Leaves pubescent beneath; fruit red or purplish. .... 3. *M. rubra.*

1. **M. álba** L. A tree 5–15 m. high; leaf-blades ovate or ovate-cordate, 5–15 cm. long, acute or acuminate, simply or doubly serrate, sometimes lobed, especially on the young shoots; staminate spikes 1–2 cm. long; pistillate spikes 5–10 mm. long; fruit 1–2 cm. long. WHITE MULBERRY. Fields and waste places: Me.—Ga.—Tex.—Neb.—Minn.; nat. from Eurasia. My.

2. **M. nìgra** L. A shrub or small tree, 3–15 m. high; twigs puberulent; leaf-blades thin, ovate, truncate, rounded, or cordate at the base, 4–15 cm. long, serrate, often 2- or 3-lobed; puberulent when young, glabrate in age; staminate spike cylindric, 1–2 cm. long, longer than the peduncles; pistillate spikes oval, 5–8 mm. long; fruit oblong, 1–2 cm. long, black when ripe. BLACK MULBERRY. Roadsides and waste places: N.Y.—Kans.—Tex.—Fla.; escaped from cultivation; native of Eurasia.

f.168.

3. **M. rùbra** L. *Fig. 168.* A tree, up to 20 m. high; with scaly bark; leaf-blades ovate or oval-ovate, 5–20 cm. long, abruptly acuminate, serrate, rounded or cordate at the base, glabrous above; staminate spikes 4–8 cm. long; pistillate spikes about 1 cm. long; fruit 3–6 cm. long. RED MULBERRY. Woods and fields: Vt.—S.D.—Tex.—Fla. *E. Temp.* Ap–My.

2. MACLÙRA Nutt. OSAGE ORANGE.

Dioecious trees or shrubs, with alternate leaves and axillary thorns. Staminate flowers green, in short spikes; sepals 4; stamens 4, opposite the sepals; anthers introrse. Pistillate flowers in dense heads, on short stout shoots of the season; sepals 4, the outer broader, accrescent. Achenes oblong, flattened, collected in a large globose syncarp. *Toxylon* Raf.

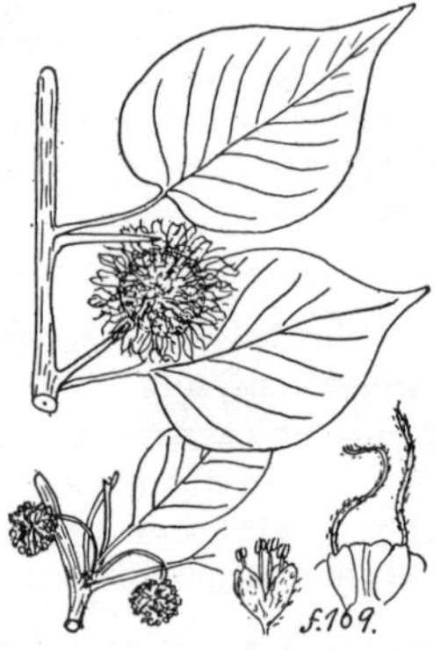
f.169.

1. **M. pomífera** (Raf.) C. K. Schneid. *Fig. 169.* A tree or large shrub, up to 20 m. high; branches zigzag; leaf-blades ovate to oblong-lanceolate, acuminate, entire, 5–30 cm. long, dark green and glossy above, paler and dull beneath; staminate spikes 1–2 cm. long; fruiting heads 5–15 cm. in diameter, globose, golden yellow. *T. pomiferum* Raf. [B]. Thickets: Va.—se Kans.—Tex.—Ga.; escaped from cultivation in Neb. and Iowa. *Carol.—Ozark.* My–Je.

FAMILY 42. **CANNABINACEAE.** HEMP FAMILY.

Herbs or vines, with mostly opposite leaves and persistent stipules. Staminate flowers in panicled racemes; sepals and stamens 5. Pistillate flowers in bracted spikes, with a cup-like calyx; pistil of 2 united carpels, but ovary 1-celled. Fruit an achene; seed solitary, pendulous.

Erect herbs; leaves digitately divided to near the base; pistillate flowers in axillary stiff spikes. 1. CANNABIS.
Twining vines; leaves merely digitately lobed; pistillate flowers in drooping ament-like spikes with imbricate bracts. 2. HUMULUS.

1. **CÁNNABIS** L. HEMP.

Erect annual dioecious herbs. Leaves alternate or opposite, digitately 5–11-divided into serrate divisions. Staminate flowers in paniculate racemes; sepals 5, imbricate; stamens 5. Pistillate flowers in leafy-bracted spikes; perianth undivided; pistil solitary. Fruit a slightly flattened achene.

1. **C. satìva** L. *Fig. 170.* Stem branched, 1–4 m. high, rough-pubescent; leaf-blades divided into 5–11 linear, serrate, acuminate divisions, 5–15 cm. long. Waste places: N.B.—Ga.—Tenn.—Colo.—S.D.; escaped from cultivation; native of Eurasia.

f. 170.

2. **HÙMULUS** L. HOPS.

Perennial twining herbaceous vines. Leaves opposite, 3–7-lobed, serrate. Stipule persistent, free. Staminate flowers in panicled racemes; sepals 5, imbricate; stamens 5; filaments short, erect. Pistillate flowers in ament-like, drooping spikes, 2 together, subtended by a bract; ovary 1-celled. Achenes a little flattened. Embryo spirally coiled.

Pistillate catkins enlarging in fruit; perennial. 1. *H. Lupulus.*
Pistillate catkins not enlarging in fruit; annual. 2. *H. japonicus.*

1. **H. Lùpulus** L. A vine 5–10 m. high; leaf-blades cordate in outline, dark green, scabrous above, glabrous beneath except the pubescent veins; lobes coarsely toothed, with ovate teeth; bracts of the pistillate flowers broadly ovate, from obtuse to short-acuminate. COMMON HOPS. Rocky banks and copses: N.S.—Ga.—Kans.—Wyo.—Mont.—Man.; Eurasia; extensively cultivated. *Canad.—Allegh.—Plain—Submont.*

2. **H. japónicus** Sieb. & Zucc. Vine 3–7 m. high; leaves similar to those of the preceding but more deeply 5–7-cleft; catkins not glandular. Escaped from cultivation: Neb.—Kans.; nat. of eastern Asia.

FAMILY 43. **URTICACEAE.** NETTLE FAMILY.

Monoecious, dioecious, or polygamous herbs (ours), often armed with stinging hairs. Leaves simple, alternate or opposite, with stipules. Flowers greenish, in axillary cymes. Calyx of 2–5 distinct or partly united sepals. Stamens 2–5, in the pistillary flowers reduced to staminodia or wanting. Pistil solitary; ovary 1-celled. Fruit an achene. Endosperm scant, oily, or wanting; embryo straight.

Plants armed with stinging hairs; sepals of the pistillate flowers distinct or nearly so.
Leaves opposite; sepals and stamens 4 in the staminate flowers; styles terminal; stigmas capitate. 1. URTICA.
Leaves alternate; sepals and stamens 5 in the staminate flowers; styles lateral; stigmas subulate. 2. LAPORTEA.

Plants without stinging hairs.
Flower-clusters not involucrate; leaves mostly opposite.
Flower-clusters in axillary cymes; sepals of the pistillate flowers nearly distinct, not enclosing the achenes. 3. PILEA.
Flower-clusters spicate, on simple branches; sepals wholly or nearly wholly united, enclosing the achenes. 4. BOEHMERIA.
Flower-clusters involucrate; leaves alternate. 5. PARIETARIA.

1. ÚRTICA (Tourn.) L. NETTLE.

Annual or perennial herbs, with stinging hairs. Leaves opposite, with membranous, toothed, 5–7-veined blades and free stipules. Plants dioecious or monoecious; flowers in axillary, paniculate cymes; sepals 4, nearly distinct, in the pistillate flowers two of them larger. Staminate flowers with 4 stamens and a rudimentary ovary, the pistillate ones with an equilateral ovary and tufted stigmas. Achenes flattened. Seeds with membranous coats, often adherent to the pericarp.

Flower-clusters simple; plant annual. 1. *U. urens.*
Flower-clusters compound; plant perennial.
Teeth of the leaves ovate, strongly directed forward.
Leaf-blades lanceolate, scarcely cordate at the base.
Stem strigose as well as bristly; leaves thicker, decidedly hairy, especially on the strong veins. 2. *U. procera.*
Stem glabrous or with a few scattered bristles, not strigose; leaves thin, not strongly veined. 3. *U. viridis.*
Leaf-blades broad, deeply cordate at the base. 4. *U. dioica.*
Teeth of the leaves broadly triangular, spreading; stem and leaves glabrous or nearly so. 5. *U. gracilenta.*

1. **U. ùrens** L. An annual herb; stem 1–4 dm. high, 4-angled; leaf-blades thin, oval or ovate, 1–4 cm. long, acute, glabrous or sparingly bristly, incised; flower-clusters from shorter than the petioles to slightly exceeding them; sepals oval to broadly oblong; achenes ovoid, 2 mm. long. Waste places: Newf.—Man.—Fla.; nat. from Eurasia. My–S.

2. **U. pròcera** Muhl. *Fig. 171.* Stem slender, 0.5–3 mm. high, sharp-angled; leaf-blades finely serrate, acuminate, 7–15 cm. long, about as wide as the length of the petioles; flower-clusters slender, but shorter than the leaves; sepals ovate, about equaling the achenes. *U. gracilis* Am. auth. [G, B, R]; not Ait. Alluvial soil and waste places: N.S.—N.C.—N.M.—Alaska. *E. Temp.—Plain—Mont.* Ap–S.

3. **U. víridis** Rydb. Stem 1–1.5 mm. high, slender; blades coarsely toothed, 4–10 cm. long, thin, light green; panicles many-flowered, often equaling the upper leaves; sepals oval or ovate, usually half longer than the achene. River bottoms: Ida.—Wyo.—S.D.—Kans.—Alta. *Submont.* Je–Au.

4. **U. dioìca** L. Stem 0.5–1.5 m. high, strongly bristly and somewhat hispidulous above, obtusely-angled; leaf-blades thin, very bristly, coarsely toothed, acute or short-acuminate, 3–10 cm. long, usually twice as broad as the length of the petioles; flower-clusters about half as long as the leaves. Waste places: N.S.—S.C.—Colo.—Man.; nat. from Eu. *E. Temp.—Plain—Submont.* Jl–S.

5. **U. gracilénta** Greene. Stem slender, 1–2 m. high, strigose or hirsutulous, as well as bristly; petioles slender, 2–8 cm. long; leaf-blades lanceolate, long-acuminate, rounded at the base, 5–15 cm. long, more or less pubescent, with salient teeth; flower-clusters slender, slightly exceeding the petioles. Along streams and in cañons: S.D.—Wyo.—Ariz.—Tex.; Mex. *Son.—Submont.* Au–S.

2. **LAPÓRTEA** Gaud. WOOD NETTLE.

Perennial herbs (in the tropics shrubs or trees), armed with stinging hairs. Leaves alternate, toothed, with distinct stipules. Flowers dioecious or monoecious, in axillary cymes. Staminate flowers with 5 sepals, 5 stamens and a rudimentary pistil. Pistillate flowers with 4 or 2 sepals, an oblique ovary, a lateral style, and a subulate stigma. Achenes with membranous or fleshy pericarp. *Urticastrum* Fabr.

1. **L. canadénsis** (L.) Gaud. Stem 2–12 dm. high; leaf-blades thin, ovate to elliptic, 5–20 cm. long, acuminate, coarsely serrate, bristly on the veins; panicles slender, the lower staminate, the upper pistillate, divergent. *U. divaricatum* (L.) Kuntze [B]. Rich woods: N.S.—N.D.—Kans.—Fla. *E. Temp.* Jl–Au.

3. **PÍLEA** Lindl. RICHWEED. CLEARWEED.

Annual or perennial herbs, with pellucid tissue, without stinging hairs; leaves opposite, 3-ribbed or nearly veinless. Flowers monoecious or dioecious, in axillary panicles or clustered cymes. Staminate flowers with 3 or 4 sepals, 3 or 4 stamens, and a rudimentary ovary. Pistillate flowers with 3 or 4 sepals, 2–4 staminodea, and an equilateral ovary. Achenes flattened, with membranous pericarp. *Adicea* Raf.

Achenes green or in age light brown; plant 2–6 dm. high; veins of the leaves conspicuous beneath. 1. *P. pumila.*
Achenes black or blackish; plant 0.5–3 dm. high; veins of the leaves not conspicuous.
 Plant 1–3 dm. high; leaves dark green, with 5–9 pairs of teeth; seeds 2 mm. long. 2. *P. opaca.*
 Plant 0.5–1.5 dm. high; leaves light green, with 3–5 pairs of teeth; seeds 1.5 mm. long. 3. *P. fontana.*

1. **P. pùmila** (L.) A. Gray. An annual herb; stem 1–6 dm. high, obtusely angled, pellucid; leaf-blades ovate to elliptic, 3–10 cm. long, short-acuminate, crenate-serrate, broadly cuneate at the base; panicles spreading, shorter than the petioles. *Adicea pumila* (L.) Raf. [B]. Damp shaded places: N.B.—Minn.—Kans.—La.—Fla. *E. Temp.* Jl–S.

2. **P. opàca** (Lunell) Rydb. Annual, simple or with 1–3 pairs of branches, 1–3 dm. high; leaf-blades appressed-pubescent, 2–6 cm. long, 1–4 cm. wide; blades ovate, cuneate at the base, 3-ribbed, not exceeding the petioles in length, the terminal tooth decidedly longer than the rest; flower-clusters short, congested; achenes roughish. *Adicea opaca* Lunell. Damp shaded soil: N.D.—Wis.—Neb. *Prairie.* Je–S.

3. **P. fontàna** (Lunell) Rydb. Annual, simple, 5–15 dm. high, leafy at the top; leaf-blades thin, appressed-pubescent, the lowest orbicular, 2–5 mm. long, the rest ovate, 1–2 cm. long, 3-ribbed, longer than the petioles; flower-clusters small, subsessile; achenes smooth. *A. fontana* Lunell. Shaded margins of brooks: N.D.—Neb. *Prairie.* Jl–Au.

4. **BOEHMÈRIA** Jacq. FALSE NETTLE.

Herbs, shrubs or trees, pubescent, but the hairs not stinging. Leaves opposite or alternate, the blades 3-ribbed, toothed or rarely lobed, the stipules free. Flowers monoecious or dioecious, in dense spikes. Staminate flowers with 4 (rarely 3 or 5) stamens and a rudimentary ovary. Pistillate flowers with a perianth of 2–4 united sepals and a single ovary with elongate stigma. Achenes flattened, included in the perianth.

Petioles much shorter than the firm, finely serrate blades. 1. *B. Drummondiana.*
Petioles equaling the thin, coarsely serrate blades or only slightly shorter. 2. *B. cylindrica.*

1. **B. Drummondiàna** Wedd. *Fig. 172.* A perennial herb; stem 2–12 dm. high, very leafy; leaf-blades firm, leathery, ovate, 2–5 cm. long, acute or short-acuminate; spike continuous, dense; sepals ovate-lanceolate, strongly hooded; achenes nearly 1.5 mm. broad. *B. cylindrica scabra* Porter [G, B]. *B. scabra* Small. Swamps: N.Y.—Kans.—Tex.—Fla. *E. Temp.* Jl–S.

2. **B. cylíndrica** (L.) Sw. A perennial herb; stem 2–10 dm. high; leaf-blades thin, ovate to lanceolate, 3–12 cm. long, acuminate, coarsely toothed; flower-clusters dense, forming interrupted spikes, the lower portion pistillate and continuous, the upper staminate and interrupted; achenes 1 mm. broad. Woods and low ground: Que.—Minn.—Kans.—Fla. *E. Temp.* Jl–S.

5. PARIETÀRIA (Tourn.) L. PELLITORY.

Annual or perennial herbs, with diffusely branched, often pellucid stems, polygamous. Leaves alternate, with 3-veined blades. Involucres of 2–6 more or less united bracts. Flowers in axillary cymes. Perianth of 4, rarely 3, more or less united sepals. Stamens 4, rarely 3, in the perfect and the staminate flowers, in the pistillate ones wanting. Pistils solitary, in the staminate flowers rudimentary; stigmas tufted. Achenes included, with a crustaceous pericarp.

1. **P. pennsylvánica** Muhl. *Fig. 173.* Annual, slender; stem weak, ascending, 1–4 dm. high, simple or branched; leaf-blades thin and flimsy, obtuse or acuminate at the apex, acute or acuminate at the base; bracts of the involucre linear, 4–5 mm. long. Shaded banks or rocks: Ont.—Fla.—Mex.—B.C. *Temp.—Plain—Submont.* My–Au.

FAMILY 44. POLYGONACEAE. BUCKWHEAT FAMILY.

Herbs or shrubs, or in the tropics trees or vines, with alternate leaves. Flowers perfect or rarely unisexual. Calyx inferior, of 2–6 more or less united sepals, often corolloid. Corolla wanting. Stamens 2–9. Pistil of 2 or 3 united carpels, but ovary 1-celled, in fruit becoming a 1-seeded, triangular or lenticular achene.

Leaves without stipules; flowers or flower-clusters subtended by involucres of partly united bracts; stamens mostly 9. — 1. ERIOGONUM.
Leaves with sheathing stipules (ocreae); flower-clusters not involucrate; stamens 4–9.
- Stigmas tufted; perianth-lobes 6.
  - Achenes winged. — 2. RHEUM.
  - Achenes not winged. — 3. RUMEX.
- Stigmas not tufted; perianth-lobes 5, rarely 4.
  - Leaf-blades jointed at the base; inner filaments dilated.
    - Branches more or less adnate to the internodes of the stem; ocreae truncate or oblique; flowers solitary. — 4. DELOPYRUM.
    - Branches not adnate to the stem; ocreae 2-lobed, becoming lacerate; flowers clustered. — 5. POLYGONUM.
  - Leaf-blades not jointed at the base; filaments slender.
    - Ocreae cylindric, truncate.
      - Perianth curved, its lobes 4; stamens 4. — 6. TOVARA.

Perianth not curved, its lobes 5, if less, the stamens more than 5. 7. PERSICARIA.
Ocreae oblique, more or less open on the side facing the leaf.
Herbs, not climbing or twining; sepals neither winged nor keeled.
Flowers in simple spike-like racemes; plants with thickened tuberous rootstocks. 8. BISTORTA.
Flowers in several racemes or panicles; rootstocks not tuberous-thickened.
Racemes in terminal corymbs; plants unarmed; embryo in the center of the endosperm. 9. FAGOPYRUM.
Racemes not in terminal corymbs; plant prickly; embryo at one side of the endosperm. 10. TRACAULON.
Herbaceous vines, with twining stems; sepals winged or keeled. 11. BILDERDYKIA.

1. **ERIÓGONUM** Michx. UMBRELLA PLANT.

Annual or perennial herbs or shrubby plants, with basal or cauline, alternate, opposite, or whorled leaves and no stipules. Blades entire. Flowers perfect or polygamo-monoecious, in involucrate clusters variously disposed. Involucres turbinate or campanulate, rarely nearly cylindric, 4–8-lobed. Perianth more or less colored, jointed to a short pedicel. Segments 6, in two series. Stamens 9; filaments filiform, often hairy at the base. Ovary 1-celled, 3-angled or 3-winged; styles 3.

Achenes 3-winged; perianth not accrescent; perennials with a thick taproot and short crown. 1. *E. alatum.*
Achenes merely 3-angled; perianth accrescent in fruit.
Perianth with a stipe-like base.
Bracts verticillate, leaf-like; perennials with a branched woody caudex; flowering branches scapiform.
Involucres in branching cymes. 2. *E. Jamesii.*
Involucres in simple or compound umbel-like or head-like clusters. 3. *E. flavum.*
Bracts not leaf-like; stem leafy, the leaves alternate. 4. *E. longifolium.*
Perianth without stipe-like base.
Ovaries and fruit pubescent; involucres few, capitate or subcymose; perennials with scapiform stems. 5. *E. lachnogynum.*
Ovaries and fruit glabrous or nearly so.
Involucres in head-like or umbellate clusters.
Perianth yellow. 6. *E. chrysocephalum.*
Perianth white or rose-colored.
Perianth glabrous; perennials with a cespitose woody caudex.
Lobes of the involucres lanceolate, not scarious-margined. 7. *E. depauperatum.*
Lobes of the involucres oval to orbicular, scarious-margined. 8. *E. pauciflorum.*
Perianth pubescent; suffruticose leafy-stemmed perennial, with decumbent branches, villous. 9. *E. multiceps.*
Involucres in open cymes; bracts scale-like.
Involucres (except those of the forks of the inflorescence) sessile, the uppermost conglomerate; cymes repeatedly dichotomous or trichotomous.
Perennials, shrubby at least at the base.
Perianth yellow. 11. *E. campanulatum.*
Perianth white or rose-colored.
Suffruticose plants, leafy only at the base; inflorescence longer than the stem. 10. *E. helichrysoides.*
Shrubby plants, with the leafy stem usually longer than the inflorescence.
Leaf-blades relatively broad, oblong to rounded-oval or obovate, obtuse. 12. *E. corymbosum.*
Leaf-blades narrow, spatulate to linear, mostly acute at the apex.
Inflorescence many times compound, copiously branched; internodes long. 13. *E. effusum.*

Inflorescence less compound; branches and internodes short, mostly spreading.
Involucres in the forks peduncled; peduncles slightly floccose. 14. *E. microthecum.*
Involucres all sessile; peduncles densely white-tomentose. 15. *E. nebraskense.*
Annuals, with a strict herbaceous stem. 16. *E. annuum.*
Involucres all peduncled, never conglomerate; scapose annuals; leaves basal, petioled.
Peduncles reflexed or at least divaricate; leaves tomentose on both sides, less so above. 17. *E. cernuum.*
Peduncles erect or ascending; leaves green, glabrate or sparingly pilose. 18. *E. Gordonii.*

1. **E. alàtum** Torr. Stems erect, 3–10 dm. high; leaves mostly basal, tufted, spatulate or oblanceolate, 3–10 cm. long, hirsute above, glabrous beneath, except the strong midrib; panicle open; involucres in small cymes; perianth campanulate, greenish yellow, 2 mm. high, glabrous; achenes 5–7 mm. long, glabrous. Plains and table-lands: Neb.—Tex.—Ariz.—Wyo. *Son.—Plain—Mont.* Je–Au.

2. **E. Jamèsii** Benth. Stems decumbent at the base, 1–3 dm. high, tomentose; leaves mostly basal, petioled, 3–8 cm. long; blades elliptic-spatulate, thick, green, densely white-tomentose beneath; involucres deeply campanulate, 5 mm. high, tomentose; perianth 4–5 mm., becoming 7–8 mm. long; inner lobes slightly longer than the outer; achenes 4 mm. long, pubescent at the base. Plains and hills: Tex.—Kans.—Colo.—Ariz. *Plain—Mont.* Jl–S.

3. **E. flàvum** Nutt. Stems 1–2 dm. high, scapiform, white-tomentose; leaves thick, petioled, 3–5 cm. long; blades oblanceolate, densely tomentose on both sides, greenish in age above, snowy-white beneath; perianth yellow; filaments villous at the base; achenes 4 mm. long, villous. *E. sericeum* Pursh. *E. crassifolium* Benth. Dry hills, mountains and cañons: Man.—Neb.—Colo.—Alta. *Plain—Subalp.* Je–Au.

4. **E. longifòlium** Nutt. Perennial, with a thick root; stem 6–12 dm. high, paniculately branched, sulcate; leaves mostly basal, the blades narrowly oblong, 5–20 cm. long, silky-strigose above, tomentose beneath, the upper sessile, the lower wing-petioled; peduncles less than 1 cm. long; involucre turbinate-campanulate, 5–6 mm. high; perianth 4–5 mm. long, densely silvery-silky; achenes 3-angled. Sandy soil: Mo.—Kans.—Tex. Je–N.

5. **E. lachnógynum** Torr. Flowering stems scapiform, 1–3 dm. high; leaves crowded, basal, oblanceolate or oblong-oblanceolate, petioled, acute, silvery-silky above, grayish tomentose beneath, with revolute margins; bracts lanceolate or subulate, rarely verticillate; involucres silky-tomentose, 3–4 mm. high, campanulate; lobes oblong; perianth campanulate, tomentose, 3 mm. high, yellow; fruit villous. Dry plains and cañons: Kans.—Colo.—Ariz.—Tex. *Son.—Plain.* Je–Au.

6. **E. chrysocéphalum** A. Gray. Scape 4–15 cm. high; leaves petioled, 3–10 cm. long; involucre turbinate, ribbed, 3 mm. long; perianth 2.5–3 mm. long, campanulate; lobes obovate-oblong, the outer usually slightly longer. *E. latifolium* (T. & G.) A. Nels. Dry hills: Neb.—Colo.—Utah—Ida. *Plain—Mont.* Jl–Au.

7. **E. depauperàtum** Small. Scapes erect, 5–10 cm. high; leaves crowded, petioled; blades thinnish, linear-spatulate, 2–6 cm. long, revolute-margined, glabrous above, tomentose beneath; bracts scale-like, lanceolate-subulate; involucres 3.5–4.5 mm. high, densely tomentose; lobes about half as long as the angled tube; perianth glabrous; lobes unequal, the outer oblong or obovate or cuneate, the inner cuneate and narrower. Dry hills: S.D. *Submont.* Je–Au.

8. **E. paucifloòrum** Pursh. Scape 1 dm. high or less; leaves crowded, short-petioled, 3–8 cm. long; blades linear or linear-oblanceolate, revolute-margined, white-tomentose beneath; bracts lance-subulate, sometimes membranous-margined; involucres 4–5 mm. high; lobes petaloid, elliptic; perianth-lobes elliptic, nearly equal. Sandy soil and dry places: Wyo.—Colo.—S.D. *Submont.—Mont.* Jl–S.

9. **E. múlticeps** Nees. Scape 5–15 cm. long, white-tomentose; leaves slender-petioled, 3–8 cm. long, densely white-tomentose on both sides, with appressed tomentum; blades from linear-oblanceolate to spatulate; inflorescence capitate or rarely umbellate; bracts subulate, or sometimes one or more of them elongate and foliaceous; involucres tubular-campanulate, tomentose, 3–4 mm. long; lobes triangular; perianth-lobes oblong or cuneate-oblong, rounded. *E. gnaphaloides* Benth. Dry plains and bad-lands: Sask.—N.D.—Neb.—Colo.—Ida. *Plain—Submont.* Jl–Au.

10. **E. helichrysoìdes** (Gand.) Rydb. Stem leafy below, 1–2 dm. high; white-floccose; leaves subsessile, linear, 4–6 cm. long, white-tomentose beneath; floccose or glabrate above, revolute on the margins; involucres campanulate, 3–4 cm. high, slightly floccose; perianth-lobes obovate, white. *E. microthecum helichrysoides* Gand. Bad-lands: Kansas. Plan. Jl–Au.

11. **E. campanulàtum** Nutt. Branches short; peduncle 1–3 dm. long; leaves basal, petioled; blades tomentose on both sides, but less densely so above; inflorescence with ascending branches; involucres campanulate, 2–2.5 mm. long, glabrous or nearly so; perianth about 2 mm. long. *E. sabulosum* M. E. Jones. Dry hills and plains: Neb.—Colo.—Utah. *Plain—Submont.* Je–Au.

E. campanulatum × multiceps. Like *E. multiceps* in habit, pubescence, and leaf-form, but the panicle is tristichous-cymose as in *E. campanulatum* and the flowers yellow. Neb.

12. **E. corymbòsum** Benth. Shrub, 3–10 dm. high, with tomentose branches; leaf-blades oval, 2–3 cm. long, 1–1.5 cm. wide, densely white-tomentose beneath, loosely floccose above; inflorescence 0.5–2 dm. high, with ascending branches; involucres sessile, tomentose, 2–2.5 mm. long, campanulate; perianth about 3 mm. long, campanulate; lobes elliptic, nearly equal. Dry plains: Kans.—Colo.—Utah—N.M. *Son.* Je–S.

13. **E. effùsum** Nutt. Shrub, 2–5 dm. high; leaves short-petioled; blades linear-oblanceolate, 2–4 cm. long, 4–8 mm. wide, white-tomentose beneath, floccose above, 3–10 cm. long, tomentulose; involucres short-peduncled, or the upper sessile, 2 mm. long, tomentulose; lobes ovate; perianth campanulate, 2 mm. long; outer lobes obovate, the inner elliptic. Dry plains and hills: Neb.—N.M.—Utah—Mont. *Son.—Plain—Submont.* Jl–S.

14. **E. microthècum** Nutt. *Fig. 174.* Low shrub, 2–6 dm. high; leaves scattered, short-petioled; blades linear, oblong, or oblanceolate, densely white-tomentose beneath, floccose above, 1–2 cm. long, 3–5 mm. wide; inflorescence 2–7 cm. long, floccose, small; branches short, spreading or ascending-spreading; involucres, at least those of the lower forks, peduncled, tomentulose, turbinate, 2.5 mm. long; lobes rounded, scarious-margined; perianth turbinate, 2.5–3 mm. long; lobes obovate, rounded, truncate or even emarginate. *E. microthecum* subsp. *myrianthum*, *intricatum*, and *sarothriforme* Gand. Plains and table-lands: Mont.—Neb.—Colo.—Ariz.—Calif.—Wash. *Son.—Plain—Mont.* Je–S.

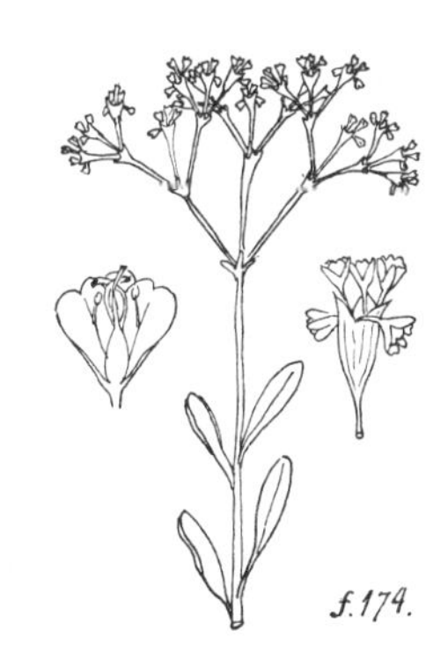
f. 174.

15. **E. nebraskénse** Rydb. A low shrub, 2–3 dm. high; branches and inflorescence tomentose;

leaves short-petioled, oblanceolate, 2–3 cm. long, densely white-tomentose on both sides; inflorescence less than 1 dm. high, trichotomously or verticillately branched; branches short, more or less spreading; involucres tomentose, 2–3 cm. high, turbinate; lobes triangular; perianth rose-colored, glabrous or slightly pubescent, 2.5 mm. long; segments oblong to obovate; fruit glabrous. Plains: w Neb. Jl.

16. **E. ánnuum** Nutt. Stem simple, leafy, 3–10 dm. high, floccose throughout; leaves petioled; blades oblong or oblanceolate, obtuse, slightly revolute, densely white-tomentose beneath, floccose above, 3–5 cm. long; inflorescence cymose, flat-topped, somewhat irregularly branched; involucres turbinate, 2–3 mm. long, white-tomentose; teeth short, obtuse; perianth white or whitish, 1.5–2 mm. long, campanulate; lobes obovate. *E. Hitchcockii* Gand. Sandy places: N.D.—Tex.—Mex.—Mont. *Submont.—Plain.* Je–O.

17. **E. cérnuum** Nutt. Scape glabrous, 2–4 dm. high, leaf-blades orbicular or oval, 1–2 cm. broad; peduncles 0.5–2 cm. long; involucres glabrous, campanulate, 1.5–2 mm. long; lobes ovate, obtuse; perianth white or pinkish, about 2 mm. long; inner lobes elliptic-cuneate, narrower, but almost as long as the outer. Bad-lands, hills and cañons: Sask.—Neb.—N.M.—Ariz.—Ida. —Alta. *Plain—Mont.* Je–S.

18. **E. Gordònii** Benth. Scape 2–4 dm. high, slender, glabrous; leaf-blades coriaceous, glabrous, orbicular, 1–3 cm. broad; peduncles 1–3 cm. long, erect; involucres turbinate-campanulate, about 1 mm. long, glabrous; lobes rounded; perianth white or pinkish, 2–2.5 mm. long, glabrous; outer lobes oblong-ovate, the inner oblong, slightly if at all shorter. *E. Visheri* A. Nels. Dry plains and bad-lands: S.D.—Colo.—Utah—Wyo. *Plain—Submont.* Je–S.

2. **RHÈUM** L. RHUBARB, PIE-PLANT.

Perennial herbs, with thick roots and large, mostly basal leaves. Stipules sheathing. Flowers perfect, greenish or white in large panicles. Perianth 6-parted. Stamens 9 or 6. Ovary 3-angled; styles 3. Achenes 3-winged.

1. **R. Rhapónticum** L. Strong perennial, often 1 m. high; petioles semi-cylindric, flat above, succulent; leaf-blades suborbicular, 5-ribbed, glabrous, flowers numerous, white. Cultivated garden plant, occasionally escaped; nat. of southern Siberia and Asia Minor.

3. **RÙMEX** L. DOCK, SORREL.

Perennial, or rarely annual, caulescent herbs, with thick roots. Leaves alternate, entire or wavy, with thin bristly ocreae. Flowers green, perfect, polygamo-monoecious or dioecious, pedicelled in distant or contiguous whorls. Sepals 6, the inner three usually developing entire, toothed, or spiny crests and one or more sometimes bearing a tubercle or callosity. Stamens 6. Ovary 1-celled, sessile; styles 3; stigmas peltate, tufted. Achenes 3-angled, usually invested by the accrescent calyx, margined or winged.

Flowers dioecious; foliage acid.
  Inner perianth-lobes not developed into wings in fruit; achenes granular. 1. *R. Acetosella.*
  Inner perianth-lobes developed into wings in fruit; achenes smooth.
    Basal leaves few, some of them sagittate, with descending basal lobes. 2. *R. Acetosa.*
    Basal leaves many, some of them hastate, with divergent basal lobes. 3. *R. hastatulus.*
Flowers perfect, or andropolygamous; foliage not acid.
  Inner perianth-lobes entire, undulate, or denticulate.
    Inner perianth-lobes in fruit without tubercles.
      Inner perianth-lobes in fruit more than 2 cm. broad; plants perennial, with deep-seated woody rootstock. 4. *R. venosus.*

Inner perianth-lobes in fruit less than 1.5 cm. in diameter; plants with taproots. — 5. *R. occidentalis.*
Inner perianth-lobes or at least one of them bearing a tubercle in fruit; perennials with a taproot.
Only one perianth-lobe tubercled.
Leaves dark green, more or less crisp; fruiting inner perianth-lobes 8–9 mm. broad, reniform. — 6. *R. Patientia.*
Leaves pale green, not crisp; fruiting inner perianth-lobes 5–6 mm. broad, deltoid-ovate. — 7. *R. altissimus.*
All three perianth-lobes bearing tubercles.
Pedicels filiform, curved or flexuose, not more than twice as long as the fruit.
Pedicels with a thickened joint.
Leaves dark green, crisp; fruiting perianth-lobes ovate to reniform, denticulate. — 8. *R. crispus.*
Leaves pale green, not crisp; fruiting perianth-lobes deltoid, entire. — 9. *R. mexicanus.*
Pedicels without a joint. — 10. *R. Britannica.*
Pedicels clavate, deflexed, straight, 2–5 times as long as the fruit. — 11. *R. verticillatus.*
Inner perianth-lobes in fruit spinulose-toothed on the margin.
Tall plants; lower leaf-blades cordate at the base; one tubercle. — 12. *R. obtusifolius.*
Low plants; lower leaf-blades narrowed at the base; three tubercles. — 13. *R. persicarioides.*

1. **R. Acetosélla** L. Perennial, with a creeping rootstock; stem erect, glabrous, 1–6 dm. high; leaf-blades hastate, 2.5–15 cm. long, obtuse or acute, with entire or 1- or 2-toothed auricles; upper stem-leaves linear; perianth green or purplish; achenes ovoid, 3-angled, 1.5 mm. long, exceeding the persistent perianth. SHEEP SORREL. Waste places, old fields, etc.: Lab.—Fla.—Calif.—Alaska; nat. from Eur. *Arct.—Temp.* Ap–Au.

2. **R. Acetòsa** L. Perennial, with a short rootstock; stem erect, 3–10 dm. high; leaf-blades oblong-hastate or ovate-sagittate, 3–12 cm. long, acute at the apex, with acute auricles, the basal ones petioled, the upper cauline ones sessile; perianth green, 2 mm. long, in fruit winged, broadly ovate or orbicular, cordate, 5 mm. long, with delicate callosities at the base. SOUR DOCK. Waste places: Lab.—N.Y.—Mont.—Alaska; adv. from Eur. Je–Au.

3. **R. hastátulus** Baldw. Perennial, with a woody base; stem 3–6 dm. long; some of the basal leaf-blades oblanceolate, some hastate, 2–12 cm. long; upper stem-leaves linear; perianth green, in fruit winged, orbicular-cordate or reniform, 2.5–4 mm. broad; achenes smooth and shining. Sandy soil: Mass.—Kans.—Tex.—Fla. *Allegh.—Austral.* Mr–Au.

4. **R. venòsus** Pursh. Stem stout, erect or decumbent at the base, 1.5–4 dm. high, somewhat fleshy; leaf-blades ovate, ovate-lanceolate, or oblong, 3–12 cm. long, fleshy, somewhat glaucous; panicle rather dense, conic; perianth red, pedicelled, about 5 mm. long; inner lobes in fruit much enlarged, orbicular, 2–5 cm. broad, venose, cordate at the base; achenes 7 mm. long. WILD BEGONIA, SOUR GREENS, WILD HYDRANGEA. Sandy soil: Man.—Mo.—Nev.—Wash.—Alta. *Prairie—Plain—Submont.* Ap–Je.

5. **R. occidentàlis** S. Wats. Stem 5–20 dm. high; blades of the basal leaves oblong-lanceolate, truncate or subcordate at the base, 1–3 dm. long; panicle narrow, elongate; perianth greenish, 3–4 mm. long; inner lobes in fruit broadly cordate, denticulate towards the base; achenes 3 mm. long. *R. polyrrhizus* Greene, in part. Wet places: Lab.—N.D.—N.M.—Calif.—B.C. *Boreal.—Plain—Mont.* Jl–Au.

6. **R. Patiéntia** L. Stem erect, grooved, 5–15 dm. high; blades of basal leaves ovate-lanceolate, 1–4 dm. long; pedicels jointed below the middle; flowers perfect; perianth green, 4–5 mm. long; inner lobes in fruit 6–7 mm. long, rounded-cordate, sinuate on the margins; achenes 3 mm. long. *R. Britannica*

of western reports; not L. PATIENCE. Waste places: Newf.—N.J.—Utah—Ida.—N.D.; cultivated and occasionally escaped; native of Eur. Je–Jl.

7. **R. altíssimus** Wood. Stem grooved, 5–12 dm. high; blades of the basal leaves lanceolate or oblong-lanceolate, 5–25 cm. long, acute at both ends, papillose; flowers perfect; pedicels jointed near the base; perianth light green, 2 mm. long; inner lobes in fruit triangular-cordate, 4–5 mm. long, reticulate, entire-margined. Along streams: Mass.—Md.—N.M.—N.D. *E. Temp.—Plain.* Ap–Je.

8. **R. críspus** L. Stem 3–10 dm. high, erect, simple; blades of the basal leaves oblong to linear-lanceolate, 1.5–3 dm. long, cordate, rounded, obtuse or acute at the base, more or less papillose; flowers perfect; pedicels jointed at the base; perianth dark green; inner lobes in fruit 3–5 mm. long. CURLED DOCK. Waste places: Newf.—Fla.—Calif—B.C.; nat. from Eur. Je–Au.

9. **R. mexicànus** Meissn. *Fig. 175.* Stem 3–6 dm. high; blades of the basal leaves lanceolate or linear-lanceolate, pale, 5–15 cm. long, 1–2.5 cm. wide; flowers perfect; perianth pale green, 2–3 mm. long; inner lobes in fruit 5–6 mm. long, truncate at the base. *R. salicifolius* Am. auth. [B]; not Weinm. Along rivers: Lab.—Me.—Mo.—N.M.—B.C.; Mex. *E. Temp.—Plain—Mont.* My–Au.

f.175

10. **R. Britannica** L. Perennial, with a taproot; stem 1–2 m. high; leaf-blades oblong-lanceolate, transversely veined, obscurely crenulate on the margins, acute at both ends, or truncate or subcordate at the base, often 3 dm. long; flowers in large panicles; inner perianth-lobes in fruit orbicular or rounded-ovate, finely reticulate, 6 mm. broad, denticulate, all 3 grain-bearing. Wet places: Newf.—Minn.—N.J.; also introduced in Utah. *Canad.—Allegh.* Jl–Au.

11. **R. verticillàtus** L. Perennial; stem 1–1.5 m. high; leaf-blades lanceolate to oblong-lanceolate, obtuse, thickish, pale green, the lower cordate at the base, 5–15 cm. long; inflorescence nearly leafless, interrupted below; pedicels verticillate; inner perianth-lobes in fruit ovate-rhombic, strongly rugose-reticulate, 4 mm. long, all three with narrowly ovoid tubercles. Swamps: Que.—Iowa—Tex.—Fla. *E. Temp.* My–Jl.

12. **R. obtusifòlius** L. Perennial, with a taproot; stem stout, 5–12 dm. high, grooved; blades of the basal leaves broadly ovate, 1.5–3 dm. long; pedicels jointed below the middle; perianth greenish or purplish, 3 mm. long; inner lobes in fruit about 5 mm. long, hastate or deltoid, strongly reticulate. Waste places: N.S.—Fla.—N.M.—Ore.—B.C.; nat. from Eur. *E. Temp.* Ap–Au.

13. **R. persicarioìdes** L. Annual; stem 3–10 dm. high; leafy; leaf-blades lanceolate or linear-lanceolate, 3–25 cm. long, undulate and crisp; pedicels jointed at the base; perianth greenish, 1–1.5 mm. long; inner lobes in fruit about 2 mm. long, with 1–3 bristle-like teeth on the margins. GOLDEN DOCK. Sandy shores: N.B.—N.C.—Calif.—B.C. *Plain—Submont.* Jl–S.

4. DELOPỲRUM Small. JOINTWEED.

Shrubby herbs, with jointed stems. Leaves alternate, the blades articulate to the truncate ocreae. Flowers perfect, solitary, the pedicels jointed. Perianth-lobes 5, petaloid, persistent, the inner 3 becoming larger than the outer 2. Stamens 8, the inner 3 filaments dilated at the base. Styles 3, stigmas capitate. Achenes 3-angled.

1. **D. articulàtum** (L.) Small. Annual; stem 1–3 dm. high; leaf-blades linear-subulate, 1–2.5 cm. long, revolute-margined; bracts imbricate; pedicels spreading or reflexed; perianth pink; achenes narrowly ovoid, brown, smooth and shining. *Polygonella articulata* Meissn. [G, B]. Sandy shores, near the coast: Me.—Fla., and along the Great Lakes to Minn.—Iowa. *E. Temp.* Jl–O.

5. **POLÝGONUM** (Tourn.) L. KNOTWEED, DOORWEED, KNOTGRASS.

Perennial or annual, sometimes somewhat shrubby herbs, with alternate leaves and somewhat fleshy or leathery leaf-blades articulate to the ocreae. Ocreae at first 2-lobed, soon lacerate, hyaline. Inflorescence of axillary small clusters. Calyx of 5 or rarely 6, partially united sepals, mostly green with white, pink, or yellowish margins. Stamens 3–8, usually 5 or 6; filaments, at least the inner ones, dilated. Ovary 1-celled, 1-ovuled; styles 3, usually distinct. Achenes 3-angled, brown or black; endosperm horny; cotlyedons incumbent.

Fruit erect.
- Perianth-lobes with yellowish green margins.
  - Plant copiously leafy throughout; the leaf-blades large, oval or elliptic. 1. *P. erectum.*
  - Plant with the upper leaves reduced; leaves narrow, oblong, spatulate, or oblanceolate.
    - Plant branched at the base, the branches spreading; lower leaves spatulate or oblong, mostly deciduous. 2. *P. latum.*
    - Plant branched above, the branches ascending; lower leaves oblanceolate, more persistent. 3. *P. ramosissimum.*
- Perianth-lobes with white or pink margins.
  - Plant copiously leafy throughout; upper leaves scarcely reduced.
    - Stem ascending; achenes obscurely punctate. 4. *P. achoreum.*
    - Stems prostrate or diffusely spreading.
      - Leaves thick, prominently veined, oblong, oval, or spatulate, obtuse or rounded at the apex, usually pale; ocreae very conspicuous; faces of the achenes granular. 5. *P. buxiforme.*
      - Leaves thin, not prominently veined, bright green; ocreae not conspicuous; faces of the achenes finely striate.
        - Perianth 2.5–3.5 mm. long; achenes 2.5–3 mm. long, acute; leaves 2–4 cm. long. 6. *P. aviculare.*
        - Perianth 2–2.5 mm. long; achenes 2–2.5 mm. long, acuminate; leaves mostly less than 2 cm. long. 7. *P. neglectum.*
  - Plants with the upper leaves more scattered and reduced; mostly erect perennials.
    - Achenes conspicuously exserted. 8. *P. exsertum.*
    - Achenes not exserted.
      - Upper bracts not subulate; achenes mostly dull.
        - Flowers hidden in the ocreae; achenes lanceolate. 9. *P. leptocarpum.*
        - Flowers not hidden in the ocreae; achenes rhombic-ovoid. 10. *P. prolificum.*
      - Upper bracts subulate; achenes smooth and shining.
        - Leaves flat, with revolute margins; achenes shining throughout. 11. *P. sawatchense.*
        - Leaves plicate, appearing as 3-ribbed; achenes glandular along the margins. 12. *P. tenue.*

Fruit reflexed.
- Perianth 1.5–2.5 mm. long; leaves narrowly linear. 13. *P. Engelmannii.*
- Perianth 3–5 mm. long; lower leaves oblanceolate to linear-oblanceolate. 14. *P. Douglasii.*

1. **P. eréctum** L. Annual; stem usually erect, 2–6 dm. high, yellowish green; leaves oval, elliptic or obovate, 1–6 cm. long; flowers in small axillary clusters; perianth 3 mm. long; achenes dark brown, granular and dull. Waste places: Me.—Ga.—Ark.—N.M.—Alta.; (? Ida.) *Plain.* Je–S.

2. **P. làtum** Small. Plants mainly yellowish green, 3–9 dm. tall, with spreading branches from the base; leaf-blades mainly 1–3 cm. long; ocreae rather inconspicuous; perianth becoming 2–2.5 mm. long; sepals usually smooth; achenes about 3 mm. long, obscurely granular. Roadsides, river-banks and shores: Tex.—S.D.—Mo. *Prairie.*

3. **P. ramosíssimum** Michx. Annual; stem 2–10 dm. high, somewhat virgate, yellowish-green; leaves lanceolate, oblong, or linear-oblong, 1–4 cm. long, yellowish-green, the upper ones much reduced in size; perianth about 3 mm. long; achenes 3 mm. long, black, not shining. River valleys and sandy soil: Man.—Ills.—N.M.—Nev.—Wash.—B.C. *Prairie—Plain—Mont.* Jl–S.

4. **P. achòreum** Blake. Annual; stem ascending, much branched, 1.5–4 dm. long, striate, glabrous; ocreae scarious, whitish with a brown base, soon lacerate, 4–9 mm. long; leaves crowded, elliptic, oval, or obovate, rounded at the apex, 7–22 mm. long, 4–10 mm. wide, blush-green; perianth bluish green, in fruit 3.5–4 mm. long, the outer lobes with or without a narrow pale margin, the inner white- or pink-margined; achenes included, obscurely punctate, dull, 2.5 mm. long. Waste places and sandy soil: Que.—Vt.—Sask.—Mont.—Kans. Mo. *Canad.—Allegh.* Jl–Au.

5. **P. buxifórme** Small. *Fig. 176.* Annual or perennial; stem usually prostrate, 3–12 dm. long; leaves oblong, elliptic or oblanceolate, 0.5–2.5 cm. long, usually obtuse, often crisp on the margin; perianth 2–2.5 mm. long, achenes 2–2.5 mm. long, dark brown, mostly dull, granular. *P. littorale* Auth. [B]; not Link. Sandy or alkaline soil: Ont.—Va.—Tex.—Nev.—B.C. *Boreal.—Plain—Mont.* Je–S.

6. **P. aviculàre** L. Annual or sometimes perennial; leaves oblong-lanceolate, acute at both ends, petioled, dull bluish green, rather thin, 1–3 cm. long; flowers pedicelled; perianth 2.5–3.5 mm. long; achenes dark, rugulose-striate, dull. Waste places: Newf.—Va.—Calif.—B.C.; nat. or adv. from Eur. *Temp.—Plain—Submont.* My–O.

f. 176.

7. **P. neglèctum** Besser. Annual or perennial; stems with short internodes, strongly striate, usually minutely roughened; leaves firm, narrowly elliptic-lanceolate, or the upper rarely linear; margins often revolute when dry; perianth 2–2.5 mm. long, venose; segments with usually purplish margins; achenes reddish brown, apiculate-acuminate. *P. aviculare angustissimum* Meissn. [G]. Waste places: Me.—Fla.—Ark.—N.M.—Alta.; nat. or adv. from Eur.; more common than *P. aviculare. Temp.—Plain—Submont.* Je–N.

8. **P. exsértum** Small. Annual; stem erect, much branched, 3–8 dm. high; leaf-blades lanceolate, light or pale green, 1–3 cm. long; perianth-lobes unequal; achene much exserted, ovate-lanceolate, chestnut-brown, smooth and shining. Brackish marshes: N.B.—N.J.—Mo.—Neb.—Minn. *Canad.—Allegh.* Au–O.

9. **P. leptocárpum** B. L. Robinson. Annual; stem glabrous, copiously branched, 3 dm. high, the branches angled; leaf-blades lanceolate or linear, the upper 6–8 mm. long, pale green; ocreae becoming lacerated, brown or reddish at the base; flowers sessile; sepals greenish, with white or reddish margins; stamens 5; achenes 3-angled, 3 mm. long, curved above. Sandy places: Kans.—S.D. *Prairie.* S–O.

10. **P. prolíficum** (Small) B. L. Robinson. Annual; stem 3–5 dm. high, much branched, strongly striate, dark green or reddish; leaves linear-oblong or linear, thick, obtuse or acute, strongly veined beneath, dark green, 1–2 cm. long; perianth about 2 mm. long; achenes brown, 2 mm. long, concave on the lanceolate-deltoid faces, shining. (?) *P. flexile* Greene. *P. ramosissimum prolificum* Small [B]. Sandy places: Me.—Va.—Colo.—Mont. *Canad.—Allegh.—Plain.* Jl–O.

11. **P. sawatchénse** Small. Annual; stems erect, striate, obscurely 4-angled, branched from the base, 5–30 cm. high; lower leaves oblanceolate, 1–2 cm. long, acute or obtuse at the apex, often more or less revolute, with a prominent midvein; perianth-segments green, only slightly lighter on the margins, 2 mm. long; achenes rather blunt at both ends, smooth and glossy. Hillsides and mountains: S.D.—N.M.—Calif.—Wash. *Submont.—Mont.—W. Temp.* Jl–Au.

12. **P. ténue** Michx. Annual; stem 1–3 dm. high, scabrous at the nodes, 4-winged below the ocreae; leaf-blades linear or lance-linear, 5–30 mm. long, acuminate, plicate, with 2 lateral impressions, appearing as ribbed; pedicels stout, 1–1.5 mm. long; sepals with white margins; stamens 8; achenes 3 mm. long, ovoid. Dry or stony soil: Me.—Man.—Neb.—Tex.—Ga. *E. Temp.* Jl–S.

13. **P. Engelmánnii** Greene. Annual; stem often diffusely branched at the base, 0.5–3 dm. high; lower leaves linear-oblanceolate, 0.5–2 cm. long, the upper reduced, bract-like, subulate; perianth-segments oblong, obtuse, with whitish margins; achenes 2–2.5 mm. long, ovoid, black, smooth and shining. *P. tenue microspermum* Engelm. Hillsides and mountains: B.C.—Mont.—Colo. *Plain—Mont.* Jl–S.

14. **P. Douglásii** Greene. Annual; stem erect, 2–4 dm. high, with ascending branches; lower leaves oblanceolate, 2–5 cm. long, mostly obtuse or acutish; the upper linear and reduced, scattered; perianth-segments white or rose-colored on the margins; achenes 3–4 mm. long, black, smooth and shining. *P. consimile* Greene. (?) *P. emaciatum* A. Nels. *P. pannosum* S. S. Sharp. Hillsides and in sandy soil: Vt.—N.Y.—N.M.—Calif.—B.C.; probably only introduced east of the Rockies. *Temp.—Plain—Mont.* Je–Au.

6. TOVÀRA Adans.

Herbaceous annuals, becoming somewhat woody at the base. Leaves alternate, membranous, acute at each end, continuous with the ocreae. Ocreae alterdric, interrupted, fringed with bristles. Flowers in slender lax racemes. Perianth more or less colored, curved. Sepals 4, the lateral ones overlapping the others. Stamens 4, alternate with the sepals or sometimes 5; filaments flattened. Styles 2, exserted, recurved. Achenes lenticular, biconvex. Embryo horny.

1. **T. virginiàna** (L.) Adans. Stem 3–13 dm. high; leaf-blades ovate or ovate-lanceolate, 3–16 cm. long, acuminate; ocreae cylindric, 5–15 mm. long; racemes 1–6 dm. long; calyx greenish or reddish, 4–5 mm. long; achenes 3.5–4 mm. long, dark brown to cream-colored, smooth and shining. *Polygonum virginianum* L. [B, G]. Woods and thickets: N.S.—Minn.—Neb.—Tex.—Fla. *E. Temp.*

7. PERSICÀRIA (C. Bauhin) Mill. SMARTWEED, LADY'S THUMB, WATER PEPPER.

Perennial or annual, caulescent herbs, not twining, with alternate leaves and entire leaf-blades, continuous with the ocreae. The latter cylindric, mostly membranous, truncate. Racemes spike-like; pedicels articulate below the calyx. Calyx more or less colored, white or greenish, glandular-punctate; perianth-segments mostly 5. Stamens 4–8, filaments not dilated. Ovary 1-celled, 1–ovuled; styles mostly 2, sometimes 3, usually partially united; stigmas capitate. Achenes mostly lenticular, sometimes 3-angular, usually black, smooth or granular. Endosperm horny; cotyledons accumbent.

Racemes spike-like, terminal only, solitary or 2, rarely 3 or 4; plant perennial, usually with both terrestrial and aquatic forms.
  Racemes broadly oblong, rounded at the apex; leaf-blades in the terrestrial form lance-linear or lance-oblong; in the aquatic form oblong, glabrous, broadest at the middle, scarcely ever subcordate at the base; ocreae of the upper leaves in the terrestrial form usually with foliaceous dilated margins.
    Peduncles glabrous or in the terrestrial form sometimes sparingly hirsute; leaves dark green and especially in the aquatic form shining above.
      Racemes 15–18 mm. thick; leaf-blades in the terrestrial form lance-oblong, in the aquatic form broadly oblong, usually ⅓–½ as broad as long, truncate or abruptly contracted at the base, and obtuse at the apex, rarely acute at each end, glabrous; upper ocreae in the aquatic form inflated, membranous. 1. *P. psycrophila.*
      Racemes 10–15 mm. thick; leaf-blades in the terrestrial form lance-linear, in the aquatic form narrowly oblong, ½–⅓ as broad as long, acute or acutish at each end; peduncles in the terrestrial form often with a few hairs; ocreae in the aquatic form not dilated. 2. *P. fluitans.*
    Peduncles in the terrestrial form distinctly hirsutulous; leaf-blades light green, in the terrestrial form lance-oblong, pubescent on both sides, in the aquatic form glabrous and dull above; flowering raceme 10–15 mm. thick. 3. *P. nebraskensis.*
  Racemes linear, tapering to the apex; peduncles in the terrestrial form pubescent or glandular, in the aquatic form sometimes slightly so; leaf-blades broadest near the base, in the terrestrial form of a lanceolate type, acuminate, in the aquatic form ovate, acute at the apex, often cordate or subcordate at the base.
    Racemes about 3 cm. long or less; flowers 3–4 mm. long; peduncles in the terrestrial form hispidulous, scarcely glandular; ocreae sometimes with a foliaceous border.
      Leaf-blades linear-lanceolate; 3cm. wide; peduncles hispidulous. 4. *P. mesochora.*
      Leaf-blades ovate, acuminate, 5–7 cm. wide; peduncles long-hirsute. 5. *P. iowensis.*
    Racemes 3–6 cm. long; flowers 4–5 mm. long; peduncles in the terrestrial form decidedly glandular and leaf-blades broadly lanceolate, usually more than 3 cm. wide; ocreae never with foliaceous border.
      Bracts glabrous or in the terrestrial form glandular at base and hispid-ciliate on the margins; leaves in the terrestrial form nearly glabrous, except the hispidulous veins and margins. 6. *P. coccinea.*
      Bracts in the terrestrial form hirsute or hispidulous all over.
        Leaves in the terrestrial form equally pubescent on both sides.
          Petioles of the middle stem-leaves about one fourth the length of the pubescent leaf-blade. 7. *P. pratincola.*
          Petioles of the middle stem-leaves half as long as the glabrous or scabrous leaf-blades. 8. *P. rigidula.*
        Leaves canescent beneath, with a finer and denser pubescence than above. 9. *P. vestita.*
Racemes axillary as well as terminal, numerous.
  Ocreae without marginal bristles.
    Racemes erect.
      Achenes strongly biconvex; racemes linear, elongate; peduncles glandless; perennials. 10. *P. portoricensis.*
      Achenes flat or nearly so, or slightly concave; peduncles glandular; annuals.
        Styles or stamens exserted; achenes slightly umbonate on the faces; racemes mostly elongate. 11. *P. longistyla.*

Styles and stamens included; achenes not umbonate; raceme oblong.
Peduncles and upper part of the stem copiously glandular with distinctly stalked glands; achenes ovate, slightly convex on one side; leaves strongly punctate beneath. 12. *P. omissa.*
Peduncles and upper part of the stem moderately glandular, with mostly subsessile or short-stalked glands; achenes suborbicular, flat or slightly concave on both sides; leaves scarcely punctate. 13. *P. pennsylvanica.*
Racemes drooping; glands on the branches and inflorescence sessile.
Styles united only at the base. 14. *P. incarnata.*
Styles united to about the middle.
Leaves deep green on both sides. 15. *P. lapathifolia.*
Leaves pale beneath. 16. *P. tomentosa.*
Ocreae bristle-fringed.
Ocreae not with spreading tips; cotyledons accumbent.
Raceme oblong.
Ocreae copiously fringed; achenes narrowly ovoid. 17. *P. maculosa.*
Ocreae inconspicuously fringed; achenes broadly ovoid. 18. *P. persicarioides.*
Racemes slender, loosely flowered, about 5 mm. thick in fruit.
Perianth not punctate; pink or white.
Ocreolae conspicuously fringed. 19. *P. opelousana.*
Ocreolae not conspicuously fringed.
Leaf-blades glabrate above; achenes pointed at the apex only. 20. *P. hydropiperoides.*
Leaf-blades strigose above; achenes pointed at each end. 21. *P. setacea.*
Perianth punctate, white or pale green.
Racemes erect; achenes smooth and shining. 22. *P. punctata.*
Racemes nodding at least in fruit; achenes granular and dull. 23. *P. Hydropiper.*
Ocreae with spreading tips; cotyledons incumbent. 24. *P. orientalis.*

1. **P. psycróphila** Greene. TERRESTRIAL FORM: Stem 3–6 dm. high, decumbent at the base, glabrous, ocreae more or less muricate and hirsute, loose, sometimes with a foliaceous spreading margin, hispid-ciliate; petioles stout, 1–2 cm. long, sometimes hispid; leaf-blades lance-oblong, acute at each end, 8–12 cm. long, 2–4 mm. wide, from glabrous to sparingly appressed-setose above and on the margins, and hirsute on the midrib beneath; peduncles terminal, 2–3 cm. long, glabrous; raceme 2–4 cm. long, 15–18 mm. thick; bracts glabrous (very rare). *P. muriculata* Greene. PALUDOSE FORM: Similar to the terrestrial one, but leaves and ocreae glabrous, the latter usually without foliaceous border (rare). AQUATIC FORM: Stem floating, 3–20 dm. long; ocreae loose, membranous, usually glabrous, with a spreading membranous rim; petioles 3–4 cm. long; leaf-blades oblong, acutish at each end or obtuse at the base, glabrous and shining, 5–12 cm. long, 2–3.5 cm. wide; spikes as in the terrestrial form, but the peduncles lateral (common). *P. psycrophila* Greene. *P. oregana* Greene. *P. coccinea* Rydb., mainly; not (Muhl.) Greene. Water and wet places: Minn.—Sask.—B.C.—Ore.—Colo.—Neb. *W. Temp.—Prairie.* Je–Au.

2. **P. flùitans** (Eaton) Greene. TERRESTRIAL FORM: Stem erect, 3–5 dm. high, glabrous, striate, ocreae more or less hirsute and somewhat glandular-muricate, closely investing the stem, the lower short, the upper more elongate, and with a spreading, foliaceous, bristly-ciliate border; leaves spreading; petioles usually not exceeding 1 cm. long; blades lance-linear, 8–15 cm. long, 2–3 cm. wide, acute at each end, glabrous or nearly so, except the hispidulous-ciliate margins, and sometimes the hirsute midrib beneath; peduncles 1–3 cm. long, glabrous or nearly so; raceme oblong, 2–3 cm. long, 10–12 mm. thick; achenes ovate-lenticular, black, shining, 2 mm. long (not common). *P. Hartwrightii* (A. Gray) Greene [B]. *Polygonum amphibium Hartwrightii* Bissell

[G]. Paludose Form: Similar, but glabrous, the ocreae usually without foliaceous margins and the leaf-blades shorter, with longer petioles. Aquatic Form: Stem floating, 3–10 dm. long, glabrous; ocreae rather loose, short, usually glabrous, wholly membranous; peduncles 3–5 cm. long; leaf-blades oblong, usually acutish at each end, 5–10 cm. long, 2–3 cm. wide, dark green, shining, glabrous; inflorescence and fruit as in the terrestrial form (common). *Polygonum amphibium* Am. auth [G, B]; not L. *P. fluitans* Eaton. Water and wet places: N.S.—Sask.—Minn.—N.J. *Canad.* Jl–O.

3. **P. nebraskénsis** Greene. *Fig. 177.* Terrestrial Form: Stem erect, 2–5 dm. long, glabrous or somewhat hirsute; ocreae usually copiously hirsute, with dilated, foliaceous, hispid-ciliate border; petioles 1 cm. long or less, hirsute; leaf-blades spreading, light green, more or less hirsute on both sides, long-hirsute on the midribs beneath, lance-linear or lance-oblong, 5–13 cm. long, 1.5–3 cm. wide; peduncles 2–4 cm. long, glabrous or sparingly hirsute; raceme oblong, 2–3 cm. long, 10–12 mm. thick; bracts sparingly hirsute. *P. nebraskensis,* proper. *P. ammophila* Nieuwl., scarcely Greene, a more robust form. Paludose Form: Similar but glabrous or nearly so, especially below, where the ocreae are shorter and without foliaceous border. *P. nebraskensis* Greene, in part. Aquatic Form: Stem floating, glabrous; plant closely resembling that of *P. fluitans,* but the leaves lighter green and rather dull (rare). Water and wet places: N.Y.—Ont.—Mont.—Utah.—Ind. *Canad.—Prairie—Plain.* Jl–S.

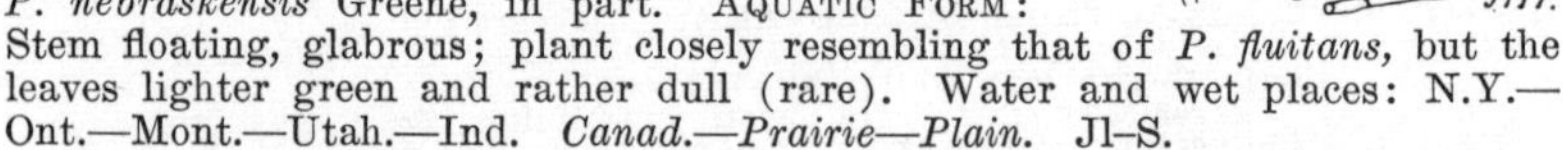

4. **P. mesochòra** Greene. Terrestrial Form: Stem decumbent at the base, rooting, 2–5 dm. high, glabrous or sparingly appressed-hirsute; ocreae more or less appressed-pubescent or the lower glabrous, rarely with a foliaceous border; petioles 1–2 cm. long, usually more or less hispidulous; leaf-blades linear-lanceolate or rarely lanceolate, 7–15 cm. long, 1.5–3 cm. wide, gradually tapering towards the apex, sparingly short-pubescent, or glabrous except on the midrib and the hispidulous-ciliolate margins; peduncles 3–5 cm. long, usually hispidulous; racemes 3–5 cm. long, about 1 cm. thick; bracts hirsute. *P. mesochora* Nieuwl. Paludose Form: Like the terrestrial, but glabrous, sometimes even to the peduncles and bracts. Aquatic Form: Stem floating and rooting at the nodes, 1–2 m. long, glabrous; petioles 3–5 cm. long, glabrous; leaf-blades lanceolate or ovate-lanceolate, acute at the apex, truncate or cordate at the base, 5–12 cm. long, 2.5–4 cm. wide, glabrous; inflorescence as in the terrestrial form, but the peduncles usually glabrous and the bracts always glabrous (less common than the terrestrial). *P. mesochora* Greene (as to type). Water and wet places: Me.—Minn.—w Neb.—D.C. *Canad.—Allegh.* Jl–S.

5. **P. iowénsis** Rydb. Terrestrial Form: Stem erect, 5–10 dm. high, hirsute throughout; ocreae inflated, densely hirsute, the margin slightly foliaceous; petioles 3–6 cm. long, hirsute; leaf-blades ovate, hirsute on both sides, acuminate, 1–2 dm. long, 5–7 cm. wide; peduncles 5–7 cm. long, densely long-hirsute, not at all glandular; racemes linear, about 3 mm. long; bracts ovate, short-hirsute, not at all glandular. Aquatic Form unknown. Perhaps of hybrid origin. Ames, Iowa. Jl.

6. **P. coccínea** (Muhl.) Greene. Terrestrial Form: Stem erect, 3–10 dm. high, striate, glabrous or somewhat hispidulous; ocreae short, truncate, short-hispidulous; leaf-blades ovate or ovate-lanceolate, 1–2 dm. long, 4–6 cm. wide, abruptly acuminate at the apex, acute to subcordate at the base, sparingly short-hispidulous or nearly glabrous, except the hispidulous margins and

veins; peduncles 1 or 2, glandular-hirsutulous, 2–5 cm. long; raceme linear, 5–8 cm. long, usually tapering upwards; bracts glandular-hispidulous at the base and hispidulous-ciliate on the margins, otherwise glabrous (common). *Polygonum coccineum terrestre* Muhl. *P. amphibium emersum* Michx. *P. Muhlenbergii* S. Wats. [G, B]. PALUDOSE FORM: Resembling the terrestrial, especially above, but often glabrous up to the inflorescence, the bract glabrous, and the lower leaves like those of the aquatic form. AQUATIC FORM: Stem floating and rooting, 1–3 m. long, glabrous, the nodes somewhat thickened; ocreae glabrous, truncate; petioles 5–7 cm. long, glabrous; leaf-blades ovate or ovate-lanceolate, acute at the apex, usually cordate at the base, 5–12 cm. long, 2–5 cm. wide, glabrous; inflorescence like that of the terrestrial form, but the peduncles sparingly glandular or glabrous and the bracts glabrous (rare). *Polygonum coccineum aquaticum* Muhl. Water and wet places: N.S.—Man. (?) —Iowa—Mo.—Va.; Utah. *Canad.—Allegh.*

7. **P. pratíncola** Greene. TERRESTRIAL FORM: Stem decumbent at the base, 3–10 dm. high, thickened at the nodes, usually more or less appressed-pubescent; sheaths truncate, appressed-hirsute; petioles 3–5 cm. long, more or less hispidulous; leaf-blades ovate or ovate-lanceolate, abruptly acuminate at the apex, acute to subcordate at the base, equally appressed-hirsutulous on both sides, 8–15 cm. long, 3–5.5 cm. wide; peduncles 1 or 2 (rarely 3 or 4), glandular-hispidulous; raceme 4–10 cm. long, 1 cm. thick or less; bracts hirsute throughout (common). *P. pratincola* Greene. *P. spectabilis, P. propinqua* and *P. Wardii* Greene. *P. rigidula* Nieuwl.; not (Sheld.) Greene. PALUDOSE FORM: like the terrestrial but glabrous or nearly so, the bracts only sometimes sparingly hirsute. AQUATIC FORM: Unknown, unless represented by *P. subcoriacea* Greene. Wet places and shallow water: Sask.—Wash.—Ore.—Utah —Mo. *W. Temp.—Plain—Prairie.* Jl–O.

8. **P. rigídula** (Sheldon) Greene. TERRESTRIAL FORM: Stem stout, 1–2 m. high, minutely scabrous with appressed hairs; sheaths truncate, the upper somewhat appressed-hirsute; petioles 5–10 cm. long, appressed-hirsutulous; leaf-blades ovate or often obcuneate, rather gradually acuminate at the apex, broadly cuneate, truncate, or rounded at the base, 1–1.5 dm. long. PALUDOSE FORM: like the terrestrial but glabrous or nearly so, the bracts only sometimes sparingly hirsute. AQUATIC FORM: Unknown, unless represented by *P. subcoriacea* Greene. Wet places and shallow water: Sask.—Wash.—Ore.—Utah—Mo. *W. Temp.—Prairie.* Jl–O.

9. **P. vestìta** Greene. TERRESTRIAL AND PALUDOSE FORMS: Stem decumbent and rooting at the base, glabrous, 5–10 dm. high; lower ocreae (on the part of the stem submerged or at some time having been submerged) glabrous, the upper ones appressed-hirsute; petioles 1.5–4 cm. long, usually more or less appressed-hirsutulous; leaf-blades ovate or ovate-lanceolate, abruptly acuminate at the apex, acute or rounded, or the lower cordate at the base, appressed-hirsutulous above, more densely pubescent with finer hairs, almost silky canescent beneath, or the lower ones growing in water glabrous; peduncles 2–4 cm. long, glandular-hispidulous; spike 4–10 cm. long; bracts densely hirsute; achenes ovate-lenticular, almost black, shining, 2 mm. long. AQUATIC FORM: Stem decumbent, rooting at the nodes; ocreae glabrous or above water somewhat hirsute; petioles 4–5 cm. long, glabrous; leaf-blades 6–10 cm. long, ovate with a cordate base, glabrous; peduncles and bracts less hairy than in the terrestrial form. *P. plattensis* Greene (transitional stage between the aquatic and the paludose form). Drying lakes and ponds: Minn.—Mont.—N.M.—Neb. *Prairie—Plain.* Jl–S.

10. **P. portoricénsis** (Bertero) Small. Perennial, glabrous, but somewhat scurfy; stem decumbent at the base, 8–13 dm. high, swollen at the nodes; leaf-blades lanceolate, 3–20 cm. long, acuminate; ocreae 2–4 cm. long, fringed with short bristles when young; racemes 2–10 cm. long, erect; bracts 3 mm.

long, with a membranous rim; perianth white, 3 mm. long; style somewhat exserted; achene broadly oblong, lenticular, black, smooth and shining. *Polygonum portoricensis* Bertero [B]. Swamps: Mo.—Kans.—Tex.—Fla.; W.Ind.; S.Am. *SE. Temp.—Trop.*

11. **P. longístyla** Small. Annual or sometimes perennial, glabrous except at the top; stem erect, 3–6 dm. high, branched; leaf-blades lanceolate or linear-lanceolate, 3–10 cm. long, acuminate, ciliate; racemes cylindric, 2–8 cm. long, 1 cm. thick; bracts 2–2.5 mm. long, hyaline-margined; flowers dimorphic, some with exserted styles, some with exserted filaments; perianth lilac; achenes lenticular, black, slightly granular, shining or dull. *Polygonium longistylum* Small [G, B]. Low ground: Mo.—Neb.—N.M.—La. *Ozark—Texan.*

12. **P. omíssa** Greene. Annual; stem 3–6 dm. high, with copious stipitate glands; ocreae short, cup-shaped; leaf-blades 5–10 cm. long, lanceolate or oblong-lanceolate, glabrous, punctate; racemes short-oblong or ellipsoid, 1–2 cm. long; perianth deep pink, about 4 mm. long; achenes round-ovate, black and shining, nearly flat on one side. *Polygonum omissum* Greene. Wet ground and dried-up ponds: Kans.—Colo.—N.M. *Plain.* Jl–S.

13. **P. pennsylvánica** (L.) Small. Annual, deep green; stem erect, 3–10 dm. high, branched; leaf-blades lanceolate, 5–20 cm. long, acuminate, ciliate, otherwise glabrous; inflorescence glandular; racemes oblong, 3–5 cm. long, dense; perianth pink or light purple, 3–4 mm. long; achene flat or slightly biconcave, orbiculate, black, smooth and shining. *Polygonum pennsylvanicum* L. [G, B]. Dry soil: N.S.—Minn.—Kans.—Fla. *E. Temp.* Jl–S.

14. **P. incarnàta** (Ell.) Small. Annual, nearly glabrous throughout; stem erect, 6–10 dm. long, more or less thickened at the nodes; leaf-blades lanceolate or linear-lanceolate, 5–20 cm. long, acuminate or acute; racemes linear, 3–8 cm. long, drooping; perianth whitish, green, or rose-colored, 2–3 mm. long; achenes lenticular, rarely 3-angled, dark brown or black, shining, smooth. *Polygonum incarnatum* Ell. [B]. Wet places: Vt.—Fla.—Calif.—Ida. *Temp.—Plain—Submont.* Jl–O.

15. **P. lapathifòlia** (L.) S. F. Gray. Annual, glabrous or nearly so; stem erect, 3–6 dm. high, thickened at the nodes; leaf-blades broadly or narrowly lanceolate, 5–20 cm. long, attenuate, punctate and ciliolate on the margins; racemes linear-oblong, 2–8 cm. long, drooping, dense; perianth flesh-colored or whitish, 2–2.5 mm. long; achenes lenticular or rarely 3-angled, black or dark brown, slightly granular, shining. *Polygonum lapathifolium* L. [B, G]. Wet places: Que.—Fla.—Calif.—B.C.; Mex., W.Ind.; Eurasia. *Temp.—Plain—Submont.* Jl–O.

16. **P. tomentòsa** (Schrank) Bickn. Annual; stem erect, 1–4 dm. high, slightly scurfy; leaf-blades lanceolate or oblong, acute or obtuse; racemes oblong, 1–3 cm. long, erect or only slightly nodding; peduncles with sessile glands; perianth 2–2.5 mm. long, greenish or pinkish-white; achenes lenticular, dark brown, shining. *Polygonum tomentosum* Schrank. [G]. *P. incanum* F. W. Schmidt. Wet places and swamps: Newf.—N.Y.—Colo.—B.C.; Eur. *Boreal.—Plain—Submont.* Je–S.

17. **P. maculòsa** S. F. Gray. Annual, glabrous or puberulent; stem erect, 2–8 dm. high, usually branched; leaf-blades lanceolate or linear-lanceolate, 2–18 cm. long, acuminate, conspicuously punctate, usually with a lunar or triangular spot in the middle; racemes oblong or ovoid, 1–3 cm. long, 0.5–1 cm. thick, mostly erect; perianth about 2.5 mm. long, pink or purple; achenes lenticular or 3-angled, smooth and shining. *P. Persicaria* (L.) Small. *Polygonum Persicaria* L. [B, G]. LADY'S THUMB. Waste places and rich soil: Newf.—Fla.—Calif.—B.C.; Mex.; Eur. *Temp.—Plain—Submont.* My–S.

18. **P. persicarioìdes** (H.B.K.) Small. Perennial, glabrous or slightly strigilose, stem erect, decumbent, or creeping, 3–7 dm. long; leaf-blades lanceolate or linear-lanceolate, 3–20 cm. long, acuminate, glabrous or sparingly pubescent, especially on the veins, ciliate; racemes erect, 2–6 cm. long, lax; perianth 2–3 mm. long, rose-colored, tinged with green; achene lenticular, biconvex, long-tipped, black, slightly granular, but shining. *P. persicarioides* H.B.K. [B]. Plains: Neb.—N.M.—Tex.; Mex. *Plain—Tex.—Trop.* Je–S.

19. **P. opelousàna** (Ridd.) Small. Perennial; stem 3–9 dm. high, glabrous, becoming woody at the base; leaf-blades linear-lanceolate or linear, 3–10 cm. long, glabrous except the ciliate midrib and margins; ocreae 1–1.5 cm. long, strigose, with long bristles; racemes linear, 1.5–4 cm. long, erect; perianth 1–1.5 m. long, white; sepals oblong, obtuse; achenes 3-angled, 2 mm. long, black, smooth and shining. *Polygonum opelousanum* Riddell. [B]. Swamps: Mo.—Kans.—Tex.—La.; Mex. *Ozark.—Tex.—Trop.* Jl–S.

20. **P. hydropiperoìdes** (Michx.) Small. Perennial, glabrous or slightly strigillose, tinged with red; stem decumbent at the base, 3–9 dm. high; leaf-blades lanceolate or lance-oblong, 4–13 cm. long, ciliate, papillose beneath; midrib short-hairy; raceme almost linear, 3–6 cm. long; perianth 2 mm. long, flesh-colored or greenish, glandular; achenes 3-angled, 3 mm. long, black, smooth and shining, pointed. *Polygonum hydropiperoides* Michx. [G, B]. Swamps: N.B.—Minn.—Calif.—Fla.; Mex. *Temp.—Trop.* Je–S.

21. **P. setàcea** (Baldw.) Small. Perennial, more or less strigose; stem erect, 5–10 dm. high, swollen above the nodes; leaf-blades oblong-lanceolate, 4–18 cm. long, acuminate, strigose on both sides; racemes erect, 1–6 cm. long, linear, loosely flowered; calyx 2 mm. long, white or pink; achene 3-angled, 2–3 mm. long, oblong, black, minutely granular, rather dull. *Polygonum setaceum* Baldw. [G]. Swamps: Mo.—Kans.—Tex.—Fla. *Austral.* Je–S.

22. **P. punctàta** (Ell.) Small. Annual or perennial, mostly glabrous throughout; stem erect, or creeping at the base, 3–10 dm. high, more or less branched; leaf-blades lanceolate to linear-lanceolate, 5–15 cm. long, acuminate, conspicuously punctate; raceme linear, erect, 1–6 cm. long, somewhat interrupted below; perianth greenish, about 2 mm. long, conspicuously glandular-punctate; achenes lenticular or 3-angled, black, smooth and shining. *Polygonum punctatum* Ell. [B]. *P. acre* H.B.K. [G]; not Lam. SMARTWEED. Swamps and wet places: Me.—Fla.—Calif.—Wash.; Mex., C.Am., W.Ind., and S.Am. *Temp.—Trop.—Plain.* Mr–D.

23. **P. Hydrópiper** (L.) Opiz. Annual, glabrous; stem erect or assurgent, 2–6 dm. high, simple or branched, sometimes reddish; leaf-blades ovate-lanceolate or lanceolate, 1.5–9 cm. long, acute, more or less papillose, punctate; racemes linear, 2–6 cm. long, interrupted and drooping; perianth greenish, 2.5–3 mm. long, glandular; achenes lenticular or 3-angled, dark brown, strongly granular and dull. *Polygonum Hydropiper* L. [G, B]. WATER PEPPER. Wet places: Newf.—Ga.—Calif.—B.C.; Mex. and C.Am.; nat. from Eur. *Temp.—Trop.* Je–S.

24. **P. orientàlis** (L.) Spach. Annual, more or less hispid; stem erect, 4–25 dm. high, woody at the base; leaf-blades ovate or broadly oblong, 5–25 cm. long, acuminate at the apex, acute to cordate at the base, ciliate, hispidulous on the veins; petiole slightly winged; ocreae with a herbaceous spreading border, ciliate; racemes oblong to linear, 3–10 cm. long, mostly drooping; perianth rose-colored, 4–4.5 mm. long; achenes lenticular, 3 mm. long, orbicular, biconcave, minutely granular and dull. *Polygonum orientale* L. [B, G]. Waste places: N.Y.—Md.—Kans.—Minn.; escaped from cultivation or nat. from India. Au–S.

8. **BISTÓRTA** (C. Bauhin) Mill. BISTORT.

Perennials, with thickened horizontal rootstocks and simple stems. Basal leaves long-petioled; stem-leaves short-petioled or sessile, narrow, entire. Ocreae cylindric, hyaline, never ciliate, oblique at the summit. Inflorescence a solitary spike-like raceme, sometimes bearing bulblets at the base. Pedicels articulate at the base of the perianth. Perianth 5-parted, not glandular-punctate. Stamens 8, exserted. Style usually 3-parted; achenes triangular or rarely lenticular. Endosperm horny; cotyledons accumbent.

1. **B. vivípara** (L.) S. F. Gray. *Fig. 178.* Blades of the basal leaves oblong or lanceolate, 2–10 cm. long, 1–2.5 cm. wide; stem-leaves lanceolate to linear; raceme narrowly cylindric, 2–10 cm. long, 0.5–1 cm. thick; perianth pale rose-colored or white; achenes dark brown, granular and dull. *Polygonum viviparum* L. [G, B]. *B. scopulina* Greene. Alpine, arctic, and subarctic swamps: Greenl.—N.H.—ne Minn.—N.M.—Alaska; Eurasia. *Arct.—Alp.—Mont.* Je–Au.

f. 178.

9. **FAGOPÝRUM** (Tourn.) Mill. BUCKWHEAT.

Caulescent glabrous annuals. Leaves alternate, petioled; blades hastate or cordate; ocreae oblique, entire. Flowers perfect, several in corymbiform cymes, with slender pedicels subtended by an ocreola. Calyx corolloid; sepals 5, equal. Stamens 8. Styles 3; stigmas capitate. Achenes 3-angled, margined or crested. Embryo S-shaped, central.

Achenes smooth and shining. — 1. *F. esculentum.*
Achenes dull and rough, pectinately sulcate. — 2. *F. tartaricum.*

1. **F. esculéntum** Moench. Stem 1–9 dm. high, branched, pubescent at the nodes; leaf-blades hastate, 2.5–8 cm. long; ocreae fugacious; perianth whitish; achenes ovoid, 5 mm. long, with pinnately striate faces, acute-angled. *F. Fagopyrum* Karst. [B, G]. Escaped from cultivation; native of Eurasia.

2. **F. tatáricum** (L.) Gaertn. Stem 1–8 dm. high; leaf-blades triangular-hastate, often broader than long, 3–10 cm. wide, short-acuminate; racemes mostly solitary in the axils, few-flowered; achenes 5 mm. long; the faces sulcate from a median groove. *P. tataricum* L. Waste places: Que.—Minn.—N.Y.; escaped from cultivation; native of Asia.

10. **TRACAÙLON** Raf. TEAR-THUMB.

Annual or rarely perennial, prickly-armed herbs, with weak 4-angled stems. Leaves alternate; blades hastate or cordate, membranous, the petioles and veins prickly. Ocreae oblique. Flowers in terminal or axillary spikes, or capitate clusters. Sepals 4 or 5, somewhat colored, neither keeled nor winged. Stamens 5–8; filaments not dilated. Ovary 1-celled; styles 2 or 3, partially united. Achenes lenticular or 3-angled, smooth and shining. Endosperm horny; cotyledons accumbent.

Leaf-blades sagittate; achenes 3-angled. — 1. *T. sagittatum.*
Leaf-blades hastate; achenes lenticular. — 2. *T. arifolium.*

1. **T. sagittàtum** (L.) Small. Annual; stem slender, decumbent or reclining, 3–15 dm. high, branched, 4-angled, armed on the angles with sharp recurved prickles; leaves lanceolate or oblong, sagittate at the base, 1–12 cm. long, 0.5–3 cm. broad, the lower petioled, the uppermost sessile; perianth white, green, or red, 4 mm. long, 5-parted; achenes 3-angled, black or brownish. *Polygonum sagittatum* L. [G, B]. Wet meadows: Newf.—Fla.—Tex.—Sask. *E. Temp.—Plain.* Jl–S.

2. **T. arifòlium** (L.) Raf. Perennial or annual; stem decumbent or reclining, 3–10 dm. long, 4-angled, with sharp, stout, recurved prickles; leaf-blades broadly hastate, 2–15 cm. long, pubescent or glabrous beneath; ocreae fringed at the summit; racemes 1–2 cm. long, interrupted; perianth greenish or rose, the lobes 4; achenes lenticular, 4 mm. long, obovoid, dark brown or variegated, smooth and shining. *Polygonum arifolium* L. [G, B]. Swamps: N.B.—Minn.—Ga. *E. Temp.* Jl–S.

11. BILDERDỲKIA Dum. FALSE BUCKWHEAT, BINDWEED.

Annual or perennial twining herbaceous vines. Leaves alternate, with cordate or hastate blades. Ocreae oblique, naked or fringed. Racemes loosely flowered, terminal or axillary, often paniculate. Sepals 5, green, white, or yellowish, the outer two sepals and the intermediate one keeled or winged. Pedicels slender, reflexed and articulate. Stamens 8; filaments short, converging. Ovary 1-celled; styles 3, short or almost wanting; stigmas capitate. Achenes 3-angled, dark brown or black. Endosperm horny. Cotyledons accumbent. *Tiniaria* Reichenb.

Outer sepals merely keeled at maturity.
  Annual; nodes naked; leaf-blades hastate; achenes rough, dull. — 1. *B. Convolvulus.*
  Perennial; nodes bristly-ciliate; leaf-blades cordate; achenes smooth and shining. — 2. *B. cilinodis.*
Outer sepals developing conspicuous wings.
  Mature calyx 5–8 mm. long; wings flat. — 3. *B. dumetorum.*
  Mature calyx 10–12 mm. long; wings crisp. — 4. *B. scandens.*

1. **B. Convòlvulus** (L.) Dum. Annual, glabrous but scurfy, pale green; stem ascending and twining, 1–12 dm. long; leaf-blades ovate-sagittate or deltoid, 2–6 cm. long, acuminate at the apex; racemes 1–6 cm. long; perianth green, 3.5–4 mm. long; segments oblong, obtuse; achenes 3-angled, 3.5 mm. long, black, granular. *Polygonum Convolvulus* L. [G, B]. *Tiniaria Convolvulus* Webb. & Moq. BINDWEED. Among bushes: N.S.—Fla.—Calif.—B.C.; nat. from Eur. *E. Temp.—Plain—Mont.* My–S.

2. **B. cilinòdis** (Michx.) Greene. Perennial, sparingly pubescent; stem twining, 5–25 dm. long, ciliate at the nodes; leaf-blades ovate, cordate at the base, minutely ciliate; raceme 1–10 cm. long, interrupted; perianth white, 3–4 mm. long, the three outer lobes keeled; achene 3-angled, broadly oblong, black, smooth and shining. *P. cilinode* Michx. [G, B]. *T. cilinodis* Small. Rocky places: N.S.—Sask.—N.C. *E. Temp.* Je–S.

3. **B. dumetòrum** (L.) Dum. Perennial, green, somewhat scurfy, weak, twining, 6–40 dm. long, much branched; leaf-blades ovate-cordate, 2–7 cm. long, acuminate, short-petioled; perianth yellowish green; segments obtuse; achenes 2–3 mm. long, black, smooth and shining. *P. dumetorum* L. [G, B]. *T. dumetorum* (L.) Opiz. Thickets: Iowa—S.D.—Mo.—Tenn. *Prairie—Ozark.*

4. **B. scândens** (L.) Greene. *Fig. 179.* Perennial, glabrous, somewhat scurfy; stem extensively twining, 5–30 dm. high; leaf-blades ovate-cordate to oblong-cordate, 1–12 cm. long, short-acuminate, long-petioled; perianth greenish yellow; segments ovate, obtuse; achenes 3.5–4.5 mm. long, black, smooth and shining. *P. scandens* L. [B, G]. *T. scandens* (L.) Small. FALSE BUCKWHEAT. Thickets: N.S.—Fla.—La.—Colo.—Mont. *E. Temp.—Plain—Submont.* Jl–O.

Family 45. **CHENOPODIACEAE.** GOOSEFOOT FAMILY.

More or less fleshy herbs or shrubs, with alternate or opposite leaves, without stipules. Flowers usually clustered in spikes, cymes, or panicles, usually greenish. Calyx of 2–5 sepals. Corolla wanting. Stamens opposite the sepals; anthers introrse. Pistils of 2–5 united carpels; ovary 1-celled; styles 2–5. Fruit a 1-seeded utricle. Embryo curved or spiral.

Embryo annular.
  Stems and branches not jointed; leaves not scale-like.
    Flowers perfect, all with perianth, not inclosed in a pair of bracts.
      Fruit inclosed in the calyx.
        Calyx in fruit not transversely winged.
          Sepals 3–5; stamens usually 2–5.
            Fruiting calyx herbaceous. 1. CHENOPODIUM.
            Fruiting calyx fleshy, red. 2. BLITUM.
          Sepals 1; stamens. 1. 3. MONOLEPIS.
        Calyx in fruit transversely winged.
          Flowers paniculate; leaves ample, sinuate, flat. 4. CYCLOLOMA.
          Flowers spicate; leaves linear, terete. 5. KOCHIA.
      Fruit laterally flattened, exserted from the marcescent calyx. 6. CORISPERMUM.
    Flowers monoecious or dioecious; the pistillate enclosed in two accrescent bractlets.
      Pericarp not hairy.
        Bracts united at least at the base; utricle not winged at the apex.
          Bracts compressed; leaves more or less farinose; testa mostly coriaceous.
            Pistillate flowers without perianth.
              Stigmas 2 or 3; bracts free at least above. 7. ATRIPLEX.
              Stigmas 4 or 5; bracts united to the apex. 8. SPINACIA.
            Pistillate flowers with 2–3 hyaline sepals shorter than the bracts. 9. ENDOLEPIS.
          Bracts obcompressed; testa membranous; pericarp hastate, with crested margins, 2-toothed apex; herbs more or less farinaceous. with toothed leaves. 10. SUCKLEYA.
        Bracts wholly free; utricle winged at the apex; perianth present in the pistillate flowers. 11. AXYRIS.
      Pericarp densely hairy, conic; low and tomentose shrubs. 12. EUROTIA.
  Stems and branches fleshy, jointed; leaves scale-like, decussately opposite; flowers sunk into the rachis of the spike; branches opposite. 13. SALICORNIA.
Embryo spirally coiled.
  Shrubs with monoecious bractless flowers; staminate flowers in spikes, without perianth; pistillate ones solitary, axillary; fruiting calyx transversely winged. 14. SARCOBATUS.
  Herbs with perfect bracteolate flowers.
    Fruiting calyx transversely winged; leaves spiny. 15. SALSOLA.
    Fruiting calyx not winged; leaves fleshy, not spiny. 16. SUAEDA.

1. **CHENOPÒDIUM** (Tourn.) L. LAMB'S QUARTERS, GOOSEFOOT, PIGWEED.

Annual (all ours) or perennial herbs, usually with mealy-coated or glandular foliage. Leaves alternate, with entire, toothed or lobed blades. Flowers perfect or rarely dioecious, in small axillary or terminal spikes or glomerules. Sepals persistent, flat or keeled. Stamens 1–5; filaments filiform. Ovary usually depressed, 1-celled; styles 2–5; stigmas filiform or subulate. Utricle containing one horizontal or vertical seed. Endosperm mealy.

Leaves more or less mealy or glabrate, never glandular or sweet-scented, sinuately lobed, dentate or entire; embryo forming a complete ring.
  Stamens 5; calyx not at all fleshy in fruit.
    Leaves entire or sinuately toothed, but not with large, acute (except in *C. murale*), divaricate teeth; seeds 1–1.5 mm. in diameter.
      Calyx-lobes carinate; at least the upper panicles exceeding the leaves.
        Pericarp easily separating from the seeds.
          Leaves linear or oblong, entire or slightly sinuately toothed.
            Whole plant almost perfectly glabrous; glomerules 1–4-flowered, in very lax spikes; leaves narrowly linear. 1. *C. subglabrum*.

- Leaves more or less mealy beneath; glomerules several-flowered; spikes denser.
  - Leaves thin; inflorescence not very dense; spikes somewhat interrupted below.
    - Leaves all narrowly linear, 1-nerved, entire. — 2. *C. leptophyllum.*
    - Lower leaves at least oblong or lanceolate, 3-nerved and often somewhat hastately toothed. — 3. *C. pratericola.*
  - Leaves thick; inflorescence dense and crowded; leaves oblong, densely mealy, yellowish. — 4. *C. desiccatum.*
- Leaves broadly ovate or triangular, more or less hastate at the base.
  - Plant densely farinose.
    - Plant low and spreading. — 5. *C. incanum.*
    - Plant tall and erect. — 6. *C. albescens.*
  - Plant sparingly farinose or glabrate, tall.
    - Leaves thin; inflorescence lax.
      - Leaf-blades about as long as broad, conspicuously hastately 3-lobed. — 7. *C. Fremontii.*
      - Leaves much longer than broad, entire or the lower sinuate-dentate. — 8. *C. Boscianum.*
    - Leaves thick, longer than broad, the upper acute; inflorescence dense. — 9. *C. atrovirens.*
- Pericarp firmly attached to the seeds.
  - Leaves subentire or merely hastately toothed.
    - Calyx open, exposing the fruit. — 10. *C. ferulatum.*
    - Calyx enclosing the fruit.
      - Seeds coarsely punctate; plant coarsely and loosely farinose, yellowish. — 11. *C. dacoticum.*
      - Seeds smooth or nearly so; plant closely and finely farinose.
        - Seeds dull; upper leaves conspicuously hastate. — 12. *C. petiolare.*
        - Seeds shining.
          - Seeds 1 mm. in diameter; upper leaves cuspidate. — 13. *C. Berlandieri.*
          - Seeds about 1.5 mm. in diameter; upper leaves long-acuminate. — 14. *C. lanceolatum.*
  - Leaves more or less sinuately dentate; inflorescence dense.
    - Leaves densely mealy. — 15. *C. album.*
    - Leaves green or nearly so.
      - Leaf-blades lanceolate in outline, acute at the base. — 16. *C. paganum.*
      - Leaf-blades deltoid in outline, more or less hastate, truncate or subcordate at the base. — 17. *C. urbicum.*
- Calyx-lobes not carinate; panicles mostly axillary, shorter than the leaves.
  - Leaves linear; calyx-lobes spreading in fruit. — 18. *C. cycloides.*
  - Leaves not linear.
    - Leaves green on both sides.
      - Seeds horizontal; leaf-blades ovate or ovate-rhombic in outline. — 19. *C. murale.*
      - Seeds vertical; leaf-blades broadly deltoid-hastate. — 20. *C. Bonus-Henricus.*
    - Leaves farinose beneath; seeds vertical.
      - Flowers in loose elongate interrupted spikes; inflorescence not villous; seeds smooth. — 21. *C. glaucum.*
      - Flowers in dense short axillary spikes; inflorescence somewhat villous; seeds finely tuberculate. — 22. *C. salinum.*
- Leaves with large divaricate acute lobes; seeds about 2 mm. in diameter. — 23. *C. hybridum.*

Stamens 1–2; calyx reddish and slightly fleshy in fruit.
Plant usually more than 1 dm. high, erect; leaves usually more or less toothed, 24. *C. rubrum.*
Plant less than 2 dm. high, prostrate; leaves entire or merely hastately toothed. 25. *C. humile.*
Leaves glandular, sweet-scented, pinnately lobed; embryo horseshoe-shaped.
Pericarp gland-dotted; inflorescence spicate-paniculate; lower leaf-blades sinuate-dentate to sinuate-pinnatifid; lobes acute or obtuse. 26. *C. ambrosioides.*
Pericarp not gland-dotted; inflorescence loosely dichotomous; lobes of the leaves rounded or broadly oblong, more or less toothed. 27. *C. Botrys.*

1. **C. subglàbrum** (S. Wats.) A. Nels. Stem 2–4 dm. high, branched, striate, obtusely angled; leaves 1-nerved, 2–10 cm. long, 1–3 mm. (rarely 4 mm.) wide, entire, light green; seeds black, shining, about 1.5 mm. in diameter. *C. leptophyllum subglabrum* S. Wats. [B]. Sandy soil: S.D.—Utah—Wash.—Ore. *Plain—Basin.* Jl–Au.

2. **C. leptophýllum** Nutt. Stem 2–5 dm. high, striately angled; leaves 1–6 dm. long, 1–6 mm. wide, green above, rather densely mealy beneath; sepals strongly carinate, scarcely covering the seed; seeds shining, fully 1 mm. broad. Sandy or dry soil: Man.—Mo.—N.M.—Ariz.—Calif.—B.C.; adv. in the East: Me.—N.J. *W. Temp.—Plain—Submont.* Je–S.

3. **C. praterícola** Rydb. Stem 3–6 dm. high, striate and angled, nearly glabrous; leaves petioled; blades 2–6 cm. long, 4–18 mm. wide; sepals scarious-margined, green on the back, slightly carinate; seeds black, shining, about 1.5 mm. in diameter. Sandy soil, fields and waste grounds: Iowa—Mo.—N.M.—Ariz.—Wash.; n Mex.; introduced in N.Y. *Plain—Prairie—Basin—Son.* Jl–S.

4. **C. desiccàtum** A. Nels. Stem 2–5 dm. high; leaves short-petioled; blades obtuse or acutish, 1–4 cm. long, 4–10 mm. wide, thick, mealy on both sides; sepals slightly carinate; seeds black, shining, 1.5 mm. broad. *C. leptophyllum oblongifolium* S. Wats. [G]. *C. oblongifolium* (S. Wats.) Rydb. *C. desiccatum* was described from a depauperate form. Arid ground: Alta.—N.D.—Mo.—Tex.—Calif.—Wash. *Plain—Submont.—W. Temp.* Jl–S.

5. **C. incànum** (S. Wats.) Heller. Stem divaricately branched, 1–3 dm. high, mealy, obtusely angled; leaf-blades rhombic or broadly ovate, hastately lobed, 3-ribbed, 1–2 cm. long and nearly as broad; flowers in dense clusters; sepals very mealy, slightly carinate; seeds black, shining. *C. leptophyllum incanum* S. Wats. [B]. Dry ground, especially in prairie-dog towns: Neb.—Kans.—N.M.—Ariz.—Utah. *Plain—Son.* My–Au.

6. **C. albéscens** Small. Stem erect, 5–12 dm. high, mealy when young, angled; leaf-blades 2–4 cm. long, sharply acute or bristle-tipped, hastately lobed or toothed, 3-ribbed; inflorescence rather lax; sepals mealy, barely keeled; seed 1–1.5 mm. broad, black, shining. Dry soil: Tex.—N.M.—Colo.—Kans. *Son.—Plain—Submont.* Je–Jl.

f. 180.

7. **C. Fremóntii** S. Wats. *Fig. 180.* Stem 2–8 dm. high, with slender branches, green; leaf-blades broadly triangular or rhombic, sinuate-dentate, 1–5 cm. long and nearly as wide, rounded and mucronate at the apex; sepals strongly carinate, nearly covering the fruit; seed black, shining, fully 1 mm. broad. In cañons and among bushes: S.D.—Kans.—N.M.—Ariz.—Ore.—Mont.; n Mex. *Plain—Mont.* Jl–S.

8. **C. Bosciànum** Moq. Stem 3–10 dm. high,

sparingly branched above, bright green; leaf-blades thin, the lower triangular-oblong, 3.5–6 cm. long, obtuse, cuneate at the base, sinuate-dentate; the upper ones lanceolate to oblong or ovate, entire bright green above, paler and finely farinaceous beneath; flowers solitary or in small clusters in interrupted spikes; sepals ovate-orbicular, white-margined, slightly carinate; seeds 1.3 mm. broad, black and shining. Woods and waste places: Iowa—Me.—Ga.—Tex.—Neb. *Temp.* Jl–S.

9. **C. atróvirens** Rydb. Stem 3–5 dm. high, striate and obtusely angled; leaf-blades broadly ovate, the upper mucronate at the apex, 3-ribbed, 1–3 cm. long, 5–15 mm. wide; flowers in short dense spikes, sparingly mealy; seeds black, shining, 1 mm. broad. Foot-hills: Mont.—S.D.—Kans.—Colo.—Wyo.—Nev.—B.C. *Submont.* Jl–Au.

10. **C. ferulàtum** Lunell. Stem 4–12 dm. high, sparingly branched above, striate, pale green; leaf-blades oblong-rhombic, 2.5–4 cm. long, obtuse or acute, shallowly 3-lobed, finely sinuate-serrate, yellowish green above and glabrate, densely and finely farinose beneath when young, glabrate in age; blades of the inflorescence lanceolate or linear, long-acuminate, entire; flowers in large glomerules in a narrow inflorescence; sepals rounded-ovate, greenish-white or yellowish; seeds 1.3 mm. broad, black, shining. Banks: N.D.—S.D. *Plain.*

11. **C. dacòticum** Standley. Ill-scented annual; stem 1.5–6 dm. high, branched throughout, with ascending branches, striate, densely farinose when young; leaf-blades broadly rhombic to round-ovate, 1.5–4 cm. long, rounded at the apex, apiculate, shallowly sinuate-dentate or sinuate, thick, pale yellowish green, densely and coarsely farinose on both sides; blades of the inflorescence ovate or oblong, cuspidate; flowers in large glomerules; sepals rounded-ovate; seeds punctate, black. Dry soil: S.D. *Plain.* Jl.

12. **C. petiolàre** H.B.K. Stem 4–10 dm. high, sparingly branched, striate, pale green, glabrous; leaf-blades rhombic or ovate-rhombic, 2.5–7 cm. long, obtuse or rounded and apiculate at the apex, shallowly 3-lobed, irregularly sinuate-dentate, with obtuse teeth, glabrous above, densely farinose beneath; upper leaves conspicuously hastate, and the uppermost entire; flowers in small glomerules, in a dense narrow inflorescence; seeds 1 mm. broad, finely punctate, black. Dry soil: Kans.—Calif.—Mex.; also Chile and Peru. *Plain—Son.*

13. **C. Berlandièri** Moq. Stem erect, 3–9 dm. high, much branched; leaf-blades lanceolate, oblong, or ovate, sometimes somewhat rhombic, 1.5–4 cm. long; inflorescence open, interrupted, lax; sepals barely keeled; seeds black, punctate. Dry soil: N.C.—Fla.—Tex.—Colo.; n Mex. *S. Temp.—Plain—Mont.* Jl–S.

14. **C. lanceolàtum** Muhl. Stem 4–12 dm. high, branched, blunt-angled; lower leaf-blades ovate or lanceolate, coarsely toothed, those of the inflorescence linear-lanceolate, entire; inflorescence open and interrupted; sepals merely keeled; seeds black, shining. *C. viride* Auth.; not L. *C. album viride* Auth. [G, B]. Waste places and fields: Mass.—Fla.—La.—Sask.; B.C. *E. Temp.—Plain.* Jl–S.

15. **C. álbum** L. Stem erect, 6–30 dm. high; leaf-blades ovate or lanceolate, more or less rhombic, 2–8 cm. long, acute or obtuse, usually hastately lobed; inflorescence dense; sepals with light margins, enclosing the fruit; seed black, shining. Fields, waste places, and thickets: Newf.—Fla.—Calif.—B.C.—Yukon; Eurasia. *Temp.—Submont.* Jl–S.

16. **C. pagànum** Reichenb. Stem erect, 0.5–3 m. high; leaf-blades ovate to broadly lanceolate, 3–15 cm. long, coarsely sinuate-dentate, acute at the apex, thin; inflorescence dense, much-branched; sepals sharply carinate, enclosing the

fruit; seeds black, rugulose-pitted. Waste places: Mass.—Va.—N.M.—Colo.—S.D.; adv. from Eur. *Submont.* Jl–S.

17. **C. úrbicum** L. Annual; stem erect, channeled, 3–10 dm. high; leaf-blades hastate or deltoid, acute at the apex, truncate or subcordate at the base, coarsely dentate or the uppermost entire, 5–13 cm. long, green; glomerules in large panicles; sepals 1 mm. long, oblong, obtuse; seeds broad, with rounded margins. Waste places: N.S.—Man.—Minn.—N.Y.; adv. from Eur. Je–S.

18. **C. cycloìdes** A. Nels. Stem 3–4 dm. high, much branched, glabrous or finely farinose, usually reddish; leaf-blades linear, 1–1.5 cm. long, 1–1.5 mm. wide, 1-nerved, thick, glabrous above, sparingly farinose beneath; flowers in large glomerules, in dense or interrupted spikes; sepals rounded on the back, scarious-margined, rounded at the apex, spreading; seeds strongly compressed, 1.3 mm. wide, black and shining. Sandhills: Kans.—N.M. *Plain.*

19. **C. muràle** L. Stem erect or decumbent, 1–6 dm. long, widely branching; leaf-blades thin, rhombic-ovate, 2–8 cm. long, acute, coarsely sinuate-dentate, cuneate or truncate at the base; flower-clusters small, shorter than the leaves; sepals oblong, obtuse; seeds firmly adherent to the pericarp, sharp-angled. Waste places: Me.—Fla.—Calif.—B.C.; Mex. and W.Ind.; adv. or nat. from Eur. *Temp.—Submont.* Mr–O.

20. **C. Bònus-Henrìcus** L. Perennial; stem 1.5–6 dm. high, simple; leaf-blades broadly deltoid-hastate, 5–12 cm. long, 4.5–9 cm. wide, entire or sinuate, bright green; flowers in dense glomerules, in naked paniculate spikes; sepals obovate or oblong, often mucronate, herbaceous; seeds 1.5 mm. broad, compressed-globose. Waste places: N.S.—Del.—Iowa; adv. from Eurasia.

21. **C. glaùcum** L. Stem prostrate, branched throughout, 1–4.5 dm. long, glabrous, often reddish; leaf-blades oblong to ovate, 1–5 cm. long, 0.5–1.5 cm. wide, obtuse, sinuately angled or sinuate-dentate; flowers in dense glomerules, in interrupted naked spikes; sepals ovate-oblong, obtuse; stamens 5, or in the lateral flowers 1–3; seeds dark, reddish brown, nearly smooth. Saline or alkaline soil: N.B.—Va.—Minn.—Neb.; adv. from the Old World.

22. **C. salìnum** Standl. Stem prostrate, decumbent or ascending, freely branched, more or less fleshy; leaf-blades broadly lanceolate to triangular-ovate or oblong, acute, 2–3 cm. long, sinuately toothed and somewhat hastate; flowers in small axillary spikes shorter than the leaves; sepals obovate, rounded at the apex; seeds finely tuberculate. *C. glaucum* Am. auth., mainly. Alkaline soils: Alta.—Neb.—Colo.—Utah—Ore. *Plain—Submont.* Jl–S.

23. **C. hỳbridum** L. Stem erect, 6–13 dm. high; leaf-blades rhombic-ovate or triangular-ovate, long-acuminate, sharply sinuate-dentate, with 1–4 large teeth on each side, or the uppermost entire, 1–2 dm. long; flowers in large open panicles; sepals oblong, slightly keeled; seeds firmly attached to the pericarp, acute-margined, punctate. Woods, thickets and waste places: Que.—Vt.—N.Y.—Ky.—N.M.—Calif.—B.C.; nat. from Eur. *Temp.—Submont.*

24. **C. rùbrum** L. Stem erect, branched, 3–8 dm. high; leaf-blades thick, triangular-hastate to lanceolate, 3–10 cm. long, coarsely sinuate-dentate or the upper entire, nearly glabrous, dark green; flower-clusters densely spicate on short branches; sepals 2–5, obtuse, rather fleshy; seeds easily separating from the pericarp, less than 1 mm. in diameter. *C. succosum* A. Nels. COAST BLITE. Alkaline or saline soil: Newf.—N.J.—Mo.—N.M.—Ariz.—B.C.; Eur. *Temp.—Submont.* Jl–S.

25. **C. hùmile** Hook. Stem decumbent or spreading, divaricately branched; leaf-blades obovate, spatulate, or lanceolate, the upper linear, 1–3 cm. long, fleshy, glabrous or nearly so; flower-clusters in small axillary spikes; sepals oblong, obtuse, somewhat fleshy; seeds less than 1 mm. broad, easily separating from the pericarp. ALKALI BLITE. Alkaline meadows: Sask.—Neb.—Colo.—Calif.—B.C. *W. Temp.—Plain—Submont.* Je–N.

26. **C. ambrosioìdes** L. Ill-scented; stem erect, 3–10 dm. high, simple or branched, glabrous or glandular-villous or tomentulose in the inflorescence; leaf-blades 2–12 cm. long, 1.5–5.5 cm. broad, oblong to ovate, or lanceolate, coarsely and irregularly sinuate-dentate, or sinuate-pinnatifid, the lobes entire or dentate, glandular-dotted, puberulent or glabrate; flowers solitary or in dense glomerules, in elongate spikes; sepals rounded-ovate; seeds 0.6–0.8 mm. broad, black. Waste places: Me.—Fla.—Calif.—Tex.; C.Am.; W.Ind. and S.Am.; nat. of the Old World. *Temp.—Trop.*

27. **C. Bòtrys** L. Stem erect, 1–6 dm. high, branched; leaf-blades 1–5 cm. long, oblong or ovate, irregularly pinnatifid; flower-clusters in small axillary cymes; seeds horizontal or vertical, 0.8 mm. broad, adherent to the pericarp. JERUSALEM OAK. Waste places: N.S.—Ga.—Tex.—Calif.—B.C.; Mex.; nat. from Eur. *Temp.—Mont.* Jl–O.

2. BLÌTUM L. STRAWBERRY BLITE, STRAWBERRY PIGWEED.

Annual fleshy herbs, with light green, glabrous, toothed leaves. Flowers small, green or reddish, aggregate in small axillary, head-like clusters, or the upper clusters forming an interrupted spike. Calyx 2–5-lobed, becoming fleshy and bright red in fruit. Stamens 1–5, mostly 2. Ovary 1-celled; styles 2–5; stigmas slender. Seed vertical, shining, separating from the pericarp. Endosperm mealy.

Inflorescence naked above; margins of the seeds acute or acutish. — 1. *B. capitatum.*
Inflorescence leafy throughout, margins of the seeds rounded. — 2. *B. virgatum.*

1. **B. capitàtum** L. Stem simple or branched from the base, 3–6 dm. high; leaves broadly triangular to lanceolate, 3–7 cm. long; the uppermost entire, rather thick; flowers in rather large clusters in the axils of the upper leaves and in a terminal spike; sepals acute or acutish. *Chenopodium capitatum* (L.) Aschers. [G]. In rocky soil: N.S.—N.J.—N.M.—Calif.—Alaska; Eurasia. *Submont.—Mont.* My–Au.

2. **B. virgàtum** L. Stem branched throughout, 1.5–8 dm. high; leaves triangular or rhombic, 2–9 cm. long, truncate or cuneate at the base, lacinate-dentate; spikes axillary; sepals rounded. Waste places: Mass.—N.Y.—Ore.—Wash.; adv. from Eurasia and n. Africa. My–Au.

3. MONÓLEPIS Schrad. POVERTY WEED.

Low branching annuals, with alternate leaves. Flowers small, perfect or polygamous, in small axillary clusters, without bracts. Calyx of a single persistent sepal. Stamen 1. Ovary 1-celled; styles 2; stigmas filiform. Seed vertical, flattened. Endosperm copious, mealy.

1. **M. Nuttalliàna** (Schultes) Greene. *Fig. 181.* Annual; stem decumbent or ascending, divaricately branched, 1–3 dm. high; leaves short-petioled or subsessile; blades thick, 2–7 cm. long, acute at both ends, sparingly sinuate-dentate or entire; sepal fleshy and foliaceous, oblanceolate or spatulate; pericarp adherent to the seed. *M. chenopodioides* Moq. Saline soils: Man.—Minn.—Tex.—N.M.—Calif.—Wash.; Sonora. *Plain—Mont.* Mr–S.

4. **CYCLOLÒMA** Moq. WINGED PIGWEED, TUMBLEWEED.

Diffusely branched coarse annuals, with alternate toothed leaves. Flowers polygamous, *i. e.*, perfect and pistillate, in paniculate spikes, without bracts. Sepals 5, keeled, each at maturity with a horizontal wing. Stamens 5. Ovary 1-celled, hairy; styles 2 or 3, partially united. Utricle depressed, enclosed in the calyx. Seed flat, horizontal. Endosperm mealy.

1. **C. atriplicifòlium** (Spreng.) Coult. *Fig. 182.* Stem erect, divaricately branched, 3–6 dm. high; leaves short-petioled or sessile; blades lanceolate or ovate, coarsely sinuately dentate, acute at the apex, cuneate at the base, 2–7 cm. long; wing of the calyx irregularly lobed and toothed, 4–5 mm. in diameter, covering the utricle. *C. platyphyllum* (Michx.) Moq. Sandy soil: Ont.—Ark.—Tex.—N.M.—Ariz.—Mont. *E. Temp.—Son.—Plain—Submont.* Je–S.

5. **KÒCHIA** Roth.

Perennial or annual herbs, or undershrubs, with alternate narrow leaves. Flowers solitary or few together in the upper axils, perfect or pistillate, sometimes bracteolate. Calyx herbaceous, 5-cleft, persistent, at length developing a horizontal wing. Stamens 3–5, usually exserted; filaments linear. Ovary ovoid, narrowed upwards; stigmas 2. Pericarp not adherent to the inverted seed. Endosperm scanty.

Plant green; leaves linear or lance-linear. — 1. *K. scoparia.*
Plant turning purple in autumn; leaves, at least the upper ones, linear-filiform. — 2. *K. trichophylla.*

1. **K. scopària** (L.) Schrad. Branches annual; stems sparingly pubescent or glabrous, 3–10 dm. high; branches strongly ascending; leaves lanceolate, or the upper linear, the lower 3-nerved, entire; flowering branches and calyces villous. Waste places and fields: Vt.—Pa.—Colo.; adv. from the Old World. *Plain.* Jl–S.

2. **K. trichophýlla** Stapf. Plant 1–1.5 m. high, much branched, ovoid, conical or globular in outline, purplish-red in autumn; leaves alternate, narrowly linear or linear-filiform, 5–7 cm. long, puberulent and with long white hairs. Waste places: N.J.—Md.—Colo.—Mich.; escaped from cult.; nat. of Asia.

6. **CORISPÉRMUM** (A. Juss.) L. BUG-SEED.

Caulescent annuals, with narrow sessile leaves and diffusely branched stems. Flowers solitary, in the axils of more or less leaf-like bracts. Sepals 1–3, unequal, scarious. Stamens 1–3, rarely 5, hypogynous, one longer than the rest; filaments dilated. Ovary 1-celled; styles 2. Utricle more or less flattened, in ours acutely margined or winged. Pericarp adherent to the vertical seed. Endosperm fleshy.

Fruit with a distinct wing about 0.5 mm. wide.
  Spike lax; lower bracts much narrower than the fruit. — 1. *C. nitidum.*
  Spikes dense; lower bracts usually overtopping, and rarely narrower than the fruit.
    Plant glabrous or nearly so. — 2. *C. marginale.*
    Plant decidedly pubescent on the upper parts. — 3. *C. simplicissimum.*
Fruit merely acute-margined, scarcely at all winged.
  Plant glabrous. — 4. *C. emarginatum.*
  Plant more or less villous. — 5. *C. villosum.*

1. **C. nítidum** Kit. Stem branched, 3–6 dm. high, glabrous; leaves linear-filiform, 2–5 cm. long, 1 mm. wide or less; lower bracts subulate, about 1 cm. long, 1–1.5 mm. broad at the base; the upper lanceolate, shorter; fruit about 2 mm. broad and 3 mm. long. *C. hyssopifolium microcarpum* S. Wats. On sand-hills and in cañons: Ill.—Tex.—N.M.—N.D.; Eurasia. *Son.—Prairie—Plain—Submont.* Jl–S.

2. **C. marginàle** Rydb. Stem glabrous, much branched, 2–5 dm. high; leaves narrowly linear, 2–5 cm. long, 1.5–2 mm. wide; lower bracts lanceolate, about 1 cm. long, the upper ovate, 5 mm. long, all with conspicuous scarious margins; fruit about 4 mm. long and 2.5 mm. wide. *C. hyssopifolium* Am. auth., in part [B, G]. *C. imbricatum* A. Nels. Sandy soil: Wyo.—N.M.—Kans.—N.D. *Plain—Submont.* Au–O.

3. **C. simplicíssimum** Lunell. Stem glabrous below, somewhat pubescent above, simple, about 3 dm. high; leaves narrowly linear, finely pubescent, or in age glabrate, 2–4 cm. long, about 1 mm. wide; bracts 2–12 mm. long, decidedly pubescent, 1–2 mm. broad, scarious-margined, the lower broader than the fruit; fruit 2.5 mm. long, wing-margined. Lake shores: Pierce Co., N.D. Au.

4. **C. emarginàtum** Rydb. Stem branched near the base, 3–4 dm. high; leaves narrowly linear, 2–4 cm. long, 1–2 mm. wide, cuspidate-pointed; bracts except the lowest ovate, 5–7 mm. long, acuminate, scarious-margined; fruit 2.5–3 mm. long and about 2 mm. wide. Sandy valleys: Alta.—Mo.—Colo.—Nev. *Plain—Submont.*

5. **C. villòsum** Rydb. *Fig. 183.* Stem 2–4 dm. high, diffusely branched from near the base; leaves linear, 2–4 cm. long, 1–3 mm. wide, cuspidate-mucronate; spikes rather dense; bracts more or less imbricate, the lower linear-lanceolate, 5–10 mm. long, the upper ovate, acuminate, 4–5 mm. long, with broad scarious margins; fruit 2.5–3 mm. long, 2 mm. wide. Sandy valleys and fields: Sask.—N.D.—N.M.—Ariz.—Ore.—Wash. *Plain—Submont.* Jl–S.

7. ÁTRIPLEX (Tourn.) L. ORACHE, SALT-BUSH, SHAD-SCALES.

Annual or perennial herbs or low shrubs, with scaly or scurfy, often silvery pubescence. Leaves alternate or some opposite. Flowers monoecious or dioecious, in axillary or terminal panicles, or congested axillary spikes. Staminate flowers without bracts; sepals 3–5; stamens 3–5; filaments distinct or united; anthers 2-celled, opening lengthwise. Pistillate flowers subtended by 2, more or less united bracts, which are entire or toothed, often crested, tuberculate, or winged on the back. Calyx wanting. Ovary 1-celled; stigmas 2, subulate or filiform. Utricle wholly or partly enclosed in the accrescent bracts. Seeds erect or nearly horizontal. Endosperm mealy.

Annuals.
  Bracts united only at the base; radicle inferior.
    Pistillate flowers of 2 kinds; some with a 3–5-lobed calyx, not enclosed in two bracts, others without perianth and enclosed in 2 bracts; bracts thin, rounded-ovate.
      Leaves green, glabrous, dull above. — 1. *A. hortensis.*
      Leaves densely white-mealy beneath, shiny above. — 2. *A. nitens.*
    Pistillate flowers all alike, bracteate, without perianth.
      Bracts rhombic-oval, mostly cuneate at the base; lower leaves rhombic-lanceolate or lanceolate or oblong, the upper ones entire. — 3. *A. patula.*

Bracts rounded-deltoid, usually truncate at the base; lower leaf-blades deltoid.
Leaves not hastate. — 4. *A. lapathifolia.*
Leaves more or less hastate. — 5. *A. hastata.*
Bracts united to about the middle; radicle superior.
Leaves thin, more or less toothed or hastate.
Bracts ovate, acute, longer than broad; branches terete or nearly so. — 6. *A. rosea.*
Bracts suborbicular, as broad as long; branches distinctly round-angled. — 7. *A. argentea.*
Leaves ovate or oblong, entire; usually less than 2 cm. long; strongly 3-ribbed; branches ascending. — 8. *A. Powellii.*
Perennials.
Bracts not winged on the back.
Bracts with entire or merely wavy (rarely slightly denticulate) margins, without appendages on the back. — 9. *A. confertifolia.*
Bracts with a distinctly toothed margin or appendaged on the back.
Bracts broadest above the middle; more or less toothed on the margin, only rarely tuberculate on the back; 3-toothed at the apex, the middle tooth the longest. — 10. *A. Gardneri.*
Bracts broadest below the middle, strongly tuberculate or appendaged.
Plant low; staminate flowers brown, in panicles; leaves short-petioled. — 11. *A. oblanceolata.*
Plant usually tall; staminate flowers yellow, in interrupted spikes; leaves sessile. — 12. *A. Nuttallii.*
Bracts broadly 4-winged. — 13. *A. canescens.*

1. **A. horténsis** L. Stem 1–1.5 m. high; leaves petioled; lower leaf-blades from cordate or hastate to ovate, sinuately toothed, 1–2 dm. long, the upper lanceolate and entire; bracts rounded-ovate, about 1 cm. in diameter, thin, reticulate. Waste places: N.Y.—Colo.—Utah—Mont.; escaped from cultivation; native of Eur. Jl–Au.

2. **A. nìtens** Sch. Stem 6–25 dm. high, with ascending or spreading branches; leaf-blades deltoid or deltoid-oval, 4–10 cm. long, acute or acuminate at the apex, subcordate to broadly cuneate at the base, subhastate, sinuate-dentate, the upper lanceolate, often entire; flowers in slender, interrupted spikes; bracts oval-rhombic, 5–17 mm. long, obtuse, entire, unappendaged, free to near the base. Waste places: N.Y.—N.J.—S.D.; Ore.; adv. from Eur.

3. **A. pátula** L. Stem 3–9 dm. high, much branched, striate, green; lowest leaves opposite, the rest alternate; blades 2.5–8 cm. long, the lower ones hastate, sinuate-dentate or entire; flower monoecious or subdioecious, in slender interrupted spikes; bracts rhombic-oval, in fruit 2–6 mm. long, subhastate, denticulate on the margins, short-tuberculate on the faces. Salt marshes: N.B.—Fla.—Ala.—N.D.; B.C.—Calif.; Eurasia and n Africa. *Temp.*

4. **A. lapathifòlia** Rydb. Stem strict, 4–6 dm. high; leaves petioled; blades somewhat fleshy, dark green, lanceolate, 3–6 cm. long, entire; bracts in fruit about 4 mm. long and 5 mm. wide, often slightly hastate, sometimes with 1 or 2 tubercles on the back, thin, veiny, acute. Alkaline flats: Mont.—Wyo.—Neb. *Plain.* Au–S.

5. **A. hastàta** L. Stem branched, 5–10 dm. high, subglabrous; leaf-blades fleshy, rarely sinuately-toothed, 3–7 cm. long; flowers numerous, in large fleshy clusters forming interrupted spikes; fruiting bracts triangular-ovate, about 5 mm. long and broad, usually with 1 or 2 small teeth on each margin, and sometimes with 1 or 2 fleshy tubercles on the back. *A. carnosa* A. Nels. *A. subspicata* Rydb., a low spreading form. Alkaline meadows or flats: Alta.—Kans.—N.M.—Nev.—Ida.; Eur. *Plain—Son.* Au–S.

6. **A. ròsea** L. Stem erect, freely branching, often 1 m. high; leaf-blades ovate, 2–5 cm. long, coarsely and irregularly toothed; flowers in axillary

clusters, staminate and pistillate mixed; fruiting bracts about 5 mm. long, hastately toothed near the base, the faces with slender green appendages. *A. spatiosa* A. Nels. Alkaline flats and railroad embankments: Wyo.—Kans.—Chihuahua—Calif.—Wash.; N.Y.—Fla.; adv. from the Old World. Au–S.

7. **A. argéntea** Nutt. *Fig. 184.* Stem divaricately branched, angled, 2–10 dm. high; leaf-blades ovate, triangular-ovate, or subrhombic, 2–5 cm. long; pistillate flowers in axillary clusters; fruiting bracts suborbicular, usually deeply toothed and with projections on the faces. *A. volutans* A. Nels. SALT-BUSH. Alkaline flats: Sask.—N.D.—Colo.—B.C. *Plain.* Jl–S.

8. **A. Powéllii** S. Wats. Stems 2–5 dm. high, freely branched; leaves silvery white, petioled; leaf-blades 0.8–2 cm. long; flowers in small clusters in the axils, the staminate and pistillate usually mixed or the staminate ones above; fruiting bracts suborbicular, about 5 mm. broad, irregularly toothed; the faces with short thick appendages. *A. philonitra* A. Nels. Alkaline flats and dry lakes: Alta.—S.D.—N.M.—Ariz.—Mont. *Plain—Son.—Submont.* Jl–S.

9. **A. confertifòlia** (Torr.) S. Wats. Shrubby dioecious perennial; stem 3–15 dm. high; branches terete, spinescent; leaf-blades ovate or obovate to orbicular, 0.5–2 cm. long, short-petioled; flowers in axillary clusters; fruiting bracts broadly ovate or suborbicular, about 1 cm. broad. SHAD-SCALES. Alkaline valleys and bluffs: N.D.—N.M.—Calif.—Ore. *Son.—Plain—Submont.* My–Au.

10. **A. Gárdneri** (Moq.) Standl. Suffruticose perennial; stems decumbent at the base, 2–4 dm. high; leaves sessile or nearly so; blades 3–6 mm. wide, 1.5–3 cm. long; flowers in axillary capitate clusters or the staminate more paniculate above; fruiting bracts 5–6 mm. long; faces smooth and reticulate; staminate flowers brown. *A. Gordoni* Hook. *A. fruticulosa* and *A. eremicola* Osterh. Alkaline flats or dry lake beds: Wyo.—Colo. *Plain.* My–Jl.

11. **A. oblanceolàta** Rydb. Suffruticose perennial, with decumbent base and ascending branches, about 2 dm. high; leaves 2–3 cm. long, obtuse or acutish; pistillate flowers in small axillary clusters; fruiting bracts ovate in outline, slightly dentate, tubercled or irregularly crested on the back. Alkaline or clayey flats: Mont.—Colo.—Neb. *Plain.* Jl–Au.

12. **A. Nuttállii** S. Wats. Suffruticose or shrubby perennial; stems 3–6 dm. high, branching near the base; leaves 2–5 cm. long; pistillate flowers in axillary clusters; fruiting bracts ovate or orbicular, 3–4 mm. long, irregularly toothed, muricate or tooth-crested on the faces. Plains, bad-lands, and arid valleys: Man.—Neb.—Colo.—Nev.—Ida. *Plain—Basin—Submont.* Je–S.

13. **A. canéscens** (Pursh) Nutt. Shrubby perennial; leaves 2–5 cm. long; pistillate flowers axillary, short-pedicelled, the staminate clusters in subterminal spikes; fruiting bracts orbicular in outline, when fully developed 12–15 mm. broad. Dry mesas and alkaline valleys: S.D.—Kans.—N.M.—Calif. Ore. *W. Temp.—Plain—Submont.* Je–Au.

8. **SPINÀCIA** L. SPINACH.

Glabrous annuals. Leaves alternate. Flowers dioecious, rarely perfect, in glomerules, or the pistillate ones axillary, the staminate ones in interrupted spikes. Staminate flowers with 4- or 5-parted perianth; stamens 4 or 5. Pistillate flowers naked, subtended by 2 bracts, united to the apex, accrescent, enclosing the utricle, with 2 spines at the base or unarmed. Seeds erect, slightly compressed; embryo annular.

1. **S. oleràcea** L. Stem 3–4.5 dm. high; leaves long-petioled, the blades triangular-ovate, hastate, or sub-hastate, 4–10 cm. long, acute, entire or sinuate-dentate; fruiting bracts 2–4 mm. long, sessile, with two divaricate spines. Escaped from cultivation.

9. ENDÓLEPIS Torr.

Monoecious or dioecious annual herbs. Staminate flowers ebracteate in glomerate terminal spikes; calyx gamosepalous, urceolate, 5-lobed, each lobe with a gibbosity at its base. Stamens 5; filaments subulate. Pistillate flowers solitary or clustered in the axils of the leaves, 2-bracted; bracts ovate, membranous, united, forming a sack enclosing the flower, or nearly free. Calyx of 2–3 distinct sepals. Utricle ovate, compressed. Radicle superior.

Leaves thin, lanceolate, 1-nerved. 1. *E. Suckleyi.*
Leaves thick, ovate or ovate-lanceolate; the lower 3-nerved. 2. *E. dioica.*

1. **E. Súckleyi** Torr. Low annual; stem erect, with ascending branches, 1–3 dm. high, almost glabrous; leaves sessile, 2–3 cm. long, entire; staminate clusters both axillary and forming short terminal spikes; fruiting bracts ovate, 2 mm. long, membranous, pubescent. *Atriplex Endolepis* S. Wats. *A. Suckleyana* Rydb. Plains: Sask.—Mont.—S.D. *Plain.* Jl.

2. **E. diòica** (Nutt.) Standl. Low annual, usually less than 1 dm. high, rarely 2–3 dm.; stem branched; leaves usually less than 1 cm. long; staminate flowers in small clusters in the axils of the upper leaves or at the ends of the branches. *Kochia dioica* Nutt. *E. ovata* Rydb. Plains: Mont.—Wyo.—N.D. *Plain.* Je–S.

10. SUCKLÈYA A. Gray.

Monoecious fleshy annuals. Flowers axillary, the staminate above. Calyx 3–4-parted to the base. Stamens 3 or 4, distinct; anther 2-celled. Fruiting bracts subhastate, obcompressed, herbaceously cristate on the margins, 2-cleft at the apex. Radicle superior.

1. **S. Suckleyàna** (Torr.) Rydb. Prostrate or ascending annual; stem diffusely branched, 1–4 dm. long, sparingly scurfy; leaves petioled; blades orbicular to rhombic-ovate, 1–3 cm. broad, acutely sinuate-dentate; fruiting bracts 5–6 mm. long, ovate-rhombic and subhastate, with crenate ridges. *Obione Suckleyana* Torr. *Suckleya petiolaris* A. Gray. Along streams: N.D.—Mont.—Colo. *Plain—Submont.* Jl–Au.

11. AXỲRIS L.

Annual monoecious herbs. Leaves alternate with entire blades. Staminate flowers on terminal glomerules or spikes; perianth hyaline, 3–5-parted; sepals obovate or elliptic; stamens 2–5. Pistillate flowers solitary in the axils or mixed with staminate ones; bracts distinct; sepals 3 or 4, hyaline, somewhat accrescent; stigmas 2, filiform. Utricle enclosed in the perianth, compressed, cuneate-obovate, short-winged at the apex; pericarp membranous. Seed erect, obovoid; embryo horseshoe-shaped.

1. **A. amaranthoìdes** L. Stem 3–7 dm. high, racemosely branched above, stellate-pilose; leaf-blades ovate or lanceolate, 3–8 cm. long, thin, finely stellate-pilose, those of the floral branches reduced, 4–6 mm. long; fruit 3 mm. long, with a short retuse wing at the apex. Waste places and fields: Man.—N.D.—Neb.—Iowa; nat. from Siberia. *Prairie—Plain.* Jl–Au.

12. **EURÒTIA** Adans. WHITE SAGE, WINTER SAGE, WINTER FAT.

Low pubescent undershrubs, with alternate, entire leaves, either polygamo-monoecious or dioecious. Flowers in small axillary clusters. Staminate flowers bractless; calyx 4-parted; stamens 4; filaments slender. Pistillate flowers bibracteate; bracts sessile, somewhat obcompressed, united to the apex, not winged, 2-horned, densely long-hairy on the sides. Ovary hairy, oblong-ovate, membranous; stigmas 2, elongate. Seeds vertical, obovate; radicle inferior.

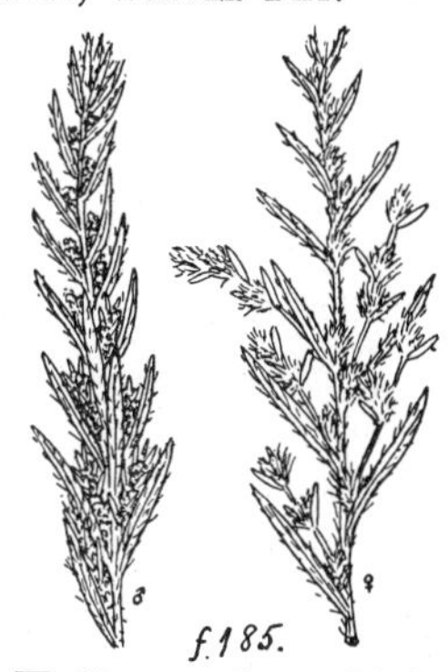

f. 185.

1. **E. lanàta** (Pursh) Moq. *Fig. 185.* Undershrub, white- or later rufous-tomentose, 2–10 dm. high; leaves linear, with revolute margins, 2–4 cm. long, 2–5 mm. wide; staminate flower-clusters above the pistillate ones, in some individuals much more numerous, in others few or none; fruiting bracts 4–6 mm. long, lanceolate; horns about 1 mm. long. Plains and hills: Sask.—Kans.—Tex.—Calif.—Wash. *W. Temp.—Son.—Plain—Mont.* Je–S.

13. **SALICÒRNIA** (Tourn.) L. GLASSWORT, SAMPHIRE.

Annual or perennial, fleshy, glabrous herbs, with scale-like leaves; branches and leaves opposite; internodes very short. Flowers perfect or polygamous, in cylindric spikes, sunk in cavities of the internodes, 3–7 together. Calyx fleshy, with a truncate or 3–4-toothed border. Stamens 2, rarely 1; filaments filiform or subulate. Utricle oblong or ovoid, included in the spongy perianth. Seeds erect; endosperm wanting; embryo conduplicate.

1. **S. rùbra** A. Nels. Stem erect, divaricately branched throughout, 1–2 dm. high; scale-like leaves short, broadly triangular, wider than long; fruiting spikes 2–4 cm. long; internodes very short, scarcely exceeding the middle flower of the nodes below. Border, of alkaline lakes: Sask.—Kans.—Colo.—Nev.—B.C. *Plain.* Au–S.

14. **SARCOBÀTUS** Nees. GREASEWOOD, CHICO.

Spinescent branched shrubs, with fleshy narrow leaves, monoecious or dioecious. Staminate flowers in terminal spikes, without calyx; stamens arranged around the base of a peltate scale. Pistillate flowers axillary, solitary, with a closed compressed calyx, margined by a narrow border, which develops into a broad membranous horizontal wing. Ovary thin and hyaline; embryo spiral; endosperm scant or none.

f. 186.

1. **S. vermiculàtus** (Hook.) Torr. *Fig. 186.* Erect shrub, 1–3 m. high, divaricately branched and spinescent; leaves linear or linear-filiform, fleshy, 1–4 cm. long; aments of the staminate flowers cylindric, 0.5–3 cm. long; stamens 3; fruiting calyx about 6 mm. long, its wing-margin 6–12 mm. broad. Alkaline flats: Sask.—Tex.—Calif.—Wash. *W. Temp.—Son.—Plain—Mont.* My–S.

15. **SÁLSOLA** L. SALTWORT, RUSSIAN THISTLE.

Annual or perennial herbs, or shrubs, bushy-branched, with narrow, entire, spine-tipped, rigid leaves. Flowers perfect, solitary or several together in the axils, with 2 bractlets. Sepals 5, appendaged with a horizontal wing at maturity. Stamens 5, rarely less; filaments linear or subulate. Ovary 1-celled, more or less depressed; styles 2. Utricle flattened; seeds erect; endosperm wanting; embryo spiral in form of a cone.

1. **S. Péstifer** A. Nels. Divaricately branched annual; stem 3–10 dm. high; leaves filiform, somewhat fleshy, 2–6 cm. long, spine-tipped; bracts subulate, 5–10 mm. long, with stout spines; calyx membranous, conspicuously veiny, 6–7 mm. in diameter. *S. Tragus* Reichenb. [B]; not L. *S. Kali tenuifolia* G. F. W. Mey. [G]. Waste places, fields, and loose sandy soil: Ont.—N.J.—Tex.—Calif.—Wash.; nat. from Eurasia. *Temp.—Submont.* Jl–S.

16. SUAÈDA Forskal. Sea Blite.

Annual or perennial herbs, or shrubby plants, more or less fleshy, with alternate, narrow, terete leaves. Flowers perfect or polygamous, solitary or clustered in the upper axils, bracteate. Sepals 5, keeled or narrowly winged at maturity. Stamens 5; filaments short. Ovary 1-celled, rounded or flat on the top; styles often 2. Utricle surrounded by the calyx; seeds horizontal or vertical; endosperm wanting or scant. Embryo coiled in a flat spiral. *Dondia* Adans.

Plant depressed, spreading. 1. *D. depressa.*
Plant erect, strict. 2. *D. erecta.*

1. **S. deprèssa** (Pursh) S. Wats. *Fig. 187.* Low and decumbent annual; stems branching at the base, 2–10 dm. long; leaves linear, subulate, 1–3 cm. long; bracts 5–10 mm. long, rather crowded on the branchlets; calyx cleft to the middle; seeds about 1 mm. broad, slightly reticulate. *Chenopodium calceolariforme* Hook. *Dondia depressa* Britton [B, R]. *Dondia calceolariformis* Rydb. Saline or alkaline soil: Man.—Kans.—Nev.—Mont. *Plain—Submont.* Jl–S.

f. 187.

2. **S. eréсta** (S. Wats) A. Nels. Erect annual; stems rather simple or with short erect branches, very leafy; leaves slender, 2–4 cm. long; bracts often more than 1 cm. long; calyx cleft below the middle; seeds smooth, nearly 1.5 mm. broad. *D. erecta* A. Nels. [R]. Alkaline or saline flats: Man.—N.D.—Colo.—Nev.—Mont. *Plain—Mont.* Jl–S.

Family 46. AMARANTHACEAE. Amaranth Family.

Coarse herbs, with alternate or opposite leaves, without stipules. Flowers inconspicuous, perfect, monoecious, dioecious, or polygamous, subtended by more or less imbricate bracts. Calyx of 2–5 scarious or herbaceous sepals. Corolla wanting. Stamens 5 or fewer, opposite the sepals. Pistil solitary, 1-celled; style 1, terminal, or wanting. Fruit a membranous utricle or pyxis.

Anthers 4-celled; green plants with alternate leaves.
  Perianth present in all flowers. 1. Amaranthus.
  Perianth wanting in the pistillate flowers. 2. Acnida.
Anthers 2-celled; leaves opposite.
  Filaments united into a short cup at the base; calyx neither crested nor spiny.
    Flowers in small glomerules in the axils of the leaves, whose bases become indurate, forming an involucre; plants stellate, diffuse. 3. Tidestromia.
    Flowers in loosely paniculate spikes; plants in ours erect, glabrate or villous. 5. Iresine.
  Filaments united into a long tube; calyx crested and tuberculate or spiny at maturity; plants woolly, erect. 4. Froelichia.

1. AMARÁNTHUS (Tourn.) L. Amaranth, Pigweed, Tumbleweed.

Annual weedy herbs, with alternate, flat, pinnately veined, entire or undulate leaves. Flowers monoecious, dioecious, or polygamous, in dense spikes or clusters, each subtended by usually 3, conspicuous, green, red, or purple bracts. Sepals 2–5, distinct; anthers 2-celled, opening lengthwise. Ovary 1-celled; styles or stigmas 2 or 3. Ovules solitary. Utricles circumscissile, irregularly splitting, or indehiscent. Seeds lenticular, shining. Embryo annular.

Sepals of the pistillate flowers clawed; flowers in terminal and axillary spikes.
  Spines present in the axils of the leaves; utricle regularly dehiscent. — 1. *A. spinosus.*
  Spines wanting.
    Dioecious; utricle indehiscent. — 2. *A. ambigens.*
    Monoecious; utricle circumscissile.
      Bracts lanceolate, not exceeding the flowers; spike not very long. — 3. *A. Torreyi.*
      Bracts subulate, pungent, exceeding the flowers; spikes very long. — 4. *A. Palmeri.*
Sepals not clawed.
  Plants tall, simple; flowers in terminal and axillary panicles; sepals 5.
    Stamens 3; sepals 1–2 mm. long; bracts 5 mm. long or more. — 5. *A. Powellii.*
    Stamens 5; sepals 2–3 mm. long; bracts 3–5 mm. long.
      Spikes stout, 8–14 mm. thick, strict; pistillate sepals obtuse or truncate; plant villous. — 6. *A. retroflexus.*
      Spikes slender, 4–6 mm. thick, usually drooping; stem glabrous; pistillate sepals acute. — 7. *A. hybridus.*
  Plants low, much branched; flowers in small axillary spike-like panicles, shorter than the leaves.
    Sepals 4–5; bracts lanceolate, a little longer than the sepals; plant prostrate. — 8. *A. blitoides.*
    Sepals 3; bracts much longer than the sepals, pungent. — 9. *A. graecizans.*

1. **A. spinòsus** L. Stem stout and succulent, 3–12 dm. high, glabrous below, pubescent above, sulcate, often reddish; leaves usually long-petioled; blades ovate or rhombic-ovate to lanceolate, 2–12 cm. long, acute at the base, obtuse or rounded at the apex, glabrous or sparingly pubescent; pistillate flowers mostly in dense axillary clusters, the staminate ones in terminal spikes; bracts lanceolate or subulate, usually shorter than the sepals; sepals 5, those of the pistillate flowers oblong, obtuse or acutish, those of the staminate ones lance-oblong, acute or short-acuminate. Waste places: Me.—Fla.—Mex.—Man.; trop. Am., Asia and Africa. *Temp.—Trop.* Je–S.

2. **A. ámbigens** Standl. Stem stout, striate, tinged with red, 5–10 dm. high; leaf-blades ovate or lanceolate-ovate; 4.5–7 cm. long, 1–2.5 cm. wide, cuneate at the base, tapering to an obtuse apex, yellowish green above, pale beneath, purple-tinged along the veins; flowers in panicles; bracts deltoid-lanceolate, acuminate, half as long as the sepals; sepals of the staminate flowers oblong or oval, acutish, scarious, tinged with red, those of the pistillate ones oblong or lance-oblong, acuminate. Wet soil: n Ill.—Wis.—se Minn. *Prairie.* Jl–Au.

3. **A. Tórreyi** (A. Gray) Benth. *Fig. 188.* Stem 3–10 dm. high, somewhat pubescent or glabrate; leaves long-petioled; blades lanceolate or oblong-lanceolate, strongly veined beneath; flowers in a rather lax panicle; sepals of the staminate plant lanceolate, spinulose-cuspidate, more or less scarious-margined; sepals of the pistillate plants obovate-spatulate, rounded at the apex. Sandy soil: Iowa—Tex.—Calif.—Nev. *Prairie—Plain—Son.* Jl–Au.

4. **A. Pálmeri** S. Wats. Stem erect, 6–10 dm. high, branching; leaves long-petioled; blades rhombic-lanceolate or obovate, 2–5 cm. long, strongly veined beneath; sepals of the staminate flowers lanceolate, spinulose-cuspidate, those of the pistillate flowers 2–3 mm. long, oblong or somewhat spatulate, distinct or nearly so. Banks and river valleys: Mo.—Tex.—Colo.—Calif.; n Mex. *Plain—Son.—Submont.*

5. **A. Powéllii** S. Wats. Stem 3–15 dm. high, glabrous, simple or with erect branches; leaves slender-petioled; blades lanceolate or ovate, 3–10 cm. long; bracts subulate, spinulose-cuspidate, 3 mm. long; sepals 1–2 mm. long, lanceolate, mucronate. Loose or sandy soil: N.M.—Wyo.—S.D.—Ore.—Calif.; n Mex. *Sun.—Mont.* Jl–S.

6. **A. retrofléxus** L. Stem erect or ascending, usually branched, 3–30 dm. high; leaves petioled; leaf-blades ovate or rhombic-ovate to lanceolate, 5–15 cm. long; bracts subulate, twice as long as the oblong, scarious sepals. PIGWEED. Waste places and fields: Vt.—Fla.—Calif.—B.C.; Mex.; nat. of Eur. *E. Temp.—Trop.—Submont.* Je–O.

7. **A. hȳbridus** L. Stem branched, 6–25 dm. high; leaves petioled; leaf-blades ovate, 4–10 cm. long, darker green above, scabrous-puberulent; bracts subulate, twice as long as the oblong acute sepals. Waste places: R.I.—Fla.—Colo.—Calif.—Alta.; Mex.; W. Ind.; nat. from Eur. *Temp.—Trop.* Mr–S.

8. **A. blitoìdes** S. Wats. Stem 3–10 dm. long, glabrous or nearly so, profusely branched; leaf-blades broadly spatulate, obovate or oblanceolate, 1–3 cm. long; bracts short-acuminate, 2–3 mm. long; sepals obtuse and mucronate or acute; utricle not rugose. Dry ground, roadsides, and waste places: Man.—La.—Calif.—B.C.; Mex.; adv. eastward to Me. and N.J. *W. Temp.—Son.—Plain—Mont.* Jl–S.

9. **A. graecìzans** L. Stems bushy-branched, whitish, 2–6 dm. high; leaf-blades oblong or spatulate, 1–4 cm. long, papillose, mucronate-cuspidate; flowers polygamous; sepals membranous. *A. albus* L. TUMBLEWEED. Waste places and cultivated ground: R.I.—Fla.—Calif.—B.C.; W. Ind.; Mex.; and Old World. *Temp.—Submont.* Jl–S.

2. **ACNÌDA** L. WATER-HEMP.

Annual coarse herbs, with branching stem and alternate, narrow, entire, pinnately veined leaves. Flowers dioecious, subtended by 1–3 bracts, in terminal or axillary, continuous or interrupted spikes. Staminate flowers with 5 scarious mucronate sepals. Stamens 5; filaments distinct, subulate; anthers 2-celled. Pistillate flowers without calyx. Ovary 1-celled; stigmas 2–5, papillose or plumose. Ovules solitary. Utricle circumscissile or opening irregularly, or indehiscent. Seeds smooth, erect, shining.

Pistillate inflorescence of slender interrupted spikes; fruit circumscissile. 1. *A. tamariscina.*
Pistillate inflorescence of closely clustered spikes; fruit indehiscent or irregularly splitting.
  Stem stout, erect; leaf-blades ovate or lanceolate, acute or acuminate, broadest near the base. 2. *A. altissima.*
  Stem slender, decumbent or prostrate; leaf-blades obovate or spatulate, rounded or obtuse at the apex, broadest above the middle. 3. *A. subnuda.*

1. **A. tamariscìna** (Nutt.) Wood. Stem erect, much branched, 1–2 m. long; leaf-blades lanceolate or ovate-lanceolate, obtuse or notched at the apex, entire or undulate; bracts lanceolate, spinulose-tipped, scarious-margined. Swamps and alluvial soil: Ill.—S.D.—Colo.—N.M.—La. *Prairie—Plain—Texan.* Jl–S.

2. **A. altíssima** Riddell. Stem 1–3 m. high, with flexuose branches; leaf-blades lanceolate to rhombic-ovate, entire; bracts rigid, acuminate; sepals of the staminate flowers lanceolate, acuminate. *A. tuberculata* Moq. [G, BB]. *A. tamariscina tuberculata* Uline & Bray. [B]. Swamps: Ont.—S.D.—Colo.—Ohio. *Plain.* Jl–S.

3. **A. subnùda** (S. Wats.) Standl. Stem 1–4 dm. high, much branched in the pistillate plant; leaf-blades 1–7 cm. long, pale and yellowish green; staminate spike slender, 3–8 cm. long; pistillate flowers in dense, usually remote clusters; bracts lanceolate to subulate, half as long as the sepals. *A. tuberculata subnuda* S. Wats. [G]. *A. tamariscina prostrata* Uline & Bray [B]. Along streams: Ont.—Tenn.—S.D. *Allegh.* Jl–S.

3. TIDESTRÒMIA Standl.

Annual or perennial herbs, with stellate pubescence and mostly opposite, entire or merely undulate, short-petioled leaves. Flowers perfect, subtended by 3 bracts, solitary or clustered in the axils. Sepals 5, unequal, pubescent. Stamens 5; filaments united at their bases; anthers 1-celled. Ovary 1-celled; styles short; stigmas capitate or 2-lobed. Ovules solitary. Utricle subglobose, indehiscent. *Cladothrix* Nutt.; not Cohn.

f. 189

1. **T. lanuginòsa** (Nutt.) Standl. *Fig. 189.* Annual; stems diffusely branched; branches ascending or prostrate, 1–6 dm. long; leaf-blades rhombic-ovate to orbicular, entire, 0.5–2 cm. long; flowers in axillary clusters; bractlets obtuse; utricle glabrous, included in the calyx. *Cladothrix lanuginosa* Nutt. [B]. Dry soil: Tex.—Kans.—Utah.—Ariz.; Mex. *St. Plain—Son.* Jl–S.

4. FROELÍCHIA Moench.

Annual or biennial herbs, with woolly or silky pubescence and opposite, entire or undulate leaves. Flowers perfect, subtended by 3 bracts, in dense spikes. Calyx 5-lobed, woolly; tube longitudinally crested or tubercled at maturity. Stamens 5, included; filaments united into a tube; anthers 1-celled. Ovary 1-celled; styles short or wanting; stigmas capitate or brush-like. Utricle indehiscent, enclosed in the tube of the filaments.

Stout, 4–12 dm. tall; crest of fruiting calyx continuous, dentate. 1. *F. campestris.*
Slender, 2–5 dm. high; crest of fruiting calyx interrupted, forming distinct spines. 2. *F. gracilis.*

1. **F. campéstris** Small. *Fig. 190.* Biennial or annual; leaves numerous and approximate below; blades spatulate to oblong or broadly linear, acute, white-wooly beneath; spikes 1–10 cm. long. *F. floridana* Coult. & Nels., not Moq. Dry or sandy soil: Minn.—Ill.—Okla.—N.M.—Colo. *Prairie—Plain—Son.* Je–S.

2. **F. grácilis** Moq. Annual or perhaps biennial; stem usually branched at the base, 2–3 dm. high; leaves numerous near the base of the plant, often clustered; blades linear-oblanceolate or linear-oblong, 1–5 cm. long, acute, white-woolly; spikes 1–3 cm. long. Sandy valleys: Iowa—Ark.—Tex.—Ariz.—Colo.; Mex. *Prairie—Plain—Son.* Jl-S.

f. 190.

5. IRESÌNE P. Br.

Annual or perennial herbs, in the tropics shrubs or small trees. Leaves opposite, entire or serrulate. Flowers perfect, polygamous, or dioecious, in paniculate spikes or heads. Perianth of 5 distinct sepals. Stamens 5, the filaments united at the base into a short tube; anthers 2-celled. Ovary compressed; stigmas 2 or 3. Ovule 1, suspended. Utricle compressed.

1. I. Celòsia L. Annual or perennial; stem 4–30 dm. high, glabrous or sparingly villous at the nodes; leaves thin, bright green, oval or rhombic-ovate, petioled, 3–14 cm. long, acute or acuminate at the apex, rounded to cuneate at the base, glabrous or somewhat villous; flowers in large, compound panicles, white or straw-colored, the pistillate ones densely woolly at the base; sepals oblong, obtuse. *I. celosioides* L. *I. paniculata* (L.) Kuntze [G, B]. Dry soil: N.C.—Fla.—Mex.—Kans.; Cent. and S.Am. *Austral.—Trop.* Au–S.

Family 47. **NYCTAGINACEAE.** FOUR-O'CLOCK FAMILY.

Annual or perennial herbs (all ours), usually with swollen nodes, and alternate or opposite leaves, without stipules. Flowers regular, perfect, often subtended by bracts forming a calyx-like involucre. Perianth simple, corolla-like, campanulate or funnelform. Stamens 1–many. Pistil solitary; ovary 1-celled, surrounded by the perianth-tube. Fruit indehiscent, angled, ribbed, or winged.

Bracts distinct.
  Wings not completely encircling the fruit, interrupted above and below. 1. ABRONIA.
  Wings completely encircling the fruit. 2. TRIPTEROCALYX.
Bracts united; fruit ribbed; involucre rotate, in fruit becoming much enlarged and membranous. 3. ALLIONIA.

1. **ABRÒNIA** Juss. SAND VERBENA, SAND PUFFS.

Annual or perennial herbs, with opposite petioled leaves. Leaf-blades entire and oblique at the base. Flowers several in heads, surrounded by five or more distinct bracts. Perianth colored and corolla-like, salver-shaped, with an elongated tube, and expanding above into a 5-lobed limb. Stamens 3–5, included. Fruit leathery, 3–5-winged, or 3–5-ribbed. Seed filling the pericarp and adherent to it.

1. A. fràgrans. Nutt. *Fig. 191.* Perennial; stems several; branches erect, 15–25 dm. high, glabrous below, puberulent and subviscid above; leaves glabrous or nearly so, petioled; blades oval, ovate, or elliptic, rounded at the apex, 3–6 cm. long; fruit about 8 mm. long, 4 mm. broad, one third from the apex. Dry soil: S.D.—Iowa—Kans.—N.M.—Ida.—Mont. *Prairie—Plain—Submont.* My–Jl.

2. **TRIPTERÓCALYX** Hook. SAND PUFFS.

Branched annuals, with opposite petioled leaves. Blades entire and usually oblique at the base. Flowers perfect, in heads surrounded by 4–6 distinct bracts. Perianth corolla-like, salverform, with elongated tube and 5-lobed limb. Stamens 5, regularly adnate to the perianth-tube. Fruit almost orbicular in outline, completely surrounded by 2–4, strongly reticulate, membranous wings. Seeds cylindric-ellipsoid.

1. T. micránthus (Torr.) Hook. Annual; stem branched at the base, decumbent or ascending, 1–3 dm. high, more or less pubescent; leaf-blades 2–4 cm. long, elliptic, about equaling the petioles; bracts ovate, acuminate, 4–5 mm. long; fruit with the wings 1–1.5 (rarely 2) cm. broad; body spongy-reticulate, not nerved. *A. micrantha* Torr. Dry plains and in sandy soil: Sask.—Kans.—N.M.—Mont. *Plain—Mont.* My–Jl.

3. **ALLIÒNIA** Loefl. UMBRELLA-WORT.

Perennial herbs, with rather fleshy, entire, opposite leaves. Flowers perfect, 1–5 in each involucre, in ours rose-colored or white. Involucre gamophyllous, 5-lobed, in fruit enlarging and membranous. Perianth campanulate or funnel-form, often oblique. Stamens 2–5, unequal; filaments very slender and united at the base. Fruit club-shaped, 5-angled or 5-ribbed, usually pubescent and minutely tuberculate. Pericarp adhering to the seed. Endosperm mealy. *Oxybaphus* L'Hér.

Leaves cordate to broadly ovate-lanceolate, all distinctly petioled.
  Leaf-blades cordate or deltoid. — 1. *A. nyctaginea.*
  Leaf-blades ovate, rounded or cuneate at the base. — 2. *A. ovata.*
Leaves ovate-lanceolate, oblong, or linear, sessile, or only the lower short-petioled.
  Stem more or less hirsute as well as viscid.
    Fruit pubescent.
      Leaves ovate or broadly oblong, as well as the stem conspicuously hirsute. — 3. *A. hirsuta.*
      Leaves linear-lanceolate, almost glabrous; stem sparingly hirsute, or glabrous except under the nodes. — 4. *A. pilosa.*
    Fruit glabrous; leaves ovate or lanceolate. — 5. *A. Carletoni.*
  Stem glabrous below, not hirsute, viscid-puberulent above.
    Flowers solitary in the involucre, on short slender pedicels; fruit nearly glabrous. — 6. *A. glabra.*
    Flowers 2–3 in the involucres, subsessile; fruit decidedly pubescent.
      Leaves of the cymes much reduced and bract-like; upper portion of the stem densely and finely puberulent. — 7. *A. bracteata.*
      Leaves of the cymes neither much reduced nor bract-like.
        Plant prostrate or diffuse; involucres and branches of the inflorescence densely viscid-hairy. — 8. *A. diffusa.*
        Plant more simple, erect or ascending; branches of the inflorescence usually merely viscid-puberulent.
          Leaves from ovate- or obovate- to linear-lanceolate, usually over 5 mm. wide. — 9. *A. decumbens.*
          Leaves narrowly linear, less than 5 mm. wide. — 10. *A. linearis.*

1. **A. nyctagínea** Michx. Stems 3–10 dm. tall, stout, dichotomously branched; leaf-blades 2–10 cm. long; inflorescence paniculate; involucres in fruit about 1.5 cm. broad, pubescent; lobes ovate or triangular; perianth red, hairy without. *Oxybaphus nyctagineus* Sweet [G]. Rich soil: Man.—Ills.—La.—N.M.—Wyo. *Prairie—Plain.* My–Au.

2. **A. ovàta** Pursh. Stems 3–9 dm. high, stoutish, swollen at the nodes; leaf-blades ovate to lanceolate, 2–5 cm. long; inflorescence paniculate, pubescent; fruiting involucres fully 1.5 cm. broad, glabrate; lobes triangular; perianth usually white; fruit obovoid, constricted near the base. *O. floribundus* Am. auth.; [G] not Choisy. Dry soil: Wis.—Mo.—Tex.—N.M.—Wyo. *Prairie—Plain.* Jl–S.

3. **A. hirsùta** Pursh. *Fig. 192.* Stem strict, 4–8 dm. high, glandular in the paniculate inflorescence; involucre glandular-pubescent, in fruit 1.5 cm. wide; lobes broadly ovate; perianth purple or pink, pubescent without. *A. aggregata* (Ortega) Speng., a form with axillary involucres. *O. hirsutus* Sweet [G]. Sandy soil: Man.—Minn.—Neb.—N.M.—Wyo. *Prairie—Plain—Submont.* Je–Au.

4. **A. pilòsa** (Nutt.) Rydb. Stem erect or decumbent at the base, 3–12 dm. high; leaf-blades lanceolate or oblong, 2–9 cm. long, thick, nearly glabrous; inflorescence cymose, glandular; fruiting involucres 1.2–2 cm. broad; lobes triangular-ovate; perianth purplish. Dry sandy soil: Sask.—Wis.—La.—Tex.—Colo.—N.D. *Prairie—Plain—Submont.* Jl–Au.

5. **A. Carletòni** Standl. Stem about 1 m. high, simple below, soft-pubescent throughout; leaves sessile or nearly so, ovate or lanceolate, very thick, acutish, wavy-margined, strongly veined, 7 cm. long, puberulent on both sides; involucres about 15 mm. wide, copiously soft-pubescent, the lobes rounded; fruit 5 mm. long, with 5 prominent smooth ribs. Plains: Kans.—Colo.—Okla. *Plain.* Je–Jl.

6. **A. glàbra** (S. Wats.) Kuntze. Stem 6–10 dm. high, erect, branched, glabrous; leaf-blades linear, 6–12 cm. long, thick, glabrous, sessile; inflorescence a large panicle; flowers solitary; fruiting involucre about 1 cm. broad, glabrous; lobes triangular, thin, veiny. Dry soil: s Utah—Kans.—N.M.—Ariz.; Chihuahua. *St. Plain—Son.* Au–S.

7. **A. bracteàta** Rydb. Stem 4–12 dm. high, erect or ascending; leaf-blades oblong to linear-lanceolate or linear, 3–9 cm. long, thick; involucres often numerous, in terminal cymes, 10–14 mm. wide; lobes broadly ovate, obtuse; perianth white or pale, about 10 mm. broad. Dry and rocky soil: Mo.—Ala.—S.D. *Prairie—Plain.*

8. **A. diffùsa** Heller. Stem 2–3 dm. long; leaves sessile, linear-lanceolate, often narrowly so, 5–6 cm. long, thick; fruiting involucres 7–10 mm. wide; lobes triangular-ovate; perianth rose-colored to white, 6 mm. long; fruit inconspicuously ribbed. *A. glandulifera* A. Nels. *A. viscida* (Eastw.) Cockerell. Gravelly hills and sandy plains: N.D.—Kans.—N.M.—Ariz.—Wyo. *Son.—Plain—Mont.* Je–Au.

9. **A. decúmbens** Nutt. Stem erect or ascending, 4–15 dm. high; leaf-blades 3–10 cm. long, very thick, obtuse or blunt at the apex; involucres numerous, usually in terminal cymes, 1–1.5 cm. wide; lobes rounded-ovate, sometimes acutish; perianth pink, about 10 mm. broad. *A. decumbens* was originally described from a form with axillary involucres. *A. lanceolata* Rydb., in part [B], represents the ordinary form. *O. albidus* Auth. [G, BB]; not Sweet. Dry sandy soil on plains and prairies: Man.—Tenn.—Tex.—N.M.—Wyo. *Plain—Submont.* My–S.

10. **A. lineàris** Pursh. Stem slender, 3–15 dm. high, terete or 4-angled below, erect or ascending; leaf-blades thick, linear, 2.5–10 cm. long, undulate, sessile, or the lower sometimes short-petioled; perianth finely pubescent, purple. *O. angustifolius* Sweet. *O. linearis* B. L. Robins. [G]. *A. montanensis* Osterh. *A. Bodini* (Holz.) Morong., a form with axillary involucres. Dry soil: Minn.—La.—Ariz.—Mont.; Mex. *Prairie—Plain—Mont.—Son.* Jl.–O.

Family 48. **PHYTOLACCACEAE.** Pokeweed Family.

Herbs, shrubs, or trees. Leaves alternate with entire blades. Flowers perfect or dioecious, in terminal or axillary racemes or spikes. Sepals 4 or 5, imbricate, distinct or partially united. Corolla wanting. Stamens 4–8, rarely more; anthers introrse. Gynoecium of a single carpel or several united carpels; ovary 1–several-celled; ovules solitary in each cavity, mostly erect. Fruit a berry or an achene. Embryo curved around a mealy endosperm.

f. 193.

1. **PHYTOLÁCCA** L. Pokeberry.

Perennial herbs or shrubs. Flowers perfect, rarely dioecious, in terminal racemes which become lateral by the production of the stem from the upper leaf axis. Sepals 4 or 5, herbaceous, or slightly colored. Stamens 5–25, inserted at the base of the ovary, reduced to staminodia in the pistillate flowers. Ovary depressed, 5–12-celled. Fruit a pulpy berry, depressed. Seeds reniform.

1. **P. decándra** L. *Fig. 193.* Perennial herb, with a large poisonous root; stem erect, 1–3 m. high, glabrous; leaf-blades ovate or elliptic, 1–3 dm. long, acute or short-acuminate, abruptly narrowed into the petiole; racemes puberulent or glabrous, 1–2 dm. long; pedicels pink; calyx white or greenish; sepals obovate, 3 mm. long; berry dark purple. Rich soil: Me.—Fla.—Tex.—Minn. *E. Temp.* Jl–S.

Family 49. **TETRAGONIACEAE.** CARPET-WEED FAMILY.

More or less succulent herbs, with opposite or whorled leaves. Flowers in ours perfect, regular. Sepals 4 or 5. Corolla wanting in our species. Stamens 4 or 5, or many, hypogynous. Gynoecium of 2 or more united carpels; ovary 2–several-celled or by reduction 1-celled, superior or partly inferior; styles as many as the cells of the ovary. Fruit a circumscissile or loculicidal capsule.

Hypanthium wanting; capsule loculicidal; leaves whorled. 1. MOLLUGO.
Hypanthium manifest; capsule circumscissile; leaves opposite. 2. SESUVIUM.

1. **MOLLÙGO** L. INDIAN CHICKWEED, CARPET-WEED.

Annual herbs, with verticillate leaves and hyaline stipules. Flowers perfect, in axillary clusters, cymes, or racemes, pedicelled, usually white. Sepals 5, persistent, with hyaline margins. Stamens 3, opposite the angles of the ovary, or 5, alternating with the sepals; filaments filiform or subulate. Ovary 3–5-celled, superior; styles 3–5, distinct; stigmas entire. Capsule 3–5-valved.

1. **M. verticillàta** L. *Fig. 194.* Annual; stem branched at the base, prostrate, 0.5–3 dm. long; leaves in whorls of 4–8, short-petioled; blades unequal, spatulate to linear-oblanceolate, 1–3 cm. long, entire; sepals 2 mm. long, oblong, with white margins; capsule oblong or oval, 4–5 mm. long. Waste places and cultivated ground: Ont.—Fla.—Tex.—Calif.—Wash.; Mex.; W.Ind., C.Am., and S.Am. *Temp.—Trop.* Ja–D.

f.194

2. **SESÙVIUM** L. SEA PURSLANE.

Annual or perennial herbs or shrubby plants, with fleshy opposite leaves, without stipules. Flowers axillary. Hypanthium in ours turbinate. Sepals 5, usually horned on the back near the tip. Stamens 1–many, perigynous. Ovary 3–5-celled, half-inferior; styles 3–5, distinct, filiform; ovules numerous. Fruit a circumscissile capsule; seeds several or many in each cavity.

1. **S. verrucòsum** Raf. *Fig. 195.* Perennial, with a thick root; stems diffuse, branched at the base, prostrate or ascending, 1–8 dm. long; leaf-blades spatulate or oblanceolate, 1–3 cm. long; sepals ovate-oblong, 8–10 mm. long; stamens many; capsule oblong, 6 mm. long. *S. sessile* B. L. Robinson. [R]; not Pers. River banks and saline plains: Wyo.—Kans.—Tex.—Calif.; Mex. and S.Am. *St. Plain—Son.* Ap–N.

Family 50. **PORTULACACEAE.** Purslane Family.

Succulent plants, with opposite or alternate leaves. Flowers perfect, regular. Sepals 2, or in *Lewisia* 6–8. Petals 4 or 5, rarely more, imbricate. Stamens as many as the petals. Gynoecium of 2–5 united carpels; ovary 1-celled, usually superior; styles 2–5, distinct or partly united. Fruit a valvate or circumscissile capsule.

Ovary wholly superior.
  Sepals usually deciduous, scarious; capsule 3-valved; plant with a fleshy rootstock or root; ours with terete leaves. 1. Talinum.
  Sepals persistent, at least in part herbaceous.
    Plants with a corm; cauline leaves opposite; ovules usually 6. 2. Claytonia.
    Plants annual, or perennial with slender rootstocks; ovules usually 3.
      Stem-leaves a single pair; stem neither rooting at the nodes nor flagelliferous. 3. Limnia.
      Stem-leaves of several pairs; stem floating and rooting at the nodes, usually flagelliferous. 4. Crunocallis.
Ovary partly inferior, circumscissile; ours low spreading leafy annuals with flat leaves. 5. Portulaca.

1. **TALÌNUM** Adans. Fame-flower.

Perennial herbs or shrubby plants, ours with fleshy rootstocks. Leaves alternate or nearly opposite, in ours terete. Flowers in terminal cymes, or in some species axillary. Sepals 2, scarious, deciduous, or rarely persistent. Petals 5 or more, early withering. Stamens usually more numerous than the petals. Ovary superior; styles 3, more or less united. Capsules 1-celled, 3-valved, parchment-like. Seeds flattened, reniform, shining.

Flowers about 1 cm. wide; stamens less than 30.
  Stamens mostly 5; capsule ellipsoid. 1. *T. parviflorum.*
  Stamens 8–30; capsule globose. 2. *T. rugospermum.*
Flowers 2–3 cm. wide; stamens 30 or more; sepals rather persistent. 3. *T. calycinum.*

1. **T. parviflòrum** Nutt. *Fig. 196.* Scape usually 1 dm. high or less; leaves numerous, basal, terete, 2–5 cm. long, glabrous; sepals ovate, 2–3 mm. long; petals rose or whitish, 5–7 mm. long; capsule ellipsoid, 3–4 mm. long. Rocky soil: Minn.—S.D.—Ariz.—Tex.—Ark. *Prairie—Son.—Submont.* Jl–Au.

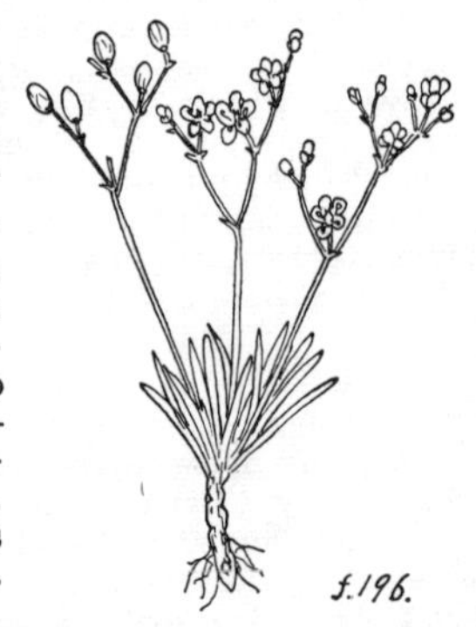

2. **T. rugospérmum** Holz. Scape 2 dm. high or less; leaves linear, terete, 2–5 cm. long; sepals ovate, early deciduous, 4 mm. long; petals rose-colored, about 8 mm. long; stigma lobes one fourth to one third as long as the style; anthers globose; capsule 4 mm. long and broad; seeds strongly roughened. It has been confused with the eastern *T. teretifolium* Pursh, which has shorter stigma lobes and oblong anthers. Prairies: *Ind.—Wis.—Minn. Prairie.* My-Au.

3. **T. calycìnum** Engelm. Scape 1–2 dm. high; leaves 1–5 cm. long, glabrous; sepals ovate-orbicular, 5 mm. long, tardily deciduous; petals 8–10, pink, 10–12 mm. long; capsule subglobose, 6–7 mm. long. Sandy soil: Ark.—Mo.—Neb.—Tex. *Prairie—Plain—Son.* Je–S.

2. **CLAYTÒNIA** (Gron.) L. Spring Beauty, Ground-nut.

Perennial fleshy herbs, with tuber-like corms or a thick fleshy root and short caudex. Basal leaves solitary or few, or clustered on the caudex; stem-leaves either 2, opposite, or a few alternate. Flowers in terminal racemes. Sepals 2, persistent. Petals 5. Stamens 5, adnate to the petals at the base. Ovary superior; styles 3, mostly united. Capsule 1-celled, 3-valved.

Leaves linear or linear-lanceolate, usually less than 1 cm. wide; sepals mostly obtuse. 1. *C. virginica.*
Leaves lanceolate, 1–3 cm. wide; sepals acute or acutish.
Plant 2–3 dm. high; leaf-blades acuminate; sepals 8–10 mm. long; petals 14–15 mm. long. 2. *C. robusta.*
Plant less than 2 dm. high; leaf-blades acute to rounded at the apex; sepals 6–7 mm. long; petal about 10 mm. long. 3. *C. caroliniana.*

1. **C. virgínica** L. *Fig. 197.* Corm globose, 1–2 cm. in diameter; stems 1–2.5 dm. high; basal leaves usually several; blades linear or narrowly linear-lanceolate, 5–12 cm. long; sepals 5–7 mm. long; petals oblong, 8–10 mm. long, white, or often pinkish, with rose-colored veins. Rich woods or among bushes: N.S.—Ga.—Tex.—Minn. *E. Temp.—Plain.* Mr–Au.

2. **C. robústa** (Somes) Rydb. Corm globose, 1–2.5 cm. thick; basal leaves petioled; blades lanceolate, 1–2 cm. wide; stem-leaves short-petioled, 5–15 cm. long, lanceolate, acuminate, 1–2.5 cm. wide; petals white, with pink or purple veins. *C. multicaulis robusta* Somes. Rich woods or copses: Ind.—Mo.—Iowa. *Ozark.*

3. **C. caroliniàna** Michx. Corm about 1 cm. thick; stem 1–2 dm. high; basal leaves petioled; blades ovate-lanceolate or lance-oblong, 3–7 cm. long, 12–18 mm. wide, obtuse; petioles of the stem-leaves 6–12 mm. long; blades 3–5 cm. long. Damp woods: N.S.—N.C.—Mo.—Neb.—Sask. *Canad.—Allegh.* Mr–My.

3. **LÍMNIA** L. Spanish Lettuce, Squaw Lettuce, Squaw Cabbage, Miners' Lettuce.

Annuals, or perennials, with slender rootstocks, rarely stoloniferous or bulbiferous in the basal leaf-axils. Basal leaves several; stem leaves opposite, a single pair, often more or less united. Inflorescence racemose. Sepals 2, somewhat unequal, persistent; petals 5, emarginate or obcordate. Stamens 5, opposite the petals. Styles 3. Capsule 3-valved from the apex; seeds 2–3.

1. **L. humifùsa** (Howell) Rydb. Annual; stem 5–10 cm. long; leaf-blades 5–15 mm. wide; stem-leaves connate, but usually less so on one side; calyx about 2 mm. long; pedicels 3–4 mm. long. *Montia humifusa* Howell. *C. parviflora depressa* A. Gray. *L. depressa* (A. Gray) Rydb. Banks and springy places: S.D.—Colo.—Ariz.—Calif.—B.C. *W. Temp.—Submont.* My–Jl.

4. **CRUNOCÁLLIS** Rydb. Water Spring-Beauty.

Floating or creeping perennials, rooting at the nodes, usually flagelliferous, bearing winter buds at the ends of the flagella. Leaves opposite, several pairs. Inflorescence racemose. Sepals 2. Petals 5, equal. Styles 3; ovules 3. Capsule 3-valved; seeds 2–3, round-reniform, muriculate.

1. **C. Chamissònis** (Ledeb.) Rydb. *Fig. 198.* Perennial; leaves spatulate or oblanceolate, 1–4 cm. long; raceme 1–9-flowered; pedicels recurved in fruit; petals pale rose-color, 6–8 mm. long; seeds densely muricate. *Claytonia Chamissoi* Ledeb. [B]. *Montia Chamissonis* Greene. *M. Chamissoi* Durand & Jacks. [G]. Wet places and in springs: Alaska—Calif.—N.M.—Minn. *W. Temp.—Prairie—Plain—Mont.* Je–Au.

5. **PORTULÁCA** (Tourn.) L. PURSLANE, PUSSLEY.

Annual or perennial fleshy herbs, mostly with diffusely branched stems. Leaves terete or flat, opposite or alternate, with bristle-like or scarious stipules. Sepals 2, deciduous. Petals 4–6, fugaceous. Stamens 8 or more. Ovary partly or wholly inferior; styles 3–8, slender. Capsule 1-celled, circumscissile. Seeds flattened, reniform.

Leaf-blades flat, glabrous in the axils or nearly so; petals yellow.
Sepals pointed in the bud; seeds obscurely granulate.
Stamens 7–12; leaves 1–3 cm. long. 1. *P. oleracea.*
Stamens 12–18; leaves 2.5–5 cm. long. 2. *P. neglecta.*
Sepals obtuse in the bud; seeds echinate-tuberculate. 3. *P. retusa.*
Leaf-blades terete or nearly so, pilose in the axils; petals red or purple.
Flowers 7–10 mm. broad. 4. *P. parvula.*
Flowers 25–50 mm. broad. 5. *P. grandiflora.*

1. **P. oleràcea** L. *Fig. 199.* Annual; stem diffusely branched at the base, prostrate, 1–6 dm. long; leaves fleshy, sessile or nearly so, cuneate or obovate, 1–3 cm. long, rounded at the apex; buds flattened; stamens 7–12. Cultivated grounds and waste places: Me.—Fla.—N.M.—Wash.; Mex. and W.Ind.; nat. from the Old World. *Temp.—Plain.* Ja–D.

f.199

2. **P. neglécta** Mack. and Bush. Annual; stem ascending, thick, forming broad tufts 1–10 dm. broad; leaves obovate-spatulate, 2.5–5 cm. long, 1–2.5 cm. wide, obtuse or retuse; sepals ovate-oblong, winged at the apex; seeds minutely tuberculate. Bottom lands: Mo.—Kans.—Ark. *Ozark.* Jl–S.

3. **P. retùsa** Engelm. Annual; stem ascending, branched near the base; leaves cuneate, 1–2.5 cm. long, often retuse, fleshy, sessile; sepals carinately winged; petals minute; stamens 10–19. Sandy soil: Ark.—Utah—Ariz.—Tex.; reported from Minn., Iowa. *Son.—Submont.* Jl–S.

4. **P. párvula** A. Gray. Annual; stem ascending, branched; leaves nearly terete, linear or linear-subulate, 1 cm. long or less, with numerous hairs in the axils; petals red, 3–4 mm. long, retuse; capsule globose. It has been confused with the tropical *P. pilosa* L. [B, G]. Sandy soil: Mo.—Colo.—Calif.; Trop. Am. *Austral—Son.*

5. **P. grandiflòra** Hook. Annual; stems ascending or spreading, 1.5–3 dm. long; leaves terete, 2–5 cm. long, 2 mm. thick; sepals obtuse, scarious-margined; petals obovate, red or purplish to yellow or white; seeds shining. Waste places: Occasionally escaped from cultivation; nat. of S. Am.

Family 51. CORRIGIOLACEAE. WHITLOW-WORT FAMILY.

Herbs, with narrow opposite leaves with scarious stipules. Flowers perfect, inconspicuous, in dichotomous cymes. Sepals 4–5, distinct or partly united. Corolla wanting. Stamens 4 or 5. Pistil solitary; ovary 1-celled. Fruit a utricle or an achene.

Stipules scarious.
Calyx sessile in a pair of bracts resembling the sepals; sepals awn-tipped. 1. PARONYCHIA.
Calyx pedicelled; bracts resembling the leaves; sepals not awn-tipped. 2. ANYCHIA.
Stipules wanting. 3. SCLERANTHUS.

1. **PARONÝCHIA** (Tourn.) Adans. WHITLOW-WORT.

Annual or perennial herbs, with stems often branched at the base. Leaves opposite, narrow. Sepals 5, narrow, concave or hooded at the awn-tipped apex. Stamens 5, included; filaments inserted at the base of the ovary, alternating with 5 small staminodia. Styles partially united. Utricle included. Radicle ascending.

Flowers solitary; leaves scarcely exceeding the bracts; plants densely pulvinate. 1. *P. sessiliflora.*
Flowers more or less clustered; leaves much longer than the bracts.
  Plants low and diffuse, less than 1 dm. high; calyx fully 3 mm. long.
    Inflorescence much contracted; bracts exceeding the flowers. 2. *P. depressa.*
    Inflorescence more open; bracts shorter than or merely equaling the flowers. 3. *P. diffusa.*
  Plants taller, 1 dm. or more high; stems erect or ascending; calyx 2–2.5 mm. long.
    Branches of the cymes ascending; calyx about 2.5 mm. long; sepals lanceolate, gradually acuminate. 4. *P. Jamesii.*
    Branches of the cymes divaricate; calyx about 2 mm. long; sepals oblong, abruptly acuminate. 5. *P. Wardii.*

1. **P. sessiliflòra** Nutt. Pulvinate-cespitose perennial, 4–10 cm. high; leaves linear or linear-subulate, spinulose-tipped, 4–6 mm. long; sepals lanceolate, brown with narrow scarious margins, spine-tipped, about 3 mm. long. Dry ridges: Man.—Tex.—Utah—Alta. *Plain—Mont.* Je–Au.

2. **P. deprέssa** Nutt. Cespitose perennial, somewhat woody at the base; stems diffuse, much branched; leaves linear, 5–8 mm. long, scabrous, spinulose-tipped; sepals linear, 2.5–3 mm. long, usually exceeded by the bracts. *P. Jamesii depressa* T. & G. [B]. Dry hills or plains: Man.—S.D.—Neb.—Wyo. *Plain—Submont.* Je–Au.

3. **P. diffùsa** A. Nels. Cespitose perennial, somewhat woody at the base; stems 5–15 cm. long; leaves linear, 7–10 mm. long, spinulose-tipped, scabrous; sepals linear, 2.5–3 mm. long; spine-tips 0.5 mm. long. Dry plains, hills, and mountains: S.D.—Kans.—Colo.—Wyo.—Alta. *Plain—Alp.* Je–S.

4. **P. Jamèsii** T. & G. Perennial, shrubby and branched at the base; stems erect, 1–3 dm. high, forking above; leaves strongly ascending, 1–2 cm. long, callous-mucronate; sepals about 2.5 mm. long, lanceolate; spine-tips strongly ascending. Dry plains and hills: Neb.—Tex.—N.M.—Wyo. *Plain—Mont.* Je–Au.

5. **P. Wárdii** Rydb. *Fig. 200.* Perennial, shrubby and branched at the base; stems erect, 1–2 dm. high; leaves narrowly linear or linear-filiform, 0.7–2 cm. long, early deciduous, mucronate, minutely scabrous. Dry stony soil: Kans.—Tex.—N.M.—Colo. *St. Plain—Plain—Submont.* Jl–O.

2. **ANÝCHIA** Michx. FORKED CHICKWEED.

Annual herbs, with forked stems; leaves opposite. Flower solitary or clustered in the forks. Calyx persistent; sepals 5, greenish, mucronate on the back. Stamens 2 or 3, rarely 5, inserted at the base of the ovary. Styles very short; stigmas spreading. Utricle as long as or longer than the sepals. Seeds erect.

Glabrate, sepals oblong; utricle longer than the calyx. 1. *A. canadensis.*
Pubescent; sepals ovate; utricle about equaling the calyx. 2. *A. polygonoides.*

1. **A. canadénsis** (L.) B. S. P. Stem 0.5–4 dm. high, simple below; leaves thin, oblong or elliptic, 0.5–2 cm. long, short-petioled; calyx 1 mm. long; sepals oblong, obtuse, slightly white-margined. Open woods and hillsides: Mass.—Ga.—Ark.—Kans.—Minn. *Canad.—Allegh.*

2. **A. polygonoìdes** Raf. Stem erect or decumbent, 0.3–3 dm. long; leaves thick, oblanceolate or linear-elliptic, 0.5–1.5 mm. long; sepals ovate, acute, green. *A. dichotoma* A. Gray [B]; not Michx. Dry soil: Me.—Fla.—Ark.—Minn. *E. Temp.*

3. SCLERÁNTHUS L. Knawel.

Annual or perennial herbs. Stems dichotomously branched. Leaves opposite, clasping, subulate or linear, without stipules. Flowers in small cymes. Hypanthium well developed. Sepals 5, or rarely 4, persistent. Stamens 1–10, included, inserted on the rim of the hypanthium. Gynoecium of 2 united carpels, 1-celled; styles 2. Ovule solitary, amphitropous. Fruit a 1-seeded utricle.

1. **S. ánnuus** L. Annual, bright green; stem branched at the base, mostly decumbent, 0.5–3 dm. high, pubescent; leaves linear-filiform, 0.5–1.5 cm. long, ciliate; flowers subsessile; hypanthium together with the calyx 3 mm. long; sepals lanceolate, obtusish, hyaline-margined. Waste places: Que.—Minn.—Fla.; nat. from Eur. Mr–O.

Family 52. ALSINACEAE. Chickweed Family.

Herbs with opposite leaves, with or without stipules. Flowers perfect, usually cymose. Sepals 4 or 5, distinct. Petals 4 or 5, clawless, rarely wanting. Stamens twice as many as the sepals or fewer. Gynoecium of 2–5 united carpels; ovary partly or completely 2–5-celled; styles 2–5. Fruit a capsule, opening with as many or twice as many valves as carpels. Seeds several to many on a central placenta. Embryo curved.

Stipules wanting.
- Capsule opening with twice as many valves or teeth as there are styles; petals deeply 2-cleft.
  - Capsule short, ovate or oblong, opening with usually 6 valves; styles usually 3. 1. Stellaria.
  - Capsule long, cylindric, often curved, opening with usually 10 teeth at the apex; styles usually 5. 2. Cerastium.
- Capsule opening with as many, entire or at length 2-cleft, valves as there are styles; petals entire or merely notched at the apex.
  - Styles fewer than the sepals, or rarely of the same number and then opposite them.
    - Seeds with a basal membranous appendage (strophiole) at the hylum. 3. Moehringia.
    - Seeds not strophiolate.
      - Valves of the capsule entire. 4. Sabulina.
      - Valves of the capsule at last 2-cleft. 5. Arenaria.
  - Styles as many as the sepals and alternate with them. 6. Sagina.

Stipules present.
- Leaves whorled; styles 5. 7. Spergula.
- Leaves opposite; styles usually 3.
  - Styles distinct; sepals not spinulose-tipped. 8. Spergularia.
  - Stigmas sessile; sepals spinulose-tipped. 9. Loeflingia.

1. STELLÀRIA L. Chickweed, Starwort.

Annual or perennial herbs, with weak spreading stems and alternate leaves. Flowers usually in open cymes. Sepals 5, rarely 4. Petals 5 or 4, deeply 2-cleft, or rarely wanting. Stamens 10 or fewer, hypogynous. Ovary 1-celled; styles 3, rarely 4 or 5, opposite the sepals, if of the same number. Capsule relatively short, opening by twice as many valves as there are styles. Seeds flattened or globose. *Alsine* L.

Leaf-blades ovate or rhombic-ovate, petioled; plant annual. — 1. *S. media.*
Leaf-blades sessile; plant perennial.
  Stem 2–3 dm. high, many-flowered.
    Upper bracts at least scarious.
      Petals minute or none; leaves linear. — 2. *S. alpestris.*
      Petals exceeding or equaling the sepals.
        Leaves broadest at about the middle, narrowed at the base. — 3. *S. longifolia.*
        Leaves broadest near the base.
          Inflorescence many-flowered, the branches spreading. — 6. *S. graminea.*
          Inflorescence few-flowered, the branches erect.
            Sepals lanceolate, very acute, nearly equaling the capsule. — 4. *S. stricta.*
            Sepals ovate or ovate-lanceolate, obtuse, scarcely more than half as long as the capsule. — 5. *S. longipes.*
    None of the bracts scarious.
      Petals equaling or exceeding the sepals; leaves less than 2 cm. long. — 7. *S. crassifolia.*
      Petals shorter than the sepals or none; leaves usually more than 2 cm. long. — 8. *S. borealis.*
  Stem 3–15 cm. high, usually 1–3-flowered, rarely 4–6-flowered. — 9. *A. laeta.*

1. **S. mèdia** (L.) Cyrill. Stem diffusely branched, prostrate or ascending, 1–4 dm. long, glabrous except the pubescent lines; leaf-blades 0.5–3.5 cm. long; sepals oblong, glandular-pubescent, about equaling the capsule; petals shorter than the sepals. *A. media* L. [R,B]. CHICKWEED. Waste places, cultivated grounds, etc.: Greenl.—Fla.—Calif.—B.C.; W.Ind.; nat. from Eur. *Temp.* Ja–D.

2. **S. alpéstris** Fries. Stem 2–4 dm. high, angled; leaves sessile, linear, 2–4 cm. long, 2–4 mm. wide; bracts lanceolate, the lower foliaceous; sepals lanceolate, about 3 mm. long, very acute; petals about half as long or less, or lacking. *A. alpestris* (Fries) Rydb. *A. borealis alpestris* Britt. [B]. Wet places: Ont.—Sask.—Colo.—Utah.—Alta.; n Eu. *Boreal.—Submont.—Mont.* Je–Au.

3. **S. longifòlia** Muhl. Stem sharply 4-angled, 2–5 dm. high; leaves sessile, lance-linear, often ciliate towards the base, 2–6 cm. long, 2–6 mm. wide; bracts lanceolate; branches of the cyme and pedicels spreading; sepals lanceolate, acute, 3-nerved, about 3 mm. long, shorter than the petals. *A. longifolia* (Muhl.) Britton [B]. Low meadows and swamps: Lab.—Newf.—Md.—N.M.—Alaska; Eurasia. *Temp.—Plain—Subalp.* My—Au.

4. **S. strícta** Richards. *Fig. 201.* Stem 2–5 dm. high, angled; leaves linear-lanceolate, 2–4 cm. long, 2–4 mm. wide, spreading; pedicels strongly ascending or almost erect; calyx 4–5 mm. long; sepals narrowly scarious-margined; petals 5–6 mm. long. *A. strictiflora* Rydb. *S. longipes* Am. auth.; not Goldie. Wet meadows: w Ont.—Colo.—Calif.—B.C. *Plain—Subalp.* My–Au.

f.201.

5. **S. lóngipes** Goldie. Stem much branched, 4-angled, 1–3 dm. high; leaves linear-lanceolate, 1–3 cm. long, glabrous, rather firm, shining; pedicels 2–5 cm. long; calyx 4–5 mm. long; sepals broadly scarious-margined; petals slightly exceeding the sepals. *A. longipes* Coville [B]. *S. valida* Goodding. Wet places: Greenl.—Que.—Colo.—Calif.—Alaska. *Arct.—Boreal—Submont.—Mont.* Jl–Au.

6. **S. gramìnea** L. Stem 4-angled, ascending or reclining, 3–5 dm. high; leaves linear-lanceolate, ciliate at the base; inflorescence terminal, diffuse; petals about as long as the sepals, lanceolate; capsule oblong, exceeding the calyx. *A. graminea* (L.) Britton [B]. Wet meadows and waste places: Newf.—Minn.—Iowa—Md.; adv. from Eur. *E. Temp.* My–Jl.

7. **S. crassifòlia** Ehrh. Stem weak, ascending, with short internodes, 2–4 dm. long; leaves small, 6–20 mm. long, lanceolate or oblong-lanceolate, acutish; cymes few-flowered; sepals ovate-lanceolate, acuminate, about 3 mm. long, exceeded by the petals. *A. crassifolia* Britton [B]. Wet ground: Lab.—Pa.—Colo.—Alta. *Boreal.—Mont.—Subalp.* Je–Au.

8. **S. boreàlis** Bigel. Stem 1.5–3 dm. high, glabrous; leaves linear-lanceolate or the lower linear-oblanceolate, 1.5–3.5 cm. long, 1-ribbed; pedicels scattered, mostly axillary, 1.5–8 mm. long; sepals ovate-lanceolate, scarious-margined, acute, petals often wanting. *A. borealis* Britton [B]. Wet places: Lab.—N.J.—N.D. *E. Boreal.* Je–Au.

9. **S. laèta** Richardson. Stem 5–15, rarely 20 cm. long, densely leafy; leaves 1–2 cm. long, 2–4 mm. wide, bluish-green, sometimes glaucous, shining; bracts foliaceous, or (in several flowered plants) the upper scarious; pedicels erect; sepals narrowly lanceolate, very acute, almost equaling the capsule; petals about 5 mm. long. *A. laeta* Rydb. [R]. *S. longipes laeta* S. Wats. [G]. Wet places in the mountains: Man.—N.M.—Calif.—Alaska—Arctic Sea. *Arct.—Subarct.—Mont.—Alp.* My–Au.

2. CERÁSTIUM L. MOUSE-EAR CHICKWEED.

Annual or perennial herbs, with pubescent, often viscid foliage. Leaves opposite. Flowers cymose. Sepals 5, rarely 4. Petals as many, white, 2-cleft. Stamens usually 10. Ovary one-celled; styles as many as the sepals and opposite them, rarely fewer. Capsule cylindric, often curved, opening by 10, rarely 8, tooth-like valves.

Annuals; pod 2–3 times as long as the calyx.
- Pedicels in fruit 1–3 times as long as the calyx, straight, or nearly so. — 1. *C. brachypodum.*
- Pedicels in fruit 5 times as long as the calyx or longer; strongly curved above. — 2. *C. nutans.*

Perennials or rarely annuals; pods 1–2 times as long as the calyx.
- Leaves oblong, oval, or ovate, mostly obtuse, or barely acutish. — 3. *C. vulgatum.*
- Leaves, at least the lower ones, linear or linear-lanceolate, acute.
  - Stem villous with reflexed hairs, glandular only on the upper part.
    - Capsule twice as long as the sepals; leaves rather thick, densely pubescent, the upper lanceolate or lance-linear. — 4. *C. velutinum.*
    - Capsule about half longer than the calyx.
      - Leaves thin and flaccid, all linear-elongate. — 5. *C. arvense.*
      - Leaves firm, with a thick midrib, rather short. — 6. *C. campestre.*
  - Stem glandular-viscid throughout.
    - Leaves rather thin, densely glandular-viscid, linear; bracts short, ovate. — 7. *C. oreophilum.*
    - Leaves firm and thick, linear-oblong, with a prominent rib; bracts lanceolate. — 8. *C. strictum.*

1. **C. brachýpodum** (Engelm.) B. L. Robinson. Annual; stem often branched at the base, finely villous-viscid, 1–3 dm. high; leaves oblanceolate or oblong, obtuse, 1–3 cm. long, viscid-villous; flowers in rather open cymes. Dry sandy soil: S.D.—Mo.—Alta.—Tex.—Ariz.—Ore.; Mex. *Prairie—Plain—Submont.* Ap–Jl.

2. **C. nùtans** Raf. Annual; stem often branched at the base, 2–5 dm. high, viscid-villous; leaves oblong or oblong-lanceolate, usually acute, or the lower oblanceolate, 2–7 cm. long; flowers in open cymes; sepals about 4 mm. long, lanceolate; petals slightly longer. *C. longipedunculatum* Muhl. [B]. *C. Bakeri* Greene. Wet places: N.S.—N.C.—Ariz.—Ore.—B.C. *Temp.—Plain—Mont.* My–Au.

3. **C. vulgàtum** L. Perennial, cespitose at the base; stems 1–4 dm. high, viscid-pubescent; leaves oblong or elliptic, obtuse, 1–3 cm. long, villous; cymes elongate; bracts foliaceous; at least the lower pedicels exceeding the sepals in fruit; sepals about 5 mm. long, obtuse, scarious-margined; petals about as long. Roadsides and fields: Lab.—Fla.—Ida.—Wash.—Alaska. *Temp.—Plain.* My–Jl.

4. **C. velùtinum** Raf. Cespitose perennial; stems decumbent at the base, densely pubescent with reflexed hairs; lower leaves linear, 2–5 cm. long, 4–7 mm. wide, densely villous, the upper broader and shorter; cymes open, 10–15-flowered; sepals 5–6 mm. long, acute, broadly scarious-margined, villous; petals about twice as long, capsule 10–12 mm. long. *C. arvense villosum* Holl. & Britt. *C. arvense velutinum* Britt. [B]. Rocky places: N.Y.—Md.—Iowa—Minn. *Allegh.* My–Jl.

5. **C. arvénse** L. Cespitose perennial; stems 2–4 dm. high, erect or ascending; primary leaves of the stem linear-lanceolate or linear, 2–4 cm. long, more or less villous; inflorescence many-flowered, somewhat glandular; sepals about 4 mm. long, lanceolate, acute, light green, only slightly viscid. *C. angustum* Greene [R]. Hillsides: Lab.—Del.—S.D.—Wyo.—Alaska; Eurasia. *Boreal—Plain—Submont.* Je–Jl.

6. **C. campéstre** Greene. *Fig. 202.* Cespitose perennial; stems decumbent at the base or ascending, 1–2 dm. high; leaves 1–2 cm. long; linear-lanceolate, ascending, or the lower ones oblanceolate; sepals lanceolate, about 5 mm. long; petals about twice as long. Plains and hills: Minn.—S.D.—Colo.—Yukon. *Plain—Submont.*

7. **C. oreóphilum** Greene. Cespitose perennial; stems decumbent at the base, 1–2 dm. high, glandular-pubescent; leaves 1–3 cm. long, viscid-pubescent; cymes many-flowered; calyx about 5 mm. long, glandular; sepals ovate-lanceolate, acute, more or less scarious-margined; petals 8–9 mm. long. Wet places in the mountains: N.D.—S.D.—Colo.—nw N.M. *Submont.—Subalp.* My–Jl.

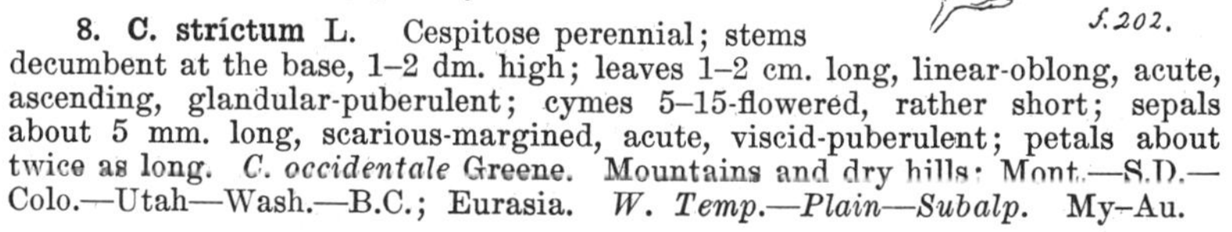

f. 202.

8. **C. stríctum** L. Cespitose perennial; stems decumbent at the base, 1–2 dm. high; leaves 1–2 cm. long, linear-oblong, acute, ascending, glandular-puberulent; cymes 5–15-flowered, rather short; sepals about 5 mm. long, scarious-margined, acute, viscid-puberulent; petals about twice as long. *C. occidentale* Greene. Mountains and dry hills: Mont.—S.D.—Colo.—Utah—Wash.—B.C.; Eurasia. *W. Temp.—Plain—Subalp.* My–Au.

3. MOEHRÍNGIA L.

Low perennials, with slender rootstocks. Leaves opposite, sessile or nearly so, soft. Flowers solitary in the axils or in terminal cymes. Sepals and petals 4 or 5. Petals entire. Stamens 8 or 10. Capsule oblong or ellipsoid, at first 3-celled, opening by 3 two-cleft valves. Seeds few, smooth and shining, appendaged at the hilum by a membranous broad strophiole.

Stem terete; leaves oval or elliptic-oblong; sepals obtuse or acutish. 1. *M. lateriflora.*

Stem angled; leaves lanceolate or oblanceolate; sepals very acute or acuminate. 2. *M. macrophylla.*

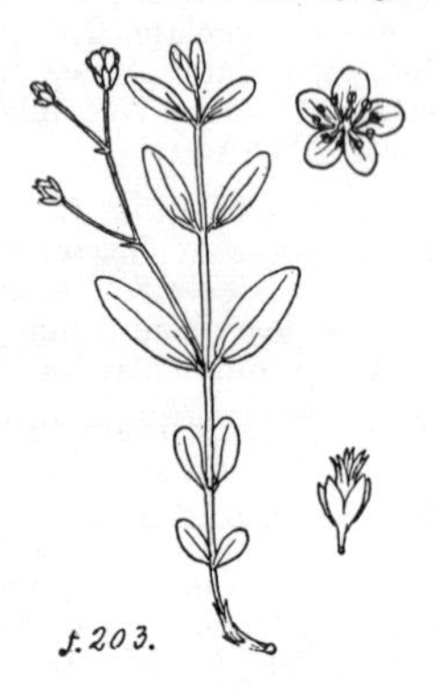

1. **M. laterifl̀ora** (L.) Fenzl. *Fig. 203.* Stems puberulent, decumbent at least at the base, 1–2 dm. high; leaves obtuse or rounded at the apex, puberulent, ciliolate on the margins and ribs, 1–3 cm. long; cymes corymbiform, 1–6-flowered; sepals ovate, 2.5 mm. long; petals obovate, 4–6 mm. long. *Arenaria lateriflora* L. [G]. Wet places among bushes: Lab.—N.J.—N.M.—Ore.—Alaska; Eurasia. *Boreal.—Submont.—Mont.* Je–Au.

2. **M. macrophýlla** (Hook.) Torr. Stems decumbent at the base, finely puberulent; leaves usually acute or acuminate at both ends, 2–7 cm. long; cymes 1–5-flowered; sepals ovate-lanceolate, about 4 mm. long, usually exceeding the petals. *A. macrophylla* Hook. [G]. Wet places among bushes: Lab.—Vt.—n N.M.—Calif.—B.C. *Boreal—Submont.—Subalp.* My–Au.

4. **SABULÌNA** Reichenb. SANDWORT.

Annual or perennial herbs, often tufted. Leaves opposite, without stipules, sometimes fascicled. Flowers in terminal cymes or rarely solitary in the axils. Sepals 5, often thick. Petals 5, white, entire or merely notched. Stamens 10. Ovary 1-celled; styles usually 3. Capsule ovoid, opening with as many entire valves as there are styles. Seeds several or many, not strophiolate. *Alsine* Gaertn.; not L. *Alsinopsis* Small.

Leaves linear-filiform or linear-subulate, 1–2 cm. long, 1-ribbed or the lateral ribs faint.
  Leaf-blades rigid, pointed, strongly ribbed beneath.
    Petals longer and capsule shorter than the sepals.
      Leaves 8–20 mm. long; inflorescence open; sepals ovate, 4 mm. long. 1. *S. stricta.*
      Leaves 5–10 mm. long; inflorescence short and dense; sepals lanceolate, 5 mm. long. 2. *S. texana.*
    Petals shorter and capsule longer than the sepals. 3. *S. litorea.*
  Leaf-blades soft, merely acute, not strongly ribbed beneath.
    Petals twice as long as the sepals, which are equaling the pod. 4. *S. patula.*
    Petals not exceeding the sepals, which are shorter than the pod. 5. *S. dawsonensis.*
Leaves short-subulate, less than 1 cm. long, distinctly 3-ribbed. 6. *S. propinqua.*

1. **S. strícta** (Michx.) Small. Annual or perennial; stem tufted, glabrous, 1.5–4 dm. high; leaves filiform, broadened at the base, 1–2 cm. long, with smaller fascicled ones in the axils; cymes open; pedicels slender, 6–35 mm. long; sepals 3-ribbed, acute; petals elliptic, 7–8 mm. long. *Arenaria stricta* Michx. [G]. *A. Michauxii* (Fenzl) Hook. [B]. Rocky places: N.H.—Va.—Mo.—Kans.—S.D. *Canad.—Allegh.* Je–Jl.

2. **S. texàna** (B. L. Robinson) Rydb. Perennial; stem branched at the base, glabrous, densely leafy below the middle, 1–2 dm. high; leaves slightly curved or straight, connate at the thickened node; pedicels slender, 5–15 mm. long; sepals long-acuminate, strongly 3-ribbed; petals 6–8 mm. long. *Arenaria stricta texana* B. L. Robinson [G]. *A. texana* Britt. [B]. Rocky hillsides; Mo.—Kans.—Tex. *Ozark—Texan.* Je–Jl.

3. **S. litòrea** (Fern.) Rydb. Perennial; stem tufted, 6–14 cm. high, leafy towards the base; leaves firm, subulate, 1–2 cm. long, with smaller tufted ones in axils; cymes open; pedicels 1–3 cm. long; sepals 3-ribbed, lanceolate, 4–4.5 mm. long; capsule 5 mm. long. *Arenaria litorea* Fern. Beeches and calcareous bluffs: Newf.—Que.—Minn.—Man. *Canad.* Je–Au.

4. **S. pátula** (Michx.) Small. Annual, with a slender root; stem branched from the base, the branches slender, 5–30 cm. long; leaves filiform, fleshy, 1–4 cm. long; branches of the cyme divergent; pedicels slender, 5–40 mm. long; sepals lanceolate, 3–3.5 mm. long, 3–5-ribbed; petals 7–10 mm. long, retuse at the apex. *Arenaria patula* Michx. [G, B]. Rocky woods and open places: Ky.—Minn.—Tex.—Tenn. *Allegh.* Ap–My.

5. **S. dawsonénsis** (Britton) Rydb. Stem branched from the base, 1–3 dm. high; leaves filiform or linear-subulate, 1–2 cm. long; cyme open, with slender pedicels; bracts lanceolate or subulate, green; sepals oblong-lanceolate, acute, 3-nerved, 4 mm. long; petals oblong. *Alsinopsis dawsonensis* (Britton) Rydb. [R]. Hillsides: Yukon—Alta.—S.D. *Arct.—Subarct.—Submont.—Subalp.* Je–Jl.

6. **S. propínqua** (Richardson) Rydb. Stems branched from the base, 3–10 cm. high; leaves linear-subulate, flat, 3-nerved, ascending, 5–10 mm. long; cyme open, branches ascending; sepals lanceolate, 3 mm. long; petals about 4 mm. long. *Arenaria propinqua* Richardson. *A. verna propinqua* Fern. *A. aequicaulis* A. Nels. *Alsinopsis propinqua* Rydb. [R]. Sandy soil: Hudson Bay—N.M.—Calif.—B.C.—Mack. *Arct.—Boreal—Submont.—Alp.* Jl–Au.

5. ARENÀRIA L. Sandwort.

More or less diffusely branched annuals or perennials, sometimes woody at the base. Leaves opposite, without stipules, often also fasciculate ones in their axils. Flowers in open or subcapitate cymes, or solitary in the leaf-axils. Sepals 5, often strongly ribbed. Petals white, entire or notched, in one species 2-cleft, sometimes wanting. Stamens normally 10. Ovary 1-celled; styles 3, rarely 2, or 4 or 5. Capsule globose or oblong, opening by 3 two-cleft valves. Seeds numerous, not strophiolate.

Annuals; leaves ovate, 4–7 mm. long. — 1. *A. serpyllifolia.*
Perennials; leaves filiform or subulate.
  Cymes open; sepals 4–5 mm. long. — 2. *A. Fendleri.*
  Cymes densely congested; sepals 6–9 mm. long. — 3. *A. Hookeri.*

1. **A. serpyllifòlia** L. Stem branched from near the base, decumbent or ascending, 1–3 dm. high, finely puberulent; leaves 4–7 mm. long, ovate, acute or acuminate, distinctly 3–5-nerved, the lowest short-petioled, the rest sessile; sepals acuminate, about 3 mm. long; petals small, scarcely 2 mm. long. Sandy soil: Que.—Fla.—Ore.—B.C.; W.Ind.; nat. from Eur. My–S.

f.204.

2. **A. Féndleri** A. Gray. Stems 1–3 dm. high, strict, glabrous beneath, glandular above; leaves filiform, glabrous, more or less glaucous, rather firm, 3–8 cm. long; pedicels rather short; sepals 4–5 mm. long. Mountains and dry hills: Wyo.—Kans.—N.M.—Ariz. *Submont.—Alp.* Je–Au.

3. **A. Hoòkeri** Nutt. *Fig. 204.* Caudex woody, tufted; stems 4–15 cm. high, terete, more or less pubescent; leaves subulate, ascending or spreading, 0.5–5 cm. long, pungent; bracts lanceolate, scarious-margined; sepals lanceolate, 6–9 mm. long; petals oblong. *A. pinetorum* A. Nels. Dry hills: Mont.—Neb.—Colo. *Plain—Submont.* Je–Au.

6. SAGÌNA L. Pearlwort.

Low matted annuals or perennials. Leaves opposite, filiform or subulate. Flowers axillary, on more or less elongated pedicels. Sepals 4 or 5, persistent.

Petals 4 or 5, white, entire or notched, or wanting. Stamens usually 5, alternate with the sepals. Capsule 4- or 5-valved, the valves opposite the sepals. Seeds usually numerous, not strophiolate.

Flowers usually 4-merous; sepals longer than the petals. 1. *S. procumbens.*
Flowers 5-merous; sepals shorter than the petals. 2. *S. saginoides.*

1. **S. procúmbens** L. Annual or perennial; stem 2–8 cm. high, depressed or spreading, glabrous; leaves glabrous, linear-filiform; fruiting peduncles often hooked at the apex; sepals broadly oval, obtuse; petals shorter or none. Springy places, mostly near the coast: Newf.—Greenl.—Ont.—Kans.—Del.; Eurasia. *Arct.—Canad.—Allegh.* My–S.

2. **S. saginoìdes** (L.) Britton. Stem 2–10 cm. high, decumbent, often rooting; leaves filiform, 5–15 mm. long; pedicels 5–20 mm. long, often curved at the summit; sepals oval, obtuse, 1.5–2 mm. long. *S. Linnaei* Presl. Wet places among rocks: Greenl.—Que.—N.M.—Calif.—Alaska; Eurasia. *Mont.—Subalp.* Je–Au.

7. SPÉRGULA L. SPURRY, CORN SPURRY.

Somewhat succulent annuals. Leaves whorled and commonly clustered, thick, narrow, with stipules. Flowers in terminal cymes. Sepals 5, persistent. Petals 5, white, entire. Stamens 10, rarely 5. Ovary 1-celled; styles 5. Capsule 5-valved, the valves opposite the sepals. Seeds narrowly winged.

1. **S. arvénsis** L. Annual; stems slender, branched from the base, 1.5–5 dm. high; leaves linear-filiform, sparingly villous or glabrate, 3–5 cm. long; cymes terminal, loose, many-flowered; pedicels reflexed in fruit; sepals ovate, 3–4 mm. long; petals equaling or slightly exceeding the sepals. Cultivated grounds and waste places: N.S.—Fla.—Calif.—Alaska; adv. or nat. from Eur. *Temp.* Je–S.

8. SPERGULÀRIA J. & C. Presl. SAND SPURRY.

Low annual, biennial, or perennial herbs, usually rather succulent. Leaves opposite, with stipules, often with secondary leaves fascicled in their axils. Flowers in terminal, racemiform cymes. Sepals 5, persistent. Petals 5, rarely fewer, or wanting, pink or whitish, entire. Stamens 2–10. Ovary 1-celled; styles 3. Capsule 3-valved to the base. Seeds often margined, smooth or tubercled. *Tissa* Adans.

Stipules lanceolate or ovate-lanceolate, longer than broad. 1. *S. rubra.*
Stipules broadly triangular, as broad as long or broader. 2. *S. salina.*

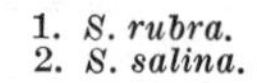

1. **S. rùbra** (L.) J. & C. Presl. *Fig. 205.* Annual or perennial; stem spreading, much branched from the base, glabrous or more or less glandular above; leaves flat or slightly grooved, linear-filiform, 8–12 mm. long; stipules attenuate, 4–6 mm. long; sepals oblong-lanceolate, scarious-margined, 3–4 mm. long; petals scarcely equaling the calyx. *T. rubra* Britton [B]. Sandy soil and waste places: Newf.—Va.—Ohio—Calif.—B.C.; Eur. *Temp.—Plain.*... My–Au.

2. **S. salìna** J. & C. Presl. Annual, stem usually diffuse, branched, more or less pubescent, 1–2 dm. high; leaves linear-filiform, fleshy, terete, 1–4 cm. long; sepals ovate, acutish or obtuse, 4–5 mm. long; petals pink, shorter than the sepals. *T. salina* Greene. *T. marina* Britton [B]. Saline soil: N.B.—Fla.—Sask.; Alaska—Calif. *Temp.* Ap–S.

9. LOEFLÍNGIA L.

Small, diffusely branched, glandular or glabrous annuals. Leaves subulate, with stipules. Flowers solitary or fascicled in the axils. Sepals 5, rigid, keeled, awn-pointed. Petals 2–5, minute, without claws, or wanting. Stamens 3–5, perigynous. Ovary 1-celled, many-ovuled; stigmas sessile. Capsule 3-valved. Cotyledons accumbent.

1. **L. texàna** Hook. Annual, branched at the base, 5–15 cm. high; leaves 4–6 mm. long; flowers on secund recurved branches; sepals straight or slightly curved; capsule shorter than the calyx; seeds obovate. Dry hills: Tex.—Neb. *Plain—Texan.* Ap–Je.

Family 53. CARYOPHYLLACEAE. PINK FAMILY.

Herbs with narrow opposite leaves, without stipules, often connate at the base. Flowers perfect, polygamous, or dioecious, cymose. Calyx of 5 or 4 united sepals. Petals 5 or 4, distinct, with a narrow claw, often with scale-like appendages (the crown) at the junction of the claw with the blade. Stamens usually twice as many as the sepals. Gynoecium of 2–5 united carpels; styles 2–5. Fruit a 1-celled or partially 2–5-celled capsule, opening by 2–5 valves at the apex.

Calyx without bractlets; seeds globular or reniform; embryo curved.
  Calyx with at least twice as many ribs (usually 10) as calyx-teeth, running both into the teeth and the sinuses.
    Styles 5, alternate with the foliaceous calyx-lobes. 1. AGROSTEMMA.
    Styles either 5 and opposite the calyx-lobes or fewer than five.
      Styles mostly 3; capsule usually septate at the base. 2. SILENE.
      Styles 5; capsule 1-celled to the base.
        Valves of the opening capsule entire; flowers perfect. 3. LYCHNIS.
        Valves of the opening capsule 2-cleft or 2-toothed.
          Plants dioecious, corolla showy; valves of the capsule deeply 2-cleft. 4. MELANDRIUM.
          Plants hermaphrodite; corolla minute or none; valves of the capsule merely 2-toothed. 5. WAHLBERGELLA.
  Calyx 5-ribbed and 5-angled; styles 2.
    Petal with a crown; calyx not strongly angled. 6. SAPONARIA.
    Petals without a crown.
      Calyx strongly angled; flowers comparatively large. 7. VACCARIA.
      Calyx not strongly angled; corolla small. 8. GYPSOPHILA.
Calyx subtended by scaly or foliaceous bractlets; seeds flattened; embryo straight. 9. DIANTHUS.

1. AGROSTÉMMA L. CORN COCKLE, CORN CAMPION.

Annual pubescent herbs, with opposite leaves. Flowers solitary at the ends of the branches. Calyx oblong, not inflated, 10-ribbed, 5-lobed; lobes linear, elongate and foliaceous. Petals 5, clawed, without appendages or crown; blade obovate or cuneate, emarginate. Stamens 10. Styles 5, opposite the petals. Capsule 1-celled. Seeds numerous.

1. **A. Githàgo** L. *Fig. 206.* Stem erect, 3–10 dm. high, densely pubescent; leaves linear-lanceolate, erect, 5–10 cm. long; calyx-lobes 2–3 cm. long; petals red; blades 1.5–2 cm. long. Waste places and grain fields: Newf.—Fla.—Calif.—B.C.; nat. from Eur. Jl–S.

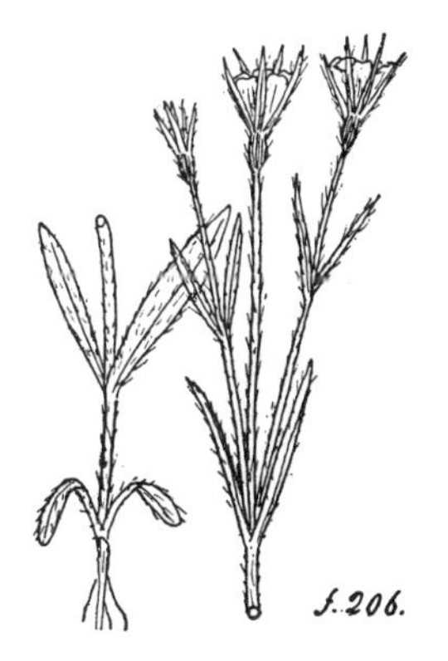
f. 206.

2. SILÈNE L. Catchfly, Campion.

Annual or perennial herbs. Leaves opposite, flat, entire. Flowers perfect, in terminal cymes, or rarely solitary. Calyx with a more or less inflated tube, 10–many-nerved, with short, erect, or spreading lobes. Petals 5, red, pink, or white, with a crown, and usually cleft or divided. Stamens 10. Ovary 1-celled, but usually with partial partitions at the base. Styles 3, seldom more. Capsule opening by twice as many tooth-like valves as there are styles. Ovary with a more or less developed stipe. Seeds tuberculate or echinate.

Annuals.
  Stems glabrous or nearly so, or the upper nodes glutinous.
    Calyx ovate, inflorescence elongate.
      Upper nodes of the stem glutinous; leaves lanceolate or oblanceolate. 1. *S. antirrhina.*
      Upper nodes not glutinous; lower leaves obovate, glaucous. 2. *S. fabaria.*
    Calyx club-shaped; leaves ovate or ovate-lanceolate; inflorescence corymbiform; stem glutinous at the nodes. 3. *S. Armeria.*
  Plant pubescent and viscid.
    Inflorescence racemiform; calyx 12–16 mm. long. 4. *S. dichotoma.*
    Inflorescence cymose; calyx 3–4 cm. long. 5. *S. noctiflora.*
Perennials.
  Calyx not strongly inflated in fruit, nor constricted at the mouth.
    Plant caulescent, rather tall, not densely tufted.
      Corolla crimson. 6. *S. virginica.*
      Corolla white. 7. *S. Menziesii.*
    Plant subacaulescent, densely cespitose-pulvinate. 8. *S. acaulis.*
  Calyx strongly inflated in fruit, more or less constricted at the mouth.
    Leaves opposite; petals emarginate or 2-cleft.
      Flowers numerous; bracts of the inflorescence much reduced. 9. *S. vulgaris.*
      Flowers few; bracts of the inflorescence leaf-like. 10. *S. nivea.*
    Leaves in whorls of 4; petals laciniate. 11. *S. stellata.*

1. S. antirrhìna L. Stem 1.5–5 dm. (rarely 6–10 dm.) high, finely puberulent below; branches strongly ascending or nearly erect, usually more or less viscid about the middle of the internodes; lower leaves oblanceolate, 5–10 cm. long, the upper linear to lanceolate; calyx in flower fusiform, 8–10 mm. long; lobes lanceolate or ovate-lanceolate, usually acute, often purple-tipped; petals purple or rose-tipped, exceeding the sepals and with a 2-cleft blade, or shorter and with a truncate blade, or wanting; fruit broadly ellipsoid, 6–8 mm. long, about 4 mm. thick. Sleepy Catchfly. Waste places, fields, etc.: Newf.—Fla.—Calif.—B.C. *Temp.—Plain—Submont.* Ap–Au.

2. S. fabària (L.) Sibth. & Sm. Annual, stems leafy, 5–10 dm. high, glabrous; leaves obovate to elliptic-lanceolate, thick, glaucous, the lower rounded at the apex, the short petioles with broadened bases; flowers fascicled; calyx ovoid, somewhat inflated, 10-nerved, glabrous; petals white; the claw dilated above; blade 2-cleft, emarginate; capsule subglobose. Waste places: N.D.; adv. from S. Eur.

3. S. Armèria L. Stem glabrous, about 3 dm. high, glutinous below the upper nodes; leaves ovate-lanceolate, 5–8 cm. long, obtuse; flowers in dense flat-topped cymes; calyx club-shaped; petals rose-colored or purplish, 10–15 mm. long, acuminate. Sweet William. Waste places: N.B.—Minn.—Iowa—N.J.; nat. from Eur. Je–Jl.

4. S. dichótoma Ehrh. Stem more or less hirsute and viscid, 3–8 dm. high; leaves oblanceolate or lanceolate, 5–8 cm. long; flowers often nodding; calyx hirsute, 12–16 mm. long, with 5 simple ribs; petals much exserted, white or pink. Fields: Me.—Minn.—Tex.—Va.; also Mont.; adv. from Eur.

5. S. noctiflòra L. Stem stout, viscid-pubescent, 3–10 dm. high; lower leaves obovate or oblanceolate, 5–12 cm. long, obtuse, viscid-hirsute, short-petioled; upper leaves lanceolate, acuminate, sessile; calyx 2–3 cm. long, at first nearly tubular; petals white or pinkish, about 3 cm. long, somewhat exceeding the calyx, 2-cleft; capsule sessile, ellipsoid. NIGHT-BLOOMING CATCHFLY. Waste places and cultivated ground: N.S.—Fla.—Utah—Wash.; nat. from Eur. *Plain—Submont.* My–Au.

6. **S. virgínica** L. Stem slender, erect, 2–6 dm. high; lower leaves spatulate, 0.5–2 dm. long, the upper oblong-lanceolate; flowers few, calyx soon obconic, 1.5–2.5 cm. long; blades of the petals oblong, 2-cleft, 2.5 cm. long. Open woods: Ont.—N.J.—Ga.—Ark.—Minn. *Canad.—Allegh.*

7. **S. Menzièsii** Hook. *Fig. 207.* Stems erect or ascending, 1–3 dm. high, dichotomously branched, very leafy, retrorsely hirsutulous and more or less glandular-villous, especially above; leaves ovate-lanceolate to linear-lanceolate, acuminate at both ends, 3–8 cm. long; calyx turbinate-obovoid, 5–8 mm. long; petals 2-cleft, white, 6–10 mm. long, without a crown (always ?); capsule sessile, ellipsoid, slightly exceeding the calyx. *S. stellarioides* Nutt., a narrow-leaved form. Among bushes: Sask.—S.D.—Mo.—N.M.—Calif.—Mont. *W. Temp.—Submont—Subalp.* My–S.

f.207.

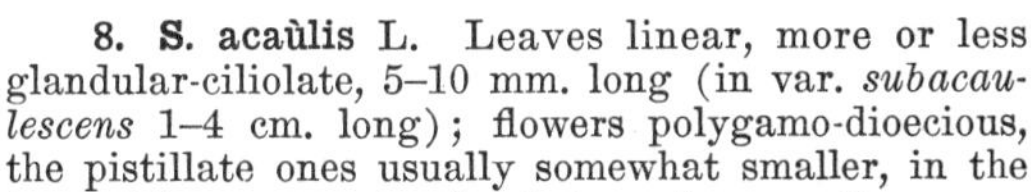

8. **S. acaùlis** L. Leaves linear, more or less glandular-ciliolate, 5–10 mm. long (in var. *subacaulescens* 1–4 cm. long); flowers polygamo-dioecious, the pistillate ones usually somewhat smaller, in the typical form on peduncles 1–4 cm. long or (in var. *exscapa*) sessile or nearly so; calyx oblong, 5–6 mm. long, rounded or sunken or rarely acute at the base; petals pink or purple, exceeding the calyx, emarginate or 2-lobed, rarely entire; capsule cylindric, exceeding the calyx, sessile, or (in the var. *exscapa*) ovoid or subglobose. MOSS CAMPION. Alpine-arctic situations: Greenl.—N.H.—N.M.—Ariz.—Alaska; Eurasia. *Arct.—Subarct.—Mont.—Alp.* Je–Au.

9. **S. vulgàris** (Moench) Garcke. Cespitose caulescent perennial; stems glabrous or nearly so, 2–4 dm. high; leaves lanceolate, acute, 3–5 cm. long, glabrous; flowers polygamo-dioecious; calyx campanulate to subglobose; petals white or pink, 2-cleft, almost without crown. *S. inflata* Smith. BLADDER CAMPION. Fields, roadsides, and waste places: N.B.—N.J.—Colo.—Alta.; Calif.—Wash.; adv. or nat. from Eur. *Boreal—Plain—Submont.* My–S.

10. **S. nívea** (Nutt.) Otth. Stem reclining, 3–8 dm. high; leaves lanceolate, oblanceolate, or linear-lanceolate, 4–15 cm. long, calyx subcylindric, viscid-pubescent, 1.5–1.8 cm. long; corolla white, 2 cm. long; blades rounded-cuneate, 2-cleft; capsule stalked. *S. alba* Muhl. [B]. River banks: Pa.—Md. Neb.—S.D.; adv. from Eur. Je–Jl.

11. **S. stellàta** (L.) Ait. f. Stem 7–10 dm. high; leaves ovate-lanceolate, acuminate, 4–12 cm. long, ciliate; flowers in large panicles; calyx campanulate, inflated above the middle, 1–1.5 cm. long; petals white, 2 cm. long, without a crown. Wooded banks: Mass.—Ga.—Tex.—Minn. *E. Temp.* Je–Au.

3. LÝCHNIS L.

Perennial herbs, with opposite leaves. Flowers cymose, perfect. Calyx ovoid, 10-ribbed, the lobes short. Petals 5, conspicuous, 2-cleft or laciniate. Stamens 10. Styles usually 5, opposite the calyx-lobes. Ovary 1-celled. Capsule opening with 5 entire valves, more or less stipitate.

1. **L. chalcedónica** L. Stem hirsute, 3–7 dm. high; leaves ovate or the upper lanceolate, acute or acuminate at the apex, rounded or subcordate at the base; flowers 2.5 cm. broad; petals scarlet, 2-cleft; stipe of the capsule nearly as long as the body. Roadsides and around dwellings: Mass.—Minn.—Iowa—N.J.; escaped from cultivation; nat. of Eur. Je–S.

4. MELÁNDRIUM Roehl. CAMPION.

Perennial herbs, with opposite leaves, dioecious. Calyx cylindric, or in fruit more or less extended by the capsule, 5-toothed, 10-nerved. Petals conspicuous, 2-cleft, with a distinct crown. Stamens 10. Styles 5, alternate with the petals. Ovary strictly 1-celled. Capsule opening with 5, 2-cleft valves at the apex.

Flowers white or pinkish. 1. *M. album.*
Flowers red. 2. *M. dioicum.*

1. **M. álbum** (Mill.) Garcke. Stem branching, 3–6 dm. high; leaves ovate-oblong or ovate-lanceolate, 2–8 cm. long; flowers fragrant, opening in the evening; calyx cylindric, becoming ovoid; corolla white or pink, 1.5–2.5 cm. broad; blades obovate, 2-cleft. *Lychnis alba* Mill. [G, B]. *L. vespertina* Sibth. Waste places: N.S.—Md.—Kans.—N.D.; adv. from Old World.

2. **M. dioìcum** (L.) Coss. & Germ. Stem branching above, 3–6 dm. high, viscid; basal leaves long-petioled, elliptic; upper stem-leaves sessile, ovate, 3–5 cm. long; flowers scentless, opening in the morning; calyx cylindric, distended in fruit; blades of the petals obovate, 2-cleft. *L. dioica* L. [B, G]. Waste places: S.C.—Va.—Minn.; adv. fr. Eur. Je–Au.

5. WAHLBERGÉLLA Fries.

Perennial herbs, with opposite leaves. Calyx ovoid, more or less inflated, 5-toothed, 10-nerved. Petals inconspicuous, often included, with small crowns and 2-cleft blades. Stamens 10. Styles 5, rarely 4, alternating with the petals. Ovary strictly 1-celled. Capsule opening by as many 2-toothed valves as there are styles. Seeds laterally attached; embryo curved. *Lychnis* Am. auth., in part.

1. **W. Drummóndii** (Hook.) Rydb. *Fig. 208.* Stems viscid-puberulent throughout, 2–4 dm. high; basal leaves oblanceolate, petioled, 5–10 cm. long, grayish puberulent, early deciduous; stem-leaves linear, attenuate; calyx oblong-cylindric, with green nerves, glandular-viscid, 10–12 mm. long; petals white or purplish; blade narrower than the claw. *Silene Drummondii* Hook. *L. Drummondii* S. Wats. [G, B]. Dry hillsides and plains: Man.—Minn.—N.M.—Ariz.—B.C. *W. Temp.—Prairie—Plain—Mont.* Je–Au.

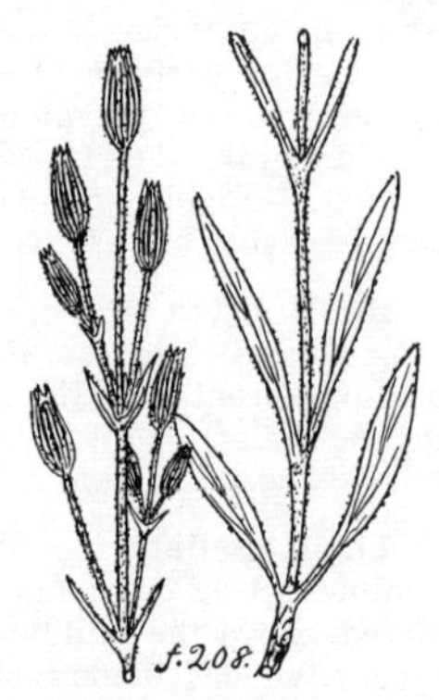

6. SAPONÀRIA L. SOAPWORT, BOUNCING BET.

Perennial herbs, with opposite broad leaves. Calyx 5-toothed, obscurely 5-nerved. Petals 5, long-clawed, with appendages (crown). Stamens 10. Ovary 1-celled or incompletely 2–4-celled; styles 2. Capsule dehiscent by 4 apical teeth or valves.

1. **S. officinàlis** L. Glabrous perennial; stem 3–6 dm. high; leaves oval or ovate, 5–8 cm. long, 3–5-ribbed; flowers in dense corymbiform cymes; calyx tubular, 1.5–2 cm. long; petals pink or white; blades obcordate; capsule oblong. Roadsides and waste places: N.S.—Fla.—N.M.—Colo.; nat. from Eur. *E. Temp.—Plain—Submont.* Jl–S.

7. **VACCÀRIA** Medic. Cow-herb, Cow Cockle.

Annual caulescent herbs, with glabrous and glaucous, opposite, often thickish, clasping leaves. Flowers in dichotomous cymes. Calyx somewhat inflated, strongly 5-angled and 5-nerved; lobes very short. Petals conspicuous, longer than the calyx, without a crown. Stamens 10. Ovary 1-celled; styles 2. Capsule opening by 4 apical, tooth-like valves. Seeds laterally attached; embryo slightly curved.

1. **V. vulgàris** Host. *Fig. 209.* Annual; stem 3–10 dm. high, branched; leaves ovate to lanceolate, connate at the base, 2–8 cm. long; calyx with the sharp angles herbaceous, and the intervening parts scarious and white, 1–1.5 cm. long; petals rose-colored, crenulate. *Saponaria Vaccaria* L. [G]. *V. Vaccaria* Britton. In waste places: Ont.—Fla.—Calif.—Alaska; nat. from Eur. *Temp.—Submont.* Je–Au.

8. **GYPSÓPHILA** L.

Slender annual or perennial herbs. Leaves narrow, glaucous. Flowers small, perfect, axillary or panicled. Calyx narrowly turbinate, 5-nerved, 5-toothed, without bractlets. Petals 5, with narrow claws, without crown. Stamens 10. Styles 2. Pod 1-celled, 4-valved at the apex, sessile. Seeds reniform.

Diffuse annuals; leaves narrowly linear; flowers axillary on long peduncles. 1. *G. muralis.*
Erect plants; lower leaves oblanceolate or oblong; flowers paniculate.
  Perennials. 2. *G. paniculata.*
  Annuals. 3. *G. elegans.*

1. **G. muràlis** L. Stem 1–2 dm. high; leaves 0.5–2 cm. long, 1 mm. wide or less; peduncles 0.5–2 cm. long; corolla purplish, 3–4 mm. broad; blades crenate or emarginate. Waste places: Me.—N.J.—Minn.; adv. from Old World. Je–S.

2. **G. paniculàta** L. Stem 3–6 dm. high; leaves 2 cm. long or more, 4–8 mm. wide, acuminate; corolla 3 mm. broad, white or pink, the petals slightly emarginate. Waste places: Man.—Neb.; adv. from Eurasia.

3. **G. élegans** Bieb. Stem 3–4 dm. high, much branched; lower leaves 5–10 cm. long, oblanceolate, the middle ones oblong; corolla white or rose, 5–6 mm. broad. Waste places: N.D.; escaped from cultivation; nat. of Caucasus and Armenia.

9. **DIÁNTHUS** L. Pink, Carnation.

Perennial or annual herbs. Leaves opposite, entire, usually narrow. Flowers perfect, in terminal cymes or solitary, involucrate. Calyx-tube elongate, finely striate; lobes 5. Petals with long claws and dentate or erose blades. Stamens 10. Ovary 1-celled, stalked; styles 2. Capsule 4-valved at the apex. Seeds flattened.

Plant annual, pubescent. 1. *D. Armeria.*
Plant perennial, glabrous. 2. *D. barbatus.*

1. **D. Armèria** L. Stem erect, 2–6 dm. high; leaves linear, or the basal ones linear-spatulate, 3–8 cm. long; flowers clustered at the ends of the branches; calyx cylindric, pilose, 1.5 cm. long; corolla *white*, usually white-spotted. Fields and roadsides: Mass.—Ga.—Iowa; nat. from Eurasia. Jl–Au.

2. **D. barbàtus** L. Stem 3–6 dm. high; leaves lanceolate or ovate-lanceolate; bracts filiform; calyx cylindric; corolla pink or whitish; blades erose. SWEET WILLIAM. Waste places: Me.—Del.—N.D.; escaped from cultivation; native of Eur.

Family 54. CERATOPHYLLACEAE. HORNWORT FAMILY.

Submerged aquatic herbs, with verticillate leaves, thrice dissected dichotomously into filiform stiff divisions. Flowers monoecious, inconspicuous, sessile, axillary. Sepals 6–12, herbaceous, valvate. Petals wanting. Stamens 10–24, with very short filaments. Pistil 1; ovary 1-celled. Fruit nut-like, with a persistent style. Seeds pendulous; embryo straight.

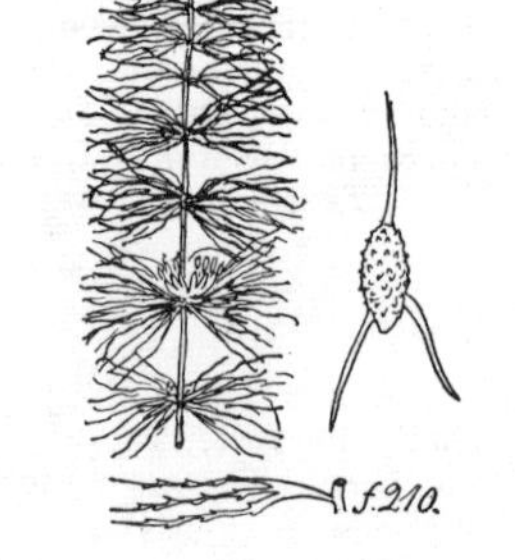

1. CERATOPHÝLLUM L. HORNWORT.

Characters of the family.

1. **C. demérsum** L. *Fig. 210.* Stems 3–12 dm. long; leaves rather rigid, 6–9 in each whorl, 1–3 cm. long, once, twice, or thrice forking, the ultimate segments spiny-toothed; achenes oblong, slightly flattened, 5 mm. long, tipped with the persistent style and armed with spreading spines at the base. Ponds and still water: Newf.—Fla.—Calif.—Wash.; W.Ind., Mex., Eurasia. *Temp.—Trop.—Plain.* Je–Jl.

Family 55. NYMPHAEACEAE. WATER LILY FAMILY.

Perennial acaulescent water plants, with stout elongate rootstocks. Leaves with elongate petioles and broad leathery floating blades cordate or sagittate at the base. Flowers solitary, long-pedicelled, perfect. Sepals 4–7, green or partly colored. Petals numerous, often passing into staminodia and stamens. Stamens numerous. Gynoecium of several more or less united carpels, forming a compound pistil; stigmas united into a disk. Ovules numerous, parietal. Fruit a leathery berry.

Petals at least as large as the sepals; stamens epigynous. 1. NYMPHAEA.
Petals small, staminodia-like; stamens hypogynous. 2. NUPHAR.

1. NYMPHAÈA (Tourn.) L. WATER LILY, POND LILY.

Flowers in ours white. Sepals 4. Petals numerous, imbricate in several series, grading into staminodia and stamens. Stamens numerous, epigynous. Stigma 12–35-rayed, with a globular projection in the middle. Fruit with a leathery pericarp, maturing under water. Seeds numerous, surrounded by a sac-like aril. *Castalia* Salish.

Flowers 5–20 cm. broad; leaf-blades orbicular in outline.
  Flowers 15–20 cm. broad, not fragrant; leaf-blades 2–4 dm. broad. 1. *N. tuberosa.*
  Flowers 7–15 cm. broad, sweet-scented; leaf-blades 5–25 cm. broad, purplish beneath. 2. *N. odorata.*
Flowers 2.5–4 cm. broad; leaf-blades oval or obovate, 4–6 cm. wide. 3. *N. Leibergii.*

1. **N. tuberòsa** Paine. Rootstock with many tubers; petals broadly oblong, obtuse; leaves strongly ribbed; seeds globular-ovoid. *C. tuberosa* (Paine) Greene [G, B]. Slow streams, ponds, and lakes: Vt.—Del.—Ark.—Neb. *Canad.—Allegh.* Je.–S.

2. **N. odoràta** Ait. *Fig. 211.* Rootstocks with few persistent, not tuberous branches; petals narrowly oblong, white or pinkish, or in f. *rosea* dark pink or red; fruit globose; seeds oblong. *C. odorata* (Ait.) Woodv. & Wood. [G, B]. Lakes, ponds, and slow streams. Newf.—Fla.—Tex.—Man. *E. Temp.* Je–Au.

3. **N. Leibérgii** Morong. Leaves broadly obovate in outline, 6–10 cm. long, 4–6 cm. wide; sinus deep, often extending more than half the length of the leaves; basal lobes acute; flowers 4–6 cm. wide; sepals green outside, elliptic, often acutish; petals 8–10, white, purple-veined, obtuse; stamens 20–30; stigma 7–8-rayed. *C. Leibergii* Morong. Ponds: Ida.—w Ont. and northward. *Mont.—Boreal.* Jl–Au.

2. **NÙPHAR** Smith. Yellow Pond Lily, Spatter-dock, Cow Lily.

Flowers yellow. Sepals 5–7, leathery, concave. Petals 10–20, small, filament-like. Filaments flattened, inserted under the ovary. Stigmas forming an 8–24-radiate disk. Fruit with a leathery pericarp; seeds pendulous. *Nymphaea* L., in part.

Sepals 5 or 6.
  Petioles decidedly flattened; blades floating, the basal sinus narrow and nearly closed.
    Flowers 1–3 cm. broad; anthers shorter than the filaments.
      Stigma-rays 6–10; flowers 1–2 cm. wide; basal lobes of the leaves acutish. — 1. *N. microphyllum.*
      Stigma-rays 8–13, usually 10–12; flowers 2.5–3, rarely 3.5 cm. broad; basal lobes of the leaves rounded. — 2. *N. rubrodiscum.*
    Flowers 3–4 cm. broad; anthers equaling the filaments; basal lobes of the leaves rounded. — 3. *N. variegatum.*
  Petioles terete; leaf-blades often erect, the basal sinus open, the basal lobes acute. — 4. *N. advenum.*
Sepals 6–9; basal sinus usually open, the lobes rounded. — 5. *N. polysepalum.*

1. **N. microphýllum** (Pers.) Fern. Leaf-blades 3–10 cm. long, oval or ovate, with a very deep sinus; sepals about 5, mostly yellow, 12–15 mm. long; petals 4 mm. long; stigma usually deep red, distinctly lobed; fruit 12–15 mm., in diameter, with a narrow neck; disk 4–5 mm. broad. *Nymphaea microphylla* Pers. [G]. *N. Kalmianum* Ait. f. [B]. Ponds: Newf.—Pa.—Sask. *Boreal.* Jl–Au.

2. **N. rubrodíscum** Morong. Leaf-blades 7–20 cm. long, oval or ovate, rounded at the apex, with a closed sinus and rounded basal lobes; sepals 5, canary-yellow, orbicular, about 2 cm. long; petals spatulate, truncate, 8–9 mm. long; stigma dark carmine; fruit ovoid, slightly constricted above, 22 mm. high, dark purplish-brown. Perhaps a hybrid of *N. microphyllum* and *N. variegatum. Nymphaea hybrida* Peck. [B]. *N. rubrodisca* Greene [G]. Water: N.S.—Minn.—N.J. *Canad.* My–S.

3. **N. variegàtum** Engelm. Leaf-blades rounded-ovate or broadly oval in outline, 1–2 dm. long, 8–15 cm. wide, with a narrow or closed sinus; outer sepals green, the rest with bright yellow edges, red or maroon towards the base within; petals oblanceolate; stigma 10–16-rayed; fruit scarcely constricted into a neck. *Nymphaea americana* Mill. & Standl. Lakes and slow streams: Lab.—N.J.—Wyo.—B.C. *Boreal—Plain—Submont.* Je–Au.

4. **N. ádvenum** Ait. Leaves erect or sometimes floating; blades ovate, 15–30 cm. long, rounded at the apex; sepals usually 6, broadly ovate, or orbicular, 3.5 cm. long; petals about 20, cuneate-oblong, truncate, 8 mm. long; fruit green, 4 cm. long and 5 mm. broad. *Nymphaea advena* Ait. [G, B]. Water: N.S.—Neb.—Kans.—Tex.—N.C. *Allegh.—Canad.* Ap–S.

5. **N. polysèpalum** Engelm. Leaf-blades oval, 2–3 dm. long, 1.5–2 dm. wide, with a narrow or closed sinus; sepals 6–12, yellow, reddish tinged; petals 12–18, cuneate, 1–1.5 cm. long; stigma 15–25-rayed, crenate; fruit about 3.5 cm. in diameter, with a constricted neck. *Nymphaea polysepala* Greene [R]. Lakes and slow streams: Mont.—S.D.—Colo.—Calif.—Alaska. *W. Temp.—Plain—Subalp.* Ap–Au.

Family 56. **NELUMBONACEAE.** Lotus Family.

Perennial aquatic herbs, with rootstocks. Leaves alternate, long-petioled; blades peltate, floating or emersed, concave on the upper side. Flowers solitary, on long scapes. Petals and sepals similar, hypogynous, numerous, imbricate, all deciduous. Stamens numerous, hypogynous; anthers narrow, introrse, the connective with an incurved appendage. Pistils many, separately immersed into an obconic, fleshy receptacle. Ovary 1-celled, with 1 or 2 suspended ovules.

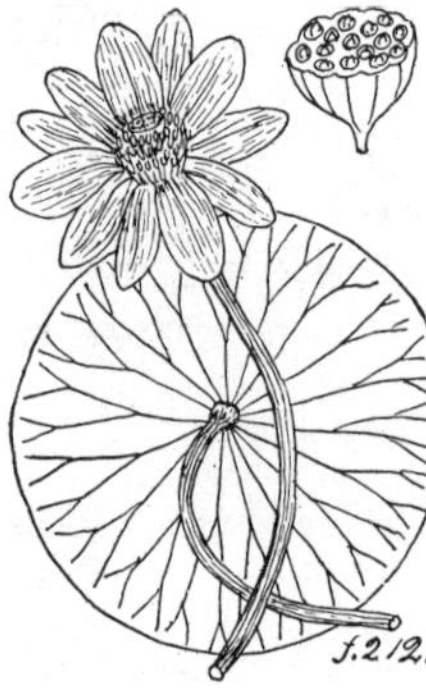

1. **NELÚMBO** Adans. Lotus, Water Chinquapin, Duck Acorn.

Characters of the family.

f. 212.

1. **N. lùtea** (Willd.) Pers. *Fig. 212.* Leaf-blades orbicular, strongly ribbed, with up-turned margins, 3–6 dm. broad; flowers pale-yellow, 1–2.5 dm. broad; petals obovate to elliptic; fruit obconic, 7–10 cm. broad; achenes subglobose. Ponds and slow streams: Mass.—Fla.—Tex.—Minn. *E. Temp.* Jl–Au.

Family 57. **CABOMBACEAE.** Water-shield Family.

Perennial aquatic plants. Leaves alternate, opposite, or whorled; blades of the submerged leaves mostly dissected, those of the floating ones peltate, mucilage-coated. Flowers perfect, inconspicuous. Sepals and petals each 3, rarely 4. Stamens 3–18, hypogynous; filaments slender; anthers extrorse. Pistils 2–18, distinct. Ovary 1-celled; stigma sessile. Fruit of several indehiscent carpels.

1. **BRASÈNIA** Schreb. Water-shield.

Aquatic branching herbs. Leaves floating; blades peltate, leathery. Flowers small, solitary on long axillary pedicels. Sepals 3, colored within. Petals 3, narrow. Stamens 12–18. Pistils 4–8; style subulate. Carpels in fruit leathery, indehiscent.

f. 213.

1. **B. Schrèberi** Gmel. *Fig. 213.* Leaf-blades oval, centrally peltate, 5–12 cm. long, green and shining above, purplish beneath; flowers purple; sepals and petals linear or linear-lanceolate, 1–1.5 cm. long. *B. peltata* Pursh. *B. purpurea* Casp. [B]. Ponds and lakes: N.S.—Fla. Tex.—Man.; Ore.—Ida.—B.C.; W.Ind. and C.Am. *Temp.—Trop.* Je–Au.

Family 58. **RANUNCULACEAE.** Crowfoot Family.

Herbs or climbing shrubs. Leaves alternate, except in *Clematis, Viorna,* and *Atragene,* simple or compound, without stipules. Flowers regular or irregular. Sepals 3–15, usually green and caducous, or in genera without petals persistent and more or less petaloid, imbricate, or in *Clematis,* etc., valvate. Petals as many as the sepals or wanting. Stamens usually many, rarely 5. Carpels numerous or few, distinct, rarely solitary, 1-celled, 1–many-ovuled. Fruit of achenes, follicles, or berries.

Carpels 1-ovuled; fruit of achenes.
- Petals wanting; sepals often petal-like.
  - Sepals imbricate in the bud; leaves all alternate, or only those subtending the inflorescence opposite.
    - Flowers subtended by opposite or verticillate leaf-like bracts.
      - Bracts more or less resembling the leaves, cleft or compound.
        - Styles present; leaf-segments sessile.
          - Styles short, not elongating in fruit. 1. Anemone.
          - Styles much elongating in fruit, plumose. 2. Pulsatilla.
        - Styles wanting; leaf-segments stalked. 4. Syndesmon.
      - Bracts entire, sepal-like, inserted closely under the flower, basal leaves reniform, 3-lobed. 3. Hepatica.
    - Flowers not subtended by opposite or verticillate bracts; leaves ternately compound. 5. Thalictrum.
  - Sepals valvate in the bud; leaves all opposite.
    - Stamens erect; sepals thickish, more or less converging; staminodia wanting. 6. Viorna.
    - Stamens spreading; sepals spreading from the base.
      - Flowers mostly paniculate; sepals white or yellow, firm, without a border; staminodia wanting. 7. Clematis.
      - Flowers solitary; sepals blue or purple, thin; staminodia usually present. 8. Atragene.
- Petals usually present.
  - Sepals spurred; small annuals with basal linear leaves; receptacle in fruit elongate-cylindrical; stamens 5. 9. Myosurus.
  - Sepals not spurred; plant usually bearing cauline as well as basal leaves; receptacle in fruit spherical, conical or short-cylindric; stamens more than 5.
    - Achenes transversely wrinkled; petals white. 10. Batrachium.
    - Achenes not transversely wrinkled.
      - Achenes not ribbed.
        - Achenes without an empty upper portion; plants not scapose. 11. Ranunculus.
        - Achenes with the lower half enclosing the seed; the upper portion empty, flattened and spongy; plants scapose. 12. Coptidium.
      - Achenes longitudinally ribbed; petals yellow; leaves simple, crenate or lobed. 13. Halerpestes.

Carpels with several ovules; fruit of follicles or berries.
- Flowers regular.
  - Petals inconspicuous or none, not spurred.
    - Fruit of follicles; flowers solitary.
      - Follicles sessile.
        - Follicles united; petals unguiculate; flowers blue. 14. Nigella.
        - Follicles distinct; petals sessile or none; flowers yellow or white.
          - Leaves simple, entire or merely toothed. 15. Caltha.
          - Leaves ternately compound or decompound. 16. Isopyrum.
      - Follicles stipitate; petals clawed; leaves compound, ternate; petals hooded and nectariferous at the summit. 17. Coptis.
    - Fruit a berry.
      - Leaves twice or thrice ternately compound; flowers racemose. 18. Actaea.
      - Leaves palmately cleft; flowers solitary. 19. Hydrastis.
  - Petals conspicuous, produced into a spur or at least saccate at the base; leaves ternately compound. 20. Aquilegia.
- Flowers irregular.
  - Posterior sepal spurred. 21. Delphinum.
  - Posterior sepal hooded, helmet-shaped or boat-shaped. 22. Aconitum.

1. **ANEMÒNE** (Tourn.) L. Wind-flower, Anemone.

Perennial herbs, usually with horizontal rootstocks and erect scapes. Leaves basal, petioled, palmately divided or dissected. Bracts verticillate, usually 3, similar to the leaves, remote from the flower. Flowers terminal, solitary or cymose. Sepals 4–20, mostly 5, petal-like. Petals wanting. Stamens numerous. Pistils usually numerous. Achenes compressed, 1-seeded. Stigmas introrse, unilateral.

Plants with tuberous roots; sepals 6–20.
  Involucre usually inserted above the middle of the stem; divisions of some of the basal leaves merely crenate. — 1. *A. decapetala.*
  Involucre inserted usually below the middle of the stem; divisions of all the basal leaves lobed or parted. — 2. *A. caroliniana.*
Plants with a rootstock; sepals 5–6, rarely more.
  Achenes densely villous.
    Styles filiform, usually deciduous; heads of fruit spherical or nearly so; involucral leaves short-petioled or subsessile.
      Leaves ternate; segments broadly cuneate or flabelliform, crenate or slightly cleft. — 3. *A. parviflora.*
      Leaves 2–4 times ternate; segments linear to oblong or oblanceolate.
        Sepals 6–12 mm. long. — 4. *A. globosa.*
        Sepals 4–7 mm. long. — 5. *A. hudsoniana.*
    Styles subulate, persistent; heads of fruit from rounded-oblong to cylindric; involucral leaves long-petioled.
      Heads elongate, cylindric; styles about 1 mm. long; flowers usually subumbellate. — 6. *A. cylindrica.*
      Heads of fruit rounded-oblong; styles about 2 mm. long; flowers truly cymose.
        Petals greenish white, 1 cm. long or less; styles in fruit divergent. — 7. *A. virginiana.*
        Petals white, 1.5 cm. long; styles suberect. — 8. *A. riparia.*
  Achenes not villous.
    Achenes wing-margined; plant stout.
      Sepals yellow; styles very long, reflexed. — 9. *A. Richardsoni.*
      Sepals white or pink; styles not reflexed; styles straight. — 10. *A. canadensis.*
    Achenes not wing-margined, pilose; styles minute; plant slender; flowers solitary. — 11. *A. quinquefolia.*

1. **A. decapétala** Ard. Stem 1–3 dm. high, simple; basal leaves few; blades 3-divided, with ovate to oblong, crenate or cleft segments, finely pubescent; involucral bracts sessile, once or twice palmately cleft into linear segments; sepals 10–20, pink or greenish-white; achenes very woolly. Prairies: Ala.—Kans.—Tex.; Mex.; C.Am. *Texan—Trop.*

2. **A. caroliniàna** Walt. Stem 1–2.5 dm. high; blades of the basal leaves 3-divided, sparingly hairy; divisions lobed or parted into narrow lobes; blades of the involucres sessile, cleft like the basal leaves; sepals 6–20, oblong, purple to white; achenes woolly, but beaks projecting. Prairies and plains: Wis.—Ga.—Tex.—S.D. *Allegh.* Ap–My.

3. **A. parviflòra** Michx. Stem 5–20 cm. high; leaves ternate, thick, and firm, glabrous above, silky-strigose beneath; divisions cuneate, 1–2 cm. long; sepals white or tinged with bluish or rose, oval, 8–18 mm. long; head of fruit spherical. In arctic-alpine situations: Lab.—Ont.—Mont.—Colo.—Alaska; Asia. *Arct.—Boreal.—Mont.—Subalp.* My–Au.

4. **A. globòsa** Nutt. Stem 2–5 dm. high, silky-villous, 1–3-flowered; leaf-blades 5–12 cm. broad, thrice cleft; sepals oval, 6–12 mm. long, pink or purplish or ochroleucous, tinged with bluish on the outside; heads of fruit globose or ovoid, 10–12 mm. in diameter. *A. multifida* Hook., in part; not Poir. Meadows and hillsides: Alaska—Calif.—Colo.—S.D.—Sask.—Mack. *W. Temp.—Plain—Subalp.* My–S.

5. **A. hudsoniàna** Richardson. Stem 2–4 dm. high, silky; leaf-blades twice cleft into narrowly linear divisions; involucral leaves similar, short-petioled; petals oblong, dark red to greenish yellow; head of fruit ovoid, 1–1.5 cm. long. *A. multifida* A. Gray [G]; not Poir. Banks and copses: N.B.—Me.—Mich.—Neb.—N.D.—Man. *Canad.* Je.

6. **A. cylíndrica** A. Gray. Stem 3–6 dm. high, 1–10-flowered; leaf-blades strigose-pubescent, 3–8 cm. wide, ternate or quinate; divisions rhombic-cuneate, 3-cleft and again cleft or toothed, with ovate or lanceolate teeth or lobes;

involucral ones similar, clustered together with those of the secondary peduncles, the lowest node of which usually is very short or obsolete, the inflorescence therefore becoming umbelliform; sepals greenish-white, oblong, 8–10 mm. long; heads in fruit 2–4 cm. long, about 1 cm. thick. Meadows, hillsides, and river banks: N.B.—N.J.—Ariz.—B.C. *Temp.—Plain—Submont.* Je–Au.

7. **A. virginiàna** L. Stem 6–10 dm. high, hirsute, 1–7-flowered; leaf-blades 10–15 cm. broad, ternate or quinate, more or less silky-hirsute; divisions rhombic-obovate, 3-cleft and coarsely toothed; inflorescence cymose; sepals greenish-white, 1 cm. or less long; heads of fruit 1–2.5 cm. long, 12–15 mm. thick. Woods and river banks: N.S.—S.C.—Kans.—Wyo.—Alta. *E. Temp.—Plain—Submont.* Je–Au.

8. **A. ripària** Fernald. *Fig. 214.* Stem 6–10 dm. high, somewhat pubescent, 1–5-flowered; leaf-blades ternate or quinate; divisions lanceolate, cuneate at base, 3-cleft and toothed; inflorescence cymose; head of fruit subcylindric, 2–3 cm. long, 1 cm. thick. River banks: Me.—Va.—Alta. *E. Temp.—Plain—Submont.* Je–Jl.

9. **A. Richardsòni** Hook. Basal leaf-blades round-reniform, deeply 5-cleft into cuneate-obovate incised divisions; involucres of 3, dilated, cuneate, 3-lobed, dentate leaves; sepals sulphur-yellow, elliptic, 8–15 mm. long; heads of fruit subglobose; achenes comparatively few. Arctic and subarctic regions: Greenl.—Man.—Alta.—Alaska. *Arct.—Subarct.—Subalp.* My–Jl.

f. 214.

10. **A. canadénsis** L. Stem strigose, 2–6 dm. high, 1–3-flowered; leaf-blades 10–12 cm. wide, 3–5-parted; divisions oblanceolate, 3-cleft and toothed; involucral leaves similar but larger, often 15 cm. long or more, sessile and more deeply cleft; sepals obovate or oval, 12–18 mm. long, white; head of fruit globose. Low ground: Lab.—Md.—N.M.—Alta. *E. Temp.—Plain—Submont.* My–Au.

11. **A. quinquefòlia** L. Stem 1–3 dm. high, simple; basal leaf solitary, appearing later than the flowering stem, long-petioled; blade 5-parted; divisions cuneate, cleft; involucral leaves 3, petioled, 3-divided into cuneate, laciniately toothed, or the lateral ones, 2-cleft divisions; flower solitary; sepals oval, white; achenes 15–20, oblong, with a hooked beak. Open woods: N.S.—Ga.—Ala.—N.D.—Man. *E. Temp.* Ap–My.

2. PULSATÍLLA Adans. PASQUE-FLOWER, BLUE TULIP, WILD CROCUS, LION'S BEARD (Fruit).

Perennial scapose herbs, with a thick taproot and short caudex. Leaves basal, palmately or rarely pinnately divided or dissected. Bracts of the scape somewhat similar to the leaves, 3 in number, verticillate, often connate. Flowers mostly solitary. Sepals 5–7, colored blue, purple, or white. Petals wanting. Stamens numerous, the outer often changed into staminodia. Pistils numerous. Fruit of numerous achenes, with much elongated, persistent, plumose styles.

f. 215.

1. **P. ludoviciàna** (Nutt.) Heller. *Fig. 215.* Leaf-blades ternate and repeatedly dissected into linear divisions, 5–10 cm. in diameter, villous-hirsute

or in age glabrate; scape 1–4 dm. high, villous; sepals ovate-oblong, 25–35 mm. long; achenes silky, their styles about 3 cm. long. *Anemone patens Nuttalliana* A. Gray. *P. patens Wolfgangiana* A. Gray [G]; not Koch. *P. hirsutissima* Britton. Prairies, plains, and hillsides: Ill.—Tex.—Utah—Wash.—Alaska. *W. Temp.—Prairie—Plain—Subalp.* Mr–S.

3. **HEPÁTICA** (Rupp.) Hill. HEPATICA, LIVER-LEAF.

Perennial acaulescent herbs, with rootstocks. Leaves basal, long-petioled, blades reniform, 3-lobed, thick, persistent over winter. Involucre of 3 small leaves, close under the flower. Sepals petal-like. Petals wanting. Stamens many. Achenes short-beaked, pubescent.

Lobes of the leaves rounded or obtuse. — 1. *H. triloba.*
Lobes of the leaves acute. — 2. *H. acutiloba.*

1. **H. triloba** Chaix. Leaf-blades 5–6 cm. broad; lobes ovate, but obtuse or rounded at the apex; bracts obtuse; sepals blue, purple, or white. *H. Hepatica* (L.) Karst [B]. Woods and copses: N.S.—Fla.—Mo.—Man.; Eur. *E. Temp.* Mr–My.

2. **H. acutíloba** DC. Resembling the preceding but leaf-lobes acute or pointed and the involucral bracts acute. *H. acuta* (Pursh) Britton [B]. Woods: Me.—Ga.—Mo.—Minn. *E. Temp.* Mr–Ap.

4. **SYNDÉSMON** Hoffmgg. RUE ANEMONE.

Perennial herbs with tuberous-thickened roots. Leaves basal, long-petioled, twice or thrice ternately compound. Bracts 3, verticillate, similar, sessile. Flowers umbellate. Sepals petal-like, white. Petals wanting. Stamens many. Stigma sessile, truncate. Achenes deeply grooved, 8–10-ribbed, ovoid. *Anemonella* Spach.

1. **S. thalictroìdes** (L.) Hoffmgg. Stem 1–2.5 dm. high, glabrous; leaflets rounded, somewhat 3-lobed; sepals oval, about 12 mm. long, white or pinkish. *A. thalictroides* Spach [G]. Open woods: N.H.—Fla.—Kans.—Minn. *E. Temp.* Mr–Je.

5. **THALÍCTRUM** L. MEADOW RUE, MAID-OF-THE-MIST.

Perennial herbs, with rootstocks. Leaves alternate, ternately decompound; petioles with dilated bases. Flowers perfect, dioecious, or polygamous, in panicles or racemes. Sepals greenish-white, 4 or 5. Petals wanting. Stamens numerous; filaments often clavate or dilated. Pistils few; stigmas elongate, unilateral. Achenes stipitate or sessile, ribbed or nerved.

Flowers perfect; achenes flat and very oblique; filaments clavate. — 1. *T. sparsiflorum.*
Flowers dioecious or polygamous; achenes not very oblique; filaments filiform.
  Stem-leaves subsessile, only the basal leaves petioled.
    Leaflets rounded-obovate in outline, scarcely paler beneath, glabrous, the lobes rounded-ovate; achenes flat, 6–7 mm. long, thin-walled, the ribs slender, separated by wide shallow grooves. — 2. *T. megacarpum.*
    Leaflets ovate or obovate in outline, the lobes at least of the upper leaves ovate, acute; achenes more terete, 4–6 mm. long, thick-walled and with narrow grooves.
      Leaves glabrous, pale beneath; achenes 4–5 mm. long, glabrous. — 3. *T. hypoglaucum.*
      Leaves viscid-pubescent beneath; achenes 5–5.5 mm. long. — 4. *T. dasycarpum.*
  Leaves all petioled.
    Achenes equally acutish at both ends, leaflets very thin, not strongly veined. — 5. *T. dioicum.*
    Achenes more acute at the upper end than at the lower.
      Leaflets firm, strongly veined, 1–1.5 cm. long. — 6. *T. venulosum.*
      Leaflets thin, not strongly veined; 2.5–4 cm. long. — 7. *T. Lunellii.*

1. **T. sparsiflòrum** Turcz. Stem glabrous, leafy; leaves bi- or tri-ternate, the lower petioled, the uppermost sessile; leaflets thin, pulverulent-glandular beneath, cordate, usually 3-cleft and the divisions 3-lobed, 1–2 cm. long; achenes 6–12, half rhombic with a straight back, glandular-pulverulent, with about three faint ribs, 5–6 mm. long, 3–4 mm. wide. *T. utahense* Greene. Moist places: Alaska—Calif.—Colo.—Man.; n Asia. *W. Temp.—Submont.—Mont.* Je–Au.

2. **T. megacárpum** Torr. Plant closely resembling *T. occidentale;* stem 3–8 dm. high; leaves 3–4 times ternate; leaflets thin, pale beneath; filaments filiform; achenes lance-oblong, acute at both ends, 6–7 mm. long, 2.5–3.5 mm. wide, with more prominent ribs than in that species. *T. propinquum* and *T. omissum* Greene. Copses, woods and river banks: Ida.—Utah—Colo.—S.D.—Mont. *Submont.—Mont.* My–Au.

f.216.

3. **T. hypoglaùcum** Rydb. *Fig. 216.* Stem 5–10 dm. high, glabrous, strict; leaves twice to thrice pinnate, the upper sessile; leaflets 2–4 cm. long, 3–5-lobed, the lobes ovate, acute or abruptly acuminate; achenes elliptic, acute at each end, 4–5 mm. long, half as wide, glabrous, plump. Banks and woods: Iowa—Kans.—Tex. *Prairie—Texan.* Je–Jl.

4. **T. dasycárpum** Fisch. & Avé-Lall. Stem tall, 5–20 dm. high; leaves 3–5 times ternate; leaflets ovate or obovate, rather thick, dark green above, paler and strongly veined beneath; achenes obliquely oblanceolate, about 5 mm. long and 2 mm. wide. *T. purpurascens* Auth. [B]; not L. *T. vegetum, T. albens, T. Nortoni* Greene. Copses and meadows: Me.—N.Y.—Ia.—n N.M.—Sask. *Prairie—Plain—Mont.* Je–Au.

5. **T. dioìcum** L. Stem 3–6 dm. high, slender; leaves 3–4 times ternate; leaflets 1–3 cm. long, thin, pale beneath, rounded, or sometimes cordate at the base, with 3–9 rounded lobes; filaments filiform; achenes oblong, about 4 mm. long. *T. Sandbergii* Greene. Woods: Lab.—Ala.—Mo.—Sask. *Boreal.* My–Je.

6. **T. venulòsum** Trelease. Stem 2–5 dm. high, glabrous; leaves 3–4 times ternate; leaflets rounded in outline, 5–15 mm. long, strongly veined, 3–5-lobed and crenate, bluish-green, paler beneath; panicle narrow and dense; filaments filiform; achenes about 5 mm. long, oblong. *T. campestre* and *T. thyrsoideum* Greene. Open mountain woods: Man.—Colo.—Utah. *Plain—Mont.* Je–Au.

7. **T. Lunéllii** Greene. Dioecious; stem about 1 m. high, light green, glabrous; upper leaves short-petioled, green above, glaucous beneath; leaflets large, 2.5–4 cm. long, 2–3.5 cm. wide, obtuse or cordate at the base, deeply 7–9-lobed above the middle, the lobes obtuse, often mucronate; achenes short, thick, about 6 mm. long, 8-ribbed. Meadows: N.D.

6. **VIÓRNA** Reichenb. VASE-VINE, LEATHER-FLOWER, OLD MAN'S WHISKERS or LION'S BEARD (Fruit).

Climbing vines or erect perennial herbs. Leaves opposite, sometimes pinnately compound or decompound. Sepals 4, rarely 5, valvate, petaloid, bluish or purplish, erect and connivent at the base or throughout. Petals and staminodia wanting. Stamens numerous, erect; anthers long and narrow, pointed; filaments hairy. Pistils many. Fruit 1-seeded achenes, with long, persistent, often plumose styles.

Climbing vines; styles pubescent but not plumose. 1. *V. Pitcheri.*
Erect herbs.

Leaves pinnately compound; styles plumose. 2. *V. Scottii.*
Leaves mostly simple; styles silky below, glabrous above. 3. *V. Fremontii.*

1. V. Pítcheri (T. & G.) Britton. A climbing vine; branches pubescent; leaves pinnate; leaflets 3–9, ovate or subcordate, entire or 3-lobed, strongly reticulate; sepals dull purple, with a narrow and slightly margined recurved point; style in fruit 2.5 cm. long. *Clematis Pitcheri* T. & G. [G]. *C. Simsii* Auth. [B]; not Sweet. Among bushes: Ind.—Mo.—Tex.—Neb. *Allegh.—Ozark—Prairie.* Je.

f.217.

2. V. Scóttii (Porter) Rydb. *Fig. 217.* Stem erect, 2–4 dm. high, more or less villous when young; leaves twice pinnately divided, spreading; peduncles about 1 dm. long; sepals purplish brown, 2.5–3.5 cm. long, more or less villous outside; achenes obovate, flattened, densely short-pubescent; their styles 4–5 cm. long. *Clematis Scottii* Porter [B]. *C. Douglasii Scottii* Coulter. Hillsides: N.M.—S.D.—Wyo. *Submont.—Mont.* My–Jl.

3. V. Fremóntii (S. Wats.) Heller. Stem erect, 2–4 dm. high, villous-pubescent, especially at the nodes; leaves sessile, simple, coriaceous, reticulate, glabrous or nearly so, broadly ovate, entire or coarsely toothed; sepals purple, 2.5 cm. long, tomentose on the margins, with recurved tips; styles in fruit 1 cm. long. *C. Fremontii* S. Wats. [G, B]. Prairies: Mo.—Kans.—Neb. *Prairie.* Ap–My.

7. CLÉMATIS L. Virgin's Bower, White Clematis, Traveler's Joy, Pipe-Stem.

Climbing vines. Leaves opposite, pinnately compound or decompound, with entire, toothed, or incised leaflets. Flowers in broad, paniculate cymes or rarely solitary, perfect, dioecious, or polygamo-dioecious. Sepals rather small, petaloid, usually white, valvate, without a border, spreading. Petals and staminodia wanting. Stamens numerous, spreading; anthers short and blunt. Pistils several or many. Fruit of 1-seeded achenes, with silky or plumose, elongate styles.

Leaves glabrate or nearly so; achenes with a thick margin.
Leaves 3-foliolate. 1. *C. virginiana.*
Leaves pinnately 5–7-foliate. 2. *C. ligusticifolia.*
Leaves silky beneath; achenes marginless. 3. *C. missouriensis.*

1. C. virginiàna L. A woody vine, 2–5 m. high; leaves 3-foliolate; leaflets ovate, acute, toothed or lobed, sometimes cordate at the base, glabrate in age; flowers polygamo-dioecious; sepals white, 8–12 mm. long, oblong-spatulate; styles in fruit 2.5–4 cm. long, plumose. Among bushes: N.S.—Ga.—Kans.—Man. *Canad.—Allegh.* Jl–S.

f.218.

2. C. ligusticifòlia Nutt. *Fig. 218.* A woody vine, 3–6 m. high; leaves pinnately 5–7-foliolate; leaflets petiolate, in the typical form lanceolate to ovate, rounded or truncate at the base, usually more or less acuminate, 3–8 cm. long, sparingly strigose; cymes with nearly erect branches; sepals white, nearly 1 cm. long, oblong-oblanceolate; filaments of the staminate flowers linear-filiform, those of the pistillate flowers broader, linear, somewhat dilated, with sterile anthers; achenes numerous, obovate, pubescent; styles plumose, 4–5 cm.

long. Very variable species. Copses and cañons among bushes: B.C.—N.D.—N.M.—Calif. *W. Temp.—Plain—Submont.* My–Au.

3. **C. missouriénsis** Rydb. A woody vine, 2 m. high or more; stem silky-puberulent; leaves pinnately 3–5-foliolate; leaflets ovate or cordate, often 3-lobed or incised; flowers polygamo-dioecious, in small panicles, the branches often subumbellate; achenes 4 mm. long; style in fruit 3 cm. long, plumose. Rich woods: Mich.—Mo.—Kans.—Neb. *Allegh.* Jl–S.

8. ATRÁGENE L. Bell Rue, Purple Virgin's Bower.

Perennial climbing or creeping vines. Leaves opposite, compound. Flowers large and showy, solitary in the axils or at the ends of the branches. Sepals large, petal-like, mostly membranous, blue or purple, spreading from the base. Petals or staminodia small, spatulate, usually present. Stamens numerous, spreading; filaments pubescent, the outer ones dilated; anthers short. Pistils many. Fruit 1-seeded achenes, with persistent, long, plumose styles.

Leaves ternate. 1. *A. americana.*
Leaves twice or thrice ternate. 2. *A. tenuiloba.*

1. **A. americàna** Sims. A trailing or climbing vine; leaves 3-foliolate; leaflets thin, ovate, acute, toothed or entire, cordate at the base; sepals purplish-blue, 2.5–5 cm. long, strongly veined; petals or staminodia spatulate, 12–18 mm. long; style plumose, in fruit 5 cm. long. *Clematis verticillaris* DC. [G]. Rocky woods: Que.—Va.—Minn.—Man. *Canad.—Allegh.* My–Je.

2. **A. tenuíloba** (A. Gray) Britton. Mostly trailing vines; leaves biternate; secondary leaflets 1–3 cm. long, usually divided to near the midrib into lanceolate or ovate divisions; these more or less toothed, acuminate; sepals 2.5–3.5 cm. long, lanceolate, acuminate, blue or purple; achenes densely pubescent; styles 4–5 cm. long. Woods: S.D.—Colo.—Ariz.—Mont. *Submont.—Subalp.* My–Jl.

9. MYOSÙRUS L. Mouse-tail.

Small acaulescent annual mud plants, with fibrous roots. Leaves basal, linear or filiform. Scapes 1-flowered. Sepals 5, rarely 6 or 7, long-spurred at the base. Petals 5–7, narrow, greenish yellow, or wanting, the claws nectariferous at the summit. Stamens 5–25. Pistils numerous, borne on a receptacle, which becomes elongated and cylindric at maturity. Achenes aristate or apiculate, numerous, 1-seeded, tardily dehiscent. Seeds suspended; embryo minute.

Achenes with a flat back, only slightly carinate, and tipped with a very short appressed beak. 1. *M. minimus.*
Achenes strongly carinate on the back, tipped with a subulate, ascending beak. 2. *M. aristatus.*

1. **M. mínimus** L. *Fig. 219.* Leaves filiform or nearly so, glabrous, 3–10 cm. long, blunt; scape 3–15 cm. long; sepals oblong, about 3 mm. long; spurs 1–2 mm. long; petals spatulate, long-clawed, sometimes lacking; fruiting spike 3–6 cm. long. In mud and shallow water: Ont.—Fla.—Calif.—Wash.; Eur. and n Africa. *Temp.—Plain—Submont.* Ap–Je.

2. **M. aristàtus** Geyer. Leaves filiform to narrowly linear-oblanceolate, blunt; scape 2–6 cm. long; sepals 2 mm. or less long; spurs about 1 mm. long; petals often lacking; carpel-spike 3–20 mm. long. (?) *M. nitidus* Eastw. In mud: B.C.—Mont.—Neb.—N.M.—Calif. *W. Temp.—Plain—Mont.* Ap–Jl.

10. BATRÀCHIUM S. F. Gray. WHITE WATER CROWFOOT.

Perennial aquatic herbs, with floating or creeping stems. Leaves alternate, palmately dissected or lobed, the submerged ones usually with filiform divisions. Flowers perfect, solitary, borne opposite the leaves. Sepals and petals usually 5, the latter white, sometimes with a yellowish base, bearing a small pit on the claw. Stamens several or numerous. Pistils many in a globular head. Achenes oblique, compressed, not margined, transversely wrinkled, short-beaked or almost beakless.

Leaves sessile, with divaricate stiff divisions. — 1. *B. divaricatum.*
Leaves petioled, with ascending, soft divisions.
  Beak of the achenes nearly 1 mm. long. — 2. *B. longirostre.*
  Beaks of the achenes minute or none.
    Petals 5–7 mm. long, broadly obovate; stamens many.
      Primary divisions of the leaves 1–1.5 cm. long, rather rigid, scarcely collapsing when withdrawn from the water. — 3. *B. trichophyllum.*
      Primary divisions of the leaves 1.5–3 cm. long, flaccid, collapsing when withdrawn from the water. — 4. *B. flaccidum.*
    Petals less than 5 mm. long, oblong-obovate; stamens 5–12; stem slender, but not capillary; leaves not very flaccid. — 5. *B. Drouetii.*

1. **B. divaricàtum** (Schrank) Wimmer. Stem 3 dm. long or more; leaves sessile, 1–2 cm. broad, with short, ovate divisions; petals broadly obovate, about 5 mm. long; head in fruit globose. *Ranunculus circinatus* Sibth. [G]. Water: Vt.—Pa.—Neb.—Man.; Eur. *Canad.—Allegh.*

2. **B. longiróstre** (Godr.) F. Schultz. Stem very leafy and branched; petioles very short; blades 2–4 cm. broad; petals broadly ovate, 5–7 mm. long; head of fruit globose; achenes many, about 1.5 mm. long. *Ranunculus circinatus* A. Gray, in part, not Sibth. Ponds and slow streams: Ont.—N.Y.—N.M.—Utah—S.D. *Boreal—Plain.* Je–Au.

3. **B. trichophýllum** (Chaix) Bosch. Stem 3 dm. long or more; petioles 1–1.5 cm. long; blades 2–3 cm. wide, usually with spreading divisions; petals about 8 mm. long; head of achenes globose, 4 mm. thick; achenes many, about 1.5 mm. long. *R. aquatilis capillaceus* DC. [G]. Ponds and slow streams: N.S.—N.C.—Calif.—B.C.; Mex.; Eurasia. *Temp.—Plain.—Mont.* Je–S.

4. **B. fláccidum** (Pers.) Rupr. Stem long and slender; leaves rather distant; leaf-blades 3–5 cm. broad, with usually ascending segments; petals 5–8 mm. long, white; head of fruit about 4 mm. wide; achenes often nearly 2 mm. long. In streams, rarely in ponds: Lab.—N.C.—L. Calif.—Wash.; Eurasia. *Temp.—Plain—Mont.* My–S.

5. **B. Drouètii** (F. Schultz) Nym. Similar in habit to *B. trichophyllum,* but more slender; leaf-blades 1.5–3 cm. wide; petals 3–5 mm. long; head of achenes about 4 mm. thick, globose; achenes about 1.5 mm. long. In streams: Vt.—R.I.—N.M.—L. Calif.—Alaska; Eurasia and Africa. *Temp.—Plain—Submont.* Je–S.

11. RANÚNCULUS (Tourn.) L. CROWFOOT, BUTTERCUP.

Annual or almost all of ours perennial herbs, with a cluster of fleshy-fibrous roots. Leaves alternate, entire, lobed, divided or dissected. Flowers solitary or cymose. Sepals mostly 5, deciduous. Petals 5 or more, in ours yellow, each with a nectariferous pit and a scale at the base of the blade. Pistils several or many, 1-ovuled. Achenes flattened, smooth, papillose, or echinate, tipped with the style, not striate.

Achenes smooth.
  Leaves linear to ovate, entire, or merely denticulate or crenate, none divided or cleft.
    Annuals; achenes beakless. — 1. *R. pusillus.*

Perennials; achenes beaked.
Stem decumbent at the base only, 3–10 dm. high; achenes with a subulate beak. — 2. *R. obtusiusculus.*
Stem trailing throughout, low; achenes with a minute beak. — 3. *R. reptans*
Some of the leaves at least cleft.
Some of the basal leaves entire; basal leaves elliptic to reniform. — 4. *R. glaberrimus.*
None of the leaves entire.
Neither floating water plants nor creeping mud plants, if rooting at the nodes, the leaves not palmately lobed or dissected.
Achenes turgid, marginless.
Petals much exceeding the sepals.
Achenes pubescent, with a short recurved beak; heads oblong to cylindric; sepals densely villous. — 5. *R. cardiophyllus.*
Achenes glabrous; beak straight, usually slender; plant more or less pubescent. — 6. *R. ovalis.*
Petals scarcely exceeding the sepals.
Basal leaves, at least some of them, merely crenate; perennials.
Basal leaf-blades cordate at the base; plant glabrous. — 7. *R. abortivus.*
Basal leaf-blades rarely cordate at the base; plant, especially the stem, with spreading hairs. — 8. *R. micranthus.*
All the leaves divided or lobed; annuals. — 9. *R. sceleratus.*
Achenes compressed, with a distinct margin.
Beak of the achenes strongly hooked; heads of fruit globose. — 10. *R. recurvatus.*
Beak of the achenes straight or nearly so; achenes in ours glabrous.
Beak short.
Leaves, at least the basal ones, pinnately ternate, the terminal division at least petioled.
Stem bulbous-thickened at the base; sepals reflexed. — 11. *R. bulbosus.*
Stem not bulbous at the base.
Petals less than 1 cm. long.
Heads of achenes decidedly oblong, about 5 mm. thick. — 12. *R. pennsylvanicus.*
Heads of achenes globose, 7–12 mm. thick.
Plants not stoloniferous; primary segments of the leaves cuneate at the base; beak half as long as the body of the achene. — 13. *R. Macounii.*
Plant producing long lateral branches, rooting at the nodes; primary segments of the lower leaves more or less truncate or subcordate at the base; beak one-third as long as the body of the achene. — 14. *R. rivularis.*
Petals more than 1 cm. long; stem creeping. — 15. *R. repens.*
Leaves palmately divided. — 16. *R. acris.*
Beak long.
Beak of the achenes broad and flat; roots slender; plant stoloniferous.
Stem glabrous or nearly so. — 17. *R. septentrionalis.*
Stem decidedly hispid. — 18. *R. caricetorum.*

Beak of the achenes subulate; roots tuberous-thickened; plant not stoloniferous.
Stem with appressed pubescence. 19. *R. fascicularis.*
Stem with spreading pubescence. 20. *R. hispidus.*
Immersed aquatics or creeping mud plants, with palmately lobed, divided or dissected leaves.
Achenes callous-margined. 21. *R. delphinifolius.*
Achenes marginless.
Stem glabrous. 22. *R. Purshii.*
Stem pubescent. 23. *R. limosus.*
Achenes prickly, annuals. 24. *R. arvensis.*

1. **R. pusíllus** Poir. Stems slender, branching, 1.5–3 dm. high; leaf-blades entire or denticulate, the lower 0.5–2 cm. long, oblong or ovate, long-petioled, the upper lanceolate or linear; petals 1–5, pale yellow, 2 mm. long, equaling the sepals; stamens 3–10; heads of fruit globose; achenes beakless, smooth or slightly papillose. Wet places: N.Y.—Fla.—Tex.—Kans. *SE. Temp.* Ap–S.

2. **R. obtusiúsculus** Raf. Stem ascending, rooting below, 3–6 dm. high, leaves lanceolate or the lowest oblong, denticulate, 4–10 cm. long, with wing-margined petioles; petals 5–7, oblong, 4–6 mm. long; head of fruit globose or slightly oblong, 10–12 mm. thick; achenes flattened, with a slender subulate beak. *R. laxicaulis* (T. & G.) Darby [G]. Ditches and muddy places: Me.—Ga.—Ark.—Minn. *E. Temp.* Je–Au.

3. **R. réptans** L. Stem filiform, 1–3 dm. long; leaves 1–5 cm. long, linear-filiform and glabrous to oblanceolate and strigose (var. *strigulosus* Freyn); peduncles 1–3 cm. long, filiform; petals obovate, 2–4 mm. long; head of fruit spherical; achenes glabrous, with a minute beak. *R. Flammula reptans* E. Meyer. [G]. Shores: Lab.—N.J.—N.M.—Ore.—Alaska. *Temp.—Plain—Mont.* My–S.

4. **R. glabérrimus** Hook. Stem 1 dm. high or less, glabrous, more or less fleshy; basal and lower cauline leaves petioled, thick, glabrous; blades reniform to rounded-oval, usually more or less 3-lobed at the apex; upper stem-leaves cuneate or oblanceolate, entire or 3-cleft; petals rounded-oval, 8–12 mm. long; heads of achenes globose; achenes glabrous, with an only slightly curved beak. *R. ellipticus* Greene, a form with thinner, more entire basal leaves. *R. Waldronii* Lunell, a form with smaller elliptic petals. Wet places: Wash.—Mont. —S.D.—N.M.—Calif. *W. Temp.—Plain—Submont.* Mr–My.

5. **R. cardiophýllus** Hook. Stem 1.5–3 dm. high, more or less villous; basal leaves petioled, thick, pubescent when young; blades 2–4 cm. wide, reniform or cordate; stem-leaves divided into linear lobes; heads of achenes usually oblong or ellipsoid. *R. affinis validus* A. Gray. Wet meadows, bogs, and along streams: Sask.—Neb.—Colo.—Wash.—Alta. *Boreal.—Plain—Mont.* My–Au.

6. **R. ovàlis** Raf. Stem 0.5–3 dm. high, more or less villous, especially when young; basal leaf-blades 1–3 cm. wide, from reniform or orbicular to rhombic-oval, crenate or round-lobed, some occasionally cleft, hairy when young; upper stem-leaves sessile, with linear or oblong divisions; sepals villous; petals obovate to oblong-cuneate, 4–6 mm. long; head of fruit globose; beak short. *R. rhomboideus* Goldie. [G]. *R. brevicaulis* Hook. Meadows: Lab.—Que.—Colo.—Alta. *Boreal.—Plain—Submont.* My–Je.

7. **R. abortìvus** L. Stem 1.5–6 dm. high, branched; basal leaf-blades 1–5 cm. wide, the later ones more or less deeply 3-cleft; upper stem-leaves sessile, with linear or oblong divisions; sepals 2–3 mm. long, oblong; petals oblong. Wet places and open woods: Lab.—Fla.—Colo.—B.C. *Temp.—Plain —Submont.* Ap–Jl.

8. **R. micránthus** Nutt. Stem 1.5–4 dm. high, branched; basal leaf-blades 1–4 cm. wide, the later 3-cleft or 3-divided, with oblanceolate-cuneate

to obovate divisions; upper stem-leaves with linear divisions, sessile; flowers and achenes as in the preceding. *R. abortivus micranthus* A. Gray. Rich woods: Que.—Pa.—Colo.—Sask. *Canad.—Allegh—Plain—Mont.* Ap–Je.

9. **R. scelerà tus** L. Stem glabrous or nearly so, stout, more or less fleshy, 1.5–6 dm. high, freely branching; blades of the basal leaves 3–10 cm. wide, reniform, 3-cleft or 3–5-parted; segments round-lobed; upper stem-leaves sessile and with linear lobes; sepals 3–4 mm. long, hairy; petals elliptic; head of achenes oblong, with numerous small glabrous achenes; beak minute. *R. eremogenes* Greene. Shallow water and swamps: N.B.—Fla.—N.M.—S.D.—Man.; Eurasia. *Temp.—Plain—Mont.* Ap–Au.

10. **R. recurvàtus** Poir. Stem erect, hirsute, 1.5–6 dm. high; leaves petioled, broadly reniform in outline, 5–7.5 cm. wide, deeply 3-cleft, the divisions broadly cuneate, lobed and toothed; petals light yellow, shorter than or equaling the reflexed sepals; head of fruit globose, 12 mm. broad; achenes compressed, margined, tipped with a recurved, hooked beak. Woods: N.S.—Man.—Mo.—Fla. *E. Temp.* Ap–Je.

11. **R. bulbòsus** L. Stem erect, pubescent, 1.5–4 dm. high; leaves ternate, the terminal leaflet stalked, the lateral ones sessile, lobed and cleft, 1–3 cm. long; petals much longer than the sepals, rounded-obovate; head of the achenes compressed, short-beaked. Fields and roadsides: Me.—Neb.—D.C.; nat. from Eur. My–Jl.

12. **R. pennsylvánicus** L. f. Mostly annual; stem erect, branching, hirsute, 3–8 dm. high; leaf-blades 5–15 cm. wide, very hairy, ternate; divisions petioled (especially the terminal one), ternately or pinnately divided, cleft and toothed, with lanceolate teeth; petals oblong, 2–4 mm. long, shorter than the reflexed sepals; beak broad, flat, less than half as long as the achene. Wet places and shallow water: N.S.—Ga.—N.M.—Wash.—B.C. *Temp.—Plain—Submont.* Je–Au.

13. **R. Macoùnii** Britton. *Fig. 220.* Stem 2–6 dm. high, usually very hirsute, at first usually erect, but later decumbent, branched and leafy; leaf-blades 5–15 cm. wide, ternate, hirsute, especially beneath; segments petiolate, especially the terminal one, rhombic-obovate, twice ternately cleft and toothed; petals obovate, 5–7 mm., equaling or somewhat exceeding the sepals. Wet meadows: Ont.—Ia.—N.M.—Utah—B.C. *Boreal—Plain—Submont.* Je–Au.

R. Macounii × pennsylvanicus. Intermediate between the parents, resembling *R. Macounii* in habit, but with smaller flowers and achenes. Devils Lake, N.D.

f. 220.

14. **R. rivulàris** Rydb. Stem hirsute, sometimes more than 1 m. long, rooting at the nodes and there producing plantlets; leaves ternate, 5–15 cm. wide; divisions petiolate, ovate, 3-cleft and coarsely toothed; petals rounded-obovate, about 4 mm. long, scarcely equaling the sepals. *R. repens* S. Wats.; not L. River banks and along ditches: Nev.—Ariz.—Neb. *Prairie—Plain—Submont.* Au.

15. **R. rèpens** L. Stem strigose or nearly glabrous, at first ascending, then decumbent and creeping; leaf-blades 3–8 cm. wide, ternate or the lower biternate, strigose or glabrate; divisions petiolate or the lateral ones sometimes subsessile, ternately or biternately cleft and toothed; petals broadly obovate or flabelliform, fully twice as long as the sepals; beak about one fourth as long as the achene. Meadows and roadsides: Newf.—Va.—Utah—Wash.; adv. or nat. from Eur. *Temp.—Plain—Submont.*

16. **R. àcris** L. Stem erect, strigose or somewhat hirsute, or nearly glabrous, 3–9 dm. high; blades of the basal leaves 3–8 cm. wide, palmately 3–5-divided to near the base, more or less hairy; divisions broadly rhombic in outline, twice cleft into lanceolate lobes; petals broadly obovate, 1 cm. long or more, more than twice as long as the sepals; beak slightly curved, about one fourth as long as the achene. Fields, meadows, and roadsides: Newf.—N.C.—Kans.—B.C.; nat. from Eur. *Temp.—Plain—Submont.*

17. **R. septentrionàlis** Poir. Stem 3–8 dm. long, villous; leaves long-petioled; blades with 3, stalked, broadly cuneate or obovate divisions, 3-cleft or parted and cut; petals obovate, twice as long as the sepals; heads of fruit subglobose, 10–13 mm. thick; achenes broadly margined. Moist places: N.B.—Ga.—Neb.—S.D.—Man. *E. Temp.* My–Au.

18. **R. caricetòrum** Greene. Stem at first erect, then reclining, 3–6 dm. high, strongly retrorsely hispid as well as the petioles; petioles of the lower leaves 2–2.5 dm. long; blades 3-divided, with stalked, 3-cleft and incised, broad divisions; petals 5, obovate, 10–12 mm. long; head of fruit globose; achenes obovate, with a thick margin; beak broad, slightly curved. *R. sicaefolius* Mack. & Bush. Wet prairies: Mo.—Minn.—Neb. (?). *Prairies.* My–Je.

19. **R. fasciculàris** Muhl. Stem low, ascending, 1–2.5 dm. high, from a cluster of fibrous-fleshy roots; basal leaves long-petioled; blades ternate, appearing pinnate; terminal division long-stalked, 3–5-divided, again 3–5-cleft into oblong or linear lobes; lateral divisions sessile; petals oblong-spatulate, twice as long as the sepals; head of fruit globose, 8–10 mm. thick; achenes scarcely margined. Hills: Mass.—Ga.—Tex.—Man. *E. Temp.* Ap–My.

20. **R. híspidus** Michx. Stem 1.5–4 dm. high, from a cluster of stout fibres; leaves ternately divided or the basal ones 3-lobed, and variously cleft or toothed; petals oblong, twice as long as the sepals; head of fruit globose, 6–10 mm. broad; achenes broad, obscurely margined. Moist woods: Ont.—Ga.—Ark.—N.D. *E. Temp.* Mr–My.

21. **R. delphinifòlius** Torr. Plant glabrous or slightly hairy when young; immersed leaves 5–10 cm. wide, finely dissected into capillary divisions, short-petioled; floating or emerged leaves 2–5 cm. wide, with longer petioles, repeatedly ternate, with linear lobes; leaves when growing in mud (var. *terrestris*) with much shorter and broader divisions, the lobes oblong or lance-ovate; petals 5–8 mm. long, obovate; head of fruit globose or slightly oblong. *R. multifidus* Pursh. Shallow water, or mud: Me.—N.C.—Kans.—Utah—B.C. *Temp.—Plain.* Je–Au.

22. **R. Púrshii** Richardson. Leaf-blades 1–3 cm. wide, cordate at the base, palmately divided into 3–7 more or less cuneate and lobed or cleft divisions, or the immersed ones dissected into filiform divisions; petals obovate, 4–5 mm. long; head of fruit globose, 5–6 mm. broad. Water or mud: N.S.—Ont.—Colo.—Ore.—Alaska. *Temp.—Mont.* Je–S.

23. **R. limòsus** Nutt. Leaf-blades 1–3 cm. wide, reniform in outline, palmately divided into more or less cuneate, cleft or lobed divisions; petals obovate, 4–5 mm. long; head of fruit globose, 4–5 mm. thick. Shallow water and mud: N.D.—Wyo.—Utah—Wash.—Alaska. *W. Boreal.—Submont.* Jl–Au.

R. limosus × sceleratus. Resembling mostly *R. limosus* in leaf-form, pubescence and the large flowers, but is an erect plant. *R. sceleratus × limosus* Lunell. Margins of swamps: Leeds, N.D.

24. **R. arvénsis** L. Stem glabrous or sparingly pubescent, branched above, 3–4 dm. high; lower leaves petioled; blades cuneate, 3–5-lobed at the apex; the rest of the leaves short-petioled, twice or thrice ternate; primary

divisions distinctly petioluled, the ultimate ones lanceolate or linear; sepals lanceolate, yellowish, membranous, long-pilose, 5 mm. long; petals obovate, strongly veined, 6–7 mm. long; achenes few, obliquely ovate, compressed, 7–10 mm. long, spiny; beak 3–4 mm. long, flat, subulate, slightly curved. Waste places: N.J.—Ohio—Utah; adv. from Eur. Je–Au.

12. **COPTÍDIUM** Beurl.

Scapose perennial herbs, with slender, stoloniferous rootstocks. Leaves basal, petioled, with reniform, 3-parted blades, the divisions of which are again lobed. Scape 1-flowered. Sepals and petals 5, the latter yellow. Pistils rather few; styles long. Achenes obliquely ovate, the lower half enclosing the seeds, the upper portion empty, flattened, somewhat spongy.

1. **C. lappónicum** (L.) Gand. Stem scapiform, about 1 dm. high, filiform, glabrous, naked or with a single small leaf; basal leaf-blades 2–5 cm. wide, glabrous, ternately divided; divisions cuneate, 3–5-lobed; petals elliptic, about 5 mm. long, equaling the sepals; achenes about 5 mm. long, 2 mm. wide, acute at both ends; beaks about 1.5 mm. long, hooked. *Ranunculus lapponicus* L. [G, B]. *Anemone nudicaulis* A. Gray. In moss: Lab.—Ont.—Minn.—Alta.—Alaska; Eur. *Subalp.—Subarctic—Boreal.* Je–Jl.

13. **HALERPÉSTES** Greene.

Perennial herbs, with runners. Leaves mainly basal, petioled, cleft or toothed. Stem more or less scapiform. Flowers small, perfect, cymose or solitary. Sepals usually 5, spreading, tardily deciduous. Petals 5–12, each with a nectariferous pit at the base. Stamens and pistils numerous. Achenes compressed, thin-walled, longitudinally striate.

1. **H. Cymbalària** (Pursh) Greene. Leaves glabrous, more or less fleshy; blades rounded-cordate or reniform, crenate, 4–18 mm. long; scapes 2–30 cm. long; flowers 1–7; sepals yellowish green, oval or oblong, 3–7 mm. long; petals oblong-spatulate, slightly shorter; heads of fruit rounded-oblong, 6–16 mm. long. *Ranunculus Cymbalaria* Pursh. [G]. *Oxygraphis Cymbalaria* Prantl. [B]. Wet sandy or alkaline soil: Lab.—N.J.—N.M.—Calif.—Alaska; Mex. and S.Am. *Temp.—Trop.—Son.—Plain—Mont.* My–Jl.

14. **NIGÉLLA** (Tourn.) L. FENNEL FLOWER. LOVE-IN-A-MIST.

Erect annuals. Leaves alternate, finely divided. Sepals 5, regular, petal-like, deciduous. Petals 5, small, with hollow claws, 2-lobed or notched at the apex. Stamens numerous. Carpels 3–10, united and fusing at the base into one cavity, dehiscent at the apex. Seeds many, black, hard.

1. **N. damascèna** L. Stem 2–6 dm. high, branched; leaves finely dissected into filiform divisions, the upper close together and forming a kind of involucre under the flower; sepals blue, or sometimes white, oval-lanceolate; follicles forming a globose fruit, the styles divergent. Waste places, near dwellings: Northeastern states, west to N.D.; escaped from cult.; nat. of s Europe.

15. **CÁLTHA** (Rupp.) L. MARSH-MARIGOLD, MEADOW-GOWAN.

Perennial herbs, somewhat fleshy. Leaves few, mostly basal, petioled; blades elliptic, oval, cordate or reniform, entire or crenate. Sepals 5–15, large, more or less petal-like, deciduous, yellow, white, or pink. Petals wanting. Stamens numerous; filaments short. Carpels numerous or several, sessile; ovules numerous, in 2 rows on the ventral suture. Fruit of follicles, dehiscent on the ventral suture.

Stem erect or ascending; flowers yellow; follicles curved. — 1. *C. palustris.*
Stem floating or creeping; flowers white or pinkish; follicles straight. — 2. *C. natans.*

1. **C. palústris** L. *Fig. 221.* Stem erect, hollow, grooved, 3–6 dm. high, glabrous; basal leaves long-petioled; blade cordate or reniform, 5–20 cm. wide, crenate or dentate; upper stem-leaves short-petioled or subsessile, truncate at the base; sepals yellow, oval, 10–15 mm. long. Swamps and wet meadows: Newf. — S.C. — Neb. — S.D. — Man.; Eurasia. *E. Temp.* Ap–Je.

2. **C. nàtans** Pall. Stem slender, floating or creeping, rooting at the nodes, 1.5–5 dm. long; lower leaves long-petioled; blades cordate-reniform, 3–5 cm. wide, entire or crenate, with a narrow sinus; upper leaves short-petioled and smaller; sepals white or pinkish, 6–8 mm. long, oval, obtuse; follicles about 4 mm. long; beak minute. In water: Man.—Minn.—Alta.—Alaska; Siberia. *Mont.—Boreal.* Jl–Au.

16. ISOPỲRUM L. FALSE RUE ANEMONE.

Perennial herbs, with 2–3-ternate leaves, the leaflets 2–3-lobed. Flowers perfect, axillary and terminal, white. Sepals 5, petal-like, deciduous. Stamens 10–40. Pistils 2–20, sessile. Pods forming a head, ovate or oblong, 2–several-seeded.

1. **I. biternàtum** (Raf.) T. & G. Stem slender from a cluster of fibrous roots, sometimes tuberiferous; lower leaves bi-ternate; segments broadly obovate, lobed; sepals 5, oblong; filaments white, clavate; pistils 3–6; follicles divaricate in fruit, 2–3-seeded. Moist woods and thickets: Ont.—Fla.—Tex.—Minn. *E. Temp.* My.

17. CÓPTIS Salisb. GOLD-THREAD.

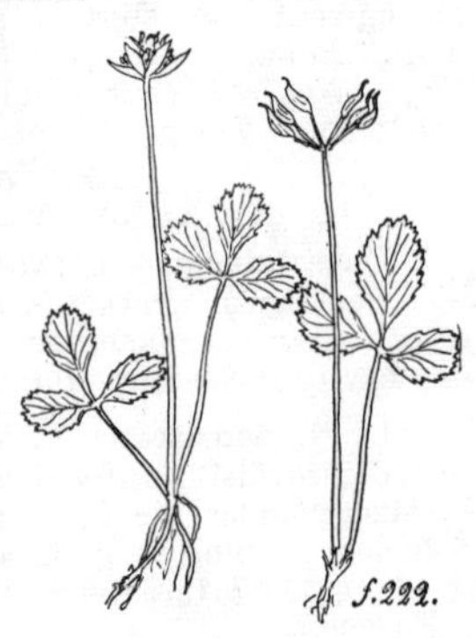

Low scapose perennial herbs, with slender rootstocks. Leaves basal, petioled, ternately compound. Scape slender, 1-flowered. Sepals 5–7, oval, petal-like, deciduous, white, with yellowish bases. Petals shorter than the stamens, club-shaped, hollowed and nectariferous at the thickened summit. Stamens numerous. Pistils 3–7. Fruit of as many stipitate follicles, forming an umbel-like cluster. Seeds smooth and shining.

1. **C. trifoliàta** (L.) Salisb. *Fig. 222.* Leaves 5–10 cm. long, ternate; leaflets evergreen, broadly obovate, slightly 3–5-lobed and crenate, shining; scape 5–12 cm. long; sepals white with yellow base; petals 2–3 mm. long; stipe of the follicle 7–8 mm. long; body about 6 mm. Woods and bogs: Greenl.—Md.—Minn.—B.C.—Alaska; n Eurasia. *Mont.—Boreal.* My–Au.

18. ACTAÈA L. BANEBERRY.

Perennial herbs, with thick rootstocks and erect stems. Leaves basal and cauline, ternately decompound. Flowers small, in terminal racemes. Sepals 3–5, petal-like, usually white. Petals 4–10, small, narrow, clawed. Stamens numerous; filaments flattened, white. Pistil solitary, sessile; style none; stigma 2-lobed, depressed. Fruit berry-like, with depressed, horizontal seeds.

Pedicels thickened in fruit, red; berries white. 1. *A. brachypoda.*
Pedicels slender.
  Fruit white, ellipsoid, 9–12 mm. long. 2. *A. alba.*
  Fruit red.

Fruit ellipsoid, 10–12 mm. long. 3. *A. rubra.*
Fruit spherical or nearly so, 5–7 mm. long. 4. *A. arguta.*

1. **A. brachýpoda** Ell. Stem 5–10 dm. high, glabrous; leaves mostly twice ternate; leaflets sharply incised; petals slender, truncate at the end; raceme ellipsoid; pedicels in fruit fully as thick as the peduncle, red; fruit globular-ovoid, white. *A. alba* Big. [G, B]; not (L.) Mill. Rich woods: N.S.—Ga.—Neb.—N.D.—Man. *E. Temp.* Ap–Je.

2. **A. álba** (L.) Mill. *Fig. 223.* Stem 6–10 dm. high, glabrate or villous-puberulent above; leaves ternate or twice ternate, the divisions pinnate; leaflets ovate, usually 3–5-lobed, and sharply serrate; pedicels slender; sepals orbicular, early deciduous; petals narrowly rhombic-spatulate, acute; fruit about 12-seeded; seeds obliquely pear-shaped, about 4 mm. long. *A. eburnea* Rydb. [R]. In rich woods: B.C.—Ore.—Utah—Colo.—Man.; n N.Y.—Newf. *Boreal.—Submont.—Mont.* My–Je.

3. **A. rùbra** (Ait.) Willd. Like the preceding, but lower, 3–6 dm. high; leaflets shorter and broader, firmer, darker, acute or short-acuminate; teeth coarser, less sharp; pedicels more divaricate; petals spatulate, acute; seeds about 3 mm. long. Rich woods: N.S.—N.J.—Ia.—Mont.—Alta. *E. Boreal.—Mont.* My–Je.

4. **A. argùta** Nutt. Very similar to *A. alba* in habit, fully as tall; leaflets very thin and light green, long-acuminate and very sharply toothed; sepals with long claws and rhombic acute blades; seeds about 10, about 3 mm. long. Woods: Alaska—Calif.—N.M.—S.D.—Alta. *W. Temp.—Submont.—Mont.* My–Je.

19. **HYDRÁSTIS** Ellis. Orange-root, Golden Seal.

Low perennials, with yellow, knotted rootstocks. Leaves one basal, long-stalked and two stem-leaves borne near the top of the stem. Flowers solitary, terminal. Sepals 3, petal-like, deciduous. Petals wanting. Stamens numerous, with white filaments. Carpels numerous, 2-ovuled, in fruit forming a head of red berries.

1. **H. canadénsis** L. Stem 2–3 dm. high, erect; basal leaf-blades reniform, 1–2 dm. wide, 5–7-lobed, doubly serrate; flowers greenish-white, 8–10 mm. broad. Rich woods: Vt.—Ga.—Ark.—Neb.—Minn. *E. Temp.* Ap.

20. **AQUILÈGIA** (Tourn.) L. Columbine.

Perennial herbs, with rootstocks and branching stems. Leaves basal and cauline, ternately decompound. Flowers perfect, usually large and showy, regular. Sepals 5, petal-like, deciduous. Petals concave, produced below into a hollow spur, or at least saccate. Stamens numerous, the inner reduced to staminodia; filaments more or less flattened. Pistils 5, sessile, many-ovuled. Fruit follicles many-seeded, with slender styles. Seeds numerous, smooth and shining, with a hard coat.

Spur of the petals hooked; sepals and spurs mostly blue or purple. 1. *A. brevistyla.*
Spur of the petals nearly straight; sepals and spurs mostly scarlet. 2. *A. latiuscula.*

1. **A. brevístyla** Hook. *Fig. 224.* Stem 4–10 dm. high, pubescent or glandular above, basal leaves biternate; stem-leaves ternate or the upper simple and 3-lobed; leaflets 3-lobed and coarsely crenate; sepals blue, lanceolate, acute, about 15 mm. long; lamina of the petals yellowish white; spur 6–8 mm.

long; follicles 2–2.5 cm. long. Meadows and open woods: Yukon—Alta.—S.D.—Minn. *Prairie—Plain—Submont.* Je–Jl.

2. **A. latiúscula** Greene. Stem 2–6 dm. high, glabrous or sparingly glandular-pubescent above; leaves biternate; leaflets obovate or cuneate, 2–4 cm. long, deeply 3-cleft and round-lobed or crenate; glaucous beneath; sepals ovate-lanceolate, acute, 10–12 mm. long, purplish-red; lamina 8–10 mm. long, yellow, rounded at the apex; spur 12–15 mm. long, abruptly narrowed at the middle, more or less incurved; follicles pubescent, about 2 cm. long. *A. vulgaris hybrida* Hook. *A. vulgaris violacea* Nutt. Closely related to *A. coccinea* Small and *A. canadensis*. Open woods: Iowa—Kans.—S.D.—N.D.—Minn. *Prairie—Plain—Submont.*

21. DELPHÍNIUM (Tourn.) L. LARKSPUR.

Annual or perennial herbs, usually with erect branching stems. Leaves alternate, mostly cauline, palmately lobed or divided. Flowers in racemes or panicles, perfect, irregular, zygomorphic. Sepals 5, the posterior produced below into a spur. Petals 2 or 4, the two posterior spurred, the lateral ones, if present, small, all in most species 2-cleft at the apex. Stamens numerous. Pistils 1–5, in ours mostly 3, many-ovuled; stigmas introrse. Fruit of many-seeded follicles.

Carpels and follicles 3, rarely 4 or 5.
- Sepals white or merely tinged with blue or with a blue spot; follicles about four times as long as broad; seeds squamellate. — 1. *D. virescens*
- Sepals dark blue; follicles only 2–3 times as long as broad; seeds not squamellate, but wing-margined.
  - Follicles erect or ascending.
    - Inflorescence few-flowered, the lower pedicels elongate; leaf-segments narrow, mostly linear.
      - Stem viscid, at least above; root thickened, but not tuber-like. — 2. *D. bicolor.*
      - Stem appressed-strigose above, not viscid; root tuber-like. — 3. *D. Nelsonii.*

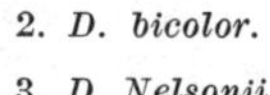

    - Inflorescence many-flowered, pedicels all short; leaf-segments lanceolate. — 4. *D. exaltatum.*
  - Follicles widely spreading. — 5. *D. tricorne.*

Carpels and follicles solitary; annuals. — 6. *D. Ajacis.*

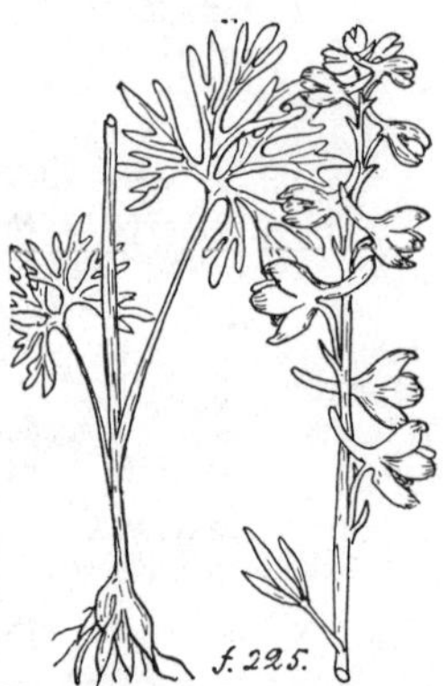

1. **D. viréscens** Nutt. *Fig. 225.* Stem 3–15 dm. high; leaves finely dissected, with linear or linear-oblong divisions; racemes long, sometimes 5–6 dm.; spur stout, tinged with blue, usually horizontal; follicles pubescent, cylindric, about 2 cm. long, nearly straight and ascending. *D. albescens* Rydb. [B]. *D. Penardi* Rob. & Fern. [G]; not Huth. Prairies and river valleys: Ill.—Man.—S.D.—Colo.—Tex.—Mo. *Prairie—Plain.* My–Jl.

2. **D. bícolor** Nutt. Stem 2–5 dm. high, few-leaved; leaf-blades 2–4 cm. broad, puberulent or glabrate, thick, dissected into linear or linear-oblong obtuse divisions; sepals ovate-lanceolate, acute or acuminate, about 15 mm. long; upper petals pale brownish-yellow, with blue veins, usually uncleft; lower petals with obtuse wavy lobes; follicles from densely viscid-pubescent to glabrous, 1.5–2 cm. long. Dry ground in the mountains: Wash.—Sask.—S.D.—Utah—Ore. *Plain—Mont.* Ap–Jl.

3. **D. Nelsònii** Greene. Closely resembling *D. bicolor* in habit, but usually more slender; stem 2–5 dm. high, finely strigose-puberulent; leaf-blades dissected into linear obtuse divisions; sepals oblong or oval, usually obtuse; petals as those of *D. bicolor* but shorter; follicles about 15 mm. long, strigose or glabrous. *D. pinetorum* Tidest. Hills and grassy slopes: N.M.—Neb.—Alta.—Wash.—Mont.—Utah. *Submont.—Mont.*

4. **D. exaltàtum** Ait. Stem 5–15 dm. high, erect; leaf-blades deeply 3–5-cleft; divisions lanceolate or oblanceolate, acuminate, cleft or toothed towards the apex; upper leaves reduced and with narrow divisions; raceme elongate, dense flowers purple or blue; spur straight; follicles erect, 8–10 mm. long. *D. urceolatum* Jacq. [B]. Woods: Pa.—N.C.—Ala.—Kans.—Minn. (?). *Allegh.* Jl–Au.

5. **D. tricórne** Michx. Stem stout, 3–10 dm. high, glabrous, from a cluster of tuberous roots; leaf-blades deeply 3–7-divided or -cleft; divisions obtuse, entire to cleft; racemes loose; flowers blue or white; spur curved, 2–3 cm. long; follicles 10–12 mm. long, short-beaked. Woods: Pa.—Ga.—Ark.—Kans.—Minn. *Allegh.* Ap–Je.

6. **D. Ajàcis** L. Annual; stem erect, 3–7 dm. high, branched, glabrous below, viscid-puberulent above; leaf-blades glabrous, finely dissected into narrowly linear lobes; flowers blue or purple, rarely white; sepals obovate, 1–1.5 cm. long, slightly pubescent; follicles 12–15 mm. long, pubescent, with a very short beak. Around dwellings: N.Y.—Va.—Tex.—Mont.; cultivated and escaped from gardens.

22. ACONÌTUM L. Aconite, Monkshood, Wolfsbane.

Perennial herbs, with rootstocks or tubers. Leaves alternate; blades palmately lobed or divided. Flowers large, perfect, irregular, zygomorphic. Sepals 5, the upper or posterior one hooded or helmet-shaped. Petals 2–5, small, the posterior two hooded, clawed, concealed in the posterior sepal, the other three when present minute. Stamens numerous, hypogynous. Pistils 3–5, sessile, many-ovuled. Fruit of many-seeded follicles.

Hood boat-shaped, slightly saccate, semicircular in outline; perennials with tubers. — 1. *A. tenue.*
Hood helmet-shaped, deeply saccate; perennials with fleshy fusiform root.
  Leaf-segments linear-lanceolate; front-line of the hood nearly straight. — 2. *A. ramosum.*
  Leaf-segments lanceolate; front-line of the hood concave.
    Follicles pubescent, almost straight. — 3. *A. porrectum.*
    Follicles glabrous, abruptly curved outward at the apex, with divaricate beaks. — 4. *A. noveboracense.*

1. **A. ténue** Rydb. *Fig. 226.* Stem very slender, about 3 dm. high; leaves few; blades glabrous, thin, shining, almost pentagonal in outline, 5–7-divided to near the base, divisions rhombic, 3-cleft and often again cleft; inflorescence racemose, 2–6-flowered; hood 15–18 mm. long; lateral sepals rounded-ovate, 10–12 mm. long; lower sepals oblong, obtuse. Damp cañons: Black Hills, S.D. *Mont.* Jl.

2. **A. ramòsum** A. Nels. Stem 3–5 dm. high; leaf-blades 5–8 cm. broad, sparsely short-pubescent, divided into 3–5 oblanceolate divisions, these variously cleft into linear-lanceolate lobes; sepals blue, short-pubescent; hood 12–16 mm. long; lateral sepals obovate, somewhat oblique; lower sepals oblong or spatulate, about three-fourths as long and one-fourth to one third as broad as the lateral ones; follicles

15–20 cm. long, more or less pubescent, straight. Open grassy ground: S.D.—Wyo. *Submont.—Mont.* Jl–Au.

3. **A. porréctum** Rydb. Stem stout, 6–7 dm. high, leafy; basal leaves glabrous; blades reniform-pentagonal in outline, 6–10 cm. wide, 5–7-divided to near the base, divisions rhombic-oblanceolate, variously cleft above; inflorescence mostly racemose; sepals dark blue, ciliate on the margins; lateral sepals broadly obovate or semi-reniform, very oblique; lower sepals lanceolate or oblong. Moist places: Wyo.—S.D.—N.M.—Ariz.—Utah. *Submont.—Mont.* Je–Au.

4. **A. noveboracénse** A. Gray. Stem erect, from a cluster of tuberous roots, pubescent above; leaf-blades deeply parted; divisions broadly cuneate, 3-cleft and incised; flowers blue; hood gibbous, obovoid, abruptly contracted above the short, descending beak. Meadows: N.Y.—Ohio—Iowa. *Allegh.* Je–Au.

Family 59. PODOPHYLLACEAE. MAY APPLE FAMILY.

Perennial herbs with rootstocks. Leaves alternate or all basal, with simple or ternately compound blades. Flowers perfect. Sepals 6, in two series, or in *Jeffersonia* 4. Petals 6–9, in two series. Stamens 6–18; anthers usually opening by valves. Pistil solitary; ovules 2–many, anatropous. Fruit a berry or capsule.

Leaf-blades simple; flowers solitary.
  Sepals 6; petals 6–9; leaves peltate; fruit a berry. 1. PODOPHYLLUM.
  Sepals 4; petals 8; leaves not peltate, 2-cleft; fruit a capsule. 2. JEFFERSONIA.
Leaves ternately compound; flowers racemose; sepals and petals each 6; fruit a capsule. 3. CAULOPHYLLUM.

1. PODOPHÝLLUM L. MAY APPLE. MANDRAKE.

Glabrous herbs, with poisonous rootstocks. Flowerless stem with a single, the flowering stems with two peltate leaves; blades 7–9-lobed. Flowers solitary, showy, nodding. Sepals 6, petal-like. Petals 6–9, larger than the sepals, white. Stamens 6–18; filaments distinct; anthers opening lengthwise. Ovary 1-celled; stigma peltate. Ovules many. Fruit a large pulpy berry.

1. **P. peltàtum** L. *Fig. 227.* Stem 2–4 dm. high; leaf-blades of the flowerless stem centrally peltate, 1–3 dm. broad; lobes 2–3-cleft; those of the flowering stems attached near the margin; petals wax-like, white, 2.5–4 cm. long; berry greenish yellow, 4–5 cm. long. Woods: Que.—Fla.—Tex.—Minn. *E. Temp.* My.

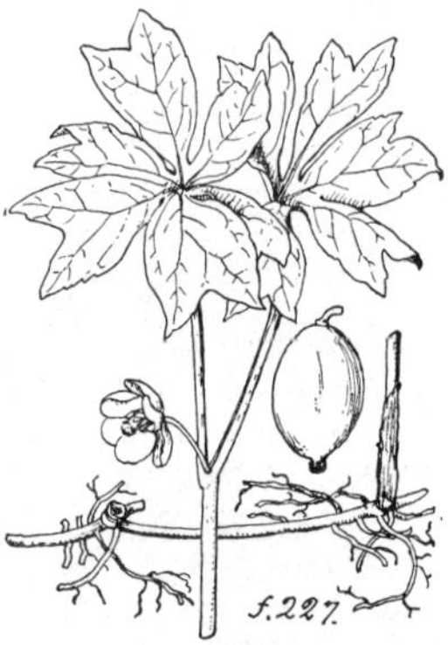

f.227.

2. JEFFERSÒNIA Barton. TWIN-LEAF.

Acaulescent herbs, with rootstocks. Leaves basal, long-petioled, with 2-cleft blade. Flowers perfect, solitary on a long scape. Sepals 4, deciduous. Petals 8, in two series, white. Stamens 8; filaments distinct; anthers opening by 2 valves. Ovary 1-celled; stigma 2-lobed. Ovules numerous. Fruit a capsule, opening at the top by a lid.

1. **J. diphýlla** (L.) Pers. Petioles 1–3 dm. long; segments of the 2-parted leaf obliquely reniform, 5–10 cm. long, crenate or wavy; petals white, about 2 cm. long; capsule obconic, 1.5–2 cm. long. Woods and thickets: N.Y.—Va.—Iowa—Wis. *Allegh.* Ap–My.

3. **CAULOPHÝLLUM** Michx. BLUE COHOSH.

Caulescent herbs. Leaves ternately compound, in ours usually near the top of the stem. Flowers perfect, in terminal racemose cymes. Sepals 6. Petals 6, hooded, gland-like, yellow. Stamens 6; filaments distinct; anthers opening by 2 valves. Ovary 1-celled; stigma one-sided. Ovules 2. Capsule opening before maturity; seeds drupe-like on a stout stalk.

1. **C. thalictroìdes** (L.) Michx. *Fig. 228.* Stem 3–10 dm. high, simple or branched above; leaves triternately compound; leaflets 2–3-lobed at the apex; seeds dark blue. Damp woods: N.B.—S.C.—e Neb.—N.D.—Man. *Canad.—Allegh.* Ap–My.

Family 60. **BERBERIDACEAE.** BARBERRY FAMILY.

Shrubs, with yellow wood and inner bark. Leaves alternate, leathery, pinnately compound or simple, usually spinose-toothed. Flowers small, yellow, racemose or paniculate. Sepals 6, in two series, subtended by 2 or 3 bractlets. Petals 6, in two series, imbricate, with glands near their bases. Stamens 6; anthers opening by 2 valves. Ovary 1-celled; stigmas peltate; ovules few, erect or ascending. Fruit a few-seeded berry, in one species rather dry. Endosperm present.

Primary leaves pinnately compound, evergreen; no secondary ones in their axils. 1. MAHONIA.
Primary leaves reduced to spines; secondary ones fascicled in their axils, simple or unifoliolate. 2. BERBERIS.

1. **MAHÒNIA** Nutt. OREGON GRAPES.

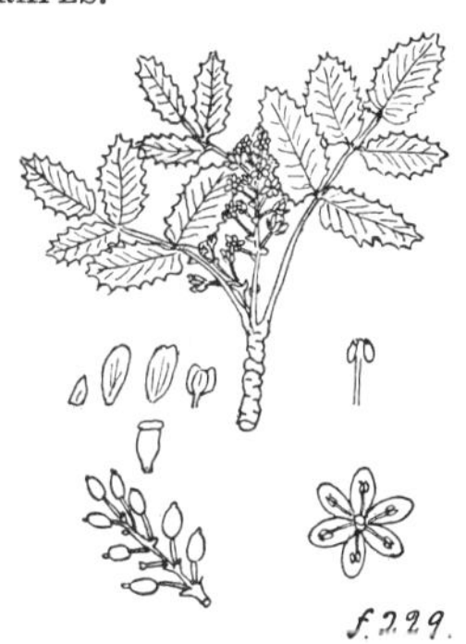

Unarmed shrubs. Leaves pinnately compound, with spinose-toothed evergreen leaflets. Flowers in erect fascicled or branched racemes. Filaments in ours with a pair of divergent teeth near the apex. Berries usually with a bloom. *Odostemon* Raf.

1. **M. Aquifòlium** (Pursh) Nutt. *Fig. 229.* Stoloniferous dwarf shrub; leaflets 3–7; oval or rarely ovate, acute to rounded at the apex, rounded or obliquely truncate at the base, coriaceous, pale and strongly reticulate beneath, sinuately 9–19-toothed, 3–9 cm. long; racemes many-flowered; berry ellipsoid-globose, 7-8 mm. long. *Berberis Aquifolium* Pursh [B]. *B. repens* Lindl. *B. nana* and *B. brevipes* Greene. *O. Aquifolium* Rydb. [BB, R]. Hills and mountainsides: Alta.—Neb.—N.M.—Calif. —B.C. *W. Temp.—Submont.—Mont.* Ap–Je.

2. **BÉRBERIS** (Tourn.) L. BARBERRY.

Spiny shrubs. Primary leaves transformed into simple or triple spines; secondary leaves fascicled in their axils, unifoliolate or apparently simple, in ours deciduous. Racemes drooping, simple. Filaments toothless. Berries red, without a bloom, acid, edible.

Flowers in elongate racemes; leaves more or less setulose-dentate. 1. *B. vulgaris.*
Flowers solitary or in small short fascicles; leaves entire. 2. *B. Thunbergii.*

1. **B. vulgáris** L. A glabrous shrub, 2–3 m. high; primary leaves reduced to 3-pronged spines; secondary leaves obovate or spatulate, 2–5 cm. long; racemes many-flowered; petals entire; berries ellipsoid, scarlet. Thickets and waste places: Me.—N.J.—Iowa—N.D.; escaped from cultivation, or naturalized; nat. of Eur. My–Je.

2. **B. Thunbérgii** DC. Low shrub, 6–15 dm. high, with spreading branches and simple spines; leaves obovate or spatulate, glaucescent beneath, entire, 1–4 cm. long; flowers 1–3, pale yellow; fruit globose, bright red. Around dwellings: Iowa—Minn.; escaped from cultivation.

Family 61. **MENISPERMACEAE.** Moonseed Family.

Twining or trailing vines. Leaves alternate, without stipules; blades in ours palmately veined. Flowers dioecious or polygamous, in racemes, panicles, or cymes. Sepals in ours 4–8, in two series. Petals as many as the sepals, also in two series, or rarely wanting. Stamens 6–12, or more numerous; anthers 2–4-celled, opening lengthwise. Pistils 3–6, rarely more. Ovules solitary. Fruit of berry-like drupes. Seeds with a crescent-like or ring-like embryo.

Petals present; anthers 4-celled; seeds incurved.
- Stamens in the staminate flowers 12–24, in the pistillate flowers represented by 6 staminodia; sepals and petals each 4–8. 1. Menispermum.
- Stamens in the staminate flowers 6, in the pistillate ones reduced or wanting; sepals and petals each 6. 2. Cocculus.

Petals wanting; anthers 2-celled; seeds saucer-shaped; stamens in the staminate flowers 12, in the pistillate ones 9, abortive. 3. Calycocarpum.

1. **MENISPÉRMUM** L. Moonseed.

Slender twining vines, woody below. Leaves alternate, petioled, with peltate, often palmately lobed blades. Flowers dioecious, in axillary cymes. Sepals 4–8, in two series. Petals of the same number, shorter than the sepals. Stamens in the staminate flowers 12–24, with 4-celled anthers, in the pistillate flowers reduced to 6 staminodia. Pistils 2–4; stigma dilated. Drupe somewhat flattened; stone broadly reniform.

1. **M. canadénse** L. *Fig. 231.* Stem 1–4 m. high, finely pubescent; leaf-blades 5–20 cm. broad, peltate (the petiole attached near the edge), round-reniform, entire or 3–7-angled or -lobed; petals fan-shaped, greenish white, clawed; drupes bluish-black, 1 cm. broad. Thickets: Que.—Ga.—Okla.—N.D.—Man. *E. Temp.* Je–Jl.

2. **CÓCCULUS** DC. Coral Bead.

Slender twining perennial vines. Leaves alternate, petioled, with ovate or cordate, entire or lobed, not peltate blades. Flowers dioecious or polygamous, in axillary cymes or panicles. Sepals and petals each 6–9, in two series each. Stamens 6, more or less reduced in the pistillate flowers; anthers 4-celled. Pistils 3–6. Drupes somewhat flattened, with a reniform or horseshoe-shaped stone. *Epibaterium* Forsk. *Cebatha* Forsk.

1. **C. carolínus** (L.) DC. Stem 0.5–8 m. long; leaf-blades broadly ovate or deltoid, 5–12 cm. long, entire or 3–5-lobed, downy beneath; flowers white, the staminate ones in panicles, the pistillate ones in simple racemes; sepals and petals erose at the apex; drupe red, 7–8 mm. broad. *E. Carolinum*

Britton. *Cebatha Carolina* Britton [B]. Woods and thickets: Va.—Fla.—Tex. Kans. *E. Temp.* Jl–Au.

3. CALYCOCÁRPUM Nutt. Cupseed.

Perennial twining vines. Leaves alternate, petioled, with palmately lobed blades. Flowers dioecious, in slender axillary panicles. Sepals 6, in two series. Petals wanting. Stamens 12, in the pistillate flowers imperfect. Pistils 3. Drupe with a thin flesh; stone cup-shaped, hollowed out on one side.

1. **C. Lŷoni** (Pursh) Nutt. High climbing vine; leaf-blades suborbicular in outline, 5–20 cm. broad, palmately 3–7-lobed; drupe black, 2.5 cm. long, round or oval. Along streams: Ky.—Fla.—La.—Kans. *Austral.* My–Je.

Family 62. LAURACEAE. Laurel Family.

Aromatic shrubs or trees. Leaves alternate, without stipules, usually glandular-punctate. Flowers in ours dioecious. Sepals 4–10, usually 6, in two series, imbricate. Stamens more numerous than the sepals, in 2–4 series, those of the pistillate flowers reduced into staminodia; anthers 2–4-celled, the sacs opening by uplifting valves. Pistil solitary; ovary 1-celled. Fruit a drupe; seed solitary.

Flowers in corymbose racemes, appearing with the leaves; anthers 4-celled. 1. Sassafras.
Flowers in axillary umbels, appearing before the leaves; anthers 2-celled. 2. Benzoin.

1. SÁSSAFRAS Nees. Sassafras, Ague Tree.

Strongly aromatic trees, with alternate, commonly lobed leaves. Flowers dioecious, rarely perfect, yellowish green, in corymbose racemes. Sepals 6, in two series. Stamens 9, in 3 series, the inner series with 2-stalked glands at the base, those of the pistillate flowers changed into staminodia; anthers 4-celled. Style elongate. Drupe subglobose, with a thin pulp.

1. **S. variifòlium** (Salisb.) Kuntze. *Fig. 232.* A tree, sometimes 30 m. high; leaf-blades usually 3-lobed, sometimes 2-lobed or entire, bright green above, glaucous beneath; drupe red or orange. *S. officinale* Nees & Eberm. *S. Sassafras* Karst. [B]. Woods and hedge-rows: Me.—Fla.—Tex.—Neb. *E. Temp.* Ap.

2. BENZÒIN Fabr. Spice-bush.

Shrubs or trees, with aromatic, spicy bark. Leaves alternate, entire. Flowers yellow, in axillary umbels, appearing before the leaves. Sepals 6, or rarely 7–9. Stamens usually 9, in 3 series, the innermost usually with glands at their bases; anthers 2-celled. Fruit a pulpy drupe.

1. **B. aestivàle** (L.) Nees. *Fig. 233.* A shrub, 1–3 m. high; leaf-blades obovate to elliptic, 5–12 cm. long, obtuse to short-acuminate, green and glabrous above, pale beneath; flowers yellow; drupe oval, 1 cm. long. *Lindera Benzoin* (L.) Blume. *B. Benzoin* Coulter [B]. Swamps and along streams: Me.—Ga.—Okla.—se Kans. *E. Temp.* Mr–Ap.

Family 63. ANONACEAE. Custard-apple Family.

Shrubs or trees, with alternate, entire leaves, without stipules. Flowers perfect, or sometimes monoecious or dioecious. Sepals usually 3, valvate or rarely imbricate. Petals 6, imbricate or valvate, in 2 series. Stamens numerous; filaments short; anthers extrorse, appendaged. Carpels few or numerous, free or slightly united; styles short or none. Fruit more or less pulpy.

ASÍMINA Adans. Papaw.

Ill-scented shrubs or trees, with deciduous leaves. Flowers perfect, solitary or in pairs. Sepals 3, valvate. Petals 6, imbricate, the outer spreading and larger. Carpels 3–5, distinct. Berry more or less elongate, several-seeded; seeds flat, enclosed in a fleshy aril.

1. **A. tríloba** (L.) Dunal. *Fig. 230.* Shrub 3–12 m. high; leaves obovate, 1–2 dm. long, abruptly acuminate; flowers axillary; pedicels recurved; petals first greenish, then purplish, the outer orbicular, 2–2.5 cm. long; fruit oblong-cylindric, 7–15 cm. long, turning yellow or brown, edible. Rich woods: N.Y.—Fla.—Tex.—Neb. *E. Temp.* Ap–My.

Family 64. PAPAVERACEAE Juss. Poppy Family.

Annual or perennial herbs, or rarely shrubby plants, with colored sap and narcotic or acrid properties. Flowers perfect, regular. Sepals 2, or rarely 3, caducous. Petals 4, 8, or 12. Stamens numerous; filaments distinct, often dilated. Gynoecium of 2 to many united carpels, forming a 1-celled ovary; placentae parietal; ovules numerous, anatropous. Fruit a capsule; seeds numerous, with a fleshy or oily endosperm.

Acaulescent herbs; petals 8–12; pod 1-celled, 2-valved. 1. Sanguinaria.
Caulescent herbs; petals 4 or 6.
  Capsule opening by valves, at least at the summit.
    Unarmed herbs; sepals neither horned nor hooded; leaves pinnately divided or lobed. 2. Chelidonium.
    Prickly-leaved leafy-stemmed herbs; sepals hooded or horned; petals deciduous. 3. Argemone.
  Capsule opening by dentiform lids under the stigma. 4. Papaver.

1. SANGUINÀRIA L. Bloodroot, Puccoon.

Acaulescent perennial herbs, with red rootstocks. Leaves basal, petioled, with 5–7-lobed blades. Flowers solitary on the scape, white. Sepals 2. Petals 8–12. Stamens numerous. Pistil 1; ovary elongate; stigmas 2. Capsule 2-valved. Seeds numerous.

1. **S. canadénsis** L. *Fig. 234.* Leaf-blades orbicular in outline, 6–25 cm. broad, cordate at the base, glaucous beneath; petals oblong, 2–3 cm. long; capsule fusiform, 3–5 cm. long. *S. mesochora* Greene. Woods: N.S.—Fla.—Ark.—Kans.—N.D.—Man. *E. Temp.* Ap–My.

2. CHELIDÒNIUM (Tourn.) L. Celandine.

Glabrous perennial herbs, with rootstocks. Leaves pinnately divided, crenately lobed, alternate. Sepals 2, yellowish-green. Petals 4, yellow, imbricate in two series. Stamens many. Pistil formed of 2 carpels; ovary linear; style present; stigmas 2-lobed. Capsule linear, torulose, glabrous, 2-valved, dehiscent from the base, separating from the persistent placentae.

1. **C. màjus** L. Stem 3–10 dm. high; leaf-segments 5–7, obovate-oblong, crenately lobed; petals 7.5–10 mm. long, rounded; capsule 2–5 cm. long, 2–3 mm. thick. Waste places: N.S.—N.C.—Utah; adv. from Eur. Ap–S.

3. ARGEMÒNE L. PRICKLY POPPY, THISTLE POPPY.

Annual or perennial herbs, or rarely shrubs, with yellow or white sap. Leaves alternate, clasping, pinnatifid or lobed, the divisions spinose-tipped. Flowers erect in bud, perfect. Sepals 2–3, hooded or horned, deciduous. Petals 4–6, showy, yellow or white. Stamens numerous; filaments slender. Ovary with 4–6 nerviform placentae; stigma sessile; ovules numerous. Capsule oblong, opening at the top by 4–6 valves. Seeds numerous, pitted; endosperm oily. *Enomegra* A. Nels.

Corolla yellow; leaves with light blotches. 1. *A. mexicana.*
Corolla white; leaves not blotched.
  Prickles of the fruit not squarrose.
    Stem unarmed, prickly or bristly, but not hispidulous. 2. *A. intermedia.*
    Stem hispidulous-pubescent as well as densely prickly. 3. *A. hispida.*
  Prickles of the fruit strong, recurved-spreading, squarrose. 4. *A. squarrosa.*

1. **A. mexicàna** L. Stem 3–6 dm. high, glabrous or sparingly prickly; leaves sinuate-pinnatifid, with spinulose dentate lobes; flowers sessile; calyx slightly bristly, its horns terete, glabrous, with stout spines; petals yellow, 1.5–2.5 cm. long; capsule ellipsoid, 2.5–3 cm. long, sparingly prickly, with stout spreading prickles. Hillsides, cultivated ground and waste places. W.Ind.—Fla.—Tex.—S.Am.; Africa, East Indies and Australia; cult. and escaped, reported from Colo., Iowa, and Kans. Ja–D.

f. 235.

2. **A. intermèdia** Sweet. *Fig. 235.* Stem stout, 3–10 dm. high, sparingly but strongly prickly; leaves bluish-green, glaucous, crisp, 1–3 dm. long; flowers usually short-peduncled; sepals sparingly prickly; horns conical, spreading, 5–8 mm. long; petals 3–4 cm. long; capsule oblong-fusiform, about 4 cm. long, with rather few strong prickles. Plains: S.D.—Mo.—Tex.—n Mex.—Wyo. *Plain—St. Plains—Submont.* Je–Au.

3. **A. híspida** A. Gray. Stem 3–6 dm. high; leaves deeply pinnately cleft, with oblong, coarsely spinulose-dentate lobes, prickly on the veins and hispidulous between; calyx densely bristly, its horns 8–10 mm. long, triangular, bristly; petals 3–4 cm. long; capsule oblong-fusiform, densely prickly with rather spreading prickles. *A. bipinnatifida* Greene. *Enomegra bipinnatifida* A. Nels. *E. hispida* A. Nels. Plains and hillsides: Kans.—N.M.—Utah—Wyo. *St. Plains—Son.—Mont.* Je–Jl.

4. **A. squarròsa** Greene. Stem 4–10 dm. high, prickly and hispid; leaves glaucescent, sinuate-pinnatifid and spinulose-dentate, hispid, 4–10 cm. long; flowers sessile; calyx densely bristly; horns broadly triangular, prickly and hispid; capsule nearly 5 cm. long. Plains: N.M.—Colo.—Kan. *St. Plains—Son.*

4. PAPÀVER (Tourn.) L. POPPY.

Annual or in ours perennial herbs, with milky sap. Leaves alternate or in ours basal, lobed or dissected. Flowers drooping in the bud, perfect. Sepals 2, rarely 3. Petals 4, rarely 6. Stamens numerous. Ovary with 4–20 septiform placentae; style none; stigma disk-like; ovules numerous. Capsule globose or pyriform, opening by 4–20 tooth-like lids under the margin of the stigma. Seeds scrobiculate, naked.

Plant glabrous and glaucous; leaves clasping, lobed. 1. *P. somniferum.*
Plant hispid, green; leaves petioled, pinnately divided. 2. *P. Rhoeas.*

1. **P. somníferum** L. Stem 3–10 dm. high; flowers white or purplish, 7–10 cm. broad; pod globose, glabrous. Near dwellings: Me.—Va.—N.D.; escaped from cultivation; nat. of Eur.

2. **P. Rhoèas** L. Stem 3–10 dm. high, hispid with spreading hairs; flowers 5–10 cm. broad, scarlet, with a darker center; pod obovoid. Waste places: Me.—Va.—Neb.—N.D.; adv. from Eur.

Family 65. FUMARIACEAE. Fumitory Family.

Herbs, with alternate, usually finely dissected leaves. Flowers perfect, irregular, in racemes or panicles. Sepals 2, scale-like. Petals 4, the outer ones spreading above, one or both spurred at the base, the inner smaller, thickened above, enclosing the stigmas. Stamens 6; filaments diadelphous. Gynoecium of 2 united carpels; ovary 1-celled, with 2 parietal placentae. Fruit a capsule, rarely indehiscent.

Both of the outer petals spurred at the base.
  Petals permanently united and enclosing the capsule; seeds crestless; herbaceous vines. 1. Adlumia.
  Petals slightly united, deciduous; seeds mostly crested; acaulescent herbs. 2. Dicentra.
One of the outer petals spurred.
  Ovules several or many; fruit an elongate capsule. 3. Corydalis.
  Ovules solitary; fruit an indehiscent nutlet. 4. Fumaria.

1. ADLÙMIA Raf. Alleghany Vine, Climbing Fumitory.

Biennial or perennial vines, with decompound alternate leaves. Flowers perfect, irregular, in panicle-like cymes terminal or opposite the leaves. Sepals 2, scale-like. Petals 4, united into a tube, the two outer gibbous below. Stamens 6, in two bundles, opposite the outer petals, the filaments united below, diadelphous above. Gynoecium of two united carpels; ovary 1-celled, with two parietal placentae; stigma 2-lobed. Fruit 2-valved. Seeds crestless.

1. **A. fungòsa** (Ait.) Greene. Slender vines, climbing by means of tendril-like petioles; leaves 2–3-pinnate; leaflets thin, ovate to cuneate, entire or lobed; cymes several- to many-flowered; corolla pink or white, 15–18 mm. long; outer petals with divergent tips, the inner with suborbicular blades; capsule 1.5–2 cm. long, few-seeded. *A. cirrhosa* Raf. Woods: N.B.—N.C.—Kans.—Mich. *E. Temp.* Je–O.

2. DICÉNTRA Bernh. Dutchman's Breeches, Bleeding Heart.

Perennial herbs, with horizontal, granular or tuberous rootstocks. Leaves basal, long-petioled, repeatedly ternately divided. Flowers perfect, irregular, in terminal racemes or panicles. Two outer petals spurred, loosely united, deciduous, their tips spreading, the inner petals crested, clawed, cohering by their tips. Stamens 6, in two bundles, opposite the outer petals; filaments diadelphous at the base. Ovary with two parietal placentae; style filiform. Capsule elongate, opening by two valves. Seeds 10–20, crested. *Bicuculla* Adans.

Corolla with two long divergent spurs; inner petals minutely crested. 1. *D. Cucullaria.*
Corolla cordate at the base, with two rounded short spurs; inner petals conspicuously crested. 2. *D. canadensis.*

1. **D. Cucullària** (L.) Bernh. *Fig. 236.* Scape 1–2.5 dm. high, as well as the petioles from a granulate-bulbous base; leaves ternately compound into linear or oblanceolate divisions; corolla white or pinkish, with yellow summit, 12–16 mm. long. *B. Cucullaria* (L.) Millsp. [B]. Rich woods: N.S.—N.C.—Neb.—Minn. *E. Temp.* Ap–My.

2. **D. canadénsis** (Goldie) Walp. Scape 1–2 dm. high, from a creeping rootstock with grain-like tubers; leaves similar to those of the preceding, glaucous; corolla greenish-white, tinged with purplish, 14–18 mm. long, fragrant. *B. canadensis* Millsp. [B]. Rich woods: N.S.—Va.—Mo.—Neb.—Minn. *E. Temp.* My–Je.

3. CORÝDALIS Vent. Corydalis.

Annual, biennial, or perennial, caulescent herbs. Leaves alternate, bipinnately dissected. Flowers perfect, irregular, in racemes. Outer petals dissimilar, distinct, one of them spurred at the base, the two inner narrower, winged or crested, coherent at the apex. Stamens 6, in two bundles, opposite the outer petals; filaments united above the middle. Ovary with two parietal placentae; style filiform. Fruit an elongate 2-valved capsule. Seeds numerous, crested. *Capnoides* Adans.

Plants low, ascending or diffuse; corolla yellow.
- Pod glabrous.
  - Flowers 6–8 mm. long; spur short; outer petals wing-margined on the back.
    - Pod drooping or spreading; crest of the petals 3-toothed; seeds reticulate. — 1. *C. flavula.*
    - Pod ascending; crests of the petals entire; seeds smooth. — 2. *C. micrantha.*
  - Flowers 8–20 mm. long; spur conspicuous; outer petals not crested on the back.
    - Pod pedulous, torulose; bracts narrow lanceolate; seeds smooth. — 3. *C. aurea.*
    - Pod erect or ascending.
      - Bracts ovate, obovate or ovate-lanceolate; seeds smooth. — 4. *C. montana.*
      - Bracts lanceolate; seeds finely pitted. — 5. *C. campestris.*
- Pod covered with translucent vesicles. — 6. *C. crystallina.*

Plants erect, tall, usually 3–6 dm. high; corolla rose or purplish with purple tips. — 7. *C. sempervirens.*

1. **C. flávula** (Raf.) DC. Slender glabrous annual, 1.5–3.5 dm. high; lower leaves petioled, the upper sessile, finely dissected; pedicels slender; bracts broadly oblong; spur 1 mm. long; pod torulose. *Capnoides flavulum* Kuntze [B]. Rocky woods: N.Y.—Va.—La.—Kans.—Minn. *E. Temp.* My–Je.

2. **C. micrántha** (Engelm.) A. Gray. Slender glabrous annual, 1–3 dm. high; pedicels short; bracts minute; pod torulose; flowers often cleistogamous, minute, spurless. *Capnoides micranthum* Britton [B]. Woods: Va.—Fla.—Tex.—Minn. *Allegh.—Carol.—Ozark.* F–Ap.

3. **C. aùrea** Willd. Glabrous and more or less glaucous annual or biennial, diffusely branched, 1–4 dm. high; racemes 1–5 cm. long; leaves thrice pinnate; corolla 12–15 mm. long, golden yellow; spur one-half to two-thirds as long as the body, curved downward; pods 2–3 cm. long, about 2 mm. thick. *Capnoides aureum* Kuntze [B, R]. *Corydalis Engelmannii, C. macrorrhiza,* and *C. Albertae* Fedde. River banks, hillsides and open woods: N.S.—Pa.—Tex.—Calif.—Alaska. *Temp.—Plain—Mont.* My–Au.

4. **C. montàna** Engelm. Glabrous and glaucous biennial, branched at the base, 2–5 dm. high; leaves as in the preceding; corolla 15–20 mm. long, golden yellow; spur usually almost as long as the body, decidedly saccate at the end and somewhat turned upward; pod usually curved at the base, 2–3

cm. long, 3 mm. thick. *Corydalis aurea occidentalis* Engelm. [G]. *C. curvisiliquaeformis, C. bilimbata* and (?) *C. hypecoiformis* Fedde. *Capnoides montanum* Britton [B, R]. River banks, cañons, and copses: S.D.—Mo.—Tex.—Ariz.—Utah; Mex. *Plain—Submont.—Son.* Mr–Au.

5. **C. campéstris** (Britton). Annual, similar to *C. aurea;* flowers spicate-racemose; corolla 8–10 mm. long; spur bright yellow, blunt, nearly straight, 4–5 mm. long; pod curved upward, short-pedicelled, stout, somewhat 4-angled; seeds sharp-margined. *Capnoides campestre* Britton [B]. Fields and woods: Ill.—Neb.—Tex.—Ark. *Prairies—Plain.* My–Jl.

6. **C. crystallìna** Engelm. Glabrous and slightly glaucous annual, branched at the base, 1–3 dm. high; leaves thrice pinnatifid, with oblong small divisions; bracts ovate or ovate-lanceolate; corolla golden yellow, 12–15 mm. long; spur nearly as long as the body; pod erect, about 2 cm. long and 3 mm. thick. *Capnoides crystallinum* Kuntze [B]. Fields and open grounds: Mo. — Ark. — Kans. *Ozark—Prairie—Plain.* Ap–Je.

f. 237.

7. **C. sempérvirens** (L.) Pers. *Fig. 237.* Glabrous and glaucous biennial, simple, 3–6 dm. high; leaves twice or thrice pinnatifid, with obovate divisions; racemes 3–10 cm. long; corolla 12–15 mm. long; spur short, one third as long as the body or less; pod ascending, 3–4 cm. long, 1.5 mm. thick. *Corydalis glauca* Pursh. *Capnoides sempervirens* Borckh. Rocky woods: N.S.—N.C.—Mont.—B.C.—Alaska. *Temp.—Plain—Boreal.* My-Au.

4. **FUMÀRIA** (Tourn.) L. FUMITORY.

Annual caulescent herbs. Leaves alternate, finely dissected. Flowers perfect, irregular, in terminal or lateral racemes. Outer petals dissimilar, one of them spurred at the base. Stamens in two bundles, opposite the outer petals. Ovary subglobose, 1-ovuled; style filiform. Fruit a 1-seeded nut. Seed not crested.

1. **F. officinàlis** L. Glabrous, diffusely branched annual, 2–10 dm. high; leaves finely dissected into linear-oblong to spatulate divisions; corolla purplish, together with the short spur 6–8 mm. long; pod spherical, 2–2.5 mm. in diameter, minutely tuberculate. Waste places: N.S.—Fla.—Utah; adv. from Eur. My–Au.

Family 66. **BRASSICACEAE.** MUSTARD FAMILY.

Herbs or rarely shrubby plants, with alternate, entire to finely dissected leaves. Flowers perfect, regular or nearly so, in spikes or racemes. Sepals 4, mostly erect. Petals 4, with spreading blades. Stamens usually 6, tetradynamous, *i.e.,* the inner 4 longer, or rarely 4 or 2. Gynoecium of 2 united carpels; ovary 2-celled or rarely 1-celled, superior or (in *Subularia* only) partly inferior. Fruit a capsule, rarely indehiscent. Seeds without endosperm. CRUCIFERAE.

I. Pod sessile, or nearly so; sepals erect, ascending, or connivent in anthesis; anthers neither twisted nor curved.
  A. Pod compressed or flattened contrary to the narrow partition.
    Pod not didymous; plants not densely stellate.
      Petals very unequal, the two lower much larger than the upper. 1. IBERIS.
      Petals equal.
        Cells of the pod 1-seeded.
          Pods ovate-cordate, acute at the apex, neither winged nor retuse. 2. CARDARIA.

Pod orbicular, elliptic or rarely ovate, retuse or notched at the apex, usually winged above. 3. LEPIDIUM.
Cells of the pods 2-seeded.
Pod more or less winged; cotyledons accumbent; hairs of the plant simple or none. 4. THLASPI.
Pod wingless; cotyledons incumbent.
Pod cuneate; plants with branched hairs. 5. CAPSELLA.
Pod elliptic; plants glabrous or minutely stellate. 6. HUTCHINSIA.
Pod more or less didymous; plants stellate. 7. PHYSARIA.
B. Pod neither compressed nor flattened contrary to the partition.
a. Pod terete or tetragonal.
†Pod short, scarcely more than twice as long as broad.
Cotyledons accumbent; valves of the pods nerveless.
Pubescence stellate or canescent with branched hairs; seeds flat.
Petals white, 2-cleft. 38. BERTEROA.
Petals yellow or tinged with red, not 2-cleft. 8. LESQUERELLA.
Pubescence not stellate; seeds terete.
Pod 1-celled; style slender; aquatics, with the submerged leaves dissected. 12. NEOBECKIA.
Pod 2-celled; style short; leaves not dissected.
Petals white; stout perennials with a thick taproot. 13. ARMORACIA.
Petals yellow or ochroleucous, rarely white; annuals, or perennials with rootstocks. 14. RORIPPA.
Cotyledons incumbent or folded transversely; valves of the pod 1-nerved; pubescence if any not densely stellate.
Submerged water plants, with subulate leaves; pods subglobose or elliptical. 9. SUBULARIA.
Land plants with ample leaves.
Capsule obovoid, dehiscent, many-seeded. 10. CAMELINA.
Capsule globose, indehiscent, 1–2-seeded. 11. NESLIA.
††Pod long, several times as long as broad.
*Pod scarcely beaked, merely tipped by a short style or a sessile stigma.
‡Pod terete or nearly so.
Pubescence of simple hairs, or none.
Seeds in two rows in each cell of the pod.
Seeds globose or oblong, turgid; valves of the pod nerveless; cotyledons accumbent.
Petals with a median nectary, usually yellow. 14. RORIPPA.
Petals without nectary, white. 15. NASTURTIUM.
Seeds flattened; valves of the pod 1-nerved; cotyledons partly incumbent. 16. TURRITIS.
Seeds in 1 row in each cell of the pod.
Pod subulate, tapering from the base to the apex. 17. ERYSIMUM.
Pod cylindric or tapering both ways.
Flowers yellowish or white; cotyledons incumbent; outer sepals not horned.
Flowers yellow; stigma 2-lobed; leaves narrow or pinnatifid with narrow lobes. 18. SISYMBRIUM.
Flowers white; stigma entire; leaves broad, dentate. 19. ALLIARIA.
Flowers violet or purple; cotyledons accumbent; outer sepals horned. 20. IODANTHUS.
Pubescence of forked hairs.
Flowers pedicelled, yellow to straw-colored, rarely white; leaves pinnatifid or lobed. 21. SOPHIA.
Flowers sessile, rose or white; leaves toothed or entire. 22. MALCOLMIA.
‡‡Pod 4-angled.
Stem-leaves clasping by a cordate base. 23. CONRINGIA.
Stem-leaves not clasping.
Corolla white. 24. ARABIDOPSIS.
Corolla yellow or purplish.
Seeds plump; leaf-blades entire or toothed.
Hairs of the stem and leaves appressed, 2-branched, attached near the middle; partition of the pod not cross-veined. 25. CHEIRINIA.

Hairs of the stem and leaves spreading, branched, attached at the end; partition of the pod cross-reticulate. 26. HESPERIS.
Seeds flat; leaf-blades pinnatifid. 27. BARBAREA.
**Pod with a long distinct beak.
Pod indehiscent, constricted between the seeds, which are separated from each other by false, spongy, transverse partitions.
Cotyledons accumbent; pod 2-seeded, 2-jointed, upper seed erect, the lower pendulous. 31. CAKILE.
Cotyledons conduplicate; pod several-seeded.
Pod few seeded, scarcely lomentaceous, but filled with a spongy tissue between the seeds. 29. RAPHANUS.
Pod constricted or articulate between the seeds into 1-seeded internodes.
Seeds in a single series; internodes without empty cells. 30. RAPHANISTRUM.
Seeds in 2 series, alternately arranged; each internode with a seed on one side of the partition and an empty space on the other. 28. CHORISPORA.
Pod opening by valves, without transverse partitions.
Beak flat and sword-like, 4-angled, or two-edged, contiguous with an internode, containing 1 seed.
Valves of the pod with 3 strong nerves. 32. SINAPIS.
Valves of the pod with 1 strong nerve. 33. ERUCA.
Beak elongate-conic.
Pod terete, seeds in a single row.
Beak-joint of the capsule containing a seed; seeds oval; cotyledons truncate. 34. ERUCASTRUM.
Beak-joint of the capsule seedless; seeds subglobose; cotyledons 2-lobed. 35. BRASSICA.
Pod flattened; seeds more or less in 2 rows. 36. DIPLOTAXIS.
b. Pod flattened parallelly to the broad partition.
Pod orbicular or short-oblong.
Pod suborbicular, with flat margins; petals entire or nearly so. 37. ALYSSUM.
Pod oblong, slightly if at all compressed; petals notched. 38. BERTEROA.
Pod oblong, ovate or linear, rarely nearly orbicular; valves flat, or if convex not with flattened margins; filaments unappendaged.
Valves nerveless.
Valves elastically dehiscent; seeds in one row; pod long.
Stem naked below, 2–3-leaved; cotyledons thick, very unequal. 39. DENTARIA.
Stem leafy; cotyledons flat, equal. 40. CARDAMINE.
Valves not elastically dehiscent; seeds in two rows; pod usually short. 41. DRABA.
Valves nerved and reticulate, not elastically dehiscent.
Pod short, from orbicular to linear-oblong. 41. DRABA.
Pod elongate-linear.
Cotyledons accumbent. 42. ARABIS.
Cotyledons incumbent. 23. CONRINGIA.
Pod stipitate.
Seeds in two rows; anthers neither twisted nor curved; sepals spreading. 43. SELENIA.
Seeds in one row; anthers twisted or curved.
Stipe short; sepals ascending; anther curved. 44. PLEUROPHRAGMA.
Stipe elongate; sepals reflexed; anther twisted. 45. STANLEYA.

1. IBÈRIS L.

Perennial, biennial or annual, glabrous or slightly pubescent herbs. Leaves entire, toothed or pinnatifid. Flowers white, lilac or purple, corymbose. Sepals equal at the base, often colored. Petals unequal, the two lower much larger. Fruit short, oval or round, compressed contrary to the septum, emarginate at the apex, more or less winged, deciduous; style filiform; locules 1-seeded.

Leaves spatulate to oblong, dentate. 1. *I. amara.*
Leaves linear or lance-linear, mostly entire. 2. *I. intermedia.*

1. I. amàra L. Annual or biennial; stem 1–4 dm. high, branched; leaves spatulate, oblong or the upper lanceolate, with 2–4 teeth or lobes on each side; flowers white or lilac; silicles suborbicular, rounded at the base, deeply notched at the apex, the wings produced into two deltoid teeth. Around dwellings: N.D.; escaped from cult.; nat. of c and s Eur.

2. **I. intermèdia** Guers. Biennial, glabrous; stem 2–6 dm. high, branched above; basal leaves slightly dentate; raceme even in fruit dense and short; pod oval with spreading deltoid lobes at the apex. Around dwellings: N.D.; adventive from w Eur. My–O.

2. CARDÀRIA Desv.

Perennial caulescent herbs. Leaves alternate, toothed, the upper clasping. Flowers perfect, in terminal panicles. Petals white. Stamens 6. Ovary sessile; styles slender, but short. Pod ovate, cordate at the base, acute at the apex, neither winged nor retuse; valves strongly convex. Seeds solitary in each cavity. Cotyledons incumbent.

1. **C. Dràba** (L.) Desv. Erect perennial, 2.5–5 dm. high, hoary-pubescent, branched above; lower leaves oblanceolate, petioled, the upper oblong, ovate, or cordate, clasping, usually dentate; pod 3 mm. long and 4 mm. broad, papillose; style 1–2 mm. long. *Lepidium Draba* L. [G, B]. Waste places and cultivated ground: N.Y.—Fla.—Calif.—Ida.; adv. or nat. from Eurasia. Ap–Je.

3. LEPÍDIUM L. PEPPERGRASS, CANARY-GRASS, BIRD-SEED.

Annual, biennial or perennial herbs. Pubescence if any of simple hairs. Leaves alternate, entire, toothed, lobed, or dissected. Flowers perfect, in racemes or panicles. Sepals 4, equal. Petals small, white or greenish, or wanting. Stamens often less than 6. Ovary sessile; styles short or wanting; ovules solitary in each cell, pendulous. Pods short, oblong to orbicular, transversely flattened, wing-margined or at least acute on the margins, notched at the apex. Seeds solitary in each cavity. Cotyledons incumbent or rarely obliquely accumbent.

Style evident, at least equaling the wing-margins; introduced annuals or biennials.
 Upper leaves dentate or entire-margined.
  Stem-leaves clasping, but not perfoliate, usually dentate. — 1. *L. campestre.*
  Stem-leaves perfoliate, entire-margined. — 2. *L. perfoliatum.*
 Leaves all pinnatifid, none clasping. — 3. *L. sativum.*
Style obsolete, or at least shorter than the width of the wing-margins; annuals or biennials.
 Wing-margins of the fruit not produced at the apex into distinct lobes or teeth.
  Petals conspicuous, at least equaling the sepals, spatulate or obovate.
   Cotyledons accumbent; petals broadly spatulate. — 4. *L. virginicum.*
   Cotyledons incumbent; petals narrowly spatulate or oblanceolate. — 5. *L. texanum.*
  Petals none or minute, scarcely more than half as long as the sepals, linear or linear-spatulate.
   Plants branched near the base; petals usually present. — 6. *L. ramosissimum.*
   Plants simple below, branched above; racemes elongate, terminal.
    Seeds wing-margined; pod suborbicular, 3 mm. broad. — 7. *L. neglectum.*
    Seeds wingless; pod broadly obovate, 2–2.5 mm. broad. — 8. *L. densiflorum.*
 Wing-margins of the fruit produced at the apex into acute lobes or teeth.
  Stem erect, 3–5 dm. high; lower leaves pinnatifid. — 9. *L. Fletcheri.*
  Stem decumbent, 3–15 cm. high; leaves linear or linear-spatulate. — 10. *L. Bourgeauanum.*

1. **L. campéstre** (L.) R. Br. Minutely soft, downy annual or biennial; stem erect, 2–4.5 dm. high; basal leaves entire or lyrate-pinnatifid, 5–8 cm. long; stem-leaves sagittate, entire or slightly dentate; raceme elongate; pod ovate, broadly winged above, rough. Fields and waste places: N.S.—Va.—Kans.—N.D.; nat. from Eur. My–Jl.

2. **L. perfoliàtum** L. Annual or biennial; stems suberect, 2–4 dm. high, more or less pilose; lower leaves bipinnatifid into linear divisions; upper stem-leaves entire, cordate or reniform, clasping, glabrous; petals yellow, fully 1 mm. long; stamens 6; pod 4 mm. long, orbicular, broadly ovate, or rhombic-elliptic, slightly emarginate; style about equaling the width of the wings. Waste places: Iowa—Neb.—Utah; nat. from Eur. My–Je.

3. **L. satìvum** L. Glabrous, bright green annual; stem 3–5 dm. high, branched; lower leaves twice pinnate, 5–20 cm. long, with obovate or oblanceolate, toothed divisions; petals spatulate, white or pinkish, 2 mm. long; stamens 6; pod obovate, 5 mm. long, glabrous, strongly winged above, emarginate; wing partly adnate to the style. Waste places: Que.—N.Y.—Mont.—B.C.—Yukon; escaped from gardens; native of Eur. My–Au.

4. **L. virgínicum** L. *Fig. 238.* Annual or biennial; stem erect, more or less pubescent, leafy, corymbosely branched above; basal leaves lyrate-pinnatifid, more or less hirsute; lower stem-leaves oblanceolate, incised, the upper narrower, remotely dentate; petals clawed, about 1.5 mm. long; stamens 2–6; pod suborbicular, 3 mm. long, winged above, with rounded lobes. Waste places, fields, and roadsides: Que.—Fla.—Tex.—Utah; Mex., W.Ind., C.Am., S.Am.; naturalized in the Old World. *E. Temp.—Trop.—Plain—Submont.* Ja–D. (in the Rockies, My–Au).

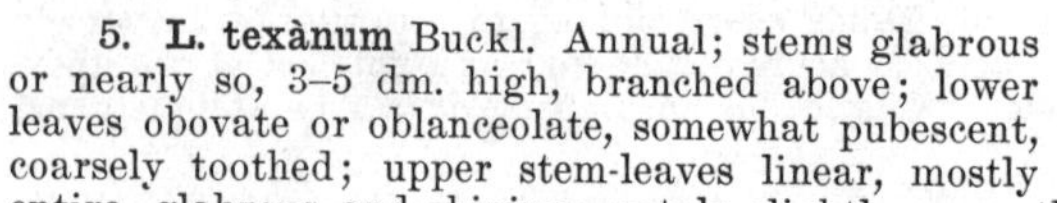

5. **L. texànum** Buckl. Annual; stems glabrous or nearly so, 3–5 dm. high, branched above; lower leaves obovate or oblanceolate, somewhat pubescent, coarsely toothed; upper stem-leaves linear, mostly entire, glabrous and shining; petals slightly more than 1 mm. long; stamens mostly 2; pod orbicular, 3 mm. long, narrowly wing-margined above. *L. intermedium* A. Gray. *L. medium* Greene. Sandy soil: Mo.—Ala.—Tex.—Calif.—B.C.; Mex. *S. Temp.—Son.* Ap–Au.

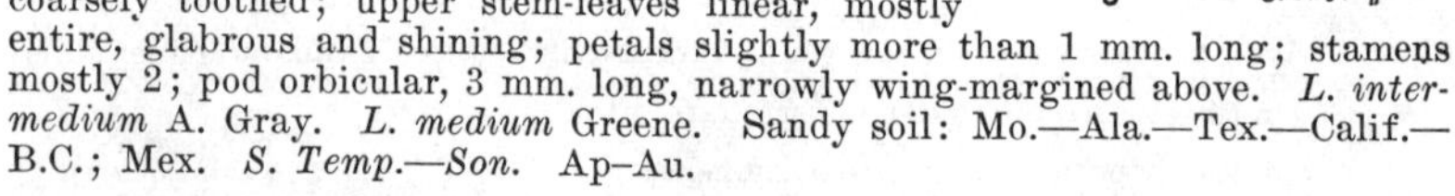

6. **L. ramosíssimum** A. Nels. Biennial; stems 2–4 dm. high, slightly puberulent, profusely branched; basal leaves pinnately lobed or toothed or incised; stem-leaves oblanceolate, all except the uppermost more or less toothed; petals white, narrowly spatulate; stamens 2; pod ovate, 3 mm. long. Mesas and mountains: Colo.—Wyo.—N.D.—Neb. *Plain—Submont.—Mont.* Je–Au.

7. **L. negléctum** Thellung. Stem erect, 2–5 dm. high; lower leaves oblanceolate, dentate; the upper ones linear, entire, with short hairs on the margins; racemes elongate; petals present but minute; pod orbicular, sometimes broader than long, rarely broadly ovate, 3 mm. wide, wing-margined, shorter than the pedicels; seeds wing-margined. Waste places: N.Y.—Ill.—Kans.—D.C. *Allegh.* Je–Au.

8. **L. densiflòrum** Schrad. Annual or biennial; stem 2–5 dm. high, puberulent, with short hairs, simple below, branched above; basal leaves oblanceolate or oblong, deeply incised-serrate or more or less pinnatifid, puberulent; stem-leaves mostly sharply toothed; petals rudimentary, linear-oblong, or none; stamens 2 or 4; pod obovate, about 3 mm. long. *L. apetalum* A. Gray [G, B]; not Willd. Plains, prairies, waste places: Vt.—Kans.—Man.—N.M.—B.C. *Temp.—Plain—Mont.* Ap–O.

9. **L. Flétcheri** Rydb. Annual or biennial; stem erect, 3–5 dm. high, puberulent with short cylindric spreading hairs; leaves narrow, pinnatifid, with linear divisions or saliently toothed, 2–5 cm. long, puberulent; petals none; stamens usually 2; pod glabrous, scarcely 3 mm. long, obovate in outline, gla-

brous; lobes of the wings nearly 0.5 mm., triangular-ovate, acutish or obtuse. Roadsides: Man.—Sask. *Plain.*

10. **L. Bourgeauànum** Thellung. Annual or biennial; stems decumbent, 1–1.5 dm. long; leaves linear or narrowly linear-spatulate, entire, pubescent with thickened hairs; petals rudimentary; stamens 2; pod obovate or elliptic, deeply notched at the apex; lobes of the wings acute and somewhat porrect. Plains: Sask. *Plain.*

4. **THLÁSPI** (Tourn.) L. PENNY CRESS, PENNY-GRASS. WILD SWEET ALYSSUM.

Glabrous annuals or perennials, with undivided and often clasping leaves. Flowers racemose, perfect. Sepals equal. Petals white, or rarely pink or purplish, obovate or oblanceolate. Stamens free, unappendaged; anthers short, oval. Styles slender or none; stigma small, entire. Capsules orbicular, oblong or cuneate; valves strongly keeled, often winged, especially towards the apex. Seeds 2–8 in each cell.

1. **T. arvénse** L. Annual, glabrous; stem 1.5–4.5 dm. high; basal leaves early deciduous, petioled; blades oblanceolate; stem-leaves oblong or elliptic, dentate, the upper auricled and clasping at the base; petals about 4 mm. long, oblanceolate or spatulate; pod 12–18 mm. long, 10–15 mm. broad. Waste places and roadsides: Lab.—Fla.—Colo.—Utah—B.C.; adv. or nat. from Eur. *Temp—Plain—Submont.* My–Jl.

5. **CAPSÉLLA** Medic. SHEPHERD'S PURSE.

Annual caulescent herbs, pubescent with branched hairs. Leaves alternate, entire, lobed, or pinnatifid. Flowers perfect, in elongate racemes. Petals white, spatulate. Stamens free and unappendaged. Styles almost none; stigma simple. Pod strongly flattened contrary to the narrow septum, cuneate or obcordate, more or less deeply notched at the apex, the valves boat-shaped. Seeds numerous in each cavity, marginless. Cotyledons accumbent. *Bursa* Weber.

1. **C. Búrsa-pastòris** (L.) Medic. *Fig. 239.* Annual or winter biennial; stem 1–5 dm. high; lower leaves usually lyrate-pinnatifid, lobed or merely dentate; stem-leaves sessile, hastate or sagittate at the base; petals white, 2 mm. long, spatulate; pods 6–8 mm. long, triangular. *B. Bursa-pastoris* Britton [B, R]. Waste places: Lab.—Fla.—Calif.—B.C.; S.Am.; nat. from Eurasia. *Temp.—Plain—Subalp.* Ja–D.

6. **HUTCHÍNSIA** R. Br.

Low spreading annuals, minutely stellate or glabrous. Leaves alternate, entire to pinnately lobed. Flowers perfect, in at last elongate racemes. Stigma sessile. Petals minute, narrow, white. Pods strongly flattened contrary to the narrow septum, elliptic, not at all obcordate or notched at the apex. Seeds numerous in each cell. Cotyledons accumbent.

1. **H. procúmbens** (L.) Desv. Stems branched at the base, slender, 5–20 cm. long; lower leaves pinnatifid, lobed, toothed, or entire, minutely stellate; upper leaves entire, oblanceolate or linear; sepals and petals about 1 mm. long; fruiting pedicels divaricate; pod elliptic, about 4 mm. long. Moist ground: Lab.—Newf.—Colo.—Calif.—B.C. *Temp.—Plain—Submont.* Mr–Je.

7. **PHYSÀRIA** A. Gray. Double Bladder-pod.

Stellate, cespitose perennials with taproots. Flowers perfect, racemose. Sepals oblong, erect or ascending, equal at the base. Petals spatulate, yellow. Stamens 6; anthers sagittate, linear. Fruit more or less distinctly didymous, inflated or obcompressed. Styles slender. Cotyledons accumbent.

Pods deeply cordate at the base; lower sinus almost as deep as the upper. 1. *P. didymocarpa.*
Pods not cordate at the base or slightly so; lower sinus none or very shallow; pod acutish at the base. 2. *P. brassicoides.*

1. P. didymocárpa (Hook.) A. Gray. *Fig. 240.* Stems numerous, 3–15 cm. long, decumbent to erect; basal leaves 1–8 cm. long, with margined petioles and broadly obovate, entire or sinuately toothed blades, finely and closely stellate; petals 8–14 mm. long, spatulate; pod 7–15 mm. in diameter. *P. macrantha* Blankinship. Sandy or rocky hills and plains: Sask.—Colo.—Utah—Alta. *Plain—Mont.* My–Jl.

2. P. brassicoìdes Rydb. Stems about 1 dm. high, ascending; basal leaves large, with winged petioles almost orbicular in outline, sinuate-dentate, 2.5–5 cm. in diameter; fruit almost obcordate, acute at the base, deeply divided above, 8–10 mm. long and 5–8 mm. in diameter. Cañons and bad-lands: Neb.—N.D. *Plain.* Je.

8. **LESQUERÉLLA** S. Wats. Bladder-pod.

Annual or tufted perennial, stellate herbs. Leaves entire or lobed, the blades linear to broadly spatulate or orbicular, the basal ones clustered. Flowers perfect, racemose. Sepals oblong, equal at the base; petals spatulate, yellow. Stamens 6; anthers linear, sagittate. Pods inflated, subglobose, ellipsoid, or ovoid; style slender; stigma entire; cells 2–16-seeded. Seeds rarely margined. Cotyledons accumbent.

Ovary and pod stellate-pubescent.
  Pod ovoid or ellipsoid.
    Pod distinctly compressed and acute at the apex.
      Basal leaves linear-oblanceolate; pod strongly compressed above. 1. *L. alpina.*
      Basal leaves spatulate; pod not strongly compressed above. 2. *L. spatulata.*
    Pod not compressed above; basal leaf-blades linear-oblanceolate; stem-leaves narrowly so. 3. *L. arenosa.*
  Pod globose.
    Plant perennial; basal leaf-blades spatulate or oblanceolate.
      Stems very slender; stem-leaves oblanceolate, 1–3 cm. long.
        Pedicels ascending. 3. *L. arenosa.*
        Pedicels in fruit recurved. 4. *L. versicolor.*
      Stems stouter, strict; stem-leaves linear, 2–6 cm. long. 5. *L. ludoviciana.*
    Plant annual; basal leaf-blades broadly oval or obovate. 6. *L. globosa.*
Ovary and pod glabrous, globose.
  Plant annual; basal leaf-blades spatulate or lyrate. 7. *L. repanda.*
  Plant perennial; basal leaf-blades broadly oval. 8. *L. ovalifolia.*

1. L. alpìna (Nutt.) S. Wats. Stem 2–10 cm. long, erect or assurgent; basal leaves 2–5 cm. long; stem-leaves narrowly linear; petals about 6 mm. long; pods usually erect, ovoid; body about 4 mm. long; style slightly longer; septum usually perforated. *L. condensata laevis* Payson. Dry hills and plains: Mont.—Utah—Colo.—N.D. *Submont.—Subalp.* Je–Jl.

2. **L. spatulàta** Rydb. Stems 3–10 cm. high; stem-leaves few, linear; petals spatulate, 7–8 mm. long; pod ovoid, about 5 mm. long; style slightly shorter; septum not perforated. (?) *L. nodosa* Greene. Dry hills: Man.—Neb. —Utah—Mont. *Plain—Submont.* Je–Jl.

3. **L. arenòsa** (Richardson) Rydb. Stems slender, decumbent or ascending, 1–3 dm. high; petals 6–8 mm. long; fruiting pedicels spreading, recurved; pod globose or slightly ellipsoid, 4–5 mm. long, shorter than the style. *L. Lunellii* A. Nels. Hills and plains: Man.—S.D.—Colo.—Mont.—Sask. *Plain—Submont.* Je–Jl.

4. **L. versícolor** Greene. Stems decumbent, 2–3 dm. high; basal leaves few and small; blades oval or oblanceolate, entire or toothed; stem-leaves oblanceolate; petals sulphur-yellow, changing into pink; pod 3 mm. thick; style 4 mm. long. Stony mountains: Man.—Sask. *Plain.* Je.

5. **L. ludoviciàna** (Nutt.) S. Wats. Stems erect or decumbent at the base, 1–4 dm. high; basal leaves linear-oblanceolate, 3–10 cm. long, entire or repand; petals 7–8 mm. long; pedicels in fruit spreading and recurved; pods 4–5 mm. long; style about the same length. *L. argentea* (Pursh) MacMill. [G, B]; not S. Wats. Plains and hills: Man.—Kans.—Colo.—Mont. *Plain—Submont.* My–Jl.

6. **L. globòsa** (Desv.) S. Wats. Stems slender, spreading or decumbent, branched; basal leaves obovate, 2–4 cm. long, entire or repand-toothed; stem-leaves linear or oblong, sessile; pedicels slender, divergent; pod globose, 2 mm. thick; style 4 mm. long. Rocky banks and open places: Ky.—Tenn.—Kans. Ap–Je.

7. **L. repánda** (Nutt.) S. Wats. Annual; stem branched; basal leaves sinuate-dentate to lyrate; stem-leaves numerous, linear-oblanceolate, repand-dentate, or entire; petals yellow; fruiting raceme elongate; pedicels horizontal to ascending; pod erect, glabrous, pyriform, 5–6 mm. long; stipe 1–2 mm. long; ovules 5–8 in each cell. *L. gracilis* Auth. [B, G]; not S. Wats. Prairies: Ark.—Neb.—Tex. *Prairie—Texan.*

8. **L. ovalifòlia** Rydb. *Fig. 241.* Stems erect, 1–2 dm. high; basal leaves petioled, 1–2 cm. long; stem-leaves sessile, linear-oblanceolate; inflorescence corymbiform; petals 8–10 mm. long, obovate; fruiting pedicels ascending; pods 5–6 mm. thick; style of about the same length. *L. ovata* Greene. *L. Engelmannii* Coult. & Nels., not S. Wats. Dry plains and hills: Neb.—Kans.—N.M.—Colo. *Plain—Son.* My–Je.

9. SUBULÀRIA L.

Small aquatic perennials, scapose and with subulate basal leaves. Flowers perfect, racemose. Sepals ovate, equal, spreading. Petals oblong or spatulate, without claws, white. Stamens 6, scarcely unequal; anthers oval. Pod short, subglobose or pear-shaped, turgid. Seeds few. Embryo folded above the radicle.

1. **S. aquática** L. Submerged perennial; leaves 10–20, erect or nearly so, 2–7 cm. long, subulate; scape 2–10 cm. long; submerged flowers cleistogamous; fruit obovate. Ponds: Newf.—N.H.—Calif.—B.C. *Boreal.—Mont.*

10. CAMELÌNA Crantz. FALSE FLAX.

Annual caulescent herbs, glabrous or with branching hairs. Leaves alternate, entire or toothed, often clasping. Flowers perfect, in elongate racemes.

Sepals equal. Petals yellowish or greenish. Pod short, inflated, nearly terete, obovoid; valves 1-nerved; style slender; stigma capitate. Seeds in two rows, usually few, marginless. Cotyledons incumbent.

Stem glabrous. 1. *C. sativa.*
Stem pubescent. 2. *C. microcarpa.*

1. **C. satìva** (L.) Crantz. Stem 3–6 dm. high, branched above; lower leaves petioled, 5–8 cm. long, lanceolate; upper stem-leaves sagittate at the base and clasping; petals light yellow, 5–6 mm. long; pod pear-shaped, 6–8 mm. long. Waste places: N.S.—N.Y.—Calif.—Mont.; adv. from Eur. *Temp.—Submont.* Je–Jl.

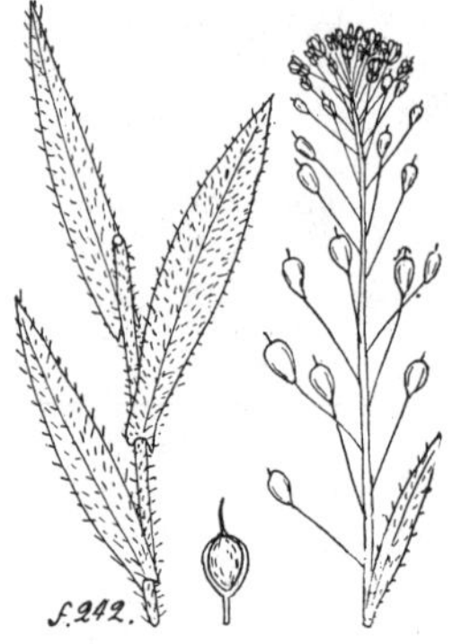

2. **C. microcárpa** Andrz. *Fig. 242.* Stem 3–6 dm. high; lower leaves lanceolate, more or less hirsute; stem-leaves sagittate at the clasping base; petals about 4 mm. long; pod 4–6 mm. long, pear-shaped, strongly margined. Waste places: R.I.—Va.—Ariz.—B.C.; nat. from Eur. *Temp.—Submont.* My–Jl.

11. NÉSLIA Desv.

Leafy-stemmed annuals. Flowers perfect. Sepals short, equal at the base. Petals spatulate, clawed. Stamens 6, free and unappendaged. Fruit globose, indehiscent, usually 1-celled by obliteration of the partition; style elongate; stigma simple. Seeds 1 or 2, neither winged nor margined. Cotyledons incumbent.

1. **N. paniculàta** (L.) Desv. Slender branched annual; stems 3–6 dm. high, rough-hispid; leaves lanceolate, sagittate at the base, 2–6 cm. long; racemes elongate in fruit; petals bright yellow, 2 mm. long; pod subglobose, reticulate and muricate, about 2 mm. thick. Waste places: Newf.—Mo.—S.D.—B.C.; adv. from Eur. *Temp.—Submont.* Jl–Au.

12. NEOBÉCKIA Greene.

Immersed perennials, with a creeping rootstock. Leaves alternate, glabrous, the submerged ones finely pinnatifid, deciduous, often rooting and producing new plants, the emersed ones simple, usually dentate. Flowers racemose, perfect. Sepals ascending. Petals white, clawed. Stamens 6, anthers sagittate at the base. Pod terete, oblong-ellipsoid, 1-celled, the partition rudimentary; style elongate, slender, stigma 2-cleft. Seeds few.

1. **N. aquática** (Eaton) Greene. Stem 3–6 dm. long, branching; submerged leaves 5–15 cm. long, the divisions filiform; emersed leaves lanceolate or oblong, 3–7 cm. long, serrate or the lower lobed; pedicels 6–8 mm. long; petals twice as long as the sepals. *Roripa americana* (A. Gray) Britt. [B]. *Radicula aquatica* (Eat.) B. L. Robinson [G]. Slow streams and lakes: N.J.—Minn.—La.—Fla. Je–Au.

13. ARMORÀCIA Gaertn. HORSE-RADISH.

Tall glabrous perennials, with thick pungent taproot and leafy stem. Leaves crenate or sinuately lobed. Flowers perfect, in dense racemes or panicles. Petals white, clawed. Stamens 6, free and unappendaged. Pod short, round-ellipsoid or subglobose; style short; stigma subcapitate.

1. **A. rusticàna** Gaertn. Stem 6–10 dm. high; basal leaves 1.5–3 dm. long, glabrous, oblong; stem-leaves lanceolate, serrate or crenate; flowers 4–8 mm. wide, showy. *Radicula Armoracia* B. L. Robinson [G]. *Nasturtium Armoracia* Fries. *A. Armoracia* (L.) Cockerell [B, R]. Around dwellings: Que.—Fla.—Tex.—Mont.; escaped from cultivation. Jl–S.

14. RORÍPPA Scop. YELLOW WATER CRESS, MARSH CRESS.

Annual or perennial herbs of wet situations. Leaves alternate, pinnately dissected or lobed. Flowers perfect, in terminal or axillary racemes. Sepals spreading during anthesis. Petals yellow, or rarely wanting, clawed. Stamens 2–6. Pods terete, from subglobose to short-cylindric; valves usually 1-nerved. Seeds in 2 rows, turgid. Cotyledons accumbent. *Nasturtium* R. Br., in part. *Radicula* Hill.

Perennials, with rootstock; petals much exceeding the sepals.
  Style very short; leaves with acute toothed or incised divisions. — 1. *R. sylvestris.*
  Style slender; leaves with obtuse, denticulate divisions.
    Pods papillose. — 2. *R. Columbiae.*
    Pods not papillose. — 3. *R. sinuata.*
Annuals or biennials.
  Pods spherical to oblong-ellipsoid, in the latter case shorter than the pedicels.
    Pedicels 4–10 mm. long.
      Stem more or less hirsute. — 4. *R. hispida.*
      Stem glabrous. — 5. *R. palustris.*
    Pedicels 2–5 mm. long, pod globose. — 6. *R. sphaerocarpa.*
  Pods oblong to linear-cylindrical, equaling or longer than the short pedicels.
    Style about 1 mm. long.
      Leaves nearly all pinnatifid, with obtuse divisions. — 7. *R. obtusa.*
      Leaves mostly sinuate, or if pinnatifid, with acute divisions
    Style minute, 0.5 mm. long or less.
      Pods fully 2 mm. thick, almost sessile; seeds pitted. — 8. *R. sessiliflora.*
      Pods 1.5 mm. thick or less, distinctly pedicelled; seeds tuberculate. — 9. *R. lyrata.*

1. **R. sylvéstris** (L.) Besser. Stem creeping with ascending branches; leaves 7–13 cm. long, pinnately parted into oblong or lanceolate divisions; pedicels slender, 6 mm. long; petals yellow, 3–4 mm. long; pod linear, 8–12 mm. long, narrow, glabrous. *Sisymbrium sylvestre* R. Br. *Radicula sylvestris* Druce [G]. Waste places and low ground: Newf.—Va.—Ill.—N.D.; nat. from Eur. Je–Au.

2. **R. Colúmbiae** (Suksd.) Howell. Stem branched, 1–3 dm. high, papillose or glabrate; leaves pinnatifid, with many oblong, often toothed divisions; petals yellow, spatulate, 4 mm. long; pedicels ascending or spreading; pod ascending or erect, curved, 5–8 mm. long; style nearly 2 mm. long. *Rorippa calycina* Rydb.; not *Nasturtium calycinum* Engelm. *Radicula Columbiae* Greene [R]. River banks and wet sandy places: Mont.—Neb.—N.M.—Wash. *Plain—Submont.* Je–Jl.

3. **R. sinuàta** (Nutt.) Hitchc. *Fig. 243.* Stems 1–4 dm. high, glabrous or nearly so; leaves glabrous, pinnately divided or cleft, with numerous oblong or lanceolate, often toothed divisions; petals yellow, 4–5 mm. long; pod curved upwards, glabrous, 1–1.5 cm. long, 2 mm. thick; style 3 mm. long. *N. sinuatum* Nutt. *Radicula sinuata* Greene [G, B, R]. River valleys: Sask.—Minn.—Ill.—Tex.—Ariz.—Wash.; Ont. *W. Temp.—Plain—Prairie.* Ap–Au.

f. 243.

4. **R. híspida** (DC.) Britton. Biennial; stem stout, branched above, 3–12 dm. high; leaves lyrate-pinnatifid, with lanceolate or ovate, toothed divisions; petals spatulate, yellow, 2 mm. long; fruiting pedicels ascending or spreading; pod 4–6 mm. long, 2 mm. thick, glabrous; style about 1 mm. long. *N. hispidum* DC. *Radicula hispida* Heller [B, R]. Water and wet places: N.B.—Fla.—N.M.—Wash.—Utah—Alaska. *Temp.—Submont.* Je–Au.

5. **R. palústris** (L.) Besser. Annual or biennial; stem branched above, 3–10 dm. high; leaves lyrate-pinnatifid, with oblong or lanceolate, toothed divisions; petals yellow, spatulate, 2 mm. long; fruiting pedicels usually spreading; pod 5–7 mm. long, 2 mm. thick; style 1 mm. long. *Sisymbrium amphibium palustre* L. *N. terrestre* R. Br. *N. palustre* DC. *Radicula palustris* Moench [G, B]. Water and wet places: Lab.—Ga.—N.M.—Alaska; Eurasia; Mex. *Temp.—Plain—Submont.* Ap–O.

6. **R. sphaerocárpa** (A. Gray) Britton. Annual or biennial; stem diffusely branched from the base, 1–3 dm. high, glabrous; leaves sinuately or lyrately lobed, with rounded lobes; petals yellow, 1.5 mm. long; pedicels 2–3, rarely 5 mm. long; pod 2–3 mm. long, nearly 2 mm. thick; style about 0.5 mm. long. *N. sphaerocarpum* A. Gray. *Radicula sphaerocarpa* Greene [G, B, R]. Along streams: Ill.—Tex.—Calif.—Wyo. *Prairie—Plain—Submont.* Jl–S.

7. **R. obtùsa** (Nutt.) Britt. Annual; stem diffusely branched at the base, 1–3 dm. high, glabrous; leaves pinnatifid, with obovate or rounded sinuately toothed divisions; petals narrowly spatulate, 1 mm. long; pedicels ascending or spreading; pod 4–8 mm. long, 1–2 mm. thick; style about 1 mm. long. *N. obtusum* Nutt. *Radicula obtusa* Greene [G, B, R]. Wet places: Mich.—Mo.—Tex.—Utah—Wash. *Temp.—Plain—Submont.* Je–S.

8. **R. sessiliflòra** (Nutt.) Hitchc. Annual or biennial; stem usually erect, branched, glabrous; leaves lobed or cleft, with rounded lobes, or merely coarsely crenate; petals 1.5–2 mm. long; pedicels 1–2 mm. long, or sometimes almost none; pod 6–12 mm. long, fully 2 mm. thick; beak short and thick. *Nasturtium sessiliflorum* Nutt. *Radicula sessiliflora* Greene [G, B, R]. Wet places: Ill.—Fla.—Tex.—Neb. *E. Temp.—Plain.* Ap–S.

9. **R. lyràta** (Nutt.) Greene. Annual; stems diffusely branched, 1–3 dm. high, glabrous; leaves pinnately divided, with oblong or ovate, coarsely toothed divisions; petals scarcely 1 mm. long; pedicels 1–3 mm. long, ascending; pod 6–15 mm. long, 1–1.5 mm. thick. *N. lyratum* Nutt. *Radicula lyrata* Greene [R]. Wet places: Mont.—Colo.—Calif.—Wash. *Plain—Submont.* Jl–Au.

15. NASTÚRTIUM R. Br. WATER CRESS.

Aquatic perennials, with floating or creeping stems, rooting at the nodes. Leaves odd-pinnate. Flowers perfect, in racemes. Sepals equal at the base, spreading during anthesis. Petals white, without nectaries. Pods cylindric; valves 1-nerved. Styles slender; stigma 2-lobed. Seeds in 2 rows, turgid. Cotyledons accumbent. *Sisymbrium* L., in part.

f.244.

1. **N. officinàle** R. Br. *Fig. 244.* Aquatic perennial; leaves pinnate, with 1–11 orbicular to oblong-lanceolate sinuate leaflets; sepals oblong, about 2 mm. long; petals white, spatulate, about 4 mm. long; fruiting pedicels divaricate; pod 1–2 cm. long, 3 mm. thick, somewhat curved; beak about 1 mm. long. *S. Nasturtium-aquaticum* L. [B,R]. In water and mud: N.S.—Va.—Calif.—B.C.; Mex.; W.Ind.; S.Am.; nat. from Eur. *Temp.—Trop.—Plain—Subalp.* My–S.

16. TÚRRITIS L. TOWER MUSTARD.

Caulescent herbs, with partly clasping, auricled stem-leaves. Flowers in elongate racemes. Sepals erect, two of them slightly saccate at the base. Petals yellowish white, small. Pods erect, linear-cylindric, terete or slightly flattened, 1-nerved and veined; stigma cupulate. Seeds flattened, partially incumbent, *i. e.*, radicle oblique.

1. **T. glàbra** L. Biennial; stem erect, 6–15 dm, high, hirsute below; basal leaves oblanceolate, sinuately toothed or somewhat lyrate, hirsute-stellate, 5–15 cm. long; stem-leaves glabrous, lanceolate, sagittately clasping, 2–10 cm. long; sepals and petals greenish white, 3 mm. long; pedicels erect, 4–12 mm. long; pod erect, 4–10 cm. long, 1.5–2 mm. wide. *Arabis perfoliata* Lam. *A. glabra* (L.) Bernh. [G, B]. Waste places: Que.—Pa.—Calif.—B.C.; nat. from Eur. *Temp.—Submont.* Ap–Au.

17. ERÝSIMUM L. HEDGE MUSTARD.

Annual caulescent herbs. Leaves alternate, pinnately lobed. Flowers perfect, in racemes. Sepals subequal, ascending. Petals small, pale yellow. Filaments unappendaged. Pod subulate, tapering from the base to the apex. Style almost none; stigma 2-lobed. Cotyledons incumbent. Seeds in one row, marginless.

f. 245.

1. **E. officinàle** L. *Fig. 245.* Annual; stem branching, hirsute, at least below; leaves hirsute, pinnatifid or the uppermost merely hastate; terminal lobes of the lower leaves rounded and toothed, those of the upper lanceolate; petals pale yellow, spatulate, 3 mm. long; pods erect, on very short pedicels, 1.5–2 cm. long, glabrous or pubescent. *Sisymbrium officinale* Scop. [B]. Waste places: N.S.—Fla.—Calif.—B.C.; nat. from Eur. *Temp.—Submont.* My–S.

18. SISÝMBRIUM L. TUMBLE MUSTARD.

Caulescent leafy annuals, with runcinately pinnatifid or toothed leaves, pubescent with simple hairs. Flowers perfect, in elongate racemes. Sepals subequal, ascending. Petals light yellow, clawed. Stamens 6; filaments unappendaged. Pod elongate, narrowly cylindric, terete; style none; stigma slightly 2-lobed. Seeds in one row, numerous, marginless. Cotyledons incumbent. *Norta* Adans.

Stem with spreading hairs; upper leaves with narrowly linear divisions; pod 7–10 cm. long, on ascending pedicels. 1. *S. altissimum.*
Stem with reflexed hairs; upper leaves with lanceolate divisions; pod 3 cm. long, on spreading pedicels. 2. *S. Loeselii.*

1. **S. altíssimum** L. Tall annual; stem freely branching, 6–12 dm. high; sparingly ciliate; leaves pinnatifid, the lower with oblong or lanceolate, the upper with narrowly linear divisions; sepals oblong, 5 mm. long; petals yellowish white, spatulate, 6–8 mm. long; pedicels and pods ascending-spreading, the latter 7–10 cm. long, 1 mm. thick, sessile. *N. altissima* Britton [B, R]. Waste places: Que.—D.C.—N.M.—Calif.—B.C.; nat. from Eur. *Temp.—Plain.* My–Au.

2. **S. Loesélii** L. Tall annual; stem 4–10 dm. long, retrorsely hirsute, at least below; lower leaves 5–20 cm. long, sparingly hirsute beneath or glabrous, lyrate-pinnatifid, with oval or oblong, toothed divisions, obtuse; middle leaves with lanceolate divisions, acuminate, the uppermost narrower, but the divisions not narrowly linear; sepals oblong, 4 mm. long; petals bright yellow, 5–6 mm. long; pedicels in fruit more or less spreading, 1–1.5 cm. long; pod

ascending, about 3 cm. long, 1 mm. thick. Waste places and roadsides: Conn. —Neb.—Ida.; adv. from Eur.

19. ALLIÀRIA Adans. HEDGE-GARLIC, GARLIC MUSTARD.

Biennial or perennial herbs, with broad leaves. Flowers white. Petals oblong, clawed. Style short-conic. Pod linear, cylindric, terete or 4-angled, valves with a strong midvein, dehiscent from below. Seeds oblong, striate, in a single row.

1. A. officinàlis Andrz. Stem erect; 3–10 dm. high, glabrous or nearly so; lower leaves petioled, the upper sessile, the blades reniform, cordate, or broadly ovate, crenate or undulate, 5–15 cm. broad; flowers white, 6–8 mm. broad; pod 3–5 cm. long, 2 mm. thick, 4-angled when dry. *A. Alliaria* (L.) Britt. [B]. Waste places: Que.—Va.—Kans.; nat. from Eurasia. My–Je.

20. IODÁNTHUS T. & G.

Glabrous erect perennials. Basal leaves lyrate-pinnatifid; stem-leaves auriculate at the base. Sepals much shorter than the petals, the inner ones gibbous at the base. Petals long-clawed, purple or white. Pod linear, slightly flattened, somewhat constricted between the seeds; style stout; stigma entire. Seeds in one row in each cell, oblong, wingless.

1. I. pinnatífidus (Michx.) Steud. Stem 3–10 dm. high; lower leaves ovate or oblong in outline, pinnatifid at the base, dentate, the upper ones ovate-oblong to lanceolate, dentate, auricled at the base; pedicels spreading; pod linear, 2–3 cm. long. River banks: Pa.—Tenn.—Tex.—Minn. *Allegh.—Prairie —Texan.* My–Au.

21. SOPHÌA Adans. TANSY MUSTARD.

Annual or biennial (all ours), or perennial herbs, or rarely suffruticose, pubescent with short branched hairs. Leaves usually twice pinnatifid to finely dissected. Flowers in elongate racemes. Sepals narrow, subequal, ascending, early deciduous. Petals yellow or yellowish-white, rarely pure white. Pod linear or club-shaped; style short; stigma minute, entire; valves one-nerved. Seeds in 1 or 2 rows in each cell. Cotyledons incumbent. *Descurainia* Webb. & Barth.

- Pedicels ascending or spreading.
  - Pods linear, 15–25 mm. long, 1 mm. wide; leaves mostly thrice pinnatifid. — 1. *S. multifida.*
  - Pods more or less clavate, 5–12 mm. long, 1–2 mm. wide; seeds usually more or less biserial.
    - Pods 8–12 mm. long, mostly erect on spreading pedicels.
      - Petals much exceeding the sepals.
        - Terminal segments of the upper leaves usually elongate, linear, entire; segments all narrow. — 2. *S. filipes.*
        - Terminal segments not greatly elongate; segments of the lower leaves obovate. — 3. *S. magna.*
      - Petals not much exceeding the sepals; segments narrow; terminal segments of the upper leaves not elongate. — 4. *S. intermedia.*
    - Pods 5–8 mm. long, spreading or ascending, on spreading pedicels.
      - Leaves nearly green. — 5. *S. brachycarpa.*
      - Leaves decidedly canescent. — 6. *S. pinnata.*
- Pedicels erect; pods 1 mm. thick; plant cinereous. — 7. *S. Richardsoniana.*

1. S. multífida Gilib. Stem much branched, 3–10 dm. high, minutely pubescent; leaves thrice pinnate, with narrowly linear divisions; petals cream-colored; pod erect, somewhat curved, glabrous, about 2 cm. long, less than 1 mm. thick, torulose. *S. Sophia* Britton [B, R]. *Sisymbrium Sophia* L. [G]. Waste places: N.B.—N.Y.—Utah—Ore.—Wash.; nat. from Eur. Je–Jl.

2. **S. fílipes** (A. Gray) Heller. Stem branched, 3–6 dm. high, almost glabrous; leaves once to twice pinnatifid, sparingly stellate or glabrous; segments linear or oblong, often lobed; petals bright yellow, 3 mm. long or more; pedicels 1–2 cm. long; pods clavate, erect, 12–15 mm. long, 1.5–2 mm. thick. Mountains, cañons, and plains: Sask.—N.D.—Colo.—Utah—Wash.—B.C. *W. Temp.—Submont.* Ap–Jl.

3. **S. mágna** Rydb. Stem branched, 5–10 dm. high, sparingly stellate-puberulent or glabrous, stout; basal leaves twice to thrice pinnatifid, 1–2 dm. long, nearly glabrous; segments obovate, often toothed; petals spatulate, nearly 3 mm. long, rather light yellow; pod glabrous, more or less clavate, 12–15 mm. long, 1.5–2 mm. thick. River bluffs: Colo.—Kans. *Mont.—Plain.*

4. **S. intermèdia** Rydb. *Fig. 246.* Stem 3–7 dm. high, often glandular above; leaves twice or thrice pinnatifid; segments linear or linear-oblong; petals spatulate, slightly if at all exceeding the sepals; pedicels in fruit 1–1.5 cm. long, usually spreading; pods clavate, glabrous, 8–12 mm. long, 1.5 mm. thick, nearly erect. Prairies, plains, and waste places: Mich.—Tenn.—Colo.—Calif.—B.C. *W. Temp.—Prairie—Plain—Mont.* My–Jl.

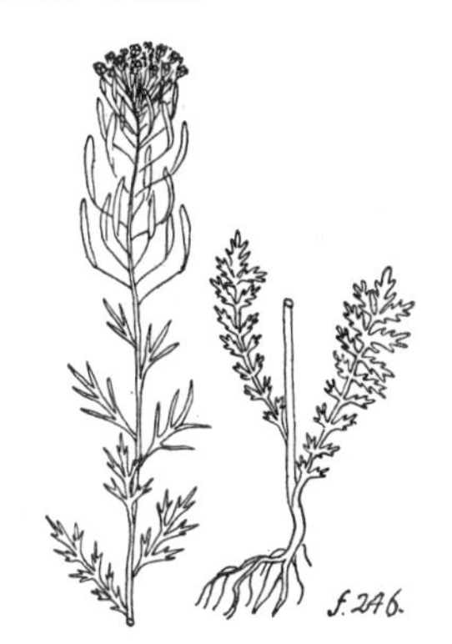

5. **S. brachycárpa** (Richardson) Rydb. Stem 2–5 dm. high, minutely stellate-canescent below, glabrous or glandular-puberulent above; leaves bi-tripinnatifid, with oblong divisions; pedicels spreading; pod short, clavate, 5–8 mm. long, with seeds in two rows; style obsolete. Waste places: Que.—Ky.—Kans.—Sask. *Prairie—Allegh.* My–Jl.

6. **S. pinnàta** (Walt.) Howell. Stem 3–6 dm. high; leaves thrice or twice pinnatifid into short, obovate, oval, or elliptic divisions; petals pale yellow; pedicels spreading, 7–12 mm. long; pod clavate, 5–8 mm. long. *Sisymbrium canescens* Nutt. Sandy or dry soil: Va.—Fla.—Ariz.—Colo. *E. Temp. —Plain.* Mr–Je.

7. **S. Richardsonìana** (Sweet) Rydb. Stem 6–20 dm. high; leaves twice pinnatifid, with lanceolate divisions; petals spatulate, light yellow, 2.5–3 mm. long; pedicels 5–8 mm. long; pod linear, 10–15 mm. long, 1 mm. thick. *Sisymbrium incisum Hartwegianum* S. Wats. [G]. *S. Richardsonianum* Sweet. *Sophia Hartwegiana* Greene, in part [B]. *S. brevipes* (Nutt.) Rydb. River banks, waste grounds, and sandy valleys: Sask.—Minn.—Colo.—Utah—B.C. *Prairie—Plain—Submont.* Jl–Au.

22. MALCÓLMIA R. Br.

Leafy-stemmed annuals, with branched hairs. Flowers perfect, racemose or spicate. Sepals erect, usually saccate at the base. Petals white or rose-colored. Longer stamens coherent. Fruit cylindric, dehiscent, with 1 or 2 rows of seeds.

Sepals alike; pubescence spreading; beak short. — 1. *M. africana.*
Sepals unlike, the outer saccate; pubescence appressed, beak of the pod elongate. — 2. *M. maritima.*

1. **M. africàna** (Willd.) R. Br. Annual; stem 2–4 dm. high, with ascending-spreading branches; leaves oblanceolate or lanceolate, 4–8 cm. long, usually sinuate-dentate, hirsute with branched hairs; flowers sessile; sepals linear, 3

mm. long; petals pinkish, veined, long-clawed; pods ascending, 5–6 cm. long, 1–1.5 mm. thick. Sandy or alkaline ground: Utah; Neb.; adv. from n Africa. Je–Jl.

2. **M. marítima** R. Br. Annual; stems 1–3 dm. high, flexuose, appressed-pubescent; leaf-blades oblong or ovate, entire or dentate, appressed, canescent; lateral sepals saccate at the base; petals violet or pink, obovate, clawed; pod ascending, on short stout peduncles, 5–8 cm. long, long-beaked. Waste places: Minn.; esc. from cult.; nat. of the Mediterranean region.

23. CONRÍNGIA (Heist.) Link. HARE'S-EAR.

Glabrous annuals, with sessile clasping leaves. Flowers perfect, in elongate racemes. Sepals subequal, long and narrow, ascending. Petals light yellow. Pod elongate-linear, more or less 4-angled; style short; stigma simple or nearly so. Seeds in a single row in each cell, oblong, thick, not margined. Cotyledons incumbent.

1. **C. orientàlis** (L.) Dum. Glabrous annual; stem 3–6 dm. high; leaves oval or elliptic, deeply cordate-clasping, 4–10 cm. long; petals oblanceolate, yellowish white, 8 mm. long; pod ascending, 8–10 cm. long, fully 2 mm. thick; beak 1.5 mm. long. Waste places and along roads: N.S.—Del.—Colo.—Utah—Ore.—B.C.; adv. from Eur. *Temp.—Plain.* My–Jl.

24. ARABIDÓPSIS (DC.) Schur. MOUSE-EAR CRESS.

Annual or perennial herbs, with branched hairs. Leaves entire or toothed. Flowers perfect, small, in terminal racemes. Sepals equal, not saccate at the base. Petals white. Style very short; stigma 2-lobed. Pod narrow, linear, with rounded, nerveless or finely nerved valves. Seeds in 1 row in each cell. Cotyledons incumbent. *Pilosella* (Thal) Kostel. *Stenophragma* Celak.

1. **A. Thalìàna** (L.) Schur. Stem simple or branched, 2–4 dm. high, sparingly pubescent; basal leaves oblanceolate or spatulate, usually petioled, 2–5 cm. long, sparingly hirsute-stellate; petals white, oblanceolate, 3–4 mm. long; pods ascending, glabrous, 1–1.5 cm. long, 0.75 mm. wide; style very short. *Arabis Thaliana* L. [G]. *Sisymbrium Thalianum* A. Gray. *Stenophragma Thalianum* Celak. [B]. Waste places: Mass.—Ga.—Kans.; Utah; nat. from Eur. Ap–Je.

25. CHEIRÍNIA Link. WILD WALL-FLOWER, PRAIRIE-ROCKET, YELLOW PHLOX.

Annual, biennial, or perennial leafy-stemmed herbs, with appressed 2-branched hairs (the hairs appearing as if attached near the middle). Flowers perfect in terminal racemes, usually rather large. Sepals erect, the outer two gibbous at the base. Petals yellow, brown, or purple, clawed, with spreading blades. Filaments free, unappendaged. Pods elongate-linear, more or less 4-angled, or at least with a strong midrib. Style short; stigma 2-lobed. Seeds in one row in each cell, numerous, marginless or margined at the apex. Cotyledons in our species incumbent or nearly so. [*Erysimum* L., in part].

Petals less than 1 cm. long.
  Petals 4–5 mm. long. — 1. *C. cheiranthoides.*
  Petals 6–10 mm. long.
    Perennials or biennials; pods ascending.
      Pods 2–5 cm. long, less than 2 mm. thick; plant green. — 2. *C. inconspicua.*
      Pods 1.5–3 cm. long, about 2 mm. thick; plant canescent. — 3. *C. syrticola.*
    Annuals; pod spreading. — 4. *C. repanda.*
Petals more than 1 cm. long.
  Basal leaves, as well as whole plant, grayish-strigose, not silvery.

Pods widely spreading, 4–8 cm. long, stout; stem-leaves usually sinuate-dentate.	5. *C. aspera.*
Pods strongly ascending or almost erect, 8–12 cm. long.	
Stem-leaves usually sinuately dentate.	6. *C. elata.*
Stem-leaves entire or rarely denticulate, linear or nearly so; pod usually twisted, slender.	7. *C. asperrima.*
Basal leaves, at least, silvery white; stem-leaves entire, narrowly linear.	8. *C. argillosa.*

1. **C. cheiranthoìdes** (L.) Link. Annual; stem 3–6 dm. high, finely strigose; leaves lanceolate or linear-lanceolate, 2–10 cm. long, entire or sinuate-denticulate; petals 4–5 mm. long; fruiting pedicels spreading or ascending; pod 2–3 cm. long, 1–1.5 mm. thick, erect. *Erysimum cheiranthoides* L. [G, B]. WORMSEED MUSTARD. Waste places and river bottoms: Newf.—N.C.—Utah.—Alaska; Eur. *Temp.—Plain—Mont.* Je–Au.

2. **C. inconspícua** (S. Wats.) Rydb. Biennial; stem 3–6 dm. high, angled; leaves linear or oblanceolate, mostly entire, canescent; petals pale yellow, 8–10 mm. long, claw shorter than the sepals; pedicels strongly ascending; pods erect, 2–5 cm. long, less than 2 mm. thick. *E. asperum inconspicuum* S. Wats. *E. parviflorum* Nutt. [G]. Dry soil: Minn.—Colo.—Nev.—B.C. *W. Temp.—Prairie—Plain—Submont.* Je–Au.

3. **C. syrtícola** (Sheldon) Rydb. Biennial or perennial; stem 3–8 dm. high; leaves linear-lanceolate, entire or denticulate; petals pale yellow, 8–9 mm. long; pedicels ascending; pods erect. *E. syrticolum* Sheldon [B]. Sandy banks: Minn. Jl–Au.

4. **C. repánda** (L.) Link. Annual; stem 3–4 dm. high, much branched; leaves lanceolate or oblanceolate, 3–8 cm. long, repand-dentate or the lower coarsely toothed; petals 6–9 mm. long, light yellow; pods 4–8 cm. long, about 2 mm. thick; style very short and stout. *E. repandum* L. [G]. Waste places: Ohio.—Kans.—Ariz.—Utah—Ore.; also as a ballast plant at sea-ports; adv. or nat. from Eur. My–Jl.

5. **C. áspera** (Nutt.) Rydb. *Fig. 247.* Biennial; stem 2–4 dm. high, mostly simple; leaves lanceolate or linear-lanceolate, usually somewhat sinuately dentate; petals light yellow, 15–18 mm. long; pedicels divergent; pods divergent, 2 mm. thick; beak about 1 mm. long, thick. *Cheiranthus asper* Nutt. [G]. *E. asperum* DC. [B]. Plains: Que.—Man.—Okla.—e N.M.—e Mont. *Plain—Submont.* Je–Au.

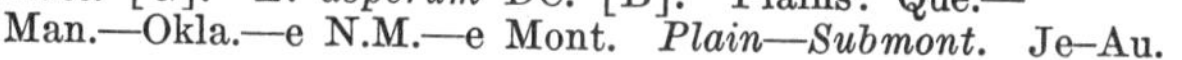

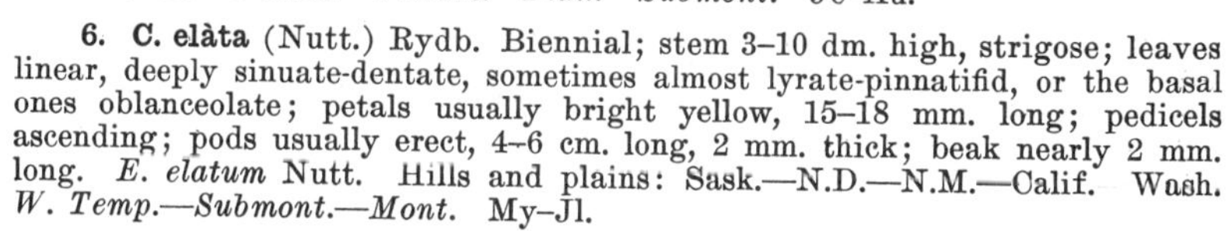

F. 247.

6. **C. elàta** (Nutt.) Rydb. Biennial; stem 3–10 dm. high, strigose; leaves linear, deeply sinuate-dentate, sometimes almost lyrate-pinnatifid, or the basal ones oblanceolate; petals usually bright yellow, 15–18 mm. long; pedicels ascending; pods usually erect, 4–6 cm. long, 2 mm. thick; beak nearly 2 mm. long. *E. elatum* Nutt. Hills and plains: Sask.—N.D.—N.M.—Calif. Wash. *W. Temp.—Submont.—Mont.* My–Jl.

7. **C. aspérrima** (Greene) Rydb. Biennial; stems 2–5 dm. high, usually simple; leaves linear, 5–8 cm. long; petals 10–15 mm. long, usually pale yellow; pedicels ascending; pods strictly erect, 5–8 cm. long, 1–1.5 mm. thick; beak 1 mm. long. *Cheiranthus asperrimus* Greene. Dry hills: S.D.—N.M.—Ariz.—Mont. *Submont.* My–Au.

8. **C. argillòsa** (Greene) Rydb. Biennial; stem usually simple, 1–3 dm. high; leaves linear-lanceolate or linear, crowded below; petals pale yellow, about 15 mm. long; pedicels ascending, 5–8 mm. long; pods erect, 5–7 cm.

long, 2 mm. thick. *Cheiranthus argillosus* Greene. Clayey bluffs and bad-lands: S.D.—Colo. *Plain.* My–Je.

26. **HÉSPERIS** (Tourn.) L. DAME'S ROCKET, DAME'S VIOLET.

Perennial or biennal herbs, with forked hairs. Leaves simple, dentate. Flowers large, racemose, perfect. Sepals equal, not saccate at base. Petals purple or white, with long claws and broad blades. Pod elongate, nearly cylindric; valves keeled and 1-nerved. Stigma with 2 erect lobes. Seeds in a single row in each cell, globose, wingless. Cotyledons incumbent.

1. **H. matronàlis** L. Stem erect, simple; branched above, 3–10 dm. high; leaves 5–20 cm. long, ovate or ovate-lanceolate, dentate, pubescent on both sides; petals purple, pink, or white; pod 5–10 cm. long, ascending or spreading, somewhat torulose when ripe. Fields and roadsides: N.S.—Pa.—Iowa—Mont.; escaped from cultivation; native of Eurasia. My–Au.

27. **BARBARÈA** R. Br. WINTER CRESS, SCURVY GRASS.

Biennial or perennial herbs with angled stems, and alternate lyrate-pinnatifid leaves. Flowers perfect in racemes or panicles. Sepals erect, the outer two slightly saccate at the base. Petals yellow, spatulate, clawed. Pod elongate, linear, somewhat 4-angled, style short; stigma more or less 2-lobed. Seeds in one row in each cell, flattish, marginless. Cotyledons accumbent. *Campe* Dulac.

Pod obtusely angled, on slender pedicels; leaves with 1–4 pairs of divisions.
  Pod ascending or spreading.
    Lateral divisions of the lower leaves 1 or 2 small pairs. — 1. *B. americana.*
    Lateral divisions of the lower leaves 3 or 4 ample pairs. — 2. *B. vulgaris.*
  Pod erect; divisions of the lower leaves 3 or 4 pairs. — 3. *B. stricta.*
Pod sharply 4-angled, on short stout pedicels; leaves with 4–8 pairs of divisions. — 4. *B. verna.*

1. **B. americàna** Rydb. *Fig. 248.* Biennial; stem 3–5 dm. high, glabrous, strict; divisions of the basal leaves broadly elliptic; stem-leaves with the petioles auriculate-clasping; petals 2–4 mm. long, light yellow; pod 2–2.5 cm. long, scarcely 2 mm. wide. *C. americana* Cockerell [R]. Wet places in the mountains: Mont.—Colo.—Utah—Wash.—B.C. *Plain—Mont.* Je–Jl.

2. **B. vulgàris** R. Br. Biennial; stem 3–6 dm. high; lower leaves petioled, pinnatifid; leaflets 5–9, oval or obovate, the terminal one much larger than the lateral ones; upper leaves sessile or nearly so, cut-toothed or lyrate-pinnatifid at the base; pods about 2.5 cm. long, obscurely 4-angled. *C. Barbarea* (L.) W. F. Wight. *B. Barbarea* Britton [B]. Waste places: Lab.—Va.—Neb.—N.D.; nat. from Eur. Ap–Je.

3. **B. stricta** Andrz. Biennial; stem erect, 3–6 dm. high, glabrous; petals light yellow; pods about 2.5 cm. long, stout, erect and appressed against the rachis. *C. stricta* W. F. Wight [R]. Waste places: Que.—Sask.—Wyo.—Fla.; Wash.—Ida.—Calif.; nat. from Eurasia. Ap–Je.

4. **B. vérna** (Mill.) Aschers. Biennial; stem branched, 3–6 dm. high; leaves with rounded or elliptic divisions; petals yellow, about 5 mm. long; fruiting pedicels ascending, stout, angled; pods 5–6 cm. long, straight or nearly so, ascending, sharply angled. *B. praecox* (Smith) R. Br. [B]. *C. verna* Heller [R]. Waste places: N.Y.—Fla.—Colo.; Calif.—Wash.; adv. from Eur. My–Je.

28. **CHORÍSPORA** DC.

Annual branched herbs. Leaves entire to pinnatifid. Flowers racemose, pedicelled, bractless. Petals clawed. Sepals erect, the lateral ones gibbous at the base. Corolla yellow or purple. Stamens free, not toothed. Pod elongate, terete, long-beaked, breaking up into one-sided, indehiscent joints. Seeds pendulous, alternately 2-serial, smooth; cotyledons accumbent, oblique.

1. **C. tenélla** DC. Sparsely glandular annual, 2–5 dm. high, branched; lower leaves runcinately lobed, the rest oblong-lanceolate, dentate, 4–10 cm. long; flowers small, purplish; pod linear, 3–4 cm. long, 2 mm. wide, tapering into an empty beak 1–1.5 cm. long, the internodes 2 mm. long. Ruderal weed, nat. of sw Asia; established at Arcadia, Neb.—w Iowa.

29. **RÁPHANUS** L. Radish.

Leafy annuals or biennials. Flowers perfect, in racemes. Sepals erect, the lateral ones somewhat saccate at the base. Petals large, white or light purple. Pod terete, tapering into a distinct beak, one-celled, few-seeded, filled with a spongy tissue between the seeds. Seeds 2-serial, globular, pendulous. Cotyledons conduplicate. Stamens unappendaged.

1. **R. satìvus** L. Annual or biennial; stem branched, 4–5 dm. high, glabrous or nearly so; lower leaves lyrate-pinnatifid with rounded, crenate divisions, more or less hairy, the uppermost lanceolate; petals 15–20 mm. long, with long claws, white and tinged with yellow; pod 4–5 cm. long; body 7–8 mm. thick, gradually tapering into a long beak. Waste places: Que.—B.C.—Calif.—Fla.; W.Ind.; escaped from cultivation. F–N.

30. **RAPHANÍSTRUM** (Tourn.) All. Jointed Charlock.

Biennial herbs. Flowers perfect, in elongate racemes. Sepals erect, the lateral ones saccate at the base. Petals large, yellow. Stamens unappendaged. Pod terete, transversely divided into several 1-seeded internodes. Seeds 1-serial, globular, pendulous. Cotyledons conduplicate. *Raphanus* L., in part.

1. **R. innócuum** Moench. Annual; stem 3–7 dm. high, sparsely pubescent or glabrous; basal leaves lyrate-pinnatifid, with 9–13 crenate or dentate divisions, rough; petals 10–15 mm. long, purple-veined; pod 2.5–4 cm. long, long-beaked. *Raphanus Raphanistrum* L. [G, B]. Waste places: Newf.—Pa.—N.D.; nat. from Eur.

31. **CAKÌLE** (Tourn.) Ludwig. Sea Rocket.

Glabrous, fleshy annual. Leaves dentate or lobed. Flowers purplish. Pod short, fleshy, indehiscent, of two internodes, each internode 1-celled, 1-seeded, or the lower one seedless; style none. Seed in the upper internode erect, that in the lower suspended.

1. **C. edéntula** (Bigel.) Hook. Stem bushy-branched, up to 3 dm. high; lower leaves obovate, sinuate and dentate, 7–13 cm. long, the upper ovate; petals light purple, twice as long as the sepals; pod 1–2 cm. long, the upper internode ovoid, flattish, longer than the lower, which is obovoid, turgid. *C. americana* Nutt. In sand, along the sea-shore: Newf.—Fla.; and along the Great Lakes: N.Y.—Minn. Jl–Au.

32. **SINÀPIS** L. Mustard, Charlock.

Annual or biennial, caulescent herbs, more or less hirsute. Flowers perfect, in racemes or panicles. Leaves alternate, runcinate-pinnatifid or lobed. Sepals more or less spreading, equal or the outer slightly saccate at the base. Petals yellow, longer than the sepals, clawed. Filaments not appendaged. Pods elongate, nearly terete, more or less constricted between the seeds, the upper portion, containing one seed, produced into a broad and sword-shaped or more or less 4-angled beak. Seeds in one row in each cell, globose, marginless and wingless. Cotyledons conduplicate.

Beak sword-shaped, constituting more than half the length of the pod. 1. *S. alba.*
Beak somewhat 4-angled, but flattened and 2-edged, constituting about one-third the length of the pod. 2. *S. arvensis.*

1. **S. álba** L. *Fig. 249.* Annual; stem 3–6 dm. high, more or less hispid; leaves more or less hirsute, the lower pinnatifid, with rounded toothed divisions, the uppermost often entire; petals yellow, about 1 cm. long; pedicels in fruit spreading; pod densely hispid, about 3 cm. long. *Brassica alba* Boiss. [G]. Waste places and fields: Me.—Fla.—Calif.—B.C.; nat. from Eur. *Temp.—Plain—Submont.* Mr–Au.

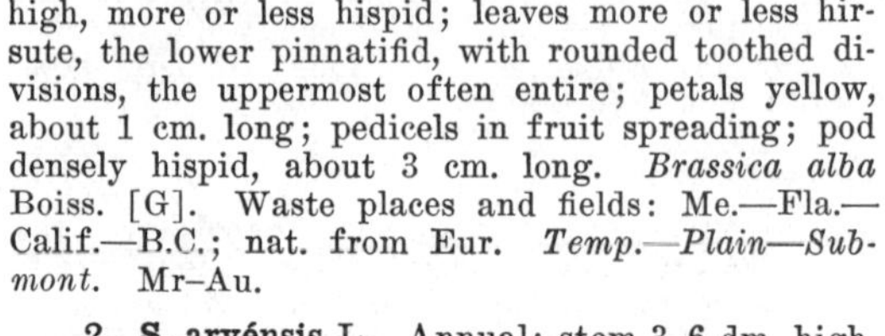

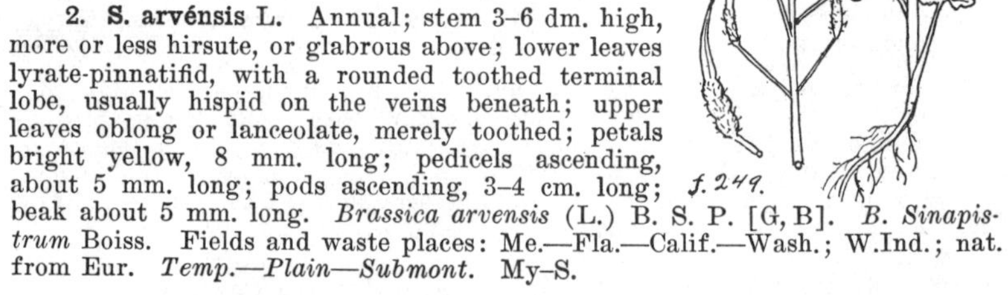

f. 249.

2. **S. arvénsis** L. Annual; stem 3–6 dm. high, more or less hirsute, or glabrous above; lower leaves lyrate-pinnatifid, with a rounded toothed terminal lobe, usually hispid on the veins beneath; upper leaves oblong or lanceolate, merely toothed; petals bright yellow, 8 mm. long; pedicels ascending, about 5 mm. long; pods ascending, 3–4 cm. long; beak about 5 mm. long. *Brassica arvensis* (L.) B. S. P. [G, B]. *B. Sinapistrum* Boiss. Fields and waste places: Me.—Fla.—Calif.—Wash.; W.Ind.; nat. from Eur. *Temp.—Plain—Submont.* My–S.

33. **ERÙCA** (Tourn.) Mill. GARDEN ROCKET.

Annual or biennial, branching herbs. Leaves pinnately lobed or toothed. Flowers perfect, large, racemose. Petals yellowish or purplish, with brown or violet veins. Pod linear, with a long sword-shaped beak; valves with a single strong and several faint nerves; seeds in two rows.

1. **E. satìva** Lam. Annual; stem glabrous, 3–4 dm. high; lower leaves pinnatifid or lobed, the upper often merely dentate; petals strongly veined; pod 1.5 cm. long or more, erect on erect pedicel. *E. Eruca* (L.) Britton. Waste places: Ont.—Pa.—Mo.—Tex.—N.M.—Mont.—Wash.; adv. from Eur. My–O.

34. **ERUCÁSTRUM** Presl.

Annual or perennial pubescent herbs. Leaves alternate, more or less divided or lobed. Flowers racemose, perfect. Sepals equal, oblong. Petals yellow, clawed. Pod elongate, cylindric, linear, the valves convex, 1-nerved; beak short, conic, gradually attenuate; seeds oval, slightly compressed, in a single row.

1. **E. Pollíchii** Spenner. Annual; stem 2–5 dm. high, pubescent; leaves pinnatifid, the cauline ones with oblong toothed divisions, decurrent, glabrous or pubescent on the veins only; petals pale yellow, small, spatulate; pod 3 cm. long, acute at each end, ascending, somewhat torulose. Fields and waste places: Mo.—S.D.—N.D.; adv. from Eur. Ap–O.

35. **BRÁSSICA** L. CABBAGE, RAPE, TURNIP, BLACK MUSTARD.

Annual, biennial, or perennial herbs, caulescent, with alternate leaves. Flowers perfect, in elongate racemes or panicles. Sepals ascending, equal or the outer somewhat saccate at the base. Petals yellow or white, clawed. Filaments free, unappendaged. Pods elongate, linear, terete or somewhat 4-angled, with an elongate-conic seedless beak, the valves 1–3-nerved; stigma truncate or

2-lobed. Seeds in one row in each cell, subglobose, marginless. Cotyledons conduplicate.

None of the leaves clasping.
Pod slender, 4-angled, erect, appressed to the stem. 1. *B. nigra.*
Pod ascending, terete, but with a strong midvein.
Leaves flat, the lower lyrate, with dentate terminal lobe. 2. *B. juncea.*
Leaves crisp, the terminal lobe laciniate or lobed. 3. *B. japonica.*
Upper stem-leaves clasping.
Leaves glaucous, all glabrous except the very earliest ones; petals cream-colored, long-clawed. 4. *B. campestris.*
Leaves not glaucous, the basal ones distinctly hairy; petals bright yellow, short-clawed. 5. *B. Rapa.*

1. **B. nìgra** (L.) Koch. *Fig. 250.* Annual; stem 5–20 dm. high, branching, glabrous or nearly so; lower leaves pinnatifid, with a large rounded or oval terminal lobe and a few small lateral ones; uppermost leaves lanceolate or oblong, entire; sepals yellowish; petals bright yellow, spatulate, 7–8 mm. long; pod 1.5–2 cm. long, a little over 1 mm. thick; beak slender. Black Mustard. Waste places and fields: Me.—Fla.—Calif.—B.C.; W.Ind.; nat. from Eur. *Temp.—Plain—Submont.* Ap–S.

f. 250.

2. **B. júncea** (L.) Cosson. Annual or biennial (?); stem erect, 3–12 dm. high, branched, glabrous or nearly so; lower leaves 1–1.5 dm. long, lyrate-pinnatifid, with a large oval dentate end-lobe, the upper lanceolate or linear; petals yellow, 8–10 mm. long, broadly spatulate; pod 3.5–5 cm. long, 2–3 mm. thick; beak 5–8 mm. long. Fields and waste places: N.S.—Va.—N.M.—Sask.; adv. or nat. from Asia. Jl–S.

3. **B. japónica** Siebold. Annual; basal leaves numerous, oblong or oblong-obovate, green and glaucous, the margins crisp and cut; stem-leaves petioled; petals small, yellow; pod small, with a slender beak, terete or nearly so, on spreading pedicel. Waste places; escaped from cultivation; nat. of Asia. Jl–S.

4. **B. campéstris** L. Annual or biennial; stem 3–10 dm. high, glabrous; basal leaves lyrate-pinnatifid, with a large toothed end-lobe; stem-leaves lanceolate, entire; petals cream-colored, spatulate, 6–7 mm. long; pod 5–7 cm. long, about 3 mm. thick; beak about 1 cm. long. Rape, Rutabaga, Swedish Turnip. Fields and waste places: N.S.—Iowa—Colo.—Mont.; escaped from cultivation. Ap–O.

5. **B. Ràpa** L. Annual or biennial, closely resembling the preceding, but greener and more hairy, and with smaller and bright yellow flowers. Turnip. Waste places: west to Wyo. and Mont.; escaped from cultivation. Ap–O.

36. DIPLOTÁXIS DC.

Annual or perennial glabrous herbs. Leaves mostly sinuate-pinnatifid. Flowers yellow, lilac or white. Sepals equal at the base. Stigma entire or emarginate. Pod linear, compressed; valves slightly convexed, 1-nerved; beak short, conic. Seeds slightly compressed in 2 rows.

1. **D. muràlis** DC. Annual or biennial; stem 2–5 dm. high, ascending or arched; leaves mostly basal, oblanceolate, petioled, sinuate-dentate or pinnatifid; flowers yellow, odorous; petals twice as long as the calyx, spatulate; pod about 2.5 cm. long, 2 mm. wide. Waste places: N.S.—Pa.—S.D.; adv. from Eur. Je–Au.

37. **ALÝSSUM** L. ALYSSUM, SWEET ALYSSUM.

Densely stellate annuals or perennials, with alternate leaves. Flowers perfect in racemes. Sepals ovate or oblong, more or less spreading. Petals whitish, obovate or cuneate to linear. Stamens 6; filaments commonly dilated at the base, in our species not toothed. Fruit orbicular, 2-celled, dehiscent; valves convex. Cotyledons accumbent.

1. **A. alyssoìdes** (L.) Gouan. Branched annual; stems 1–3 dm. high; leaves linear-oblong or spatulate, 1–3 cm. long, densely stellate-canescent, entire; petals white, or at first yellowish, 3.5 mm. long; pod orbicular in outline, 3 mm. broad, notched at the apex. *A. calycinum* L. Fields and waste places: N.H.—N.J.—Utah—Calif.—B.C.; nat. from Eur. *Temp.—Plain—Submont.* My–Je.

38. **BERTERÒA** DC. HOARY ALYSSUM.

Annual or perennial herbs, stellate or canescent with forked hairs. Leaves alternate, entire. Flowers racemose, perfect. Sepals ascending in anthesis, equal. Petals white, notched. Stamens 6; filaments 2-toothed at the base. Pod oblong (in ours) to nearly globose, only slightly compressed. Seeds winged, several. Cotyledons accumbent.

1. **B. incàna** (L.) DC. Stem 3–6 dm. high, branched; leaves numerous; entire, lanceolate or lance-elliptic, pale green, or the lower oblanceolate, petioled; petals 2–3 mm. long, white; pod oblong, 6–8 mm. long, 2.5–3.5 mm. thick, canescent. Waste places and around dwellings: Me.—N.J.—Mo.—Mont.; adv. or nat. from Eu. Je–S.

39. **DENTÀRIA** L. TOOTHWORT, MILKMAIDS.

Erect scapose perennial herbs, with often thickened rootstocks. Leaves basal; blades palmately cleft or divided. Bracts leaf-like, subopposite or subverticillate. Flowers perfect, in terminal racemes or corymbs. Sepals equal at the base, erect or nearly so. Petals white or purplish, with slender claws and spreading blades. Pods elongate, linear, flattish or nearly terete, the valves with faint or no midnerve, elastically dehiscent from the base; style slender; stigma entire or rarely 2-lobed. Seeds in one row in each cell, somewhat flattened, but not margined. Cotyledons accumbent.

Leaves pedately 3–5-divided, the segments linear, lanceolate, or oblong, the margins lobed, cleft, or coarsely dentate.
  Segments narrowly linear, dentate. — 1. *D. furcata.*
  Segments lanceolate or oblong, more deeply divided. — 2. *D. laciniata.*
Leaves ternate, the segments rhombic, coarsely crenate. — 3. *D. diphylla.*

1. **D. furcàta** Small. Perennial, with a moniliform rootstock; stem scapiform, 1–2 dm. high; basal leaves long-petioled, with 3 linear, entire or dentate segments; stem-leaves 3, verticillate, with 3–5 divisions; petals white or pink, 12–14 mm. long; pod narrowly fusiform, 3 cm. long, with a slender beak. Rich woods: Ohio—Tenn.—Ala.—Iowa. *Allegh.* Mr–Ap.

2. **D. laciniàta** Muhl. Perennial, with a deep-seated moniliform rootstock; stem scapiform, 2–4 dm. high; basal leaves long-petioled, appearing after the flowers; blades 3–5-parted to near the base; divisions lanceolate or oblong, laciniate; stem-leaves 3, verticillate, short-petioled, similar; flowers pink or white; pod linear, 2–4 cm. long. Rich woods: Que.—Fla.—La.—Neb.—Minn. *E. Temp.* Ap–Je.

3. **D. diphýlla** Michx. Perennial, with an elongate rootstock; stem glabrous, 2–4 dm. high; basal leaves long-petioled, ternate; divisions rhombic-ovate, coarsely crenate; stem-leaves usually 2, opposite or nearly so, similar, short-petioled; flowers white; pod 2.5–3 cm. long, rarely maturing. Woods: N.S.—S.C.—Ky.—Minn. *E. Temp.* My.

40. **CARDÁMINE** L. Bitter Cress.

Annual or perennial herbs, with alternate, entire or pinnate leaves. Flowers perfect, in racemes or panicles. Sepals equal at the base, erect or ascending. Petals white or purple, obovate to narrowly spatulate. Pods elongate, many-seeded, narrow, flattened, the valves nerveless or nearly so, elastically dehiscent at maturity. Seeds in one row in each cell, not margined. Cotyledons accumbent.

Leaves all entire.
  Petals white; stem-leaves 5–8, scattered; stem glabrous. 1. *C. bulbosa.*
  Petals rose-purple; stem-leaves 2–6, the upper approximate; stem pubescent when young. 2. *C. Douglasii.*
Leaves, at least some of them, pinnatifid.
  Petals 5–8 mm. long. 3. *C. pratensis.*
  Petals 2–4 mm. long.
    Stem leafy; leaves glabrous; stamens 6.
      Leaflets of the upper leaves elliptic or oblong. 4. *C. pennsylvanica.*
      Leaflets of the upper leaves narrowly linear. 5. *C. parviflora.*
    Stem-leaves few; leaves pubescent above; stamens 4. 6. *C. hirsuta.*

1. **C. bulbòsa** (Schreber) B.S.P. Perennial with a tuberiferous base; stem 2–5 dm. high; basal leaves oval or nearly orbicular or ovate-cordate, 2–4 cm. long, angled or entire; stem-leaves ovate or oblong, dentate or entire; petals 7–12 mm. long; pod erect, 2.5 cm. long; beak 2–4 mm. long. Wet places: N.S.—Fla.—Tex.—Minn. *E. Temp.* Ap–Je.

2. **C. Douglásii** (Torr.) Britton. Perennial, with a slender tuberiferous rootstock; stem more or less pubescent, 1.5–4 dm. high; leaves simple, the lower blades cordate, 1–3 cm. broad, sinuate, the upper ovate, usually coarsely dentate; petals obovate, pink or purple, 8–12 mm. long; pod 2–3 cm. long; beak 3–4 mm. long. *C. rhomboidea purpurea* Torr. *C. purpurea* (Torr.) Britt.; not Cham. & Schlecht. Springy places: Que.—Md.—Mo.—Alta. *E. Boreal.* Ap–My.

3. **C. praténsis** L. Perennial, with a short rootstock; stem 2–5 dm. high, simple; leaves pinnatifid; divisions 7–15, those of the lower leaves rounded, those of the upper linear; petals thrice as long as the sepals; pod 2–3 cm. long, 2 mm. wide; beak short. Wet places: Lab.—N.J.—Minn.—Man.; Eur. Ap–My.

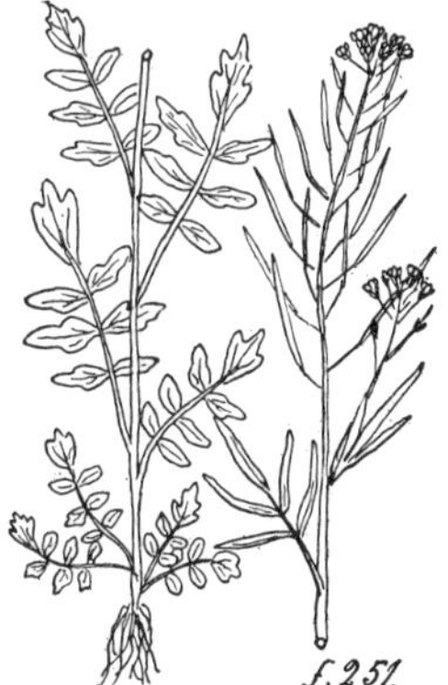
f. 251.

4. **C. pennsylvánica** Muhl. *Fig. 251.* Perennial, with a horizontal rootstock; stem glabrous or nearly so, branched, 2–8 dm. high; leaves with 5–17 oblong, oval, or obovate divisions, or the terminal one orbicular; pod erect or ascending, 2–3 cm. long, 1 mm. wide. Wet places: Newf.—Fla.—Colo.—Ore.—B.C. *Temp.—Plain—Mont.* Ap–Au.

5. **C. parviflòra** L. Stem very slender, glabrous or slightly pubescent, 0.5–3 dm. high; leaflets of the basal leaves oblong or the terminal one suborbicular; pod erect on ascending pedicel, 1.5–2.5 cm. long, less than 1 mm. wide; petals 1 mm. long. Rocky soil: Me.—w Ont.—Iowa—Kans.—Ga.; Eur. *Allegh.* Ap–My.

6. **C. hirsùta** L. Stem 1–2 dm. high; basal leaves many, 2–10 cm. long, pinnatifid, the terminal segment orbicular; stem-leaves few, with linear segments; pods linear, 2.5 cm. long, 1 mm. wide, on erect pedicels, 4–8 mm. long. Wet places: Pa.—N.C.—Neb.—Mich.; Eurasia. *E. Temp.* Mr–My.

41. **DRÀBA** L. Whitlow Grass, Draba.

Annual or perennial herbs, mostly low, often with stellate or branched pubescence. Flowers perfect, in racemes. Sepals equal at the base. Petals yellow or white, longer than the sepals. Pods elliptic to linear, flat; valves nerveless; stigma capitate. Seeds in 2 rows in each cell, neither margined nor winged. Cotyledons usually accumbent.

Annuals; style obsolete.
Petals deeply 2-cleft. 1. *D. verna.*
Petals entire, or nearly so.
Pod 6–15 mm. long.
Petals white or wanting.
Pod pubescent.
Pod linear, appressed-hirsute; leaves entire.
Inflorescence even in fruit corymbiform; petals minute or none. 2. *D. micrantha.*
Inflorescence in fruit elongate; petals conspicuous. 3. *D. coloradensis.*
Pod oblong or elliptic, with branched pubescence, often stellate; leaves usually toothed. 4. *D. cuneifolia.*
Pod glabrous. 5. *D. caroliniana.*
Petals yellow, sometimes turning whitish in age.
Pod hairy. 6. *D. nemorosa.*
Pod glabrous. 7. *D. lutea.*
Pod 2–5 mm. long; stem-leaves thick. 8. *D. brachycarpa.*
Perennials; style evident.
Plants acaulescent or nearly so; petals white.
Leaves finely stellate, spatulate or obovate. 9. *D. nivalis.*
Leaves hirsute-ciliate on the margins, oblanceolate. 10. *D. fladnizensis.*
Plants leafy-stemmed.
Petals white.
Pod densely stellate, linear or linear-oblong; petals about 3 mm. long; pedicels in fruit nearly erect. 11. *D. cana.*
Pod glabrous or sparingly and finely pubescent, broadly oblong or elliptic; petals about 5 mm. long; pedicels in fruit ascending.
Style in fruit 0.75–1 mm. long; pod usually twisted. 12. *D. arabisans.*
Style in fruit less than 0.5 mm. long; pod not twisted. 13. *D. hirta.*
Petals yellow.
Petals 5 mm. long or more, bright yellow. 14. *D. surculifera.*
Petals 3–4 mm. long, pale yellow. 15. *D. aureiformis.*

1. **D. vérna** L. Subacaulescent annual; scape 2.5–10 cm. long; leaves basal, oblong or oblanceolate, 1–2.5 cm. long, stellate-pubescent; racemes elongate in fruit; flowers cleistogamous; petals white, 2-cleft; pod oblong, glabrous, 6–8 mm. long, 2 mm. wide, shorter than the pedicels. Sandy soil and roadsides: Mass.—Va.—Neb.—Minn.; B.C.; nat. from Eur. Ap–My.

2. **D. micrántha** Nutt. Annual; stems 5–10 cm. high, branched at the base, more or less pubescent; leaves 5–15 mm. long, ovate or elliptic, entire, stellate; petals shorter than the hairy sepals, or none, sometimes in the earlier flowers 2.5–3 mm. long, 2-cleft at the apex; pedicels ascending; pod hispid, 6–12 mm. long, 4 mm. wide. *D. caroliniana micrantha* A. Gray [G, B]. Dry plains: Ill.—Tex.—Ariz.—Wash. *Temp.—Plain.* Mr–Au.

3. **D. coloradénsis** Rydb. Annual, leafy below, often branched, 5–10 cm. high; leaves obovate, entire, 1–2 cm. long, 0.5–1 cm. wide, subsessile, coarsely hirsute; petals white, 3–3.5 mm. long, spatulate, emarginate; pods linear-oblong, 10–12 mm. long and nearly 2 mm. wide. Plains and hillsides: N.M.—Colo.—Kans.—S.D. *Plain—Submont.* Ap–My.

4. **D. cuneifòlia** Nutt. Annual; stem 5–20 cm. high, branched; leaves obovate or cuneate, 1–4 cm. long, usually coarsely toothed above the middle, stellate; racemes elongate in fruit; petals white, 3–4 mm. long; pod 7–15 mm. long, 2 mm. wide, rarely glabrate. Grassy places and plains: Kans.—Ill.—Fla.—Calif.—Ida. *Temp.—Plain.* F–My.

5. **D. caroliniàna** Walt. Annual, leafy at the base, 3–13 cm. high; leaves oblong to obovate, pubescent, sessile, 1–2 cm. long, mostly entire; racemes short and corymbiform, 1–2.5 cm. long; pod linear, 8–12 mm. long,

glabrous, longer than the pedicel. Sandy places: Mass.—Ga.—Ark.—Neb.—Minn. *E. Temp.* Mr–My.

6. **D. nemoròsa** L. *Fig. 252.* Winter annuals; stem leafy, 0.5–3 dm. high, branched below; leaves often more or less dentate; petals light yellow, 2–3 mm. long; fruiting pedicels divaricate, 1–2 cm. long; pods oblong, 5–10 mm. long, 2 mm. wide. Hills, prairies, and plains: Mich.—Colo.—Nev.—Ore.—B.C.; Eurasia. *Temp.—Plain—Mont.* Ap–Au.

7. **D. lùtea** Gilib. Winter annual of the habit of *D. nemoralis;* stems 1–4 dm. high, branched; basal leaves obovate or oblanceolate, densely and coarsely stellate; stem-leaves ovate, usually coarsely toothed; petals yellow, or in age paler, 2–3 mm. long; fruiting pedicels 15–30 mm. long, ascending or spreading; pod oblong, 5–10 mm. long, 2 mm. wide. *D. nemorosa leiocarpa* Lindb. *D. dictyota* Greene. Hills and plains: Man.—Mich.—Colo.—Ore.—Alaska; Eurasia. *W. Temp.—Plain—Mont.* My–Jl.

8. **D. brachycárpa** Nutt. Annual; stem usually much branched, stellate; leaves ovate to lanceolate, entire, 1 cm. long or less, thick, finely stellate; petals yellow or whitish, 2 mm. long; pedicels 3–4 mm. long; pod oblong, 2–5 mm. long. Dry hills and fields: Va.—Ga.—La.—Colo.; Ore.—Mont. *Austral.—Plain.* Mr–My.

9. **D. nivàlis** Liljebl. Cespitose tufted perennial; leaves 5–8 mm. long, densely stellate; scape 1–5 cm. long; petals white, obovate, 3 mm. long; pedicels ascending; pod 6–8 mm. long, 2 mm. wide. Arctic-alpine situations: Greenl.—Lab.—Utah—Alaska; Eur. *Arct.—Subalp.—Alp.* Je–Au.

10. **D. fladnizénsis** Wulfen. Cespitose acaulescent perennial; leaves spatulate or oblanceolate, 1 cm. or less long, rosulate, hirsute-ciliate on the margins or nearly glabrous; peduncles 1–5 cm. long; petals white, 3 mm. long; pod oblong, glabrous, 4–5 mm. long, 2 mm. wide. Alpine-arctic situations: Greenl.—Que.—Colo.—Utah—B.C.; Eur. *Arct.—Subalp.—Alp.* Jl–Au.

11. **D. càna** Rydb. Perennial, with a taproot and short caudex, densely grayish stellate; stem 1–2 dm. high; basal leaves numerous, oblanceolate or spatulate, 1–1.5 cm. long, densely stellate; stem-leaves lanceolate to ovate, about 1 cm. long; pedicels short, nearly erect; petals white, about 3 mm. long; pod linear-oblong, 6–8 mm. long, densely pubescent. *D. valida* Goodding. Foothills and mountains: w Ont.—Alta.—N.M.—B.C. *Submont.—Mont.* Je–Jl.

12. **D. arábisans** Michx. Perennial, with a branched caudex; stems mostly simple, 1.5–4.5 dm. high, sparingly pubescent; basal leaves spatulate or oblanceolate, entire or denticulate, thin, sparingly stellate, 2–7 cm. long; stem-leaves serrate-dentate; pedicels ascending or divergent; pod elliptic, 1–1.5 cm. long, glabrous, strongly twisted; style 1 mm. long. *D. incana arabisans* S. Watson [B]. Rocky banks: Newf.—N.Y.—Minn.—Man. *Canad.* My–Jl.

13. **D. hírta** L. Perennial, with a branched caudex; stems 1–2 dm. high, glabrous or nearly so; basal leaves numerous, oblanceolate, entire or denticulate, 1–4 cm. long, minutely stellate; stem-leaves oblong to ovate, sessile, often dentate; pedicels ascending, in fruit about 5–10 mm. long; petals white, 5 mm. long, emarginate; pod elliptic or elliptic-ovate, acute, 6–12 mm. long. *D. Henneana* O. E. Schultz, in part; not Schlecht. Rocky banks: Greenl.—Newf.—Man.—Kew. *Arct.* Jl–Au.

14. D. surculífera A. Nelson. Cespitose perennial; stems 2–4 dm. high, stellate and villous; basal leaves 3–5 cm. long, oblanceolate, finely and rather sparingly stellate; petals 5 mm. long, golden yellow; pedicels 3–4 mm. long; pod lanceolate, 8–12 mm. long, finely stellate, straight or somewhat twisted. Mountains: Wyo.—S.D. *Submont.—Mont.* Jl–Au.

15. D. aureifórmis Rydb. A slender, cespitose, grayish stellate plant; stem strict, 2–4 dm. high; basal leaves spatulate, 2–3 cm. long, acute, entire; stem-leaves closely sessile, oblong-lanceolate; pedicels ascending; pod erect, linear-oblong to lanceolate, 10–15 mm. long, about 3 mm. wide, slightly if at all twisted, stellate. *D. Bakeri* Greene. Mountains: Utah—Colo.—S.D. *Mont.—Alp.* Je–Au.

42. ÁRABIS L. Rock Cress, Wild Candytuft.

Annual or perennial herbs, glabrous, or pubescent with branched hairs. Leaves alternate, in ours mostly toothed. Flowers perfect, in terminal or axillary racemes. Sepals equal or nearly so, sometimes saccate at the base. Petals white, rarely yellowish, pink, or purple, mostly clawed. Stamens 6; filaments free, unappendaged. Pods sessile, elongate, flat, the valves sometimes nerved, not elastically dehiscent; stigma capitate or 2-lobed. Seeds in 1–2 rows in each cell, marginless, margined, or winged. Cotyledons accumbent or nearly so.

Basal leaves pinnatifid.
  Stem-leaves also pinnatifid; seeds winged. — 1. *A. virginica.*
  Stem-leaves entire-margined or dentate; seeds wingless. — 2. *A. lyrata.*
Basal leaves entire or dentate, not pinnatifid.
  Pod erect or strongly ascending.
    Leaves coarsely hirsute; stem-leaves auricled at the base. — 3. *A. ovata.*
    Leaves not coarsely hirsute; pubescence, if any, consisting of 2- or 3-forked hairs.
      Stem 3 dm. high or more, solitary or 2 or 3 from the root.
        Pod about 1.5 mm. wide. — 4. *A. Drummondii.*
        Pod 3 mm. wide. — 5. *A. connexa.*
      Stems less than 3 dm. high, several from the branched caudex. — 6. *A. albertina.*
  Pod spreading or reflexed.
    Pod spreading, or ascending on a spreading pedicel.
      Plant more or less pubescent.
        Stem-leaves narrowly lanceolate or linear-lanceolate.
          Calyx more or less stellate. — 7. *A. Bourgovii.*
          Calyx glabrous.
            Basal leaves finely stellate; pod straight or slightly curved upwards.
              Pod 5–8 cm. long; stem-leaves conspicuously sagittate at the base. — 8. *A. confinis.*
              Pod 2–5 cm. long; stem-leaves not conspicuously sagittate at the base. — 9. *A. brachycarpa.*
            Basal leaves coarsely hirsute with branched hairs; pod downwardly curved. — 10. *A. Fendleri.*
        Stem-leaves broadly lanceolate, oblong or obovate.
          Style obsolete; petals scarcely exceeding the sepals; seeds not margined. — 11. *A. dentata.*
          Style 1 mm. long; petals much exceeding the sepals; seeds wing-margined. — 12. *A. patens.*
      Plant perfectly glabrous and glaucous; pod arcuate. — 13. *A. laevigata.*
    Pod pendulous on a recurved or reflexed pedicel.
      Plant not stellate.
        Plant low, 0.5–1.5 dm. high, decumbent; leaves small, entire. — 14. *A. arenicola.*
        Plant tall, 3–10 dm. high, erect; leaves mostly sinuate-dentate. — 15. *A. canadensis.*
      Plant stellate; stem 3–5 dm. high.
        Petals 6–10 mm. long; stem-leaves usually ample. — 16. *A. retrofracta.*
        Petals about 5 mm. long; stem-leaves narrow, serrulate. — 17. *A. lignipes.*

1. **A. virgínica** (L.) Trelease. Annual or biennial; sparingly pubescent or glabrous; stem branched at the base, 1–4 dm. high; leaves pinnatifid, 2–7 cm. long, the divisions linear or oblong, entire or 1- or 2-toothed; pedicels short, spreading; corolla small, white; pod linear, 2–2.5 cm. long, ascending; seeds in a single row, winged. *A. ludoviciana* C. A. Meyer. Open ground in sandy soil: Va.—Fla.—Tex.—Kans. *Austral.* Mr–My.

2. **A. lyràta** L. Stem 1–3 dm. high, glabrous or sparingly pubescent below; basal leaves 2–5 cm. long, sparingly hairy or glabrate; stem-leaves spatulate to linear, 1–2.5 cm. long; petals 6–8 mm. long, white; fruiting pedicels ascending or spreading, 6–8 mm. long; pods 1.5–3 cm. long. Rocky places: Conn.—S.C.—Mo.—Sask. *Canad.—Allegh.* Ap–Jl.

3. **A. ovàta** (Pursh) Poir. Stem simple or branched above; basal leaves oblanceolate or spatulate, 1–8 cm. long, slightly toothed or entire; stem-leaves lanceolate, cordately or sagittately clasping, usually sinuate-dentate, 2–8 cm. long; petals 4–5 mm. long; pedicels 5–7 mm. long; pods 3–4 mm. long, 1 mm. wide; style very short. *A. hirsuta* Hook., not Scop. [B]. Among rocks and in waste places: N.B.—Ga.—Calif.—Alaska. *Temp.—Mont.* My–Au.

4. **A. Drummóndii** A. Gray. *Fig. 253.* Stems simple and strict, 3–6 dm. high; basal leaves oblanceolate, 2–4 cm. long, glabrous or somewhat pubescent; stem-leaves lanceolate, sagittate-clasping, glabrous and glaucous; petals white or pink, 6–10 mm. long; pedicels 5–10 mm. long; pod 5–8 cm. long, 1.5–2 mm. wide; seeds winged, in two rows. *Streptanthus angustifolius* Nutt. *Turritis stricta* Grah. *Arabis philonipha* A. Nels. *A. oxyphylla* Greene. Mountains: Que.—Conn.—N.M.—Utah—Alta. *Boreal—Submont.—Subalp.* My–Au.

f. 253.

5. **A. connéxa** Greene. Closely resembling the preceding and perhaps not distinct, differing in the stronger and more branched caudex, spatulate basal leaves, shorter, less acuminate stem-leaves, and broader pod, which is 5–8 cm. long. Mountains: Que.—Alta.—Colo.—Utah. *Boreal—Mont.* Je–Jl.

6. **A. albertìna** Greene. Stem 1–2 dm. high, glabrous, simple; basal leaves oblanceolate, 1–2 cm. long; stem-leaves narrowly lanceolate, auriculate-clasping; inflorescence short; petals purple or rose-colored, about 8 mm. long; pedicels 5–10 mm. long, erect; pod 2–6 cm. long, 2 mm. wide. *A. Drummondii Wardii* A. Gray. Mountains: Alta.—S.D.—Utah—Wash. *Submont.—Mont.* Jl–Au.

7. **A. Bourgòvii** Rydb. Stem 3–6 dm. high; basal leaves oblanceolate, finely stellate, almost entire, 2–4 cm. long; stem-leaves lanceolate, sagittate at the base; petals pink, 8 mm. long; pedicels spreading, 5–7 mm. long; pods sometimes pendulous, 4–8 cm. long, 2 mm. wide; seeds more or less in two rows. *Turritis patula* Grah. *A. dacotica* Greene, small form. Meadows and hills: Man.—Wyo.—Ida.—Alta. *Boreal—Plain—Submont.* Ap–Jl.

8. **A. confìnis** S. Wats. Stem 3–10 dm. high; basal leaves lanceolate or obovate, 2–5 cm. long, usually sharply serrate, finely stellate-pubescent; stem-leaves mostly glabrous, pale, strongly ascending, 2–4 cm. long; pedicels 5–12 mm. long, ascending-spreading, 5–8 cm. long, glabrous; pod more or less spreading, 5–8 cm. long, 2 mm. wide; seeds winged. River banks: Man.—Minn.—Ill.—Ont. *Canad.* Je–Jl.

9. **A. brachycárpa** (T. & G.) Britton. Stem 4–6 dm. high; basal leaves oblanceolate, usually dentate, 2–5 cm. long; stem-leaves lanceolate, sagittate at

the base, glabrous; petals pink, about 6 mm. long; pedicels about 5 mm. long, spreading or ascending; pods usually spreading, 2–5 cm. long, 2 mm. wide; seeds in two rows. Dry ledges: Que.—Mich.—Colo.—Utah—Alta. *Boreal.—Plain.* My–Au.

10. **A. Féndleri** (S. Wats.) Greene. Biennial or short-lived perennial; stem 2–5 dm. high; basal leaves oblanceolate, 1–3 cm. long; stem-leaves glabrous, linear-lanceolate; petals 6–7 mm. long, pink or whitish; pedicels 1–2 cm. long, ascending to somewhat recurved; pods somewhat arcuate, 4–6 cm. long, 2 mm. wide; seeds 2-rowed. Mountains: N.M.—S.D.—Wyo. *Submont.—Mont.* My–Je.

11. **A. dentàta** T. & G. Biennial; stem erect, 2–6 dm. high, pubescent; basal leaves spatulate, 5–12 cm. long, dentate, short-petioled; stem-leaves ample, clasping, dentate, oblanceolate or oblong, auricled at the base, stellate-pubescent; pedicels 1–3 mm. long, finally spreading; petals 3 mm. long, white; pod 2–2.5 cm. long, straight, spreading; seeds in a single row, marginless. Moist soil: N.Y.—Va.—Neb.—Minn. *Allegh.* My–Je.

12. **A. pàtens** Sulliv. Biennial; stem more or less hirsute, 3–7 dm. high; basal leaves oblanceolate or spatulate, with winged petioles; stem-leaves oblanceolate to ovate or broadly oblanceolate, sessile, serrate; pedicels spreading, 1–2.5 cm. long; petals white, 8–9 mm. long; pod narrowly linear, spreading or ascending-spreading, 2.5–4 cm. long; seeds in a single row, narrowly winged. River banks: Pa.—Ala.—Mo.—Minn. *Allegh.* Ap–My.

13. **A. laevigàta** (Muhl.) Poir. Biennial, glabrous and glaucous; stem erect, 3–10 dm. high; basal leaves spatulate, dentate; stem-leaves 3–15 cm. long, oblong, lanceolate, or linear, entire or toothed, clasping and auricled at the base; pedicels ascending or spreading, 8–10 mm. long; petals about 1 cm. long; pod linear, 6–10 cm. long, arcuate-recurved; seeds in a single row, broadly winged. Rocky places: Que.—Me.—Ga.—Ark.—S.D. *E. Temp.*

14. **A. arenícola** (Richardson) Gelert. Stem glabrous, or pubescent below, ascending, 7–15 cm. long; leaves spatulate or oblong, nearly entire, 8–16 mm. long, 2–4 mm. wide, the lower petioled, the upper sessile; corolla purplish or white, 6 mm. broad; pod at last drooping, 1.5–2.5 cm. long, 1 mm. wide; seeds oblong, wingless. Arctic regions: Greenl.—Lab.—Man.—Mack. *Arct.—Subarct.* Jl.

15. **A. canadénsis** L. Annual or biennial, more or less pubescent; stem erect, 2–10 dm. high; lower leaves short-petioled, toothed or lyrate-pinnatifid; upper leaves 3–12 cm. long, lanceolate or oblanceolate, entire or nearly so, acute; pedicels spreading, at last recurved, 5–12 mm. long; petals greenish white, twice as long as the sepals; pod narrowly linear, falcate, 5–7 cm. long, drooping; seeds in one row, winged. Rocky woods: Vt.—Ont.—Minn.—Tex.—Ga. *E. Temp.* Je–Au.

16. **A. retrofrácta** Grah. Biennial or perennial; stems 3–5 dm. high, usually stellate throughout; basal leaves oblanceolate, 1–3 cm. long, densely stellate, usually entire; stem-leaves lanceolate, sagittate at the base; sepals stellate; petals 6–8 mm. long, white or pinkish; pedicels 1 cm. long, abruptly reflexed; pods 3–6 cm. long, 2 mm. wide. *A. Holboellii* Am. auth. [G, B]. *A. Kochii* Blankinship. (?) *A. demissa* Greene. Plains, hills, and mountains: Mack.—Man.—Neb.—Utah—Calif.—B.C. *Plain—Mont.* My–Jl.

17. **A. lígnipes** A. Nels. Short-lived perennial; stems 3–5 dm. high, finely stellate below, glabrous above; basal leaves entire, narrowly oblanceolate, densely stellate-pubescent, 1–2 cm. long; stem-leaves almost linear, auriculate, revolute; sepals stellate; petals 5–6 mm. long, white or purplish; pedicels abruptly reflexed; pod 6–8 cm. long, 2 mm. wide. Sandy or stony ground: S.D.—Wyo.—Mont.—Ida. *Submont.—Mont.* Je–Jl.

43. SELÈNIA Nutt.

Glabrous tufted annuals, with pinnatisect leaves. Flowers in leafy-bracted racemes. Sepals spreading. Petals narrow, with 10 glands at their bases. Pod flat, stalked, narrowed at each end; valves ribless, veined; style long, slender. Seeds flat, broadly winged.

1. S. aùrea Nutt. Stems 0.5–2 dm. high; basal leaves 2–5 cm. long, once or twice pinnatifid, with many oblong dentate segments; stem-leaves and bracts similar, but smaller; flowers 6–8 mm. high; pod 1–2 cm. long, 4–6 mm. broad. Open places: Mo.—Tex.—Kans.—Iowa. Mr–Ap.

44. PLEUROPHRÁGMA Rydb.

Tall glabrous perennials or biennials, with paniculate inflorescence and thick entire leaves. Basal leaves oblanceolate or spatulate; stem-leaves linear-lanceolate, sessile, but not clasping. Sepals ascending, thin, more or less petaloid. Petals white or purplish, with slender claws. Filaments subulate, white, with conspicuous glands at their bases; anthers linear, sagittate at the base, curved. Receptacle dilated. Pod slender, terete, torulose; stipe short; style short, slender; stigma minute, entire; septum with a strong midrib.

1. P. lilacìnum (Greene) Rydb. *Fig. 254.* Perennial, with a taproot; stem simple, strict or branched above, 6–20 dm. high; basal leaves 5–30 cm. long; stem-leaves 4–10 cm. long; inflorescence paniculate, at first corymbiform; pedicels in fruit spreading; petals 6–7 mm. long; pod 2.5–3 cm. long, 1–1.5 mm. thick. *Thelypodium integrifolium* Endl., in part [B]. *T. lilacinum* Greene. *P. integrifolium* Rydb. [R]. Dry plains: N.D.—Neb.—N.M.—Utah—Mont. *Plain—Submont.* Jl–Au.

45. STANLÈYA Nutt. PRINCE'S PLUME.

Leafy perennials, usually more or less glaucous. Flowers perfect, in elongate, many-flowered racemes. Sepals oblong, widely spreading in anthesis. Petals with long claw and narrow elongate blades. Stamens 6; filaments elongate, spreading; anthers linear, curved and spirally coiled. Pods long-stipitate, elongate-linear, terete; valves 1-nerved. Stigma sessile, small, simple. Seeds in one row, oblong; cotyledons incumbent.

Blades of the petals one third to two thirds as long as the claws; lower leaves pinnatifid, with lanceolate or oblong lobes.	
Pod arcuate, not tortuose.	
Leaves sparingly hirsute.	1. *S. pinnata.*
Leaves glabrous.	2. *S. bipinnata.*
Pod decidedly tortuose.	3. *S. glauca.*
Blades of the petals as long as the claws; leaves broadly lanceolate, the lower with short broad lobes.	4. *S. integrifolia.*

1. S. pinnàta (Pursh) Britton. Stem sparingly hairy or glabrate, 3–10 dm. high, terete; leaves thick, commonly pinnatifid, with oblong or lanceolate divisions, or the upper entire, more or less pubescent; petals 10–15 mm. long; pod 6–7 cm. long, spreading; stipe 10–15 mm. long, twice as long as the pedicels. *S. pinnatifida* Nutt. Plains: S.D.—Neb.—Utah. *Plain—Submont.* My–Au.

2. **S. bipinnàta** Greene. Stem flexuose, 2–6 dm. high, terete, glabrous; leaves except the uppermost pinnatifid, with oblong or linear, often lobed divisions; petals yellow, 8–10 mm. long; pod 4–5 cm. long, torulose and tortuose; stipe 7–8 mm. long, about equaling the pedicel. Plains and dry draws: Wyo. — Colo. — Kans. *Plain — Submont.* Jl–Au.

3. **S. glaùca** Rydb. Stem 4–6 dm. high, bluish green; lower leaves 1–1.5 dm. long, more or less pinnatifid, with linear-lanceolate or oblong lobes, the terminal usually much longer than the rest; upper leaves usually entire, linear-lanceolate; petals lemon-yellow, about 1 cm. long; claws pubescent; pedicels about 1 cm. long, more or less spreading; stipes about 1.5 cm. long; pod proper about 5 cm. long and 1.5 mm. thick, somewhat arcuate. Dry hills and table-lands: Colo.—Kans. —N.D.—Wyo. *Plain—Submont.*

4. **S. integrifòlia** James. *Fig. 255.* Stem glabrous, 3–6 dm. high; leaves 5–10 cm. long, thick, glabrous, the upper entire, the lower with triangular to oblong lobes; petals sulphur-yellow, about 12 mm. long; pod 5–6 cm. long; stipe 12–15 mm. long, about equaling the pedicel. Dry plains and hills: S.D.—Kans.—Colo.—Wyo. *Plain—Submont.* Je–Jl.

Family 67. **CAPPARIDACEAE.** Caper Family.

Herbs with alternate, digitately compound or rarely simple leaves. Flowers perfect, regular or irregular, usually in racemes. Sepals 4, distinct or united at the base. Petals 4, rarely more, clawed. Receptacle often thickened or produced between the stamens and the petals. Stamens 6 or more. Gynoecium of 2 united carpels, 1-celled, with 2 parietal placentae, sessile or stalked. Fruit in ours dehiscent, the valves separating from the frame-like placenta (replum).

Pods elongate, linear; receptacle with an appendage or gland.
  Appendages tubular; petals cuneate-flabelliform, laciniate, very unequal. — 1. Cristatella.
  Appendages solid; petals entire, emarginate or 3-toothed, but not laciniate.
    Stamens 12–32; capsule sessile or short-stipitate. — 2. Polanisia.
    Stamens 6; capsule long-stipitate. — 3. Peritoma.
Pods short, broader than long, more or less flattened contrary to the replum; receptacle without appendages. — 4. Cleomella.

1. **CRISTATÉLLA** Nutt.

Slender glandular annuals. Leaves alternate, without stipules, palmately trifoliolate, with narrow entire leaflets. Flowers perfect, small, white or yellowish, in terminal racemes. Sepals 4, united at the base, deciduous. Receptacle produced between the posterior petals and the ovary into a tubular appendage, as long as the smaller petals. Petals 4, more or less fan-shaped, clawed, laciniate at the apex, the posterior larger. Stamens 6–14, declined; filaments filiform, distinct; anthers sagittate. Capsule elongate, flattened. Seeds numerous.

1. **C. Jamèsii** T. & G. Erect annual; stem 1–4 dm. high, glandular-puberulent, branched above; leaflets 3, linear or linear-oblong, 8–25 mm. long; petals cream-colored, 3–4 mm. long; stipe about 4 mm. long; body of the capsule about 2 cm. long and 4 mm. wide, glandular, oblanceolate or oblong. Sandy soil: Iowa—Ark.—Colo.—S.D. *Plain.* Je–Au.

2. POLANÍSIA Raf. CLAMMY-WEED.

Glandular, clammy annuals. Leaves alternate, without stipules, palmately trifoliolate or sometimes simple. Flowers perfect, somewhat irregular, in terminal racemes. Receptacle inconspicuous, with a solid gland on the upper side. Sepals 4, imbricate, deciduous. Petals 4, imbricate, entire or notched at the apex, with slender claws. Stamens 12–32; filaments purple, more or less declined; anthers introrse. Capsule elongate, flattened, sessile, 2-valved at the apex. Seeds numerous. *Jacksonia* Raf.

Stamens about 11, barely exceeding the petals; capsule slightly stipitate. 1. *P. graveolens.*
Stamens 12–16, much exceeding the petals; pod sessile. 2. *P. trachysperma.*

1. **P. gravèolens** Raf. Stem branched, 1–8 dm. high; leaflets 3, oblong, 1.5–3.5 cm. long; petals white or pink, 4–5 mm. long; claws shorter than the blades; pod linear-oblong, 3–5 cm. long. Sandy soil: Que.—Md.—Tenn.—Neb.—S.D.—Man. *Canad. Allegh.* Je–Au.

2. **P. trachyspérma** T. & G. *Fig. 256.* Stem 3–8 dm. high, usually branched; leaflets 3, oblanceolate, elliptic, or oval, 2–5 cm. long; petals long-clawed, 8–12 mm. long; stamens about 16; filaments purple; pods sessile, linear, 4–5 cm. long, 6–7 mm. wide. *J. trachysperma* Greene [R]. Sandy soil and cañons: Sask.—Iowa—Mo.—Tex.—Ariz.—B.C. *W. Temp.—Plain—Submont.* Je–Au.

3. PERÍTOMA DC. BEE FLOWER, INDIAN PINK, STINK FLOWER.

Tall glabrous or pubescent annuals. Leaves palmately trifoliolate, petioled, alternate. Flowers perfect, in terminal racemes, nearly regular. Sepals 4, united below, valvate; calyx deciduous by circumscission at base. Receptacle with a solid appendage opposite the upper sepal. Petals 4, nearly equal, convolute, scarcely clawed. Stamens 6; filaments filiform, unequal, more or less declined; anthers spirally coiled. Capsule elongate, stalked, 1-celled, 2-valved. Seeds numerous, pendulous, conduplicate. *Cleome* L., in part.

Petals yellow. 1. *P. luteum.*
Petals purple, pink, or white. 2. *P. serrulatum.*

1. **P. lùteum** (Hook.) Raf. Stem 3–6 dm. high; leaflets 3–7, linear-oblanceolate to oblong, entire, 3–5 cm. long; petals yellow, oblanceolate, 6–8 mm. long; pedicels about 1 cm. long, spreading; stipe 1–2 cm. long; body of the pod 1–5 cm. long. *Cleome lutea* Hook. [G, B]. River bottoms and banks: Neb.—N.M.—Ariz.—Ore.—Wash. *W. Temp.—Plain—Submont.* My–Au.

2. **P. serrulàtum** (Pursh) DC. *Fig. 257.* Stem 6–15 dm. high; leaves 3-foliolate; leaflets lanceolate to obovate-oblong, entire or rarely denticulate, 3–10 cm. long, minutely pubescent or glabrous; petals fully 1 cm. long, purplish, rose-colored, or white; stipe of the pod 1.5–2 cm. long; body 2.5–5 cm. long, 4–6 mm. thick. *Cleome serrulata* Pursh [G, B]. *C. integrifolia* T. & G. Prairies, sand-draws, and waste places: Sask.—Kans.—N.M.—Ariz.—Ore. *Plain—Mont.* My–Au.

4. CLEOMÉLLA DC.

Glabrous annuals. Leaves alternate, without stipules, digitately trifoliolate, petioled. Flowers small, yellow, in terminal bracted racemes, nearly regular. Receptacle elongate, without appendage. Sepals 4, distinct, thin, imbricate, deciduous. Petals 4, entire, sessile, deciduous. Stamens 6; filaments filiform, equal, distinct. Capsule short, broader than long, inflated, with two helmet-like valves. Seeds usually 2 on each placenta, pendulous.

1. C. angustifòlia Torr. Stem glabrous, branched, 3–6 dm. high; leaflets acuminate or acute, 3–4 cm. long; sepals lanceolate, 1 mm. long; petals yellow, 5 mm. long, oblanceolate; pedicels ascending, 10–15 mm. long; stipes spreading, 5–6 mm. long; body of the pod rhombic, about 5 mm. long, 6 mm. broad, with rounded angles. River valleys and plains: Neb.—Tex.—Utah. *St. Plains—Plain—Son.* Jl–S.

Family 68. DROSERACEAE. SUNDEW FAMILY.

Mostly subacaulescent herbs, with glandular-viscid leaves, which are either filiform and circinate in the bud or with broadened sensitive blades. Flowers perfect, regular, racemose or spicate. Sepals 4–8, imbricate. Petals as many, convolute. Stamens 5–8, mostly 5. Gynoecium of 2–5 united carpels; ovary 1-celled, with 2–5 parietal placentae, rarely 2–5-celled, loculicidally 2–5-valved.

1. DRÓSERA L. SUNDEW.

Perennials, in ours scapose, with basal leaves. Sepals, petals, and stamens 4–8. Petals white or pink, spatulate or oblanceolate. Filaments subulate or filiform; anthers extrorse. Ovary 1-celled, superior; styles 2–5, distinct. Capsule 2–5-valved.

Leaf-blades suborbicular or broader than long. — 1. *D. rotundifolia.*
Leaf-blades elongate.
  Leaf-blades spatulate or oblanceolate.
    Stipules adnate; seeds appendiculate at the ends, fusiform. — 2. *D. longifolia.*
    Stipules free; seeds not appendiculate, oblong. — 3. *D. intermedia.*
  Leaf-blades linear or linear-oblanceolate; seeds muricate. — 4. *D. linearis.*

1. D. rotundifòlia L. *Fig. 258.* Rosulate perennial; leaf-blades 6–10 mm. wide; scape glabrous, 1–3 dm. high; raceme secund; calyx about 3 mm. long; lobes ovate-oblong; petals about 4 mm. long, white; capsule erect, about 5 mm. long; seeds smooth. Sphagnum bogs: Lab.—N.J.—Minn.—Ida.—Calif.—Alaska; Eurasia. *Boreal.—Mont.* Je–S.

f. 258.

2. D. longifòlia L. Rosulate perennial; leaf-blades spatulate or oblanceolate, 1.5–3 cm. long, about 4–5 mm. wide; scape 1–3 dm. high; calyx about 4 mm. long; lobes oblong; petals white, 5 mm. long; pods 7–8 mm. long. *D. anglica* Huds. [G]. Bogs: Newf.—Ont.—Ida.—Calif.—B.C.; Eurasia. *Boreal.—Mont.* My–Au.

3. D. intermèdia Hayne. Stem often evident below the rosette; leaf-blades spatulate, long-petioled; scape 2–20 cm. long; corolla white; seeds reddish-brown, papillose. Bogs: Newf.—Fla.—La.—Minn.; W.Ind.; Eur. *E. Temp.*

4. **D. lineàris** Goldie. Rosulate perennial; leaf-blades linear or narrowly linear-oblanceolate, 1.5–4 cm. long, 2–3 mm. wide; scape 3–10 cm. high, 1–4-flowered; calyx 3–4 mm. long, lobes ovate; petals white, 5–6 mm. long; capsule 5–6 mm. long. Bogs: Que.—Mich.—Minn.—Alta. *Boreal.* Jl–Au.

Family 69. SARRACENIACEAE. PITCHER-PLANT FAMILY.

Acaulescent bog plants, with tubular or pitcher-like leaves. Flowers perfect, on a long scape from the rootstock. Sepals 4 or 5, imbricate, persistent. Petals 5, imbricate, or none. Stamens many; anther versatile. Gynoecium of 3–5 united carpels; ovary 3–5-celled; style single; stigma peltate; ovules numerous.

1. SARRACÈNIA L. PITCHER-PLANT, SIDE-SADDLE FLOWER.

Leaves hollow, with a lateral wing and a terminal hood-like blade. Flowers nodding, solitary at the end of the scape. Sepals 5, subtended by 3 bractlets. Petals 5, oblong, incurved. Ovary 5-celled; style short; stigma umbrella-like. Fruit a loculicidal capsule.

f. 259.

1. **S. purpùrea** L. *Fig. 259.* Leaves pitcher-shaped, 1–3 dm. high, broadly winged, the hood heart-shaped; scape 3–6 cm. high; flowers globose, deep purple, 5 cm. broad; petals fiddle-shaped. Peat-bogs: Lab.—Fla.—Iowa—Mack. *E. Temp.* My–Je.

Family 70. PODOSTEMONACEAE. RIVER WEED FAMILY.

Small aquatics, with poorly differentiated stems and leaves, the latter more or less decompound. Flowers naked or at first included in a spathe. Perianth none, or consisting of a membranous 3–5-cleft calyx. Stamens 1–many, hypogynous; anthers 2-celled. Gynoecium of 2 or 3 united carpels; ovary 1–3-celled; styles 2 or 3, short; ovules numerous in each cavity. Capsule septicidal.

1. PODÓSTEMON Michx. RIVER WEED. THREAD-FOOT.

Aquatic herbs, with distichous filiform or dichotomously branched leaves. Flowers at first enclosed in a spathe, with rudimentary perianth. Stamens 2, filaments united to near the apex. Ovary 2-celled; stigmas 2, subulate. Capsule 2-valved, 6–10-ribbed.

1. **P. Ceratophýllum** Michx. Stem less than 2 dm. long, usually less than 5 cm. long; leaves usually dichotomously divided into filiform divisions; flowers solitary or scattered; capsule 2.5–3 mm. long, ellipsoid, stipitate. On rocks in shallow streams: Me.—Ga.—Ala.—Minn. *E. Temp.* Jl–S.

Family 71. CRASSULACEAE. STONE-CROP FAMILY.

Herbs or rarely shrubby plants, mostly fleshy or succulent, without stipules. Flowers mostly cymose, perfect, regular. Calyx persistent;

sepals 4 or 5, usually free, rarely united. Petals 4 or 5, free or more or less united, rarely wanting. Stamens as many or twice as many as the petals. Pistils as many as the sepals, distinct or united at the base. Fruit of follicles, dehiscent on their ventral suture.

Stamens twice as many as the sepals; land plants. 1. SEDUM.
Stamens as many as the sepals; minute annual mud or water plants. 2. TILLAEASTRUM.

1. SÈDUM L. STONE-CROP, ORPINE.

Annual or perennial, fleshy herbs. Leaves mostly alternate, often imbricate, entire or dentate. Flowers perfect, in terminal, often one-sided cymes. Sepals distinct or somewhat united, 4 or 5. Petals 4 or 5, distinct, or rarely slightly united at the base. Stamens 8–10, the alternate ones often adnate to the base of the petals. Carpels 4 or 5, distinct, or united at the very base, more or less spreading or at least ascending, usually many-seeded.

Annuals.
  Petals purple, pink, or white; follicles ascending. 1. *S. pulchellum.*
  Petals yellow; follicles divergent. 2. *S. Nuttallianum.*
Perennials.
  Petals yellow; flowers cymose; leaves terete or nearly so.
    Leaves linear. 3. *S. stenopetalum.*
    Leaves short, obovate, thick, imbricate. 4. *S. acre.*
  Petals purple; flowers corymbose; leaves flat. 5. *S. triphyllum.*

1. **S. pulchéllum** Michx. Glabrous; stem 1–3 dm. long, ascending or trailing; leaves crowded, terete, linear, obtuse, 6–25 mm. long; cyme 4–7-forked, with spreading or recurved branches; petals linear-lanceolate, twice as long as the sepals; follicles 4–6 mm. long. Rocks: Va.—Ga.—Tex.—Kans. *Carol.—Ozark.* My–Jl.

2. **S. Nuttalliànum** Raf. Glabrous; stem 5–8 cm. high; leaves alternate, linear-oblong, terete, obtuse, 4–15 mm. long; cyme 2–5-forked; petals lanceolate, slightly exceeding the sepals; follicles 2.5–3 mm. long. *S. Torreyi* Don. [B]. Dry ground: Mo.—Ark.—Tex.—Kans. *Ozark.* My.

3. **S. stenopétalum** Pursh. *Fig. 260.* Perennial, tufted; flowering stems 8–18 cm. high; leaves crowded, but scarcely imbricate, except on the sterile shoots, sessile, linear, 6–16 mm. long; cymes forked; petals yellow, narrowly lanceolate; follicles about 4 mm. long, ascending with divergent tips. *S. subalpinum* Blankinship. On rocks: Sask.—Neb.—N.M.—Calif.—Alta. *Plain—Mont.* Je–Au.

f. 260.

4. **S. àcre** L. Perennial; stems prostrate, tufted; floral branches erect, 2–8 cm. long; leaves sessile, alternate, imbricate, thick, ovate, 3 mm. long; flowers sessile, 8 mm. broad; petals yellow, lanceolate; follicles divergent. Rocky places and roadsides: N.B.—Va.—Minn.; escaped from cultivation; nat. of Eur. Je–Au.

5. **S. triphýllum** (Haw.) S. F. Gray. Perennial, stem stout, 3–5 dm. high; leaves alternate, ovate, obtuse, 2–5 cm. long, dentate; cyme dense, compound, corymbiform, 5–8 cm. broad; petals purple, twice as long as the sepals; follicles 4 mm. long. *S. Fabaria* Koch. *S. Telephium* Torr. [B]; not L. Fields and roadsides: Que.—Minn.—Iowa—Md.; escaped from cult.; Eur.

2. **TILLAEÁSTRUM** Britton.

Small glabrous aquatic annuals. Leaves opposite, entire. Flowers perfect, small, solitary, axillary. Sepals 4, distinct. Petals usually 4, distinct or united at the base. Stamens 4. Carpels 4, distinct, several-seeded; styles short.

1. **T. aquáticum** (L.) Britton. Stem 1–8 cm. high; leaves linear-oblong, connate at the base, 4–6 mm. long; flowers subsessile; petals greenish; about twice as long as the sepals; follicles ovoid, 8–10-seeded. *Tillaea aquatica* L. [B, G]. In mud or water: N.S.—Md.—La.—Tex.—L.Calif.—Wash.; Eurasia. *Temp.—Plain—Mont.*

Family 72. PENTHORACEAE. Ditch Stonecrop Family.

Perennial herbs, with rootstocks and alternate serrate leaves. Flowers perfect, in terminal cymes, secund on spreading branches. Sepals 5 or 6, greenish, distinct. Petals as many, inconspicuous, or wanting. Stamens twice as many as the sepals; anthers oblong. Gynoecium of 5 or 6 carpels, united to the middle; styles short, abruptly pointed; ovules numerous in each carpel, on axial placentae. Fruit depressed, of 5 or 6 follicles, inserted obliquely, each follicle circumscissile just above its union.

1. **PENTHÒRUM** (Gron.) L.
Ditch Stonecrop.

Characters of the family.

1. **P. sedoìdes** L. *Fig. 261.* Stem 1–6 dm. high; leaves shining, petioled; blades elliptic-lanceolate or elliptic, 3–15 cm. long, acuminate at each end, finely serrate; cymes 2- or 3-branched, the branches 2–8 cm. long, spreading; sepals triangular-ovate, light green; petals linear or linear-spatulate, or often wanting. Ditches and swamps: N.B.—Fla.—Tex.—Minn. *E. Temp.* Jl–S.

F. 261.

Family 73. PARNASSIACEAE. Grass of Parnassus Family.

Scapose perennials, with rootstocks and basal petioled entire leaves. Flowers solitary on the scape, perfect, regular. Sepals 5, imbricate. Petals 5, imbricate, white or nearly so, conspicuously veined. Stamens 5, alternate with the petals and with 5 clusters of united gland-tipped staminodia. Gynoecium of 3 or 4 united carpels; ovary superior or half inferior, with 3 or 4 parietal placentae; style obsolete; stigmas 3 or 4. Fruit a 1-celled capsule, loculicidal at the apex.

1. **PARNÁSSIA** L. Grass of Parnassus.

Characters of the family.

Staminodia 3–5 in each fascicle, nearly free, not forming an obovate scale. 1. *P. americana.*
Staminodia united below into an obovate scale fringed above by the free portion of the filaments.
  Staminodia 7–15 in each fascicle; basal leaf-blades cordate or rounded at the base.
    Petals nearly twice as long as the sepals; hypanthium inconspicuous; staminodia usually 9–15 in each fascicle. 2. *P. palustris.*

Petals only slightly exceeding the sepals; hypanthium conspicuous, obconic, fully half as long as the sepals and nearly as broad as high; staminodia 7–9 in each fascicle. 3. *P. montanensis.*
Staminodia 5–7 in each fascicle; basal leaf-blades acute at the base. 4. *P. parviflora.*

1. P. americàna Muhl. Leaf-blades broadly ovate to suborbicular, sometimes subcordate at the base, paler beneath, 7–9-ribbed, 1.5–6 cm. long; scape 2–4 dm. high, bearing the bracts on the lower part; sepals elliptic, 3-nerved, 4–5 mm. long; petals broadly ovate or oval, 10–18 mm. long, strongly 9-veined; staminodia usually 3 in each set, slightly shorter than the stamens. *P. caroliniana* Auth. [G]; not Michx. Swamps: N.B.—Va.—Iowa—S.D.—Man. *Canad.—Allegh.* Je–S.

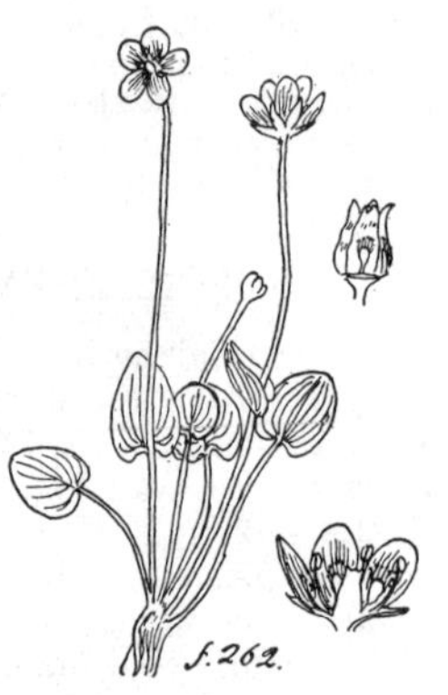

2. P. palústris L. *Fig. 262.* Leaf-blades cordate, 1–3 cm. wide; scape 1–3 dm. high; sepals oblong or elliptic, 4–6 mm. long; petals oval, 8–12 mm. long; capsule ovoid, about 1 cm. long. Wet places: Lab.—Que.—Mich.—Wyo.—Alaska; Eurasia. *Boreal.—Mont.* Jl–Au.

3. P. montanénsis Fern. & Rydb. Leaf-blades ovate, with a subcordate or rounded base, 1–2 cm. long, 8–18 mm. wide; scape about 2 dm. high; sepals oblong to lanceolate, acute, 7–9 mm. long; petals oval to elliptic; capsule rounded-ovoid, about 1 cm. long. Wet river banks: Mont.—B.C.—Sask. *Plain—Mont.* Jl–Au.

4. P. parviflòra DC. Leaf-blades oval or ovate, 1–3 cm. long, 0.75–2 cm. wide; scape slender, 1–3 dm. high; sepals oblong, 5–8 mm. long; petals elliptic or oval, 6–10 mm. long; capsule ovoid, about 1 cm. long. Wet places and swamps: Lab.—Que.—S.D.—Colo.—Utah—Alta. *E. Boreal.—Plain—Mont.* Jl–S.

Family 74. SAXIFRAGACEAE. Saxifrage Family.

Herbs, usually without stipules. Leaves alternate, rarely opposite, often all basal. Flowers perfect, regular or nearly so. Hypanthium often well developed, from flat to cylindric, often more or less adnate to the ovary. Sepals 5, rarely 4, distinct. Petals as many or wanting. Stamens as many or twice as many as the sepals, in one genus only 3. Gynoecium of 2, rarely of 3 or 4, carpels, more or less united, rarely wholly distinct; placentae parietal, basal, or axial. Fruit a capsule or follicles.

Placentae parietal, sometimes nearly basal.
  Flowers solitary and axillary to leaf-like bracts, or 2–4 in small corymbs, each subtended by a leaf-like bract; sepals 4; petals wanting. 1. Chrysosplenium.
  Flowers in more or less elongate racemes or panicles.
    Gynoecium of 2 or 3 equal or essentially equal carpels.
      Flower-stalk axial, from a slender bulbiferous rootstock; gynoecium 3-carpellary. 2. Lithophragma.
      Flower-stalk a lateral shoot from a stout scaly rootstock; gynoecium 2-carpellary.
        Inflorescence racemose; stamens 10; petals pinnately cleft or pinnatifid. 3. Mitella.
        Inflorescence paniculate; stamens 5; petals broadened upward. 4. Heuchera.
    Gynoecium of 2 very unequal carpels. 5. Tiarella.

Placentae axial.
Hypanthium well-developed and accrescent, at maturity longer than the sepals,
Stamens 5; sepals imbricate; petals marcescent; seeds winged.
6. SULLIVANTIA.
Stamens 10.
Petals clawed; styles partially united; plants with thick rootstocks.
7. TELESONIX.
Petals clawless; styles distinct; plants with slender rootstocks, often with offsets.
Plants without caudices, producing only annual flowering stems.
8. SAXIFRAGA.
Plants with perennial very leafy caudices, often with offsets, the flowering stems very different from the caudices; leaves of the caudex with serrate blades, each tooth with an encrusted pore.
11. CHONDROSEA.
Hypanthium only slightly developed, unchanged at maturity, or if slightly accrescent, then flat and plants acaulescent.
Leaves alternate, sometimes all basal.
Plants acaulescent; corolla essentially regular, the petals about equal in shape and length. 9. MICRANTHES.
Plants caulescent; leaf-blades entire or rarely with 3 apical tooth-like lobes. 10. LEPTASEA.
Leaves opposite, except sometimes on the flower-stalks.
12. ANTIPHYLLA.

1. CHRYSOSPLÈNIUM (Tourn.) L. GOLDEN SAXIFRAGE, WATER CARPET.

Low, somewhat succulent herbs, mainly semi-aquatic; all North American species with perennial rootstocks. Flowers small, axillary or terminal, solitary or in small corymbs. Hypanthium saucer-shaped or campanulate, adnate to the lower portion of the gynoecium and usually lined with a disk. Sepals normally 4. Petals none. Stamens 4–8 (rarely 10), inserted on the margin of the disk. Gynoecium 1-celled, 2-lobed; styles 2, usually short; placentae parietal, many-ovuled.

Leaves all alternate; flowers clustered near the ends of the branches.
1. *C. iowense.*
Lower leaves opposite; flowers solitary, axillary. 2. *C. americanum.*

1. C. iowénse Rydb. *Fig. 263.* Perennial, with a slender rootstock; stem 3–5 cm. high; leaf-blades reniform, 5–15 mm. wide, crenate, with 4–7 broad rounded teeth; stamens 6–8. Among moss: Iowa. My–Je.

2. C. americànum Schw. Perennial, with a slender, stoloniferous root-stock; stem decumbent, forked above, glabrous or nearly so, 7–20 cm. long; leaf-blades broadly obovate, orbicular or rarely reniform, 4–20 mm. wide. Wet shady places: N.S.—Ga.—Minn.—Sask. *E. Temp.—Plain.* Ap–Au.

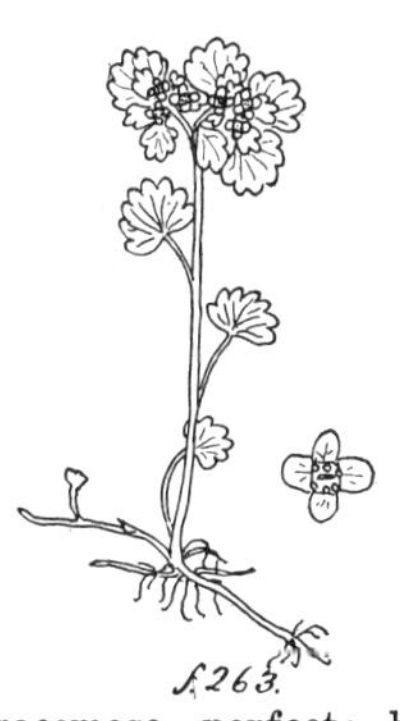
f. 263.

2. LITHOPHRÁGMA Nutt. STAR-FLOWER, PRAIRIE STAR, WOODLAND STAR.

Perennial herbs, with slender bulblet-bearing rootstock and axial leafy flowering shoots. Flowers racemose, perfect; hypanthium campanulate or hemispheric and adnate to the very base of the ovary, to elongate-turbinate and adnate to the lower half thereof. Sepals 5, valvate, rounded to triangular. Petals white or rose-colored, clawed, digitately or pinnately divided, toothed, or entire, much exceeding the sepals. Stamens 10, included; filaments short; anthers cordate. Gynoecium 1-celled with 3 parietal placentae, 3-valved at the apex; styles 3, short. Seeds many.

Hypanthium campanulate or hemispheric, mostly rounded at the base, adnate only to the base of the ovary; stem leaves usually bulbiferous in their axils. 1. *L. bulbiferum.*

Hypanthium deeply obconic, adnate to the lower half of the ovary; plant not bulbiferous. 2. *L. parviflorum.*

1. **L. bulbíferum** Rydb. *Fig. 264.* Stem 1–2 dm. high, glandular-puberulent; leaf-blades ternately divided to the base; divisions 0.5–1 cm. long, cuneate or obovate, 3-cleft and toothed; pedicels 3–5 mm. long, or in fruit 1 cm. long; hypanthium campanulate, together with the sepals 3–4 mm. long; petals 4–7 mm. long, 3–5-cleft, white or sometimes rose-colored. Dry hills: S.D.—Colo.—Calif.—B.C. *Submont.—Subalp.* Ap–Jl.

2. **L. parviflòrum** (Hook.) Nutt. Stem 1–3 dm. high, glandular-puberulent and somewhat hirsutulous; leaf-blades more or less hirsutulous, divided to the base into 3–5 divisions, which are 1–3 cm. long, twice ternately cleft into oblong or linear lobes; pedicels 2–4 mm. long; hypanthium together with the sepals 5–8 mm. long; petals deeply 3–5-cleft into narrowly oblong divisions. Rocky or gravelly places: Alta.—S.D.—Colo.—Calif.—B.C. *Submont.—Mont.* Ap–Jl.

3. MITÉLLA (Tourn.) L. Miterwort, Bishop's Cap.

Low perennials, with scaly rootstocks and lateral flowering branches. Inflorescence racemose. Hypanthium saucer-shaped or open-campanulate, adnate to the base of the gynoecium. Sepals triangular or ovate, valvate in aestivation. Petals 5, pectinately pinnatifid. Stamens 10, included; filaments short, subulate. Gynoecium mostly superior, 1-celled, with 2 parietal placentae, ovuliferous only at the base; styles distinct, short; stigma obtuse, simple. Capsule 2-valved at the summit, few-seeded.

Stem-leaves alternate, those of the flowering stems solitary or wanting. 1. *M. nuda.*

Stem-leaves opposite, sessile. 2. *M. diphylla.*

1. **M. nùda** L. *Fig. 265.* Perennial, producing long runners later in the season; flowering branches scapiform, 0.5–2 dm. high, sparingly hairy; leaf-blades reniform, 2–5 cm. in diameter, rounded-crenate or slightly round-lobed; calyx about 6 mm. wide; sepals yellowish green, triangular-ovate; petals about twice as long as the sepals. Cold woods and peat-bogs: Lab. — Newf. — B.C. — Mont.; e Asia. *Boreal.—Mont.* My–Jl.

2. **M. diphýlla** L. Perennial, with a branched rootstock; flowering branches 2–4.5 dm. high, puberulent; basal leaves petioled; blades broadly cordate, 3–5-lobed and dentate, 3–7 cm. long; stem-leaves sessile, ovate or cordate, 3-lobed, dentate; sepals whitish; divisions of the petals subulate. Woods and copses: Que.—N.C.—Mo.—Man. *E. Temp.* Ap–My.

4. **HEÙCHERA** L. ALUM-ROOT, KALISPELL.

Perennials, with scaly rootstocks and lateral flowering branches, which are either scapiform or leafy. Basal leaves petioled, with palmately veined, broad blades. Inflorescence paniculate. Hypanthium cylindric to saucer-shaped, most often campanulate, adnate to the lower part of the gynoecium, often more or less oblique. Sepals 5, often very unequal. Petals 5, often more or less clawed. Stamens 5, opposite to the sepals; filaments usually filiform. Gynoecium partly inferior, 1-celled, with 2 parietal, many-ovuled placentae; styles 2, distinct, elongate. Capsule opening between the two more or less divergent beaks.

Hypanthium short-campanulate; filaments twice as long as the sepals. 1. *H. americana.*
Hypanthium deeply campanulate, strongly oblique; filaments only slightly exceeding the sepals.
 Hypanthium with the sepals less than 1 cm. long; petioles long-hispid. 2. *H. hispida.*
 Hypanthium with the sepals more than 1 cm. long; petioles glabrous or sparingly short-hairy. 3. *H. Richardsonii.*

1. **H. americàna** L. Flowering branches scapiform, pubescent with short hairs or glabrous, 6–9 dm. high, naked; leaf-blades rounded-cordate, hirsute beneath, hirsutulous above, 9–11-lobed, with rounded lobes and broad mucronate teeth; petals spatulate, white, about equaling the sepals. Dry and rocky woods: Conn.—Ala.—La.—Minn. *E. Temp.* My–Au.

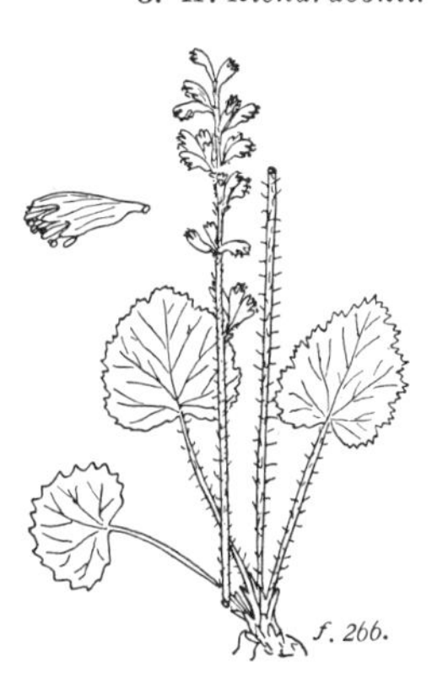
f. 266.

2. **H. híspida** Pursh. *Fig. 266.* Flowering branches scapiform, usually densely hispid, 6–12 dm. high; leaf-blades orbicular-cordate or reniform, 4–7 cm. wide, 5–9-lobed, with shallow rounded lobes and broad teeth, hirsute beneath and ciliate on the margins; petals spatulate, equaling or shorter than the sepals. *H. ciliata* Rydb. Woods and hillsides: Ont.—Va.—Kans.—Wyo.—Mont. *Canad.—Allegh.—Plain—Mont.* My–Jl.

3. **H. Richardsònii** R. Br. Flowering branches 3–4 dm. high, scapiform, sparingly short-hirsute and glandular-puberulent above; leaf-blades rounded-cordate, 3–6 cm. broad, with rounded lobes and broadly ovate teeth, sparingly hispidulous, ciliate on the margins; petals spatulate, purplish, slightly exceeding the oblong sepals. Hills: Man.—S.D.—B.C. *Plain—Submont.* Je.

5. **TIARÉLLA** L. FALSE MITERWORT.

Perennial herbs, with scaly rootstocks and lateral leafy flowering branches. Leaves petioled; blades cordate in outline. Stipules present. Inflorescence racemose or paniculate. Hypanthium small, short-campanulate, almost free from the gynoecium. Sepals 5, ovate or lanceolate. Petals clawed, with oblong or elliptic blades, or else clawless and filiform. Stamens 10, exserted with long filiform filaments. Gynoecium 1-celled, with 2 parietal nearly basal placentae. Capsule membranaceous, with 2 very unequal valves. Seeds few.

1. **T. cordifòlia** L. Acaulescent herb, with sarmentose runners later in the season; flowering branches 1–4 dm. high, with white retrorse hairs, sometimes with a single small leaf; leaf-blades broadly cordate in outline, 3–7-lobed, acute, sparingly hirsute, 4–10 cm. broad; lobes and teeth rounded, ovate; flowers in simple racemes; sepals white, oblong, 2 mm. long; petals white, oblong or oblanceolate, half longer than the sepals. Rich woods: N.S.—Ga.—Ala.—Minn. *E. Temp.* Ap–My.

6. SULLIVÁNTIA T. & G.

Perennial acaulescent slender herbs, with small rootstocks and scape-like stems. Leaves alternate, mostly basal, only one on the lower part of the stem; blades reniform to orbicular in outline, shallowly lobed and often coarsely toothed, cordate at the base, long-petioled. Flowers in panicled cymes. Hypanthium campanulate, longer than the calyx. Sepals 5, erect. Corolla white or whitish, regular. Petals 5, persistent, clawed. Stamens 5; filaments subulate. Ovary fully half-inferior, the carpels united up to the beaks. Follicles erect, well included in the drooping hypanthium. Seeds winged.

1. **S. renifòlia** Rosendahl. *Fig. 267.* Perennial, with a short rootstock; stem 1–2 dm. high, few-leaved; petioles of the basal leaves long, the blades reniform, 3–9 cm. wide, 7–13-lobed and dentate, glabrous, deeply cordate at the base; inflorescence cymose-paniculate, glandular-pubescent; flowers secund on the branches; sepals broadly ovate, 3-nerved; petals rhombic-ovate, about 2.5 mm. long. Shaded cliffs: Wis.—Minn.—Iowa—nw Ills. *Allegh.* Je–Au.

7. TELESÒNIX Raf.

Low glandular-pubescent perennials, with thick scaly rootstocks. Leaf-blades reniform, deeply and doubly crenate. Flowers in a contracted, leafy, somewhat secund panicle. Hypanthium turbinate-campanulate, adnate to the lower half of the ovary. Sepals 5, ovate-lanceolate. Petals 5, red or purple, with long claws. Stamens 10; filaments subulate. Ovary 3-celled; styles more or less united. Fruit dehiscent between the beaks, which are not divergent. Seeds numerous.

1. **T. heucheraefórmis** Rydb. *Fig. 268.* Cespitose perennial, with a scaly caudex; stem 1–2 dm. high, glandular-hirsute; basal leaf-blades round-reniform, deeply and doubly crenate, 2–3.5 cm. wide; hypanthium 6–7 mm. long, hirsute; petals dark violet, scarcely exceeding the sepals, the blades obovate-spatulate. Mountains: Alta. —B.C. — Wyo. — S.D. *Mont.—Subalp.* Jl–S.

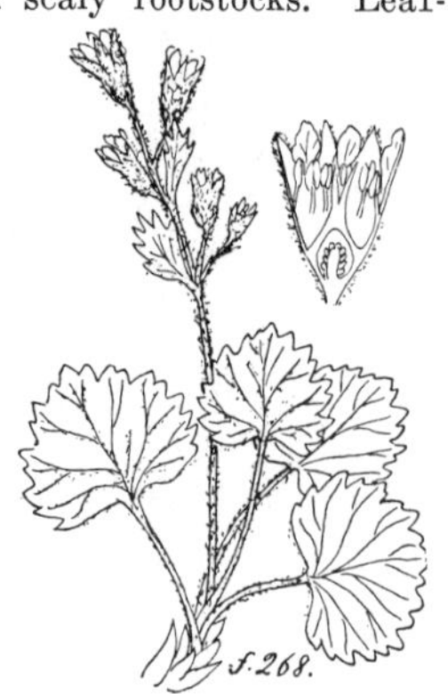

8. SAXÍFRAGA (Tourn.) L. SAXIFRAGE.

Perennial caulescent herbs, from a small rootstock. Leaves alternate, not crowded, but sometimes tufted at the base of the plants; blades more or less 3-lobed, or in diminutive plants nearly entire, mostly petioled. Flowers solitary or in terminal cymes. Hypanthium turbinate or campanulate, longer than the calyx, at least at maturity. Sepals 5, erect, usually with a terminal gland. Petals 5, white, somewhat narrowed at the base, but usually clawless. Stamens 10; filaments subulate. Ovary about half-inferior, the carpels united to above the middle. Follicles well united, erect except the more or less spreading tips.

Lobes of the stem-leaves linear to triangular-lanceolate; petals cuneate. 1. *S. cernua.*

Lobes of the stem-leaves broad and rounded, as broad as long or broader; petals fiddle-shaped. 2. *S. simulata.*

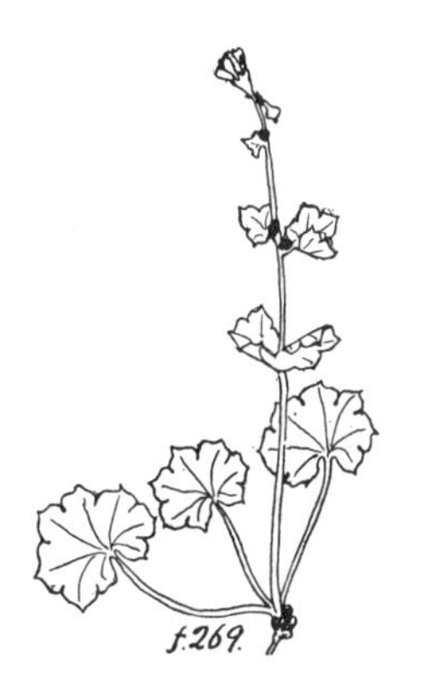

1. **S. cérnua** L. Stems 8–19 cm. tall, somewhat glandular-pubescent; leaf-blades reniform in outline, 9–23 mm. wide, prominently 5–7-lobed, the lobes linear to triangular-lanceolate; upper stem-leaves with 3–5-lobed blades; hypanthium 3.5–5 mm. long at maturity; sepals oblong to oblong-lanceolate, 3–4.5 mm. long; petals cuneate, 6–8 mm. long. Wet rocks: Greenl.—Lab.—N.M.—Utah—Alaska; Eurasia. *Arct.—Alp.—Subalp.* Je–Jl.

2. **S. simulàta** Small. *Fig. 269.* Stems 9–22 cm. tall, sparingly glandular-pubescent; leaf-blades reniform or orbicular-reniform in outline, 8–21 mm. wide, 5–7-lobed, the lobes much broader than long; upper stem-leaves with 3–5-lobed blades; sepals oblong-ovate, 3–3.5 mm. long; petals fiddle-shaped, 6–7 mm. long. Wet rocks: S.D.—Colo. *Mont.—Alp.* Jl.

9. **MICRÁNTHES** Haw. SAXIFRAGE.

Perennial acaulescent herbs, with very short caudices and solitary or tufted scapes, sometimes bulblet-bearing at the base. Leaves basal; blades fleshy, entire or toothed. Flowers in terminal cymes disposed in narrow or broad panicles. Hypanthium rather flat, shorter than the calyx. Sepals 5, erect or reflexed. Corolla essentially regular. Petals 5, mostly white. Stamens 10. Ovary slightly inferior, the carpels partially united, often only at the base. Follicles spreading or with spreading tips.

Cymules of flowers dense, on the slender branches of the panicle.
 Petals greenish; basal leaves oblong to spatulate, without distinct petiole. 1. *M. pennsylvanica.*
 Petals white; basal leaves ovate, distinctly petioled. 2. *M. texana.*
Cymules of the flowers becoming open and lax, forming raceme-like branches of the panicle. 3. *M. virginiensis.*

1. **M. pennsylvánica** (L.) Haw. *Fig. 270.* Leaves erect or nearly so, narrowly oblong to spatulate, 5–35 cm. long, pubescent, shaggy at the base, glandular-dentate or crenate; scape stout, 3–11 dm. high, finely glandular-pubescent, paniculate above; sepals ovate, 1–1.5 mm. long; petals greenish, 1.5–3 mm. long, lanceolate or oblong-lanceolate, clawless. *Saxifraga pennsylvanica* L. [G, B]. Wet places: Me.—Va.—Mo.—Minn. My.

2. **M. texàna** (Buckley) Small. Leaves spreading, 2–4.5 cm. long, the blade ovate to suborbicular, glabrous except the ciliate margins, sinuate-crenate; scape 7–16 cm. high, glandular-pubescent; panicle dense, the cymules aggregated in 1–4 heads terminating the branches; sepals oblong to ovate, 1.5–2 mm. long, obtuse; petals oval or obovate, 2.5–3 mm. long, clawed. Wet places: Ark.—Kans.—Tex. *Texan.*

3. **M. virginiénsis** (Michx.) Small. Leaves spreading or ascending, 2–15 cm. long, ovate to oblong, sinuate, dentate, or crenate, sparingly pubescent or glabrous on both sides, ciliolate, narrowed into a petiole-like base; scape 8–40 cm. long, glandular-pubescent, paniculately branched above; sepals 1.5–2 mm. long, obtuse; petals white, 4–5.5 mm. long, oblong or elliptic. *S. virginiensis* Michx. [G, B]. Rocky woodlands: N.B.—Ga.—Tenn.—Minn. *E. Temp.* Mr–My.

10. LEPTÀSEA Haw. YELLOW SAXIFRAGE, SPOTTED SAXIFRAGE.

Perennial matted caulescent herbs, with copiously leafy caudices and offsets. Leaves alternate, fleshy or parchment-like; blades entire, ciliate or rarely with 3 apical tooth-like lobes, sessile. Flowers solitary or in terminal simple or compound cymes. Hypanthium small, much shorter than the calyx. Sepals 5, often spreading or reflexed. Corolla white or yellow, regular. Petals 5, with claws. Stamens 10; filaments subulate (in all ours) or club-shaped. Ovary mostly superior, the carpels united to above the middle. Follicles erect, with spreading tips.

Leaves entire, ciliate or denticulate on the margins.
  Flowers cymose; petals 3–4.5 mm. long. — 1. *L. aizoides.*
  Flowers solitary; petals 9–13 mm. long. — 2. *L. Hirculus.*
Leaves 3-toothed at the apex, the teeth spinulose-tipped. — 3. *L. tricuspidata.*

1. **L. aizoìdes** (L.) Haw. Leaves of the rosettes 7–13 mm. long, thick, flat, sparingly ciliate or glabrate; flower-stalk 4–11 cm. tall; stem-leaves longer than those of the rosettes; sepals ovate, 2.5–4 mm. long; petals yellow, oblong, clawless. *Saxifraga aizoides* L. Subarctic or subalpine rocks: Greenl.—Vt.—Mich.—Man.; Europe. *Arctic—Subarctic.* Je–Jl.

2. **L. Hírculus** (L.) Small. Leaves of the caudices 9–41 mm. long; blades flat, elliptic or elliptic-spatulate, glabrous, obtuse, often long-petioled; flower-stalks 6–30 cm. tall, more or less pubescent, mostly 1-flowered; sepals oblong to oblong-ovate, 3–4 mm. long, copiously ciliate, obtuse, reflexed at maturity; petals yellow. *S. Hirculus* L. [B]. In wet places: Greenl.—Colo. B.C.—Alaska; Eur. *Arct.—Alp.—Subalp.* Jl–Au.

3. **L. tricuspidàta** (Retz.) Haw. Leaves of the caudices 7.5–21 mm. long, parchment-like, crowded, spreading, the blades linear to cuneate-linear, 3-toothed, ciliate; flower-stalks 5.5–18 cm. tall, sometimes sparingly glandular-pubescent, sepals ovate, 1.5–3 mm. long, obtuse, ciliate; petals white, oblong-elliptic to oval; 6–7 mm. long. *S. tricuspidata* Retz. [B]. Arctic regions or high mountains: Greenl.—Lab.—Mich.—Alta.—Alaska. *Arctic.—Alp.* Jl–Au.

11. CHONDRÒSEA Haw.

Perennial caulescent herbs, with a densely leafy short caudex and offsets, and sparingly leafy flower-stalks. Leaves alternate; blades serrate, each tooth terminating in an encrusted pore. Flowers in terminal compound cymes. Hypanthium turbinate to campanulate, at maturity longer than the calyx. Sepals 5, erect. Corolla white, regular. Petals 5, clawless. Stamens 10; filaments lanceolate or lanceolate-subulate. Ovary about half-inferior, the carpels united to above the middle. Follicles united to the spreading tips.

1. **C. Aizòon** (Jacq.) Haw. Stems 1–4 dm. tall; basal leaves linear-oblong to spatulate, 1.5–4 cm. long; stem-leaves few, remote, more abruptly broadened than those on the caudex; sepals ovate, 1.5–2 mm. long; petals oval, oval-obovate or oval-oblong, 3.5–4 mm. long, sessile. *Saxifraga Aizoon* Jacq. [B, G]. Wet rocks: Lab.—N.S.—Vt. Sask.; Eurasia. *Boreal—Subarctic.* Je–Jl.

12. ANTIPHÝLLA Haw. PURPLE OR MOUNTAIN SAXIFRAGE.

Perennial caulescent densely matted herbs, with copiously leafy stems and sterile branches, and naked or sparingly leafy flower-stalks. Leaves opposite, except sometimes on the flower-stalks, often 4-ranked, imbricate; blades relatively broad, conspicuously ciliate, mostly with an almost apical pore, keeled, sessile. Flowers solitary, erect. Hypanthium shorter than the calyx. Sepals 5, strongly ciliate. Corolla regular; petals 5, blue or purple, much longer than the sepals, narrowed into claw-like bases. Filaments subulate to triangular. Ovary more than half superior, the carpels united to above the middle. Follicles erect, with more or less spreading tips.

1. **A. oppositifòlia** (L.) Fourr. Plants in dense mats; leaves densely imbricate, 4-ranked, the blades obovate to spatulate, 3–5 mm. long, ciliate, keeled; flower-stalks 1–3 cm. long, leafy; sepals oblong to ovate, 2.5–3 mm. long, ciliate all around; petals elliptic to oval, 8–9 mm. long. *Saxifraga oppositifolia* L. [B, G]. Rocks and stony places: Greenl.—Vt.—Wyo.—B.C.—Alaska; Eurasia. *Arct.—Mont.—Alp.* Je–Au.

Family 75. HYDRANGEACEAE. Hydrangea Family.

Shrubs or trees, or rarely vines. Leaves opposite, without stipules. Flowers commonly perfect, in cymes. Hypanthium well developed, usually ribbed. Sepals 4–10. Petals as many. Stamens from 8 to many. Gynoecium of 2–5, rarely 10, united carpels. Ovary partly or wholly inferior; styles distinct or united, sometimes wanting. Fruit a capsule.

Petals small, valvate; stamens 8–10; hypanthium in fruit ribbed. 1. Hydrangea.
Petals large, convolute or imbricate; stamens 12–60; hypanthium not ribbed. 2. Philadelphus.

1. HYDRANGEA (Gron.) L. Hydrangea.

Shrubs with opposite leaves. Flowers perfect, except some of the marginal ones, in corymbiform or thyrsoid cymes. Sepals 4 or 5, often minute, or in the marginal sterile flowers enlarged and petal-like. Petals 4 or 5, valvate. Stamens 8 or 10. Gynoecium of 2–4 united carpels; ovary inferior, 2–4-celled; styles wanting; stigmas 2–4, introrse. Capsule membranous-walled, opening at the top. Seeds numerous.

Leaves glabrous or with scattered hairs. 1. *H. arborescens.*
Leaves densely grayish-tomentose beneath. 2. *H. cinerea.*

1. **H. arborèscens** L. *Fig. 271.* A branching shrub, 1–3 m. high; leaf-blades ovate, oval, or elliptic, rounded or cordate at the base, 5–20 cm. long; cymes 5–20 cm. broad, usually with a few neutral ray-flowers; capsule prominently ribbed, 2–2.5 mm. high and still broader. Along streams: N.Y.—Fla.—La.—Iowa. *E. Temp.* Je–Jl.

2. **H. cinèrea** Small. A spreading shrub, 1–2 m. high; leaf-blades from rounded-ovate to elliptic, rounded or cordate at the base, acuminate, serrate; cymes round-topped, 5–20 cm. broad; ray-flowers usually present; capsule 2–2.5 mm. high, scarcely as broad. Mountains: N.C.—Ga.—Ala.—Mo. *Canad.—Ozark.* Je–Jl.

2. PHILADÉLPHUS L. Mock Orange, Syringa.

Branching shrubs. Leaves toothed or entire. Flowers perfect, borne at the ends of short leafy branches, solitary or in cymes. Sepals 4, rarely 5, valvate, persistent, more or less tomentose within. Petals 4, rarely 5, convolute, white or ochroleucous. Stamens usually many, 25–60; filaments flat, subulate, free or more or less united below; anthers short, didymous. Ovary at least two thirds inferior and adnate to the hypanthium, 4-celled; styles elongate, distinct or more or less united; stigmas distinct or more or less united. Ovules numerous, pendulous. Capsule more or less woody, loculicidal, the septa separating in the center. Seeds numerous; coat reticulate; albumen fleshy.

1. **P. coronàrius** L. A shrub 2–4 m. high; bark of the twigs brown, glabrous, exfoliating the second season into flakes; leaf-blades ovate, elliptic-ovate, or oval, denticulate, 3–10 cm. long, glabrous; cymes 5–9-flowered, the flowers fragrant; sepals lance-ovate, 7 mm. long, acute; petals cream-colored, obovate, 12–15 mm. long. Around dwellings: Northeastern states—Iowa—S.D.; escaped from cultivation.

Family 76. **GROSSULARIACEAE.** GOOSEBERRY FAMILY.

Shrubs, with palmately veined leaf-blades; stipules none or adnate to the petioles. Inflorescence racemose. Flowers regular, perfect. Hypanthium more or less developed, from saucer-shaped to tubular. Sepals 5, rarely 4. Petals as many as the sepals, small. Stamens as many as and alternate with the petals, inserted in the mouth of the hypanthium. Gynoecium of 2 united carpels; ovary 1-celled, with 2 parietal placentae, inferior; styles 2. Fruit a pulpy berry.

Leaf-blades plicate (folded like a fan) in vernation; flowers regular; style not inclined.
- Plant armed with nodal spines and usually also bristly.
  - Pedicels not jointed near the flowers; bractlets if present enclosed in the bract; hypanthium-tube well developed, deeply campanulate to cylindric. 1. GROSSULARIA.
  - Pedicels jointed below the short stipe-like base of the flower, bearing the bractlets just below the node; hypanthium-tube short, saucer-shaped. 2. LIMNOBOTRYA.
- Plant unarmed; pedicels jointed just under the flowers; bractlets if present borne just under the node. 3. RIBES.

Leaf-blades convolute (rolled in) in vernation; flowers slightly irregular; styles somewhat declined. 4. CHRYSOBOTRYA.

1. **GROSSULÀRIA** (Tourn.) Mill. GOOSEBERRY.

Shrubs, normally armed with simple or 3-forked nodal spines. Leaves in ours rounded or reniform, or more or less pentagonal in outline, 3–5-cleft and crenate or dentate. Flowers bracteate in few-flowered racemes; pedicels not jointed. Hypanthium distinctly produced beyond the ovary; tube campanulate to cylindric. Fruit a berry, with rather tough skin, in most smooth, or somewhat glandular-hispid or spiny.

Hypanthium bristly, at least the larger bristles not gland-tipped. 1. *G. Cynosbati.*

Hypanthium smooth or with some stalked glands.
- Sepals white; filaments more than twice as long as the petals; style pubescent below; filaments and anthers glabrous. 2. *G. missouriensis.*
- Sepals green or purplish; filaments not more than twice as long as the petals.
  - Hypanthium-tube cylindric, usually longer than the sepals. 3. *G. setosa.*
  - Hypanthium-tube campanulate or turbinate, not longer than the sepals.
    - Stamens equaling the petals. 4. *G. oxyacanthoides.*
    - Stamens about twice as long as the petals.
      - Leaf-blades truncate or cordate at the base. 5. *G. inermis.*
      - Leaf-blades wedge-shaped at the base. 6. *G. hirtella.*

1. **G. Cynósbati** (L.) Mill. Shrub less than 1.5 m. high; nodal spines 1–3, slender, 6–10 mm. long; leaf-blades almost orbicular, 3–5 cm. broad, deeply 3–5-lobed, crenate-dentate, somewhat pubescent, at least when young; peduncles 3-flowered; sepals green, oblong, shorter than the hypanthium; petals obovate, shorter than the sepals; stamens a little longer than the petals; berry globose or ellipsoid, 8–12 mm. thick, wine-colored. *Ribes Cynosbati* L. [G, B]. Rocky woods: N.B.—N.C.—Ala.—S.D.—Man. *E. Temp.* Ap–Je.

2. **G. missouriénsis** (Nutt.) Cov. & Britt. A shrub 1–2 m. high, with glabrous branches; spines reddish brown, 1–2 cm. long, straight; leaf-blades 2–6 cm. broad, thin, sparingly pubescent or glabrous above, densely pubescent beneath; peduncles slender, longer than the petioles; hypanthium cylindric, pubescent, about 2.5 mm. long, greenish; sepals linear, 2–3 times as long as the hypanthium, greenish white; petals much shorter than the sepals, erose; berry purple or brownish, about 1 cm. in diameter. *Ribes missouriense* Nutt. [B]. *R. gracile* A. Gray [G]; not Michx. River banks: Ills.—Tenn.—Kans.—S.D. *Prairie—Plain.* Ap–Je.

3. **G. setòsa** (Lindl.) Cov. & Britt. *Fig. 272.* A shrub, usually less than 1 m. high, with bristly branches; spines subulate, usually less than 1 cm. long; leaf-blades thin, cordate or truncate at the base, 1–4 cm. wide, finely pubescent; peduncles shorter than the leaves; hypanthium-tube 5–8 mm. long, about twice as long as the white sepals; petals one half to two thirds as long as the sepals and as long as the stamens; berry red to black, somewhat bristly or smooth, 8–12 mm. in diameter. *R. setosum* Lindl. [B]. *R. saximontanum* E. Nels. Plains and hills: Man.—Minn.—Neb.—Colo.—Ida.—Alta. *Plain—Submont.* Ap–Je.

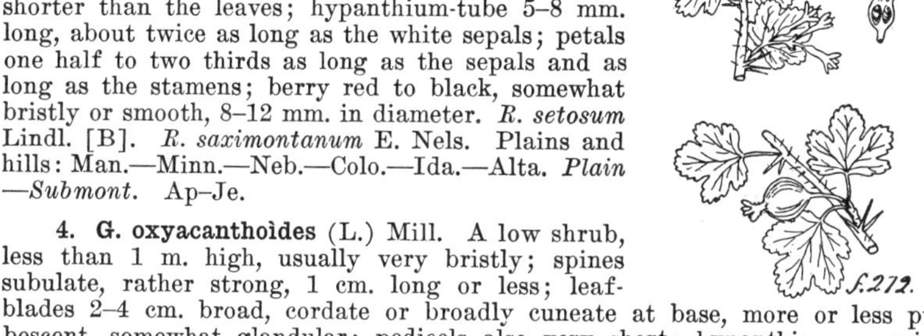

4. **G. oxyacanthoìdes** (L.) Mill. A low shrub, less than 1 m. high, usually very bristly; spines subulate, rather strong, 1 cm. long or less; leaf-blades 2–4 cm. broad, cordate or broadly cuneate at base, more or less pubescent, somewhat glandular; pedicels also very short; hypanthium greenish white, glabrous; sepals white, glabrous, 2.5–4 mm. long, a little longer than the hypanthium; petals obovate, two thirds as long as the sepals; berry smooth, about 1 cm. in diameter. *R. oxyacanthoides* L. [G, B]. Ont.—Mich.—Mont.—B.C.—Yukon. *Boreal.—Submont.* My–Je.

5. **G. inérmis** (Rydb.) Cov. & Britt. Shrub with glabrous stem; nodal spines few, 1 cm. long or less, or none; leaves truncate or cordate at the base, glabrous or somewhat puberulent and glandular; peduncles exserted but short, glabrous, 1–4-flowered; bract small and glabrous; hypanthium glabrous; tube 2.5–3.5 mm. long, a little longer than the glabrous sepals; berry red-purple, glabrous, about 8 mm. thick. *R. inerme* Rydb. *R. vallicola* Greene, a more pubescent form. Mountains: B.C.—Alta.—N.D.—N.M.—Utah—Calif. *Submont.—Mont.* My–Jl.

6. **G. hirtélla** (Michx.) Spach. A shrub 6–12 dm. high, the young shoots gray and glabrous; nodal spines usually wanting, when present subulate, 5–12 mm. long; leaf-blades suborbicular in outline, 2–6 cm. wide, glabrous or sparingly pubescent; peduncles 1–3-flowered, shorter than the petioles; hypanthium-tube not longer than the green or purplish sepals; petals obovate, half as long as the stamens; berry purple or black, globose, 8–10 mm. thick. *R. hirtellum* Michx. Thickets: Newf.—N.J.—W.Va.—S.D.—Man. *Canad.—Allegh.* My–Jl.

2. **LIMNOBÒTRYA** Rydb. Swamp Currants.

Shrubs, armed with pectinately divided nodal spines, usually also bristly. Leaves more or less pentagonal in outline, palmately veined and quinately-cleft and incised or toothed. Flowers perfect, usually several, in bracted racemes; pedicels jointed just under the short stipe-like base of the flower, usually with 2 minute bractlets just below the node. Hypanthium slightly produced beyond the ovary; tube saucer-shaped. Sepals 5, ascending, oval or rounded. Petals reniform-flabellate, clawed, erect. Stamens not exceeding the petals. Fruit a currant-like berry.

1. **L. lacústris** (Pers.) Rydb. Shrub 1–2 m. high; spines and bristles slender, 2–4 mm. long; petioles ciliate; leaf-blades 2–5 cm. long, cleft about three-fourths their length; divisions rhombic, incised into oblong acute teeth; flowers 4–10, light green or purplish; stamens shorter than the petals; berry 6–8 mm. in diameter, densely glandular-hispid. *Ribes lacustre* (Pers.) Poir. [B, G]. Swamps: Newf.—Mass.—Pa.—S.D.—Mack. *E. Boreal.—Submont.* My–Jl.

3. RÌBES L. CURRANTS.

Unarmed shrubs. Leaves alternate, palmately veined and usually also palmately lobed, plicate in vernation, mostly deciduous. Inflorescence several- to many-flowered; pedicels jointed beneath the flowers, a pair of small bractlets often present at the node. Flowers perfect or in some exotic species dioecious. Hypanthium-tube more or less developed, from saucer-shaped to cylindric. Fruit a berry, with rather thin skin, with or without glands, never spiny, disarticulating from the pedicel.

Hypanthium-tube poorly developed, saucer-shaped or short campanulate.
  Fruit not glandular-hispid.
    Berry red, without glands; racemes drooping; bracts short-ovate.
      Pedicels usually glandless; sepals and petals green. 1. *R. vulgare.*
      Pedicels usually with glands; sepals mottled with purple or wholly purple. 2. *R. triste.*
    Berry black, with sessile glands; racemes erect; bracts subulate-linear. 3. *R. hudsonianum.*
  Fruit glandular-hispid, red. 4. *R. glandulosum.*
Hypanthium-tube well developed.
  Fruit not glandular-hispid, black; hypanthium campanulate or urn-shaped.
    Fruit with sessile glands, glabrous. 5. *R. nigrum.*
    Fruit without glands, pubescent. 6. *R. americanum.*
  Fruit glandular-hispid, red; hypanthium cylindric. 7. *R. inebrians.*

1. **R. vulgàre** Lam. A shrub 0.5–1.5 m. high; twigs and petioles pubescent; leaves thin, suborbicular in outline, subcordate at the base, dark green above, pale and pubescent beneath, 3–5-lobed, 7 cm. or less broad; racemes several-flowered, sepals obtuse, spreading, 2–3 mm. long; petals yellowish green; berry globose, red, shining, 1 cm. in diameter. RED CURRANT. Around dwellings: Mass.—Ont.—Minn.—Mo.—Va.; escaped from cult.; Eur.

2. **R. tríste** Pall. A shrub about 1 m. high; leaf-blades thin, reniform-orbicular, 6–10 cm. wide, usually 3-lobed, rarely 5-lobed, dark green and glabrous above, pale and pubescent or glabrate beneath, with conspicuous veins and coarsely dentate-serrate lobes, cordate or rarely truncate at the base; racemes somewhat glandular, usually shorter than the leaves; flowers more or less purplish; sepals obtuse, spreading; berry smooth, glabrous, red, 6–8 mm. in diameter. *R. albinervium* Michx. *R. rubrum* A. Gray [B]; not L. Woods: Newf.—N.J.—S.D.—Ore.—Alaska. *Boreal.—Submont.* My–Je.

3. **R. hudsoniànum** Richards. A shrub 1–1.5 m. high; leaf-blades reniform, broader than long, 3–10 cm. wide, more or less pubescent and resinous-dotted, their lobes ovate, obtuse or acutish, coarsely dentate; racemes 3–6 cm. long, loosely flowered; bracts setaceous, nearly equaling the pedicels; sepals oval, obtuse, spreading; fruit globose, glabrous, 5–10 mm. in diameter. Wooded places: Hudson Bay—Ont.—Minn.—Mont.—B.C.—Alaska. *Boreal.—Mont.* My–Je.

4. **R. glandulòsum** Weber. A shrub, with decumbent or spreading branches, 1 m. long or less; leaf-blades thin, fetid, glabrous above, puberulent on the veins beneath, orbicular in outline, cordate at the base, 3–8 cm. wide, deeply 5–7-lobed, the lobes ovate, acute, incised-serrate; racemes 8–12-flowered, ascending, shorter than the leaves, puberulent; pedicels much longer than the narrow, glandular bracts; tube of the hypanthium short-campanulate; sepals 2–2.5 mm. long, pubescent without; fruit 6–8 mm. in diameter. *R. prostratum* L'Hér. [G]. Woods: Newf.—N.C.—Wis.—B.C.—Mack. *Boreal.* My–Je.

5. **R. nìgrum** L. Shrub 1 m. high or less; leaf-blades 3–5-lobed, thin, sparingly pubescent and resinous-dotted, cordate or truncate at the base, the lobes ovate, sharply serrate; racemes few-flowered, 4–10 mm. long; sepals oblong, obtuse, green, recurved; petals oblong, reddish; berry globose, 8–10 mm. broad. Around dwellings: Mass.—Va.—Minn.; escaped from cultivation; nat. of Eur. My–Je.

6. **R. americànum** Mill. *Fig. 273.* A shrub 1–1.5 m. high; young shoots somewhat pubescent and glandular-dotted; leaves suborbicular in outline, cordate at the base, with an open sinus, 3–8 cm. broad, glabrate above, more or less pubescent (at least on the veins) and glandular-dotted beneath, with ovate serrate-dentate lobes; petioles long-ciliate towards the base; bracts linear or linear-lanceolate; hypanthium urceolate-cylindric, 3–4 mm. long, greenish, sparingly pubescent; sepals obtuse, a little longer than the hypanthium; petals oblong, greenish, two thirds as long; fruit 6–10 mm. in diameter. *R. floridum* L'Hér. [G, B]. Thickets and damp woods: N.S.—Va.—N.M.—Alta. *E. Temp.—Plain —Submont.* My–Je.

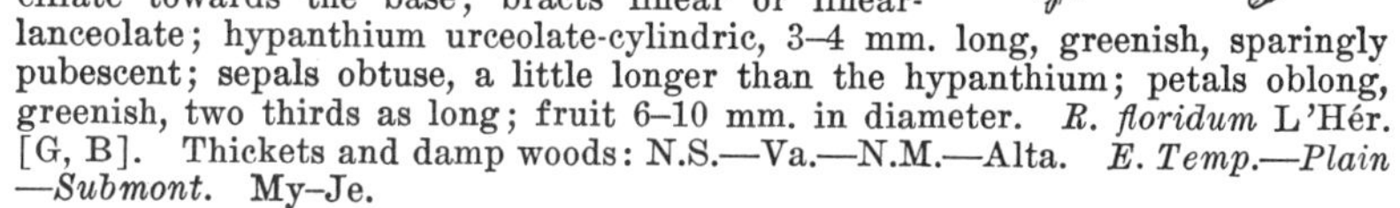

7. **R. inèbrians** Lindl. A low scraggling shrub, 0.5–1 m. high; leaf-blades round-reniform, 1–3 cm. wide, the lobes rounded, crenate; racemes few-flowered, pendulous, puberulent and usually glandular; tube of the hypanthium 5–8 mm. long, usually pink or pinkish; sepals ovate, acutish; fruit 6–8 mm. in diameter. *R. pumilum* Nutt. *R. cereum* Auth. [B]; not Dougl. Dry hills: S.D.—w Neb.—N.M.—Ariz.—Ida. *Plain—Submont.* My–Jl.

4. CHRYSOBÒTRYA Spach. FLOWERING, GOLDEN, MISSOURI, OR BUFFALO CURRANTS.

Unarmed erect shrubs. Leaves palmately veined, palmately 3–5-lobed, convolute in vernation. Inflorescence racemose, several-flowered; pedicels jointed beneath the flower. Flowers perfect, slightly irregular. Hypanthium well developed, yellow, tubular, slightly oblique. Petals erose or denticulate at the apex. Fruit a berry, disarticulating from the pedicel.

Sepals less than half as long as the hypanthium-tube.	1. *C. odorata.*
Sepals more than half as long as the hypanthium-tube.	2. *C. aurea.*

1. **C. odoràta** (Wendl.) Rydb. A shrub 1–2 m. high, with pubescent young shoots and gray bark on the old stems; leaves obovate or suborbicular in outline, cuneate at the base, 2–5 cm. broad, deeply 3–5-lobed and coarsely dentate, glabrous on both sides; racemes 4–8-flowered; bracts ovate or oval, longer than the pubescent pedicels; hypanthium 10–15 mm. long; petals 2–3 mm. long, erose; berry black, globose, about 8 mm. in diameter. *Ribes odoratum* Wendl. *R. aureum* Auth. [G, B]; not Pursh. *R. longiflorum* Nutt. *Chrysobotrya revoluta* Spach. Hillsides: S.D.—Tex. *Plain.*

2. **C. aùrea** (Pursh) Rydb. Erect shrub, 1–2 m. high; leaves reniform-orbicular to obovate in outline, firm, 3–5-lobed and dentate, from cuneate to subcordate at the base, 3–5 cm. broad, glabrous or essentially so; racemes 5–15-flowered; bracts oblong or obovate, 5–12 mm. long; tube of the hypanthium 6–10 mm. long; petals oblong, erose, yellow, often tipped with red; fruit black, red, or amber-yellow, 6–8 mm. in diameter. *Ribes aureum* Pursh. *Chrysobotrya Lindleyana* and *C. intermedia* Spach. Hills and river banks: Wash.—Sask.—S.D.—N.M.—Calif. *W. Temp.—Submont.*

Family 77. PLATANACEAE. Plane-tree Family.

Trees or shrubs, with exfoliating bark. Leaves broad, alternate, long-petioled, palmately 3–5-ribbed and lobed, with membranous, connate, deciduous stipules. Flowers monoecious, in globose heads, arranged in racemes or spikes. Sepals 3 or 4, rarely 6. Petals alternating with the sepals, spatulate. Stamens alternate with the sepals; filaments short; anthers linear, with the connective enlarged at the apex. Pistillate flowers with caducous staminodia. Pistils 3 or 4, opposite the petals; ovary linear; styles recurved at the tip. Fruits 4-sided, linear achenes, truncate and tipped with persistent styles, crowded in dense heads.

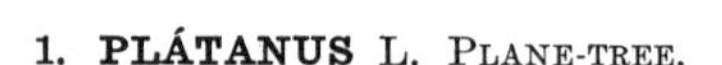

1. **PLÁTANUS** L. Plane-tree.

Characters of the family.

1. P. occidentàlis L. *Fig. 274.* A tall tree, reaching a height of 50 m.; leaf-blades 15–25 cm. long and wide, broadly deltoid, 3–5-ribbed, 3-lobed, coarsely and sharply serrate, the lobes abruptly acuminate; fruiting heads 1 or 2 on the peduncle, 2.5–3 cm. in diameter; achenes 7–8 mm. long. Woods and along streams: N.H.—Fla.—Tex.—Neb.—Minn. *E. Temp.* My.

Family 78. HAMAMELIDACEAE. Witch-hazel Family.

Shrubs or trees, with alternate simple leaves. Flowers perfect, polygamous, or monoecious. Hypanthium campanulate. Sepals 4 or 5, imbricate. Petals 4 or 5, linear or spatulate, or wanting. Stamens 4 to many; filaments distinct; anthers introrse. Gynoecium of 2 united carpels; ovary 2-celled, partly inferior; styles 2; ovules solitary in each carpel, pendulous. Fruit a woody or coriaceous capsule, elastically dehiscent.

1. **HAMAMÈLIS** L. Witch-Hazel.

Shrubs or small trees. Leaves alternate, slightly oblique, with small deciduous stipules. Flowers perfect or polygamous, borne 3 together, bracteate. Sepals 4, reflexed. Petals 4, ribbon-like or wanting in the fertile flowers. Stamens 4, opposite the sepals, alternating with 4 strap-shaped staminodia. Ovary of 2 carpels, united at the base. Capsule woody, 2-beaked.

1. H. virginiàna L. *Fig. 275.* A shrub or tree, 1–10 m. high; leaf-blades firm, obovate, oval, or suborbicular, 5–15 cm. long, coarsely crenate, at least above the middle, more or less stellate-pubescent; flowers opening in the autumn; petals linear, bright yellow, up to 2 cm. long; capsule 12–15 mm. long, nearly black. Low woods and copses: N.S.—Fla.—Minn.—Tex. *E. Temp.* Au–D.

Family 79. **ROSACEAE.** Rose Family.

Herbs, shrubs or trees, usually with alternate leaves and stipules. Flowers perfect, rarely dioecious or monoecious, regular or nearly so. Hypanthium usually well developed, from flat or saucer-shaped to urn-shaped, salverform, or tubular. Sepals normally 5, rarely 4 or 6–9, often subtended by as many bractlets. Petals as many as the sepals or rarely wanting. Stamens 1–many, most commonly 20, in three series. Gynoecium of 1–many usually wholly distinct carpels. Fruit of achenes, follicles, or drupelets.

Fruit consisting of 1–5 dehiscent follicles.
Carpels alternate with the sepals, or less in number; stipules none or deciduous.
Carpels 1–5, if more than one more or less united below; seeds shining, pear-shaped, with a bony coat; endosperm present. 1. Physocarpa.
Carpels usually 5, distinct; seeds dull, linear or linear-lanceolate, with a thin coat; endosperm lacking.
Stamens inserted on the margin of the hypanthium; flowers perfect; shrubs or undershrubs with simple leaves.
Carpels dehiscent on the ventral suture; shrubs with deciduous leaves. 2. Spiraea.
Carpels dehiscent on both sutures; cespitose tufted undershrubs with persistent leaves. 3. Petrophytum.
Stamens inserted on the inside of the hypanthium; flowers dioecious; tall herbs with twice or thrice compound leaves. 4. Aruncus.
Carpels opposite the sepals; stipules present, persistent; leaves ternate. 5. Porteranthus.
Fruits consisting of indehiscent achenes or drupelets.
Carpels not enclosed in a fleshy hypanthium.
Fruits of dry achenes.
Ovules 2, one above the other; achenes usually 1-seeded and more or less flattened, arranged in a single circle; perennial herbs, with pinnately dissected leaves. 6. Filipendula.
Ovules and seeds solitary; achenes usually turgid and, if many, spirally arranged.
Seeds inserted at the distal end of the ovary, *i. e.*, opposite the origin of the style, pendulous or in genera with basal styles ascending; radicle superior.
Style articulate to the ovary; hypanthium from campanulate to almost flat, neither contracted at the throat nor closely investing the achenes; flowers cymose or solitary.
Style not basal.
Style terminal or nearly so; ovules pendulous and anatropous. 7. Potentilla.
Style lateral; ovules ascending and amphitropous.
Achenes glabrous; herbs.
Achenes numerous; stamens usually about 20.
Leaves odd-pinnate.
Receptacle not enlarged in fruit; petals yellow, obtuse or retuse; leaves interruptedly pinnate. 8. Argentina.
Receptacle somewhat enlarged in fruit, becoming spongy; petals red, acute or acuminate; leaves regularly pinnate. 9. Comarum.
Leaves trifoliolate; receptacle much enlarged and usually red in fruit, pulpy; petals white or pinkish. 10. Fragaria.
Achenes 10–15; stamens 5, leaves trifoliolate. 11. Sibbaldia.
Achenes hairy; shrubs or plants shrubby at the base.
Style filiform; leaves 3-foliolate; petals white. 12. Sibbaldiopsis.
Style club-shaped; leaves pinnate; petals in ours yellow. 13. Dasiphora.
Style nearly basal; ovules ascending or nearly erect, orthotropous.
Stamens and pistils numerous; bractlets present; leaves pinnate. 14. Drymocallis.
Stamens 5; bractlets wanting; leaves twice or thrice ternate. 15. Chamaerhodos.

Style not articulate to the ovary; inflorescence spicate, racemose, or paniculate; hypanthium indurate, closely investing the achenes in fruit.
Hypanthium not prickly; petals lacking.
Stamens 2–4, not declined; pistil 1.
Perennials, with rootstocks; stigmas muricate-papillose; leaflets toothed. 16. SANGUISORBA.
Annuals or biennials, with taproots; stigmas brush-like; leaflets pectinate-pinnatifid. 17. POTERIDIUM.
Stamens in the staminate flowers numerous and declined; pistils 2. 18. POTERIUM.
Hypanthium prickly; petals present; prickles of the hypanthium hooked. 19. AGRIMONIA.
Seeds inserted at the proximal end of the ovary, *i. e.,* perfectly basal; radicle inferior.
Styles wholly deciduous. 20. WALDSTEINIA.
Styles partly or wholly persistent.
Hypanthium hemispheric, campanulate, or turbinate, persistent.
Flowers 8–10-merous; low depressed undershrubs with crenate or entire leaf-blades. 21. DRYAS.
Flower usually 5-merous.
Style conspicuously bent and distinctly geniculate above, the upper hairy portion readily deciduous. 22. GEUM.
Style neither conspicuously bent nor distinctly geniculate, the upper glabrous portion persistent or tardily deciduous. 23. SIEVERSIA.
Hypanthium salver-shaped, the limb deciduous; the tube persistent and closely investing the fruit; shrubs. 24. CERCOCARPUS.
Fruits of more or less fleshy drupelets; ovules 2, collateral.
Styles club-shaped; stigmas slightly 2-lobed; receptacle flat; unarmed shrubs, with exfoliating bark and simply digitately ribbed and lobed leaves; drupelets capped by a hard pubescent cushion. 25. RUBACER.
Styles filiform; stigmas capitate; receptacle convex, hemispheric or nipple-shaped; drupelets without cushion; leaves in most species compound and stem prickly or bristly. 26. RUBUS.
Carpels enclosed in the hypanthium which becomes fleshy in fruit. 27. ROSA.

1. PHYSOCÁRPA Raf. NINE-BARK.

Shrubs with exfoliating bark. Leaves alternate, 3–5-ribbed, more or less lobed and usually with more or less stellate hairs. Flowers in terminal corymbs. Hypanthium hemispheric or nearly so. Sepals 5, persistent. Petals 5, white or rarely pinkish, spreading. Stamens 20–40 on a disk, clothing the mouth of the hypanthium. Pistils 1–5, more or less united at the base; styles filiform, terminal; stigmas capitate; ovules 2–4. Follicles more or less inflated, opening along both sutures; seeds obliquely pear-shaped, shining with a bony coat; endosperm copious. *Physocarpus* Maxim. *Opulaster* Medic.

Carpels 3–5, united only at the base, turgid.
Carpels usually 5, glabrous; leaves ovate in outline. 1. *P. opulifolia.*
Carpels 3 or 4, finely pubescent; leaves suborbicular in outline. 2. *P. intermedia.*
Carpels 2, united half their length, somewhat compressed, pubescent. 3. *P. monogyna.*

1. **P. opulifòlia** (L.) Raf. Shrub 1–3 m. high; leaves ovate, generally acutely 3-lobed, cordate, or truncate at the base, glabrous, 3–5 cm. long; sepals ovate, nearly glabrous within; petals 4 mm. long; carpels 5, glabrous, shining, 8–10 mm. long, acute. *O. opulifolius* Kuntze [B, R]. River banks and rocky places: Que.—Ky.—Ga.—ne Minn.; often cultivated and escaped. *Canad.—Allegh.* Je.

2. **P. intermèdia** (Rydb.) C. K. Schneider. *Fig. 276.* Shrub 1–1.5 m. high; branches glabrous or nearly so; blades ovate to orbicular in outline, usually shallowly 3-lobed, 2–6 cm. long, doubly crenate, glabrous or nearly so above, usually sparingly hairy beneath; inflorescence dense; bracts linear-oblanceolate,

caducous; sepals ovate, acute, densely stellate on both sides; petals 4 mm. long; carpels united at the base, rounded-ellipsoid, abruptly acuminate, 7–8 mm. long. *O. Ramaleyi* A. Nels., mainly. *O. intermedius* Rydb. [B, R]. Creek banks and hillsides: w N. Y.—Mo.—Colo.—S.D. *Allegh.—Prairie—Plain—Submont.* Je–Jl.

3. **P. monógyna** (Torr.) Coult. Shrub, usually less than 1 m. high, with decumbent stems; leaves 1–3 cm. long, rounded-ovate to reniform in outline, rather deeply 3–5-lobed, incised; sepals lance-ovate to elliptic, usually obtuse; petals about 3 mm. long; follicles densely stellate, 3–4 mm. long, with ascending-spreading beaks. *P. Torreyi* Maxim. *O. monogynus* [R] Kuntze. Mountains: S.D.—Wyo.—N.M—Nev. *Submont.—Mont.* My–Jl.

2. **SPIRAÈA** L. MEADOW-SWEET.

Shrubs, with simple feather-veined leaves, which are usually serrate, and no stipules. Inflorescence racemose, corymbose or paniculate. Flowers in ours perfect. Hypanthium hemispherical, campanulate, or turbinate. Sepals 5, valvate or slightly imbricate. Petals 5, imbricate, white to red. Stamens 15–70, inserted in 1–several series under the margin of a more or less developed disk. Pistils 5, rarely 3–8, distinct, inserted in the bottom of the hypanthium; styles terminal; stigmas capitate or discoid; ovules 2–several, anatropous, pendulous. Follicles leathery, not inflated, opening along the ventral suture; seeds several, in ours usually 4, pendulous, linear-lanceolate or oblong, tapering at both ends; endosperm none or almost none.

Panicle flat-topped; sepals erect or merely spreading.
  Petals pink or rose-colored; corymb small. — 1. *S. densiflora.*
  Petals white, seldom tinged with rose. — 2. *S. lucida.*
Panicle elongate, conical or ovoid, or lanceolate.
  Sepals merely spreading; disk obsolete; leaves not tomentose beneath.
    Petals rose-colored. — 3. *S. salicifolia.*
    Petals white or slightly pinkish in bud.
      Inflorescence glabrous; leaf-blades broadly oblanceolate to obovate. — 4. *S. latifolia.*
      Inflorescence pubescent; leaf-blades narrowly oblanceolate. — 5. *S. alba.*
  Sepals soon reflexed; disk usually rather conspicuous; leaves tomentose beneath. — 6. *S. tomentosa.*

1. **S. densiflòra** Nutt. Low shrub, 2–6 dm. high, with ascending branches; bark of the twigs dark reddish brown, glabrous; leaf-blades oval or elliptic, rounded at both ends, glabrous or essentially so, bright green above, somewhat paler beneath, crenate or serrate above the middle, 1.5–3 cm. long; sepals ovate, obtuse; petals nearly 1.5 mm. long, obovate; follicles glabrous and shining. Mountains: B.C.—Ore.—Wyo.—S.D.—Mont. *Submont.—Mont.* Jl–Au.

2. **S. lùcida** Dougl. *Fig. 277.* Low shrub, with creeping rootstock; stems and branches erect, often dying down annually to near the base, 3–10 dm. high; blades of the lower leaves obovate, of the upper oval, often acutish, 2–6 cm. long, thin, glabrous, shining above, paler beneath, coarsely serrate or incised; sepals triangular, acute; petals white, orbicular, about 2 mm. long; follicles glabrous, shining. Mountains: B.C.—Ore.—Wyo.—Sask. *Submont.—Mont.* My–Jl.

3. **S. salicifòlia** L. A shrub, 1–2 m. high; bark of the young twigs brown, finely puberulent; leaf-blades lance-oblong, broadest at the middle, acute, sharply serrate, 3–7 cm. long, puberulent on the midrib and margins; inflorescence narrow, 5–10 cm. long, 2–3 cm. wide, puberulent; petals pink, rounded-ovate; follicles glabrous. Around dwellings: Me.—Minn.—Kans.—Va.; escaped from cultivation; nat. of Siberia.

4. **S. latifòlia** (Ait.) Borkh. A shrub, 1–20 dm. high, branched; bark of the twigs glabrous, chestnut; leaf-blades broadly oblanceolate or obovate, thin, 2–8 cm. long, acute, obtuse or rounded at the apex, sharply and sometimes doubly serrate; inflorescence 0.5–2 dm. long and about half as wide; bracts sometimes ciliate; sepals triangular, acute; petals white or a little pinkish in bud; follicles glabrous and shining. Meadows: Newf.—N.C.—w Pa.—Neb.—Sask. *E. Temp.* Je–S.

5. **S. álba** DuRoi. A shrub, 1–2 m. high; bark of the twigs brown; leaf-blades rather firm, more or less puberulent on the veins, sharply serrate, acute at both ends, 3–6 cm. long; inflorescence 5–20 cm. long; sepals triangular, acute; petals white, about 2 mm. long; follicles glabrous, shining. *S. simplex* Greene. *S. salicifolia* Am. auth. [G]. Swamps and wet ground: Newf.—N.C.—Ia.—N.D.—Sask. *E. Temp.* Je–Au.

6. **S. tomentòsa** L. A shrub, 3–12 dm. high; bark of the twigs purplish or brown, tomentose; leaf-blades oval to oblong, usually acute at the apex, coarsely crenate-serrate, dark green above, densely white- or rusty-tomentose beneath, 3–6 cm. long; panicle narrow, 1–2 dm. long, 3–5 cm. broad; sepals triangular, reflexed in fruit; petals pink, purplish, or white, orbicular, 1.5 mm. long; follicles tomentose, 2.5 mm. long. Wet places: N.S.—Man.—Kans.—Ga.

3. **PETRÓPHYTUM** (Nutt.) Rydb.

Densely cespitose and depressed undershrubs, with prostrate branches, growing on rocks. Leaves oblanceolate or spatulate, entire, coriaceous, evergreen, crowded on the short branches. Inflorescence racemose, rarely compound. Flowers perfect. Hypanthium hemispheric. Sepals 5, valvate. Petals 5, imbricate, white. Stamens about 20; filaments filiform, distinct, about twice as long as the sepals. Disk evident, entire, margined. Pistils 3–5; ovary and lower part of the style very hairy; style filiform, terminal; ovules 2–4, pendulous. Follicles leathery, dehiscent along both sutures; seeds linear.

1. **P. caespitòsum** (Nutt.) Rydb. Season's shoots very short; leaves spatulate, 5–12 mm. long, 2–4 mm. wide, densely silky, 1-ribbed, obtuse or mucronate; sepals ovate-lanceolate, acute, 1.5 mm. long; petals spatulate, obtuse, 1.5 mm. long; follicles 3–5, 2 mm. long. *Spiraea caespitosa* Nutt. Rocks: S.D.—N.M.—Ariz.—Calif.—Ida.—Mont. *Plain—Mont.* Jl–S.

4. **ARÚNCUS** (L.) Adans. GOAT'S BEARD.

Perennial herbs, with thick rhizomes, twice or thrice ternate-pinnatisect leaves, and no stipules. Inflorescence a large panicle. Flowers dioecious. Hypanthium of the staminate flowers flat or saucer-shaped, pentagonal. Sepals triangular, valvate. Petals 5, orbicular to spatulate, white. Stamens 15–30; filaments subulate, distinct, long-exserted; anthers didymous. Disk 5-lobed. Pistils 3–5, rudimentary, styleless. Pistillate flowers much smaller. Petals elliptic to oblanceolate. Stamens rudimentary. Pistils 3–5, distinct; styles short, obliquely terminal; stigmas capitate; ovules 8–12, pendulous. Follicles oblong, cartilaginous, dehiscent along the ventral suture and then splitting dorsally at the apex. Seeds few, attentuate at both ends.

1. **A. pubéscens** Rydb. Stem 1–2 m. high; lower leaves pinnate, with 5 leaflets, the terminal one trifoliolate or ternate, with the divisions pinnately 5-foliolate; leaflets 3–10 cm. long, decidedly hairy beneath, ovate, short-

acuminate, doubly serrate; petals of the staminate flowers obovate, hardly 1 mm. long, those of the pistillate ones still smaller; follicles about 2 mm. long. Rich woods: Ills.—Iowa.—Okla.—Ark. *Ozark.*

5. **PORTERÁNTHUS** Britton. AMERICAN IPECAC.

Perennial herbs, with horizontal rootstocks. Leaves alternate, trifoliolate, with stipules. Hypanthium cylindro-campanulate. Sepals 5, imbricate. Petals 5, convolute, linear-oblong, clawed. Stamens 20–25, in three series. Pistils 5, opposite the sepals; style terminal; ovules 2–8, ascending. Follicles free, dehiscent along the ventral suture and dorsally at the apex. *Gillenia* Moench; not Adans.

1. **P. stipulàtus** (Muhl.) Britton. *Fig. 278.* Stem 5–10 dm. high, more or less pubescent; stipules ovate to orbicular, double-toothed to laciniate, 1–2 cm. long; leaflets oblanceolate, those of the lower leaves incise, those of the upper ones sharply serrate; petals white or pinkish, 12–15 mm. long. *G. stipulacea* Barton. [G]. Woods: N.Y.—Kans.—La.—Ga. *Allegh.—Austral.* Je–Jl.

6. **FILIPÉNDULA** (Tourn.) Mill. MEADOW QUEEN, QUEEN-OF-THE PRAIRIE.

Perennial herbs, with rootstocks. Leaves alternate, pinnatisect, with ample stipules. Flowers in corymbiform cymes. Hypanthium flat. Disk obsolete. Stamens 20–40, of which 10 are shorter; filaments subclavate. Pistils 5–15, distinct, opposite to the petals or more numerous, in a single series; stigmas large, capitate. Carpels indehiscent, stipitate, 1-seeded. *Ulmaria* Hill.

1. **F. rùbra** (Hill) B. L. Robinson. *Fig. 279.* Stem 0.5–2 m. high, red-tinged; leaves dark green and glabrous above, pubescent beneath; segments 3–9, the terminal one reniform, 1–2 dm. broad, 5–9-cleft and doubly serrate, the lateral ones obovate, 3–5-lobed, 5–12 cm. long; petals pink or purple, orbicular, 3–3.5 mm. long; achenes about 10, straight, fixed by the base. *Spiraea lobata* Gronov. *Ulmaria rubra* Hill [B]. Moist ground: Vt.—Iowa—Ky.—Ga. *Canad.—Allegh.* Je–Jl.

7. **POTENTÍLLA** L. CINQUEFOIL, FIVE-FINGER.

Annuals, or perennial herbs, with rootstocks. Leaves pinnately or digitately compound. Inflorescence usually cymose-paniculate. Hypanthium concave, mostly hemispheric. Bractlets, sepals, and petals 5 (sometimes 4). Petals deciduous, obcordate, obovate, rotund, or cuneate, as a rule not unguiculate, yellow, white, or dark purple. Stamens most commonly 20, in three series, viz., 10, 5, and 5, respectively, sometimes more, sometimes fewer; filaments filiform or subulate, neither flattened nor dilated. Receptacle hemispheric or conic, bearing usually numerous pistils. Style usually long and filiform, attached near the apex of the ovary, deciduous. Seeds inserted near the base of the style, pendulous and anatropous.

Flowers solitary, axillary, on long peduncles. I. TORMENTILLAE.
Flowers cymose.
Stamens about 30; mature achenes reticulate; petals sulphur yellow. XII. RECTAE.
Stamens about 20 or less; mature achenes smooth; petals darker yellow.
Cymes very leafy, many-flowered.
Annuals or biennials, rarely short-lived perennials; styles fusiform and glandular at the base. II. SUPINAE.
Perennials; styles filiform but short, not glandular; leaves silvery beneath. III. ARGENTEAE.
Cymes not very leafy, generally rather few-flowered; perennials with rootstocks.
Leaves digitate.
Plants more than 2 dm. high.
Leaves green on both sides, sparingly hairy, not at all tomentose. IV. NUTTALLIANAE.
Leaves either densely hairy or more or less tomentose beneath.
Leaflets merely crenate or cut-toothed, the toothing not extending half-way to the midrib. VI. GRACILES.
Leaflets cleft more than half-way to the midrib into linear, oblong, or lance-oblong divisions. V. PECTINISECTAE.
Plants less than 2 dm. high.
Leaves neither tomentose nor silky, green. VII. AUREAE.
Leaves tomentose or silky, at least beneath. VIII. CONCINNAE.
Leaves odd-pinnate.
Style much longer than the mature achene, filiform.
Leaves with 1–3, more or less approximate pairs of leaflets; plants low. VII. AUREAE.
Leaves with 3–13 pairs of leaflets; plants usually comparatively tall.
Leaves grayish or whitish, silky or tomentose, at least beneath. IX. LEUCOPHYLLAE.
Leaves green on both sides or merely strigose. X. MULTIJUGAE.
Style not longer than the mature achene, thickened and glandular below; leaves usually more or less tomentose beneath. XI. MULTIFIDAE.

I. TORMENTILLAE.

Pubescence of the stem, petioles, and pedicels appressed. 1. *P. simplex.*
Pubescence of the stem, petioles, and pedicels spreading. 2. *P. canadensis.*

II. SUPINAE.

Achenes with a corky gibbosity on the upper suture.
Leaves all pinnate, with 3–5 pairs of leaflets; inflorescence truly cymose. 3. *P. paradoxa.*
Lower leaves pinnate, with 2 or 3 pairs of leaflets, the upper ternate; inflorescence falsely racemose. 4. *P. Nicolletii.*
Achenes not gibbous.
Lower leaves pinnate with 2 approximate pairs of leaflets, the upper ternate; stem strict; inflorescence cymose. 5. *P. rivalis.*
Leaves all digitate; the basal ones rarely pinnate in *P. monspeliensis.*
Leaves ternate.
Petals about half as long as the sepals, cuneate or obovate; achenes whitish, smooth; hypanthium in fruit 5 mm. wide or less; stamens 10.
Stem with diffuse, spreading branches; leaflets cuneate; inflorescence cymose. 6. *P. millegrana.*
Stem with erect or strongly ascending branches; inflorescence inclined to be falsely racemose; leaflets broadly obovate; plant decidedly glandular-puberulent. 7. *P. biennis.*
Petals equaling the sepals or nearly so; hypanthium in fruit about 7 mm. wide; stamens 15–20. 8. *P. monspeliensis.*
Lower leaves digitately 5-foliolate, or ternate with the lateral leaflets 2-cleft to near the base; stamens 5. 9. *P. pentandra.*

III. ARGENTEAE.

Teeth of the leaflets ovate, the margins not revolute; fruiting calyx 5–6 mm. broad. 10. *P. collina.*
Teeth of the leaflets oblong to linear, the margins revolute; fruiting calyx 5 mm. broad or less. 11. *P. argentea.*

IV. Nuttallianae.

One species. — 12. *P. Nuttallii.*

V. Pectinisectae.

One species. — 13. *P. flabelliformis.*

VI. Graciles.

Leaflets cut-toothed or cleft, with lanceolate teeth.
  Petioles with appressed or ascending hairs, leaflets only sparingly tomentose beneath. — 14. *P. viridescens.*
  Petioles with spreading hairs; leaflets densely white-tomentose beneath. — 15. *P. camporum.*
Leaflets merely crenate, with rounded teeth; leaflets white-tomentose beneath. — 16. *P. filipes.*

VII. Aureae.

Leaves rather densely silky-strigose, the basal ones often inclined to be pinnate. — 17. *P. diversifolia.*
Leaves slightly pubescent, in age glabrate and somewhat glaucous, all digitate. — 18. *P. glaucophylla.*

VIII. Concinnae.

Leaflets merely crenate or serrate. — 19. *P. concinna.*
Leaflets deeply dissected. — 20. *P. divisa.*

IX. Leucophyllae.

Leaves almost equally white-pubescent on both sides.
  Bractlets much shorter than the acuminate sepals; leaflets finely tomentose. — 21. *P. effusa.*
  Bractlets at least three fourths as long as the acute sepals.
    Inflorescence open; leaves silky and shining. — 22. *P. Hippiana.*
    Inflorescence dense; leaves loosely tomentose and rather dull. — 23. *P. argyrea.*
Leaves green and merely silky above.
  Leaflets usually 9, all usually ascending, the upper pair decurrent. — 24. *P. propinqua.*
  Leaflets 5–7, the lower spreading or reflexed, the upper pair not decurrent. — 25. *P. pulcherrima.*

X. Multijugae.

One species. — 26. *P. plattensis.*

XI. Multifidae.

Pubescence of the under side of the leaves silvery-white.
  Segments of the leaves acute; style not thickened. — 27. *P. multifida.*
  Segments of the leaves obtuse; style decidedly thickened at the base.
    Segments of the leaflets linear or linear-oblong; branches of the inflorescence usually short and ascending. — 28. *P. bipinnatifida.*
    Segments of the leaflets lanceolate or oblong, usually green above; branches of the inflorescence long and erect. — 29. *P. platyloba.*
Pubescence of the under side of the leaves grayish.
  Petioles and stem appressed-pubescent.
      Leaves glandular-pruinose above and sometimes with scattered long hairs. — 30. *P. glabrella.*
      Leaves short-strigose above. — 31. *P. pennsylvanica.*
  Petioles and stem with spreading hairs.
    Flowers distinctly pedicelled; upper leaflets of the basal leaves with 9–13 segments; pubescence of the stem and petioles long.
      Plant dark green; leaves scarcely paler beneath. — 32. *P. atrovirens.*
      Plant not dark green; leaves decidedly paler beneath. — 33. *P. strigosa.*
    Flowers usually subsessile, densely conglomerate; upper leaflets of the basal leaves with 15–21 segments; pubescence of the stem short and dense. — 34. *P. lasiodonta.*

XII. Rectae.

One species. — 35. *P. sulphurea.*

1. P. símplex Michx. Stem slender, at first erect, soon decumbent, 4–10 dm. long, appressed-pubescent; basal leaves digitately 5-foliolate; leaflets 2–6 cm. long, oblanceolate, rarely obovate, coarsely toothed; stem-leaves short-petioled; peduncles about 3.5 cm. long; petals broadly obcordate, 5–6 mm. long. Open woods and among grass: N.S.—Minn.—Mo.—N.C. *Canad.—Allegh.* Ap–Au.

2. **P. canadénsis** L. Stem slender, silky-villous, at first erect or assurgent, soon decumbent or prostrate; leaves digitately 5-foliolate, or 3-foliolate, with the lateral leaflets 2-cleft; leaflets 2–5 cm. long, obovate-oblanceolate, coarsely serrate, silky on both sides; pedicels 3–10 cm. long; petals yellow, obcordate, 5–6 mm. long; stamens 20. Dry ground: N.B.—Minn.—Tex.—N.C. *E. Temp.*

3. **P. paradóxa** Nutt. Stem spreading or ascending, seldom erect, about 2–5 dm. high, leafy; leaves in age nearly glabrous, light green in color; leaflets obovate-cuneate, deeply crenate or cleft, 1–2 cm. long; flowers about 7 mm. in diameter; bractlets and sepals oblong-ovate, acute or mucronate, about equal in length; petals yellow, obovate-cuneate, truncate or slightly emarginate; stamens 15–20. Low ground: N.Y.—Ont.—N.M.—Utah—Wash.; Mex., e Asia. *Temp. —Submont.* Je–Au.

4. **P. Nicollétii** (S. Wats.) Sheldon. Stem spreading, branched, 2–5 dm. high; lower leaves pinnate, the upper trifoliolate; leaflets obovate-cuneate, with sharper teeth than in the preceding, 5–25 mm. long; flowers 5 mm. broad; bractlets and sepals oblong-ovate, mucronate; petals obovate-cuneate, equaling the sepals; stamens 10–15. Low ground: N.D.—Kans.—Mo. *Prairie.* Je–Au.

5. **P. rivàlis** Nutt. Stem erect and simple, finely villous-hirsute, leafy; leaflets 2–5 cm. long, obovate, with coarse ovate teeth; flowers less than 5 mm. in diameter; bractlets oblong, obtuse or acute, rather shorter than the ovate acute sepals; petals cuneate, much shorter than the sepals; stamens about 10. River valleys: B.C.—Sask.—Mex. *W. Temp.—Plain—Submont.* Je–Au.

6. **P. millegràna** Engelm. Stem slender, 4–8 dm. high, divaricate, softly pubescent, sometimes nearly glabrous; leaves ternate, finely pubescent; leaflets oblong-cuneate, deeply serrate, 1–6 cm. long, the middle one often petiolulate; flowers 3–4 mm. in diameter; bractlets and sepals oblong-ovate, acute, of about the same length; petals oblong-cuneate, about half as long as the sepals, light yellow; stamens generally 10; achenes smooth, light-colored, small. *P. leucocarpa* Rydb. River valleys: Man.—Ills.—N.M.—Calif.—Wash. *Temp.—Submont.* My–S.

7. **P. biénnis** Greene. Annual or biennial; stems often several from the root, 3–5 dm. high, finely and rather densely pubescent, often tinged with red or purple; leaves all ternate; leaflets broadly obovate, coarsely crenate, 2–4 cm. long; flowers small, about 5 mm. broad; bractlets ovate-lanceolate or oblong, acute, a little shorter than the ovate acute sepals; petals yellow, obovate-cuneate, much shorter than the sepals; ripe achenes whitish, smooth. Waste places: B.C.—Sask.—S.D.—Colo.—Ariz.—Calif. *W. Temp.—Plain—Submont.* My–S.

8. **P. monspeliénsis** L. Stems stout and very leafy, 3–8 dm. high, often several from the annual or biennial root, hirsute; leaves digitately 3-foliolate, or in luxuriant forms the lower sometimes digitately or pinnately 5-foliolate; leaflets usually obovate, 3–10 cm. long, serrate; flowers about 1 cm. in diameter; bractlets and sepals oblong-lanceolate, acute, nearly of the same length; petals light yellow, obovate or cuneate, truncate, nearly equaling the sepals; stamens generally 20, sometimes only 15; achenes usually rugulose when ripe. *P. norvegica hirsuta* Michx. Rich soil and waste places: Lab.—D.C.—Kans.—Mex.—Calif.—Alaska; Eurasia. *Temp.—Submont.* My–S.

9. **P. pentándra** Engelm. Annual; stems stout, very leafy, 3–7 dm. high, finely hirsute and much branched above; leaflets 2–10 cm. long, oblong to oblanceolate or cuneate, deeply serrate, pubescent on both sides; flowers less than 5 mm. in diameter; bractlets oblong, acute, about 3 mm. long, nearly as long as the ovate acute sepals; petals pale yellow, obovate, scarcely half as long as the sepals; achenes smooth, brownish. Bottom-land: Mo.—Ark.—Sask. —Alta.—Neb. *Prairie—Plain.* Je–Au.

10. **P. collìna** Wibel. Perennial; stems several, ascending or spreading, glabrous or tomentose; basal leaves digitate, long-petioled, glabrous or puberulent above, grayish-tomentose beneath; leaflets 5, broadly cuneate, 1.5–2.5 cm. long; upper stem-leaves ternate, short-petioled; bractlets and sepals equal, oblong, obtuse, 4 mm. long; petals obcordate, a little longer than the sepals; stamens 20. Roadsides: Mass.—Minn.; adv. from Eur.

11. **P. argéntea** L. Perennial; stems many, branched and leafy, 1–5 dm. long, grayish-tomentose; leaves digitate, all except the uppermost 5-foliolate, glabrous and green above, white-tomentose beneath; leaflets obovate or cuneate, 1–3 cm. long, deeply divided into narrow divisions, with revolute margins; bractlets oblong, nearly equaling the ovate sepals, 3 mm. long; petals obovate-cuneate, emarginate, scarcely longer than the sepals; stamens 20. Roadsides and waste places: N.S.—Man.—N.D.—Kans.—D.C.; nat. from Eur. My–S.

12. **P. Nuttállii** Lehm. Stem 6–8 dm. high, stout, sparingly hirsute; basal leaves digitate, sparingly hirsute, not at all tomentose beneath; petioles with appressed hairs; leaflets 5–10 cm. long, oblanceolate, toothed or divided nearly halfway to the midrib; hypanthium hirsute, in fruit about 1 cm. in diameter; bractlets linear to lanceolate, generally shorter than the ovate-lanceolate sepals; petals yellow, obovate, emarginate, 6–8 mm. long. Mountain valleys: Sask.—Minn.—Colo.—Ore.—B.C. *Prairie—Plain—Mont.* My–Au.

13. **P. flabellifórmis** Lehm. Stem slender, but strict, 4–6 dm. high, silky-strigose; leaves densely silky above, white-tomentose beneath; leaflets 3–5 cm. long, pectinately divided into narrowly linear lobes; hypanthium silky-villous; bractlets linear-lanceolate, much shorter than the triangular-lanceolate acuminate sepals; corolla 10–12 mm. in diameter; petals obcordate, a little longer than the sepals. Meadows: B.C.—Sask.—S.D.—Wyo.—n Calif. *Son.—Submont.* Je–Au.

14. **P. viridéscens** Rydb. Stem 5–7 dm. high, sparingly silky with appressed or slightly spreading hairs; leaves silky, but green above, silky and somewhat tomentulose beneath; leaflets oblanceolate or spatulate, deeply toothed or divided nearly halfway to the midrib, with lanceolate often somewhat divergent teeth; hypanthium silky; bractlets lanceolate, a little shorter than the ovate-lanceolate sepals; corolla 10–15 mm. in diameter; petals obcordate, equaling or slightly exceeding the sepals. Valleys: Man.—Alta.—Ida.—N.D. *Plain—Submont.—Mont.* My–Au.

15. **P. campòrum** Rydb. Stem simple, strict, 3–4 dm. high, hirsute with spreading or reflexed hairs; leaflets obovate or broadly oblanceolate, 2–5 cm. long, cut-toothed, densely hirsute-strigose above, white-tomentose and silky beneath; bractlets lanceolate, 3–4 mm. long; sepals ovate-lanceolate, 4–5 mm. long; petals obcordate, 5–6 mm. long. Meadows: S.D.—Man. *Plain.* Je–Jl.

16. **P. fílipes** Rydb. Stems 3–5 dm. high, slender, silky-hirsute or silky-strigose; leaflets obovate to oblanceolate, obtuse, coarsely crenate, 2–6 cm. long, green and sparingly strigose above, white-tomentose beneath; bractlets linear-lanceolate, a little shorter than the lanceolate acute sepals; petals yellow, broadly obcordate, 6–8 mm. long, exceeding the sepals. *P. Hippiana pulcherrima* S. Wats. *P. gracilis* Porter & Coulter; not Dougl. Mountains, valleys: Man.—N.M.—Utah—Alta. *Submont.—Mont.* Je–Au.

17. **P. diversifòlia** Lehm. Stem erect, few-leaved, smooth or strigose, 1–2 dm. (rarely 3 dm.) high; basal leaves digitate, or often pinnate with approximate leaflets, or both in the same plant, more or less silky-strigose, especially beneath; leaflets most commonly 7, oblanceolate, cuneate, or sometimes obovate, more or less toothed; bractlets and sepals lanceolate, acute, 3.5–5 mm. long; petals obcordate or obovate and emarginate, 6–7 mm. long, about one third longer than the sepals. *P. dissecta* Nutt., not Pursh. Mountain valleys: Sask.—B.C.—Calif.—Colo.—S.D. *Submont.—Alp.* Je–Au.

18. **P. glaucophýlla** Lehm. Stems 1.5–4 dm. high, glabrous below, sparingly strigose above; leaflets oblanceolate-cuneate, 1–5 cm. long, coarsely serrate, with acute teeth directed forward, strigose when young, soon glabrate, glaucous; cyme 3–12-flowered; bractlets lanceolate, 3–5 mm. long, acute; sepals ovate, acuminate, 5–7 mm. long; petals obcordate, 6–10 mm. long. *P. diversifolia glaucophylla* Lehm. *P. dissecta glaucophylla* S. Wats. Mountain valleys: B.C.—Sask.—N.M.—Utah. *Submont.—Subalp.* Je–Au.

19. **P. concínna** Richardson. Stems many, spreading, more or less tomentose; leaves densely white-tomentose beneath, silky and slightly tomentose above when young, digitate (sometimes approximately pinnate); leaflets obovate or cuneate, more or less deeply toothed, 1–3 cm. long; sepals ovate, acute, 4–5 mm. long; petals obcordate, a little exceeding the sepals, about 6 mm. long. Dry hills and mountains: Colo.—S.D.—Sask.—Alta.—Ida. *Plain—Mont.* My–Jl.

20. **P. divìsa** Rydb. Stems spreading or diffuse, 5–10 cm. long, pubescent with long white hairs; basal leaves 5-foliolate, digitate, or pinnate with approximate leaflets; leaflets 1–2 cm. long, obovate or oblanceolate in outline, deeply cleft, with oblong or lanceolate divisions, pubescent above with long white hairs, densely white-tomentose beneath; sepals ovate, about 4 mm. long; petals broadly cuneate, emarginate, about 5 mm. long. *P. nivea dissecta* S. Wats. *P. concinna divisa* Rydb. Dry hills: S.D.—Sask.—Alta.—Colo. *Submont.—Alp.* Je–Jl.

21. **P. effùsa** Dougl. Stems many, 2–4 dm. high, slightly silky, ascending or diffuse; basal leaves usually interruptedly pinnate, grayish tomentose on both sides; leaflets 5–11, cuneate-oblong or oblanceolate, the upper often confluent, crenate, with broad usually ovate teeth; sepals lanceolate-acuminate, 4–5 mm. long; petals obovate, retuse, a little longer than the calyx. Plains and hills: Alta.—Sask.—Minn.—Neb.—N.M. *Plain—Mont.* Je–Au.

f. 280.

22. **P. Hippiàna** Lehm. *Fig. 280.* Stem erect, 3–5 dm. high, more or less silky-canescent or white with appressed hairs, leafy; basal leaves several; leaflets 7–11, gradually diminishing downward, white- or grayish-silky on both sides and tomentose beneath, obovate or cuneate-oblong, 2–5 cm. long, deeply obtusely toothed, the upper often confluent; sepals ovate-lanceolate, 5–7 mm. long; petals 6–8 mm. long, obovate, retuse. Plains and hills: Alta.—Sask.—Minn.—Neb.—N.M.—Ariz. *Prairie—Plain—Mont.* Jl–Au.

23. **P. argýrea** Rydb. Stems erect or nearly so, white-tomentose and silky; basal leaves pinnate, with 7–9 leaflets; leaflets oblanceolate or oblong-obovate, 1–3 cm. long, white-tomentose on both sides, coarsely crenate; sepals ovate, acute, 4–5 mm. long; petals obcordate, 5 mm. long. Plains:—Man.—Sask.—N.D. *Plain.* Je–Jl.

24. **P. propínqua** Rydb. Stems decumbent, ascending, or erect, finely silky-strigose, 2–5 dm. high; basal leaves pinnate, with 9–11 leaflets; leaflets oblanceolate, 1–6 cm. long, coarsely crenate, green and silky above, white-tomentose beneath, the upper two pairs usually more or less decurrent on the rachis; sepals ovate-lanceolate, 5–6 mm. long; petals obcordate, 6–8 mm. long. *P. diffusa* A. Gray; not Willd. *P. Hippiana pulcherrima* S. Wats., in part. Mountain meadows: Alta.—S.D.—N.M.—Ariz. *Submont.—Mont.* Je–Au.

25. **P. pulchérrima** Lehm. Stems erect or ascending, 4–6 dm. high, pubescent with ascending hairs; basal leaves pinnate, with 5–7 approximate leaf-

lets; leaflets oblanceolate, 2–7 cm. long, crenate, green and silky above, densely white-tomentose beneath, the lower two pairs spreading or the lowest reflexed, usually very close together; sepals ovate-lanceolate, acuminate, 5–7 mm. long; petals obcordate, 6–8 mm. long. *P. Hippiana pulcherrima* S. Wats. *P. Wardii* Greene. Mountains: Alta.—Sask.—S.D.—N.M.—Utah. *Submont.—Mont.* Je–Au.

26. **P. platténsis** Nutt. Stems 1–2 dm. high, decumbent or spreading; basal leaves many, pinnate; leaflets 9–17, light green in color, appressed-strigose or glabrate, obovate-oblong in outline, deeply pinnatifid, with oblong, obtuse to nearly linear, acutish lobes, 5–8 mm. long; flowers in few-flowered, rather open cymes; sepals and bractlets lanceolate, long-acuminate; sepals 3–4 mm. long; petals yellow, obovate, slightly retuse, longer than the sepals. Valleys: S.D.—N.M.—Utah—Sask. *Submont.—Mont.* Je–Au.

27. **P. multífida** L. Cespitose; stem at last spreading, 1–2 dm. high, silky-strigose; basal leaves pinnate; leaflets 5–7, grayish-tomentose beneath, glabrous above, pectinately divided into linear, acute, revolute divisions; sepals ovate-lanceolate, 3–4 mm. long; petals cuneate, emarginate, 4–5 mm. long. Subarctic regions: Greenl.—Lab.—Man.—Mack. *Subarctic—Arct.*

28. **P. bipinnatífida** Dougl. Stems several, erect or ascending, finely white-silky-villous, 3–5 dm. high; basal leaves many, pinnate, with 7–9 approximate leaflets, densely and finely silky above, white-tomentose beneath; leaflets 2–4 cm. long, obovate in outline, pectinately divided to near the midrib into almost linear, mostly obtuse segments; sepals ovate, 4–5 mm. long; petals obovate, cuneate, truncate, about equaling or slightly exceeding the sepals. Plains and hills: Man.—Minn.—Colo.—Alta. *Prairie—Plain—Mont.* Jl–Au.

29. **P. platýloba** Rydb. Stems several, erect or ascending, 3–5 dm. high, pubescent with white appressed hairs; leaves pinnate, usually with but two pairs of leaflets; leaflets obovate, 2–6 cm. long, greenish and silky-strigose above, white-tomentose beneath, cleft into broad lanceolate or oblong divisions; stem-leaves short-petioled and often subdigitate; sepals ovate, acute; petals about equaling the sepals, obovate. Plains: Man.—Colo.—Alta. *Plain—Mont.* Je–S.

30. **P. glabrélla** Rydb. Stem erect and strict, rather low, 1–2 dm. high, glabrate or sparingly pubescent; leaves pinnate, with 5–11 rather approximate leaflets, glabrate or glandular-pruinose, sparingly hairy on the veins and margins; leaflets obovate to oblanceolate, the lower ones smaller, very deeply dissected into oblong, acute or obtuse segments; sepals broadly ovate, rather strongly veined, about 5 mm. long; petals obovate, about equaling the sepals. *P. pennsylvanica glabrata* S. Wats. Mountains: Minn.—Man.—Alta.—Wyo. *Boreal.—Submont.* Jl–Au.

31. **P. pennsylvánica** L. Stem generally erect, strict, 4–8 dm. high, more or less appressed-pubescent and tomentulose; leaves pinnate, with 7–15 leaflets, grayish-tomentose and veiny beneath, short-strigose and green above; leaflets gradually reduced downward, often somewhat decurrent, oblong or oblanceolate in outline, divided halfway to the midrib into oblong divisions; sepals ovate-triangular, acute, usually not prominently veiny, about 5 mm. long; petals obovate, cuneate, slightly emarginate or truncate, about 6 mm. long. Plains: Ont.—Yukon—Neb.—Colo. *Prairie—Plain—Mont.* Jl.

32. **P. atróvirens** Rydb. Stem stout, 2–3 dm. high, with erect branches, dark-colored, densely pilose; leaves very dark green, densely pilose on both sides, strongly veined, pinnate, with 7–9 leaflets; these obovate to oblanceolate, coarsely dissected about halfway to the midrib into oblong segments; sepals and bractlets subequal, about 5 mm. long, ovate, exceeded by the cuneate-obovate bright yellow petals. Plains: Minn.—Wyo.—Colo.—Sask. *Prairie—Plain.* Je–Jl.

33. **P. strigòsa** Pall. Stems 1–4 dm. high, densely puberulent and with long spreading hairs; basal leaves pinnate, with 7–11 leaflets; leaflets obovate or oblanceolate, 1–5 cm. long, densely silky-pubescent, but yellowish green above, densely grayish-tomentose and silky beneath, deeply cleft into lanceolate or linear, revolute divisions; sepals ovate, slightly longer, strongly ribbed; petals obovate or orbicular, about equaling the sepals. Plains: Man.—Kans. —N.M.—B.C. *Plain—Submont.* Je–Au.

34. **P. lasiodónta** Rydb. Stem strict, 3–4 dm. high, densely pubescent, with short spreading pubescence; basal leaves pinnate, with 11–15 leaflets; leaflets oblong, 1–4 cm. long, green and densely short-pubescent, almost velvety above, grayish tomentose beneath, deeply serrate; sepals ovate, acute, strongly ribbed, about 6 mm. long; petals rounded-obovate, slightly if at all exceeding the sepals. Plains: Sask.—S.D.—Alta. *Plain.* Jl–Au.

35. **P. sulphùrea** Lam. Stem stout, 4–7 cm. high, leafy, finely pubescent and with long scattered hairs, pale green; leaves digitate, long-petioled; leaflets 5–7, or on the upper leaves 3; narrowly oblanceolate, deeply toothed, hirsute, light green; petals obovate, deeply emarginate, about 1 cm. long, sulphur-yellow. Around dwellings: Vt.—Minn.—Ills.—D.C.; escaped from cultivation; nat. of Eurasia.

8. ARGENTÌNA Lam. SILVER-WEED, GOOSE TANSY.

Perennial herbs, with slender prostrate stolons. Leaves interruptedly pinnate, with many leaflets. Flowers solitary in the axils of small leaves or scales on the stolons. Hypanthium almost flat. Bractlets, sepals, and petals normally 5. Petals yellow, broadly elliptic or almost orbicular, not at all unguiculate. Stamens 20–25, in three series, inserted as in *Potentilla;* filaments filiform; anthers didymous, dehiscent by a longitudinal slit. Receptacle hemispheric, bearing very numerous pistils. Styles filiform, lateral, scarcely deciduous. Mature achenes with thick pericarp. Seeds ascending and amphitropous.

Achenes corky, with a deep groove; stem and petiole and rachis of the leaves densely pubescent, with at first ascending and later spreading hairs.
  Leaves silvery on both sides. 1. *A. argentea.*
  Leaves green and glabrate above. 2. *A. Anserina.*
Achenes not corky, without a groove; stem and petiole and the rachis of the leaves glabrous or slightly appressed-hairy and glabrate; hypanthium acute at the base. 3. *A. subarctica.*

1. **A. argéntea** Rydb. Basal leaves 1–2 dm. long, pinnate, with 11–25 larger leaflets and smaller ones interposed; larger leaflets 1–3 cm. long, obovate, rounded at the apex, serrate, white-silky on both sides or slightly greener above; bractlets usually entire, about equaling the ovate or ovate-lanceolate sepals; petals obovate or broadly oval, 6–9 mm. long. *Potentilla Anserina concolor* Am. auth. [G]; not Ser. Wet places: Mack.—S.D.—N.M.—Ariz.—Ore.—B.C. *Plain—Mont.* My–Au.

2. **A. Anserìna** (L.) Rydb. Leaves 1–2 dm. long, interruptedly pinnate, with 9–31 larger leaflets and smaller ones interposed, spreading or flat on the ground, slightly silky and green above, white-silky and tomentose beneath; larger leaflets 1–4 cm. long, oblong or oblanceolate, usually acute, deeply and sharply serrate; bractlets simple and lanceolate, or ovate-lanceolate, toothed or divided, generally a little longer than the broadly ovate sepals; petals oval, 7–10 mm. long. *Potentilla Anserina* L. [G]. Wet places: Newf.—N.Y.—Calif. —Alaska; Eurasia. *Boreal.—Mont.* My–Au.

3. **A. subárctica** Rydb. Basal leaves 7–15 cm. long, ascending; rachis silky, with appressed or ascending hairs; larger leaflets 15–19, dark green and sparingly silky above, densely white-silky and tomentose beneath, deeply serrate,

the upper leaflet 1.5–2 cm. long; hypanthium about 5 mm. wide, silky and tomentulose, acute at the base; bractlets linear-lanceolate, 3–6 mm. long; sepals ovate, 5–6 mm. long; petals oval or suborbicular, about 8 mm. long. Wet places: Alaska—Yukon—Ida.—Mont.—Man. *Subarct.—Mont.—Subalp.* Je–Au.

9. CÓMARUM L. Marsh Cinquefoil, Purple Marshlocks, Cowberry.

Aquatic perennials, with long creeping rootstocks and pinnate leaves. Inflorescence cymose. Hypanthium almost flat or slightly saucer-shaped, enlarging in fruit, more or less tinged with red. Bractlets, sepals, and petals 5. Petals red, ovate, acuminate. Stamens 20–25, inserted near the base of the receptacle; filaments filiform, stout; anthers flat, cordate at the base, attached by the back, and opening by longitudinal marginal slits. Receptacle hemispheric, enlarging in fruit and becoming ellipsoid or hemispheric and spongy. Pistils numerous. Styles lateral, filiform. Seeds amphitropous.

1. C. palústre L. Stem ascending, 2–5 dm. high, more or less hirsute, with short spreading hairs, somewhat glandular on the upper portion; leaves pinnate, with 5–7 more or less approximate leaflets, green above, paler and purple-veined beneath, sparingly hairy, in age glabrate; leaflets elliptic or oval, mostly acute at both ends, to linear-oblong, 5–8 cm. long and only 1–2 cm. wide, obtuse or rounded at the apex; hypanthium in flower 7–8 mm., in fruit about 15 mm. in diameter, short-pilose and glandular-pubescent; sepals ovate to lanceolate, acuminate, about 1 cm. long in flower and 1.5 cm. in fruit; petals spatulate or ovate, acuminate or acute, scarcely half as long as the sepals. *Potentilla palustris* (L.) Scop. [G]. Wet bogs: Greenl.—Lab.—N.H.—Wyo.—Calif.—Alaska; Eurasia. *Mont.* Jl–S.

10. FRAGÀRIA L. Strawberry.

Perennial acaulescent herbs, with scaly rootstock, and producing runners which root and form new plants. Leaves (in all ours) ternate, basal. Hypanthium almost flat. Bractlets, sepals, and petals normally 5. Petals white, or in one species reddish, broadly obovate, elliptic, or almost orbicular, obtuse, never emarginate. Stamens about 20, in three series as in *Potentilla,* closely surrounding the base of the receptacle; filaments short; anthers dehiscent by a longitudinal slit. Receptacle hemispheric or conic, bearing very numerous pistils, in fruit becoming enlarged, very juicy, and delicious in taste. Styles filiform, but rather short, attached near the middle of the ovaries. Seeds ascending and amphitropous.

Pubescence of the scape and the petioles divaricate.
  Achenes superficial on the receptacle; leaflets sessile. — 1. *F. americana.*
  Achenes set in deep pits in the receptacle; leaflets petioluled.
    Fruit globose.
      Pedicels with appressed hairs. — 2. *F. virginiana.*
      Pedicels with spreading hairs. — 3. *F. Grayana.*
    Fruit oblong-conic. — 4. *F. canadensis.*
Pubescence of the scape and the petioles appressed-ascending; achenes in pits.
  Plant not glaucous; leaflets small, elliptic or narrowly obovate. — 5. *F. pumila.*
  Plant more or less glaucous.
    Leaflets obovate; scape several-flowered. — 6. *F. glauca.*
    Leaflets oblong-cuneate; scape 1–4-flowered. — 7. *F. pauciflora.*

1. F. americàna (Porter) Britton. Leaflets 3–8 cm. long, rhombic-obovate, mostly acute, sharply and deeply serrate, very thin, very soon glabrate on both sides; scape slender, seldom over 1.5 dm. high, seldom much exceeding the leaves, very rarely leafy-bracteate; fruit elongate-ovoid, 5–7.5 mm. in diameter and 1–1.5 cm. long. *Fragaria vesca americana* Porter [G]. Woods: Newf.—Man.—N.M.—Va. *Boreal.—Submont.—Mont.* Je–Au.

2. **F. virginiàna** Duchesne. Leaflets 3–10 cm. long, dark green, all petioluled, silky when young, glabrate in age; petioles with spreading hairs; scape more or less villous, 1–2 dm. high; flowers 1–2 cm. broad; fruit 1–1.5 cm. broad, subglobose; achenes set in deep pits. Grassy places: N.S.—Minn.—Okla.—Ga. *Canad.—Allegh.* Ap–Je.

3. **F. Grayàna** E. Vilm. Like the last but more hirsute; leaflets 4–10 cm. long, coarsely serrate; scape densely hirsute, 1–1.5 dm. high; petals orbicular, 5–7 mm. long; fruit subglobose, 1.5 cm. broad; achenes set in pits. *F. virginiana illinoensis* (Prince) A. Gray [G]. Open places: Ind.—N.C.—Kans.—La.—Ala. *Prairie—Ozark.* Ap–Je.

4. **F. canadénsis** Michx. Leaflets thin, often reddish, glabrate above, silky beneath, obovate or cuneate, 2–4 cm. long; scape shorter than the leaves, 4–10 cm. high, with spreading hairs, 2–4-flowered; corolla 1–1.5 cm. broad; petals broadly obovate, half longer than the sepals; fruit 6–7 mm. broad, 1–1.25 cm. long. Open woods: Newf.—N.Y.—Mich.—Man.—Mack. *Boreal.* Je–Jl.

5. **F. pùmila** Rydb. Leaflets oblong or narrowly obovate, 1–3 cm. long, crenate, small and thick, dark green and silky above when young, soon glabrate, silky-strigose and finely puberulent beneath; scape few-flowered, usually about 5 cm. long; flowers 1–1.5 cm. in diameter; petals obovate, exceeding the sepals by about a half; fruit scarcely 1 cm. in diameter. Dry hills: N.D.—Colo.—Wyo. *Submont.—Subalp.* My–Jl.

6. **F. glaùca** (S. Wats.) Rydb. Leaflets broadly obovate, 3–5 cm. long, coarsely toothed, thin, glaucous and almost glabrous above, silky or at last glabrate beneath; scape slender, rarely exceeding the leaves; flowers 1.5–2 cm. in diameter; petals obovate, exceeding the sepals by about a half; fruit subglobose, 1–1.25 cm. in diameter. *Fragaria virginiana* (?) *glauca* S. Wats. Mountains: B.C.—Nev.—N.M.—S.D.—Mack. *W. Temp.—Plain—Subalp.* My–Au.

7. **F. pauciflòra** Rydb. *Fig. 281.* Leaflets cuneate, subsessile, coarsely toothed above the middle, the lateral ones scarcely at all oblique at the base, thin and glaucous, almost glabrous above and soon glabrate beneath; scape 2–4-flowered, 5–15 cm. long, seldom exceeding the leaves; flowers 1–1.5 cm. in diameter; petals obovate; fruit subglobose, about 1 cm. in diameter; achenes set in very deep pits. Hills: Man.—Colo.—Alta. *Plain—Mont.* Je–Au.

11. SIBBÁLDIA L.

Low, tufted perennial herbs, with short cespitose caudices or rootstocks and ternate leaves. Hypanthium saucer-shaped or cup-shaped, small. Bractlets, sepals, and petals 5. Petals yellow, obovate, cuneate, or oblanceolate, scarcely equaling the sepals. Stamens 5, inserted not very close to the small receptacle; filaments filiform, but short, inclined. Pistils 5–20; styles lateral. Ovule and seed attached near the base of the style, ascending and amphitropous.

1. **S. procúmbens** L. Stems less than 1 dm. high, more or less hirsute-strigose; leaflets sparingly appressed-pilose, 1–2 cm. long, broadly cuneate, 3–5-toothed at the apex; flowers few in rather dense cymes; bractlets and sepals subequal, broadly oblong or ovate; petals yellow, spatulate, shorter than the sepals. Arctic and alpine regions: Greenl.—N.H.—Man.—Colo.—Calif.—Alaska; arctic and alpine Eurasia. *Arct.—Alp.—Subalp.* Je–Au.

12. **SIBBALDIÓPSIS** Rydb.

Cespitose undershrubs, with a branched caudex. Leaves trifoliolate, leathery. Bractlets, sepals, and petals 5. Petals white, obovate or elliptic, sessile. Stamens 20, in 3 series; filaments filiform; anthers cordate. Receptacle hemispheric. Pistils many; styles slender, lateral; stigmas truncate. Achenes turgid, villous.

1. **S. tridentàta** (Solander) Rydb. Stem 1–2 dm. high, appressed-silky; leaflets obovate-cuneate, subcoriaceous, green and shining above, pale beneath; bractlets oblong, shorter than the lance-ovate sepals; petals white, obovate or elliptic, half longer than the sepals. *Potentilla tridentata* Ait. [G]. Mountains and hills: Greenl.—Mass.—Minn.—Iowa—Ga.—N.J. *Arct.—E. Boreal.* Je–Au.

13. **DASÍPHORA** Raf. SHRUBBY CINQUEFOIL, YELLOW ROSE.

Shrubs, with scarious sheathing stipules, pinnate leathery leaves, and axillary flowers. Hypanthium saucer-shaped. Bractlets, sepals, and petals 5. Petals in ours yellow, nearly orbicular, neither unguiculate nor emarginate. Stamens about 25, in 5 festoons on a pentagonal disk; filaments filiform; anthers oblong, flat, not didymous, dehiscent by a longitudinal slit along the margin. Receptacle hemispheric, with numerous pistils; styles club-shaped, thick and glandular upward, inserted near or below the middle of the ovaries; stigmas large and evidently four-lobed; achenes densely covered with long straight hairs. Seeds ascending and amphitropous.

1. **D. fruticòsa** (L.) Rydb. *Fig. 282.* Shrub 3–15 dm. high; branches with brownish shreddy bark; leaves pinnate, with 3–7 approximate leaflets, silky above, silky and whitish beneath; leaflets oblong or linear-oblong; flowers in small loose cymes or solitary, large, 1.5–3 cm. in diameter; petals yellow, orbicular, often twice as long as the sepals. *Potentilla fruticosa* L. [G]. Cold valleys and among rocks: Lab.—Alaska—Calif.—N.M.—N.J.; w Eur.; e Asia. *Boreal—Submont.—Subalp.* Je–Au.

14. **DRYMOCÁLLIS** Fourr.

Perennial herbs, with scaly rootstocks, glandular foliage, pinnate leaves, and cymose inflorescence. Hypanthium saucer-shaped or hemispheric. Bractlets, sepals, and petals 5. Petals obovate, elliptic, or nearly orbicular, neither unguiculate nor emarginate, yellow or white. Stamens 20–30, in 5 festoons on the much thickened pentagonal disk around the receptacle; filaments filiform; anthers oblong, truncate at each end or cordate at the base, flat and dehiscent by longitudinal marginal slits. Receptacle hemispheric or semi-ellipsoid, with very numerous pistils. Style nearly basal, in all our species somewhat thickened and glandular a little below the middle and tapering to each end; stigma minute. Seed attached near the base of the ovary, ascending and orthotropous.

Petals white or cream-colored.
  Petals slightly if at all exceeding the sepals; inflorescence many-flowered, dense. — 1. *D. agrimonioides.*
  Petals much exceeding the sepals; cyme open, few-flowered; stem slender. — 2. *D. pseudorupestris.*
Petals yellow in anthesis.
  Petals much exceeding the sepals in length; plant low, leafy up to the inflorescence. — 3. *D. fissa.*
  Petals slightly if at all exceeding the sepals; plant tall, the upper leaves reduced. — 4. *D. glandulosa.*

1. **D. agrimonioìdes** (Pursh) Rydb. *Fig. 283.* Stem stout and erect, 3–10 dm. high, striate, generally densely glandular- or viscid-hirsute; basal leaves many; leaflets 7–11, strongly veined, densely pubescent on both sides; upper three leaflets larger than the others, usually 4–5 cm. long, doubly serrate, the odd one commonly rhomboid, the others usually obliquely ovate; flowers 12–18 mm. in diameter; bractlets lanceolate, much smaller than the oblong-ovate, acute or mucronate sepals. *Potentilla arguta* Pursh [G]. Meadows and rocky places: N.B.—D.C.—Colo.—Mack. *Canad.—Allegh.—Plain—Submont.* Je–Au.

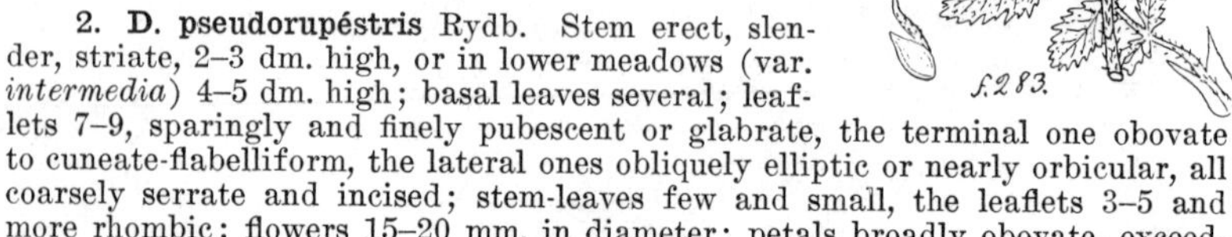

2. **D. pseudorupéstris** Rydb. Stem erect, slender, striate, 2–3 dm. high, or in lower meadows (var. *intermedia*) 4–5 dm. high; basal leaves several; leaflets 7–9, sparingly and finely pubescent or glabrate, the terminal one obovate to cuneate-flabelliform, the lateral ones obliquely elliptic or nearly orbicular, all coarsely serrate and incised; stem-leaves few and small, the leaflets 3–5 and more rhombic; flowers 15–20 mm. in diameter; petals broadly obovate, exceeding the sepals by a third; sepals ovate-lanceolate, pointed. Mountains: Alta.—S.D.—n Wyo.—Ida. *Submont.—Subalp.* Je–Au.

3. **D. físsa** (Nutt.) Rydb. Stem low, 2–3 dm., or sometimes 4 dm. high, very leafy; basal leaves with short petioles, pinnate; leaflets, as a rule, 9, nearly orbicular, except the upper ones, which are somewhat rhombic, all deeply incised and doubly serrate; stem-leaves similar, only the upper ones slightly reduced; flowers 15–20 mm. in diameter; sepals triangular-lanceolate, long-acuminate; petals orbicular, very concave, much exceeding the sepals. Mountains: S.D.—Alta.—Utah—Colo. *Submont.—Subalp.* My–Au.

4. **D. glandulòsa** (Lindl.) Rydb. Stem strict, 3–6 dm. high; basal leaves pinnate, sometimes interruptedly so; leaflets 7–9, sparingly hairy, obovate, obtuse, simply or doubly serrate, with broad teeth, the upper 1–3 cm. long; stem-leaves smaller, short-petioled, 3–7-foliolate; flowers in an open many-flowered cyme, 10–15 mm. in diameter; sepals oblong or ovate-lanceolate; petals obovate, about equaling the sepals. Mountains: S.D.—B.C.—Calif.—N.M. *W. Temp.—Submont.—Mont.* Je–Au.

15. CHAMAÈRHODOS Bunge.

Perennial or biennial herbs, with taproots, 2–4 times ternately divided leaves, and small cymose flowers. Hypanthium cup-shaped, small. Petals and sepals 5; bractlets wanting. Petals obovate or cuneate, somewhat clawed. Stamens 5, opposite the petals; filaments subulate, short, persistent; anthers didymous, opening by a slit. Pistils 5–10, or more; styles basal, filiform. Seed inserted near the base of the style, ascending and nearly orthotropous.

1. **C. Nuttállii** Pickering. Plant 1–3 dm. high, erect, branching and leafy, hirsute and glandular; basal leaves numerous and rosulate, 2–4-ternately divided into linear or oblong divisions; inflorescence many-flowered; hypanthium 2–3 mm. in diameter, hispid; sepals narrowly lanceolate, equaling or somewhat shorter than the white obovate-cuneate petals. *C. erecta* Hook.; not Bunge. Plains: Man.—S.D.—Colo.—Alaska. *Plain—Mont.* Jl–Au.

16. SANGUISÓRBA L. Burnet.

Leafy perennial herbs, with thick rootstocks. Leaves odd-pinnate, with adnate stipules and toothed leaflets. Flowers perfect or some of them pistillate, in dense spikes. Hypanthium urn-shaped, contracted at the mouth, angled, and usually winged. Sepals 4, petaloid, deciduous, very concave. Petals none.

Stamens in ours 4, opposite the sepals. Pistils solitary; styles terminal; stigmas muricate-papillose; ovule solitary, suspended. Achenes dry, enclosed in the indurate 4-angled smooth hypanthium.

1. **S. officinàlis** L. A glabrous perennial; stem 1–2 m. high; lower leaves with 9–13 leaflets, the upper ones reduced; leaflets lanceolate-oblong or sometimes ovate, serrate, 2–8 cm. long, slightly cordate at the base; spike rounded to oblong, peduncled, 1–3 cm. long, about 1 cm. thick; sepals dark purple, 2–2.5 mm. long; filament filiform, not exceeding the sepals; fruiting hypanthium obovoid, 4-angled. Waste places and around dwellings: Me.—Minn. —Calif.; escaped from cultivation; native of Eurasia.

17. **POTERÍDIUM** Spach.

Leafy branched annuals or biennials, with taproots. Leaves odd-pinnate, with adnate stipules and pectinately pinnatifid leaflets. Flowers greenish, perfect, in dense, oblong spikes. Hypanthium urn-shaped, contracted at the mouth, 4-winged. Sepals 4, green, with white-scarious margins. Stamens 2 or 4, opposite all or the inner two sepals; filaments short. Pistils solitary; styles terminal; stigmas brush-like. Ovules solitary, suspended. Achenes enclosed in the 4-winged, indurate, dry hypanthium.

1. **P. ánnuum** (Nutt.) Spach. Stem branched, leafy, 1–4 dm. high; leaves glabrous, odd-pinnate; leaflets of the lower leaves 7–15, obovate, pectinately pinnatifid into 9–15, linear divisions; spike globose or oblong, 0.5–3 cm. long, 7 mm. thick; sepals ovate, white-margined; stamens 4; fruiting hypanthium ovoid, 4-angled, winged, not reticulate on the faces. *Poterium annuum* Nutt. Moist places: Ark.—Kans.—Tex. *Ozark.* Je–Au.

18. **POTÈRIUM** L. Salad Burnet.

Leafy perennial herbs, with rootstocks. Leaves odd-pinnate, with adnate stipules and toothed leaflets. Flowers polygamo-monoecious, in dense spikes. Hypanthium urn-shaped, contracted at the mouth, 4-angled. Sepals 4, petaloid, deciduous, concave. Petals none. Stamens in the staminate flowers numerous, with filiform, exserted and declined filaments, in the perfect flowers fewer. Pistils 2, styles terminal; stigmas many-cleft and brush-like; ovules solitary, suspended. Achenes dry, enclosed in the indurate, 4-angled and rugose hypanthium.

1. **P. Sanguisórba** L. Stem branched, 2–5 dm. high; leaflets 7–9, oval to orbicular, coarsely serrate, 1–2 cm. long; spike globose, 10–12 mm. broad; lower flowers staminate, the upper perfect or pistillate; sepals oval, greenish within, purplish without, 3.5–4 mm. long; fruiting hypanthium 4 mm. long, 4-angled, with thick ridges and alveolate-favose on the faces. *Sanguisorba minor* Scop. [G]. Around dwellings: Me.—D.C.—Neb.—Minn.; escaped from cultivation in the East; nat. from Eur. Jl-S.

19. **AGRIMÒNIA** L. Agrimony.

Perennial herbs, with rootstocks. Leaves odd-pinnate, with smaller leaflets interposed between the larger ones. Flowers in narrow racemes, regular, perfect. Hypanthium hemispheric to obconic, constricted at the throat and enclosing the achene in fruit, usually 10-grooved longitudinally, with a ring of hooked bristles above. Sepals 5, after anthesis more or less connivent, forming a nipple-shaped beak on the fruit. Petals 5, small, yellow, clawless. Stamens 5–15; filaments slender. Pistils 2; styles terminal; stigmas 2-lobed; ovules suspended.

Fruiting hypanthium with more than 4 series of hooked bristles, the outer bristles reflexed; sepals hooked at the apex. 1. *A. gryposepala.*
Fruiting hypanthium with 2–4 series of hooked bristles; the latter all erect or merely spreading; sepals not hooked.

Racemes and leaves glabrous or nearly so, merely glandular-granuliferous; roots tuberous. — 2. *A. rostellata.*
Racemes and lower surface of the leaves decidedly hairy.
Leaflets not glandular-granuliferous beneath.
Hypanthium in fruit 2.5–3 mm. long, with shallow grooves; leaflets 5–7; roots tuberous-thickened. — 3. *A. pubescens.*
Hypanthium in fruit 5 mm. long, with deep grooves; leaflets 7–13; roots not tuberous. — 4. *A. Eupatoria.*
Leaflets glandular-granuliferous beneath.
Leaflets 5–9, broadly lanceolate to oblanceolate, with few teeth. — 5. *A. striata.*
Leaflets 9–23, narrowly lanceolate, with numerous teeth. — 6. *A. parviflora.*

1. **A. gryposépala** Wallr. A perennial, with short thick rootstocks; stem 5–8 dm. high, sparingly hirsute and glandular-granuliferous; principal leaflets 5–9, oval or obovate, or those of the upper leaves narrower, coarsely serrate, acute, glabrous or sparingly strigose above, hirsute and glandular-granuliferous beneath, 4–12 cm. long; petals 2.5–3 mm. long, bright yellow; fruiting hypanthium 4–5 mm. long, abruptly contracted at the base, strigose, strongly grooved. *A. hirsuta* Bickn. Thickets: N.S.—Man.—N.D.—N.M.—S.C.; Calif.; Mex. *Temp.—Plain—Submont.* Je–Au.

2. **A. rostellàta** Wallr. A perennial, with short rootstock and tuberous-thickened roots; stem 2–10 dm. high; principal leaflets 2–9, obovate or oval, thin, glabrous or nearly so; petals 2–3 mm. long, pale yellow; fruiting hypanthium 2 mm. long, hemispheric, glandular-granuliferous, with broad ribs and shallow grooves; rim obsolete; bristles in 3 series, only the lowest spreading. Hillsides: Conn.—Kans.—Ala.—Ga. *Allegh.* Jl–S.

3. **A. pubéscens** Wallr. A perennial, with tuberous-thickened roots; stem 3–15 dm. high, finely and densely crisp-hairy and slightly hirsute; principal leaflets 5–7, dark green and hirtellous or glabrous above, velvety-pubescent beneath, 2–10 cm. long, the terminal one obovate, the lateral ones more elliptic, coarsely toothed; petals 1.5 mm. long, deep yellow; fruiting hypanthium turbinate, 2.5–3 mm. long, deeply grooved, glabrous, with a small and obsolete rim. *A. mollis* (T. & G.) Britton [G, B]. Open woods: N.Y.—Iowa—Kans.—Ark.—Ga. *Allegh.* Jl–O.

4. **A. Eupatòria** L. A perennial, with a stout rootstock; stem 3–6 dm. high, hirsute; principal leaflets 9–13, or on the upper part of the stem 7, elliptic or lance-elliptic, 2–6 cm. long, coarsely serrate, acute at each end, dark green and appressed-pubescent above, finely pubescent and hirsute beneath; petals golden-yellow, 4–6 mm. long; fruiting hypanthium turbinate, strongly grooved, with a distinct rim; bristles in 4 series, ascending. Waste places: Wis.—Minn.; adv. from Eur. Jl–S.

5. **A. striàta** Michx. *Fig. 284.* A perennial, with stout rootstock and fibrous roots; stem 3–20 dm. high, hirsute and glandular-papillose; principal leaflets 7–13, hispidulous or scabrous above, copiously glandular-granuliferous and somewhat pubescent beneath, lanceolate, elliptic, or oblanceolate, or rhombic-obovate, acuminate, sharply serrate, 3–10 cm. long; petals deep yellow, obovate, 3 mm. long; fruiting hypanthium 5 mm. long, turbinate, strongly ribbed, glandular-granuliferous. *A. Brittoniana* Bickn. [B]. Roadsides and copses: N.S.—B.C.—N.M.—W.Va. *Temp.—Plain—Submont.* Jl–Au.

6. **A. parviflòra** Ait. A perennial, with a stout rootstock; stem 3–20 dm. high, densely hirsute, finely pubescent and glandular-granuliferous; principal leaflets 11–23, strongly veined, 1.5–12 cm. long, subsessile, strongly serrate, hispidulous or glabrate above, pubescent and glandular-granuliferous beneath; petals pale yellow, 2 mm. long; fruiting hypanthium turbinate, minutely glandu-

lar-granuliferous, with shallow grooves; rim obsolete; bristles in 2 series, the outer spreading. Damp ground: N.Y.—Neb.—Miss.—Fla.; Hispaniola; Mex. *Allegh.—S. Temp.—Trop.* Jl–O.

20. WALDSTEÌNIA Willd. BARREN STRAWBERRY.

Herbs, with perennial rootstocks and membranous stipules. Leaves basal, palmately 3–5-foliolate or 3–7-lobed. Stem in our species scapiform. Hypanthium mostly obconic. Sepals 5, valvate, spreading. Petals 5, yellow, obovate. Stamens numerous; filaments subulate, flattened below. Pistils 1–6, inserted on a short villous receptacle; styles filiform, terminal, deciduous. Achenes obliquely ovoid, pubescent. Seeds erect.

1. **W. fragarioìdes** (Michx.) Tratt. A scapose perennial; leaves basal, 3–5-foliolate; leaflets obovate, rounded at the apex, cuneate at the base, lobed and dentate, 2–10 cm. long, pubescent on the veins or glabrate; scape 1–2 dm. high; hypanthium turbinate, glabrous, 3–4 mm. long; petals bright yellow, obovate, 6–10 mm. long. Woods: N.S.—ne Minn.—Ind.—Ga. *Canad.—Allegh.* My–Je.

21. DRỲAS L. DRYAD, MOUNTAIN AVENS.

Low depressed undershrubs, strongly cespitose, with short horizontal branches. Leaves alternate, petioled, simple, crenate or entire, white-tomentose beneath. Flowers solitary on naked peduncles. Hypanthium saucer-shaped, little developed. Bractlets wanting. Sepals 8–10, persistent. Petals 8–10, obovate or oval. Stamens numerous, inserted in the mouth of the hypanthium; filaments subulate. Pistils numerous, sessile; style terminal, persistent, elongate and plumose in fruit. Fruit indehiscent, 1-seeded. Seeds basal, ascending.

Sepals linear or linear-lanceolate; petals white, spreading.
  Leaf-blades entire or with a few acutish teeth, shining and not conspicuously rugose above, broadest near the base. — 1. *D. integrifolia.*
  Leaf-blades coarsely round-crenate, dull, strongly rugose and impressed-reticulate above, broadest near the middle. — 2. *D. octopetala.*
Sepals ovate or ovate-lanceolate; petals yellow, strongly ascending. — 3. *D. Drummondii.*

1. **D. integrifòlia** Vahl. Leaf-blades lanceolate or elliptic-lanceolate, rounded, truncate, or subcordate at the base, acutish at the apex, 6–25 mm. long, usually with entire revolute margins, densely white-tomentose beneath, glabrate and shining above; scapes 5–15 cm. long, tomentose; hypanthium and calyx tomentose and usually more or less densely black- or brown-hairy; sepals linear or linear-lanceolate, 5 mm. long; petals white, elliptic, 8–12 mm. long. Ridges: Greenl.—Lab.—N.H.—Man.—B.C.—Alaska. *Arct.—Alp.* Je–Jl.

2. **D. octopétala** L. *Fig. 285.* Leaf-blades oblong, elliptic or oval, white-tomentose beneath, glabrate and dull above, 6–25 mm. long, rounded or obtuse at both ends, or subcordate at the base; scape 5–20 cm. long, tomentose and more or less black-hairy; hypanthium black-hairy and tomentose; petals white, elliptic or obovate-elliptic, 1–1.5 cm. long. Ridges: Greenl.—Colo.—Wash.—Alaska; alpine and arctic Eurasia. *Arct.—Alp.—Subalp.* Je–Au.

3. **D. Drummòndii** Richardson. Leaf-blades elliptic or obovate, 1–3 cm. long, white-tomentose beneath, dark green, dull, slightly tomentose or glabrous above, veiny and somewhat rugose, coarsely crenate, acute at the base and rounded at the apex; scape 5–20 cm. long, tomentose and somewhat black-hairy above; petals yellow, about 1 cm. long, elliptic-spatulate or obovate. Ridges: Que.—Mont.—Ore.—B.C.—Mack. *Mont.—Subalp.* Je–Au.

22. GÈUM L. Avens.

Perennial herbs, with rootstocks. Basal leaves lyrately odd-pinnate, with large terminal divisions; stem-leaves few, reduced, most often 3-fid or the upper bract-like. Stipules persistent, adnate to the clasping bases of the petioles. Flowers cymosely corymbose. Hypanthium turbinate, hemispheric, or saucer-shaped. Bractlets 5, sometimes reduced. Sepals 5, imbricate or valvate. Petals 5, orbicular, obovate, or cuneate, yellow, white, or red. Stamens numerous, in several series; filaments free, filiform. Pistils numerous, inserted on a clavate receptacle; styles filiform, terminal, strongly curved and geniculate above, the upper portion deciduous. Achenes small, dry, with a hooked beak. Seed erect.

Sepals reflexed; upper internode of the style less than one third as long as the lower..
  Receptacle stalked; bractlets none or rudimentary. 1. *G. vernum.*
  Receptacle sessile; bractlets present, seldom rudimentary.
    Petals white or ochroleucous.
      Receptacle glabrous; achenes sparingly hispid towards the apex, otherwise glabrous. 2. *G. virginianum.*
      Receptacle long-hairy; achenes pubescent throughout.
        Carpels 50–80, in heads 15–18 mm. broad; leaves thick; petals broadly obovate or suborbicular, ochroleucous. 3. *G. camporum.*
        Carpels 30–50, in heads 10–15 mm. broad; leaves thin; petals oblong or cuneate, white or cream-colored. 4. *G. canadense.*
    Petals bright or golden yellow; receptacle short-hairy.
      Upper internode of the style hirsute, the hairs about 1 mm. long; lower internode hirsute below, not glandular; petals orbicular, rounded at the base. 5. *G. strictum.*
      Upper internodes of the style short-hispidulous at the base or glabrous; petals usually cuneate at the base, except in *G. perincisum;* lower internode of the style glandular, puberulent.
        Divisions of the upper stem-leaves ample, rhombic or cuneate; inflorescence corymbiform, usually with short branches. 6. *G. macrophyllum.*
        Divisions of the upper small stem-leaves oblanceolate or rarely obovate; inflorescence open, with slender branches.
          Leaflets doubly serrate or dentate; terminal leaflet of the basal leaves rounded or reniform. 7. *G. oregonense.*
          Leaflets, especially those of the upper leaves, deeply incised; terminal leaflet of the basal leaves rhombic, ovate, or subcordate. 8. *G. perincisum.*
Sepals ascending or merely spreading, upper internode of the style at least half as long as lower internode, hirsute. 9. *G. rivale.*

1. **G. vérnum** (Raf.) T. & G. Stem 3–6 dm. high, slightly pubescent; basal leaves 3–9-foliolate; leaflets oval or obovate, 1–3 cm. long, or those of the new shoots simple and reniform, 3–6 cm. wide; stem-leaves pinnate, with double-serrate divisions, the uppermost ternate; sepals triangular; petals oblong, yellow or ochroleucous, equaling the sepals; fruiting heads 1 cm. broad; body of the achenes finely strigose; style glabrous, the upper internode 1 mm. long. Meadows and woods: Ont.—Kans.—Tex.—Tenn.—N.J. *Allegh.—Carol.—Ozark.* Ap–Je.

2. **G. virginiànum** L. A perennial; stem 3–10 dm. high, hirsute; earlier basal leaves often simple, with rounded-reniform to cuneate blades, the rest pinnate, with obovate, dissected leaflets; upper stem-leaves 3-divided or 3-lobed; petals white, oblong or cuneate, shorter than the sepals; fruiting heads 1.5–2 cm. broad; body of the achenes 3–4 mm. long, hispid at the apex; lower internode of the styles 4–5 mm. long, glabrous, the upper 1 mm. long, sparingly pubescent. Thickets: N.B.—Minn.—Kans.—N.C. *Canad.—Allegh.* My–Jl.

3. **G. campòrum** Rydb. A perennial; stem 5–10 dm. high, short-hirsute; basal leaves either simple, with rounded-cordate blades 3–6 cm. long, or ter-

nate, with rhombic-ovate divisions, somewhat lobed and dentate, sparingly pubescent; upper stem-leaves simple, rhombic-ovate, 3-lobed; petals ochroleucous, 4–5 mm. long, about equaling the sepals, obovate or orbicular; fruiting heads 15–18 mm. broad; body of the achenes 3–3.5 mm. long, hispid; lower internodes of the style glabrous, the upper 1–1.5 mm. long, with a few hairs. Moist woods: Minn.—S.D.—Tex.—Ark. *Prairie—Plain—Ozark.* Je–Au.

4. **G. canadénse** Jacq. Stems 3–10 dm. high, finely pubescent or glabrate; blades of the earlier basal leaves reniform or rounded-cordate, round-lobed and dentate, those of the later usually ternate, with rhombic or obovate leaflets; upper stem-leaves often simple and subsessile, ovate, acute, and 3-lobed; petals white, elliptic or oblong, seldom exceeding the sepals; fruiting head 10–12 mm. in diameter. *Geum album* J. F. Gmel. Banks and among bushes: N.S.—Ga.—Tex.—S.D.; Mex. *E. Temp.—Plain.* Je–S.

5. **G. stríctum** Soland. Stem 5–15 dm. high, strongly hirsute; basal leaves interruptedly lyrate-pinnate; principal leaflets 5–9, cuneate, obovate, or on the young shoots the terminal leaflet rounded or reniform, 2–10 cm. long, cleft or divided and doubly toothed; uppermost stem-leaves 3-foliolate and short-petioled; petals orbicular, 5–8 mm. long, bright yellow; fruiting head 15–18 mm. in diameter. *G. scopulorum* Greene. Low meadows: Newf.—Pa.—Mo.—Ariz.—B.C.; Mex. *Temp.—Submont.* Je–Au.

6. **G. macrophýllum** Willd. Stem very hirsute, especially below, with yellowish hairs; basal leaves interruptedly lyrate-pinnate; principal leaflets 5–15, the terminal one reniform or rounded, often 3-lobed, 5–10 cm. broad, the lateral ones obovate or rounded, 1–5 cm. long, hirsute on both sides, doubly toothed; petals bright yellow, obovate or obcordate, 4–7 mm. long; fruiting head 15–18 mm. in diameter. Wet meadows: Newf.—N.H.—S.D.—Mont.—Calif.—Alaska; e Siberia. *Boreal.—Submont.—Mont.* My–Au.

7. **G. oregonénse** (Scheutz) Rydb. Stem 4–7 dm. high, densely hirsute, with yellowish hairs; basal leaves interruptedly lyrate-pinnate; principal leaflets 5–9; terminal leaflet reniform or rounded, 5–10 cm. broad, doubly dentate; lateral leaflets obovate or cuneate, 1–5 cm. long, more or less hirsute on both sides; upper stem-leaves small, usually ternate or 3-cleft, with obovate or oblanceolate divisions, or simple; petals pale yellow, 4–5 mm. long, obovate; fruiting head about 15 mm. in diameter. *G. urbanum oregonense* Scheutz. Meadows: Mack.—Minn.—N.M.—Calif.—B.C. *W. Temp.—Submont.—Mont.* My–S.

8. **G. perincìsum** Rydb. Stem strict, 4–6 dm. high, conspicuously hirsute, with divaricate hairs; basal leaves interruptedly pinnate; principal leaflets 5–11, 1–6 cm. long, obovate or rhombic-obovate, or the terminal one ovate or cordate and then deeply 3-cleft, all cleft and incised; upper stem-leaves 3-foliolate; leaflets oblanceolate; petals pale yellow, oval or obovate, about 5 mm. long. Meadows: Mack.—Mich.—Alta.—Yukon. *Boreal.—Subalp.—Mont.* Je–Jl.

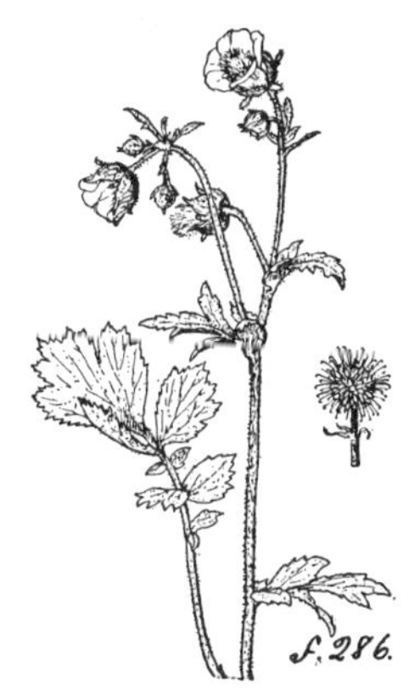

9. **G. rivàle** L. *Fig. 286.* Stems 3–6 dm. high, simple, more or less hirsute; basal leaves lyrate-pinnate; leaflets obovate or cuneate or the terminal one rounded or reniform, 2–10 cm. long, bidentate or biserrate; stem-leaves ternate, with oblanceolate or obovate leaflets; petals flesh-colored or sometimes tinged with yellow, purple-veined, clawed, flabelli-

form, emarginate, 7–10 mm. long. In swamps and low ground: Lab.—Newf.—N.J.—Mo.—N.M.—B.C.; Eurasia. *Boreal.—Plain—Subalp.* My–Au.

G. macrophyllum × rivale. Resembling *G. rivale* in habit and foliage, but the flowers smaller and more numerous, the sepals widely spreading or somewhat reflexed, and the petals golden yellow, obovate, and less clawed. *G. pulchrum* Fernald [G]. Que.—Vt.—Alta.

G. rivale × strictum. With the habit of *G. strictum,* but differing in the petals, which are golden yellow, obcordate, clawed at the base and emarginate at the apex, longer than the reflexed sepals; the basal leaves are those of *G. rivale. G. aurantiacum* Fries. B.C.—Alta.—n N.Y.

23. SIEVÉRSIA Willd. OLD MAN'S WHISKERS.

Perennial herbs, with rootstocks or suffruticose bases. Basal leaves usually numerous, lyrately or odd-pinnately divided, usually with smaller segments interposed between the larger ones; stem-leaves usually reduced; stipules usually large, adnate to the petioles. Flowers solitary or corymbose-cymose. Hypanthium turbinate or hemispheric. Bractlets usually present. Sepals 5, valvate. Petals 5, white, light yellow, pinkish, or purplish. Stamens many, inserted on a disk in the mouth of the hypanthium. Pistils numerous, inserted on a short carpophore. Styles not articulate or obsoletely so, the lower portion plumose, elongate in fruit, not at all hooked or rarely slightly so, the upper portion glabrous, persistent, or withering and then deciduous, but not falling off early at a distinct node. Fruit a hairy achene. Seed erect, basal. *Erythrocoma* Greene.

Leaflets cuneate, toothed at the apex. 1. *S. triflora.*
Leaflets dissected at least half-way into linear or linear-lanceolate divisions. 2. *S. ciliata.*

1. **S. triflòra** (Pursh) R. Br. Stem finely pilose and somewhat hirsute, 2–4 dm. high, cymosely 3-flowered; basal leaves pinnate; principal leaflets 7–15, at the apex 2- or 3-cleft less than half their length, finely puberulent; hypanthium densely pilose, acute at the base; bractlets linear-subulate, 12–18 mm. long, longer than the lanceolate sepals; petals yellowish or flesh-colored. *Geum triflorum* Pursh [G]. *S. ciliata* Rydb., in part [B]. *Erythrocoma triflora, E. cinerascens,* and *E. affinis* Greene. Hills: Newf.—N.Y.—Ill.—Neb.—Mont.—Alta. *Canad.—Plain—Submont.* My–Jl.

2. **S. ciliàta** (Pursh) G. Don. Stem 2–5 dm. high, finely pilose and silky-hirsute, cymosely 3–15-flowered; basal leaves 1–2 dm. long; principal leaflets 9–19, obovate in outline, finely pilose-puberulent, and more or less densely silky-ciliate on the veins and margins; bractlets linear-subulate; sepals finely pilose, ovate-lanceolate; petals broadly elliptic or oval, flesh-colored or yellowish. *Geum ciliatum* Pursh. *E. dissecta* and *E. ciliata* Greene. Mountains and hills: Alta.—S.D.—N.M.—Utah—Wash.—B.C. *Submont.—Subalp.* My–Au.

24. CERCOCÁRPUS H.B.K. MOUNTAIN MAHOGANY.

Shrubs or trees with hard wood. Leaves alternate, simple, toothed or entire-margined. Flowers axillary, solitary or fasciculate. Hypanthium with a cylindric persistent tube, abruptly widening into a turbinate or campanulate, deciduous limb. Sepals 5, from broadly triangular to nearly subulate. Corolla wanting. Stamens 15 or more, inserted at different heights on the limb of the hypanthium. Pistil solitary, inserted in the bottom of the hypanthium; ovary cylindric-fusiform, sessile; style terminal, elongate, plumose; stigma undivided, terminal. Seed cylindric, solitary, basal.

f. 287.

1. **C. montànus** Raf. *Fig. 287.* Shrub 1–2 m. (rarely 3 m.) high; leaf-blades 2–5 cm. long, coarsely serrate, with ovate teeth, pale and finely tomentose beneath; tube of the hypanthium about 1

cm. long, pilose; limb campanulate or somewhat turbinate, together with the sepals 6–7 mm. wide. *C. parvifolius* Nutt. [B]. Hills and mountain sides: S.D.—w Kans.—N.M.—Utah—Mont. *Plain—Mont.* My–Jl.

25. **RUBÀCER** Rydb. FLOWERING RASPBERRY, THIMBLEBERRY, SALMON BERRY.

Unarmed shrubs, with more or less shreddy bark, more or less glandular-hispid above. Leaves petioled, simple, digitately ribbed and lobed, with acute or acuminate lobes and cordate at the base, resembling the leaves of certain maples. Flowers showy, in small panicles. Sepals 5, ovate, with long slender acuminations. Petals 5, pink or white. Stamens numerous, in several series; filaments slender. Pistils numerous; styles glabrous, club-shaped; stigmas slightly 2-lobed. Drupelets numerous, coherent with each other and separating from the flat receptacle, capped with a dry, rather hard, finely and densely pubescent cushion, the style inserted under its margin.

Petals white; middle lobe of the leaves not decidedly longer than broad. 1. *R. parviflorum.*
Petals rose-colored; middle lobe of the leaves decidedly longer than the rest. 2. *R. odoratum.*

1. **R. parviflòrum** (Nutt.) Rydb. Stem erect, shrubby, 5–20 dm. high, with flaky bark; leaf-blades reniform in outline, 5–20 cm. broad, 3–5-lobed, the lobes triangular; sepals broadly ovate, abruptly caudate-acuminate, about 1.5 cm. long, densely glandular on the back; petals white, broadly oval, or ovate, 15–30 mm. long; fruit convex, red, 15–20 mm. wide. *Rubus parviflorus* Nutt. [G, B]. *Rubus nutkanus* Moc. *Bossekia parviflora* Greene. Open woods and among bushes: Mich.—w Ont.—S.D.—N.M.—Calif.—s Alaska; n Mex. *Submont.—Mont.* My–Jl.

2. **R. odoràtum** (L.) Rydb. Stem 1–1.5 m. high, the young parts villous and glandular-hispid; leaf-blades broadly cordate in outline, 1–3 dm. broad, 3–5-lobed, the lobes triangular or ovate, abruptly acuminate; sepals broadly ovate, densely glandular without; petals suborbicular, 15–25 mm. long; fruit depressed-hemispheric. *Rubus odoratus* L. [G, B]. Rocky woods: N.S.—Mich. —Tenn.—Ga.; cult. and escaped: Kans. *E. Temp.*

26. **RÙBUS** (Tourn.) L. RASPBERRY, BLACKBERRY, CLOUDBERRY.

Perennial plants, ours either shrubs with biennial, prickly or bristly stems, or herbaceous plants, with rootstocks or creeping stems, unarmed. Leaves alternate, usually pinnately or pedately compound or rarely simple. Flowers racemose, paniculate or rarely solitary, perfect, dioecious or polygamo-dioecious. Hypanthium flat or saucer-shaped. Sepals 5, rarely 6–8, without bractlets. Petals as many, deciduous. Stamens numerous, inserted on a disk at the margin of the hypanthium. Pistils many, inserted on a convex or nipple-shaped, dry or fleshy receptacle. Fruit of several or many fleshy drupelets. Ovules 2, collateral, one abortive.

Stems herbaceous, never prickly, rarely bristly; floral branches directly from the root-stock or from stolons; stipules broad, free or nearly so.
  Petals spreading or ascending; filaments not dilated; plants dioecious; drupelets many, coherent; leaves merely lobed. 1. *R. Chamaemorus.*
  Petals erect; filaments dilated.
    Petals pink or rose-color; plants not stoloniferous; floral branches directly from the rootstocks.
      Petals oblanceolate or obovate, 10–15 mm. long, distinctly clawed; stem less than 1 dm. high, usually 1-flowered; leaflets rounded at the apex. 2. *R. acaulis.*
      Petals broadly obovate, rounded or emarginate, scarcely clawed, 10 mm. long or less; stem leafy, 1–3 dm. high, 1–5-flowered; stem-leaves with rhombic acute leaflets. 3. *R. arcticus.*
    Petals white; plant stoloniferous, wholly unarmed; flowers solitary or racemose. 4. *R. pubescens.*

Stems more or less woody, biennial or perennial, usually prickly or bristly, if unarmed or merely bristly, then the stipules narrow and more or less adnate.
- Carpels united into a thimble-shaped aggregate fruit, separating from the dry receptacle (RASPBERRIES).
  - Inflorescence corymbiform; fruit black or purple; stem prickly; leaves, when having more than 3 leaflets, pedately compound. — 5. *R. occidentalis.*
  - Inflorescence racemose; fruit red; leaves, when having more than 3 leaflets, pinnately compound.
    - Plant not at all glandular-hispid; inflorescence with weak prickles, finely villous. — 6. *R. idaeus.*
    - Plant more or less glandular-hispid, especially in the inflorescence, which is not prickly.
      - Turions (new canes), petioles and inflorescence neither densely tomentose nor puberulent.
        - Turions sparingly bristly, scarcely glandular, leaves dark green above. — 7. *R. strigosus.*
        - Turions densely bristly, more or less glandular, leaves light green above. — 8. *R. melanolasius.*
      - Turions, petioles and inflorescence densely villous, densely bristly. — 9. *R. subarcticus.*
- Carpels coherent to the fleshy receptacle (BLACKBERRIES).
  - Stems prickly or rarely unarmed, the prickles comparatively few and confined to the angles of the stem.
    - Stems erect or merely arching.
      - Inflorescence densely glandular with long-stalked glands.
        - Inflorescence elongate-racemose; terminal leaflets of the turions elongate-cordate or ovate, caudate-accuminate. — 10. *R. nigrobaccus.*
        - Inflorescence corymbiform; terminal leaflets of the turions very broadly cordate. — 11. *R. sativus.*
      - Inflorescence not at all glandular or but slightly so and the glands subsessile or sessile.
        - Leaves glabrous or nearly so beneath. — 12. *R. canadensis.*
        - Leaves densely pubescent beneath.
          - Inflorescence elongate-racemose; young branches angled. — 13. *R. argutus.*
          - Inflorescence short, corymbiform; young branches terete. — 14. *R. frondosus.*
    - Stems prostrate, only the floral branches erect.
      - Leaves decidedly pubescent beneath.
        - Inflorescence conspicuously leafy-bracted, with unifoliolate, broadly ovate or cordate leaves, terminal; leaflets on the turions cordate at the base. — 15. *R. Baileyanus.*
        - Inflorescence not conspicuously leafy-bracted, the unifoliate leaves, if any, cuneate at the base; terminal leaflets on the turions ovate or lanceolate.
          - Leaflets more or less plicate, sharply serrate with lanceolate teeth; those of the turions abruptly acuminate. — 16. *R. plicatifolius.*
          - Leaflets not plicate, with broad ovate teeth; those of the turions acute. — 17. *R. aboriginum.*
      - Leaves glabrous or nearly so, except on the veins beneath.
        - Leaflets merely dentate. — 18. *R. flagellaris.*
        - Leaflets distinctly incised. — 19. *R. heterophyllus.*
  - Stems bristly, prostrate at least in age; bristles numerous, not confined to the angles of the stem.
    - Stem at first ascending; leaflets of the turions oblanceolate or lanceolate, acute or acuminate, not evergreen. — 20. *R. nigricans.*
    - Stem prostrate; leaflets of the turions obovate, usually obtuse or rounded at the apex, evergreen. — 21. *R. hispidus.*

1. R. Chamaemòrus L. A herbaceous perennial, with creeping rootstock; stem 1–3 dm. high, 1–3-leaved; leaf-blades reniform, with 3–7 rounded lobes, palmately veined; flowers solitary; petals white, obovate, 8–12 mm. long; stamens numerous, in the pistillate flowers without anthers; fruit at first red, when ripe yellow or golden. CLOUDBERRY. BAKED-APPLE BERRY. Arctic and subarctic sphagnum swamps. Greenl.—N.H.—B.C.—Alaska; Eurasia. *Subarct.* Jl–Au.

2. **R. acaùlis** Michx. An unarmed herbaceous, subdioecious perennial, with creeping rootstock; stem 3–12 cm. high, 2–4-leaved; leaflets 3, broadly obovate to flabelliform, unevenly serrate; petals rose-colored, with yellowish claws; fruit red; drupelets 20–30, coherent. Arctic and subarctic swamps and tundras: Lab.—Newf.—Minn.—Alaska. *Arctic—Boreal.—Subalp.* Je–Au.

3. **R. árcticus** L. An unarmed herbaceous perennial, with a rootstock; stem 0.5–2 dm. high, 2–6-leaved, more or less flexuose; leaflets 3, obovate to rhombic, coarsely dentate, 1.5–4 cm. long; petals rose-colored, clawless, sometimes denticulate; fruit red, globose; drupelets 20–40, coherent. Bogs: Lab.—Que.—Man.—Alta.; Eur. *E. Boreal.—Subalp.* Jl–Au.

4. **R. pubéscens** Raf. An unarmed perennial, with flagelliform, procumbent shoots 1–10 dm. long; leaves ternate, rarely quinate; leaflets 3–10 cm. long, green on both sides, sharply and doubly serrate, the lateral ones obliquely ovate, the terminal one rhomboid; petals elliptic or oblanceolate; fruit red, globose; drupelets rather few, large, slightly coherent. *R. triflorus* Richardson. [G]. *R. americanus* (Pers.) Britt. [B]. Damp woods and swamps: Newf.—N.J.—Ia.—Mont.—B.C.; Colo. *Canad.—Allegh.—Mont.* My–Jl.

5. **R. occidentàlis** L. Stems biennial, 1–3 m. long, recurved, at length rooting at the tips, usually purplish or bluish; leaves of the shoots 3-foliolate, or some of them pedately 5-foliolate; median leaflet ovate to broadly cordate, sometimes more or less lobed, doubly serrate, acuminate at the apex, rounded to cordate at the base, 5–8 cm. long, white-tomentose beneath; lateral leaflets in the 5-foliolate leaves similar, but narrower; outer leaflets subsessile, often oblique and sometimes 2-lobed; floral branches green, their leaves 3-foliolate; leaflets similar, but rarely cordate at the base; corymbs few-flowered; sepals ovate-lanceolate, long-acuminate, 7–8 mm. long; petals white, elliptic, shorter than the sepals; fruit dark bluish purple, with a bloom, hemispheric. BLACK RASPBERRY, BLACK-CAP. Copses: N.B.—Que.—Ga.—Colo.—Minn. *E. Temp.—Plain—Submont.* Je–Jl.

6. **R. idaèus** L. Stem biennial, erect, light-colored, finely tomentose when young, armed with bristles or weak prickles; leaves of the turions pinnately 5-foliolate, those of the floral branches 3-foliolate; terminal leaflet broadly ovate, rounded or cordate at the base, double-serrate, short-acuminate, 5–10 cm. long, dark green above, white-tomentose beneath, the lateral ones ovate, somewhat smaller; inflorescence short-racemose, terminal and often with branches from the upper axils; peduncle and pedicels finely tomentulose and with small recurved prickles; fruit red, thimble-shaped. EUROPEAN RED RASPBERRY. Around dwellings: Me.—R.I.—Minn.; escaped from cultivation. My–Je.

7. **R. strigòsus** Michx. Stem biennial, brownish or reddish, more or less bristly, but not tomentulose, 1–2 m. high; leaves of the new shoots pinnately 5-foliolate, those of the floral branches 3-foliolate, the terminal leaflet ovate, sometimes 3-lobed, doubly serrate, abruptly acuminate at the apex, rounded or cordate at the base, dark green and short-hairy or glabrate above, white-tomentose beneath, 5–10 cm. long, the lateral leaflets obliquely ovate; peduncles and pedicels bristly and glandular-hispid; petals white, elliptic, 5–6 mm. long; fruit hemispheric, 1 cm. broad, light red. AMERICAN RED RASPBERRY. Copses and roadsides: Newf.—N.D.—Neb.—Va. *Canad.—Allegh.* My–Jl.

R. occidentalis × strigosus. Like *R. occidentalis* in habit, but leaves pinnate and plant more or less glandular-hispid. *R. neglectus* Peck. Me.—Minn.—Pa.

R. idaeus × strigosus. Intermediate between the parents; inflorescence with short-stalked glands and slightly tomentulose. Often cultivated and rarely escaped: Minn.

8. **R. melanolàsius** Focke. *Fig. 288.* Stems biennial, erect, 0.5–1 m. high, purple or yellow, and often glaucous, densely bristly and more or less glandular; leaves of the shoots pinnately 3–5-foliolate; leaflets ovate or lanceolate, short-acuminate or acute at the apex, usually rounded or cuneate at the base, sharply double-serrate, light green and sparingly and minutely pubescent above, finely and densely white-tomentose beneath; leaves of the floral branches similar, but always 3-foliolate; flowers in small terminal and axillary racemes; peduncles, pedicels, hypanthium, and calyx densely bristly and glandular-hispid; petals white, elliptic, 5–6 mm. long, erect; fruit red or purplish red, hemispheric, very sour. *R. strigosus* of Western reports. *Batidaea laetissima, B. dacotica, B. unicolor,* and *B. Sandbergii* Greene. ROCKY MOUNTAIN RED RASPBERRY. Mountains: B.C.—Alta.—S.D.—Colo.—Utah—Ore. *W. Temp.—Submont.—Mont.* Je–Jl.

9. **R. subárcticus** (Greene) Rydb. Stem biennial, 3–10 dm. high, densely and finely villous when young, subtomentose as well as bristly; leaves of the new shoots 3–5-foliolate; terminal leaflet petioluled, broadly ovate, rounded or subcordate at the base, sometimes 3-lobed, 5–8 cm. long, double-serrate; lateral leaflets obliquely ovate; inflorescence corymbiform, the peduncle, pedicels and hypanthia glandular-hispid; petals white, elliptic, 4.5 mm. long; fruit hemispheric, tomentose. *R. idaeus canadensis* Richardson. Open woods and copses: Lab.—N.S.—Minn.; B.C.; Alaska. *Boreal.—Subarct.*

10. **R. nigrobáccus** Bailey. Stem biennial, furrowed, 5–30 dm. high, erect or the branches recurving, armed with short, somewhat recurved prickles, the younger parts villous and somewhat glandular; leaves of the young shoots pedately 5-foliolate; petioles, petiolules, and midribs villous and glandular; leaflets petioluled, ovate, doubly serrate, abruptly acuminate, pilose above, softly pubescent beneath; leaves of the floral branches 3-foliolate, the leaflets 2–10 cm. long, less acuminate; inflorescence racemose; petals oval or elliptic, 10–15 mm. long; fruit hemispheric to more elongate, black, 1–1.5 cm. long. *R. alleghenensis* Blanchard [G]; not Porter. Open woods: N.B.—Minn.—Neb.—Ark.—N.C. *E. Temp.* My–Jl.

R. argutus × nigrobaccus. Differs from *R. nigrobaccus* in the sharply angled stem, stronger prickles, and short-stipitate glands. Mass.—Conn.—Pa.—Minn.

11. **R. satìvus** (Bailey) Brainerd. Stem biennial, stout, 1–2 m. high, angled; prickles stout, straight; leaves of the turions pedately 5-foliolate; petioles, petiolules and midribs villous and prickly; leaflets pilose above, villous beneath, doubly serrate, 10–15 cm. long; leaflets of the floral branches ovate, 4–8 cm. long; inflorescence corymbiform, densely villous, and glandular-hirsute; petals white, broadly obovate or orbicular, 12–15 cm. long; fruit globose, 15 mm. broad, juicy. Open fields: Que.—Conn.—Pa.—Minn. *Canad.* My–Je.

12. **R. canadénsis** L. Stem erect, 1–4 m. high, grooved and round-angled, glabrous, unarmed or with a few weak straight prickles; leaves on the new shoots 5-foliolate; leaflets thin, dark green, glabrous or nearly so, sharply serrate, rounded or subcordate at the base, abruptly long-acuminate, 5–15 cm. long; leaves on the floral branches 3-foliolate, the leaflets oval, less acuminate, 3–10 cm. long; raceme lax, 8–15 cm. long; petals oval, 10–15 mm. long; fruit round, 1–1.5 cm. long. Woods: Newf.—ne Minn.—N.C. *Canad.—Allegh.* My–Je.

13. **R. argùtus** Link. Stem biennial, 1–2 m. high, erect or recurving, strongly angled and furrowed; prickles mostly straight, stout, flattened; leaves of the new shoots digitately 5-foliolate; leaflets ovate, double-serrate, short-acuminate, rounded or the terminal one cordate at the base, sparingly pubescent or glabrate above, softly and densely pubescent beneath, mostly 6–10 cm. long; leaves of the floral branches trifoliolate or unifoliolate in the inflorescence, 3–6 cm. long, oval, acuminate; petals 1 cm. long; fruit black, 10–12 mm. long, nearly as broad. *R. floricomus* [G], *R. Andrewsianus* [G], and *R. amnicola* Blanchard. Open places: N.S.—Iowa—Kans.—N.C. *E. Temp.* My–Jl.

14. **R. frondòsus** Bigel. Stems biennial, erect or recurved, terete, 1–2 m. high, glabrous; prickles less stout than in the preceding, straight or slightly recurved, nearly terete; leaflets on the turions 5, rather thick, sparingly pubescent or glabrate above, softly and densely pubescent beneath, double-serrate, abruptly acuminate, the median leaflet rounded-cordate, 8–14 cm. long, the lateral leaflets broadly ovate; leaflets on the floral branches 3, or on the upper leaves 1, ovate, oval, or obovate, acute, 3–7 cm. long; raceme short and leafy, corymbiform; petals orbicular, 1 cm. long; fruit black, 15 mm. long, 1 cm. broad. *R. philadelphicus* Blanchard. Thickets: R.I.—Ont.—se Minn.—Kans. —Va. *Allegh.*

15. **R. Baileyànus** Britton. Stem biennial, decumbent, trailing, 1–2 m. long; prickles small, recurved; leaflets on the turions 3, rarely 5, abruptly acuminate, softly pubescent on both sides, serrate with broad teeth, the terminal one broadly ovate, subcordate at the base, 5–7 cm. long, 4–6 cm. broad, the lateral ones oblique, broadly ovate, somewhat rhombic; leaflets on the floral branches 3, oval or ovate, 3–5 cm. long, or on the upper leaves only 1, rounded-ovate or cordate; flowers 1–3 at the ends of the branches or solitary in the upper axils; petals elliptic, 10–14 mm. long; fruit black, hemispheric, 10–15 mm. broad. *R. villosus humifusus* T. & G. [G]. Open woods: Mass.—Ont.—Minn.—Okla.—N.C. *Allegh.* My–Je.

16. **R. plicatifòlius** Blanchard. Stem biennial; prostrate, 1–2 m. long, terete, glabrous, not glandular; prickles straight, weak, retrorse; leaflets on the turions 3 or 5, oval or ovate, abruptly long-acuminate, rugose beneath, 7–10 cm. long; leaflets on the floral branches 1–3, oval or obovate, 3–5 cm. long, densely pubescent beneath; inflorescence 3–8-flowered; petals 8–10 mm. long; fruit oblong, 12–15 mm. long. *R. villosus* Ait. [G]. Open places: Me.—Ont.—Minn.—Kans.—Pa. Je–Jl.

17. **R. aboríginum** Rydb. Stem biennial, decumbent, straw-colored; prickles short, curved; leaflets on the turions 3–5, ovate or oval, 4–10 cm. long, softly pilose on both sides, acute at each end, or rounded at the base; leaflets of the floral branches smaller, only 3, 1–5 cm. long; inflorescence 1–4-flowered; petals obovate, 15 mm. long. Open prairies: Kans.—Tex. *Texan.*

18. **R. flagellàris** Willd. Stem biennial, prostrate, 3–20 dm. long, glabrous; prickles rather weak, recurved, scattered; leaflets on the turions 3–5, rather firm, 4–10 cm. long, green and glabrous on both sides, or slightly pubescent on the veins beneath, abruptly acuminate, the terminal one inclined to be rhombic-ovate, the lateral ones ovate; leaflets on the floral branches 3, obovate or oblanceolate, 3–5 cm. long; flowers 1–4, often solitary in the upper axils; petals 10–15 mm. long; fruit black, hemispheric, 1–1.5 cm. long. *R. canadensis* T. & G. *R. villosus* Bailey. *R. procumbens* Muhl. *R. subuniflorus* Rydb. Open places, especially in sandy soil: Me.—Minn.—Mo.—Va. *Canad.—Allegh.* My–Jl.

19. **R. heterophýllus** Willd. Stem biennial, 2–4 m. long, prostrate, glaucous, armed with recurved prickles; leaves of the turions 3–5-foliolate; leaflets yellowish green, sparingly pubescent, short-acuminate, coarsely double-serrate, 5–10 cm. long; leaves of the floral branches 3-foliolate or 1-foliolate;

leaflets oval, the terminal one short-petiolulate; petals white, elliptic, 10–14 mm. long; fruit short-oblong or globose, 16–18 mm. long. Dry ground: Me.—Conn.—Mich.—Minn. *Canad.* Je.

20. **R. nìgricans** Rydb. Stem biennial, retrorsely hispid and somewhat glandular; leaves of the turions 5-foliolate, the petioles and ribs bristly; leaflets glabrous or nearly so, dark green and shining above, doubly serrate, abruptly acuminate, 4–10 cm. long; leaves of the floral branches 3-foliolate, rarely 5-foliolate, or in the inflorescence 1-foliolate; leaflets obovate; inflorescence densely glandular-hispid and bristly; petals elliptic, 10–12 mm. long; fruit hemispheric, at first red, black when ripe, 1 cm. broad. Swamps: Newf.—R.I.—Pa.—Minn. *Canad.* Je–Jl.

R. nigricans × plicatifolius. Resembles *R. plicatifolius* in habit and leaf-form, but the stem and petioles more or less glandular-hispid as well as prickly; inflorescence more or less corymbiform and glandular-hispid. N.S.—Me.—Minn.

R. flagellaris × nigricans. With the inflorescence, fruit and bristly stem of *R. nigricans*, but the stem harsher, the bristles fewer and stronger, and the leaflets broader and shining as in *R. flagellaris*. Vt.—Conn.—Minn.

21. **R. híspidus** L. Stem biennial, prostrate and trailing, 5–15 dm. long, usually densely retrorsely bristly; leaflets 3, more or less persistent, rather firm, dark green and shining above, paler and duller beneath, obovate or rhombic-obovate, acute or short-acuminate, or those of the floral branches rounded at the apex, 3–6 cm. long; inflorescence corymbiform, terminal, or with also 1 or 2 flowers in the upper axils; petals 8 mm. long, obovate; fruit dark red or purple, with few drupelets. Damp woods and swampy ground: N.S.—Minn.—Kans.—Ga. *E. Temp.* Je–Jl.

27. RÒSA (Tourn.) L. Rose.

Shrubs or vines, usually prickly. Leaves alternate, pinnate, with more or less adnate stipules, and serrate leaflets. Flowers perfect, solitary or corymbose. Hypanthium well-developed, urceolate, globose, ellipsoid or turbinate, contracted at the mouth, enclosing the achenes, becoming fleshy in fruit. Sepals 5 (rarely 4), with or without bractlets. Petals normally 5 (rarely 4), or by the transformation of the stamens numerous, spreading, usually obcordate. Stamens numerous, inserted on the thickened margin of the hypanthium. Pistils numerous or several, inserted in bottom of the hypanthium or also on the inside walls of the same. Styles ventral, reaching the mouth of the hypanthium or long-exserted, sometimes united into a column. Achenes bony.

Styles much exserted, about equaling the stamens; sepals reflexed, deciduous.
  Leaves glabrous or slightly pubescent on the veins beneath, dark green above. — 1. *R. setigera.*
  Leaves velutinous-pubescent beneath, rather dull above. — 2. *R. rubifolia.*
Styles not exserted or slightly so, the stigmas forming a head closing the mouth of the hypanthium.
  Sepals reflexed after flowering and soon deciduous.
    Hypanthium and pedicel glandular-hispid.
      Achenes inserted both on the inner walls and in the bottom of the hypanthium; prickles stout, scattered, rarely infrastipular. — 3. *R. rubiginosa.*
      Achenes inserted only in the bottom of the hypanthium; stem with infrastipular prickles and usually with numerous bristles on the young shoots.
        Infrastipular prickles decidedly curved. — 4. *R. palustris.*
        Infrastipular prickles straight or nearly so.
          Leaves decidedly pubescent beneath.
            Leaflets comparatively thin, not subcoriaceous. — 5. *R. Lyoni.*
            Leaflets firmer, subcoriaceous. — 8. *R. rudiuscula.*
          Leaves glabrous or pubescent only on the veins beneath.
            Leaflets not glandular-dentate. — 6. *R. Carolina.*
            Leaflets glandular-dentate; rachis usually glandular-hispid. — 7. *R. serrulata.*

Hypanthium and pedicel glabrous. — 14. *R. conjuncta.*
Sepals after flowering erect, connivent, long-persistent on the fruit; hypanthium glabrous.
Infrastipular prickles not present; branches mostly unarmed and turions bristly but not prickly.
Flowers solitary and bractless. — 27. *R. pimpinellifolia.*
Flowers either corymbose or, if solitary, supported by a broad bract.
Inflorescence corymbose, terminating the stems (or rarely the branches); plant suffruticose; stem usually dying back to near the ground; leaflets 9–11.
Leaves glabrous or nearly so.
Leaflets obovate, pale and glaucous on both sides. — 9. *R. subglauca.*
Leaflets elliptic or oval, dark green above.
Leaflets mostly 2–5 cm. long, not glaucous, decidedly acute; plant 3 dm. high or more; flowers corymbose. — 10. *R. arkansana.*
Leaflets rarely more than 2 cm. long, mostly rounded at the apex, somewhat glaucous beneath; plant 1–2 dm. high; flowers 1–3. — 11. *R. Lunellii.*
Leaves densely pubescent, especially beneath.
Plant 3–6 dm. high, usually dying back in the fall to near the ground.
Sepals erect in fruit; plant not glaucous.
Upper stipules and bracts not densely glandular; leaflets 1.5–4 cm. long; plant 3–5 dm. high. — 12. *R. suffulta.*
Upper stipules and bracts densely glandular; leaflets seldom more than 1.5 cm. long; plant 1–3 dm. high. — 13. *R. alcea.*
Sepals reflexed in fruit; plant glaucous. — 14. *R. conjuncta.*
Plant shrubby, not dying back; branches from the upper axils soon overtopping the inflorescence. — 15. *R. polyanthema.*
Inflorescence of solitary or few corymbose flowers at the ends of the branches; plant shrubby; leaflets 5 or 7.
Stem densely bristly even in age; flowers solitary.
Hypanthium decidedly pear-shaped or elliptic, acute at the base, with a distinct neck at the top.
Leaflets conspicuously glandular-granuliferous, but scarcely at all hairy beneath, rounded-oval, double-toothed. — 16. *R. Engelmannii.*
Leaflets densely pubescent, but rarely slightly if at all glandular-granuliferous beneath, mostly elliptic, simple-toothed. — 17. *R. acicularis.*
Hypanthium subglobose, almost without a neck.
Hypanthium in fruit 1.5 cm. broad; leaflets glandular-granuliferous as well as somewhat puberulent beneath. — 18. *R. Bourgeauiana.*
Hypanthium in fruit about 1 cm. broad; leaflets villous beneath. — 19. *R. acicularioides.*
Stem unarmed, or when young armed with a few deciduous bristles.
Leaves glabrous on both sides, shining. — 20. *R. subblanda.*
Leaves finely pubescent beneath. — 21. *R. blanda.*
Infrastipular prickles present.
Hypanthium globose; neck usually obsolete.
Stipules, petioles, and rachis conspicuously glandular. — 22. *R. Fendleri.*

Stipules, petioles, and rachis not conspicuously glandular, usually glandless.
Leaflets glabrous or nearly so. 23. *R. Woodsii.*
Leaflets more or less pubescent beneath.
Turions armed with stout flat prickles; fruit about 15 mm. broad. 24. *R. terrens.*
Turions merely bristly; fruit rarely more than 10 mm. broad. 25. *R. Macounii.*
Hypanthium ellipsoid with a distinct neck. 26. *R. pyrifera.*

1. **R. setígera** Michx. Stem 2–5 m. high, climbing, terete; prickles scattered, curved, 4–8 mm. long, flattened below; petiole and rachis glandular-hispid; leaflets 3, or on the turions sometimes 5, the lateral ones sessile, lanceolate or ovate, the terminal one petioluled, rounded or subcordate at the base, 4–9 cm. long, dark green; flowers corymbose; pedicels and the globose hypanthium slightly glandular-hispid; petals 2–3 cm. long, rose-colored. Copses: Ohio—N.C.—Fla.—Kans.—Neb. *Allegh.—Carol.—Ozark.* Je–Jl.

2. **R. rubifòlia** R. Br. Closely resembling the preceding; stem 2–4 m. high, climbing, glabrous; prickles 4–10 mm. long, flattened below, curved; leaflets 3, rarely 5, ovate or ovate-lanceolate, serrate with gland-tipped teeth, pale and short-villous beneath, 3–7 cm. long; pedicel and hypanthium glandular-hispid; petals 1.5–2.5 cm. long. Copses: Ont.—Ga.—Tex.—Neb. *Allegh.—Carol.—Ozark.* Je–Jl.

3. **R. rubiginòsa** L. Stem branched, often 2 m. high, somewhat climbing, sometimes bristly; prickles scattered, flattened, curved, 5–10 mm. long; leaflets 5–7, broadly oval or suborbicular, rounded at each end, 1–3 cm. long, doubly serrate, glandular-pruinose beneath; flowers 1–4 together, the pedicels glandular-hispid; hypanthium broadly ellipsoid or somewhat obovoid, in fruit 10–12 mm. thick, 12–15 mm. long, orange or scarlet; sepals usually lobed, glandular on the back; petals 15–20 mm. long. SWEET BRIER. Roadsides and around dwellings: N.S.—Kans.—Minn.—Ga.; B.C.—Calif.; escaped from cultivation and nat. from Eur. Je–Jl.

4. **R. palústris** Marsh. *Fig. 289.* Stem 0.3–2 m. high, erect, glabrous, terete; prickles infrastipular, more or less curved, 4–6 mm. long; floral branches unarmed; leaflets 7, rarely 9, dull and dark green, glabrate above, finely appressed beneath, lance-elliptic or oblanceolate, 2–6 cm. long, finely crenate; flowers usually corymbose; hypanthium subglobose, glandular-hispid, in fruit 10–12 mm. broad; sepals glandular on the back; petals 1.5–2 cm. long. *R. Carolina* L. (1762, not 1753) [G, B]. Damp places: N.S.—Minn.—Miss.—Fla. *E. Temp.* Je–Au.

f.289.

5. **R. Lỳoni** Pursh. Stem low, 3–15 dm. high, terete, glabrous or more or less bristly when young; infrastipular prickles straight, slender, 4–8 cm. long; leaflets 5–7, oval to lance-elliptic, acute at each end, 1.5–5 cm. long, regularly serrate, sparingly pubescent or glabrate above, villous beneath; flowers 1–4; pedicels glandular-hispid; hypanthium depressed-globose, more or less glandular-hispid, in fruit 8–10 mm. broad; petals obcordate, 2–2.5 cm. long. *R. humilis villosa* Best. [B]. Hillsides: Mass.—Minn.—se Kans.—Ark.—Ga. *Canad.—Carol.—Ozark.* Je–Au.

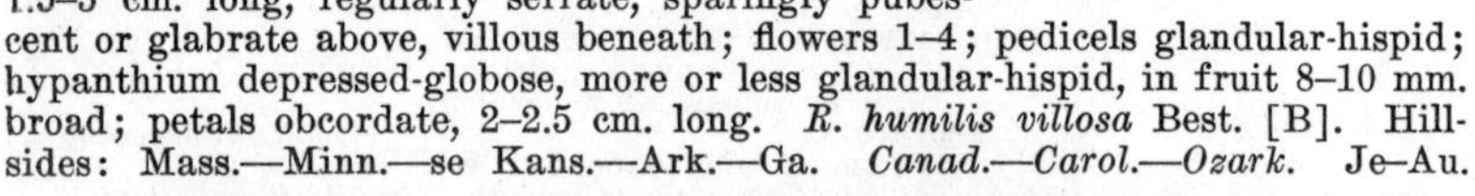

6. **R. Carolìna** L. Stem low, slender, 3–6 dm. high, terete, bristly when young; infrastipular prickles straight, terete, 5–10 mm. long; leaflets 5–7, elliptic or lance-elliptic, rarely oval or oblanceolate, 1–3 cm. long, glabrous above, somewhat paler beneath, glabrous or pubescent only on the veins, sharply serrate; flowers usually solitary; hypanthium globose, glandular-hispid, in fruit about 8 mm. broad; some of the sepals usually lobed; petals 2–2.5 cm. long. *R. humilis* Marsh. [G, B]. Banks and hillsides: Me.—Minn.—se Kans.—Tex.—Fla. *E. Temp.* My–Au.

7. **R. serrulàta** Raf. Stem slender, 3–10 dm. high; infrastipular prickles straight, terete, 4–8 mm. long; leaflets 3–7, usually 5, lance-elliptic, rarely oval, glabrous on both sides, or slightly pubescent on the veins beneath, paler beneath, sharply serrate with gland-tipped, often double teeth; flowers solitary; hypanthium globose or slightly depressed, glandular-hispid, in fruit 8–12 mm. broad; petals 1.5–2.5 cm. long. Highlands: Mass.—Ont.—Iowa—Tex.—Fla.; n Mex. *E. Temp.* My–Au.

8. **R. rudiúscula** Greene. Stem stout, mostly simple, about 1 m. high, usually bristly; prickles slender, 3–5 mm. long; petioles and rachis densely pubescent and glandular; leaflets 5–7, subcoriaceous, elliptic or oval, 2–3 cm. long, coarsely serrate, dark green and shining above, paler and densely pubescent beneath; flowers corymbose; hypanthium glandular-hispid, in fruit 12 mm. broad; sepals glandular on the back. Banks: Wis.—Mo.—Iowa—Okla. Je–Jl.

9. **R. subglaùca** Rydb. Stem erect, about 5 dm. high, dying back in the winter to the base, densely bristly, otherwise glabrous; leaflets 7–9, obovate, rounded at the apex, coarsely serrate, glabrous and glaucescent on both sides, 2–4 cm. long; flowers corymbose at the ends of the stem and the branches; hypanthium globose, glabrous, in fruit 12–15 mm. broad; sepals sparingly glandular on the back; petals about 2 cm. long. Prairies: Sask.—Iowa. *Prairie.*

10. **R. arkansàna** Porter. Stem erect, 2–4 dm. high, copiously bristly, mostly simple, killed back to near the root; glabrous or nearly so, glandular-toothed; rachis and petiole glabrous or sparingly pilose, sometimes glandular; leaflets elliptic, 1.5–6 cm. long, glabrous and shining, coarsely and sharply serrate; hypanthium subglobose, glabrous, without a neck, in fruit 12–15 mm. broad; sepals somewhat glandular on the back; petals obcordate, 2–2.5 cm. long. Plains and cañons: Colo.—Kans.—Wis.—Minn. *Prairie—Plain—Submont.* Je–Jl.

11. **R. Lunéllii** Greene. Stem erect, simple, 1–3 dm. high, densely bristly; sepals glabrous or nearly so, sometimes glandular, entire or gland-toothed on the margins; petiole and rachis glabrous or sparingly puberulent, sometimes glandular; leaflets elliptic or oval, serrate, 1–2.5 cm. long, glabrous on both sides or sparingly pubescent on the veins beneath; hypanthium globose, glabrous, in fruit orange-red and 10–12 mm. thick; sepals glandular on the back; petals rarely 2 cm. long, obcordate. Plains and hills: Man.—N.D.—S.D. *Plain—Submont.* Je–Jl.

12. **R. suffúlta** Greene. Stem erect, 3–5 dm. high, usually simple, dying back to near the ground, densely bristly, green; stipules densely and finely pubescent, glandular-dentate on the margins; leaflets obovate, 1.5–4 cm. long, light green, finely and rather densely pubescent on both sides or in age glabrate above, densely serrate; rachis and petiole finely pubescent; hypanthium globose, glabrous or rarely bristly, in fruit about 1 cm. broad; sepals somewhat glandular on the back; petals obcordate, 2–2.5 cm. long. *R. arkansana* S. Wats.; not Porter [B]. *R. dulcissima* Lunell. *R. pratincola* [G] and *R. heliophila* Greene. Prairies and plains: Ill.—Tex.—N.M.—Wyo.—Alta.—Man. *Prairie—Plain—Submont.* Je–Au.

13. **R. álcea** Greene. Stem low, 1–3 dm. high, terete, densely bristly, even the floral branches; stipules puberulent, glandular-hispid, and even bristly on the back, glandular-ciliate on the margins, rather broad; rachis and petiole puberulent and glandular; leaflets obovate, 1–2 cm. long, serrate above the middle, glabrous above, finely and densely pubescent and strongly veined beneath; hypanthium globose, glandular-bristly; sepals glandular-hispid on the back; petals obcordate, 15 mm. long. Plains: Sask.—N.D.—Man.—Colo. *Plain.*

14. **R. conjúncta** Rydb. Stem erect, about 5 dm. high, densely bristly, glabrous; leaflets mostly 9, oblanceolate or elliptic, sharply serrate, 2–5 cm. long, glabrous and glaucous above, densely pubescent beneath; flowers corymbose; hypanthium subglobose, about 1.5 cm. broad in fruit; sepals glandular-hispid, persistent, but reflexed in fruit. Prairies: Mo.—Iowa. *Prairie.*

15. **R. polyánthema** Lunell. Stem 5–15 dm. high, very bristly and prickly; infrastipular prickles usually wanting; leaflets 7–11, mostly 9, oval or obovate, 2.5–5 cm. long, glabrate above, short-pubescent beneath; flowers corymbose; hypanthium globose, glabrous, in fruit 15–20 mm. broad; sepals pubescent and glandular on the back. Banks: N.D. *Prairie.*

16. **R. Engelmánnii** S. Wats. Stem low, 3–5 dm. high, densely bristly, but rarely prickly; stipules rarely pubescent, conspicuously glandular-granuliferous and glandular-ciliate; petiole and rachis more or less glandular; leaflets 5 or 7, oval, 1–3 cm. long, doubly glandular-serrate, shining above, paler beneath; hypanthium decidedly ellipsoid or pear-shaped, in fruit 8–10 mm. thick, and 12–15 mm. long, with a distinct neck; sepals slightly if at all glandular; petals broadly obcordate, about 2.5 cm. long. Woods: Colo.—s Mont.—N.D. *Submont.—Mont.* Je–Jl.

17. **R. aciculàris** Lindl. Stem low, about 1 m. high, densely covered with straight bristles or weak prickles; stipules pubescent, more or less glandular-granuliferous and glandular-ciliate; leaflets 3–7, usually 5, elliptic or oval, 1.5–5 cm. long, simply and regularly serrate, dull and glabrous above, pale and finely and densely pubescent beneath; hypanthium glabrous, pear-shaped or elliptic, in fruit 1–1.5 cm. broad and 1.5–2 cm. long, usually with a neck; sepals pubescent and more or less glandular on the back; petals obcordate, 2–2.5 cm. long. Woods: Alaska—Wyo.—Mich.—N.Y. *Subarct.—Boreal.—Mont.* Je–Au.

R. acicularis × **Woodsii.** Shrub 9–12 dm. high, with straight infrastipular prickles; leaflets 7, obovate, rather thin, serrate, 2–4 cm. long, sparingly villous beneath; stipules slightly glandular; fruit pyriform, small, with erect sepals. Dumfries, Zumbro Valley, Minn.

18. **R. Bourgeauiàna** Crép. Stem usually low, 3–10 dm. high, densely covered with straight bristles or weak prickles; stipules slightly, if at all, pubescent, glandular-dentate; petiole and rachis somewhat pubescent and glandular; leaflets usually 5 or 7, broadly elliptic, 1.5–5 cm. long, usually rounded or obtuse at both ends, more or less double-serrate, thin, glabrous or nearly so above, slightly pubescent and glandular-granuliferous beneath; hypanthium glabrous, in fruit 12–15 mm. broad; sepals slightly, if at all, glandular; petals broadly obcordate, 2–2.5 cm. long, rose-colored. *R. Sayi* S. Wats [B]; not Schw. *R. acicularis Bourgeauiana* Crép. [G]. Woods: Ont.—Colo.—Mont. —Mack. *Boreal—Submont.—Mont.* Je–Jl.

19. **R. acicularioìdes** Schuette. Stem up to 5 dm. high, densely bristly; floral branches usually unarmed; leaflets 7–9, elliptic or oval, obtuse or acutish, 1.5–3 cm. long, glandular-serrate, finely puberulent above, villous and pale beneath; hypanthium globose, glabrous, with a short neck; flowers corymbose, bracts conspicuous; sepals 1.5 cm. long, caudate, densely glandular on the back. Damp woods: Wis.—Minn. *Canad.*

20. **R. subblánda** Rydb. Stem 1–2 m. high, bristly when young, in age unarmed; leaflets 7 or 9, oval or obovate, mostly acute at each end, glabrous on both sides, dull green above, paler beneath, sharply serrate, 1.5–5 cm. long; flowers solitary or in few-flowered corymbs; sepals somewhat glandular on the back; petals 2.5–3 cm. long. Thickets: N.H.—Man.—Iowa. *Canad.* Je–Jl.

21. **R. blánda** Ait. Stem 1–2 m. high, unarmed, or bristly when young, the branches unarmed; leaflets 5 or 7, rarely 9, oval or obovate, coarsely toothed, often obtuse, dull and glabrous above, paler and finely pubescent beneath, 2–6 cm. long; flowers usually solitary; hypanthium glabrous, sub-

globose, without a neck, in fruit 12 mm. broad; sepals glandular on the back; petals 2.5–3 cm. long. *R. gratiosa* Lunell. Thickets: Que.—Man.—N.D.—Kans.—Pa. *Canad.—Allegh.* Je–Jl.

22. **R. Féndleri** Crép. Stem low, 1 m. high or less, terete, armed with a few prickles, which are straight, slender, 5 mm. long or less; stipules glandular-pruinose on the back, more or less glandular-dentate; rachis and petiole densely glandular-puberulent and often somewhat glandular-hispid or with weak prickles; leaflets 5–7, elliptic, oval, or obovate, 1–3 cm. long, rather thin, green and glabrous above, slightly paler, puberulent and glandular-pruinose beneath, often double-serrate; hypanthium globose, glabrous, in fruit 8–10 mm. broad; sepals sparingly glandular on the back; petals rose-colored, obcordate, about 1.5 cm. long. *R. poetica* Lunell. Woods and copses: N.M.—Ariz.—S.D.—Mont. *Plain—Submont.* My–Jl.

23. **R. Wòodsii** Lindl. Stem terete, glabrous, 0.5–2 m. high, armed with rather numerous, straight or slightly curved prickles, 4–8 mm. long; stipules glabrous, usually glandless, entire or somewhat toothed; rachis and petiole glabrous, occasionally with a few prickles or stalked glands; leaflets 5–7, obovate, cuneate at the base, serrate above, glabrous on both sides, glaucous beneath, 1–2 cm., rarely 3 cm. long; hypanthium globose, glabrous, in fruit 8–10 mm. thick; sepals glabrous or slightly glandular on the back. *R. Macounii* Rydb. (Fl. Colo.); not Greene. *R. Sandbergii*, and (?) *R. fimbriatula* Greene. *R. deserta* Lunell. River banks and copses: Man.—N.D.—Kans.—Colo.—Utah—B.C. *Plain—Mont.* Je–Jl.

24. **R. térrens** Lunell. Stem 1–1.5 m. high, copiously armed with strong, more or less flattened prickles, both infrastipular and scattered ones; leaflets 5–11, usually 7, dark green, obovate or oval, obtuse, serrate towards the apex, cuneate and subentire at the base, 2–3 cm. long, softly pubescent beneath, glabrous above; flowers solitary or few; hypanthium subglobose, in fruit 1 cm. broad; sepals glandular-hispid on the back. Plains: N.D. *Plain.*

25. **R. Macoùnii** Greene. Stem 0.5–2 m. high, usually armed with straight terete prickles, 5–8 mm. long; new shoots bristly, stipules entire or glandular-denticulate, puberulent on the back; rachis and petiole finely puberulent, unarmed; leaflets obovate, serrate, 1–3 cm. long, green and glabrous above, pale or glaucous, finely puberulent, and sometimes pruinose beneath; hypanthium globose, without a neck, glabrous, 8–10 mm. thick; sepals glabrous or sparingly pubescent on the back; petals about 2 cm. long, obcordate, rose-colored. *R. Woodsii* S. Wats. [B, G]; not Lindl. *R. grosseserrata* E. Nels. *R. Maximiliani* Rydb; not Nees. Banks and copses: Sask.—Neb.—w Tex.—N.M.—Utah—e Wash. *Plain—Submont.* My–Jl.

26. **R. pyrífera** Rydb. Stem 1 m. high or more, armed with slender, straight prickles 5–8 mm. long; stipules finely puberulent and usually glandular-granuliferous on the back; petiole and rachis puberulent and often glandular; leaflets about 7, oval, 2–4 cm. long, coarsely serrate, dark green and glabrous above, finely puberulent and more or less glandular-granuliferous beneath; hypanthium pyriform or ellipsoid, acute at the base, with a distinct neck at the apex, in fruit 10–12 mm. thick and 15–20 mm. long; sepals glandular on the back; petals obcordate, about 2 cm. long. Banks: Mont.—S.D.—Neb.—Utah—Calif.—Wash. *Submont.* Je–Jl.

R. pyrifera × Woodsii. Like *R. Woodsii* in habit, leaf-form, and the lack of pubescence, but with an ellipsoid fruit, approaching that of *R. pyrifera*. Valentine, Neb., and Colorado Springs, Colo.

27. **R. pimpinellifòlia** L. Stem 1–4 dm. high, with spreading branches, densely armed with straight terete prickles and bristles; leaflets 5–11, small, orbicular to broadly elliptic, sharply serrate, with gland-tipped teeth, glabrous and green, 1–2 cm. long; flowers solitary; pedicels glandular-hispid or naked; fruit globose, 15–20 mm. long. Waste places: N.H.—Ont.—Ill.—Kans.; escaped from cultivation.

Family 80. **MALACEAE.** APPLE FAMILY.

Trees or shrubs, with simple or pinnate, alternate leaves, with stipules. Flowers perfect, regular. Hypanthium well developed, adnate to the ovary, becoming fleshy, and constituting a part of the fruit. Sepals and petals mostly 5. Stamens numerous, distinct, inserted on the margins of the receptacle. Gynoecium of 1–5 united carpels; cells of the ovary and styles as many, the latter distinct or partly united. Fruit a more or less fleshy pome.

Cells of the fruit by false partitions twice as many as the styles. 1. AMELANCHIER.
Cells of the fruit as many as the styles.
  Endocarp of the carpels leathery or papery.
    Fruit berry-like, with a thin pulp; endocarp of the carpels leathery.
      Leaves odd-pinnate; styles distinct. 2. SORBUS.
      Leaves simple; styles united below. 3. ARONIA.
    Fruit large, fleshy; endocarp of the carpels papery; leaves simple.
      Fruit globose or subglobose, without grit-cells; mouth of the hypanthium open. 4. MALUS.
      Fruit pear-shaped, the flesh containing grit-cells; mouth of the hypanthium more or less closed by a fleshy disk. 5. PYRUS.
  Endocarp of the carpels bony; leaves simple. 6. CRATAEGUS.

1. **AMELÁNCHIER** Medic. SERVICE-BERRY, JUNE-BERRY, SHADBERRY.

Shrubs or small trees. Leaves alternate, simple, petioled, toothed or entire-margined. Flowers in ours racemose. Hypanthium campanulate, becoming globose, adnate to the ovary. Sepals 5, reflexed, persistent. Petals 5, white. Stamens many; filaments subulate. Styles 3–5, rarely less. Ovary inferior or nearly so, the cells becoming twice as many as the styles, by false partitions intruding from the back. Ovules solitary in each cell, erect. Pome berry-like, 6–10-celled.

Flowers racemose; leaves conduplicate in the bud; fruit globose (except in *A. oblongifolia*).
  Leaves short-acuminate, finely toothed; summit of the ovary glabrous; sepals reflexed.
    Leaves white-tomentose when young, glabrate in age. 1. *A. canadensis.*
    Leaves glabrous from the beginning. 2. *A. laevis.*
  Leaves obtuse or merely acute.
    Leaves oblong or elliptic, finely and densely serrate; sepals erect or ascending; top of the ovary glabrous. 3. *A. oblongifolia.*
    Leaves ovate, oval, or suborbicular, coarsely serrate; top of the ovary pubescent; sepals reflexed or spreading.
      Leaves mostly acute or acutish, serrate to near the base.
        Petals 11–20 mm. long, narrow; racemes drooping; sepals 4 mm. long. 4. *A. sanguinea.*
        Petals 7–10 mm. long, broad; racemes erect; sepals 2–3 mm. long. 5. *A. humilis.*
      Leaves truncate or rounded at the apex, few-toothed at end above the middle.
        Sepals lance-deltoid, decidedly longer than broad; leaf-blades serrate above, usually suborbicular, somewhat glaucous beneath in age. 6. *A. alnifolia.*
        Sepals deltoid, scarcely longer than broad.
          Leaf-blades coarsely serrate above, rounded-oval, firm, not glaucous; fruit 10 mm. broad. 7. *A. macrocarpa.*
          Leaf-blades subentire or with a few short teeth at the apex; fruit 6–7 mm. broad. 8. *A. Carrii.*
Flowers solitary, or 2 or 3 together, not racemose; leaves flat and imbricate in the bud; fruit ellipsoid; sepals erect. 9. *A. Bartramiana.*

1. A. canadénsis (L.) Medic. A bushy tree, 5–10 m. high, or sometimes a tall shrub; leaves white-tomentose when young; glabrate in age, rounded or cordate at base, usually acuminate, sharply and finely and often doubly serrate, the teeth ascending, 50–60 on each margin, the lateral veins 11–17 pairs; racemes 3–5 cm. long; sepals broadly oblong-lanceolate, abruptly pointed, tomentose; petals oblong, 10–14 mm. long; fruit purple, not very juicy. *A. canadensis Botryapium* (L.) T. & G. [G]. *A. Botryapium* DC. [B]. Copses: Me.—S.C.—Ark.—Kans.—Minn.—Alta. *E. Temp.—Mont.*

2. **A. laèvis** Wieg. Tree up to 13 m. high; leaves glabrous from the beginning, ovate or oval, subcordate to rounded at the base, acuminate at the apex, sharply serrate to the base, the teeth 35–45 on each margin, the veins 12–17 pairs; hypanthium glabrous; sepals deltoid-lanceolate; petals oblong-linear, 10–18 mm. long; fruit purple, glaucous. *A. canadensis* A. Gray [B, G]; not Medic. Damp soil: Newf.—Wis.—Iowa—Kans.—N.C. *E. Temp.* My–Je.

3. **A. oblongifòlia** (T. & G.) M. Roemer. Shrub 2–8 m. high, growing in clumps; leaves nearly glabrous above, densely white-tomentose beneath when young, oblong or elliptic, 3–5.5 cm. long, 1–3 cm. wide, rounded to acutish at the apex, rounded at the base, finely serrate, with 25–40 teeth on each margin; veins 10–15 pairs; hypanthium tomentose at the base or throughout; sepals deltoid, tomentose within; petals ovate-oblong to linear, 7–10 mm. long; fruit nearly black, glaucous. *A. spicata* (Lam.) DC. [B]. Low ground: Me.—Minn.—Ala.—S.C. *Canad.—Allegh.* My–Je.

4. **A. sanguínea** (Pursh) DC. Shrub 1–3 m. high; leaves green above, densely tomentose beneath when young, glabrate in age, oval or elliptic, rarely suborbicular, rounded or subcordate at the base, acutish at the apex, coarsely serrate, with 20–25 short spreading teeth on each margin, the veins 13–15 pairs; hypanthium and sepals glabrous without, the latter ovate-lanceolate, hairy within; petals linear, 11–15 mm. long; fruit almost black, glaucous. *A. rotundifolia* (Michx.) M. Roemer [B]. *A. spicata* Robins. & Fern. [G]. Cliffs and ravines: Me.—Minn.—Ala.—N.C. *E. Temp.* My–Jl.

5. **A. hùmilis** Wieg. A stoloniferous shrub, 0.5–1.5 m. high, with grayish brown twigs; leaves green above, tomentose beneath when young, becoming glabrate and glaucous, oval-oblong or oval, 2.5–5 cm. long, subcordate or rounded at the base, the teeth often double, ascending, about 20 on each margin; veins 9–13 pairs on each side; racemes 4–5 cm. long; sepals triangular-lanceolate; petals obovate-oblong, 7–10 mm. long; fruit almost black, glaucous, juicy. Banks and hills: Vt. — Neb. — Alta. — (? Mack.). *Boreal.—Plain—Mont.* Je.

f. 290.

6. **A. alnifòlia** Nutt. *Fig. 290.* A low shrub, 1–2 m. high, with dark gray twigs; leaves suborbicular or rounded-oval, 2–5 cm. long, glabrous above, somewhat tomentose beneath, the veins about 12 pairs; raceme 2–4 cm. long; sepals lanceolate; petals oblanceolate-oblong, about 1 cm. long; fruit purple, juicy, with a bloom, 8–9 mm. thick. Banks and hills: Sask.—Neb.—Colo.—Ida.—Yukon; Mich. *Plain—Mont.* My–Jl.

7. **A. macrocárpa** Lunell. Shrub 5–20 dm. high; leaf-blades rounded-oval, rounded or subcordate at the base, 2–6 cm. long, 1.5–4.5 cm. wide, firm, coarsely serrate at and above the middle, densely tomentose beneath when young, becoming glabrate, but some tomentum long-remaining on the peduncles, petioles and ribs of the leaves beneath; sepals deltoid, nearly as broad as long, tomentose on both sides when young; petals oblanceolate, 8 mm. long; fruit black or black-purple, 10 mm. in diameter. Woods and rocky places: N.D.—Minn. *Plains.* Ap–Je.

8. **A. Cárrii** Rydb. A low shrub with dull brown branches; leaves suborbicular to oval, 2–4 cm. long, 1.5–2.5 cm. wide, thin, entire or with few

short teeth at the rounded apex, probably glabrous from the beginning; sepals deltoid, as broad as long, obtuse, soon glabrous, 1–1.5 mm. long; petals unknown; ovary tomentose at the apex; fruit purple, fleshy, 6–7 mm. in diameter. Hillsides: Deadwood, S.D.

9. **A. Bartramiàna** (Tausch) M. Roemer. Shrub 0.5–3 mm. high; cespitose; leaves glabrous or nearly so, firm, elliptic, 3–6 cm. long, 1.5–3 cm. wide, rounded at the apex, acute or acuminate at the base, sharply and often doubly serrate, with 15–50 teeth on each margin, the veins 12–17 pairs; flowers 1–3, terminal, not racemose; sepals deltoid-subulate, erect or slightly spreading; petals broad and short, 6–9 mm. long; fruit 13 mm. broad, dark purple. *A. oligocarpa* (Michx.) M. Roemer [G, B]. Damp uplands: Lab.—Minn.—Pa. *Canad.* Je–Jl.

2. SÓRBUS L. Mountain Ash.

Trees or shrubs. Leaves alternate, pinnate; leaflets several, more or less toothed; stipules deciduous. Flowers perfect, regular, in terminal compound cymes. Hypanthium urn-shaped. Sepals 5, erect or spreading, in ours deciduous. Petals white, spreading. Stamens many. Ovary inferior, adnate to the hypanthium; styles usually 3, distinct; cells of the ovary as many, with 2 ovules each. Fruit small, berry-like, in ours red and very acid; carpels cartilaginous.

Leaves glabrous above, at least in age.
  Leaflets long-acuminate; fruit 4–6 mm. in diameter. — 1. *S. americana.*
  Leaflets obtuse or acute, rarely short-acuminate; fruit about 8 mm. in diameter.
    Leaflets 11–13, bright green on both sides, glabrous, or rarely slightly short-hairy on the veins beneath. — 2. *S. scopulina.*
    Leaflets 13–17, dark green and rather dull above, paler and more or less long-hairy beneath, at least when young. — 3. *S. subvestita.*
Leaves pubescent on both sides. — 4. *S. Aucuparia.*

1. **S. americàna** Marsh. A small tree, up to 10 m. high, with smooth bark; leaflets 11–17, lanceolate, long-acuminate, glabrous or slightly pubescent when young, bright green above, paler beneath, 3.5–10 cm. long, sharply serrate; cymes compound, 8–15 cm. broad; corolla 4–6 mm. broad; fruit globose, 4–6 mm. broad. *Pyrus americana* DC. [G]. Woods: Greenl.—Man.—Minn.—N.C. *Arct.—E. Temp.* My–Je.

f. 291.

2. **S. scopulina** Greene. *Fig. 291.* A shrub 1–4 m. high; young twigs, petioles and inflorescence sparingly villous; leaflets 11–13, elliptic or elliptic-lanceolate, rounded at the base, abruptly acute or acuminate, 3–6 cm. long, sharply and doubly serrate, glabrate, pale beneath; fruit red, subglobose, 7–8 mm. broad. *Pyrus sambucifolia* Porter; not Cham. & Schlecht. Hillsides and ravines: Alta.—S.D.—N.M.—Ariz.—Ore.—B.C. *Submont.—Mont.* Je–Jl.

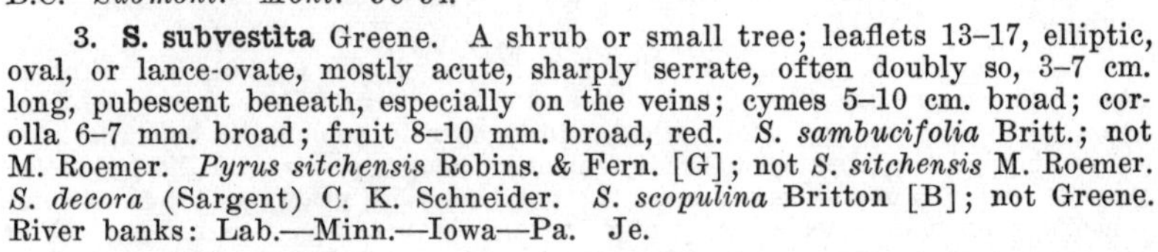
3. **S. subvestìta** Greene. A shrub or small tree; leaflets 13–17, elliptic, oval, or lance-ovate, mostly acute, sharply serrate, often doubly so, 3–7 cm. long, pubescent beneath, especially on the veins; cymes 5–10 cm. broad; corolla 6–7 mm. broad; fruit 8–10 mm. broad, red. *S. sambucifolia* Britt.; not M. Roemer. *Pyrus sitchensis* Robins. & Fern. [G]; not *S. sitchensis* M. Roemer. *S. decora* (Sargent) C. K. Schneider. *S. scopulina* Britton [B]; not Greene. River banks: Lab.—Minn.—Iowa—Pa. Je.

4. **S. Aucupària** L. A tree 8–12 m. high; leaflets 11–15, narrowly oblong, mostly obtuse, finely serrate, pale beneath, pubescent, especially beneath when young. Cultivated and occasionally escaped: Me.—Pa.—N.D.; native of Eur.

3. **ARÒNIA** Medic. CHOKEBERRY, CHOKEPEAR.

Low shrubs, with simple leaves. Flowers small, in terminal compound cymes, perfect. Hypanthium urnshaped. Sepals 5, distinct. Petals 5, concave, spreading, white or pink. Stamens numerous, born on a disk. Styles 3–5, united below. Fruit small, globose, sunken at the base; the carpels with a leathery endosperm.

1. **A. melanocárpa** (Michx.) Britt. A shrub 1–4 m. high; leaf-blades obovate or oval, obtuse to abruptly short-acuminate, crenulate, glabrous or nearly so on both sides; calyx-lobes glandular-toothed; petals white or tinged with purple, 6–10 mm. long; fruit globose, nearly black, 6–8 mm. broad. *A. nigra* Britt. *Pyrus melanocarpa* Willd. [G]. Low woods: N.S.—Minn.—Fla. *E. Temp.* My—Je.

4. **MÀLUS** Juss. APPLE.

Trees or shrubs, with alternate, toothed or lobed leaves. Flowers in simple terminal cymes. Hypanthium urn-shaped, becoming fleshy in fruit. Sepals 5. Petals 5, clawed, white or pink. Styles 2–5, mostly 5, usually united at the base. Endocarp of the carpels papery. Fruit a pome, usually depressed-globose, hollowed at the base and apex, the flesh without grit-cells. Carpels 2-seeded.

Leaves glabrous, at least at maturity; so also the sepals without.
  Leaves oblong or oblong-lanceolate; styles nearly distinct; sepals deciduous in fruit. — 2. *M. coronaria.*
  Leaves ovate; styles united below; sepals persistent on the fruit. — 1. *M. angustifolia.*
Leaves persistently pubescent, often tomentose beneath; so also the sepals without.
  Leaves narrowed at the base; fruit 2–4.5 cm. broad; pedicels slender. — 3. *M. ioensis.*
  Leaves rounded or cordate at the base; fruit 5–10 cm. broad; pedicel stout. — 4. *M. sylvestris.*

1. **M. angustifòlia** (Ait.) Michx. A small tree, often with spinescent branches, rarely 6 m. high; leaves thick, shining and dark green above, narrowed at the base, dentate or entire-margined, acute or obtuse at the apex, narrowed at the base, those of the shoots often lobed; mid-vein glandular above; cymes 3–7-flowered; corolla pink, less than 2.5 cm. broad; fruit about 2.5 cm. *Pyrus angustifolia* Ait. [G]; *M. coronaria* Britton [B, B]; not Mill. Thickets: N.J.—Kans.—La.—Fla. *Austral.* Mr–My.

2. **M. coronària** (L.) Mill. A small tree, up to 10 m. high, often spinescent; leaves ovate, sharply serrate and often lobed, acute at the apex, rounded or cordate at the base, sparingly pubescent along the veins beneath when young; corolla rose-colored, 2.5–5 cm. broad; fruit 2.5–3.5 cm. broad, greenish yellow. *P. coronaria* L. [G]. *M. glaucescens* Britton [BB]; scarcely Rehder. Thickets: N.J.—Ont.—Kans.—S.C. *Allegh.* Ap–My.

f. 292.

3. **M. ioénsis** (Wood) Britt. *Fig. 292.* A small tree; leaves ovate, oval, or oblong, dentate or crenate, or few-lobed, obtuse at the apex, white-tomentose beneath, 2.5–5 cm. long; corolla rose-colored; fruit globose, 1–2 cm. in diameter; pedicels slender, 2.5–3.5 cm. long, pubescent. *P. ioensis* Bailey [G]. Woods: Ill. — Minn. — Okla. — La. *Allegh.—Ozark.* Ap–My.

4. **M. sylvéstris** (L.) Mill. A tree with spreading branches; leaves broadly oval or ovate, obtuse to short-acuminate, 2.5–7.5 cm. long, pubescent,

usually wooly beneath, serrate; pedicels and calyx tomentose, the former 2.5–5 cm. long, stout; corolla pink or white, 3–5 cm. broad; fruit depressed-globose, rarely elongate. *Pyrus Malus* L. [G]. *M. Malus* Britt. [B]. COMMON APPLE. Woods: Me.—Minn.—Kans.—Ga.; cult., escaped, and nat. from Eur. Ap–My.

M. ioensis × sylvestris. Intermediate between the two parents, the pedicels shorter and stouter than in *M. ioensis,* the fruit larger, 2.5–6 cm. broad. *M. Soulardi* (Bailey) Britton. *P. Soulardi* Bailey [G, B]. Woods: Mo.—Minn.—Tex.

5. PỲRUS L. PEAR.

Trees or shrubs, with simple alternate leaves. Flowers in terminal cymes. Hypanthium urn-shaped, becoming fleshy. Sepals 5. Petals 5, white or pink, short-clawed, rounded. Stamens numerous. Styles mostly 5, distinct or nearly so. Fruit fleshy, obovoid, tapering at the base, with grit-cells in the flesh; carpels 2-seeded, the endocarp papery.

1. P. commùnis L. A tree, up to 17 m. high and a trunk 6 dm. in diameter, often thorny; leaf-blades ovate, elliptic, or obovate, usually finely serrulate, 3–8 cm. long, downy when young, glabrate in age; corolla white, about 2.5 cm. broad; fruit in the wild form 5 cm. long or less. Woods and around dwellings: Me.—Minn.—Fla.; rarely escaped from cult.; nat. of Eur.

6. CRATAÈGUS L.* HAW OR HAWTHORN.

Small trees or shrubs, usually armed with thorns or spines. Leaves alternate, petioled, simple, toothed and usually more or less lobed. Flowers in terminal clusters, cymose or corymbose. Hypanthium cup-shaped or campanulate, adnate to the ovary. Sepals 5, reflexed after anthesis. Petals 5, white or pink, spreading, rounded. Stamen 5–25, inserted on the margin of the hypanthium; filaments slender, incurved. Ovary inferior, of 1–5 carpels; styles 1–5, distinct. Pome globose, pear-shaped, or ellipsoid, yellow, red, blue, or black, containing 1–5, bony and 1-seeded carpels.

Leaves not deeply cut; nutlets 2–5.
- Leaves broadest towards the apex; fruit red or yellow.
  - Leaves not impressed-veiny above, shining. — 1. *C. Crus-galli.*
  - Leaves impressed-veiny above, dull.
    - Fruit ellipsoid; nutlets 3 or 4. — 2. *C. punctata.*
    - Fruit globose; nutlets 2 or 3. — 3. *C. Margaretta.*
- Leaves broadest at or below the middle.
  - Leaves pubescent beneath, at least on the veins (sometimes glabrous on both sides in *C. viridis*); fruit red or yellow.
    - Leaves broadest at or just below the middle.
      - Leaves impressed-veined.
        - Calyx-lobes deeply cut; nutlets deeply pitted on the ventral faces.
          - Leaves dark green, glabrous and shining above, coriaceous.
            - Fruit about 15 mm. thick, stamens 10. — 4. *C. succulenta.*
            - Fruit about 12 mm. thick; stamens 15–20. — 5. *C. neofluviatilis.*
          - Leaves gray-green, pubescent and dull above, subcoriaceous.
            - Leaves rhombic-ovate, lobed. — 6. *C. Calpodendron.*
            - Leaves oval, mostly unlobed. — 7. *C. globosa.*
        - Calyx-lobes less deeply cut; nutlets with shallow pits on the ventral faces.
          - Fruits and corymbs pubescent; calyx-lobes sharply glandular-serrate. — 8. *C. pertomentosa.*
          - Fruits and corymbs glabrous; calyx-lobes remotely glandular-serrate. — 9. *C. Brainerdii.*
      - Leaves not impressed-veined.
        - Calyx-lobes glandular-margined; fruit more than 8 mm. broad; leaves not 3-lobed.
          - Leaf-blades suborbicular. — 10. *C. chrysocarpa.*
          - Leaf-blades oval. — 11. *C. columbiana.*

* The treatment of this genus has been adapted, more or less, from Mr. Eggleston's various publications.

Calyx-lobes not glandular-margined; fruit 5–6 mm. broad; leaves often 3-lobed at the apex. — 12. *C. viridis.*
Leaves broadest near the base.
Calyx-lobes usually entire; fruit glabrous. — 13. *C. macrosperma.*
Calyx-lobes usually serrate; fruit pubescent.
Leaves usually glabrate above in age; anthers pink. — 14. *C. coccinoides.*
Leaves tomentose above, on vegetative shoots cordate; anthers yellow. — 15. *C. mollis.*
Leaves glabrous beneath, pubescent above; fruit black when ripe. — 16. *C. Douglasii.*
Leaves deeply cut; nutlets usually solitary. — 17. *C. monogyna.*

1. **C. Crùs-gálli** L. A small tree, up to 9 m. high; spines numerous, straight, 2–17 cm. long; leaf-blades obovate, 2–10 cm. long, 1–4 cm. wide, sharply serrate, except at the cuneate base, dark green and shining above, glabrous or nearly so, coriaceous; corymbs glabrous; calyx-lobes usually entire, lanceolate; stamens 10–20; anthers pink; fruit about 1 cm. thick, greenish or red; nutlets 2. Conn.—Ont.—Kans.—Ga. *Allegh.—Carol.—Ozark.* My–Je; fr. O.

2. **C. punctàta** Jacq. A flat-topped tree, up to 9 m. high; with spreading branches; spines gray, 2–5 cm. long; leaf-blades obovate, 2–7 cm. long, 1–5 cm. wide, impressed-veined, dull and gray-green above, pubescent, doubly serrate at the apex, entire at the cuneate base; corymbs tomentose; stamens 20; anthers white or pink; fruit yellow or red, 12–15 mm. broad. Que.—Pa.—Ky.—Minn. *Canad.* Je; fr. O.

3. **C. Margarétta** Ashe. A shrub or small tree, up to 7 m. high, with ascending branches; spines 2–4 cm. long; leaf-blades oblong-ovate or obovate, 2–6 cm. long, 2–4 cm. wide, cuneate at the base, serrate or doubly serrate, sometimes lobed at the apex, glabrous when mature; corymb slightly pubescent; calyx-lobes reflexed; stamens 20; fruit rusty green, yellow, or red, angular, compressed, 8–15 mm. thick. *C. Brownii* Britton. Thickets: Ont.—Iowa—Mo.—Va. *Canad.—Allegh.* My; fr. O.

4. **C. succulénta** Schrad. *Fig. 293.* A small tree, 4–7.5 m. high, with ascending branches; branches glabrous or hairy; spines numerous, 3–10 cm. long; leaves rhombic-ovate to obovate, 3–9 cm. long, acute at the apex, broadly cuneate at base, doubly serrate, dark green and shining above; corymb slightly villous; sepals lanceolate, acuminate, glandular-laciniate, villous; stamens 10–20; anthers pink, rarely yellow or white; fruit dark red, villous, about 1 cm. thick. *C. macracantha* (Lindl.) Lodd. [G]. *C. occidentalis* Britton. *C. coloradensis* A. Nels. *C. Colorado* Ashe and *C. coloradoides* Ramaley, the variety with hairy twigs. Hillsides and cañons: N.S.—N.C.—Colo.—Sask. *Temp.—Plain—Submont.* My–Je.

5. **C. neofluviátilis** Ashe. A tree up to 9 m. high, with ascending branches; spines numerous, 2–7 cm. long; leaf-blades elliptic-ovate to obovate, 2–7 cm. long, 2–6 cm. wide, sharply and doubly serrate, slightly lobed at the apex, coriaceous, dark green and shining above, pubescent on the veins beneath; calyx-lobes reflexed, glandular-laciniate; stamens 15–20; anthers pink; fruit globose or short-ellipsoid, 6–12 mm. thick, glabrous or pubescent; nutlets 2 or 3. Woods: Vt.—Wis.—Iowa—N.C. *Canad.—Allegh.* My; fr. S.

6. **C. Calpodéndron** (Ehrh.) Medic. A small tree or shrub, up to 6 m. high, with ascending branches; spines sometimes present, 2–5 cm. long; leaf-

blades rhombic-obovate, 4–11 cm. long, 3–7 cm. wide, finely and doubly serrate, pubescent on both sides, scabrous above, subcoriaceous, dull green; corymb tomentose; stamens 20, rarely 10; anthers pink, calyx-lobes laciniate, reflexed; fruit pyriform or ellipsoid, orange-red, 8–10 mm. thick; nutlets 2 or 3. *C. tomentosa* DuRoi [B, G]; not L. N.Y.—Minn.—Mo.—Ga. *Canad.—Allegh.* My–Je; fr. S.

7. **C. globòsa** Sargent. A slender shrub or tree, up to 7.5 m. high; spines 2–6 cm. long; leaf-blades oval or obovate, 4–10 cm. long, 3–7 mm. wide, coarsely serrate, broadly cuneate at the base, scabrous above, pubescent beneath; corymb villous; calyx-lobes glandular-laciniate; stamens 20; anthers pink; fruit globose or short-ellipsoid, orange-red, shining, about 8 mm. thick; nutlets 2. Mo.—Kans. *Ozark.* My; fr. O.

8. **C. pertomentòsa** Ashe. A small tree, up to 6 m. high, with spreading branches; spines many, 2–8 mm. long; leaf-blades oval or obovate, 3–7 cm. long, 2–5 cm. wide, finely and doubly serrate, often lobed, slightly villous, or glabrate above, villous beneath, dark green, subcoriaceous; corymb densely villous; stamens 10–15; calyx-lobes deeply serrate; fruit globose, 10–12 mm. thick, cherry-red, villous when young. *C. campestris* Britton [B]. Rocky barrens: Iowa—Kans.—Mo. *Ozark.* My; fr. S.

9. **C. Brainérdii** Sargent. A shrub or tree, up to 6 m. high, with ascending branches; spines 2.5–6 cm. long; leaf-blades elliptic, oval, or ovate, finely and doubly serrate, or lobed, 3–8 cm. long, 2–4 cm. wide; bright green above, pubescent on the veins beneath; corymb glabrous; stamens 5–20; anthers pink; fruit short-ellipsoid to globose, cherry-red or scarlet, about 1 cm. thick; nutlets 2–4. *C. scabrida, C. Egglestonii,* and *C. asperifolia* Sargent. *C. Schuettei* Ashe. Woods: Mass.—Iowa—Pa. *Allegh.* My; fr. S.

10. **C. chrysocárpa** Ashe. A round-topped shrub or tree, 2–7 m. high, with spines 2–7 cm. long, numerous; leaves orbicular or rounded-obovate, 3–5 cm. long, acute at the apex, broadly cuneate at the base, with 3 or 4 pairs of triangular lobes and doubly serrate; sepals lanceolate, acuminate, glandular-margined; stamens 5–10; fruit depressed-globose, about 1 cm. thick, red. *Crataegus rotundifolia* (Ehrh.) Borckh.; not Lam. *C. Sheridana* A. Nels. *C. Doddsii* Ramaley, a form with glandular petioles and dark fruit. Cañons and banks: N.B.—N.C.—N.M.—Wyo.—Sask. *E. Temp.—Plain—Submont.* My; fr. Au–S.

11. **C. columbiàna** Howell. A shrub or small tree, much branched; leaves rather thin, obovate or ovate, 2–6 cm. long, cuneate at the base, with 3 or 4 pairs of sharp lobes, sharply and doubly toothed, sparingly long-hairy, especially along the veins; corymb sparingly long-hairy; sepals triangular, long-pointed, slightly long-hairy within, glandular-toothed; stamens about 10; fruit pear-shaped or ellipsoid, 8–12 mm. long, scarlet. *C. Piperi* Britton. River banks: B.C.—N.D.—Ida.—Ore. *W. Temp.—Submont.*

12. **C. víridis** L. A tree often 12 m. high; spines rare, 2–5 cm. long; leaf-blades oblong-ovate, doubly serrate, often lobed at the apex, 2–8 cm. long, 1–5 cm. wide, dark green, shining, often pubescent along the veins beneath, otherwise glabrous; corymb glabrous; stamens 20; anthers usually yellow; fruit globose, red or orange, glaucous, 4–6 mm. broad; nutlets 4 or 5. *C. arborescens* Ell. Alluvial soil, along water: Va.—Kans.—Tex.—Fla. *Allegh. —Austral.* Mr–Ap; fr. O.

13. **C. macrospérma** Ashe. A shrub or tree, up to 7.5 m. high, with ascending branches; spines numerous, curved, 2–7 cm. long; leaf-blades elliptic-ovate to broadly ovate, truncate or rounded at the base, doubly serrate, 2–7 cm. long and as wide, slightly villous, becoming glabrate, dark yellowish green above; corymb glabrous or nearly so; stamens 5–10, rarely up to 20; fruit

ellipsoid or pyriform, scarlet or crimson, 10–18 mm. thick, glaucous. Thickets: N.S.—Minn.—Tenn.—N.C. *Canad.—Allegh.* My; fr. Au–S.

14. **C. coccinoìdes** Ashe. A shrub or tree, up to 6 m. high, with spreading branches; spines 2–4 cm. long; leaf-blades broadly ovate, rounded or truncate at the base, doubly serrate and broadly lobed, 3–8 cm. long, 3–7 cm. wide, dark green above, paler beneath, tomentose along the veins; corymb glabrous; calyx-lobes ovate, spreading, glandular-serrate; stamens 20; fruit globose, 15–20 mm. broad. *C. Eggersii* Britton. *C. dilatata* and *C. speciosa* Sargent. Thickets: Que.—Kans.—Mo. *Canad.—Allegh.* My; fr. S.

15. **C. móllis** (T. & G.) Scheele. A tree, up to 12 m. high, with spreading branches; spines 3–5 cm. long; leaf-blades broadly ovate, cordate or truncate at the base, 4–12 cm. long, 3–10 cm. wide, doubly serrate and with acute lobes, densely tomentose beneath, tomentose, but becoming scabrate above; corymb tomentose; stamens 20; anthers light yellow; fruit subglobose, scarlet, 15–25 mm. broad; nutlets 4 or 5. Ont.—S.D.—Ark.—Tenn. *Allegh.—Ozark.* My; fr. S.

16. **C. Douglásii** Lindl. A tree 5–12 m. high, with ascending branches; spines red, stout, 1–3 cm. long; leaf-blades ovate or obovate to broadly oval, 2–7 cm. long, short-acuminate, doubly toothed and somewhat lobed above; sepals acuminate, long-hairy above; anthers 10–20, light yellow; fruit black. *C. brevispina* (Dougl.) Farwell. River banks: B.C.—Mich.—Wyo.—Calif.—N.M. *W. Temp.—Submont.* My–Je.

17. **C. monógyna** Jacq. A shrub or tree, with ascending branches; spines numerous; leaf-blades ovate in outline, deeply 3–15-lobed or cleft, broadly cuneate or truncate at the base, 1–4 cm. long; 1–5 cm. wide, dark green and glabrous above when mature, paler beneath and slightly pubescent; corymb glabrous; calyx-lobes entire, obtuse; stamens 20; anthers pink; fruit subglobose, red, 6 mm. thick; nutlets solitary. Around dwellings: Me.—Minn. —Neb.—Va.; escaped from cult.; intr. from Eurasia. My–Je; fr. S.

E. J. Palmer has lately reported the following species from Kansas: *C. bracteata*, *C. dasyphylla*, *C. munita*, *C. Munsoniana*, *C. paradoxa*, and *C. Stevensiana*, *C. McGeeae* Ashe was from Iowa. They are all unknown to the author.

Family 81. **AMYGDALACEAE.** PLUM FAMILY.

Trees or shrubs, with alternate simple leaves and deciduous stipules. Flowers perfect, in ours regular. Hypanthium well developed, mostly cup-shaped, with an angular disk at the mouth bearing the stamens. Sepals and petals 5. Stamens in ours 10 or more. Pistils usually solitary. Fruit a drupe.

1. **PRÙNUS** L. PLUMS, CHERRIES.

Shrubs or trees. Leaves alternate, deciduous, usually toothed. Flowers perfect, either solitary, umbellate, or corymbose from scaly buds, or corymbose or racemose at the ends of leafy branches. Hypanthium in ours campanulate or turbinate. Sepals 5, imbricate. Petals 5, imbricate, inserted with the stamens on a disc in the throat of the hypanthium. Stamens 15–30; filaments filiform, distinct. Drupe with a fleshy exocarp, often with a bloom; stone bony, smooth.

Flowers in small umbels or corymbs, from scaly buds, usually naked or rarely subtended by one or two small leaves.
  Stone usually more or less flattened, with a ventral groove (PLUMS).
    Leaves abruptly acuminate; petals 8–16 mm. long.
      Sepals entire, not glandular-serrate.
        Leaves and twigs glabrate in age. — 1. *P. americana.*
        Leaves beneath and twigs permanently copiously soft-pubescent. — 2. *P. lanata.*

Sepals glandular-serulate.
Sepals glabrous within; leaves oval or obovate. — 3. *P. nigra.*
Sepals pubescent on both sides; leaves lanceolate. — 4. *P. hortulana.*
Leaves acute or rarely gradually acuminate; petals 4–8 mm. long.
Leaves pubescent, at least beneath; twigs soft-pubescent when young.
Leaves sharply serrate; fruit rounded-ellipsoid. — 5. *P. gracilis.*
Leaves crenate; fruit ellipsoid, yellow. — 6. *P. rugosa.*
Leaves and twigs glabrous or nearly so.
Leaves ovate-lanceolate, or oblong; fruit with thick skin. — 7. *P. Watsonii.*
Leaves lanceolate; fruit with thin skin. — 8. *P. angustifolia.*
Stone of the fruit turgid, usually without a groove (CHERRIES).
Flowers in sessile naked umbels.
Low shrubs; flowers 6–12 mm. broad.
Leaves oblanceolate or spatulate. — 9. *P. pumila.*
Leaves oval or obovate.
Petioles 8–20 mm. long; drupe 8–10 mm. thick. — 10. *P. cuneata.*
Petioles 4–6 mm. long; drupe 12–16 mm. thick. — 11. *P. Besseyi.*
Trees; flowers 1–3 cm. broad.
Leaves glabrous; fruit sour. — 12. *P. Cerasus.*
Leaves pubescent beneath; fruit sweet. — 13. *P. avium.*
Flowers in short-peduncled corymbs.
Leaves ovate or lanceolate, acute at the base. — 14. *P. pennsylvanica.*
Leaves rounded-ovate, subcordate or rounded at the base. — 15. *P. Mahaleb.*
Flowers in elongate racemes, terminating leafy branches.
Sepals deciduous; leaves serrate with spreading teeth.
Leaves glabrous on both sides.
Fruit red or purplish, astringent; leaves thin. — 16. *P. nana.*
Fruit black, sweet; leaves firmer. — 17. *P. melanocarpa.*
Leaves pubescent beneath. — 18. *P. demissa.*
Sepals persistent; leaves crenulate-serrate, with incurved teeth. — 19. *P. virginiana.*

1. P. americàna Marsh. *Fig. 294.* Tree, 3–10 m. high, spinescent; leaves narrowly obovate, abruptly acuminate, sharply and doubly serrate, 4–10 cm. long, pubescent near the veins beneath, the teeth not gland-tipped; petioles 0.5–2 cm. long, usually glandless; flowers 2–5; sepals pubescent within; petals about 1 cm. long; fruit rounded-ellipsoid, 2 cm. thick, red or purplish; stone oval, not crested on the upper suture. River banks and thickets: Mass.—Fla.—N.M.—Utah—Mont. *E. Temp.—Submont.* Ap–My; fr. Au–O.

2. P. lanàta Mack. & Bush. Small tree, forming thickets; leaves obovate to oblong-obovate, 6–12 cm. long, acuminate, sharply and often doubly serrate, pubescent beneath; petioles usually glandless; flowers 2–5; pedicels and calyx pubescent, the lobes entire; fruit globose, 2–3 cm. broad, red or yellowish; stone flattened. *P. Americana mollis* T. & G. *P. Palmeri* Sarg. Thickets: Ill.—Iowa—Tex.—Mo. *Prairie—Ozark—Texan.*

3. P. nìgra Ait. Tree or shrub, 2–10 m. high; leaves thin, broadly obovate or oval, acuminate, sharply doubly serrate, 6–12 cm. long, glabrous above, sparingly pubescent beneath, the teeth gland-tipped and the petioles with 2 glands; flowers 3–5; petals obovate, 12–14 mm. long; fruit oval-ellipsoid to subglobose, 2.5 cm. long, orange-red to yellow, without a bloom; stone oval, acutely crested on the upper suture. Woods and thickets: Newf.—N.Y.—Ga.—Minn.—Man. *Canad.—Allegh.* My; fr. Au.

4. **P. hortulàna** Bailey. Small tree, rarely spinescent; leaves lanceolate, rarely ovate, abruptly long-acuminate, glandular-serrate; petioles 2.5 cm. long or less, with 2 glands near the base of the blade; flowers 2–4; petals obovate, 8–10 mm. long; fruit subglobose or rounded-ellipsoid, 2–2.3 cm. long, yellow or red, thin-skinned; stone less flattened, rough, neither margined nor crested. Thickets: Ills.—Iowa—Kans.—Tex.—Ala.—Ky. *Carol.—Ozark.—Texan.* Ap–My; fr. S.

5. **P. grácilis** Engelm. & Gray. Shrub, 3–13 dm. high, with spreading unarmed branches; leaves oblong or elliptic, densely pubescent beneath when young, sharply serrate, 3–7 cm. long, short-petioled; flowers 2–4, the pedicels slender; sepals ovate, obtuse, finely pubescent; petals 4–5 mm. long; fruit subglobose or rounded-ellipsoid, 8–10 mm. long; stone suborbicular, slightly flattened. *P. normalis* (T. & G.) Small. Sandy soil: Tenn.—Kans.—Tex.

6. **P. rugòsa** Rydb. Shrub, 5–15 dm. high, with spreading, almost black branches; petioles 3–5 mm. long; leaf-blades rather thick, elliptic, 2–3.5 cm. long, acute at the base, acute or obtuse at the apex, crenate, strongly rugose-reticulate, finely villous on both sides, densely so beneath; flowers unknown; pedicels about 1 cm. long; fruit ellipsoid, fully 1.5 cm. long, about 1 cm. thick, yellow, the skin rather thin. Sandhills: w Kans. fr. Jl.

7. **P. Watsònii** Sarg. Sparingly spinescent shrub, 1–3.5 m. high; leaves thick, oblong or oblong-lanceolate, acute, crenate-serrate, shining above, pale and dull beneath; flowers 2–3; pedicels short, red; sepals ciliate, glandless; petals oblong-obovate; petals 6–8 mm. long; fruit subglobose, 2–2.2 cm. long, red-orange, without a bloom. *P. angustifolia Watsonii* Waugh [G]. Sandy soil: Neb.—Kans.—Ark. Ap–My.

8. **P. angustifòlia** Marsh. Shrub, 2–8 m. high, scarcely spiny; leaves thin, 3–10 cm. long, elliptic-lanceolate, acute, serrulate, short-petioled, glabrous; flowers 2–4; petals obovate, 6–8 mm. long; fruit globose, 12–16 mm. thick, red, without a bloom, thin-skinned; stone ovoid, turgid, rounded on both sutures, the upper slightly grooved. CHICKASAW PLUM. Thickets: N.J.—Neb.—Tex.—Fla. *Allegh.—Carol.* Ap; fr. Jl.

9. **P. pùmila** L. Prostrate, spreading shrub; stems 1–2 m. long; leaves oblanceolate or linear-spatulate, acute or acutish, pale beneath, entire or serrate above the middle; flowers 2–4; petals 5–6 mm. long; fruit globose, pendulous, dark brown-purple, without a bloom, 1 cm. thick; stone turgid. SAND CHERRY. Sandy shores: N.B.—Pa.—Wis.—Man. *Canad.* Ap–My; fr. Au.

10. **P. cuneàta** Raf. Low, erect shrub, 3–12 dm. high; leaves spatulate-oblong or obovate, mostly rounded at the apex, 3–8 cm. long, bright green above, glaucous beneath, serrate above the middle; flowers 2–4; petals 6–7 mm. long; fruit globose, 1 cm. thick, nearly black, without a bloom; stone turgid. DWARF CHERRY. Thickets, in sandy soil: N.H.—Minn.—N.C. *Canad.—Allegh.*

11. **P. Bésseyi** Bailey. Prostrate shrub, 3–15 dm. long; leaves elliptic or oblong-oblanceolate, often acute at each end, serrate-crenate, flowers 2–4, short-pedicelled; petals 5–6 mm. long; fruit globose, 12–16 mm. thick, black, often mottled. *P. prunella* Daniels. WESTERN SAND CHERRY. Sand hills: Can.—Minn.—Kans.—N.D. *Prairie—Plain.* Ap–My.

12. **P. Cérasus** L. Tree, up to 10 m. high; bark gray; leaves ovate, 4–10 cm. long, abruptly acute, serrate-dentate, glabrous; inner bud-scales small; petals 12–15 mm. long; fruit depressed-globose, red, sour, about 1 cm. thick; stone globose. SOUR or MORELLO CHERRY. Woods and hedge-rows: N.H.—Pa.—Ga.—Minn.; escaped from cultivation; nat. of Eur. Ap–My; fr. Je–Jl.

13. **P. àvium** L. Tree, sometimes more than 20 m. high, with reddish brown bark; leaves oval or ovate, abruptly short-acuminate, serrate, glabrous,

resinous when young; 5–10 cm. long; petals 12–15 mm. long; pedicels 2.5–3.5 cm. long; fruit yellow or red, sweet, 8–10 mm. thick; stone globose. SWEET CHERRY or BIRD CHERRY. Woods and hedge-rows: Conn.—Ga.—Kans.—Minn.; escaped from cultivation; native of Eur. Ap–My.

14. **P. pennsylvánica** L. f. Small tree, 5–10 m. high; leaves oblong-lanceolate, acuminate, finely and doubly serrate, 8–15 cm. long, glabrous; flowers in small, peduncled corymbs, usually naked at the base; petals 6–7 mm. long; fruit globose, 4–6 mm. thick, sour; stone globose. WILD RED CHERRY. Rocky places and open woods: Newf.—Ga.—Colo.—N.D. *E. Temp.* Ap–Je.

15. **P. Màhaleb** L. Small tree or shrub, up to 7 m. high, glabrous; leaves rounded-ovate, crenulate-dentate, glandular between the teeth, 3–6 cm. long; flowers in small peduncled corymbs, usually subtended by a small leaf; fruit globose, black, 8 mm. long; stone slightly flattened. PERFUMED CHERRY. Waste places and roadsides: Conn.—Del.—Kans.; escaped from cult.; nat. of Eur.

16. **P. nàna** DuRoi. Shrub or small tree, up to 12 m. high; leaves thin, oval or obovate, 5–10 cm. long, short-acuminate, sharply serrate, glabrous; petioles gland-bearing; racemes lax, 8–15 cm. long; sepals reflexed; petals 6 mm. long; fruit globose, red, very astringent, 8–10 mm. thick; stone globose. CHOKE-CHERRY. *P. virginiana* Auth.; not L. River banks and rocky woods: Newf.—Ga.—Tex.—Man. *E. Temp.* Ap–My; fr. Jl–Au.

17. **P. melanocárpa** (A. Nels.) Rydb. *Fig. 295.* Tree or shrub up to 10 m. high; twigs reddish brown, glabrous; leaves obovate or oval, abruptly acuminate at the apex, mostly rounded at the base, rather firm, paler beneath, glabrous on both sides; petioles glandless; peduncles slender, glabrous or rarely minutely puberulent, many-flowered; petals about 6 mm. long; fruit dark purple or black, 6–8 mm. thick, sweet, slightly astringent, shorter than the pedicels. *Cerasus demissa melanocarpa* A. Nels. Hills and river banks: Alta.—N.D.—Kans.—N.M.—Calif.—B.C. *W. Temp.—Plain—Submont.* My–Je.

f.295.

18. **P. demíssa** (Nutt.) Walp. Tree up to 15 m. high; twigs densely pubescent when young; leaves obovate or oblong-obovate, 5–10 cm. long, rounded or subcordate at the base, sharply toothed, paler and pubescent at least on the veins beneath; petioles gland-bearing; peduncles 7–10 cm. long, more or less pubescent; petals about 5 mm. long; fruit globose, 8–10 mm. in diameter, purplish or red. *Cerasus demissa* Nutt. River banks: B.C.—Ida.—Calif.; apparently also Black Hills of S.D., but there a mere low shrub (f. *Rydbergii* Koehne). *Submont.* My–Je.

19. **P. virginiàna** L. Tall tree, up to 30 m. high, with aromatic bark; leaves oblong, elliptic, or lanceolate, acuminate, 5–15 cm. long, glabrous or nearly so; racemes elongate, 1–3 dm. long, dense; petals obovate, 6–8 mm. long; fruit globose, 8–10 mm. thick, black or purple, sweet, slightly astringent. WILD BLACK CHERRY. *P. serotina* Ehrh. *Padus serotina* Agardh. Woods: N.S.—Fla.—Tex.—N.D. My; fr. Au–S.

Family 82. **MIMOSACEAE.** MIMOSA FAMILY.

Herbs, shrubs, or trees, usually with twice or thrice pinnately compound leaves, with stipules, the latter often modified into spines. Flowers mostly perfect, regular, in heads or spikes. Sepals 3–6, partly united, valvate. Petals as many, valvate, distinct or partly united. Stamens of the same number, twice as many, or numerous. Pistil solitary, in fruit becoming a legume.

Stamens numerous; ovary stipitate; trees or shrubs. 1. ACACIELLA.
Stamens 10 or less; ovary sessile.
  Anthers without glandular appendages at the apex; herbs or low shrubs.
    Legumes flat; valves not separating from the continuous margins. 2. DESMANTHUS.
    Legumes 4-angled; valves separating from the continuous margins. 3. LEPTOGLOTTIS.
  Anthers tipped with glandular appendages; trees or high shrubs. 4. NELTUMA.

1. ACACIÉLLA Britton & Rose. ACACIA.

Shrubs or trees, rarely herbs, usually unarmed. Leaves usually twice pinnate; leaflets often numerous. Flowers perfect or polygamous, in dense globose or cylindric spikes on axillary peduncles. Calyx mostly campanulate, 4- or 5-lobed. Petals 4 or 5, distinct, or somewhat united below. Stamens numerous; filaments distinct, or the inner somewhat united below. Ovary often stipitate. Pod flat, 2-valved.

1. **A. hírta** (Nutt.) Britton & Rose. An unarmed herb, with angled stem; leaves 5–15 cm. long, with 8–13 pairs of pinnae, hirsute; leaflets numerous, oblong or linear, 4–5 mm. long; inflorescence globose, 1 cm. thick; peduncles 1–2 cm. long, hirsute; flowers pedicelled; pod linear-oblong, 5–7 cm. long, 6–9 mm. wide. *Acacia hirta* Nutt. *A. angustissima hirta* (Nutt.) B. L. Robinson [G]. Sandy prairies: Mo.—Kans.—Ariz.—Tex.; Fla.; Mex.

2. DESMÁNTHUS Willd. PRAIRIE MIMOSA.

Perennial unarmed herbs, rarely somewhat shrubby. Leaves bipinnate, with numerous leaflets. Flowers in peduncled heads or head-like spikes, perfect or the lower ones staminate, sometimes without petals. Calyx 5-lobed. Petals 5, distinct, or slightly united at the base. Stamens 5 or 10, exserted; filaments distinct or nearly so. Ovules numerous. Pod elongate, unarmed. *Acuan* Medic.

Legume oblong, strongly falcate, 1.5–2.5 cm. long, 5 mm. wide. 1. *D. illinoensis.*
Legume linear, straight or nearly so, 3–7 cm. long, 2–3 mm. wide.
  Legume constricted between the seeds; stamens 5. 2. *D. leptolobus.*
  Legume not constricted between the seeds; stamens 10. 3. *D. Cooleyi.*

1. **D. illinoénsis** (Michx.) MacMill. Erect or ascending herb, 3–10 dm. high, glabrous or nearly so; leaves bipinnate, with 12–30 pinnae; leaflets numerous, linear, 2–3.5 mm. long; peduncles 2.5–8 cm. long, longer than the pods; stamens 5; pods 4–6 mm. wide, in compact heads, slightly spirally twisted. *Desmanthus brachylobus* (Willd.) Benth. *A. illinoensis* (Michx.) Kuntze [B, R]. River banks: Tenn.—Fla.—Tex.—N.M.—S.D. *Carol.—Tex.—Son.—Plain.* My–Je.

2. **D. leptólobus** T. & G. Stem ascending, 6–10 dm. high, scabrous on the angles; leaves bipinnate, with 10–20 pinnae; leaflets numerous, linear-lanceolate or linear-oblong, 3–4 mm. long, acute; peduncles 1–2 cm. long; pods linear, 5–8 cm. long, straight or nearly so, acuminate, glabrous, 4–7 cm. long. *A. leptolobum* (T. & G.) Kuntze [B]. Prairies: Mo.—Kans.—Tex. *Texan.* Je–S.

3. **D. Coòleyi** (Eaton). Shrubby, decumbent below, 2–5 dm. high; leaves bipinnate, 3–4 cm. long, with 16–30 pinnae; leaflets numerous, oblong or linear-oblong, 3–4 mm. long, mucronate, ciliate, peduncles 2 cm. long or less; legumes linear, 3–7 cm. long, 3 mm. wide. *Acacia Cooleyi* Eaton. *D. Jamesii* T. & G. *A. Cooleyi* Britt. & Rose. Plains: Neb.—Tex.—Ariz.; Mex. *Plain—Son.*

3. **LEPTOGLÓTTIS** DC. SENSITIVE BRIER.

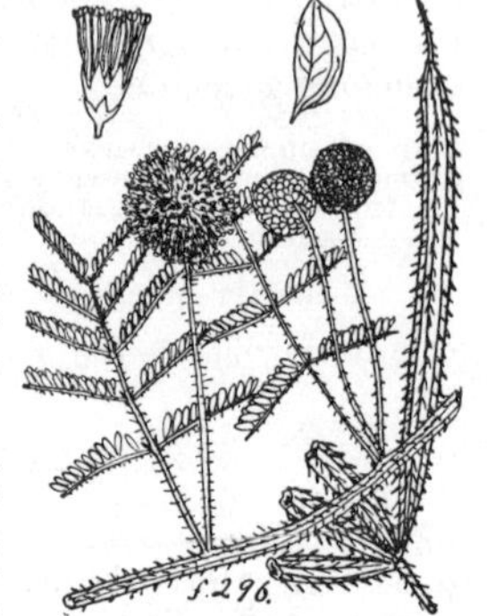

Perennial herbs or shrubs, with spreading prickly stems. Leaves bipinnate, usually sensitive; leaflets numerous, usually small. Flowers perfect or polygamous, in axillary peduncled heads. Calyx 4- or 5-lobed. Petals 4 or 5, united to about the middle. Stamens 8–10, exserted; filaments distinct or nearly so. Ovules numerous. Pod narrow, 4-angled, prickly all over, finally 4-valved, the valves separating from the margins. *Schrankia* Willd.; not Medic. *Morongia* Britt.

1. **L. Nuttállii** DC. *Fig. 296.* Herbs; stem 3–15 dm. long, armed with recurved prickles; leaves bipinnate; pinnae 8–16; leaflets numerous, oblong, 4–8 mm. long, apiculate, prominently veined beneath; peduncles prickly; pod 3–10 cm. long, more or less densely prickly, beaked. *Schrankia uncinata* Am. auth. [G]; not Willd. *M. uncinata* Britton. Prairies and dry soil: Va.—Fla.—Tex.—Colo. *Plain.* Je–S.

4. **NELTÙMA** Raf. MESQUITE.

Shrubs or trees, with axillary spines. Leaves alternate, bipinnate, with relatively few pinnae and leaflets. Flowers perfect, in cylindric spikes. Calyx 5-lobed. Petals 5, distinct or slightly united, yellowish or greenish. Stamens 10; filaments distinct; anthers with glandular appendages. Pods flat and straight or slightly curved, indehiscent, spongy within.

1. **N. glandulòsa** (Torr.) Britton & Rose. A shrub or tree, sometimes 15 m. high, with zigzag branches and thorn-like spines; leaves long-petioled; pinnae 2–4; leaflets 12–60, linear, 1.5–4 cm. long; spikes 4–8 cm. long; petals linear, 4–5 times as long as the calyx, pubescent within; pod linear, 1–2 dm. long. *Prosopis glandulosa* Torr. Plains and prairies: Kans.—Ariz.—Tex.; Mex.

Family 83. CAESALPINIACEAE. SENNA FAMILY.

Trees, shrubs or herbs. Leaves alternate, usually with stipules, simple or compound. Flowers mostly perfect, regular or irregular. Calyx of 5, more or less united sepals. Petals usually 5, imbricate, the upper one enclosed by the lateral ones in the bud. Stamens 10 or fewer, filaments distinct or monadelphous. Ovary 1-celled. Fruit a 2-valved legume or indehiscent.

Corolla regular or somewhat irregular, but not zygomorphic; leaves pinnately compound.
- Leaves bipinnate.
  - Flowers dioecious or polygamous; trees.
    - Receptacle small; stamens 6–10, longer than the petals; pod leathery. 1. GLEDITSIA.
    - Receptacle long; stamens 10, shorter than the petals; pod woody. 2. GYMNOCLADUS.
  - Flowers perfect; herbs. 3. LARREA.
- Leaves once pinnate.
  - Corolla essentially regular, the petals nearly equal; calyx-lobes obtuse. 4. CASSIA.
  - Corolla very irregular, one of the lateral petals (representing the banner) and the lowest petal large, the other three small; calyx-lobes acuminate. 5. CHAMAECRISTA.

Corolla zygomorphic; leaves simple. 6. CERCIS.

1. GLEDÍTSIA L. Honey Locust.

Trees usually armed with simple or branched thorns. Leaves equally bipinnate, or some of them merely pinnate; leaflets often crenulate. Flowers small, polygamous, in axillary or lateral racemes or panicles. Calyx 3–5-lobed, the tube campanulate-turbinate, the lobes nearly equal. Petals 3–5, nearly equal. Stamens 6–10, longer than the petals; filaments distinct; anthers uniform, opening lengthwise. Pod elliptic or elongate, flattened, more or less curved and twisted, indehiscent or tardily dehiscent, the valves leathery. Seeds flattened.

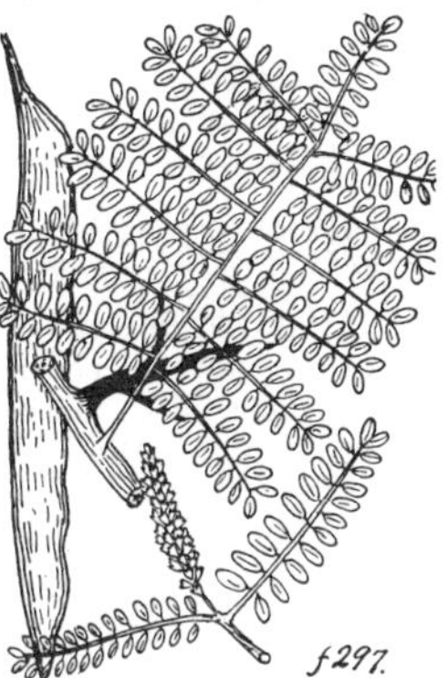

1. **G. triacánthos** L. *Fig. 297.* A tree, often 40 m. high; leaflets 12–28, ovate-lanceolate to elliptic, 1–2.5 cm. long, inequilateral, crenulate, obtuse, pubescent beneath; staminate racemes short-peduncled, 3–12 cm. long; petals greenish, longer than the calyx-lobes; pod linear, 2–3 cm. long, curved and twisted, black. Rich woods: N.Y.—S.D.—Tex.—Fla. *Allegh.—Austral.* My–Jl.

2. GYMNÓCLADUS Lam. Kentucky Coffee-tree.

Unarmed trees, with furrowed bark. Leaves alternate, bipinnate, the leaflets entire. Flowers polygamous, regular, in terminal racemes or thyrsoid panicles. Hypanthium elongate. Calyx-lobes 5, equal. Petals 5, greenish-white, inserted on the edge of a disk. Stamens 10; filaments subulate, those opposite the petals shorter; anthers opening lengthwise. Pod oblong, flattened, woody. Seeds flattened, on slender panicles.

1. **G. dioìca** (L.) Koch. A tree, 20–35 m. high; leaves bipinnate, the pinnae 5–9; each with 3–7 pairs of leaflets, the terminal one present or absent; leaflets ovate or oval, 2–7 cm. long, acuminate, entire; panicles 1–3 dm. long, the pistillate ones longer than the staminate ones; calyx pubescent within and without; hypanthium 1 cm. long; petals oblong, nearly white, longer than the calyx-lobes; pod 1–2 dm. long, curved. *G. canadensis* Lam. Woods: N.Y.—S.D.—Okla.—Tenn. *Allegh.—Carol.—Ozark.* My–Je.

3. LÁRREA Ort.

Herbs, sometimes somewhat woody at the base. Leaves bipinnate; leaflets several to many, often glandular-punctate. Flowers perfect, racemose. Calyx-lobes equaling or exceeding the tube. Petals 5, yellow or yellowish, imbricate. Stamens 10, slightly declined; filaments distinct, glandular at the base; anthers opening lengthwise. Pot flat, often more or less curved, few-seeded. *Hoffmanseggia* Cav.

Leaflets not glandular-punctate; pod narrow, strongly falcate. 1. *L. densiflora.*
Leaflets glandular-punctate; pod broad and short, scarcely falcate. 2. *L. Jamesii.*

1. **L. densiflòra** (Benth.) Britt. Stem branched at the base, ascending, 5–30 cm. high, pubescent or puberulent; leaves with 5–9 pinnae; leaflets 10–20, oblong, 4–6 mm. long, obtuse; calyx-lobes linear or linear-oblong, 7–8 mm. long; petals spatulate; pod 3–4 cm. long. *H. densiflora* Benth. Plains and prairies: Kans.—Calif.—Tex.; Mex. *Texan—Trop.* Ap–Je.

2. **L. Jamèsii** (T. & G.) Britt. Herb, with fusiform root; stem branched at the woody base, 1–3.5 dm. high, finely pubescent and more or less glandular; leaves with 5–7 pinnae, each bearing 10–20 leaflets, which are oblong or oblong-ovate, 3–5 mm. long, obtuse; calyx-lobes oblong or linear-oblong, 7–9 mm long; petals about 1 cm. long, clawed; pod 2.25 cm. long. *H. Jamesii* T. & G Plains: Kans.—Tex.—Calif.—Colo. *St. Plains—Son.* My–Au.

4. CÁSSIA L. SENNA.

Herbs, shrubs, or trees with abruptly pinnate leaves. Flowers perfect, mostly yellow, in terminal or axillary racemes or panicles. Petiolar glands often present. Calyx-lobes 5, obtuse. Petals 5, nearly equal, imbricate, spreading. Stamens 5 or 10, all perfect, or the upper 3 reduced to staminodia; anthers opening by apical pores. Pod turgid or somewhat compressed, elongate, often with cross-partitions within. Seeds transverse or nearly so.

Petiolar glands at the base of the rachis. 1. *C. Medsgeri.*
Petiolar glands between the lowest pair of leaflets. 2. *C. Tora.*

1. **C. Médsgeri** Shafer. Plant biennial, about 1 dm. high; stipules linear-lanceolate; petiolar gland cylindric or conic-obovoid; leaflets 14–20, oblong-lanceolate, obtuse; racemes short, many-flowered; calyx-lobes oval, somewhat petal-like; pod 5–9 cm. long, thick, its segments shorter than broad; seeds twice as high as broad. *C. marilandica* A. Gray, in part [B]. Dry soil: Pa.—Iowa—Kans.—Tex.—Ga. *Allegh.—Carol.—Ozark.* Au.

2. **C. Tòra** L. Plant annual, 5–15 dm. high, glabrous; leaflets 4–6, glabrous or pubescent beneath, elliptic to obovate-cuneate, slightly cuspidate, the uppermost the largest, 3–5 cm. long; racemes loose, few-flowered; flowers large, pod linear, falcate, acute, 4-angled, glabrous, 15–18 cm. long; seeds quadrate. Waste ground: Va.—Ind.—Kans.—Tex.—Fla.; Mex. *Allegh.—Carol.—Tex.* Jl–S.

5. CHAMAECRÍSTA Moench. PARTRIDGE PEA, SENSITIVE PEA, WILD SENSITIVE PLANT.

Herbs or shrubs. Leaves pinnate; rachis with 1 or 2 glands at the base. Flowers solitary or clustered, axillary or supra-axillary; pedicels twisted. Calyx 5-lobed. Corolla irregular; petals 5, one of the lateral ones (banner) and the lower one (one of the wings) larger than the rest. Stamens 10, all perfect or some of them reduced to staminodia; anthers unequal in size, opening by terminal pores. Pod linear, flat, dehiscent with elastic valves.

Corolla 2.5–4 cm. broad; anthers 10. 1. *C. fasciculata.*
Corolla less than 10 mm. broad; anthers 5. 2. *C. nictitans.*

1. **C. fasciculàta** (Michx.) Greene. *Fig. 298.* Erect herb; stem glabrous, 3–6 dm. high; leaflets 16–28, oblong-elliptic, about 1.5 cm. long, 3 mm. wide; petiolar gland sessile; peduncles supra-axillary, 3–5-flowered; sepals lanceolate, fully equaling the petals; pods linear, pubescent, 5 cm. long, 5 mm. wide. *Cassia Chamaecrista* Am. auth. [G, B]. Meadows: Me.—Fla.—Mex.—Colo.—Minn. *E. Temp. —Plain.* Je–S.

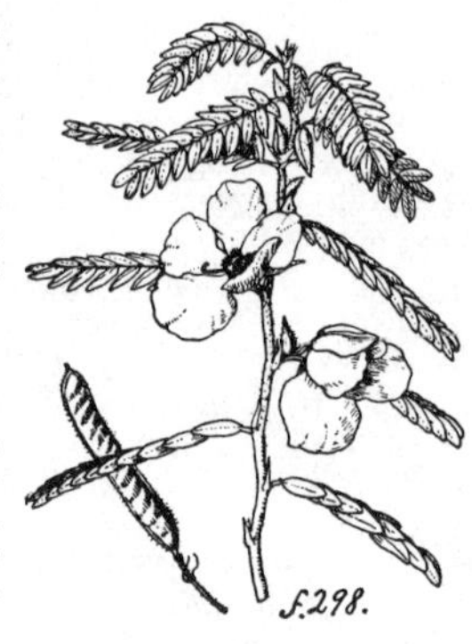
f.298.

2. **C. níctitans** (L.) Moench. Erect annual; stem 1.5–4 dm. high, pubescent; leaflets 12–28, glabrous, linear-elliptic, acute or mucronate, 1–1.5 cm. long, 2–3 mm. wide; petiolar gland urceolate, subsessile; peduncles 1-flowered; flowers sessile; pod pubescent, 3.5 cm. long, 6 mm. wide. *Cassia nictitans* L. [G, B]. Woods and fields: Me.—Kans.—Ariz.—Tex.—Fla. *E. Temp.* Jl–O.

6. CÉRCIS L. RED BUD. JUDAS TREE.

Unarmed shrubs or trees, with scaly bark. Leaves alternate, simple;

flowers perfect, in axillary clusters, or on branches of the preceding year, appearing before the leaves. Calyx campanulate, 5-lobed, swollen on one side and lined within with a disk. Corolla pink or rose; petals 5, the uppermost smallest and enclosed by the lateral ones, the lowest 2 auricled on the upper side. Stamens 10, in two series, the inner shorter; anthers opening lengthwise. Pod narrow, flat, 2-valved, the central suture with 2 narrow wings. Seeds flat.

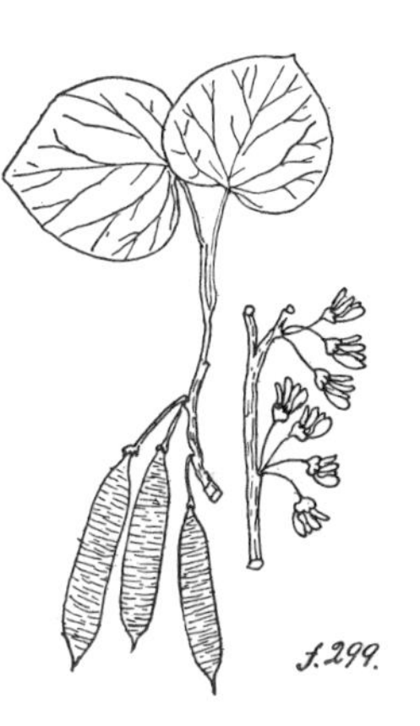
f.299.

1. **C. canadénsis** L. *Fig. 299.* Shrub or small tree, 15 m. high or less; leaves petioled, the blade suborbicular in outline or reniform, cordate at the base, acuminate, 6–12 cm. broad; corolla pink-purple; pod linear-oblong, 5–10 cm. long, acuminate at both ends, short-stipitate. Rich soil: N.Y.—Minn. —Neb.—Tex.—Fla. *E. Temp.* Ap.

Family 84. **KRAMERIACEAE.** KRAMERIA FAMILY.

Shrubs or perennial herbs. Leaves alternate, without stipules, simple or trifoliolate. Flowers perfect, irregular. Sepals 4 or 5, colored, deciduous, unequal. Petals 4 or 5, shorter than the sepals, the 2 or 3 posterior with long, often united claws, thin, the 2 anterior thick, sessile. Stamens 3 or 4, filaments united below; anthers 2-celled, opening by terminal pores. Pistil 1, simple; ovary 1-celled or 2-celled by the intrusion of the placenta. Fruit indehiscent, spiny. Ovules 2, pendulous, anatropous.

1. **KRAMÈRIA** Loefl.

Characters of the family.

1. **K. secundiflòra** DC. *Fig. 300.* An herb, with a thick rootstock; stem ascending or spreading, 1–4 dm. high; leaves simple; blades oblanceolate or linear, 1–3 cm. long, spine-tipped; peduncles axillary, 2–3 cm. long, 1-flowered; sepals ovate-lanceolate, 8–10 mm. long; petals 5; stamens 4; fruit subglobose, 7–9 mm. in diameter; spines stout, retrorsely scabrous. Sandy soil: Fla.—Kans.—N.M.—Mex. *Austral.*

f.300.

Family 85. **FABACEAE.** PEA FAMILY.

Herbs, shrubs, or trees, with alternate, mostly pinnately or palmately compound leaves with stipules. Flowers perfect, rarely polygamo-dioecious, irregular, zygomorphic. Calyx of 5, or rarely 4, more or less united sepals, or sometimes by the complete union of some of the sepals becoming 2-lipped. Corolla papilionaceous, *i.e.*, consisting of 5 more or less united or distinct petals, of which the upper (banner or standard) is broader and surrounds the rest, the 2 lateral ones (wings) are curved upwards, and the 2 lowest (keel) more or less united, forming a boat-shaped organ. Stamens usually 10, sometimes 9 or 5; filaments diadelphous (9 and 1), monadelphous, or rarely distinct. Pistil 1, superior. Fruit a 2-valved or indehiscent legume, 1-celled or by the intrusion of the sutures 2-celled, or a loment, *i.e.*, divided by false cross-partitions into several cells.

Filaments distinct.
  Leaves pinnate. Tribe I. SOPHOREAE.
  Leaves digitately 3-foliolate. Tribe II. PODALYRIEAE.
Filaments monadelphous or diadelphous.
  Rachis of the leaves not produced into a tendril or bristle-like appendage, representing the terminal leaflet.
    Fruit 2-valved or indehiscent, not a loment.
      Foliage not glandular-dotted.
        Anthers of 2 kinds; filaments monadelphous; leaves digitate; calyx 2-lipped. Tribe III. GENISTEAE.
        Anthers all alike; filaments diadelphous; calyx 5-toothed.
          Leaflets toothed. Tribe IV. TRIFOLIEAE.
          Leaflets not toothed.
            Filaments all or at least 5 of them (alternately) dilated above; flowers umbellate or solitary. Tribe V. LOTEAE.
            Filaments all filiform; flowers racemose or capitate.
              Trees, shrubs, or not twining herbs, if trailing, the leaves not 3-foliolate.
                Connective of the anthers produced or appendaged; plants with hairs attached at the middle; wings with a lateral spur. Tribe VII. INDIGOFEREAE.
                Connective of the anthers neither produced nor appendaged; hairs rarely attached at the middle; wings not spurred.
                  Keel not spirally twisted. Tribe VIII. GALEGEAE.
                  Keel spirally twisted (*Apios*). Tribe XI. PHASEOLEAE.
              Twining or trailing herbs with 3-foliolate leaves. Tribe XI. PHASEOLEAE.
      Foliage glandular-dotted.
        Pod prickly. Tribe IX. GLYCYRRHIZEAE.
        Pod not prickly. Tribe VI. PSORALEAE.
    Fruit a loment, *i. e.*, breaking up transversely into 1-seeded indehiscent reticulate internodes. Tribe X. HEDYSAREAE.
  Rachis of the leaves produced into a tendril or bristle-like appendage. Tribe XII. VICIEAE.

TRIBE I. SOPHOREAE.

One genus. 1. SOPHORA.

TRIBE II. PODALYRIEAE.

Ovary sessile; pod flat. 2. THERMOPSIS.
Ovary stipitate; pod turgid or inflated. 3. BAPTISIA.

TRIBE III. GENISTEAE.

Pod inflated; stipules decurrent; leaves in ours uni-foliolate. 4. CROTALARIA.
Pod flat; stipules not decurrent; leaves in ours digitately 5–11-foliolate. 5. LUPINUS.

TRIBE IV. TRIFOLIEAE.

Leaves digitate, in most 3-foliolate. 6. TRIFOLIUM.
Leaves pinnately 3-foliolate.
  Pod curved or coiled. 7. MEDICAGO.
  Pod straight. 8. MELILOTUS.

TRIBE V. LOTEAE.

Filaments diadelphous; pod dehiscent. 9. ACMISPON.
Filaments monadelphous; pod indehiscent or nearly so. 10. ANTHYLLIS.

TRIBE VI. PSORALEAE.

Petal 1 (banner); leaves odd-pinnate. 11. AMORPHA.
Petals 5.
  Wings and keel free from the filament-tube; ovules solitary.
    Pod indehiscent, the beak short.
      Leaves pinnately 1–3-foliolate; pod bony, cross-wrinkled. 13. ORBEXILUM.
      Leaves digitately 3–7-foliolate; pod not cross-wrinkled. 12. PSORALIDIUM.
    Pod circumscissile, or bursting irregularly, the beak long; leaves digitately 3–7-foliolate. 14. PEDIOMELUM.
  Wings and keel more or less adnate to the filament-tube; ovules 2–6; leaves odd-pinnate.
    Stamens 9 or 10. 15. PAROSELA.
    Stamens 5. 16. PETALOSTEMON.

TRIBE VII. INDIGOFEREAE.

One genus. 17. INDIGOFERA.

TRIBE VIII. GALEGEAE.

Leaflets with stipels; pods margined on one edge; trees or shrubs. 18. ROBINIA.
Leaflets without stipels; pods not margined; herbs or undershrubs.
  Racemes terminal or subopposite the leaves; leaflets obliquely striate.
    Upper filament wholly united with the staminal sheath, forming a closed tube; banner glabrous. 19. GALEGA.
    Upper filament free, at least at the base; banner silky-strigose on the back. 20. CRACCA.
  Racemes axillary; leaflets not obliquely striate.
    Keel (lower petals) not produced into a beak.
      Fruit wholly 1-celled, the lower suture rarely sulcate, but never inflexed and forming a partition.
        Pod more or less compressed laterally, both sutures prominent.
          Leaflets spine-tipped; pod 1–4-seeded. 21. KENTROPHYTA.
          Leaflets not spine-tipped; pod several- to many-seeded. 22. HOMALOBUS.
        Pod not laterally compressed, or if slightly so, the lower suture not prominent.
          Pod usually leathery or woody in texture.
            Leaves pinnate.
              Pod stipitate, the body with two grooves on the upper side. 23. DIHOLCOS.
              Pod sessile, without two grooves on the upper side.
                Calyx campanulate; stem usually elongate.
                  Pod erect, not cross-ribbed; corolla middle-sized, the banner narrow, moderately arched.
                    Pod rounded at the base, ovoid or elliptic, terete in cross-section or nearly so. 24. CNEMIDOPHACOS.
                    Pod mostly acute at each end, more or less lunate, ovate or cordate in cross-section. 30. BATIDOPHACA.
                  Pod spreading, cross-ribbed; corolla small, the banner broad, strongly arched. 25. MICROPHACOS.
                Calyx cylindric; flowers comparatively large; plant mostly low, with decumbent stems; pod horizontal. 26. XYLOPHACOS.
            Leaves digitately 3-foliolate; pod ovoid, few-seeded, indehiscent; plants mostly pulvinate. 27. OROPHACA.
          Pod membranous or papery.
            Pod linear- or oblong-cylindric, not inflated, the valves rather close-fitting to the seeds.
              Legume sulcate on both sutures, more or less link-shaped in cross-section. 28. HOLCOPHACOS.
              Legume not sulcate on either suture; the cavity more or less traversed by fibers. 29. PISOPHACA.
            Pod inflated, the valves extended from the seeds.
              Legume membranous, lunate or obliquely ovoid in side-view, the upper suture less curved than the lower. 30. BATIDOPHACA.
              Legume papery, if membranous, ellipsoid or ovoid. 31. PHACA.
      Fruit perfectly or imperfectly 2-celled, the lower suture inflexed, forming a false partition (septum).
        Septum incomplete, not meeting the upper suture.
          Pod membranous, papery, more or less inflated, the body oblong or oval. 32. ATELOPHRAGMA.
          Pod leathery or woody, the body linear in outline. 33. TIUM.
        Septum complete, meeting the upper suture, not inflated, firm or fleshy.
          Leaves odd-pinnate, *i. e.,* with a terminal leaflet; herbs.
            Fruit dry, dehiscent.
              Body of the fruit elongate, linear, in ours laterally compressed. 34. HAMOSA.
              Body of the fruit short or oblong, turgid. 35. ASTRAGALUS.
            Fruit fleshy, indehiscent or very tardily dehiscent. 36. GEOPRUMNON.
          Leaves abruptly pinnate, *i. e.,* the terminal leaflet lacking. 37. CARAGANA.
    Keel (lower petals) produced into a beak. 38. OXYTROPIS.

TRIBE IX. GLYCYRRHIZEAE.

One genus. 39. GLYCYRRHIZA.

TRIBE X. HEDYSAREAE.

Leaves odd-pinnate, with several pairs of leaflets, without stipels.
  Pods 4–several-seeded, neither spiny nor dorsally toothed.
    Pod terete or 4-angled; segments subcylindric; flowers umbellate. 40. CORONILLA.
    Pod flat; segments suborbicular in outline; flowers racemose. 41. HEDYSARUM.
  Pods 1–2-seeded, more or less spiny or toothed. 42. ONOBRYCHIS.
Leaves 3-foliolate.
  Corolla purple or white.
    Pod with several internodes; leaflets stipellate. 43. MEIBOMIA.
    Pod with 1 or 2 internodes; leaflets not stipellate. 44. LESPEDEZA.
  Corolla yellow; pod with 1 or 2 internodes, the lower internode empty. 45. STYLOSANTHES.

TRIBE XI. PHASEOLEAE.

Style glabrous throughout or slightly pubescent below.
  Keel spirally twisted; leaves 5–7-foliolate. 46. APIOS.
  Keel not twisted; leaves 3-foliolate.
    Calyx subtended by 2 bractlets. 47. GALACTIA.
    Calyx without bractlets. 48. AMPHICARPA.
Style bearded along the upper suture.
  Keel of the corolla not coiled nor incurved. 49. CLITORIA.
  Keel of the corolla coiled or incurved.
    Keel merely incurved; inflorescence capitate. 51. STROPHOSTYLES.
    Keel spirally twisted; inflorescence racemose or paniculate. 50. PHASEOLUS.

TRIBE XII. VICIEAE.

Style scarcely dilated.
  Style filiform, hairy all around and below the apex; stamen-tube usually oblique at the summit. 52. VICIA.
  Style flattened towards the apex, hairy on the inner side; stamen-tube usually truncate or nearly so. 53. LATHYRUS.
Style strongly dilated, the edges reflexed, forming a deep groove. 54. PISUM.

1. SOPHÒRA L.

Trees or shrubs, or (ours) perennial herbs, with rootstocks. Leaves odd-pinnate, with bristle-like stipules. Flowers perfect, in racemes. Calyx campanulate, 5-lobed. Petals clawed, white, yellow, or violet, nearly equal in length; banner suborbicular, notched at the apex; keel 2-beaked at the apex. Stamens 10; filaments distinct or nearly so. Pod nearly terete, usually moniliform, indehiscent or tardily dehiscent. Seeds globose.

1. **S. serícea** Nutt. *Fig. 301.* Perennial, with a creeping rootstock; stem 1–3 dm. high, silky-canescent, leafy; leaflets 15–23, elliptic or obovate, canescent; calyx gibbous on the upper side; lobes lanceolate to triangular; corolla about 15 mm. long, ochroleucous; keel with two subulate beaks; pod 3–5 cm. long, constricted between the seeds. Dry prairies, plains, and hills: S.D.—Tex.—Ariz.—Wyo. *Plain—Submont.* Ap–Je.

f 301.

2. THERMÓPSIS R. Br. YELLOW PEA, GOLDEN PEA, PRAIRIE BEAN.

Perennial herbs, with creeping rootstocks. Leaves alternate, with foliaceous stipules, digitately 3-foliolate, with entire leaflets. Flowers perfect, in racemes. Calyx campanulate, 5-lobed, the lobes nearly equal, or the upper two united. Corolla in ours yellow; banner with a broad spreading blade and reflexed margins; keel as long as the wings or longer. Stamens 10, distinct; anthers nearly alike. Pod narrow, flat, 2-valved, many-seeded.

Fruit reflexed, curved into half a circle or more. 1. *T. rhombifolia.*
Fruit mostly horizontal, merely arcuate. 2. *T. arenosa.*

1. **T. rhombifòlia** (Nutt.) Richards. *Fig. 302.* Stems several, glabrous or strigose above, 1–3 dm. high; stipules ovate to suborbicular, 1–2 cm. long, often obtuse; leaflets rhombic-obovate to oblong-oblanceolate, 2–3 cm. long; corolla 15–20 mm. long; pod 5–6 cm. long, grayish-strigose, 10–13-seeded. Sandy places, especially in draws: Man.—Neb.—Colo.—Mont. *Plain—Submont.* Ap–Je.

2. **T. arenòsa** A. Nels. Stems several, 2–4 dm. high, glabrous or slightly pubescent above; stipules ovate to suborbicular, 2–4 cm. long; leaflets oblong, oblanceolate, or obovate, 3–4 cm. long, glabrous on both sides, or sparingly hairy beneath; pod spreading, 4–7 cm. long, more or less torulose, 8–13-seeded. Sandy or gravelly places: Sask.—Colo.—Mont. *Plain—Mont.*

3. **BAPTÍSIA** Vent. WILD INDIGO.

Perennial herbs, with large rootstocks. Leaves alternate, digitately 3-foliolate or rarely simple, the leaflets entire, broadest above the middle. Flowers perfect, in simple or branched racemes, or axillary. Calyx-lobes 5, nearly equal, or the upper two wholly united. Corolla yellow, white, or blue; banner broad, spreading, the margins reflexed; wings oblong or broader upward; keel somewhat curved, as long as the wings. Stamens 10, distinct, the anthers uniform. Ovary stipitate, the style incurved. Pod more or less inflated, stipitate, cylindric to globose.

Racemes terminal, numerous, short, often leafy at the base; corolla yellow. — 1. *B. tinctoria.*
Racemes opposite the leaves, few.
  Corolla white or cream-colored.
    Racemes naked or with minute bracts; plants glabrous or nearly so. — 2. *B. leucantha.*
    Racemes distinctly bracted, the bracts sometimes foliaceous; plant very pubescent. — 3. *B. leucophaea.*
  Corolla blue. — 4. *B. vespertina.*

1. **B. tinctòria** (L.) R. Br. Stem 3–12 dm. high, much branched; leaves glabrous, turning black in drying, the leaflets obovate to cuneate, 1–3 cm. long, rounded or retuse at the apex; calyx campanulate, about 3 mm. long, the lower lobes triangular, shorter than the tube; corolla yellow, 15 mm. long or less; pod long-stipitate, the body oval or oblong, 9–12 mm. long. Dry soil: Me.—Minn.—La.—Fla. *E. Temp.* Je–S.

2. **B. leucántha** T. & G. *Fig. 303.* Stem 4–10 dm. high, widely branching; leaves glabrous, glaucous, blackening in drying, the leaflets obovate or cuneate, rarely oblong-oblanceolate, 2–5 cm. long, obtuse or retuse; racemes 2–5 dm. long; calyx 7–9 mm. long, the three lower lobes triangular, acute; corolla 2 cm. long, white; stipe of the pod equaling the calyx, the body oblong, 2–3 cm. long. Rich alluvial soil: Ont.—Minn.—Tex.—Fla. *Allegh.—Austral.* Je–Jl.

3. **B. leucophaèa** Nutt. Stem 3–8 dm. high, widely branching; leaves villous, the leaflets oblong or oblanceolate, 3–10 cm. long, obtuse or retuse, strongly reticulate; racemes 1–2 dm. long; bracts lanceolate or ovate-lanceolate; calyx strigillose; corolla white or cream-colored, 2–2.5 cm. long; the tube

7–8 mm. long, the lower lobes triangular, shorter than the tube; pod short-stipitate, the body ovoid or ellipsoid, 4–5 cm. long. *B. bracteata* A. Gray; not Ell. [G, B]. Prairies: Mich.—Minn.—Tex.—La. *E. Temp.—Prairie.* Ap–My.

4. **B. vespertìna** Small. Stem 6–12 dm. high, widely dichotomously branched; leaves glabrous, glaucous, turning black in drying, the leaflets leathery, cuneate or oblanceolate, about 2 cm. long, mucronate; racemes 1–3 dm. long; bracts lanceolate or oblong-lanceolate; calyx 1 cm. long, the lobes triangular or ovate, shorter than the tube; corolla 25–30 mm. long, blue; stipe of the pod longer than the body, the latter oblong, 3–4 cm. long. Rich soil: Mo.—Kans.—Tex. *Texan.* Je–Au.

4. CROTALÀRIA L. RATTLE-BOX.

Annual or perennial herbs, rarely shrubby plants. Leaves alternate, often with decurrent stipules, odd-pinnate, sometimes 1-foliolate. Flowers perfect, solitary or in terminal racemes. Calyx 5-lobed, the upper 2 and the lower 3 often partially united and forming an upper and a lower lip. Corolla yellow or rarely blue, the banner broad, the keel scythe-shaped. Stamens monadelphous, the anthers alternately smaller and versatile, and larger and erect. Style curved. Pod globular or cylindric, inflated, 2-valved. Seeds becoming loose in the pod at maturity.

1. **C. sagittàlis** L. Annual or in warmer regions perennial; stem 3–10 dm. high, pubescent with loose, spreading hairs; leaflet solitary, oblong to lanceolate or linear-lanceolate, 2–6 cm. long, abruptly acute, sessile or nearly so; stipules conspicuous, decurrent; calyx shaggy-pubescent, the lobes of the lower lip about 1 cm. long, acuminate; corolla yellow, 8–12 mm. long; pod ellipsoid, 2.5–4 cm. long. Dry soil: Mass.—S.D.—Tex.—Fla.; Mex. *E. Temp. Trop.* Je–S.

5. LUPÌNUS (Tourn.) L. LUPINE, WOLF'S BEAN, BLUE BONNET, QUAKER BONNET.

Annual or perennial herbs, or rarely shrubs. Leaves alternate, with narrow stipules, digitately 5–15-foliolate, rarely 1- or 3-foliolate. Flowers perfect, in terminal racemes. Calyx 2-lipped, the upper lip of 2 partially, and the lower of 3 partially or usually wholly united sepals. Corolla in most of our species blue, purple, or white, in a few yellow; banner broad, with reflexed margins; wings curved; keel sickle-shaped. Stamens monadelphous; anthers alternately larger and smaller. Pods flat, 2-valved, dehiscent; ovules and seeds 2–many.

Perennials with rootstocks; cotyledons petioled after germination.	
Stem with appressed hairs.	
Leaves glabrous above, at least in age.	
Flowers large, more than 12 mm. long.	
Banner with a dark spot.	1. *L. plattensis.*
Banner without dark spot.	2. *L. perennis.*
Flowers small, less than 12 mm. long.	
Corolla about 10 mm. long; banner pubescent on the back; plant silvery.	3. *L. argenteus.*
Corolla about 6 mm. long; banner glabrous; plant green.	
Leaflets narrowly oblanceolate.	4. *L. floribundus.*
Leaflets broadly oblanceolate.	5. *L. parviflorus.*
Leaves permanently pubescent above.	
Corolla less than 12 mm. long; plant short-hairy.	
Calyx produced backwards into a distinct spur; corolla dark blue, about 10 mm. long.	6. *L. aduncus.*
Calyx merely gibbous at the base; corolla about 10 mm. long.	3. *L. argenteus.*
Corolla 12–14 mm. long; plant long-hairy.	7. *L. leucopsis.*
Stem with spreading hairs; leaves hairy above; corolla 12–14 mm. long.	8. *L. sericeus.*
Annuals; cotyledons sessile, persistent, clasping; pod 2-seeded.	9. *L. pusillus.*

1. **L. platténsis** S. Wats. Stem short, densely leafy, 2–5 dm. high; leaflets 7–9, oblanceolate, obtuse or acutish, thick, somewhat glaucous, grayish silky-strigose beneath, 2–4 cm. long; racemes 1–1.5 dm. long; bracts lanceolate, equaling the calyx, deciduous; calyx only slightly gibbous at the base; lips lanceolate, about 8 mm. long; corolla pale blue; banner with a dark spot, about 12 mm. long; legume 2.5–3.5 cm. long. *L. crassus* Payson. Plains and hills: Neb.—Wyo.—Colo.—Kans. *Plain—Submont.* Je–Jl.

2. **L. perénnis** L. Perennial, more or less pubescent; stem 2–6 dm. high; leaves spreading; leaflets 7–11, oblanceolate to elliptic, 1.5–5 cm. long, obtuse or mucronate, appressed-pubescent or glabrate; raceme terminal, 1.5–3 dm. long; calyx 6–8 mm. long; corolla blue, rarely white or pink, 15–18 mm. long; pod linear-oblong, 3–4 cm. long, pubescent, 4–5-seeded. Sandy soil: Me.—Minn. —La.—Fla. *E. Temp.* My–Je.

3. **L. argénteus** Pursh. *Fig. 304.* Stems diffusely branched, 5–10 dm. high; leaflets 7–8, narrowly oblanceolate and usually conduplicate, or in var. *argentatus* broader and flat, 2–5 cm. long; bracts linear-lanceolate, about equaling the flower-buds; lower lip of the calyx longer than the upper; corolla light violet, rose or white, about 1 cm. long; legume about 2 cm. long, 5–6-seeded. *L. decumbens* Torr. *L. leptostachys* Greene. Plains and hills: N.D. — Neb. — Colo. — Calif. — Ore. *Plain—Mont.* Je–Au.

f. 304.

4. **L. floribúndus** Greene. Stems branched, leafy; leaflets 7–9, narrowly oblanceolate, acute, 3–6 cm. long, strigose beneath; bracts lanceolate, shorter than the calyx; upper lip of the calyx 3 mm. long, lance-ovate; lower lip lanceolate, 4 mm. long, corolla darker or lighter, blue or violet, about 6 mm. long; petals subequal; pod 1.5–2 cm. long; 3–4-seeded. *L. myrianthus* Greene. (?) *L. fulvomaculatus* Payson. Mountain meadows: S.D.—Colo.—Wyo.—Utah. *Mont.* Jl–Au.

5. **L. parviflòrus** Nutt. Stems slender, 5–10 dm. high, branching; leaflets 5–11, broadly oblanceolate to obovate, obtuse, 3–5 cm. long, slightly strigose beneath, green; racemes 1–3 dm. long; bracts linear-subulate, equaling the calyx, deciduous; lips of the calyx nearly equal, 6 mm. long, broadly lanceolate, corolla light blue, sometimes white, about 6 mm. long; banner glabrous; legume 2–2.5 cm. long, 3–4-seeded. Meadows and among bushes: S.D.—Colo.—Utah—Mont. *Submont.—Alp.* Je–S.

6. **L. adúncus** Greene. Stems 3–10 dm. high, silky-canescent; leaflets 7–9, narrowly oblanceolate, 2–3 cm. long, acute, appressed-silky on both sides, often conduplicate; raceme 1–2.5 dm. long; bracts ovate-lanceolate, shorter than the calyx; calyx densely appressed-silky; lower lip lanceolate, 7 mm. long, the upper ovate-lanceolate, nearly as long; corolla dark blue, about 10 mm. long; petals subequal; pod 2–3 cm. long, 5–6-seeded. *L. decumbens argophyllus* A. Gray. *L. Helleri* Greene. *L. argophyllus* Cockerell. Plains, cañons, and river banks: Neb.—N.M.—Utah—Wyo. *Submont.—Mont.* My–Jl.

7. **L. leucópsis** Agardh. Stems 4–8 dm. high; leaflets 6–12, densely appressed silky-canescent, oblanceolate, 3–6 cm. long; raceme dense, 1–2 dm. long; bracts lanceolate, slightly exceeding the flower-buds; calyx densely canescent; lower lip lanceolate, 7–8 mm. long, the upper ovate-lanceolate, about 6 mm. long; corolla blue; banner with a light spot, densely hairy outside; pod 2.5–3 cm. long, about 5-seeded. *L. sericeus* Wats., in part. *L. axillaris* Blankinship. Plains: Wash.—Ore.—Nev.—Wyo.—Sask.—S.D. *Plain—Mont.* My–Au.

8. **L. seríceus** Pursh. Stem erect, 3–6 dm. high, densely silky-hirsute, leafy; leaflets 5–10, narrowly oblanceolate, densely silky, acute, 3–7 cm. long; raceme 1–2 dm. long, rather dense; calyx silky-hirsute; lower lip narrowly lanceolate, 7–9 mm. long; upper lip ovate, 6–7 mm. long; corolla dark blue; banner with a light spot, pubescent on the back; legume about 3 cm. long, 4–6-seeded. Plains and hills: Ore.—Ida.—Wyo.—S.D.—Mont. *Submont.* My–Au.

9. **L. pusíllus** Pursh. Stem with branches decumbent near the base, stout, 0.5–2.5 dm. high, hirsute; leaflets 5–8, oblong-oblanceolate, glabrous above, sparingly hirsute beneath, 2–3 cm. long, rounded at the apex; racemes short-peduncled, 3–8 cm. long; calyx hirsute-ciliate; lower lip broadly lanceolate, 6 mm. long, the upper 4–5 mm. long; corolla tinged with rose or purple, or almost white, 8–10 mm. long; legume nearly 2 cm. long. Plains and sandy places: Wash.—Nev.—N.M.—Kans.—Sask. *Son.—Plain.* My–Jl.

6. TRIFÒLIUM (Tourn.) L. CLOVER, TREFOIL.

Annual or perennial herbs. Leaves alternate, digitately 3-foliolate, rarely 5–7-foliolate, with toothed (in ours mostly finely denticulate) or rarely entire leaves. Flowers perfect, in heads or short dense racemes, bracted at the base. Calyx pedicelled, campanulate; lobes 5, elongate, often subequal, or the upper broader and more or less united. Corolla in ours purple, pink, or white; banner in ours straight; wings narrow and longer than the keel. Stamens 10, diadelphous; anthers all alike. Pod flattish or terete, included in the persistent corolla, often indehiscent; seeds few or solitary.

Flowers yellow.
Terminal leaflet not longer-stalked than the lateral ones; stipules linear; head 12–18 mm. broad. 1. *T. aureum.*
Terminal leaflet longer-stalked than the lateral ones; stipules ovate; head 8–12 mm. broad.
Heads 20–40-flowered; banner conspicuously striate. 2. *T. procumbens.*
Heads 8–12-flowered; banner scarcely striate. 3. *T. dubium.*
Flowers red, purple, pink, or white.
Flowers sessile or subsessile, not reflexed in fruit.
Head cylindric; corolla white, shorter than the plumose-pubescent calyx. 4. *T. arvense.*
Head subglobose or short-ovoid; corolla red or purple, longer than the calyx.
Calyx pubescent.
Heads cylindric; corolla scarlet to deep-red. 5. *T. incarnatum.*
Heads globose or ovoid; corolla magenta or purple.
Calyx soft-hairy; heads sessile. 6. *T. pratense.*
Calyx nearly glabrous; heads peduncled. 7. *T. medium.*
Calyx glabrous; head peduncled. 8. *T. Beckwithii.*
Flowers stalked, reflexed in fruit.
Head 2.5 cm. broad or more.
Annual or biennial; leaves pubescent. 9. *T. reflexum.*
Perennial; leaves glabrous. 10. *T. stoloniferum.*
Head less than 2 cm. broad.
Calyx barely half as long as the white or rose-colored corolla.
Plant ascending; corolla rose or pink, rarely white. 11. *T. hybridum.*
Plant creeping; corolla white or tinged with pink. 12. *T. repens.*
Calyx nearly as long as the purple corolla. 13. *T. carolinianum.*

1. **T. aùreum** Poll. Stem diffusely branched, 2–8 dm. long, strigillose; stipules acuminate; leaflets broadest above the middle, denticulate, 12–18 mm. long, truncate or retuse at the apex; peduncles 1–5 cm. long, finely pubescent; calyx 3.5–4 mm. long, the lobes subulate, longer than the ribbed tube; corolla deep yellow, 4–5 mm. long, turning brown in age. *T. agrarium* L., in part [G]. Fields and waste places: Newf.—Minn.—Kans.—N.C.; nat. from Eur. My–S.

2. **T. procúmbens** L. Stem erect, simple or with decumbent branches at the base, 1–7 dm. long; leaflets cuneate or obovate, 6–12 mm. long, rounded or notched at the apex, denticulate, pilose or strigillose; calyx sparingly pubescent, the teeth lanceolate or lance-subulate, the longer exceeding the tube in length; corolla 3–4 mm. long. Fields and waste places: N.S.—N.D.—Miss.—S.C.; Wash.—Calif.; nat. from Eur. My–S.

3. **T. dùbium** Sibth. Stem with spreading branches at the base, 0.3–5 dm. long; leaflets cuneate or obovate, truncate or notched at the apex, denticulate, glabrous; calyx glabrous, otherwise as in the preceding; corolla pale yellow, turning brown, 3 mm. long. Waste places: N.S.—Wis.—Kans.—Tex.—Fla.; Ore.—Wash.; nat. from Eur.

4. **T. arvénse** L. Annual; stem 1–3 dm. high, with ascending branches, villous; leaflets linear or oblanceolate, 1–2.5 cm. long, entire or few-toothed, villous; heads dense, 1–3 cm. long; calyx gray, the teeth subulate, longer than the tube; corolla white, shorter than the calyx-teeth. Roadsides and dry fields: N.H.—Que.—Minn.—Iowa—S.C.; Ore.—Wash.; nat. from Eur. My–S.

5. **T. incarnàtum** L. Annual; stem 2–9 dm. high, suberect, soft-pubescent; leaves long-petioled; leaflets sessile, obovate or obcordate, 1–2.5 cm. long, denticulate; heads often 5 cm. long; flowers sessile; calyx-lobes plumose; corolla 7–12 mm. long. CRIMSON CLOVER. Fields and waste places: Me.—Minn.—Iowa—D.C.; escaped from cult.; nat. of Eur.

6. **T. praténse** L. Stem erect or decumbent, 2–6 dm. high, more or less pubescent; leaflets oval or obovate, 2–3 cm. long, rounded or retuse at the apex, often with a purplish spot in the middle; heads globose or nearly so, subsessile; calyx-tube about 3 mm. long, sparingly hairy; teeth subulate, 3 mm. long; corolla about 1.5 cm. long. Meadows, roadsides, and fields: Newf.—Fla.—Calif.—Alaska; W.Ind.; cult. and nat.; native of Eur. My–S.

7. **T. mèdium** L. Perennial; stem often zigzag, glabrate, 3–6 dm. high; leaflets oblong or lanceolate, not spotted; flowers deep purple; calyx-teeth slightly pubescent; corolla 12–14 mm. long, bright-purple. ZIGZAG CLOVER. Dry hills: Mass.—Que.—Kans.—N.Y.; adv. from Eur. Jl–Au.

8. **T. Beckwíthii** Brewer. *Fig. 305.* Stem stout, erect, glabrous, 3–5 dm. high; leaflets elliptic or oval, the upper acute, the lower rounded at the apex, strongly veiny, glabrous, 2–4 cm. long; peduncles 1–2 dm. long; heads globose; flowers reflexed in age; calyx glabrous; tube nearly 3 mm. long; teeth subulate, 3–4 mm. long; corolla reddish purple. Wet meadows: S.D.—Mont.—Ore.—Calif. *Plain—Mont.* Je–Jl.

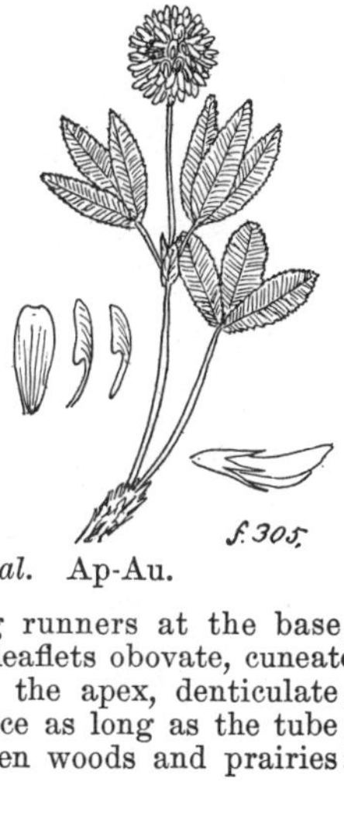

f. 305.

9. **T. refléxum** L. Annual or biennial, without runners; stem branched at the base, 1–5 dm. long, villous; leaflets oblong to suborbicular, 1.5–3 cm. long, rounded or emarginate at the apex, denticulate; calyx glabrous, the teeth much longer than the tube, lance-subulate; corolla 6–10 mm. long; banner red or purple, longer than the white wings and keel. Waste places: N.Y.—S.D.—Tex.—Fla. *Allegh.—Austral.* Ap-Au.

10. **T. stoloníferum** Muhl. Perennial, producing runners at the base; stem with spreading branches, 1–3 dm. long, glabrous; leaflets obovate, cuneate or suborbicular, 1–3 cm. long, rounded or notched at the apex, denticulate; calyx glabrous, ribbed, the teeth subulate, twice or thrice as long as the tube; corolla white, tinged with purple, 8–12 mm. long. Open woods and prairies: Ohio—S.D.—Kans.—Ky. *Allegh.* My–Au.

11. **T. hýbridum** L. Stems several, glabrous, ascending or erect, 3–6 dm. high; leaflets broadly obovate, 1–3 cm. long, glabrous, rounded or rarely emarginate at the apex; peduncles 5–10 cm. long; heads globose; flowers reflexed in fruit; calyx glabrous; tube 1.5–2 mm. long; teeth subulate, 2–3 mm. long. Pastures and waste places: Newf.—Fla.—Calif.—Wash.—Alaska; W.Ind.; cultivated and escaped, native of Eur. Je–S.

12. **T. rèpens** L. Stems creeping, 1–6 dm. long, glabrous; leaflets obovate, emarginate or obcordate, 0.5–2 cm. long; glabrous; peduncle 1–3 dm. long; flowers reflexed in age; calyx glabrous or essentially so; tube about 2 mm. long; teeth lance-subulate, 1.5 mm. long; corolla 7–8 mm. long. Waste places, roadsides and pastures: N.S.—Fla.—Calif.—Alaska; W.Ind.; cult. and naturalized; native of Eurasia. Ap–O.

13. **T. caroliniànum** Michx. Perennial; stem branched at the base, the branches ascending or decumbent, 1–3 dm. long, pubescent; leaflets obovate or cuneate, 5–12 cm. long, rounded or emarginate at the apex, denticulate; calyx pilose, reticulate, the teeth lanceolate, longer than the tube; corolla purplish, the banner pointed. Sandy soil and waste places: Pa.—Kans.—Tex. —Fla. *Allegh.—Austral.* Mr–O.

7. MEDICÀGO (Tourn.) L. MEDICK, ALFALFA, LUCERNE, NONESUCH.

Mostly perennial herbs, sometimes shrubby plants. Leaves alternate, pinnately trifoliate, with commonly toothed leaflets. Flowers perfect, in elongate or head-like racemes. Calyx pedicelled, campanulate; lobes slender, nearly equal. Corolla purplish or yellow; banner oblong, subsessile; wings nearly oblong; keel obtuse, shorter than the wings. Stamens 10, diadelphous; anthers all alike. Stigma oblique. Pods curved or spirally coiled, often spiny reticulate, indehiscent.

Pod unarmed.
  Perennials, erect; corolla 6–10 mm. long.
    Corolla violet or blue; pod spirally coiled. — 1. *M. sativa.*
    Corolla yellow; pod merely curved. — 2. *M. falcata.*
  Annuals, prostrate; corolla yellow, 3 mm. long; pod merely curved. — 3. *M. lupulina.*
Pod prickly, spirally coiled, corolla yellow. — 4. *M. hispida.*

1. **M. satìva** L. Perennial; stem decumbent or ascending at the base, sparingly hairy, in age glabrous, branched, 3–10 dm. high; leaflets oblong, oblanceolate, or cuneate-obovate, 1–3 cm. long, sharply denticulate towards the apex, obtuse or truncate, mucronate; racemes oblong, 1–5 cm. long; corolla 8–10 mm. long; pods pubescent, twisted into 2–3 spires. ALFALFA, LUCERNE. Fields and fence rows: Me.—Va.—Calif.—B.C.; Eur.; cult. and escaped. My–S.

2. **M. falcàta** L. Perennial, stem erect, glabrous, branched, 3–5 dm. high; leaflets linear, denticulate towards the apex, mucronate; corolla yellow, 6–8 mm. long; pod more or less falcate. Waste ground: N.Y.—Iowa—Man.; adv. from Eur. My–S.

3. **M. lupulìna** L. Annual; stem prostrate, branched at the base, 3–6 dm. long, sparingly pubescent; leaflets usually broadly obovate or obcordate, denticulate above the middle, emarginate and mucronate, sparingly hairy, 4–15 mm. long; raceme dense, short and head-like, less than 1 cm. long; corolla about 3 mm. long; pod pubescent, strongly reticulate. NONESUCH. Fields and waste places: N.S.—Fla.—Mex.—Calif.—Wash.; Mex.; nat. from Eurasia. Mr–D.

4. **M. híspida** Gaertn. Annual; stem branched at the base, with ascending or spreading branches, glabrous or sparingly strigose; leaflets obovate, rounded to obcordate at the apex, crenulate above the middle, 1–1.5 cm. long; flowers in few-flowered heads; corolla yellow; pods several-seeded, reticulate, armed on the edges with curved prickles. *M. denticulata* Willd. [B]. BUR

Clover. Waste places: N.S.—Fla.—Tex.—Mont.; Calif.—Wash.; Mex.; adv. from Eur. Je–Au.

8. **MELILÒTUS** (Tourn.) Hill. Sweet Clover, Honey Clover.

Annual or perennial herbs, sweet-scented. Leaves alternate, pinnately trifoliolate, with toothed leaflets. Flowers perfect, in elongate lax racemes. Calyx pedicelled, campanulate, teeth 5, nearly equal. Corolla yellow or white, free from the filaments; banner abruptly contracted at the base, subsessile; wings narrow, cohering with the short obtuse keel. Stamens 10, diadelphous; anthers equal. Stigma terminal. Pods short and thick, straight, often subglobose, indehiscent or nearly so, in ours reticulate.

Corolla white; banner a little longer than the wings. 1. *M. alba.*
Corolla yellow; banner about equaling the wings.
  Corolla 5–6 mm. long. 2. *M. officinalis.*
  Corolla 2–2.5 mm. long. 3. *M. indica.*

1. **M. álba** Desv. Stem 1–3 m. high, erect, branched, glabrous, or puberulent when young; leaflets obovate, oblanceolate or oblong, denticulate, except at the base, 1.5–2.5 cm. long, usually truncate at the apex; corolla white, 4–6 mm. long. Waste places and roadsides: N.S.—D.C.—N.M.—Calif. —Wash.; adv. or nat. from Eurasia, or escaped from cultivation. My–S.

2. **M. officinàlis** (L.) Lam. Stem erect, 1–3 m. high, glabrous, or slightly pubescent when young; leaflets from broadly obovate to oblong, sharply denticulate, except at the base, about 2 cm. long, rounded or obtuse at the apex; raceme lax and slender; corolla light yellow, 5–7 mm. long. Waste places and roadsides: N.S.—Fla.—Colo.—Utah—Ida.—Mont.; adv. or nat. from Eur., or escaped from cultivation. Je–S.

3. **M. índica** (L.) All. Stem erect, 2–7 dm. high, branched; leaflets oblong, oval, or cuneate, 0.5–2 cm. long, pubescent when young, rounded or retuse at the apex; raceme lax and slender, 2–5 cm. long; corolla yellow; pod subglobose or oval, 2 mm. long, wrinkled. Adventive along the seacoasts; escaped from cult.; inland, west to N.D.; nat. of Eur. Je–S.

9. **ACMÍSPON** Raf.

Leafy-stemmed annuals. Leaves alternate, pinnately 1–5-foliolate, with small gland-like stipules. Flowers perfect, usually solitary, on bracted axillary peduncles. Petals usually pinkish, slightly exceeding the calyx; claws equally approximate to each other; banner rounded; wings oblong; keel narrowed above into a rather short, acute, incurved beak, equaling or exceeding the wings. Stamens diadelphous; filaments alternately dilated under the subequal anthers. Pods linear, straight or nearly so, somewhat compressed, readily dehiscent, many-seeded.

f. 306.

1. **A. americànus** (Nutt.) Rydb. *Fig. 306.* Stem 3–5 dm. high, silky when young, much branched, with strongly ascending branches; leaves 1–3-foliolate; leaflets lanceolate or linear-lanceolate, acute, silky-villous, in age glabrate, 1–2 cm. long; peduncles 7–25 mm. long; bracts linear-lanceolate; calyx-teeth subulate, subequal, 4 mm. long, nearly equaling the pinkish corolla. *Hosackia Purshiana* Benth. *H. americana* (Nutt.) Piper [G]. *Lotus americanus* (Nutt.) Bisch. [B]. Prairies, especially in sandy soil: Man.—Ark.—Tex.—Sonora—Ida. *Prairie—Plain—Submont.* Je–Au.

10. ANTHÝLLIS (Riv.) L.

Perennial herbs, with odd-pinnate leaves. Flowers perfect, in loose head-like spikes. Calyx 5-toothed, persistent, loose, somewhat inflated in fruit. Corolla yellow or crimson, the keel blunt or short-pointed. Pod stipitate, included in the calyx, usually indehiscent, few-seeded.

1. **A. Vulnerària** L. Stem pubescent, 2–3 dm. high; leaflets of the lower leaves 1–3, those of the upper leaves 5–13; spikes ovoid or subglobose, involucrate; corolla yellow, 10–12 mm. long. Clover fields: Ont.—N.D.—Iowa—Pa.; adv. from Eur.

11. AMÓRPHA L. FALSE INDIGO, SHOE-STRINGS.

Shrubs, with glandular-punctate foliage. Leaves alternate, odd-pinnate. Flowers perfect, incomplete, in spike-like racemes. Calyx subcampanulate; lobes 5, short. Corolla incomplete; banner erect, clawed, folded around the stamens, blue, purple or white; wings and keel lacking. Stamens 10; filaments united at the base only. Ovary 2-ovuled. Pods short, 1–2-seeded, nearly indehiscent.

Tall shrubs; leaflets 2–5 cm. long; pods usually 2-seeded; leaves petioled; pod curved.
- Pubescence closely appressed; leaflets mostly cuneate at the base. — 1. *A. fragrans.*
- Pubescence spreading; leaflets mostly rounded at the base. — 2. *A. fruticosa.*

Low shrubs; leaflets 0.5–1.5 cm. long; pods 1-seeded.
- Glabrous or nearly so; spikes usually solitary at the ends of the branches; leaves petioled; pod straight. — 3. *A. nana.*
- Densely canescent; spikes usually clustered; leaves subsessile; pod curved. — 4. *A. canescens.*

1. **A. fràgrans** Sweet. Branched shrub, 1.5–5 m. high; young growth finely strigose-canescent; leaflets 9–27, oval to oblong, 2–4 cm. long, mostly acute at both ends, minutely puberulent; racemes 5–20 cm. long; calyx about 3.5 mm. long; upper two lobes rounded, the others triangular, pointed, the lowest almost twice as long as the rest; banner broadly obovate, short-clawed, 4.5–5 mm. long; pod 6–7 mm. long. *A. fruticosa* Coult.; not L. *A. angustifolia* (Pursh) Boynton. Banks of streams: Man.—Ill.—Ark.—Tex.—Ariz.—Mont. *Prairie—Plain.* My–Jl.

2. **A. fruticòsa** L. *Fig. 307.* A shrub, 2–6 m. high; twigs sparingly pilose or glabrate; leaflets 11–25, oval or elliptic, 1.2–4 cm. long, 7–20 mm. wide, rounded at each end, finely pubescent or glabrate; racemes 7–15 cm. long; calyx 3 mm. long; upper two lobes rounded, the lower three triangular, the lowest one slightly longer than the rest; banner rounded-obovate, emarginate, 6 mm. long, blue; pod 7–9 mm. long. Banks: Conn.—Minn. (?)—La.—Fla. *Allegh.—Austral.* My–Jl.

f.307

3. **A. nàna** Nutt. Undershrub, 1–4 dm. high, glabrate or minutely strigose; leaflets 15–31, elliptic or oval, rounded at both ends, mucronate, 5–10 mm. long; racemes solitary, 3–6 cm. long; calyx about 3 mm. long, glabrous; lobes lanceolate, acuminate, subequal; banner broadly obovate, purple; pod about 5 mm. long. *A. microphylla* Pursh [G]. Hills and prairies: Man.—Sask.—Ida.—N.M.—Iowa. *Prairie—Plain—Submont.* My–Je.

4. **A. canéscens** Pursh. Undershrub, 3–10 dm. high, white-canescent; leaflets 15–47, crowded, oval or ovate-lanceolate, 9–18 mm. long, rounded at the base, acute or obtuse at the apex; racemes clustered, dense, 5–18 cm. long;

calyx about 5 mm. long; lobes lance-subulate; banner obovate, 5 mm. long, blue; pod about 4 mm. long. Dry prairies and plains: Man.—Mich.—La.—N.M.—Mont. *Prairie—Plain.* Je–Au.

12. PSORALÍDIUM Rydb.

Perennial herbs, with glandular-punctate foliage. Leaves alternate, 3–5-foliolate, with entire leaflets. Flowers small, perfect, in racemes or interrupted spikes. Calyx campanulate, lobes equal, or the lower longer. Corolla white or purplish; banner usually broad and auricled at the base; wings clawed, scythe-shaped; keel shorter than the other petals, incurved. Stamens diadelphous or monadelphous; anthers alike. Pods broad, indehiscent, mostly 1-seeded; ovules 1 or 2.

Flowers in racemes, *i. e.*, distinctly pedicelled.
  Pods globose; corolla white, only the keel tipped with purple.
    Leaflets lance-oblong to linear. — 1. *P. lanceolatum.*
    Leaflets narrowly linear. — 2. *P. micranthum.*
  Pods ovoid; corolla blue or purplish; raceme lax.
    Leaflets from linear-oblanceolate to obovate; pods with short beaks.
      Stem, peduncles, and calyces pilose with spreading hairs. — 3. *P. Batesii.*
      Stem, peduncles, and calyces strigose.
        Racemes short, few-flowered; flowers 5 mm. long, 1 or 2 at each node; leaves 3-foliolate, rarely the lower 5-foliolate. — 4. *P. tenuiflorum.*
        Racemes elongate, many-flowered; flowers 6 mm. long, usually 2–4 at each node; stem-leaves 5-foliolate, only those of the branches 3-foliolate. — 5. *P. floribundum.*
    Leaflets narrowly linear; pods with longer beaks. — 6. *P. linearifolium.*
Flowers in interrupted spikes.
  Leaves not silvery below; flowers about 6 mm. long; bracts minute. — 7. *P. collinum.*
  Leaves silvery-white, at least below; flowers 7–8 mm. long; bracts at least half as long as the calyx.
    Leaflets oval; bracts lanceolate, acuminate; calyx not inflated in fruit. — 8. *P. argophyllum.*
    Leaflets linear; bracts obovate; calyx inflated in fruit. — 9. *P. digitatum.*

1. **P. lanceolàtum** (Pursh) Rydb. Stem very glandular-punctate, glabrous or sparingly strigose, 1.5–4 dm. high; leaflets 1.5–4 cm. long, strigose with scattered hairs or glabrate; peduncles 1–3 cm. long; calyx-tube 1.5 mm. long, campanulate; lobes ovate, obtuse, 0.5 mm. long; corolla 5–6 mm. long; pod 4 mm. thick. *Psoralea lanceolata* Pursh [B, G]. Sandy plains: Sask.—Iowa—Kans.—N.M.—Ariz.—Mont. *Plain—Submont.* My–Au.

2. **P. micrânthum** (A. Gray) Rydb. Stem very glandular-dotted, glabrous or strigose; leaflets 1–5 cm. long, 1–3 mm. wide, glabrous or nearly so; peduncles 2–4 cm. long; calyx-tube 1 mm. long; lobes rounded-oval, 0.5 mm. long; corolla scarcely 5 mm. long. *Psoralea micrantha* A. Gray. Sandy plains: Kans.—Tex.—Ariz.—s Utah. *Son.* Ap–Jl.

3. **P. Batèsii** Rydb. Stem about 5 dm. high, erect, branched, pilose or glabrate below; leaves 5-foliolate, or those of the floral branches 3-foliolate; petioles 5–10 mm. long; leaflets obovate, 1–2 cm. long, glabrous and strongly glandular-dotted above, pilose beneath; racemes 3–5 cm. long, the rachis conspicuously pilose with spreading hairs; flowers mostly 2 at each node; calyx pilose, the tube about 1 mm. long, the lobes lanceolate, 1.5 mm. long; corolla apparently white, 5–6 mm. long; pod oval, 6 mm. long, with a short stout beak, densely glandular; seeds about 5 mm. long, 4 mm. broad, dark brown. Prairies: Wymore, Neb.

4. **P. tenuiflòrum** (Pursh) Rydb. Stems erect, diffusely branched, striate and strigose; leaflets 1–4 cm. long, glabrous above, strigose beneath; racemes slender, 2–10 cm. long, rather few-flowered; calyx strigose, usually

purplish; tube 1.5 mm. long; calyx-lobes lanceolate, 1 mm. long, acute; corolla light blue, 4–5 mm. long; pod ovoid, 8 mm. long. *Psoralea tenuiflora* Pursh [B, G]. Prairies and plains: S.D.—Tex.—Ariz.—Mont. *Plain—Submont.* Je–O.

5. **P. floribúndum** (Nutt.) Rydb. Stem erect, branched above, 5–12 cm. high, canescent-strigose; leaflets 1–5 cm. long, 3–10 mm. wide, elliptic or oblanceolate, strigose beneath; racemes 10–20 cm. long, many-flowered; calyx 3 mm. long, strigose; lobes subulate, as long as the corolla; corolla 6–7 mm. long, blue or white; pod ovoid, 8 mm. long. *Psoralea floribunda* Nutt. [B]. Prairies and valleys: Minn.—Ill.—Neb.—Tex.—Ark. *Prairie—Tex.* Je–O.

6. **P. linearifòlium** (T. & G.) Rydb. Stem 4–10 dm. high, widely branching, sparingly strigose; leaflets 2–6 cm. long, 2–3 mm. wide, sparingly strigose beneath, glabrous above; calyx-tube campanulate, 2–2.5 mm. long; teeth lanceolate, acute, 1–1.5 mm. long; corolla blue, 7–8 mm. long; pod ovoid, 8 mm. long. *Psoralea linearifolia* T. & G. [B]. Plains and hills: Neb.—Ark.—Tex.—Colo. *Plain.* Je–S.

7. **P. collìnum** Rydb. Stem 4–6 dm. high, branched, strigose; leaflets oblong-oblanceolate, obtuse, 1.5–2.5 cm. long, glabrous or nearly so above, grayish-strigose beneath; bracts lanceolate, 2–4 mm. long; calyx densely silky-strigose; tube and the lanceolate lobes each 2 mm. long; corolla blue, about 6 mm. long. *Psoralea collina* Rydb. Hills: w Neb. *Plain.* Jl.

8. **P. argophýllum** (Pursh) Rydb. *Fig. 308.* Stem erect, 3–6 dm. high, branched above, strigose-canescent; leaflets 1.5–4 cm. long, 6–20 mm. wide, obovate or oval, obtuse, densely white-silky on both sides, or grayish-strigose above; spike rather short; calyx silvery; tube about 2 mm. long; upper teeth 2–3 mm. long, lanceolate, the lower one narrower, 6 mm. long, in fruit over 1 cm. long. *Psoralea argophylla* Pursh [B, G]. Prairies and plains: Man.—Wis.—Mo.—N.M.—Alta. *Prairie—Plain.* Je–Au.

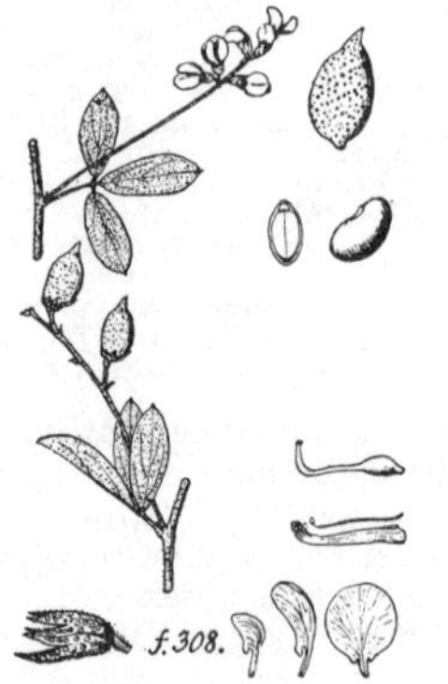

f.308.

9. **P. digitàtum** (Nutt.) Rydb. Stem erect, simple below, branched above, canescent, 3–6 dm. high; leaflets 2–3.5 cm. long, 3–5 mm. wide, densely strigose-canescent beneath, sparingly strigose when young, soon glabrate above; spike lax; bracts 5 mm. long; calyx silky; tube in flower 3–4 mm. long; lobes lanceolate, abruptly acuminate, the upper 2 mm., the lowest one 4 mm. long; corolla 7–8 mm. long, blue; pod ovoid, 6–7 mm. long. *P. campestris* Nutt. *Psoralea digitata* Nutt. [B, G]. Sandy plains: S.D.—Ark.—Tex.—Colo. *Prairie—Plain.* My–Jl.

13. ORBÉXILUM Raf.

Perennial herbs, with rootstocks or fusiform roots. Leaves alternate, glandular-dotted, pinnately 1–3-foliolate, rarely 5-foliolate. Flowers perfect, in long-peduncled axillary racemes. Calyx campanulate, glandular-dotted, 5-lobed, the lowest lobe the longest. Corolla purple; banner broadly obovate or suborbicular; wings as long as the banner, the blades obliquely lunate, with a small basal auricle; keel-petals shorter, lunate, with a dark tip. Stamens monadelphous or diadelphous. Pod ovate or orbicular in outline, flattened, indehiscent, with a short incurved beak, cross-wrinkled, 1-seeded.

Pod obliquely ovate; leaves rhombic-ovate to rhombic-lanceolate; perennials with rootstocks. 1. *O. Onobrychis.*
Pod suborbicular; leaflets linear-lanceolate to elliptic; perennials with fusiform roots. 2. *O. pedunculatum.*

1. **O. Onóbrychis** (Nutt.) Rydb. *Fig. 310.* Rootstock stoloniferous; stem erect, 1–1.5 m. high, finely pubescent; leaves 3-foliolate; leaflets 5–10 cm. long, 2.5–6 cm. wide, glabrous above, pubescent beneath, ciliate on the margins; peduncles 6–12 cm. long; raceme 1 cm. long; bracts subulate; calyx glabrate; lobes triangular; corolla 6–7 mm. long, pale with purple veins; pod 1 cm. long, rugose-reticulate. *Psoralea Onobrychis* Nutt. [B, G]. Banks: Ohio—Iowa—Mo.—Tenn. *Allegh.* Je–Jl.

2. **O. pedunculàtum** (Mill.) Rydb. Root fusiform, 2 cm. thick; stem 3–8 dm. high, glabrous or nearly so; leaves 3-foliolate; leaflets 4–7 cm. long, 1–2 cm. wide, glabrous or nearly so; peduncles 8–15 cm. long; racemes 4–10 cm. long; bracts broadly ovate, deciduous; calyx strigose, 3 mm. long; lobes lanceolate, the lowest slightly longer than the rest; corolla 5–6 mm. long, pale purple; pod obliquely orbicular, flat, 4 mm. long. *Psoralea pedunculata* (Mill.) Vail [B, G]. Prairies: Va.—Kans.—Tex.—Fla. *Austral.*

14. **PEDIOMÈLUM** Rydb. POMME DE PRAIRIE, POMME BLANCHE, BREAD-ROOT, INDIAN TURNIP.

Perennial herbs, with glandular-punctate foliage and tuberous farinaceous roots. Leaves alternate, 3–5-foliolate, with entire leaflets. Flowers middle-sized or large, perfect, in peduncled spikes. Calyx campanulate, the lower lobes longer. Corolla white or purplish; banner usually broad and auricled at the base; wings clawed, scythe-shaped; keel shorter than the other petals, incurved. Stamens diadelphous or monadelphous; anthers alike. Pods broad, circumscissile or bursting irregularly, mostly 1-seeded; ovules 1 or 2.

Tall, leafy and branched, usually over 4 dm. high; strigose throughout; lower calyx-teeth cuspidate-acuminate. 1. *P. cuspidatum.*
Low and more simple, 1–3 dm. high; lower calyx-teeth not cuspidate-acuminate.
  Stem and peduncles hirsute, with spreading pubescence; lowest calyx-lobe not much larger than the rest. 2. *P. esculentum.*
  Stem, short peduncles, and pedicels with appressed pubescence; lowest calyx-lobe much broader than the rest. 3. *P. hypogaeum.*

1. **P. cuspidàtum** (Pursh) Rydb. Stems strigose or glabrate, erect, 4–6 dm. high, with spreading branches; leaflets 5, 2–4 cm. long, obovate to elliptic, glabrate above, grayish-strigose beneath; spikes head-like, 2–5 cm. long; calyx strigose, enlarged in fruit; tube in flowers 4–5 mm. long, gibbous above; lobes ovate to lanceolate-acuminate, the upper two united half their length, 4 mm. long, the lower long-cuspidate, 6–9 mm.; corolla blue, 15–20 mm. long; pod strigose, about 8 mm. long, ovoid. *Psoralea cuspidata* Pursh [B]. Sandy soil, hillsides: Minn.—Ark.—Tex.—N.M.—Mont. *Prairie—Plain.* Ap–Jl.

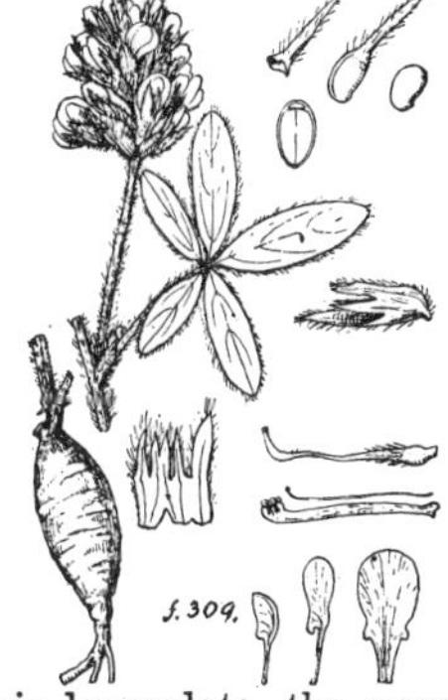

2. **P. esculéntum** (Pursh) Rydb. *Fig. 309.* Stem erect, 1–3 dm. high, hirsute; leaflets 5, 2–6 cm. long, obovate to oblong, glabrous above, strigose beneath; spike short and dense, 2–8 cm. long; calyx-tube about 5 mm. long, gibbous above, lobes in anthesis lanceolate, the upper four 6 mm. long, the lowest 8 mm.; corolla about 15 mm. long; body of the pod ovoid, 6 mm. long; beak 15 mm. long. *Psoralea esculenta* Pursh [B, G].

Prairies and plains: Man.—Wis.—Mo.—Okla.—Alta. *Prairie—Plain—Submont.* My–Jl.

3. **P. hypogaèum** (Nutt.) Rydb. Stem very short, strigose; leaflets 5–7, 2–5 cm. long, linear or linear-lanceolate, glabrous or nearly so above, strigose beneath; peduncles 1–5 cm. long, strigose; spike head-like; calyx-tube 4 mm. long, gibbous above; upper 4 lobes lance-subulate, 5 mm. long, the lowest broadly-lanceolate, in flower 7–10 mm. long; corolla about 12 mm. long. *Psoralea hypogaea* Nutt. [B]. Dry plains: Neb.—Tex.—N.M.—Mont. *Plain.* Je.

15. PARÓSELA Cav.

Perennial or annual herbs or shrubby plants, with glandular-punctate foliage. Leaves alternate, odd-pinnate, with small leaflets. Flowers perfect, in terminal or lateral spikes. Calyx campanulate, its lobes nearly equal. Corolla white, purple, or yellow; banner broad, long-clawed, auricled at the base; wings and keel usually longer than the banner, their claws adnate to the staminal tube. Stamens 10 or 9, monadelphous; anthers alike. Pod usually included in the calyx, indehiscent, 2–3-ovuled, but usually 1-seeded. *Dalea* Willd.

Stem and leaves glabrous.	
Annual; spike dense, cylindric.	1. *P. alopecuroides.*
Perennial; spike rather lax, with scattered flowers.	2. *P. enneandra.*
Stem and leaves pubescent.	
Calyx-lobes short, ovate, not subulate-tipped; spike lax; prostrate or decumbent villous herbs.	3. *P. lanata.*
Calyx-lobes subulate-tipped, longer than the tube; plants erect or decumbent, with a short cespitose caudex; spike dense.	
Leaves 3-foliolate.	4. *P. Jamesii.*
Leaves 5-foliolate.	
Stem low, 1–2 dm. high, decumbent at the base; spike sessile or short-peduncled.	5. *P. nana.*
Stem tall, 3–5 dm. high, with the upper leaves reduced and spike rather long-peduncled.	6. *P. aurea.*

1. **P. alopecuroìdes** (Willd.) Rydb. Annual; stem erect, glabrous, with ascending branches, 2–6 dm. high; leaflets 15–41, oblong, cuneate, or oblanceolate, 3–8 mm. long, obtuse at the apex; spike 2–7 cm. long; bracts ovate to lanceolate, acuminate, hyaline-margined; calyx-lobes linear-lanceolate, long-acuminate, plumose; corolla 2.5–3 mm. long, white, tinged with rose or lilac. *Dalea alopecuroides* Willd. *P. Dalea* Britton [B]. Prairies: Minn.—Ill.—Ala.—Tex.—N.M.—S.D. *Prairie—Texan—Plain.* Je–S.

2. **P. enneándra** (Nutt.) Britton. Perennial, with a taproot; stem simple below, branched above, 3–10 dm. high, flat-topped; leaflets 5–11, linear, linear-oblong, or oblanceolate, 5–10 mm. long, glabrous, conspicuously glandular-dotted; spikes 2–11 cm. long; calyx silky; lobes filiform-subulate, plumose, about 6 mm. long; corolla white; banner about 12 mm., keel 15 mm. long. *D. laxiflora* Pursh. *D. enneandra* Nutt. [G]. Prairies: N.D.—Ia.—Mo.—Tex.—Colo. *Prairie—Plain.* My–Au.

3. **P. lanàta** (Spreng.) Britton. Stems branched at the base, decumbent, 2–6 dm. long, densely short-villous; leaflets 9–13, cuneate, 4–12 mm. long, obtuse, truncate or retuse, densely short-villous; spikes slender, 2–10 cm. long, many-flowered; bracts ovate, acuminate, 3–4 mm. long; calyx 3–3.5 mm. long, velutinous; lobes lanceolate, shorter than the tube; corolla purple, rarely white; keel nearly 5 mm. long; banner 3 mm. long. Dry soil: Kans.—Tex.—N.M.—Colo.; Mex. *St. Plain—Son.* Je–Au.

4. **P. Jamèsii** (T. & G.) Vail. Stems silky-strigose, ascending or decumbent, 5–20 cm. high; leaflets obovate, 5–15 mm. long, appressed silky-canescent;

spike short-peduncled, 2–3 cm. long, dense; bracts ovate, acuminate, 8–10 mm. long; corolla yellow, in age turning purplish; keel 10–12 mm. long; banner about half as long. *P. Porteri* A. Nels. Dry plains and hills: Colo.—Kans.—Tex.—N.M. *St. Plain—Son.—Submont.* My–Au.

5. **P. nàna** (Torr.) Heller. Perennial, branched at the base; stems 1–2 dm. long, decumbent, silky-canescent; leaves numerous, spreading, 2–3 cm. long; leaflets 5, obovate, 5–15 mm. long, sericeous on both sides; spikes short-peduncled, dense, 1–3 cm. long, 1 cm. thick; bracts ovate, short-acuminate; corolla yellow, fading rose; banner reniform, shorter than the other petals. *Dalea nana* Torr. [G]. Dry soil: Kans.—Ariz.—Mex. *St. Plain—Son.*

6. **P. aùrea** (Nutt.) Britton. Stems few, strigose-canescent; leaflets 5–9, oblong, obovate, or oblanceolate, 6–18 mm. long, obtuse or mucronate, silvery-canescent; spikes 2–7 cm. long; bracts broadly ovate, short-acuminate; corolla yellow, not turning red or purple; keel about 12 mm. long, about twice as long as the banner. Plains and hills: S.D.—Mo.—Tex.—Colo. *Plain.* Je–Au.

16. PETALOSTÈMON Michx. Prairie Clover.

Perennial (all ours), or annual caulescent herbs, with glandular-dotted foliage. Leaves alternate, odd-pinnate. Flowers perfect, in spikes, not subtended by involucres. Calyx campanulate; teeth short and broad. Corolla white, purple, pink, or yellowish; banner broad, clawed; wing- and keel-petals similar to each other, narrower, distinct, inserted at the mouth of the staminal tube, between the stamens. Stamens 5, monadelphous. Pods short, indehiscent, mostly 2-ovuled, but 1-seeded. *Kunistera* Kuntze; not Lam. [B].

Calyx glabrous except the teeth; corolla white.
  Spikes numerous, globose, corymbosely arranged; petals elliptic. — 1. *P. multiflorus.*
  Spikes oblong or cylindrical, solitary at the ends of the stem and long leafy branches; petals oval.
    Spikes compact; bracts longer than the buds; leaflets 1.5–2.5 cm. long. — 2. *P. candidus.*
    Spikes looser; bracts shorter than the buds; leaflets usually less than 1.5 cm. long. — 3. *P. oligophyllus.*
Calyx pubescent.
  Corolla white or ochroleucous; spike long and compact; leaves glabrous. — 4. *P. compactus.*
  Corolla rose or purple, very rarely white.
    Leaflets mostly 5, linear or nearly so.
      Stem and leaves glabrous or slightly pubescent.
        Calyx villous, with spreading hairs.
          Bracts with glabrous or merely puberulent tips.
            Spike dense, about 1 cm. thick; calyx not ribbed. — 5. *P. purpureus.*
            Spike slender, about 8 mm. thick; calyx ribbed. — 6. *P. Porterianus.*
          Bracts with pubescent tips; spike slender, about 8 mm. thick. — 7. *P. tenuifolius.*
        Calyx silky with appressed pubescence.
          Spike globose; leaflets mostly 3. — 8. *P. tenuis.*
          Spike oblong or cylindric; leaflets usually more than 3.
            Spike oblong, 12–15 mm. thick. — 9. *P. pulcherrimus.*
            Spike cylindric, 7–8 mm. thick. — 10. *P. Stansfieldii.*
      Stem and leaves densely villous. — 11. *P. mollis.*
    Leaflets 7–17, oblong. — 12. *P. villosus.*

1. **P. multiflòrus** Nutt. Stems several, 3–6 dm. high, glabrous, sulcate, much branched, the upper branches short; leaflets 3–9, linear or linear-oblanceolate, obtuse, mucronate, glabrous, 6–12 mm. long; spikes subglobose, 8 mm. thick; bracts rhombic-lanceolate, subulate-tipped; calyx 3 mm. long, strongly 10-ribbed, glabrous; lobes lance-triangular, ciliolate; corolla white. Plains: Mo.—Kans.—Tex.—Ark. *Texan.* Jl–Au.

2. **P. cándidus** (Willd.) Michx. *Fig. 311.* Stems several, erect, 3–7 dm. high, glabrous; leaflets 7–9, linear, oblong, or oblanceolate, 1–3 cm. long, acute or mucronate, glabrous; spike 2–10 cm. long, compact; bracts subulate-tipped; calyx 3 mm. long; lobes triangular to lanceolate, shorter than the tube; corolla white. Prairies: Man.—Ind.—La.—Tex.—Sask. *Prairie—Plain.* Je–Jl.

3. **P. oligophýllus** (Torr.) Rydb. Stems erect or decumbent, 3–10 dm. high, branched, glabrous; leaflets 5–9, firm, linear, oblanceolate, or elliptic, acute or mucronate, glabrous; spike oblong, becoming more or less cylindric and looser, 1–7 cm. long; bracts with filiform tips; calyx-lobes 3 mm. long, triangular, acute; corolla white. *P. gracilis oligophyllus* Torr. Plains, prairies, and cañons: Minn.—Iowa—Tex.—Ariz.—Alta. *Plain—Submont.* Je–Au.

4. **P. compáctus** (Spreng.) Swezey. Stems few, glabrous, 3–8 dm. high; leaflets 5–7, oblong-oblanceolate or linear-oblong, 8–25 mm. long; spike long-peduncled, cylindric, compact, 4–15 cm. long, about 13 mm. thick; calyx silky-villous, 4 mm. long; teeth lanceolate, acute; corolla ochroleucous. *P. macrostachyus* Torr. Sand hills and plains: Neb.—Colo. *Plain.* Jl–Au.

5. **P. purpùreus** (Vent.) Rydb. Stems several, erect or ascending, 3–10 dm. high; leaflets 3–5, narrowly linear, 8–20 mm. long, strongly involute; spikes oblong or cylindric, 1–5 cm. long; bracts oblanceolate, abruptly acuminate; calyx densely silky-velutinous, 3 mm. long; teeth triangular; corolla violet or purple, rarely white. *P. violaceus* Michx. Prairies, plains or hills: Man.—Ind.—Ark.—N.M.—Sask. *Prairie—Plain—Submont.* Je–Jl.

6. **P. Porteriànus** Small. Stems 2–4 dm. high, sparingly pubescent; leaves 3–5 cm. long; leaflets 3–5, linear, 8–15 mm. long, involute, conspicuously punctate; spike oblong-cylindric, 1.5–4 cm. long, 8 mm. thick; calyx villous, 10-ribbed; lobes lanceolate, shorter than the tube; corolla rose-purple. Plains: Kans.—Tex. *St. Plains.*

7. **P. tenuifòlius** A. Gray. Stems several, slender, 1.5–3 dm. high, glabrous, striate; leaflets 3–5, linear or narrowly oblanceolate, 8–18 mm. long, involute, glabrous, sparingly glandular-punctate; spikes cylindric, in fruit 2–5 cm. long, 8 mm. thick; bracts silky-villous throughout, ovate, abruptly subulate-caudate; calyx 4 mm. long, silky-villous; corolla rose-purple. Plains: Ark.—Kans.—Tex. *Texan—St. Plains.* Jl–Au.

8. **P. ténuis** (Coult.) A. Heller. Stems several, branched, glabrous, about 3 dm. high; leaflets mostly 3, linear, strongly revolute, 5–12 mm. long, glabrous; spike subglobose or short-oblong, 1–1.5 cm. long, 1 cm. thick; bracts ovate-lanceolate; calyx 4 mm. long, the lobes deltoid-ovate; corolla pink-purple; pod obliquely obovate, finely pubescent. Dry soil: Ark.—Kans.—Tex. *St. Plains.* Jl–Au.

9. **P. pulchérrimus** A. Heller. Stems 3–7 dm. high, glabrous or nearly so; leaves 4–6 cm. long, numerous, leaflets 3–7, linear, 2–4 cm. long, 1–3 mm. wide; spike long-peduncled, oblong, 2–4 cm. long, 12–15 mm. thick; calyx silvery pubescent on the ribs; corolla deep rose-purple. Prairies: Mo.—Kans.—Tex.—Ark. *Texan.*

10. **P. Stansfièldii** Small. Stems 1–3 dm. high, glabrous, branched; leaves 2–4 cm. long; leaflets 3–5, linear-filiform, 5–12 mm. long, involute; spike long-peduncled, 1–4 cm. long; calyx silky-puberulent; corolla rose-purple. Plains: Kans.—Tex. *St. Plains.*

11. **P. móllis** Rydb. Stems several, erect, densely villous, 3–4 dm. high; leaflets 5, densely short-villous, linear, 10–15 mm. long, 1.5–2 mm. wide, obtuse; spike cylindric or oblong, 2–4 cm. long; calyx densely silky-villous, yellowish or ferruginous; teeth lanceolate; corolla rose-purple. Dry plains and hills: Mont.—S.D.—Colo. *Plains.* Je–Au.

12. **P. villòsus** Nutt. Perennial, with a taproot; stems several, ascending or decumbent, 3–6 dm. high, densely villous; leaflets 7–17, approximate, linear to oblong, silky-villous, 6–12 mm. long; spikes cylindric, 2.5–8 cm. long; calyx densely villous; teeth subulate; corolla rose-purple, pink, or rarely white. Sandy hills: Man.—Mo.—Tex.—Colo.—Sask. *Prairie—Plain.* Jl–S.

17. INDIGÓFERA L. Indigo Plant.

Perennial herbs or shrubs, more or less covered by malpighiaceous hairs, *i. e.*, appressed hairs which have two opposite branches and therefore appear as if attached by their middle. Leaves alternate, usually odd-pinnately compound, sometimes 3- or 1-foliolate. Flowers in axillary racemes. Calyx 5-lobed. Corolla often pink or purple or orange; banner broad, sessile or short-clawed; wings elongate, slightly adnate to the keel; keel-petals furnished with a lateral oblique spur or pouch and a basal auricle. Stamens 10, monadelphous, or the tenth stamen united with the sheath at the middle; anthers all alike. Ovary sessile; style glabrous; ovules many or few. Pod linear to subglobose; seeds more or less prismatic.

1. **I. leptosépala** Nutt. Stem branched from the base, 2–10 dm. long, spreading or decumbent, strigose; leaflets 7–9, oblong to cuneate or obovate, 1–3 cm. long, sessile; racemes often surpassing the leaves; calyx strigose, the lobes subulate, longer than the tube; corolla pale scarlet, 8–10 mm. long; pod linear, 3–3.5 cm. long, straight. Dry soil. Kans.—Tex.; n Mex. *St. Plain.* My–N.

18. ROBÍNIA L. Locust-tree.

Shrubs or trees, often with spine-like stipules. Leaves alternate, odd-pinnate, with entire leaflets. Flowers in axillary or terminal racemes. Calyx campanulate, 5-lobed, the two upper lobes more united than the rest. Corolla white, pink, or purplish; banner with a broad reflexed blade; wings curved, free; keel-petals incurved, united at the base. Stamens 10, diadelphous, or the upper filament united with the tube up to the middle. Pods narrow, flat, short-stalked, many-seeded; seeds oblique, with a crustaceous coat.

Flowers white, drooping; pod glabrous, wing-margined; stipular spines stout; twigs not bristly. — 1. *R. Pseudo-Acacia.*
Flowers rose-purple, on erect pedicels; pods and twigs glandular-hispid; stipular spines none or weak. — 2. *R. hispida.*

1. **R. Pseùdo-Acàcia** L. *Fig. 312.* Tree 6–35 m. high, with firm brown bark, and spreading branches; stipular spines 3–10 mm. long; leaflets 9–19, thin, 2.5–4.5 cm. long, elliptic or oval, glabrous; calyx finely silky-strigose; teeth triangular, 2 mm. long; corolla 15–20 mm. long, white, except a yellow spot on the banner; pods 5–10 cm. long, 10–15 mm. wide. (?) *R. coloradensis* Dode. Woods: Pa.—Minn.—Okla.; cult. and nat. as far as Ont.—Ida.—Utah. *Allegh.* My–Jl.

f. 312.

2. **R. híspida** L. Straggling shrub, 3–20 dm. high; branches usually densely hispid; spines, if present, 3–5 mm. long; leaflets 7–13, oval or suborbicular, rounded at both ends; racemes 3–5-flowered; calyx-lobes deltoid-lanceolate, acuminate; corolla 22–25 mm. long; pod 5–8 cm. long, densely

hispid. Woods: Va.—Ill.—Ala.—Ga.; escaped from cult. west to Minn.—Kans. *Allegh.* My–Je.

19. GALÈGA L.

Perennial herbs. Leaves odd-pinnate, with numerous entire leaflets with parallel oblique lateral veins and semi-sagittate stipules. Racemes terminal and axillary. Flowers without bractlets. Calyx campanulate, with 5 equal teeth. Banner oblong-obovate, short-clawed. Wings oblong, slightly coherent with the incurved keel. Stamens monadelphous. Ovary sessile; style filiform, with a terminal stigma. Legume linear, 2-valved, the valves obliquely striate.

1. **G. officinàlis** L. Plant bright green; stem 4–6 dm. high, striate; leaves short-petioled; leaflets 11–17, oblong to lanceolate, obtuse, acuminate-mucronate; racemes many-flowered; bracts subulate; corolla blue, rarely white; legume glabrous, linear. Escaped from cultivation: Utah—Kans.; native of c and s Eur. and the Orient. Jl–Au.

20. CRÁCCA L. CAT-GUT. GOAT'S RUE.

Perennial herbs or shrubs. Leaves alternate, odd-pinnately compound, the leaflets usually numerous, entire, obliquely striate-veined. Flowers perfect, in terminal or terminal and axillary racemes, or by the development of the branches in their axils, the racemes appearing as if opposite to the leaves. Calyx 5-lobed, but the upper two lobes often more or less united and shorter than the three lower. Corolla varying from white to purple; petals clawed, the banner suborbicular; wings oblique, slightly adhering to the incurved keel. Stamens monadelphous or diadelphous, or the tenth stamen united to the sheath at the middle; anthers alike. Ovules numerous or few. Pod elongate, 2-valved, flat. Seeds variegated. *Tephrosia* Pers.

Pubescence spreading; calyx-lobes narrowly lanceolate. 1. *C. virginiana.*
Pubescence appressed, silky; calyx-lobes ovate-lanceolate. 2. *C. leucosericea.*

1. **C. virginiàna** L. *Fig. 313.* Root long and tough; stem more or less densely pilose with spreading hairs, tufted at the base, 3–6 dm. high, erect; leaves 6–10 cm. long; leaflets 11–21, 1–2 cm. long, oblong or elliptic, glabrous above and silky-villous beneath, or villous on both sides in var. *holosericea* (Nutt.) Vail; racemes terminal, many-flowered, or with 1–2 smaller racemes in the uppermost axils; calyx silky-villous, the lobes long-attenuate, longer than the tube; corolla cream-colored, tinged with purple or pink, 15–17 mm. long; banner pubescent without; pod 3–5 cm. long, slightly curved, villous. *T. virginiana* (L.) Pers. [G]. Sandy soil: Me.—Man.—Tex.—Fla. *Canad.—Allegh.* Je–Jl.

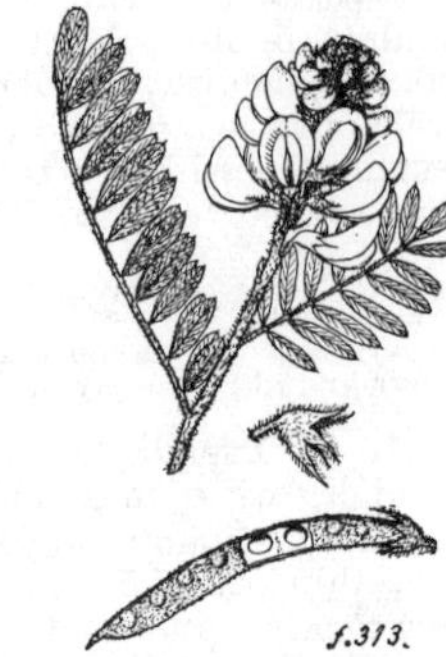

f.313.

2. **C. leucosericea** Rydb. Stem 3–5 dm. high, appressed-silky, striate; leaves 5–10 cm. long; leaflets 11–25, linear-oblong to lance-elliptic, those of the lower leaves obtuse, 1.5–2.5 cm. long, 5–6 mm. wide, glabrous above, silky beneath; racemes terminal; calyx silvery-silky, the lobes ovate-lanceolate, short-acuminate, 4 mm. long; corolla 15–18 mm. long; pod 4–5 mm. long, white-silky. Sandy places: Kans.—Tex. *Texan.*

21. KENTRÓPHYTA Nutt.

Perennial, diffusely branched herbs. Leaves pinnately 3–7-foliolate; leaflets linear or subulate, inarticulate, rigid, spinulose-tipped; stipules more or less united. Flowers small, 2 or 3 together in the axils. Calyx campanulate, deeply 5-cleft; teeth equal, subulate. Corolla ochroleucous or purplish; banner oblong, reflexed; keel obtuse, shorter than the wings. Legume ovate, one-celled, flat, with both sutures prominent, 1- or 2-seeded.

Stipules united only at the base, more or less herbaceous. 1. *K. viridis.*
Stipules united for about half their length, scarious. 2. *K. montana.*

1. **K. víridis** Nutt. Stems diffusely spreading and much branched, 1–4 dm. long; leaflets usually 5, subulate, pungent, 8–18 mm. long; calyx-tube 1.5–2 mm. long; teeth 1.5 mm. long; corolla ochroleucous, 5 mm. long. *Astragalus Kentrophyta* A. Gray in part. Plains, bad-lands, and hills: S.D.—Wyo.—Colo. *Plain—Submont.* My–Je.

2. **K. montàna** Nutt. *Fig. 314.* Stems diffusely spreading and intricately branched; stipules ovate, 1–3 mm. long; leaflets 5–7, subulate, pungent, 8–15 mm. long; calyx-tube 1.5–2 mm. long; teeth 1 mm. long; corolla ochroleucous, 4–5 mm. long. *A. Kentrophyta* A. Gray, in part. *Homalobus montanus* Britt. [B]. Bad-lands, cañons, and sandy hills: Sask.—Neb.—Utah—Alta. Plain—Submont. Je–Au.

22. HOMÁLOBUS Nutt.

Perennial herbs, with rootstocks or woody caudices. Leaves alternate, odd-pinnate, or sometimes simple. Flowers racemose, purple or ochroleucous. Calyx campanulate. Keel obtuse. Pod several- or many-seeded, usually narrow, perfectly 1-celled, without trace of a partition, with both sutures prominent, usually more or less compressed, often flat.

Leaves simple or some of them rarely 3-foliolate; plants pulvinate-cespitose; pod sessile. 1. *H. caespitosus.*
Leaves several-foliolate.
  Legume oval, ovate, broadly oblong, or elliptic.
    Pod without long black hairs.
      Pod sessile. 2. *H. vexilliflexus.*
      Pod stipitate.
        Stipe rarely exceeding the tube of the calyx; leaves usually spreading in age; leaflets oblong to oval, obtuse and mucronate. 3. *H. dispar.*
        Stipe of the mature legume usually exceeding the calyx-teeth; leaves strongly ascending; leaflets narrowly linear to linear-oblong, mostly acute.
          Stipe 3–4 mm. long; leaflets linear or linear-oblong; stem usually conspicuously strigose; calyx-teeth half as long as the tube. 4. *H. tenellus.*
          Stipe 5–7 mm. long; leaflets narrowly linear; stem glabrous or nearly so; calyx-teeth usually nearly equaling the tube. 5. *H. stipitatus.*
    Pod covered with black hairs. 6. *H. Bourgovii.*
  Legume linear, 1.5 cm. long or more. 7. *H. hylophilus.*

1. **H. caespitòsus** Nutt. Leaves 2–5 cm. long, appressed silky-canescent, either simple, linear-oblanceolate, or some of them with 3–5 linear-oblanceolate leaflets; peduncles 3–7 cm. long; racemes short, 2–10-flowered; calyx strigose; tube 2 mm. long; teeth of about the same length; corolla about 8 mm. long; pod lance-oblong, about 1 cm. long, acute, finely strigose. *A. caespitosus* A. Gray. *A. spatulatus* Sheld. Dry hills and plains: Sask.—Neb.—Colo.—Utah—Mont. *Plain—Submont.* My–Je.

2. **H. vexilliféxus** (Sheld.) Rydb. Stem strigose-canescent, decumbent, branched, 1–3 dm. high; leaflets 7–11, strigose-canescent, oblong or oblanceolate, acute, 5–18 mm. long; calyx strigose; tube 1.5 mm. long; teeth subulate, 1.5 mm. long; corolla purplish, 6–8 mm. long; pod oblong, strigose, abruptly acute, 8 mm. long, 3 mm. wide. *A. pauciflorus* Hook. *Astragalus vexilliflexus* Sheld. Hills and mountains: Sask.—S.D.—Wyo.—B.C. *Plain—Submont.* My–Jl.

3. **H. díspar** Nutt. *Fig. 315.* Stem 3–4 dm. high, sparingly strigose or glabrate; leaflets oblong to oval, rarely linear-oblong, obtuse, often mucronate, glabrous above, sparingly strigose beneath, 8–20 mm. long; raceme 5–15-flowered; calyx strigose or glabrous; tube 2 mm. long; teeth lance-subulate, 1 mm. long; corolla 8–10 mm. long; legume short-stipitate; body oblong, 10–13 mm. long, abruptly contracted into the stipe. Hills and mountains: w Neb.—Colo. —Utah. *Prairie—Plain—Mont.* Je–Jl.

4. **H. tenéllus** (Pursh) Britton. Stems erect, 2–4 dm. high, more or less strigose, branched; leaflets 6–10 pairs, linear or linear-oblong, glabrous above, more or less strigose beneath, 8–15 mm. long, 1–4 mm. wide; racemes 4–20-flowered; calyx strigose; tube nearly 2 mm. long; teeth triangular-subulate, 1 mm. long or less; corolla 8–9 mm. long; body of the pod oblong, 8–12 mm. long, 3–5 mm. wide, gradually tapering into the stipe. *A. tenellus* Pursh [G]. *A. multiflorus* (Pursh) A. Gray. Hills and mountain-sides: Man. —Neb.—Colo.—Utah—Yukon. *Prairie—Plain—Mont.* Je–Jl.

5. **H. stipitàtus** Rydb. Stems slender, 3–4 dm. high, erect; leaflets 9–15, narrowly linear, 1–2 cm. long, 1–2.5 mm. wide, glabrous or sparingly strigose beneath; racemes lax, 5–20-flowered; calyx strigose; tube about 1.5 mm. long; teeth subulate, about as long; corolla 8–10 mm. long; body of the pod oblong, about 1 cm. long, 3 mm. wide, acute, gradually tapering into the stipe. Dry hills: Sask.—Minn.—S.D. *Prairie—Plain.*

6. **H. Bourgòvii** (A. Gray) Rydb. Stems numerous, decumbent or ascending, 0.5–3 dm. long, minutely strigose; leaflets oblong or elliptic to lance-linear, 4–12 mm. long, strigose; raceme lax, 3–10-flowered; calyx strigose, with black hairs; tube about 3 mm. long; teeth 1–2 mm. long, subulate; corolla 8–10 mm. long, dark purple; pod lance-oblong, slightly stipitate, about 15 mm. long, black-hairy, with a round ridge on each side. *A. Bourgovii* A. Gray. Mountains: Alta.—S.D.—Mont.—B.C. *Submont.—Mont.* Jl–Au.

7. **H. hylóphilus** Rydb. Stems slender, leafy, sparingly strigose; leaflets 13–25, elliptic to lance-oblong, 1–2 cm. long, glabrous above, sparingly strigose beneath; raceme short, 6–12-flowered; calyx strigose, with dark hairs; tube 3 mm. long; teeth subulate, 1–1.5 mm. long; corolla about 1 cm. long, white, tinged with purple at the tip; keel purple-tipped; pod linear, 2–2.5 cm. long, 3–4 mm. wide. Open woods: Mont.—S.D.—Wyo.—Utah—Ida. *Submont.—Mont.* Je–Au.

23. DIHÓLCOS Rydb.

Stout erect perennial herbs, forming clumps. Leaves alternate, odd-pinnate, with numerous thick leaflets, strigose beneath, glabrous above; stipules lanceolate, free. Raceme many-flowered, dense, spikelike; bracts subulate. Calyx campanulate, gibbous at the base on the upper side; teeth subulate-setaceous. Corolla purple, white, or ochroleucous. Pod coriaceous, stipitate, oblong, straight, rounded on the lower suture, deeply 2-grooved above and with a prominent upper suture, 1-celled.

1. **D. bisulcàtus** (Hook.) Rydb. *Fig. 316.* Stems erect or decumbent at the base, sparingly strigose or glabrate, 3–7 dm. high, leaflets 17–27, elliptic or rarely oval, 1–2.5 cm. long; racemes many-flowered; calyx strigose, tinged with purple; tube 4 mm. long; teeth subulate or lance-subulate, the upper 1.5–2 mm. long, the lower 2–3 mm.; corolla 12–15 mm. long, purple; legume strigose. *A. bisulcatus* A. Gray [B]. Plains, hills, and valleys: Man.—Neb.—Colo.—Alta. *Plain—Submont.* Je–Au.

24. CNEMIDOPHÀCOS Rydb.

Stout perennials, with often several stems rising from the same root, and often more or less stoloniferous. Leaves pinnate, with 11–19 narrow leaflets, stipules ovate, free from the petioles, but united back of the stem. Peduncles usually equaling the leaves; raceme short. Calyx cylindro-campanulate, very gibbous at the base on the upper side. Corolla ochroleucous, with narrow petals; keel almost straight. Pod oblong, terete, coriaceous, usually with thick and prominent sutures, neither inflexed. *Ctenophyllum* Rydb.

1. **C. pectinàtus** (Hook.) Rydb. Stems 3–5 dm. high, erect or decumbent at the base; leaflets 15–21, minutely strigose when young, soon glabrate; raceme short and dense, 5–20-flowered; calyx sparingly black-strigose; tube narrowly linear, not jointed to the rachis, 5–7 mm. long; teeth subulate, 2 mm. long; corolla ochroleucous, about 2 cm. long; pod oblong-ellipsoid, 1.5–2 cm. long, terete, 8 mm. in diameter. *Astragalus pectinatus* Dougl. [B]. Dry plains: Man.—Kans.—Colo.—Alta. *Plain—Submont.* My–Je.

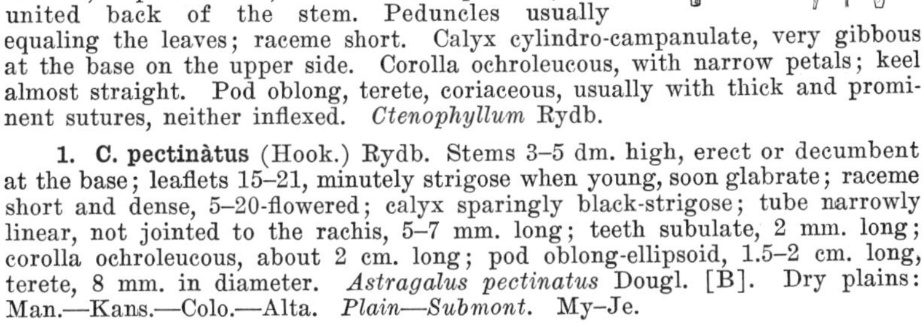

25. MICROPHÀCOS Rydb.

Subcinereous perennials, with very slender stems. Leaves pinnate, with 7–21 linear to oblong leaflets; stipules very broad, triangular, free from the petioles and only the lower more or less united with each other. Racemes many-flowered. Calyx short-campanulate, about 2 mm. long, with very short and broad teeth. Corolla purple, 5–6 mm. long, all petals strongly curved. Pod 6–8 mm. long, coriaceous, cross-wrinkled, wholly 1-celled, 6–7 ovuled, boat-shaped, *i. e.*, upper suture prominent, strongly curved, lower suture nearly straight or curved upwards, flat or slightly sulcate.

Pod slightly sulcate on the lower suture, transversely veined; leaflets narrowly linear-obtuse. 1. *M. parviflorus.*
Pod merely flattened on the lower suture, cross-wrinkled; leaflets linear-oblong or linear-cuneate, truncate or emarginate. 2. *M. gracilis.*

1. **M. parviflòrus** (Pursh) Rydb. *Fig. 317.* Stem minutely strigose, 3–6 dm. high; leaflets 11–21; racemes very slender, lax; calyx-tube 1 mm. long; teeth triangular, 0.5 mm. long; corolla purplish, 5–6 mm. long; pod pendent, straight, ovoid, 5–7 mm. long. *Astragalus gracilis* A. Gray [B]; not Nutt. *A. parviflorus* MacMill. [G]. Plains and hills: N.D.—Tex.—Wyo. *Plain.* Je–Jl.

2. **M. grácilis** (Nutt.) Rydb. Stem branched, 3–4 dm. high, finely strigose; leaflets 7–17, minutely strigose; raceme lax, short; calyx-tube 1.5–2 mm. long; teeth triangular, 0.5 mm. long; corolla about 7 mm. long, purplish; pod 6–10 mm. long, ovoid. *A. gracilis* Nutt. *A. microlobus* A. Gray [B]. Plains and hills: S.D.—Kans.—Colo.—Mont. *Plain—Submont.* Je–Jl.

26. XYLOPHÀCOS Rydb. SHEEP-POD.

Perennial herbs, mostly low, usually copiously hairy, often canescent, with short stems. Leaves alternate, with nearly free and distinct stipules; leaflets few or more numerous, entire. Flowers in short, often subcapitate racemes. Calyx cylindric; lobes much shorter than the tube. Corolla purple, rarely ochroleucous or in one species crimson; banner narrow, slightly longer than the wings. Stamens 10; filaments diadelphous; anthers alike. Pods fusiform, or ovoid, leathery or woody, 1-celled, without partition; the lower suture sometimes slightly sulcate. Seeds numerous.

Corolla purple; pod strigose.
  Pod more or less curved. 1. *X. Shortianus.*
  Pod straight. 2. *X. missouriensis.*
Corolla yellow; pod villous. 3. *X. Purshii.*

1. **X. Shortiànus** (Nutt.) Rydb. *Fig. 318.* Stem 1–10 cm. long; leaflets 11–15, obovate or oval, 1–2.5 cm. long; peduncles shorter than the leaves; raceme short, 5–15-flowered; calyx strigose, usually with white hairs; tube 8–10 mm. long; teeth lance-subulate, about 3 mm. long; corolla violet or purple, 18–25 mm. long; pod elongate-ovoid, arcuate, 3–5 cm. long, 12–14 mm. wide, sulcate on the lower suture, strigose. *A. Shortianus* Nutt. [B]. Plains and mesas: Neb.—Ariz.—Wyo. *Plain—Mont.* My–Je.

f.318.

2. **X. missouriénsis** (Nutt.) Rydb. Stems 5–10 cm. long; leaflets 11–21, elliptic to obovate, 5–15 mm. long; raceme short, 5–15-flowered; calyx-tube 7–8 mm. long; teeth subulate, 2–2.5 mm. long; corolla purple, 15–20 mm. long; pod oblong, 1.5–2.5 cm. long, 7–8 mm. broad, neither suture sulcate, rather prominent in age. *A. missouriensis* Nutt. [B]. Plains and hills: Sask.—Minn.—Kans.—N.M.—Mont. *Plain—Mont.* My–Je.

3. **X. Púrshii** (Dougl.) Rydb. Stem usually less than 5 cm. long, villous; leaflets 9–13, oblanceolate or oblong, acute, 8–15 mm. long, villous; calyx densely villous; tube about 1 cm. long; teeth subulate, about 3 mm. long; corolla 2–2.5 cm. long, ochroleucous; keel tipped with purple; pod ovoid, densely woolly, about 2 cm. long, slightly curved, scarcely sulcate. *A. Purshii* Dougl. *A. Booneanus* A. Nels., in part. Dry plains and hills: Sask.—Neb.—Colo.—Calif.—B.C. *Son.—Submont.* My–Je.

27. OROPHÀCA (T. & G.) Britton.

Cespitose, silvery or villous perennials, with branched woody caudices and deep roots. Leaves crowded, digitately 3-foliolate (rarely 5-foliolate), with scarious sheathing stipules. Flowers few, capitate or racemose. Calyx campanulate to cylindric. Corolla yellowish or purplish. Keel blunt. Pod coriaceous, 1-celled, without partition, ovoid or elliptic, few-seeded, villous or puberulent, partly included in the calyx.

Flowers sessile in the axils of the leaves, 12 mm. or more long; corolla ochroleucous; calyx-tube longer than the teeth; plant pulvinate-cespitose.
  Corolla 2 cm. or more long, glabrous; calyx-tube cylindrical, more than thrice as long as the teeth. 1. *O. caespitosa.*
  Corolla 15–18 mm. long, pubescent on the outside; calyx-tube short-cylindric, about twice as long as the teeth. 2. *O. argophylla.*
Flowers in small, peduncled, 1–3-flowered racemes, purple or turning yellowish in age, less than 10 mm. long; calyx-tube campanulate, not longer than the teeth. 3. *O. sericea.*

1. **O. caespitòsa** (Nutt.) Britton. *Fig. 319.* Stems 5 cm. or less, densely covered with leaves and scarious stipules; leaflets elliptic to narrowly oblanceolate, 0.5–2 cm. long, appressed silvery-silky on both sides; calyx silky; tube cylindric, 10–15 mm. long; teeth 3–4 mm. long; corolla 2–3 cm. long; legume oblong-ovate, terete, silky-villous. *Phaca caespitosa* Nutt. *Astragalus triphyllus* Pursh. Dry gravelly hills: Man.—Neb.—Wyo.—Mont. *Plain.* My–Je.

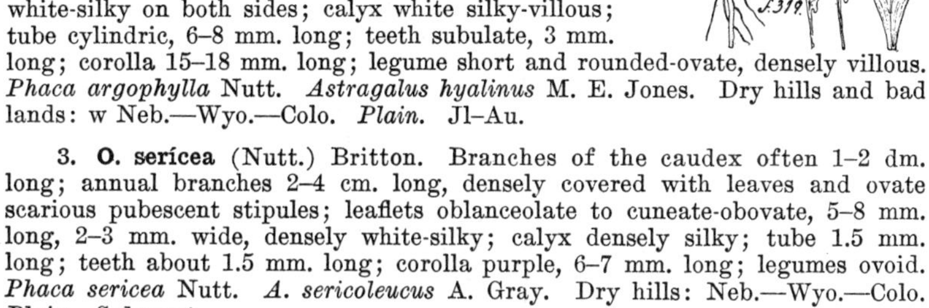

2. **O. argophýlla** (Nutt.) Rydb. Stems less than 5 cm. high, densely covered with scarious, ovate stipules; leaflets elliptic or obovate to oblanceolate, 5–15 mm. long, 3–5 mm. wide, appressed white-silky on both sides; calyx white silky-villous; tube cylindric, 6–8 mm. long; teeth subulate, 3 mm. long; corolla 15–18 mm. long; legume short and rounded-ovate, densely villous. *Phaca argophylla* Nutt. *Astragalus hyalinus* M. E. Jones. Dry hills and bad lands: w Neb.—Wyo.—Colo. *Plain.* Jl–Au.

3. **O. serícea** (Nutt.) Britton. Branches of the caudex often 1–2 dm. long; annual branches 2–4 cm. long, densely covered with leaves and ovate scarious pubescent stipules; leaflets oblanceolate to cuneate-obovate, 5–8 mm. long, 2–3 mm. wide, densely white-silky; calyx densely silky; tube 1.5 mm. long; teeth about 1.5 mm. long; corolla purple, 6–7 mm. long; legumes ovoid. *Phaca sericea* Nutt. *A. sericoleucus* A. Gray. Dry hills: Neb.—Wyo.—Colo. *Plain—Submont.*

28. HOLCOPHÀCOS Rydb.

Perennial herbs. Leaves alternate, odd-pinnate, the stipules distinct. Flowers perfect, in short few-flowered racemes. Calyx short-campanulate, the lobes short and broad. Corolla pink or purplish; banner broad, deeply notched; keel shorter than the wings. Stamens diadelphous; anthers uniform. Pod linear or oblong, sessile, leathery, turgid, sulcate on both sutures, 1-celled, without a septum. Seeds numerous.

1. **H. distórtus** (T. & G.) Rydb. *Fig. 320.* A glabrous or slightly pubescent herb; stem branched at the base, the branches prostrate or ascending, 1–4 dm. long; leaflets 17–23, ovate, obovate or cuneate, rounded or notched at the apex, 4–10 mm. long; calyx minutely pubescent; corolla purplish-blue, 8–10 mm. long; pod lunate, 2–2.5 cm. long, short-beaked. *Astragalus distortus* T. & G. [B, G]. Dry soil: W.Va.—Iowa—se Kans.—Tex. *E. Temp.* Mr–Jl.

29. PISOPHÀCA Rydb.

Perennial herbs. Leaves odd-pinnate. Flowers perfect, racemose. Calyx-tube campanulate, the teeth subulate. Corolla purple or white; banner obovate, broad, strongly arched. Ovary sessile or nearly so; stigma minute. Pod leathery, oblong or linear to elliptic or obovoid, terete in cross-section, when young filled with a spongy tissue, which becomes fibrous.

Pod straight or nearly so. 1. *P. flexuosa.*
Pod distinctly arcuate. 2. *P. elongata.*

1. **P. flexuòsa** (Dougl.) Rydb. *Fig. 321.* Perennial, with a rootstock; stem flexuose, leafy, slender; leaves 4–10 cm. long, spreading; leaflets 15–25, oblong or obovate, obtuse or retuse at the apex, 5–18 mm. long, 1–4 mm. wide, strigose beneath, glabrous above; racemes elongate; corolla whitish or tinged with purple, 8–10 mm. long; pod linear, terete, 1.5–2 cm. long, 3–4 mm. wide, strigose. *Astragalus flexuosus* Dougl. *Phaca flexuosa* Hook. Dry plains: Man.—Kans.—N.M.—Alta. *Prairies—Plain.* Je–Jl.

2. **P. elongàta** (Hook.) Rydb. Perennial with a rootstock; stem 3–5 dm. high, grayish-canescent; leaves 4–8 cm. long, spreading; leaflets 15–19, linear or those of the lower leaves oblong, 5–15 mm. long, 1–2 mm. wide, strigose beneath; racemes lax; corolla white; pod linear, 12–18 mm. long, 2–2.5 mm. wide, curved upwards. *Phaca elongata* Hook. Prairies: Minn.—Mont.—Alta. *Prairies—Plain.* Je.

30. BATIDOPHÀCA Rydb.

Perennial herbs. Leaves odd-pinnate. Flowers perfect, spicate or racemose. Calyx campanulate, the teeth subulate. Banner broadly obovate, strongly arched. Ovary sessile, style arched, stigma minute. Pod sessile, membranous or leathery, more or less lunate or obliquely ovate in outline, tapering at each end. Seeds round-reniform.

Pubescence silky, appressed or ascending.
  Corolla 8–9 mm. long; pod 1.5–2 cm. long, equally acute at each end. — 1. *B. lotiflora.*
  Corolla 10–11 mm. long; pod 2.5–3 cm. long, more abruptly tapering at the apex than at the base. — 2. *B. cretacea.*
Pubescence villous, spreading. — 3. *B. nebraskensis.*

1. **B. lotiflòra** (Hook.) Rydb. Perennial with a caudex but apparently blooming the first season; stems branched at the base, 2–10 cm. high, densely covered by the stipules; leaves 4–10 cm. long; leaflets 9–13, or on the earlier leaves fewer, oblong or elliptic, 5–15 mm. long; flowers of two kinds, the earlier in short 4–12-flowered racemes on elongated peduncles; corolla ochroleucous or yellow, 8–9 mm. long; the later flowers subsessile in the axils setting fruit without fully opening; fruit 7–8 mm. wide, semi-ovoid, lunate, upcurved at the apex. *Astragalus lotiflorus* Hook. *Phaca lotiflora* Nutt. Plains: Man.—Neb.—Colo.—Sask. *Plain.* My–Je.

2. **B. cretàcea** (Buckl.) Rydb. Perennial of the habit of the preceding; leaflets oblong or oval, 5–20 mm. long, 2–6 mm. wide; peduncles of the early flowers 4–15 cm. long; corolla ochroleucous, tinged or streaked with purple, fully 1 cm. long; pod lunate, 7–8 mm. wide. *Phaca cretacea* Buckl. *A. Reverchoni* A. Gray. *A. elatiocarpus* Sheldon (the form with subsessile flowers). *A. Batesii* A. Nels. Prairies: Minn.—Mont.—N.M.—Tex. *Prairie—Plain—St. Plain.* My–Je.

3. **B. nebraskénsis** (Bates) Rydb. A perennial with a caudex; stems branched at the base, less than 1 dm. high; leaves 5–7 cm. long; leaflets oblong or elliptic, 1–2 cm. long, hirsute on both sides; flowers 1–3 in the axils (peduncled early flowers not seen); corolla ochroleucous; pod lance-ovoid, slightly lunate, 2 cm. long, 8 mm. wide, densely hirsute. *A. lotiflorus nebraskensis* Bates. Dry plains: S.D.—Neb. *Plain.* Je.

31. PHÀCA L. Rattle-weed, Rattle-pod.

Perennial or annual herbs, caulescent or rarely almost stemless. Leaves alternate; stipules nearly free from the petioles, often partially united with each other; leaflets many or few, entire, or in one species often none. Flowers perfect, racemose, or in one species subsessile in the axils of the leaves. Calyx campanulate. Corolla ochroleucous or purple; banner rather broad; wings longer than the keel; the latter beakless. Stamens 10, diadelphous; anthers alike. Pod membranous or papery, inflated, sessile or stipitate, 1-celled, with neither suture intruded, without even a partial partition. Seeds numerous.

Pod stipitate.
  Pod long-stipitate, pendent, not mottled; leaflets 7–17, well developed. — 1. *P. americana.*
  Pod short-stipitate, mottled; leaflets linear or none, the terminal one represented by the much produced rachis. — 2. *P. longifolia.*
Pod sessile.
  Pod 12–18 mm. long, glabrous; flowers white. — 3. *P. neglecta.*
  Pod about 8 mm. long, black-strigose; flowers purple. — 4. *P. Bodinii.*

1. **P. americàna** (Hook.) Rydb. *Fig. 322.* Perennial; stems single or two or three, erect, 3–10 dm. high; leaflets 7–17, oval, elliptic, or oblong, obtuse, thin, 2–4 cm. long, glabrous above, sparingly pubescent beneath; raceme short; calyx glabrous or nearly so; tube campanulate, 4 mm. long; teeth obsolete; corolla ochroleucous, about 12 mm. long; pod glabrous; stipe about 5 mm. long; body ellipsoid, acute, 2 cm. long, 7–8 mm. wide. *Astragalus frigidus americanus* S. Wats. Along streams and in wet copses: Que.—Wyo.—Yukon. *Subarct.—Submont.—Mont.* Jl–Au.

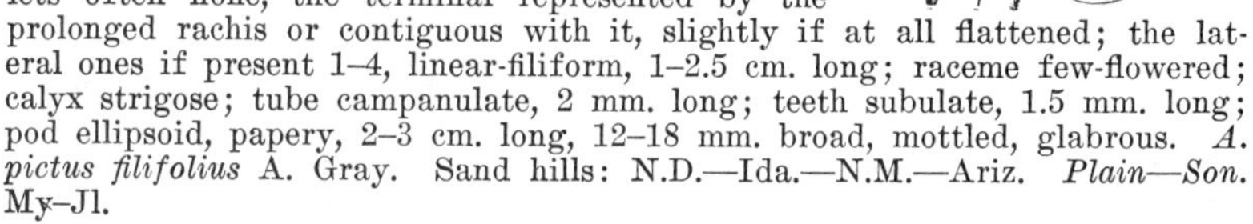
f. 322.

2. **P. longifòlia** (Pursh) Nutt. Perennial, with horizontal rootstock; stem erect, 1.5–4 dm. high; leaves 5–15 cm. long, grayish-canescent; leaflets often none, the terminal represented by the prolonged rachis or contiguous with it, slightly if at all flattened; the lateral ones if present 1–4, linear-filiform, 1–2.5 cm. long; raceme few-flowered; calyx strigose; tube campanulate, 2 mm. long; teeth subulate, 1.5 mm. long; pod ellipsoid, papery, 2–3 cm. long, 12–18 mm. broad, mottled, glabrous. *A. pictus filifolius* A. Gray. Sand hills: N.D.—Ida.—N.M.—Ariz. *Plain—Son.* My–Jl.

3. **P. neglécta** T. & G. Stem erect, 3–6 dm. high, glabrous or nearly so; leaflets 11–21, elliptic or oblong, 15–25 mm. long, minutely puberulent beneath; calyx more or less black-hairy, the teeth subulate; corolla white, 10–15 mm. long; flowers in a short spike; pod inflated, rather firm, broadly ellipsoid, sulcate on both sutures, 1.5–2 cm. long, glabrous. *Astragalus Cooperi* A. Gray. *A. neglectus* Sheldon [G]. Banks and shores: Que.—Minn.—Iowa—N.Y. *Allegh.* Je–Jl.

4. **P. Bodìnii** (Sheld.) Rydb. Perennial, scarcely cespitose; stems slender, glabrous, 3–6 dm. high, decumbent, branched; leaflets 11–17, oval to lanceolate, acute, 7–18 mm. long, glabrous or sparingly strigose beneath; raceme lax, 4–16-flowered; calyx strigose, with black hairs; tube 3 mm. long; teeth subulate, 2 mm. long; corolla purple, 10–12 mm. long; pod ovoid, black-strigose, 8 mm. long, 4 mm. wide. *A. Bodinii* Sheld. [B]. Meadows: Neb.—Wyo.—Colo. Jl–Au.

32. ATELOPHRÁGMA Rydb.

Caulescent leafy slender perennial herbs, with rootstocks. Stipules nearly free both from each other and from the petioles. Leaves pinnate, usually with numerous leaflets. Inflorescence a spike-like raceme. Calyx short, campanulate; lobes slender, subulate. Corolla white or purplish. Pod membranous or papery, more or less compressed laterally, in the typical species decidedly flattened and stipitate; dorsal suture intruding in the pod and forming a partial partition.

Pod stipitate, usually somewhat compressed.
Pod sulcate on the lower suture, black-hairy. 1. *A. alpinum.*
Pod not sulcate on either suture.
Pod long-stipitate, glabrous at least in age.
Plant loosely short-villous. 2. *A. aboriginum.*
Plant appressed-silky or glabrous.
Leaflets oblong. 3. *A. Forwoodii.*
Leaflets linear.
Plant grayish-silky. 4. *A. Herriotii.*
Plant glabrous or with a few scattered hairs. 5. *A. glabriusculum.*
Pod short-stipitate, black-hairy. 6. *A. Macounii.*
Pod sessile, more turgid. 7. *A. elegans.*

1. **A. alpìnum** (L.) Rydb. Stems decumbent or ascending, 1–3 dm. (rarely 4 dm.) high, glabrous or nearly so; leaflets 13–25, oval or elliptic, obtuse or retuse, 6–12 mm. long, pilose on both sides; peduncles 5–10 cm. long; raceme short, 1–4 cm. long; calyx black-hairy; tube about 3 mm. long; teeth subulate, 1.5 mm. long; corolla 8–12 mm. long; stipe of the pod 3 mm. long; body oblong, about 1 cm. long, 3 mm. thick. *A. alpinus* L. [B]. Alpine-arctic regions: Lab.—Vt.—Colo.—Ida.—Alaska; Eurasia. *Arct.—Mont.—Alp.* Je–Au.

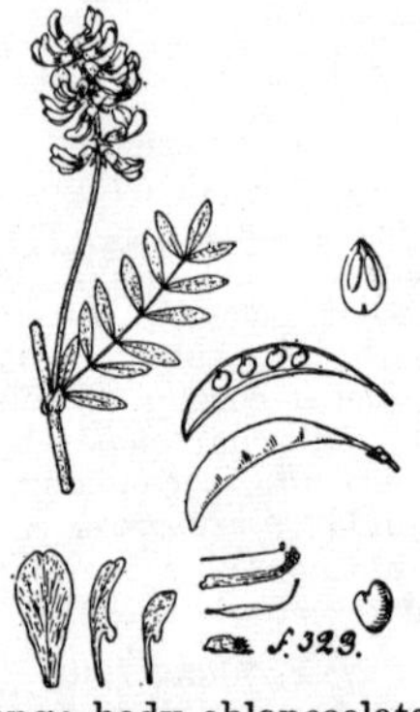

2. **A. aboríginum** (Richards.) Rydb. *Fig. 323.* Stems 2–4 dm. high, short-villous; leaflets 9–15, linear-oblong to elliptic or lanceolate, 1–2 cm. long; raceme short; calyx more or less black-hairy; tube 3–4 mm. long; teeth 2–3 mm. long; corolla ochroleucous, 9–11 mm. long; pod glabrous; stipe 5–8 mm. long; body oblanceolate, acute at both ends, 1.5–2 cm. long; lower suture almost straight. *Astragalus aboriginorum* Richards. *Homalobus aboriginum* Rydb. [B]. Mountains: Man.—S.D.—Colo.—Nev.—Yukon. *Submont.—Mont.* My–Je.

3. **A. Forwoòdii** (S. Wats.) Rydb. Stems ascending, minutely strigose, 1–3 dm. high; leaflets 9–17, oblong or oblanceolate, strigose on both sides or glabrous above, 1–1.5 cm. long; raceme short; calyx more or less black-hairy; tube 3 mm. long; teeth 2 mm. long; legume glabrous, flat; stipe 3–5 mm. long; body 2–2.5 cm. long, lunate-oblanceolate; lower suture usually somewhat concavely curved. *Astragalus Forwoodii* S. Wats. Hills: S.D.—Wyo.—Mont. *Submont.—Mont.* Je–Jl.

4. **A. Herriótii** Rydb. Stem grayish-strigose, 2–4 dm. high; leaflets 7–11, linear, 1.5–2.5 cm. long, 1.5–3 mm. wide, grayish-strigose beneath, glabrous above; racemes short; calyx black-hairy, the tube 2.5–3 mm. long, the teeth 1.5 mm. long; corolla about 8 mm. long, ochroleucous, the keel tipped with purple; legume strigillose, stipitate, the stipe 4 mm. long, the body 15 mm. long, the upper suture arched, the lower straight. Slopes: Man.—Alta. *Plains.* Je.

5. **A. glabriúsculum** (Hook.) Rydb. Stem 2–4 dm. high; leaflets linear or nearly so, 15–25 mm. long, glabrous or nearly so; raceme short, in fruit elongate; calyx black-hairy; tube 3 mm. long; corolla ochroleucous, 8 mm. long; keel tipped with dark purple; pod glabrous; stipe 4–5 mm. long; body 2 cm. long; lower suture straight or nearly so. *Phaca glabriuscula* Hook. *A. glabriusculus* A. Gray. Gravelly slopes: Alta.—S.D.—Wyo. *Submont.—Mont.* Je.

6. **A. Macoùnii** Rydb. Stem 5–6 dm. high, finely and sparingly strigose or glabrate; leaflets 9–17, very thin, oblong or oval, 15–25 mm. long, obtuse, glabrous above, sparingly strigose beneath; calyx black-hairy; tube about 3 mm. long; teeth 1 mm. long; corolla about 8 mm. long, cream-colored or white, tinged with purple; stipe of the pod 2–3 mm. long; body acute at both ends,

15–18 mm. long, 5 mm. wide. *Astragalus Macounii* Rydb. Mountains: B.C.—Alta.—Que.—Colo. *Boreal.—Submont.—Mont.* My–Jl.

7. **A. élegans** (Hook.) Rydb. Stem finely strigose or glabrous, erect or decumbent at the base, 2.5–6 dm. high; leaflets 9–17, oblong to linear, 15–20 mm. long, obtuse; raceme lax, 5–10 cm. long, in fruit even 15–20 cm. long; calyx black-hairy; tube 2.5 mm. long; teeth 1 mm. long; corolla 6–8 mm. long, purple, or ochroleucous and tinged with purple; pod ellipsoid, pendent, slightly inflated, sessile, black-hairy, 8–15 mm. long. *Phaca elegans* Hook. *Astragalus oroboides americanus* A. Gray. *A. elegans* Britton [B]. *A. minor* M. E. Jones. *A. eucomus* B. L. Robins. [G]. Mountains and woods: Lab.—Que.—S.D.—Colo.—Ida.—Alta. *Boreal.—Submont.—Alp.* Je–Jl.

33. TÌUM Medic.

Perennial caulescent herbs. Leaves alternate, with nearly free and distinct stipules; leaflets numerous, entire. Flowers perfect, in racemes. Calyx campanulate; teeth nearly equal, usually rather long. Corolla purple, white or ochroleucous; banner with a rather broad, erect, often notched blade. Stamens 10, diadelphous; anthers alike. Pod narrow, tapering at each end, more or less stipitate, usually membranous, 1-celled, obcordate, or inverted V-shaped in cross-section, the lower suture strongly sulcate, rarely with a narrow partial partition. Seeds numerous.

Plant villous; pod cordate in cross-section. 1. *T. Drummondii.*
Plant appressed-pubescent or glabrous; pod more or less triangular or inverted V-shaped in cross-section.
  Calyx not black-hairy; pod straight, its septum very narrow. 2. *T. racemosum.*
  Calyx black-hairy; pod arcuate, its septum broad, almost reaching the upper suture. 3. *T. scopulorum.*

1. **T. Drummóndii** (Dougl.) Rydb. Stems 3–7 dm. high, villous; leaflets 25–31, elliptic to linear-oblong, obtuse, about 1 cm. long, glabrous or nearly so above, softly long-villous beneath; peduncles 5–10 cm. long; raceme 3–10 cm. long, at first dense, soon elongate; calyx hirsute-villous, with mixed black and white hairs; tube 5–6 mm. long; teeth subulate, 1.5–2 mm. long; corolla 15–20 mm. long, cream-colored; stipe of the pod about 8 mm. long; body linear, 2–2.5 cm. long, glabrous, 3–4 mm. thick. *Astragalus Drummondii* Dougl. [B]. Plains and hills: Sask.—Neb.—Colo.—Alta. *Plain—Submont.* My.

2. **T. racemòsum** (Pursh) Rydb. Stems 3–6 dm. high, strigose; leaflets 15–31, pale green, linear or oblong, obtuse, 1.5–3 cm. long, glabrous above, strigose beneath; peduncles about 1 dm. long; raceme lax, 1–2 dm. long, many-flowered; calyx strigose; tube 5–8 mm. long; teeth subulate, 4–6 mm. long; corolla white, 15–20 mm. long; stipe of the pod about 5 mm. long; body linear, about 2.5 cm. long, 4 mm. broad, glabrous. *A. racemosus* Pursh [B, G]. Plains and hills: Minn.—N.D.—N.M.—Okla. *Plain.*

3. **T. scopulòrum** (Porter) Rydb. *Fig. 324.* Stem 3–5 dm. high, strigose, angled; leaflets 15–27, oblong to obovate or oval, glabrous above, sparingly strigose beneath; peduncles 5–8 cm. long; raceme 5–10 cm. long; calyx-tube about 6 mm. long; teeth subulate, 3 mm. long; corolla ochroleucous, 15–20 mm. long; body of the pod linear, curved, glabrous, 2–5 cm. long, 3 mm. wide. *A. scopulorum* Porter. *A. rasus* Sheld. *A. subcompressus* A. Gray. Mountains, in open woods: S.D. (?)—Colo.—N.Mex.—Utah. *Submont.—Mont.* My–Je.

34. **HAMÒSA** Medic.

Annual, biennial, or perennial herbs, with decumbent or tufted stems. Leaves alternate, odd-pinnate; stipules almost free and distinct; leaflets numerous, entire, usually truncate or notched at the apex. Flowers perfect, in short racemes, few. Calyx short-campanulate, 5-lobed; lobes subulate, nearly equal. Corolla usually purplish; banner with rather broad blade, erect; wings usually longer than the unbeaked keel. Stamens 10; filaments diadelphous; anthers alike. Pods linear, laterally flattened, dehiscent, membranous, completely 2-celled by the intrusion of the lower suture. Seeds numerous.

1. **H. leptocárpa** (T. & G.) Rydb. Annual; stem erect, simple or branched at the base, with ascending branches, 1–3 dm. high; leaflets 9–19, glabrous or nearly so, oblong to cuneate, 3–8 mm. long, notched at the apex; peduncles 3–7 cm. long; raceme few-flowered; calyx-tube and the lanceolate teeth each 2 mm. long; corolla purplish, 8–10 mm. long; legume linear, 2.5–3 cm. long, straight or nearly so, glabrous. *A. leptocarpus* T. & G. Dry soil: Kans. (?)—Tex.—Colo.—Ark. *Son.—Texan.* Mr.–Ap.

35. **ASTRÁGALUS** (Tourn.) L. LOCO-WEED, MILK VETCH.

Perennial or annual herbs. Leaves alternate, odd-pinnate. Flowers perfect, racemose. Calyx campanulate to cylindric, 5-lobed; lobes usually distinctly unequal. Corolla purplish or white, or rarely ochroleucous; banner with an erect blade, usually longer than the wings and the keel; keel not beaked, its petals wholly united. Stamens 10; filaments diadelphous; anthers alike. Pods ovoid to oblong, leathery or woody, turgid, 2-celled, dehiscent; partition formed by the intruded lower suture. Seeds usually many.

Plants cespitose, subscapose, villous-pubescent; pods sulcate on both sutures. — 1. *A. mollissimus.*
Plants with elongate leafy stems.
  Pods not sulcate or slightly so on the lower suture, round or nearly so in cross-section.
    Bracts linear-lanceolate, long-attenuate, the lower almost as long as the calyces; calyx-teeth all narrow, subulate, fully half as long as the tube; pod glabrous. — 2. *A. canadensis.*
    Bracts ovate to lanceolate, scarcely half as long as the calyces; calyx-teeth short, less than half as long as the tube, the upper broader; pod more or less hairy. — 3. *A. pachystachys.*
  Pod deeply sulcate on the low suture, cordate or triangular in cross-section.
    Pod with appressed gray or black pubescence.
      Corolla purple or pink, seldom white; calyx-teeth much shorter than the tube. — 4. *A. striatus.*
      Corolla sulphur-yellow or white; calyx-teeth almost equaling the tube. — 5. *A. Chandonetti.*
    Pod villous with long spreading hairs.
      Calyx black-hairy; teeth decidedly shorter than the tube. — 6. *A. goniatus.*
      Calyx not black-hairy; teeth about equaling the tube. — 7. *A. agrestis.*

1. **A. mollíssimus** Torr. Stems less than 1 dm. long, decumbent, densely villous; leaves mostly basal; leaflets 23–29, obovate to oval, 6–25 mm. long, silky-villous; racemes short; calyx-tube cylindric, silky, 7–9 mm. long; teeth subulate, about 3 mm. long; corolla bright purple, about 18 mm. long; pod cylindric, 2 cm. long, curved upwards. WOOLLY LOCO-WEED. Prairies and plains: Neb.—Tex.—N.M.—Wyo.—(? Mont.). *Plain.* My–Jl.

2. **A. canadénsis** L. Stem erect or ascending, 3–12 dm. high, sparingly strigose; leaflets 15–25, elliptic, oblong, or oblong-lanceolate, 2–4 cm. long, usually obtuse at the apex, strigose on both sides or glabrate above; peduncles 5–10 cm. long; raceme 5–20 cm. long; calyx-tube strigose, 5 mm. long; teeth subulate, the upper 1.5 mm., the lower 2 mm. long; corolla ochroleucous, about 12 mm. long; pod oblong, 1–1.5 cm. long. *A. carolinianus* MacMill.; not L.

[B]. River banks and hillsides: Que.—Fla.—Utah—B.C. *Temp.—Plain—Submont.* Je–Au.

3. **A. pachystàchys** Rydb. Stems 2–4 dm. high, strigose; leaflets oblong to oval, obtuse or rounded at the apex, 1–2.5 cm. long, strigose on both sides or glabrate above; peduncles 1–1.5 dm. long; flowers often reflexed; calyx-tube 6 mm. long; lower teeth subulate, 2 mm. long, the upper ones triangular, 1.5 mm. long; corolla ochroleucous, 12–15 mm. long; pod cylindric, 1 cm. long, 4–5 mm. thick. *A. spicatus* Nutt.; not Pall. *A. Mortoni* Coult.; not Nutt. Meadows: Mont.—S.D.—Wyo.—Nev.—Ida. *Submont.—Mont.* Je–Au.

4. **A. striàtus** Nutt. Stems 1–4 dm. high, strigose, decumbent or ascending; leaflets 9–19, oblong or elliptic, acute or obtuse, 1–2 cm. long, grayish-strigose; peduncles 7–10 cm. long; spike dense, 2–6 cm. long; calyx-tube 5 mm. long, strigose, with mixed black and white hairs; teeth subulate, 2–3 mm. long; corolla 12–14 mm. long; pod ovoid, 8–19 mm. long, strigose. *A. adsurgens* Hook. [B, G]; not Pall. *A. nitidus* Dougl. *A. Crandallii* Gand. Plains and hills: Man.—Minn.—Colo.—Ore.—B.C. *Prairie—Plain—Mont.* Je–Jl.

5. **A. Chandonétti** Greene. Stems ascending, about 4 dm. high, angled, strigose-canescent; leaflets 13–19, elliptic, obtuse or acutish, mucronulate, silky-canescent on both sides, 12–30 mm. long; spike dense and elongate; calyx white-strigose, sometimes with scattered black hairs; tube about 5 mm. long; lobes almost filiform, 5 mm. long; corolla light yellow, white, or pinkish; banner 15–18 mm. long, much exceeding the wings and keel; pod about 1 cm. long, 3–4 mm. wide, strigose. Mountains: Minn. —Mont. *Prairie—Mont.* Je–Au.

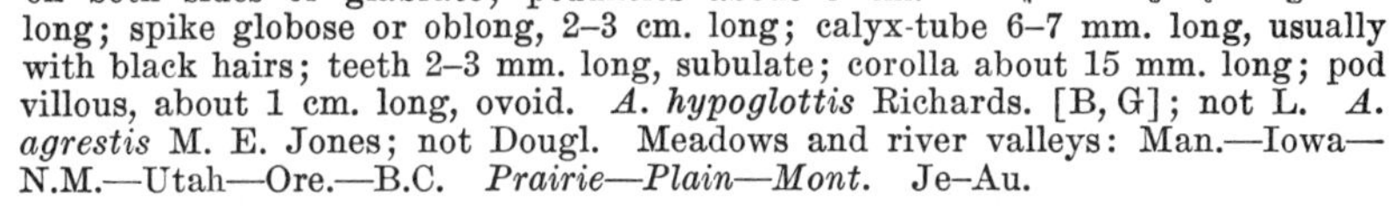

6. **A. goniàtus** Nutt. *Fig. 325.* Stems 1–2 dm. (rarely 3 dm.) high, strigose, decumbent or ascending, zigzag; leaflets 15–21, linear-oblong to elliptic, obtuse or retuse, 5–10 mm. long, sparingly strigose on both sides or glabrate; peduncles about 5 cm. long; spike globose or oblong, 2–3 cm. long; calyx-tube 6–7 mm. long, usually with black hairs; teeth 2–3 mm. long, subulate; corolla about 15 mm. long; pod villous, about 1 cm. long, ovoid. *A. hypoglottis* Richards. [B, G]; not L. *A. agrestis* M. E. Jones; not Dougl. Meadows and river valleys: Man.—Iowa—N.M.—Utah—Ore.—B.C. *Prairie—Plain—Mont.* Je–Au.

7. **A. agréstis** Dougl. Stems decumbent or ascending, about 2 dm. high, glabrous or nearly so; leaflets oblong or oblong-lanceolate, 7–15 mm. long, obtuse or retuse, sparingly strigose or glabrate; peduncles 5–10 cm. long; spikes oblong or globose, 2–3 cm. long; calyx-tube 5–6 mm. long; teeth subulate, 4–5 mm. long; corolla purple, 15–18 mm. long; pod as in the preceding. Meadows: Man.—Sask.—Mont. *Prairie.* Je.

36. GEOPRÚMNON Rydb. Buffalo Beans, Ground Plums, Buffalo Peas, Prairie Apples.

Perennial tufted herbs, with decumbent or ascending stems. Leaves alternate, odd-pinnate; leaflets numerous, entire, not notched. Flowers perfect, in dense racemes; calyx deeply campanulate or nearly cylindric, 5-lobed; lobes nearly equal. Corolla white or purplish, rarely ochroleucous; banner with a rather narrow, erect blade, longer than the wings and keel; keel-petals wholly united, not produced into a beak. Stamens 10; filaments diadelphous. Pods globose or ellipsoid, fleshy, becoming spongy, 2-celled, the partition formed by the intruded lower suture, which meets the upper suture. Seeds numerous.

Pods pubescent, abruptly pointed. 1. *G. plattense.*
Pods glabrous, pointless.
Calyx strigose, its lobes subulate or lanceolate.
Corolla purple; leaflets oblong to linear. 2. *G. crassicarpum.*
Corolla yellowish-white, with purple keel; leaflets oval or obovate. 3. *G. succulentum.*
Calyx villous, its lobes triangular or deltoid. 4. *G. trichocalyx.*

1. **G. platténse** (Nutt.) Rydb. Stems diffusely branched, 1–4 dm. long, decumbent, strigose; leaflets 13–23, oblong, elliptic, or cuneate, 5–15 mm. long, acute, rounded or retuse at the apex; racemes few-flowered, head-like; calyx-tube strigose, 5 mm. long; teeth subulate, 2 mm. long; corolla about 15 mm. long, yellowish white, tipped with purple; pod ovoid, 1–1.5 cm. in diameter. *Astragalus plattensis* Nutt. [B, G]. Prairies: Ind.—Ala.—Tex.—N.D. *Plain.* Ap–Je.

2. **G. crassicárpum** (Nutt.) Rydb. *Fig. 326.* Stems strigose, branched, prostrate, 1–4 dm. long; leaflets 13–27, oblong to linear, 8–20 mm. long, acute or obtuse, strigose beneath, glabrous above; raceme few-flowered; calyx-tube about 5 mm. long, strigose, at least some of the hairs black; corolla violet-purple, 1.5–2 cm. long; pod subglobose, 1.5–2 cm. thick. *A. caryocarpus* Ker [G]. *A. crassicarpus* Nutt. *A. mexicanus* of (Fl. Neb.); not DC. Prairies: Man.—Mo.—Tex.—Mont. *Plain.* Ap–My.

f.326.

3. **G. succuléntum** (Richards.) Rydb. Stems prostrate, spreading, branched, 2–4 dm. long, finely strigose; leaflets 17–25, elliptic or broadly oblong, obtuse, truncate or emarginate, 1–1.5 cm. long, strigose beneath, glabrous above; raceme short; calyx strigose, sometimes with dark hairs; tube 7–8 mm. long; teeth subulate, 2 mm. long; corolla 2 cm. long; pod rounded-ellipsoid, over 1.5–2 cm. in diameter. *A. succulentus* Richards. *A. prunifer* Rydb. Plains and hills: Sask.—Colo.—Mont. *Plain—Submont.* My–Je.

4. **G. trichócalyx** (Nutt.) Rydb. Stem branched at the base, villous, decumbent; leaflets 17–33, obovate or oblong, rounded or retuse at the apex, 5–15 mm. long; peduncles usually longer than the leaves; racemes 2–8 cm. long; calyx densely villous; corolla cream-colored, sometimes purple-tipped, 15–25 mm. long; pod subglobose, glabrous, 2.5–3 cm. in diameter. *A. mexicanus* A. Gray [B, G]; not DC. *A. trichocalyx* Nutt. Prairies: Ill.—Kans.—Tex.—Ark. *Prairie—Texan.* My.

37. **CARAGÀNA** Lam. PEA TREE.

Shrubs or trees. Leaves often fascicled, abruptly pinnated, the leaflets entire, the rachis sometimes produced into a spiny point. Flowers perfect, axillary, solitary or fascicled. Calyx campanulate or tubular, 5-toothed. Corolla yellow, rarely white or pinkish; banner erect, long-clawed; wings also long-clawed; keel-petals obtuse, straight. Stamens diadelphous (9–1). Pod sessile or stipitate, terete, 2-valved, 2-celled, several-seeded

1. **C. arboréscens** Lam. Small tree or shrub, 3–6 m. high; leaflets 8–12, obovate or oblong, rounded at the apex, sparingly pubescent beneath or glabrate, 1–2.5 cm. long, flowers in fascicles of 2–4, yellow, 1.5–2 cm. long; pod about 2 cm. long. Fence rows and wood lots: N.D.—Kans.; escaped from cult.; nat. of Siberia.

38. **OXÝTROPIS** DC. LOCO-WEED.

Perennial herbs, mostly acaulescent. Leaves alternate, odd-pinnate. Flowers racemose or spicate. Calyx campanulate, its teeth almost equal. Petals clawed; banner erect, ovate or oblong; wings oblong; keel produced at the apex into a porrect point. Stamens 10, diadelphous; anthers all alike. Pods sessile or stipitate, coriaceous, often more or less 2-celled by the intrusion of the upper suture. *Aragallus* Necker, in part.

Stipules conspicuously adnate to the petioles; pods not pendent; plant acaulescent.
  Fruiting calyx inflated, enclosing the fruit; plant tufted; scape less than 1.5 cm. high. — 1. *O. multiceps.*
  Fruiting calyx not inflated, much exceeded by the fruit; scape more than 1.5 cm. high.
    Leaves strictly pinnate, the leaflets opposite.
      Corolla purple, rarely white.
        Pod villous, almost completely 2-celled; caudices conspicuously white-hairy. — 2. *O. plattensis.*
        Pod appressed-silky or short-hairy; caudices not conspicuously white-hairy.
          Leaves white-silky on both sides; corolla 2 cm. long.
            Corolla purple or in albino forms pure white. — 3. *O. sericea.*
            Corolla white, with a purple spot on the keel petals. — 4. *O. pinetorum.*
          Leaves not silky on both sides; corolla less than 2 cm. long
            Leaves white-silky beneath; corolla 15 mm. long. — 5. *O. albertina.*
            Leaves green on both sides, only slightly silky.
              Plant usually less than 2 dm. high; leaves, leaflets, flowers and pods more or less spreading; spike short and rather dense.
                Leaflets narrowly linear-lanceolate. — 6. *O. Hookeriana.*
                Leaflets lanceolate. — 7. *O. dispar.*
              Plant usually 3–4 dm. high; leaves, leaflets, flowers, and fruit usually strongly ascending.
                Leaflets linear or narrowly linear-lanceolate. — 8. *O. involuta.*
                Leaflets broader, lanceolate or narrowly lanceolate. — 9. *O. Lambertii.*
      Corolla yellow or ochroleucous, only the keel purple-tipped.
        Plant, including the pod, with long, loose hairs; pod not black-hairy. — 10. *O. villosa.*
        Plant, including the pod, with short, appressed hairs.
          Corolla 12–15 mm. long, usually without a spot on the keel; plant not sericeous.
            Calyx and pod white-hairy. — 11. *O. gracilis.*
            Calyx and pod with mixed black and white hairs. — 12. *O. glabrata.*
          Corolla 18–20 mm. long.
            Corolla yellow; banner and wings narrower than in the next, slightly notched; legume nearly 2-celled. — 13. *O. Macounii.*
            Corolla white or straw-colored; banner and wings fully 1 cm. wide, deeply notched; legume half 2-celled. — 4. *O. pinetorum.*
    Leaflets verticillate.
      Leaflets elliptic or oval; plant densely villous. — 14. *O. splendens.*
      Leaflets lanceolate, appressed-silky. — 15. *O. Richardsonii.*
Stipules adnate to the petioles only slightly at the base; pod pendulous; plant often caulescent.
  Stem well developed, leafy, with several axillary peduncles; corolla 7–9 mm. long, merely tipped with purple or wholly white; pod oblong. — 16. *O. deflexa.*
  Stem very short, with short internodes, the leaves therefore collected near the base; corolla dark purple, 10 mm. long; pod ellipsoid. — 17. *O. foliolosa.*

1. O. múlticeps Nutt. Pulvinate perennial; leaflets 7–9, elliptic to lanceolate, densely canescent, with appressed silky hairs, 5–10 mm. long, acute;

peduncles 3–5 cm. long, 1–3-flowered; calyx silky-villous; tube 8 mm. long; teeth lanceolate, 3 mm. long; corolla reddish purple; keel with a very short porrect beak; legume ellipsoid, silky-strigose, 1 cm. long, with a very narrow partial partition. *A. multiceps* Heller [R]. Hills and mountains: Neb.—Wyo.—Colo. *Plain—Submont.* My–Je.

2. **O. platténsis** Nutt. Leaflets 11–15, elliptic or oblong-lanceolate, 4–20 mm. long, densely silky-villous, with rather loose hairs, acute; scape 4–8 cm. long, with rather loose hairs; raceme 3–5 cm. long; calyx-tube 5 mm. long; teeth subulate, 2 mm. long; corolla purple, 12–14 mm. long; legume villous, about 15 mm. long, coriaceous, half 2-celled. *A. plattensis* Rydb. [R]. Dry hills: Wyo.—Neb.—N.D.—Utah. *Plain—Submont.* Je.

3. **O. serícea** Nutt. Leaflets 9–19, elliptic or oblong, 1–3 cm. long, shining, with white silky-appressed hairs; scape 1–2.5 dm. high, stout; raceme 5–10 cm. long; flowers rather spreading; calyx about 7 mm. long; teeth lanceolate, about 3 mm. long; corolla light purple; wings very broad and deeply emarginate; pod cylindric, erect, about 2 cm. long, long-acuminate, nearly 2-celled. *A. sericeus* Greene [R, B]. Hills: Neb.—Colo.—Wyo. *Plain.* My–Jl.

4. **O. pinetòrum** (Heller) Rydb. Leaflets 13–21, oblong, elliptic, or lanceolate, 1.5–2.5 cm. long, grayish-strigose; scape 1.5–3 dm. long; spike elongate, 5–10 cm. long; calyx silky, with intermixed black hairs; tube 7–8 mm. long; teeth lanceolate, 3 mm. long; corolla 2–2.5 cm. long; keel with a deep purple blotch; pod densely pubescent, oblong, about 2 cm. long, rather abruptly acuminate, about half 2-celled. *A. saximontanus* A. Nels. *A. albiflorus* A. Nels. [R]. *A. pinetorum* Heller. *A. majusculus* Greene. Plains: Mont.—S.D.—Okla.—N.M.—Utah.—Colo. *Plains—Submont.—Mont.* My–Jl.

5. **O. albertìna** (Greene) Rydb. Leaflets 15–30, lanceolate, broadest near the base, 15–20 mm. long, sparingly silky and light green above, densely silky beneath, the margins somewhat involute; scape slender, 1–2 dm. high; raceme about 1 dm. long; calyx silky-villous, with rather loose hairs; tube 6 mm. long; teeth 3 mm. long; corolla about 15 mm. long, white below, sky-blue at the end; pod ovoid, erect, dark, long-acuminate, thin, half 2-celled. *A. albertinus* Greene [R]. Exposed rocky places: Man.—Sask. *Plain.* Au.

6. **O. Hookeriàna** Nutt. Leaflets 7–13, narrowly linear-lanceolate to narrowly linear, 1–2 cm. long; scape about 1 dm. high; raceme 3–5 cm. long; calyx finely silky-strigose; tube 5–6 mm. long; teeth subulate; wings broad and slightly emarginate; legume ovoid, coriaceous, half 2-celled. *A. angustatus* Rydb. [R]. *A. Aven-Nelsonii* Lunell. Plains and hills: N.D.—Neb.—Colo. *Plain—Submont.* Jl–Au.

7. **O. díspar** (A. Nels.) K. Schum. Leaves spreading or ascending; leaflets 9–23, elliptic or oblong, acutish at both ends, 1–2 cm. long, 4–6 mm. wide, those of the lower leaves often shorter and broader, somewhat silvery, with appressed hairs; scape 1–1.5 dm. high; raceme short; flowers usually spreading; calyx sparingly appressed-silky with short hairs, often somewhat tinged with purple above; tube 5–7 mm. long; teeth subulate, 2 mm. long; corolla dark bluish purple to rarely cream-white, about 15 mm. long; keel with a very dark purple spot; pod ascending-spreading, thin-coriaceous, nearly straight, less than 2 cm. long. *A. dispar* A. Nels. *A. patens* Rydb. [R]. *A. formosus* Greene, a coarse form. Plains and hills: N.D.—Okla.—Colo.—Wyo. *Plain—Submont.* My–S.

8. **O. involùta** (A. Nelson) K. Schum. Leaflets linear, 1–3 cm. long, 1–3 mm. wide, often involute, falcate, sparingly silky-hirsute, acute at each end; peduncles 1.5–2.5 cm. long; spike about 5 cm. long, in age often 1 dm. long; calyx-tube 6–7 mm. long; corolla purple, in age blue; pod 1.5–2 cm. long, firm, white-strigose. Prairies: Minn.—N.D.—Okla.—Mo. *Prairie.* Jl.

9. **O. Lambértii** Pursh. *Fig. 327.* Leaflets 11–17, narrowly lanceolate, 1.5–4 cm. long, 3–5 mm. wide, acute at both ends, green, silky on both sides; scape 1.5–2 dm. high; racemes elongate, 5–10 cm. long; flowers usually somewhat ascending; calyx-tube about 7 mm. long; teeth subulate, 2 mm. long; corolla usually dark bluish-purple; wings slightly emarginate; legume almost erect, lance-oblong, half 2-celled. *A. Lambertii* (Pursh) Greene [B, R]. *A. falcatus* Greene. Dry plains and prairies: Man.—Mo.—Colo.—Mont. *Prairie—Plain—Submont.* My–Au.

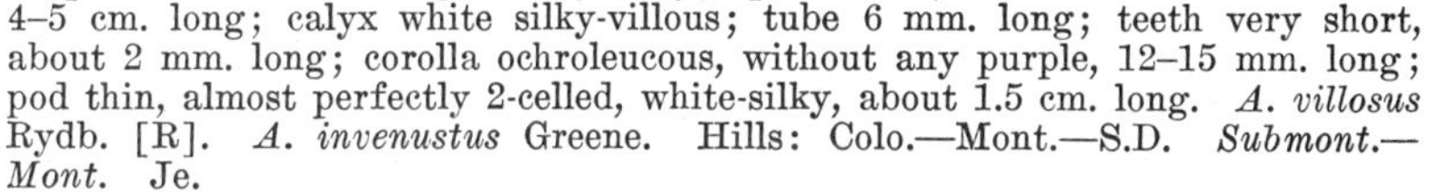

10. **O. villòsa** (Rydb.) K. Schum. Leaflets 25–31, rather crowded, 1–1.5 cm. long, lanceolate, very acute, densely but somewhat loosely silky; scape about 1.5 dm. long, loosely silky; spike dense, 4–5 cm. long; calyx white silky-villous; tube 6 mm. long; teeth very short, about 2 mm. long; corolla ochroleucous, without any purple, 12–15 mm. long; pod thin, almost perfectly 2-celled, white-silky, about 1.5 cm. long. *A. villosus* Rydb. [R]. *A. invenustus* Greene. Hills: Colo.—Mont.—S.D. *Submont.—Mont.* Je.

11. **O. grácilis** (A. Nels.) K. Schum. Leaves erect; leaflets 21–31, oblong-lanceolate, appressed-silky, 1–2.5 cm. long, acute; scape 2–4 dm. high; spike elongate, 5–10 cm. long; calyx densely silky; tube 6–7 mm. long; teeth linear-subulate, 3–4 mm. long; corolla ochroleucous, about 15 mm. long; wings slightly emarginate; pod oblong-ovate, about 1.5 cm. long, semi-membranous, nearly 2-celled. *A. gracilis* A. Nels. [R]. Open woodlands: Man.—S.D.—Ida. —Alta. *Submont.—Mont.* Je–Jl.

12. **O. glabràta** (Hook.) A. Nels. Leaves ascending, 8–12 cm. long; leaflets 13–25, elliptic or oblong, 5–25 mm. long, 2–5 mm. wide, silky-pubescent on both sides; peduncles 1.5–2.5 dm. high; spike 10–15-flowered; calyx with black and white hairs; tube 5–6 mm. long; teeth subulate, 2 mm. long; corolla 10–12 mm. long, ochroleucous; pod lance-elliptic, 2 cm. long, black-hairy. Open places: Man.—Mack.—Alaska. *Subarctic.*

13. **O. Macoùnii** (Greene) Rydb. Leaflets oval to oblong, 8–25 mm. long, 4–8 mm. wide, usually obtuse, silky-strigose; scape 1–2 dm. long; spike 4–10 cm. long; calyx silky-strigose, often with intermixed black hairs; tube 6–7 mm. long; teeth lanceolate, 3 mm. long; corolla 18–20 mm. long; pod short-hairy, often with some black hairs, oblong, about 2 cm. long. *O. campestris spicata* Hook. *A. spicatus* Rydb. [R]. *A. Macounii* Greene. Plains and hills: Sask.—S.D.—Wyo.—Alta. *Plain—Mont.* My–Jl.

14. **O. spléndens** Dougl. Leaflets very numerous, linear-lanceolate, 1–2.5 cm. long, very acute, white, silky-villous; scape 2–3 dm. high; spike dense or in age laxer, 4–10 cm. long; bracts narrowly linear to subulate, silky-villous, 1–2 cm. long; calyx densely white villous; tube about 6 mm. long; teeth about 5 mm. wide; corolla 12–15 mm. long, dark blue; legume densely long-villous, 1 cm. long, ovoid, short-beaked, nearly 2-celled. *A. splendens* (Dougl.) Greene [R]. *A. caudatus* Greene. Plains and hills: Man.—Minn.—Mont.—B.C. *Boreal. —Mont.* Jl.

15. **O. Richardsònii** (Hook.) K. Schum. Leaves erect, 1–2 cm. long, appressed-silky; leaflets in more than 10 verticils, lanceolate, acute or acuminate, 1–2 cm. long, 2–4 mm. wide; peduncles 1–2 dm. high; spike 4–10 cm. long; bracts subulate, 5–20 mm. long; corolla dark purple, 15 mm. long; pod 1 cm. long, short-villous. Hills and plains: Man.—Sask.—Alta. *Boreal.*

16. **O. defléxa** (Pall.) DC. Stem 1–4 dm. high, villous, flexuose; leaflets 25–41, lanceolate or ovate, green, loosely villous, 5–20 mm. long, rounded at the base, acute at the apex; raceme usually lax, 5–10 cm. long; calyx black-hairy; tube 4 mm. long; teeth subulate, 3 mm. long; corolla 6–9 mm. long, dirty white below, blue at the apex; pod oblong, black-hairy. *A. deflexus* Heller [R]. Damp places: Man.—S.D.—N.M.—Ida.—Alaska. *Mont.—Subalp.* Je–Au.

17. **O. foliolòsa** Hook. Subacaulescent; leaflets 13–31, ovate or elliptic, 3–8 mm. long, villous, green, acutish or obtuse; calyx black-villous; tube 5–6 mm. long; teeth subulate, 3 mm. long; corolla bluish purple, 8–10 mm. long; legume ellipsoid, black-hairy. *A. foliolosus* (Hook.) Rydb. [R]. High mountains: Alta.—Colo.—Yukon. *Subalp.* Jl–Au.

39. GLYCYRRHÌZA (Tourn.) L. LIQUORICE.

Perennial herbs, with thick sweet roots and glandular-dotted foliage. Leaves odd-pinnate. Flowers perfect, in spikes or heads. Calyx with the two upper teeth somewhat shorter and partly united. Banner oblong or narrowly ovate, clawed; keel acute or obtuse. Stamens diadelphous (9 and 1); anthers alternately smaller and larger. Pod sessile, indehiscent, covered with prickles or in exotic species with glands.

1. **G. lepidòta** Nutt. Tall leafy perennial, 3–10 dm. high; leaflets 11–19, lanceolate or oblong, entire, 2–3.5 cm. long, mucronate-pointed, scaly when young; spikes many-flowered; corolla yellowish white, 12–13 mm. long; pod 12–15 mm. long, with hooked prickles. Copses and rich meadows: Ont.—N.Y.—Ariz.—Calif.—Wash.; Mex. *Son.—Plain—Submont.* My–Au.

40. CORONÍLLA L.

Glabrous herbs or shrubs. Leaves odd-pinnate. Flowers perfect, in axillary peduncled umbels. Calyx 5-toothed. Banner orbicular. Keel-petals incurved. Stamens diadelphous (9 and 1). Pod terete or 4-angled, jointed, the joints subcylindric or prismatic.

1. **C. vària** L. A perennial herb, with ascending stems, 3–4 dm. high; leaves sessile: leaflets 15–25, oblong; flowers rose-colored; pod coriaceous, 3–7-jointed, 4-angled, the joints 6–8 mm. long. Roadsides and waste places: Mass.—S.D.—Mo.—Md.; nat. from Eur.

41. HEDÝSARUM (Tourn.) L. HEDYSARUM.

Perennial herbs, rarely shrubby, with odd-pinnate leaves, with several leaflets and no stipels. Flowers showy, in axillary, peduncled racemes, perfect. Calyx bracteolate, campanulate, nearly equally 5-toothed. Banner obovate or orbicular, clawed; wings oblong, shorter than the other petals; keel obliquely truncate, obtuse. Stamens diadelphous (9 and 1). Legume flat, divided transversely into rounded or subrhombic indehiscent internodes.

Calyx-teeth shorter than the tube; reticulations of the pods polygonal.
  Fruit glabrous, except as to a few hairs on the margins; internodes 6–8 mm. long. — 1. *H. americanum.*
  Fruit distinctly appressed-pubescent; internodes 8–15 mm. long. — 2. *H. boreale.*
Calyx-teeth subulate, longer than the tube; reticulations of the pods transversely elongate, usually reaching from the middle to the margins, without cross-veins.
  Flowers 18–20 mm. long; leaves glabrous above. — 3. *H. Mackenzii.*
  Flowers 10–15 mm. long; leaves canescent on both sides. — 4. *H. cinerascens.*

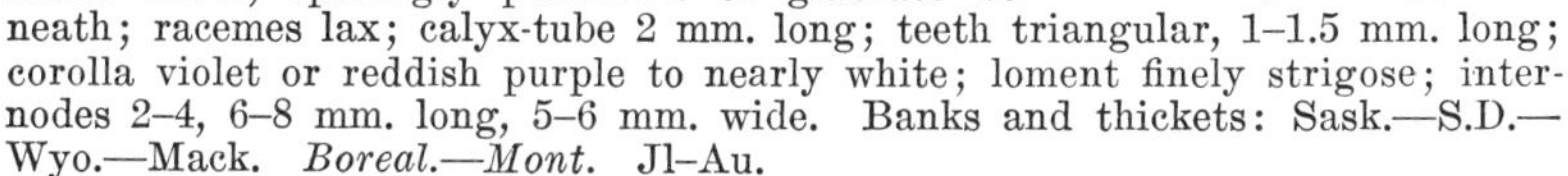

1. **H. americànum** (Michx.) Britton. *Fig. 328.* Stem 2–7.5 dm. high, glabrous or nearly so; leaflets 11–21, oblong, 1.5–3 cm. long, rounded at both ends, glabrous above, sparingly hairy beneath; raceme lax; calyx-tube 2 mm. long; teeth triangular, 1–2 mm. long; corolla violet, rarely white, 12–18 mm. long; loment glabrous or nearly so; internodes 3–5, about 5 mm. wide. *H. boreale* A. Gray [G]; not Nutt. *H. alpinum americanum* Michx. *H. philoscia* A. Nels., a white-flowered form. Rocky places and open woods: Lab. — Vt. — S.D. — Wyo. — Alaska. *Boreal.—Mont.* Je–Au.

2. **H. boreàle** Nutt. Stem 3–7 dm. high, strigose, especially above; leaflets oblong or lanceolate, rounded to acute at the apex; 1.5–3 cm. long, glabrous above, sparingly pubescent or glabrate beneath; racemes lax; calyx-tube 2 mm. long; teeth triangular, 1–1.5 mm. long; corolla violet or reddish purple to nearly white; loment finely strigose; internodes 2–4, 6–8 mm. long, 5–6 mm. wide. Banks and thickets: Sask.—S.D.—Wyo.—Mack. *Boreal.—Mont.* Jl–Au.

3. **H. Mackénzii** Richardson. Stems 2–4 dm. high, strigose above; leaflets 7–15, elliptic, 1–2.5 cm. long; finely grayish-strigose beneath, glabrate above; racemes short; calyx-tube 3 mm. long, the teeth 4–5 mm. long; corolla 18–20 mm. long, rose-purple; loments minutely strigose; internodes 5–7 mm. long, about 5 mm. wide. Meadows: Man.—Alta.—Yukon. *Boreal.—Subarct.* Je.

4. **H. cineráscens** Rydb. Stems 3–5 dm. high, strigose-canescent or glabrate below; leaflets 9–15, oblong or elliptic, 1–2 cm. long, silvery canescent-strigose on both sides; calyx-tube about 3 mm. long; teeth 5 mm. long; corolla reddish purple, about 15 mm. long; loment strigose; internodes 3–4, 6–8 mm. long, about 6 mm. wide. *H. canescens* Nutt.; not L. Dry hills: Sask.—N.D.—Wyo.—Utah—Ida.—Alta. *Submont.* Je–Jl.

42. ONÓBRYCHIS Scop. SANFOIN.

Herbs or undershrubs. Leaves odd-pinnate, without stipels. Flowers perfect, in spikes or racemes. Calyx-tube campanulate; teeth subulate, subequal, the lower smaller. Banner obcordate or obovate, sessile; wings shorter; keel equaling the banner. Stamens partly diadelphous, *i.e.*, the tenth stamen united with the rest at the middle, but free at the base. Legume flat, 1–2-seeded, half-round, or spirally coiled, not jointed, pectinately prickly.

1. **O. satìva** Lam. Perennial herb; stem erect, 3–5 dm. high; leaflets 11–23, elliptic; calyx-lobes subulate, subequal; corolla rose-colored, striate; wings about half as long as the banner; legume brown, obliquely semi-orbicular, keeled above, the lower suture and apex with short broad prickle-points. *O. Onobrychis* (L.) Rydb. Cultivated and occasionally escaped: N.Y.—Pa.—Colo.—Utah—B.C.

43. MEIBÒMIA Heister. TICK-TREFOIL, BEGGAR-TICKS.

Herbs (all ours), shrubs, or vines. Leaves alternate, pinnately 3-foliolate, rarely 1- or 5-foliolate; leaflets stipellate. Flowers perfect, in racemes or panicles. Calyx with bractlets at the base, campanulate; lobes unequal, the two upper ones more or less united. Corolla in ours reddish purple; banner more or less clawed; wings oblique, more or less coherent with the keel. Stamens monadelphous or partly diadelphous; anthers alike. Pod divided transversely into several indehiscent internodes, flat, in ours retrorsely hispid. *Desmodium* Desv.

Loment long-stipitate, straight or nearly so on the upper, deeply constricted on the lower suture.
  Peduncles arising from the base of the plant. 1. *M. nudiflora.*
  Peduncles terminating the leafy stems.
    Leaves clustered at the base of the peduncle. 2. *M. acuminata.*
    Leaves scattered on the stem. 3. *M. pauciflora.*
Loment constricted on both sutures.
  Stem trailing or reclining; leaflets suborbicular. 4. *M. Michauxii.*
  Stem erect or ascending; leaflets ovate to oblong.
    Leaves sessile or subsessile. 5. *M. sessilifolia.*
    Leaves, at least the lower ones, distinctly petioled.
      Internodes of the loments decidedly longer than broad, nearly semi-rhombic.
        Leaflets coriaceous, pubescent, strongly reticulate and whitish beneath. 6. *M. canescens.*
        Leaflets thin, not reticulate.
          Leaves glabrous. 7. *M. bracteosa.*
          Leaves pubescent. 8. *M. longifolia.*
      Internodes of the loment scarcely longer than broad.
        Loment long-stipitate, the stipe longer than the calyx, the internodes obliquely rhombic or subtriangular.
          Stem and lower surface of the leaflets neither densely soft-pubescent nor villous.
            Stem and leaves glabrous. 9. *M. paniculata.*
            Stem puberulent; leaflets slightly pubescent beneath. 10. *M. pubens.*
          Stem and lower surface of the leaflets densely soft-pubescent or villous. 11. *M. Dillenii.*
        Loment sessile or short-stipitate, the stipe shorter than the calyx, the internodes more or less oval.
          Internodes of the loment 3–7; corolla 1 cm. long; bracts conspicuous.
            Leaflets subcoriaceous, strongly reticulate; stipules large. 12. *M. illinoensis.*
            Leaflets membranous, not reticulate; stipules inconspicuous. 13. *M. canadensis.*
          Internodes of the loment 1–3; corolla 5–7 mm. long.
            Leaflets scabrous above, pubescent beneath, ciliate on the margins. 14. *M. rigida.*
            Leaflets glabrous. 15. *M. marilandica.*

1. **M. nudiflòra** (L.) Kuntze. Sterile stem 2–3 dm. high, with a cluster of leaves at the top; scape 3–8 dm. long, mostly naked; stipules setaceous, caducous; petioles 3–10 cm. long; leaflets oval or ovate, 2–8 cm. long, acuminate, with a blunt apex, glabrous or sparingly pubescent; corolla purplish; loment 2–5 cm. long; internodes 2–4, straight or slightly concave on the upper suture, 10–13 mm. long, glabrous. *D. nudiflorum* (L.) DC. [G]. Woods: Me.—Minn.—Kans.—La.—Fla. *E. Temp.* Jl–Au.

2. **M. acuminàta** (Michx.) Blake. Stem erect, 3 dm. high, not including the long peduncle; stipules setaceous, persistent; petioles 7–15 cm. long; leaflets 3, ovate, 5–13 cm. long, acuminate, softly pubescent with scattered hairs on both sides; bracts caducous; corolla rose-purple, 6–7 mm. long; loment long-stipitate, its internodes 2 or 3, 1 cm. long, concave above, rounded or semi-rhombic on the lower suture, uncinate-pubescent on the faces. *D. grandiflorum* [G]. *M. grandiflora* (Walt.) Kuntze [B]. Rocky woods: Me.—Sask.—Tex.—Fla. *E. Temp.* Je–S.

3. **M. pauciflòra** (Nutt.) Kuntze. Stem ascending or decumbent, 3–5 dm. long, sparingly pubescent; stipules caducous; petioles 6–10 cm. long; leaflets 3, ovate or subrhombic, 2.5–8 cm. long, acuminate or blunt, pubescent, paler beneath; racemes few-flowered; corolla white, 5–6 mm. long; loment long-stipitate, its internodes 1–3, rarely 4, rounded on the lower suture, 10–12 mm.

long, uncinate-pubescent. *D. pauciflorum* DC. [G]. Woods: N.Y.—Ont.—Mich. —Kans.—Tex.—Fla. *E. Temp.* Jl–Au.

4. **M. Michaùxii** Vail. Stem branched at the base, the branches prostrate, 5–10 dm. long, villous above; stipules ovate-cordate, striate, persistent, ciliate; petioles 3–5 cm. long; leaflets 3, nearly orbicular, 3–6 cm. long, sparingly pubescent or glabrate above, pubescent beneath; bracts ovate, caducous; corolla purplish, 8–10 mm. long; internodes of the loment 3–5, rounded on the upper, semi-rhombic on the lower suture, uncinate-pubescent. *D. rotundifolium* (Michx.) DC. [G]. Rocky woods: Me.—Minn.—La.—Fla. *E. Temp.* Jl–S.

5. **M. sessilifòlia** (Torr.) Kuntze. Stem 3–8 dm. long, striate, uncinate-pubescent; stipules linear, mostly caducous; leaves sessile; leaflets 3, linear or oblong, obtuse at each end, thick and reticulate, 2.5 cm. long, downy-pubescent beneath; bracts caducous; corolla purplish, turning greenish, 4–5 mm. long: loment rather short-stipitate, its internodes 1–3, obliquely obovate, hispid, 5 mm. long. *D. sessilifolium* T. & G. [G]. Copses: Mich.—Kans.—Tex.—Miss. *Allegh.—Carol.—Ozark.* Jl–S.

6. **M. canéscens** (L.) Kuntze. Stem 6–10 dm. high, hirsute and densely pubescent with hooked hairs; stipules ovate-cordate, 8–15 mm. long, persistent; petioles 3–10 cm. long; leaflets 3, broadly ovate or subrhombic, rough on both sides, pale and reticulate beneath; bracts caducous; corolla violet-purple, 1 cm. long; loment long-stipitate, its internodes 4–6, elongately obliquely subrhomboid, 8–15 mm. long, uncinate-pubescent. *D. canescens* (L.) DC. [G]. Moist ground and borders of woods: Ont.—Minn.—Tex.—Fla. *E. Temp.* Jl–S.

7. **M. bracteòsa** (Michx.) Kuntze. Stem 1 m. high or more, glabrous or nearly so; stipules lanceolate, 8–15 mm. long, caducous; petioles 3–8 cm. long; leaflets 3, ovate or ovate-lanceolate, 5–15 cm. long, acuminate or cuspidate, glabrous or nearly so; bracts glabrous, caducous; corolla lilac, 1 cm. long; internodes of the loment 3–7, elongate, semi-rhombic, 1 cm. long, reticulate. *D. bracteosum* (Michx.) DC. [G]. Thickets: Me.—Minn.—Tex.—Fla. *E. Temp.* Au–S.

8. **M. longifòlia** (T. & G.) Vail. Stem erect, stout, 6–8 dm. high, angled, striate, mostly pubescent in lines; stipules ovate-lanceolate, 5 mm. long; persistent; leaflets lanceolate or lance-ovate, acuminate, 2.5–5 cm. long, scabrous or glabrate above, appressed-pubescent beneath; flowers numerous;, calyx-lobes attenuate; corolla lilac-purple, 1 cm. long; loment 4–6-jointed, pubescent with hooked hairs; internodes longer than broad, rounded on the upper suture, more triangular on the lower, 8–10 mm. long. *Desmodium bracteosum longifolium* (T. & G.) Robins. [G]. Plains and prairies: Ill.—Kans.—La.—Ala. *Prairie—Canad.*

9. **M. paniculàta** (L.) Kuntze. Stem erect, 5–10 dm. high, nearly glabrous; stipules setaceous, caducous; petioles 1–3.5 cm. long; leaflets 3, oval to lanceolate, obtuse, glabrous, paler beneath, 3–5 cm. long; bracts caducous; corolla purplish, turning greenish, 5–6 mm. long; internodes of the loment obliquely obovate or subrhombic, 5–8 mm. long, minutely pubescent. *D. paniculatum* (L.) DC. [G]. Copses: Me.—Minn.—Neb.—Tex.—Fla. *E. Temp.* Jl–S.

10. **M. pùbens** (T. & G.) Rydb. Stem erect, 5–10 dm. high, puberulent; stipules setaceous, caducous; leaflets 3, oblong-oblanceolate, coriaceous in age, scabrous above, sparingly pubescent beneath; flowers as in the preceding; internodes of the loment more rhombic. *D. paniculatum pubens* T. & G. [B, G]. Sandy soil: N.J.—Kans.—Tex.—Fla. *E. Temp.* Jl–S.

11. M. Dillénii (Darl.) Kuntze. Stem erect, about 6 dm. high, sulcate, glabrous or nearly so; stipules subulate, caducous; petioles 2.5–5 cm. long; leaflets 3, oblong to ovate, obtuse, 4–10 cm. long, thin, scabrous or sparingly hairy above, softly and finely pubescent beneath; bracts small, caducous; corolla purple, 6–8 mm. long; internodes of the loment 3–6, semirhombic, 5–12 mm. long. *D. Dillenii* Darl. [G]. Woodlands: Me.—Minn. (?)—Kans.—Tex.—Fla. *E. Temp.* Je–S.

12. M. illinoénsis (A. Gray) Kuntze. Stem 5–12 dm. high, uncinate-pubescent; leaflets lanceolate or ovate-lanceolate, hispidulous above, cinereous beneath, 2.5–9 cm. long; flowers paniculate, 6–9 mm. long, purple; loments 1–2.5 cm. long; internodes 3–6, oval or orbicular, densely uncinate. *D. illinoense* A. Gray [G]. Prairies: Ill.—Kans.—S.D. *Prairie—Plain.* Je–S.

13. M. canadénsis (L.) Kuntze. *Fig. 329.* Stem erect, 5–20 dm. high, pubescent; leaflets elliptic or oblong-lanceolate, appressed-pubescent and paler beneath, hispidulous or glabrate above; flowers paniculate, 11–17 mm. long, purple; loments about 2.5 cm. long; internodes 3–5, somewhat triangular, straight or nearly so on the upper suture, uncinate-pubescent. *Desmodium canadense* (L.) DC. [G]. Thickets and river banks: N.B.—N.C.—Okla.—S.D.—Man. *E. Temp.—Plain—Submont.* Jl–S.

14. M. rígida (Ell.) Kuntze. Stem erect, rigid, striate, minutely pubescent, 6–9 dm. high; stipules small, caducous; upper leaves nearly sessile; leaflets 3, ovate-oblong or elliptic, obtuse, 2.5–8 cm. long, yellowish green and scabrous above, pubescent and reticulate beneath; bracts very small, caducous; corolla purplish, 5–6 mm. long; internodes of the loment 2 or 3, rarely 4, obliquely oval, 3–5 mm. long. *D. rigidum* (Ell.) DC. [G]. Hills and sandy fields: N.H.—Mich.—Neb.—Tex.—Fla.; Mex. *E. Temp.—Trop.* Jl–O.

f.329.

15. M. marilándica (L.) Kuntze. Stem erect, slender, striate, glabrous or nearly so, 6–9 dm. high; leaves crowded; stipules subulate, caducous; leaflets 3, ovate to suborbicular, obtuse, 1–2.5 cm. long, glabrous or nearly so, glaucous beneath; corolla purplish, 2–4 mm. long; internodes of the loment 1–3, obliquely oval, 5–10 mm. long. *D. marilandicum* (L.) DC. [G]. Copses: Mass.—Minn. (?)—La.—Fla. *E. Temp.* Jl–S.

44. LESPEDÈZA Michx. BUSH-CLOVER.

Herbs or shrubs. Leaves pinnately 3-foliolate, rarely 1-foliolate, with often veiny leaflets, mucronate stipules, and no stipels. Flowers perfect, petaliferous or some of them apetalous. Calyx deeply 5-cleft, the lobes nearly equal, or the two upper ones more united. Corolla white, purplish, or yellow; the banner broad, clawed; wings and keel-petals curved. Stamens 10, diadelphous or the upper stamen partially united with the sheath. Anthers alike. Style filiform, incurved. Pod of 1 or 2, lenticular, indehiscent internodes, prominently veined.

Perennials; stipules subulate, bracts minute; calyx-lobes narrow.
- Flowers both petaliferous and apetalous; corolla purplish.
  - Flower-clusters slender-peduncled, surpassing the leaves.
    - Stems branched at the base, the branches trailing or procumbent. — 1. *L. repens.*
    - Stems erect.
      - Petaliferous flowers in spike-like or head-like clusters.
        - Calyx-lobes 5–7 mm. long, as long as the pod or longer. — 2. *L. Manniana.*

Calyx-lobes 3–4 mm. long, shorter than the pod. — 3. *L. Nuttallii.*
Petaliferous flowers paniculate.
Inflorescence close, short-peduncled; calyx-lobes half as long as the pod. — 4. *L. violacea.*
Inflorescence lax, long-peduncled; calyx-lobes one-fourth as long as the pod. — 5. *L. prairea.*
Flower-clusters sessile or nearly so.
Leaflets densely woolly beneath. — 6. *L. Stuvei.*
Leaflets glabrous or strigose beneath.
Leaflets suborbicular to elliptic. — 7. *L. frutescens.*
Leaflets linear or nearly so. — 8. *L. virginica.*
Flowers all petaliferous; calyx-lobes fully as long as the pod; corolla whitish.
Leaflets suborbicular to broadly oblong.
Peduncles longer than the leaves; spike cylindric. — 9. *L. hirta.*
Peduncles shorter than the leaves; spike subglobose. — 10. *L. capitata.*
Leaflets linear or linear-oblong.
Spike short, oblong-capitate. — 11. *L. longifolia.*
Spike elongate, lax. — 12. *L. leptostachya.*
Annuals; stipules and bracts ovate, scarious; calyx-lobes as broad as long. — 13. *L. striata.*

1. **L. rèpens** (L.) Bart. Stem procumbent or trailing, 2–7 dm. long; leaflets oval or obovate, 6–15 mm. long, obtuse or retuse; inflorescence lax, corolla violet-purple, 4–6 mm. long; pod oval-orbicular, 3–4 mm. long, finely pubescent. Sandy soil: Conn.—Minn.—Tex.—Fla. *E. Temp.* Au–S.

2. **L. Manniàna** Mack. & Bush. Stem 4–9 dm. high, erect, densely pubescent; leaflets oblong to cuneate, 1–4 cm. long, mucronulate, sparingly appressed-pubescent above, densely so beneath; corolla 5–6 mm. long; pod oval, 5–6 mm. long, minutely pubescent. Barrens or open woods: Mich.—Kans.—Ark. *Allegh.—Ozark.*

3. **L. Nuttállii** Darl. Stem erect, 5–10 dm. high, pubescent; leaflets leathery, obovate to suborbicular, 1–2 cm. long, obtuse or notched at the apex, dark green and glabrate above, closely pubescent beneath; inflorescence capitate or short-spicate; corolla 5–7 mm. long; pod oval, acute at each end, 5–6 mm. long, densely pubescent. Dry soil: Mass.—Mich.—Kans.—Fla. *E. Temp.* Au–S.

4. **L. violàcea** (L.) Pers. Stem usually branched at the base, the branches erect, ascending, or spreading, 3–10 dm. long; leaflets rather thin, oblong-elliptic to oval, 1–2 cm. long, obtuse or retuse at the apex, appressed-pubescent beneath; corolla violet-purple, 6–10 mm. long; pod ovate, 4–6 mm. long, acute, sparingly finely pubescent. Dry soil: N.H.—Minn.—Kans.—La.—Fla.; Mex. *E. Temp.—Trop.* Au–S.

5. **L. praìrea** (Mack. & Bush) Britton. Stem ascending, 3–8 dm. high, branched, appressed-pubescent above; leaflets obovate, sometimes obcordate, 6–18 mm. long, glabrous above, finely appressed-pubescent beneath; corolla purple, 6–8 mm. long; pod broadly oval, 3–4 mm. long, prominently veined. *L. violacea prairea* Mack. & Bush [G]. Dry prairies: Iowa—Neb.—Tex.—Ark. *Prairie—Texan.* Jl–S.

6. **L. Stùvei** Nutt. Stem upright or spreading, 3–12 dm. high, very leafy, downy, with spreading pubescence, simple or with a few branches; leaves short-petioled; leaflets firm, elliptic to orbicular, 1–3 cm. long; peduncles very short; flower-clusters crowded; calyx 3–5 mm. long, shorter than the pod; corolla violet-purple, 4–6 mm. long. Dry soil: Mass.—Mich.—Kans.—Okla.—Va. *Allegh.* Au–S.

7. **L. frutéscens** (L.) Britton. Stem erect or ascending, 3–10 dm. high, appressed-pubescent or glabrate; leaflets oval to oblong, 1–3.5 cm. long, dark green and glabrous above, paler and slightly pubescent beneath; flower-clusters

usually crowded near the summit of the plant; corolla violet-purple, 4–6 mm. long; pod ovate-oblong, about 4 mm. long, acute, pubescent. Dry soil: Me.—Minn.—Tex.—Fla. *E. Temp.* Au–S.

8. **L. virgínica** (L.) Britton. Stem erect or ascending, 4–10 dm. high, slender; leaflets linear or oblong-linear, 1–3.5 cm. long, truncate or obtuse, rarely acute at the apex, glabrous on both sides, or finely pubescent beneath; flower-clusters crowded in the upper axils; corolla violet, 4–6 mm. long; pod ovate to suborbicular, 4 mm. long, obtuse or acute, pubescent or glabrous. *L. angustifolia* Ell. Dry soil: N.H.—Minn.—Kans.—Tex.—Fla. *E. Temp.* Au–S.

9. **L. hírta** (L.) Ell. Stem erect or ascending, 5–12 dm. high, branched above; leaflets oval or suborbicular, 1–4 cm. long, rounded at each end or notched at the apex, pubescent; corolla yellowish, 6 mm. long; the banner sometimes purple-spotted; pod oval or obovate, 7–8 mm. long, acute. Dry soil: Me.—Minn.—La.—Fla. *E. Temp.* Au–O.

10. **L. capitàta** Michx. Stem rigid, 5–15 dm. high, simple to the inflorescence, velutinous, usually with spreading hairs; leaflets oblong or oval, 2.5–3.5 cm. long, obtuse or acute at both ends, many times as long as the petioles, silky or silvery-pubescent; corolla yellowish white, 6 mm. long, the banner sometimes with a purple spot; pod oblong-ovate, 5–5.5 mm. long, acute, pubescent. Dry soil: Me.—S.D.—Kans.—La.—Fla. Au–S.

11. **L. longifòlia** DC. Stem stout, 5–10 dm. high, densely velutinous, with ascending hairs; leaves subsessile; leaflets linear or linear-oblong, 2–6 cm. long, 4–8 mm. wide, glabrous and reticulate above, densely appressed-silky beneath, rather thick; racemes dense, short-peduncled or subsessile; calyx silky, the tube 3 mm. long, the teeth linear-lanceolate, attenuate, 5–7 mm. long, exceeding the white corolla. Open woods and fields: Ky.—Iowa—La.—Ala. *Ozark.* Au–S.

12. **L. leptostàchya** Engelm. Stem erect, 3–10 dm. high, silvery-pubescent; leaflets linear, 2.5–3 cm. long, silvery-pubescent, longer than the short petioles; peduncles equaling or exceeding the leaves; corolla as in *L. capitata;* pod ovate, about 3 mm. long, pubescent. Prairies: Ill.—Minn.—Iowa. *Prairie.* Au–S.

13. **L. striàta** (Thunb.) H. & A. Stem erect, diffusely branched, 1–4 dm. high, sparingly pubescent, slender; leaflets oblong or oblong-obovate, 0.8–2.5 cm. long, ciliate, longer than the petioles; flowers solitary or 2 or 3 together, subsessile in the axils; corolla white, pink or purple, 6–8 mm. long; pod suborbicular, acute, surpassing the sepals. Roadsides and fields: Va.—Ill.—Kans.—La.—Fla.; nat. from Asia. Jl–Au.

45. STYLOSÁNTHES Sw. PENCIL-FLOWER.

Rigid perennial herbs, with bristly or sticky pubescence. Leaves alternate, pinnately 3-foliolate, the leaflets prominently veined. Flowers of two kinds, the one petaliferous and sterile, the other without corolla and stamens but fertile, in axillary or terminal heads or spikes. Calyx with a slender tube, 5-lobed, but disposed to be 2-lipped by the union of the upper 4 lobes. Corolla yellow or orange; banner with spreading blade, the keel incurved. Stamens monadelphous, the anthers alternately longer and attached basally, and shorter and versatile. Loment with 1 or 2 internodes, the lower one, when present, empty, the upper opening at the apex.

1. **S. biflòra** (L.) B. S. P. Stem tufted, 2–5 dm. high, flexuose, villous below the stipules; leaflets elliptic, lanceolate or oblanceolate, 1–3.5 cm. long; corolla orange, 8 mm. long; internodes of the pod 2, the terminal one with a subbasal beak. *S. elatior* Sw. Dry sandy soil: N.Y.—Kans.—Tex.—Fla. *E. Temp.* Je–S.

46. **ÀPIOS** Moench. GROUND-NUT.

Twining vines, with large tuberous roots. Leaves alternate, pinnate, with small stipules. Flowers perfect, in racemes. Upper two calyx-teeth united, the lateral ones very small, and the lowest one longer. Corolla red, purple, or brown; banner reflexed; keel elongate, spirally twisted. Stamens diadelphous. Style glabrous. Pod elongate, curved, flattened. *Glycine* L.

1. **A. tuberòsa** Moench. *Fig. 330.* Stem 3–20 dm. long; leaflets 5–9, ovate to lanceolate, 3–10 cm. long, acute or acuminate; calyx-tube 2–4 mm. long; corolla brownish-purple; banner with a broad blade, about 1 cm. long; pod linear, 6–12 cm. long. *A. Apios* (L.) MacMill. [B]. *Glycine Apios* L. [B, B]. Thickets: N.B.—Fla.—Tex.—Colo.—Minn. *E. Temp.* —*Plain*—*Submont.*

47. **GALÁCTIA** P. Br. MILK PEA.

Perennial herbs, with postrate or twining stems. Leaves pinnately trifoliolate, with deciduous stipules; leaflets entire, firm. Flowers perfect, in axillary racemes or panicles. Calyx with unequal lobes, the upper two united, the lowest one the longest. Corolla white, violet, or red; banner broad, spreading; wings oblong or obovate; keel-petals shorter, coherent to the wings. Stamens diadelphous or monadelphous; anthers uniform. Pod elongate, flat. Plants sometimes with underground cleistogamous flowers.

Stem puberulent or nearly glabrous, prostrate; pods slightly hairy. 1. *G. regularis.*
Stem climbing, retrorsely hirsute; pod very hirsute. 2. *G. mississippiensis.*

1. **G. regulàris** (L.) B. S. P. Stem 4–10 dm. long, retrosely puberulent or glabrate; leaflets oblong or ovate-oblong, 2–4.5 cm. long, often emarginate, glabrous above, appressed-pubescent beneath; panicle 3–8 cm. long, rigid, erect; calyx glabrate, 6–9 mm. long; corolla reddish purple, 12–17 mm. long; pod 2.5–3 cm. long, appressed-pubescent, but glabrate; seeds brown, oval. *G. glabella* Michx. Sandy soil: N.Y.—Kans.—Okla.—Miss.—Fla. *E. Temp.* Jl–Au.

2. **G. mississippiénsis** (Vail) Rydb. Stem 4–12 dm. long, retrorsely hirsute; leaflets oval or elliptic, 2–4 cm. long, rounded or slightly cordate at the base, rounded or retuse, rather firm, reticulate, equally pubescent on both sides, paler beneath; panicle 2–4 dm. long; calyx 5–6 mm. long, densely pubescent; corolla 8–10 mm. long, lilac or pink-purple; pod linear, 3–4 cm. long; seeds truncate at each end. *G. volubilis mississippiensis* Vail. Sandy dry soil: Ill.—Kans.—Tex.—La. Je–Jl.

48. **AMPHICÁRPA** Ell. HOG-PEANUT.

Perennial twining herbs, the rootstock bearing small tubers. Leaves alternate, pinnately 3-foliolate, with stipules, the leaflets broadest below the middle. Flowers of two kinds, the upper in axillary racemes, perfect. Calyx oblique, 5-lobed. Corolla white, blue, or violet; banner spreading; wings and keel-petals falcate, the latter obtuse, incurved. Stamens 10, diadelphous; anthers uniform. Pod narrow, curved; style slender, incurved, glabrous. Flowers of the filiform branches near the base with no or rudimentary corollas and few stamens, but fertile. Fruit subterranean, fleshy. *Falcata* Gmel.

Plant glabrous or slightly pubescent; bracts shorter than the pedicels; pod pubescent on the margins. 1. *A. comosa.*
Plant densely villous; bracts longer than the pedicels; pod pubescent throughout. 2. *A. Pitcheri.*

1. **A. comòsa** (L.) Nieuwl. & Lunell. *Fig. 331.* Stem 1–3 m. high; leaflets ovate or ovate-lanceolate, 3–8 cm. long, acute or apiculate; bracts lanceolate; calyx 8–10 mm. long; corolla white or purplish, 10–13 mm. long; pod of the upper flowers 2.5–3 cm. long, abruptly beaked. *A. monoica* (L.) Ell. [G]. *F. comosa* Kuntze [B]. Thickets: N.B.—Man.—Neb.—La.—Fla. *E. Temp.* Au–S.

2. **A. Pítcheri** T. & G. Stem 2–3 m. high, brown-hirsute; ovate or rhombic, 3–8 cm. long, acute or acuminate; bracts ovate; calyx 1 cm. long; corolla purple, 12–15 mm. long; pod 3–3.5 cm. long, gradually acuminate. *F. Pitcheri* Kuntze [B]. Thickets: Mass.—Minn.—S.D.—Tex.—Tenn. *Allegh.—Ozark.* Au–S.

49. CLITÒRIA L. Butterfly Pea.

Perennial herbs or shrubs, with erect or twining stems. Leaves alternate, pinnately 3-foliolate; leaflets entire, broadest below the middle. Flowers perfect, axillary or racemose. Corolla white, blue, purple, or red; banner large, notched; wings falcate; keel-petals shorter, incurved, acute. Stamens diadelphous or partly monadelphous. Ovary stipitate; style slender, incurved, flattened at the apex. Pod narrow, flattened.

1. **C. Mariàna** L. Stem 3–12 dm. high, branched; leaflets ovate-lanceolate, glabrous, mucronate at the apex, rounded or cordate at the base; peduncles shorter than the petioles; calyx 10–15 mm. long; lobes minute, ovate, acuminate; corolla lilac or pale blue, 5–6 cm. long; banner striped with magenta at the center; pod linear or linear-oblong, 2.5–3 cm. long. Sandy soil: N.Y.—Iowa—Tex.—Fla. *E. Temp.* Je–Jl.

50. PHASÈOLUS L. Bean.

Perennial herbs, with prostrate or twining stems. Leaves alternate, pinnately 3-foliolate, with persistent stipules; leaflets entire or lobed. Flowers in axillary racemes or panicles. Calyx 5-lobed, the lobes nearly equal, the upper two partly united. Corolla often variegated; banner suborbicular, spreading or reflexed; wings broadest above the middle; keel-petals spirally twisted, with long obtuse beak. Stamens diadelphous; anthers uniform. Style bearded; stigma oblique or lateral. Pod slightly flattened. Seeds reniform or suborbicular.

1. **P. polystàchyus** (L.) B. S. P. Stem climbing, 1–4 m. high; leaflets minutely pubescent, ovate or rounded-ovate, 4–10 cm. long, acuminate, rounded or subcordate at the base; racemes simple or compound, exceeding the leaves; corolla purple or whitish, long; pod falcate, 4–8 cm. long, glaucous. *P. perennis* Walt. Thickets: Conn.—Que.—Minn.—Neb.—La.—Fla. *E. Temp.* Jl–S.

51. STROPHOSTÝLES Ell. Wild Bean.

Herbaceous vines, twining or trailing. Leaves alternate, pinnately 3-foliolate, with stipules. Flowers perfect, in dense clusters, on long axillary peduncles. Calyx subequally 5-toothed, or the upper two teeth partially united. Corolla white or purplish; banner with a broad blade; keel curved. Stamens diadelphous. Style bearded, bent. Pod narrow, flat, straight or nearly so. Seeds truncate.

Leaflets less than twice as long as broad.
  Stem 1–2 m. long, mostly trailing; leaflets, at least some of them, lobed. 1. *S. helvola.*
  Stem 3–9 m. long, high-climbing; leaflets all entire. 2. *S. missouriensis.*
Leaflets at least twice as long as broad, entire. 3. *S. leiosperma.*

1. **S. hélvola** (L.) Ell. Annual, often villous; stem branching, 3–12 dm. long; leaflets rhombic-oval to 3-lobed, 2–4 cm. long; peduncles twice as long as the leaves; calyx-lobes abruptly pointed; corolla purple, becoming greenish; pod linear, subterete, 5–7 cm. long. *S. angulosa* Ell. Sandy soil: Que.—Minn.—Tex.—Fla. *E. Temp.* Jl–O.

2. **S. missouriénsis** (S. Wats.) Small. Annual; stem 3–10 dm. high, climbing, retrorsely hispid; leaflets rhombic-ovate, 3–8 cm. long, shining above; peduncles longer than the leaves; calyx sparingly pubescent; corolla pink or bluish; pod 7–9 cm. long. *S. angulosa missouriensis* S. Wats. [G]. Alluvial soil: Mo.—Kans.—Ark. *Ozark.* Au–S.

3. **S. leiospérma** (T. & G.) Piper. *Fig. 332.* Annual, stem 3–10 dm. long; leaflets linear to lanceolate, 1.5–3.5 cm. long, rarely sinuately lobed; peduncles longer than the leaves; calyx 1–1.5 mm. long; lobes subequal; corolla pale purple; pod 2–3 cm. long, strigose. *Phaseolus pauciflorus* Benth; not Sessé & Moc. River banks: Minn.—Ind.—Iowa—Tex.—Colo. *Prairie—Plain.*

f.332.

52. VÍCIA L. Vetch, Wild Pea.

Perennial or annual herbaceous vines. Leaves alternate, abruptly pinnate, usually with tendrils or these represented by a tip. Flowers axillary, racemose or sessile. Calyx somewhat oblique and gibbous at the base, the upper two teeth shorter. Banner obovate or oblong, emarginate; wings obliquely oblong, adherent to the curved keel. Stamens diadelphous (9 and 1); tube oblique at the summit; anthers all alike. Style slender, with a tuft or ring of hairs at the summit. Pod flat, dehiscent, 2-valved, few- or several-seeded.

Peduncles very short or none; flowers 1–3 in the axils, subsessile; introduced annuals.
  Flowers 2–3 cm. long; plant pubescent, glabrate only in age; leaflets cuneate or oblong. — 1. *V. sativa.*
  Flowers 1–1.8 mm. long; plant glabrous or nearly so; leaflets of the upper leaves usually linear. — 2. *V. angustifolia.*
Peduncles evident.
  Flowers 5–40; corolla usually 1 cm. long or more; perennials, except *V. villosa.*
    Racemes dense, 1-sided, 15–40-flowered; introduced species.
      Plant finely pubescent or glabrate, perennial. — 3. *V. Cracca.*
      Plant villous, annual. — 4. *V. villosa.*
    Racemes lax, 3–20-flowered; native perennials.
      Corolla 1.5–2 cm. long; flowers 5–10.
        Leaflets linear to oblong; stipules usually narrow, semi-sagittate and often entire.
          Leaves decidedly pubescent, rather thick and strongly veined.
            Stem low; leaflets linear to oblong. — 5. *V. trifida.*
            Stem tall; leaflets oblong or elliptic, only those of the lower leaves sometimes linear. — 6. *V. oregana.*
          Leaves glabrous or slightly pubescent when young.
            Leaflets elongate, narrowly linear; plant low and erect. — 7. *V. sparsifolia.*
            Leaflets, at least the upper ones, oblong or linear-oblong; plant tall and climbing. — 8. *V. dissitifolia.*
        Leaflets broad; stipules broadly semi-sagittate or semi-orbicular in outline, sharply toothed. — 9. *V. americana.*
      Corolla about 1 cm. long; flowers 8–20. — 10. *V. caroliniana.*
  Flowers 1–6, 3–8 mm. long.
    Corolla 6–8 mm. long; native perennial. — 11. *V. ludoviciana.*
    Corolla 3–4 mm. long; introduced annual. — 12. *V. hirsuta.*

1. **V. satìva** L. Annual, pubescent, in age glabrate; leaflets 8–16; oblong to obovate, truncate to emarginate and mucronate at the apex, 1.5–3 cm. long, 5–13 mm. wide; flowers usually 2 in each axil; corolla purple or rose-colored; pod pubescent when young, 4–8 cm. long, torulose. Waste places and cultivated ground: Minn.; escaped from cultivation; native of Eur. Jl–Au.

2. **V. angustifòlia** (L.) Reichard. Annual; stem glabrous, 3–5 dm. high; leaflets 5–11, those of the lower leaves oblong, truncate, those of the upper linear, mucronate, 1.5–3 cm. long; corolla purple or rose-colored; pod 4–5.5 cm. long, 5–7 mm. wide. Fields, meadows, and waste places: N.S.—Fla.—Mo.—Minn.; Ida.; nat. from Eur. Ap–Jl.

3. **V. Crácca** L. Stem weak, 3–12 dm. high; stipules narrowly semi-sagittate, entire; leaflets 18–24, thin, linear or linear-oblong, mucronate, 1.5–2 cm. long, finely pubescent; corolla bluish-purple, 11–13 mm. long; pod 1.8–2.5 cm. long. Dry soil: Newf.—N.J.—Kans.—Ida.—B.C.; nat. from Eur. Je–Au.

4. **V. villòsa** Roth. Annual; stem 3–10 dm. high, villous; leaflets 8–12, elliptic or oblong, villous, 1.5–2 cm. long, mucronate; corolla about 15 mm. long, claw of the banner half as long as the blade. Waste places and fields: Me.—Sask.—Kans.—Pa.; rarely escaped from cultivation; nat. of Eur. My–S.

5. **V. trífida** Dietr. Stem low, 1–4 dm. high, more or less pubescent when young; stipules broadly semi-sagittate, often toothed; leaflets 8–12, linear or narrowly oblong, acute at both ends or truncate at the apex, mucronate or cuspidate, dark green; racemes 3–6-flowered; corolla bluish purple, 15–18 mm. long; pod 2 cm. long, puberulent or glabrate. *V. tridentata* Schw. *V. caespitosa* A. Nels., a depauperate form. *V. callianthema* Greene. Wet meadows: w Ont.—Kans.—Tex.—Nev.—B.C. *Prairie—Plain—Mont.* My–Au.

6. **V. oregàna** Nutt. Stem 3–8 dm. high, somewhat pubescent when young, or glabrate, angled; stipules usually narrowly semi-sagittate, mostly toothed; leaflets 8–12, elliptic or those of the lower leaves linear, often truncate and somewhat toothed at the apex, 1–2.5 cm. long, dark green, appressed-pubescent, at least below; raceme 3–8-flowered; corolla pinkish or bluish-purple, 15–20 mm. long; pod about 3 cm. long. *V. vexillaris* Greene. *V. americana truncata* Britt. [B]. Meadows: Sask.—N.M.—Calif.—B.C. *W. Temp.—Plain—Mont.* My–Au.

7. **V. sparsifòlia** Nutt. Stem low, 2–5 dm. high, glabrous; leaflets 8–12, narrowly linear, firm, strongly veined, acute at both ends, mucronate, glabrous or nearly so, 2–4 cm. long, 1–3 mm. wide; raceme 2–6-flowered; corolla bluish purple, about 18 mm. long; pod 3 cm. long, 6–7 mm. broad. *V. linearis* (Nutt.) Greene [B]. Prairies and plains: Man.—Okla.—N.M.—Calif.—B.C. *W. Temp.—Plain—Mont.* My–Au.

8. **V. dissitifòlia** (Nutt.) Rydb. Stem tall, slender, 3–7 dm. high, glabrous; leaflets 8–12, linear or narrowly oblong, 1.5–5 cm. long, 2–5 mm. wide, rather thin, not strongly veined, acute at both ends, mucronate; raceme 3–6-flowered; corolla violet-purple, 15–18 mm. long. *Lathyrus dissitifolius* Nutt. Confused with and grading into *V. oregana*. Valleys: w Neb.—Colo.—Utah. *Plain—Mont.* Je–Au.

9. **V. americàna** Muhl. *Fig. 333.* Stem glabrous or nearly so, 3–10 dm. high; leaflets 8–12, oval or elliptic, or those of the lower leaves linear-oblong, thin, glabrous, usually rounded at both ends or retuse at the apex, mucronate, 1.5–3.5 cm.

long, not strongly veined; racemes shorter than the leaves, 3–9-flowered; corolla bluish purple, 15–20 mm. long; pod glabrous, 2.5–3 cm. long. Meadows: N.B.—Va.—Ariz.—B.C. *Temp.—Plain—Mont.* My–Au.

10. **V. caroliniàna** Walt. Stem branched at the base, strigillose or glabrous, spreading or climbing, 4–10 dm. long; leaflets 8–16, oblong or elliptic, mucronate, 1–2 cm. long; peduncles longer than the leaves; calyx-teeth very short; corolla about 1 cm. long, white, the keel tipped with blue; pod oblong, 2.5–3 cm. long. Open woods and river banks: Ont.—Minn.—Kans.—Ga. *E. Temp.* Ap–Au.

11. **V. ludoviciàna** Nutt. Stem branched at the base, the branches decumbent or climbing, 4–10 dm. long; leaflets 6–12, oblong or oval, 7–25 mm. long, glabrous or sparingly pubescent when young, notched or rounded at the apex; racemes shorter than the leaves; calyx-teeth as long as the whitish tube; corolla bluish; pod oblong, 2.5–3 cm. long. River banks and dry soil: Fla.—Mo.—Kans.—Tex. *Austral.* Ap–My.

12. **V. hirsùta** (L.) Koch. Sparingly pubescent annual; stem branched at the base, 3–7 dm. long; leaflets 6–12, linear or oblong, 6–15 mm. long, truncate or notched at the apex; peduncles shorter than the leaves; calyx-lobes subulate, as long as the tube; corolla whitish or pale purplish blue; pod oblong, 6–8 mm. long, usually 2-seeded, pubescent. Waste places: N.B.—Alta.—Ohio—Ga.; nat. from Eur. My–S.

53. LÁTHYRUS (Tourn.) L. VETCHLING.

Herbaceous vines, rarely erect herbs, mostly perennials, with horizontal rootstocks. Leaves alternate, abruptly pinnate, the rachis usually tendril-bearing at the apex, or the tendril represented by a tip or an appendage. Flowers racemose (raceme sometimes 1-flowered), perfect. Calyx obliquely campanulate, gibbous at the base, its teeth nearly equal, or the upper shorter. Banner obovate, emarginate, clawed; wings obliquely oblong, adherent to the shorter curved keel. Stamens diadelphous (9 and 1), monadelphous below; tube usually truncate at the apex. Style curved, flattened, hairy along its inner side. Legumes linear, more or less flattened, dehiscent, 2-valved. Seeds usually several.

Perennials.
  Tendrils much reduced; stem mostly erect.
    Plant glabrous or nearly so.
      Flowers ochroleucous. — 1. *L. Hapemanii.*
      Flowers purple. — 2. *L. stipulaceus.*
    Plant decidedly villous-pubescent. — 3. *L. incanus.*
  Tendrils usually well developed; stem more or less climbing.
    Corolla purple.
      Stipules small, less than half as long as the leaflets.
        Flowers 2–8; leaflets linear to elliptic, rarely oval, 2–5 pairs.
          Stem distinctly winged.
            Leaflets linear or linear-lanceolate; corolla 12–16 mm. long; plant glabrous. — 4. *L. palustris.*
            Leaflets elliptic or lance-elliptic; corolla 18–22 mm. long; young parts of the plant pubescent. — 5. *L. macranthus.*
          Stem not winged or slightly so; leaflets oblong or broader.
            Corolla 1–1.5 cm. long; plant climbing; leaflets 4–6. — 6. *L. myrtifolius.*
            Corolla 2.5–4 cm. long; plant mostly erect; leaflets mostly 6–10. — 7. *L. decaphyllus.*
        Flowers 8–25; leaflets broadly oval, 4–6 pairs. — 8. *L. venosus.*
      Stipules large, foliaceous, almost as large as the leaflets. — 9. *L. maritimus.*
    Corolla ochroleucous or yellow.
      Leaflets 6–8, oval. — 10. *L. ochroleucus.*
      Leaflets 2, lanceolate or linear-lanceolate. — 11. *L. pratensis.*
Annual; leaflets 2; flowers purplish. — 12. *L. pusillus.*

1. **L. Hapemánii** A. Nels. Stem 1–2 dm. high, from a slender rootstock, glabrous; leaflets 6–12, linear, nerved, glabrous, 1.5–3 cm. long, 2–4 mm. wide, peduncles 2–4 cm. long; flowers 3–6; calyx about 5 mm. long; corolla lemon-yellow or ochroleucous, 2.5 cm. long; pod oblanceolate, 4–5 cm. long, glabrous. Dry prairies: Neb. My–Je.

2. **L. stipulàceus** (Pursh) Butters & St. John. *Fig. 334.* Stem erect, glabrous or nearly so, 2–4 dm. high; leaflets linear or lance-linear, 1.5–4 cm. long, 3–5 mm. wide, glabrous; peduncles about 5 cm. long; raceme 3–5-flowered; corolla purple, 2–2.5 cm. long; pod 4–5 cm. long. *L. ornatus* Nutt. [B, R]. Plains and prairies: S.D.—Okla.—Colo.—Wyo. *Plain—Submont.* My–Jl.

f.334.

3. **L. incànus** (Smith & Rydb.) Rydb. Stem erect, 1–3 dm. high, more or less densely villous with short hairs; leaflets 3–4 pairs, linear or linear-oblong, 1–4 cm. long, 3–5 mm. wide, usually densely villous with short hairs; corolla as in the preceding; pod 3–4 cm. long. *L. ornatus incanus* Smith & Rydb. Dry sandy plains: Neb.—Colo.—Wyo. *Plain—Submont.* My–Jl.

4. **L. palústris** L. Stem glabrous or nearly so, 3–6 dm. high; leaflets 2–4 pairs, 2.5–6 cm. long; tendrils usually branched; inflorescence 7–15 cm. long, 2–6-flowered; corolla purple, 12–15 mm. long; pod linear, 4–5 cm. long. Wet places: Lab.—N.Y.—S.D.—Ore.—Alaska; Eurasia. *Boreal.—Plain—Submont.* My–Au.

5. **L. macránthus** (White) Rydb. Stem 4–10 dm. high, more or less winged; leaflets mostly 3 pairs, elliptic or lance-elliptic, rarely linear-oblong, acute or obtuse at the ends, 3–5 cm. long, 8–16 mm. wide; stipules obliquely ovate, half-sagittate; peduncles 5–8 cm. long; racemes 4–6-flowered; lower calyx-teeth longer than the tube; corolla purple; pod about 8 mm. broad, abruptly acute at each end. Wet places: Me.—N.D.—Iowa—N.Y. *Canad.* Je–Jl.

6. **L. myrtifòlius** Muhl. Stem slender, glabrous, angled, 3–10 dm. high; stipules obliquely ovate or half-sagittate; leaflets mostly 3 pairs, oblong to narrowly oval, 2–5 cm. long, rather thin; flowers 3–9; corolla purple; pod linear, 2.5–5 cm. long, glabrous. *L. palustris myrtifolius* A. Gray [G]. Wet ground: N.B.—Man.—Tenn.—N.C. *Canad.—Allegh.* My–Jl.

7. **L. decaphýllus** Pursh. Stem erect, 2–5 dm. high, angled; leaflets 2–5 pairs, thick and veiny, elliptic or lance-oblong, acute, somewhat glaucous, glabrous, or finely pubescent beneath; stipules of the lower leaves reduced; peduncles 5–10 cm. long; racemes short, 3–5-flowered; corolla purple, 2.5–3 cm. long; pod 4–5 cm. long. Plains and prairies: Neb.—N.M.—Ariz. *Plain—Submont.* My–Jl.

8. **L. venòsus** Muhl. Stem 6–10 dm. high, erect, usually finely villous, 4-angled; leaflets 4–7 pairs, oval, often very veiny, glabrous above, finely pubescent beneath; peduncles 5–10 cm. long; raceme short and dense, 12–16-flowered; corolla purple, 12–15 mm. long; pod 4–5 cm. long, glabrous. River banks and wet places: Ont.—Ga.—La.—Kans.—Mont.—Sask. *Canad.—Allegh.—Plain—Submont.* My–Jl.

9. **L. marítimus** (L.) Bigel. Stem stout, trailing or climbing, glabrous; stipules broadly ovate, half-halberd-shaped, nearly as large as the leaflets; leaflets 6–10, thick, ovate-oblong to oval; racemes 6–10-flowered; corolla 18–25 mm. long, purple; pod 4–8 cm. long, glabrous, veiny. Beaches: Arctic coast—N.J. and Ore.; lake shores: N.Y.—Minn.—Man.; Eur. Je–S.

10. L. ochroleùcus Hook. Stem 4–10 dm. high, terete, glabrous; leaflets 3–4 pairs, broadly oval or ovate, 2–5 cm. long, thin, glabrous, somewhat glaucous beneath; peduncles 2–5 cm. long; racemes 5–10-flowered; corolla ochroleucous, about 15 mm. long; pod about 4 cm. long, glabrous. Woods and river banks: Man.—Que.—N.J.—Wyo.—B.C.—Mack. *Boreal.—Plain—Submont.* My–Au.

11. L. praténsis L. Perennial; stem 3–10 dm. long, angled, slightly wing-margined, glabrous; leaves glabrous with a single pair of leaflets and well-developed tendrils; leaflets lanceolate, 3–5 cm. long, usually with 3 stronger ribs; peduncles 5–10 cm. long; raceme short, 5–10-flowered; corolla yellow, 12–15 mm. long. Roadsides and pastures: Que.—N.J.—Iowa—Minn.; adv. from Eur. Au.

12. L. pusíllus Ell. Stem branched at the base, decumbent or climbing, 1–6 dm. long, winged; leaflets 2; tendrils simple or forking; stipules semisagittate; leaflets 2, linear to linear-elliptic, 1–5 cm. long, acute; raceme 1- or 2-flowered; calyx glabrous, the teeth subulate-lanceolate, longer than the tube; corolla purplish, 6–7 mm. long; pod linear, 1.5–4 cm. long. Sandy soil: N.C.—Kans.—Tex.—Fla. *Austral.* Ap–My.

54. PÌSUM L. Pea.

Annual herbs. Leaves pinnate, with tendrils and foliaceous stipules. Flowers in axillary racemes. Calyx 5-toothed, the teeth unequal, the upper two shorter and broader. Corolla rather large, the banner orbicular, spreading. Stamens diadelphous, the sheath truncate at right angles. Style laterally compressed at the apex. Pod oblong, obliquely truncate, many-seeded.

Seeds globose, separate, white or green; hilum linear-oblong. 1. *P. sativum.*
Seeds angled, compressed to each other, gray or brown; hilum oval. 2. *P. arvense.*

1. P. satìvum L. Stem 5–10 dm. high; leaflets usually 3 pairs, oval, 2–3 cm. long; stipules usually toothed at the base and larger than the leaflets; racemes 1- or 2-flowered; corolla white. Garden Pea. Occasionally escaped from cultivation; introd. from Eur.

2. P. arvénse L. Stem 3–8 dm. high; leaflets 1 or 2 pairs, oval, entire, 2–3 cm. long; racemes 1–3-flowered; banner usually rose-colored or rarely white, the wings dark-purple, and the keel white. Field Pea. Occasionally escaped from cultivation; nat. of Eur.

Family 86. GERANIACEAE. Geranium Family.

Herbs, with opposite leaves with stipules; blades usually palmately lobed or divided. Flowers perfect, regular, cymose or subumbellate. Sepals 5, imbricate. Petals 5, deciduous. Stamens 10, rarely 5; filaments monadelphous. Gynoecium of 5 carpels, the styles of which are adnate to an elongate central column, from which they separate at maturity. Carpels 2-ovuled, but 1-seeded.

Carpels rounded, their tails (styles) glabrous within, merely recoiling at maturity; leaves in ours palmately veined and lobed. 1. Geranium.
Carpels spindle-shaped, their tails (styles) pubescent within, spirally twisted at maturity; leaf-blades in ours pinnately veined, pinnately lobed or dissected. 2. Erodium.

1. GERÀNIUM (Tourn.) L. Cranebill, Wild Geranium.

Annual or perennial herbs, with opposite leaves. Leaf-blades palmately lobed or parted. Flowers cymose, perfect, regular. Sepals 5, usually awn-tipped. Petals 5, often pubescent near the base, not clawed. Stamens 10, rarely only 5; filaments ciliate at the base, more or less united, monadelphous. Style column usually beaked, the styles glabrous within, recoiled but not spirally twisted, when freed from the axis. Carpels turgid, permanently attached to the styles. Seeds smooth, reticulate, or pitted.

Leaf-blades ternately divided to near the base, the divisions 1–2-pinnatifid; carpel-bodies deciduous. — 1. *G. Robertianum.*
Leaf-blades palmately lobed, cleft, or parted; carpel-bodies permanently attached to the recoiled styles.
  Petals 2–7 mm. long, at most slightly exceeding the calyx; plants annual or biennial.
    Sepals not bristle-tipped; seeds smooth.
      Carpels wrinkled; stamens 10. — 2. *G. molle.*
      Carpels finely pubescent; stamens 5. — 3. *G. pusillum.*
    Sepals awn-tipped; seeds reticulate or pitted.
      Awn-tips less than 1 mm. long. — 4. *G. rotundifolium.*
      Awn-tips 1–2 mm. long.
        Beak and branches of the style less than 3 mm. long. — 5. *G. carolinianum.*
        Beak and branches of the style more than 4 mm. long; inflorescence open. — 6. *G. Bicknellii.*
  Petals 1–3 cm. long; plants perennial, with rootstocks or caudices; seeds reticulate.
    Petals white; style-column and carpels glandular; upper petioles retrorsely hirsute or glabrous; sepals slightly if at all glandular. — 7. *G. Richardsonii.*
    Petals purple or rose.
      Style-column finely pubescent; petals glabrous within. — 8. *G. maculatum.*
      Style-column densely glandular-puberulent; petals white-hairy within. — 9. *G. viscosissimum.*

1. **G. Robertiànum** L. Stem 1.5–5 dm. high, branched from near the base, finely glandular-pubescent; primary divisions of the leaves ovate in outline, 1.5–6 cm. long, the terminal one with a petiole-like base; sepals lanceolate, 6–8 mm. long, with filiform tips; petals red-purple, 8–11 mm. long, cuneate or obovate. Woods: N.S.—Man.—Neb.—N.J.; nat. or adv. from Eur. Je.–O.

2. **G. mólle** L. Stem spreading or decumbent, branched, 1–5 dm. long, finely pubescent; leaf-blades reniform or rounded-reniform, 2–6 cm. broad, the lower 6–9-cleft, the upper 3–5-cleft or parted, the divisions toothed or lobed at the apex; sepals elliptic or ovate, 3–4 mm. long; petals deep purple, obovate, truncate or emarginate; style-column short-beaked. Waste places and around dwellings: Me.—B.C.—Calif.—N.C.; nat. from Eur. My–Jl.

3. **G. pusíllum** Burm. Stem decumbent or prostrate, branched, 1–5.5 dm. long, puberulent; leaf-blades reniform, 1.5–5 cm. broad, 5–7-parted, the divisions toothed or lobed at the apex; sepals elliptic, 2.5–4 mm. long; petals violet, often pale, cuneate, notched; style-column short-beaked. Waste places: Ont.—B.C.—Utah—Va.; nat. from Eur. *Plain.* My–Au.

4. **G. rotundifòlium** L. Stem decumbent, branched, 1–6 dm. long, softly pubescent with gland-tipped hairs; leaf-blades rounded-reniform in outline, 1.5–4 cm. broad, cleft to about the middle; divisions 3–5-toothed at the apex; sepals 4–5.5 mm. long, oblong or elliptic; tips minute; petals purple, longer than the sepals; style-column 12–15 mm. long, minutely glandular; carpels glandular-pubescent. Waste places: N.Y.—Mich.—Colo.; adv. from Eur. *Plain—Submont.* Mr–S.

5. **G. caroliniànum** L. Stem branched above, more or less glandular-pubescent; leaf-blades 3–6 cm. broad, reniform in outline; divisions cleft or parted, with oblong or linear-oblong lobes; inflorescence congested; sepals ovate, with long awn-tips; petals pink or whitish; style-column with spreading glandular hairs; carpels pilose; seeds reticulate. Meadows and waste places: Newf.—Alaska—Calif.—Fla.; Mex. and W.Ind. *Temp.—Trop.—Plain—Submont.* Ap–Jl.

6. G. **Bicknéllii** Britton. Stem with ascending branches and loose spreading hairs; leaf-blades pentagonal or the lower rounded, 2–7 cm. broad, the division incised or cleft into oblong or lanceolate segments; petioles, peduncles, and pedicels glandular-pubescent; sepals lanceolate or oblong-lanceolate; petals rose-purple, about as long as the sepals; style-column glandular-hirsute; carpels sparingly hirsute. *G. carolinianum longipes* S. Wats. *G. longipes* Goodding. Meadows: N.S.—N.Y.—Utah—Wash.—B.C.—Yukon. *Boreal.—Plain—Submont.* My–Au.

7. G. **Richardsònii** Fisch. & Trautv. *Fig. 335.* Stem erect, 4–9 dm. high, pubescent with spreading hairs or glabrous; leaf-blades thin, pentagonal in outline, 3–5 parted, the divisions incised or toothed; pedicels glandular-pubescent; sepals awn-tipped, 8.5–11.5 mm. long, elliptic, pubescent and sometimes glandular-ciliate; petals white, 12–20 mm. long; style-column 19–23 mm. long. *G. albiflorum* Hook. *G. gracilentum* Greene. Mountain valleys: Sask.—S.D.—N.M.—Calif.—B.C. *Mont.—Submont.* My–Au.

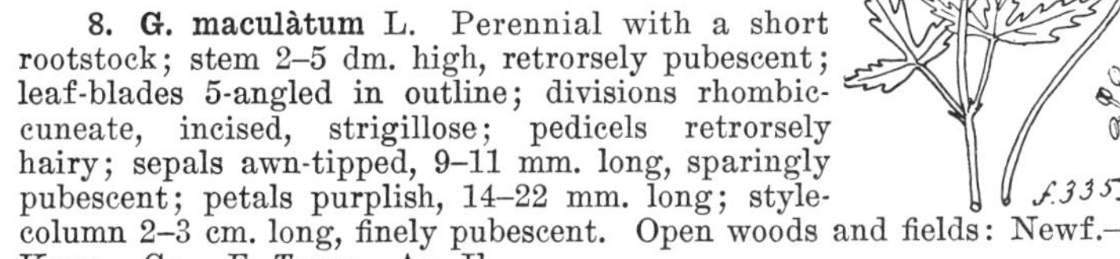

8. G. **maculàtum** L. Perennial with a short rootstock; stem 2–5 dm. high, retrorsely pubescent; leaf-blades 5-angled in outline; divisions rhombic-cuneate, incised, strigillose; pedicels retrorsely hairy; sepals awn-tipped, 9–11 mm. long, sparingly pubescent; petals purplish, 14–22 mm. long; style-column 2–3 cm. long, finely pubescent. Open woods and fields: Newf.—Man.—Kans.—Ga. *E. Temp.* Ap–Jl.

9. G. **viscosíssimum** F. & M. Stem 2.5–6 dm. high; leaf-blades pentagonal in outline, 4–11 cm. wide, hirsute, 3–5-parted, the divisions sharply and irregularly incised; sepals 12–15 mm. long, oblong-lanceolate, finely hirsute, awn-tipped; petals pink-purple, 17–20 mm. long; carpels glandular-pilose. Woods and among bushes: S.D.—Colo.—Calif.—B.C. *Mont.—Submont.* My–S.

2. **ERÒDIUM** L'Hér. STORKBILL, ALFILARIA, PIN CLOVER, FILAREE.

Annual or perennial herbs, with opposite leaves. Leaf-blades pinnately divided or lobed, or if merely toothed, pinnately veined, long-petioled. Sepals 5, usually awn-tipped. Petals 5, those of the later flowers often reduced. Stamens 5, alternating with 5 staminodia. Style-column very elongate, the styles pubescent within, and spirally twisted, when freed from the central axis. Carpels narrow, spindle-shaped. Seeds smooth.

1. E. **cicutàrium** (L.) L'Hér. An annual, with low spreading stem; leaf-blades pinnately divided; segments oblong to ovate, pinnatifid or incised; peduncles and pedicels more or less hirsute, rarely glandular, sepals oblong, 6–7 mm. long, the short tips with 1 or 2 bristle-like appendages; petals slightly longer than the sepals, pink, with darker veins; anther-bearing filaments toothless; style-column 3–4 cm. long. Waste places and fields: N.S.—N.J.—N.M.—Calif.—B.C.—Alta.; Mex. and C. Am.; nat. from Eur. *Plain—Submont.* My–Au.

Family 87. **OXALIDACEAE.** WOOD-SORREL FAMILY.

Mostly herbs, with elongate or bulb-like rootstocks. Leaves alternate, sometimes all basal, compound, in ours palmately trifoliolate. Flowers perfect, regular or nearly so, cymose. Sepals and petals 5. Stamens 10, in 2 series, monadelphous. Gynoecium of 5 united carpels; styles distinct or coherent. Fruit a valvate capsule.

Plants acaulescent, with rootstocks.
  Rootstock elongate; sepals without apical tubercles; flowers white or pink. 1. OXALIS.
  Rootstock short and bulb-like; sepals with apical tubercles; flowers purple or violet. 2. IONOXALIS.
Plants caulescent; flowers yellow. 3. XANTHOXALIS.

1. ÓXALIS L. WOOD-SORREL.

Perennial acaulescent herbs, with slender rootstocks. Leaves basal, with long petioles, palmately 3-foliolate. Flowers of two kinds, solitary on long peduncles. Sepals 5, thick. Petals of the petaliferous flowers 5, cuneate or obovate, white or pinkish. Stamens 10; filaments united at the base. Ovary as broad as long, each carpel 1- or 2-ovuled. Capsule subglobose. Cleistogenous flowers borne on peduncles lower than the petaliferous ones, nodding, without petals.

1. **O. Acetosélla** L. Scape 3–10 cm. high; petioles sparingly villous; leaflets 1.5–2.5 cm. broad, pubescent on both sides; petaliferous flowers erect; sepals oblong-lanceolate, 4–5 mm. long, ciliate; petals 10–16 mm. long, white or pink, striped with purple. Woods: N.S.—Sask.—Tenn.—N.C.; Eur. *Boreal.*

2. IONÓXALIS Small. VIOLET WOOD-SORREL.

Perennial succulent herbs, with scaly bulbs, scapose. Leaves basal, with elongate petioles; blades palmately 3–9-foliolate. Flowers in umbel-like cymes, subtended by several bracts. Sepals 5, each with one or two pairs of apical tubercles. Petals 5, violet, blue, red, or white. Stamens 10; filaments united at the base, the longer sometimes appendaged on the back. Capsules erect.

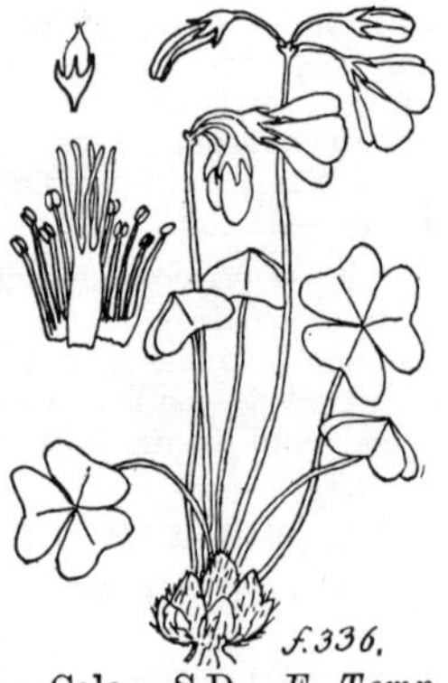

1. **I. violàcea** (L.) Small. *Fig. 336.* Plant 1–4 dm. high; bulb-scales 3-ribbed; petioles glabrous; leaflets 3, obreniform, 1–2.5 cm. broad, glabrous, bright green above, slightly paler beneath; scape and pedicels glabrous, the former 4–19-flowered; sepals oblong to ovate-oblong, 4–6 mm. long, glabrous; petals violet, 14–20 mm. long; filaments pubescent; capsule globose-ovoid, 4–5 mm. long. *Oxalis violacea* L. [G, B]. Rich soil: Me.—Fla.—Tex.—Colo.—S.D. *E. Temp.—Plain—Submont.* Ap–S.

f. 336.

3. XANTHÓXALIS Small. YELLOW WOOD-SORREL.

Herbs, perennial, with horizontal rootstocks or annual taproots, or rarely shrubs. Stem leafy. Leaves alternate, sometimes clustered at the nodes. Leaves palmately 3-foliolate; leaflets mostly obcordate (in our species), or rarely broadest at the base, sessile. Flowers in umbel-like or dichotomous cymes. Sepals 5, herbaceous or petaloid. Petals 5, yellow or purplish, deciduous. Stamens 10; filaments united into a tube at the base. Ovary elongate; styles filiform or subulate. Capsule columnar or oblong, prismatic, erect.

Stem creeping, often rooting; pod finely strigose. 1. *X. corniculata.*
Stem erect or decumbent at the base only, not rooting.
  Pedicels appressed-pubescent.
    Stem appressed-pubescent; capsule pubescent. 2. *X. stricta.*
    Stem loosely pubescent with spreading hairs, or nearly glabrous; capsule glabrous.
      Cyme open in fruit; leaves bright green. 3. *X. Bushii.*
      Cyme cluster-like in fruit; leaves reddish or purplish. 4. *X. rufa.*
  Pedicels with spreading hairs; cymes dichotomous; capsules glabrous. 5. *X. cymosa.*

1. **X. corniculàta** (L.) Small. Stem branched at the base, creeping, 0.5–4 dm. long, loosely but sparsely pubescent; leaflets 5–12 mm. broad, ciliate and with scattered hairs on both sides; peduncles fully as long as the petioles; pedicels short, strigulose; petals pale yellow, 7–10 mm. long; capsule columnar, 8–18 mm. long, finely pubescent, abruptly acute. *Oxalis corniculata* L. [B]. Waste places and around dwellings: Man.—Ont.—Fla.—Tex.—Calif.; escaped from cultivation; Eur. and the Tropics.

2. **X. strícta** (L.) Small. Stem erect or decumbent in age, 0.5–5 dm. long, strigillose; leaflets 15–20 mm. wide; peduncles longer than the petioles; sepals 4–5 mm. long; petals pale yellow, 5–10 mm. long; capsule columnar, 16–30 mm. long, rather abruptly pointed. *Oxalis stricta* L. [G, B]. Woods, roadsides, and cultivated grounds: N.S.—Fla.—N.M.—Wyo.—N.D.—Man.; Mex. *E. Temp.—Plain—Submont.* Ap–N.

3. **X. Bûshii** Small. *Fig. 337.* Stem slender, erect, 1–2 dm. high, villous; leaflets broadly obcordate, glabrous, 8–20 mm. long; peduncles very slender, about 5 cm. long; inflorescence cymose, but often reduced to 2–3 flowers and umbel-like; sepals 4–5 mm. long, oblong; petals 6–7 mm. long; capsule columnar, 12–15 mm. long. *Oxalis Bushii* Small [B]. *Oxalis* and *Xanthoxalis coloradensis* Rydb. River bottoms: N.S.—Ga.—Colo.—N.D. *E. Temp.—Plain—Mont.* My–Au.

4. **X. rùfa** Small. Stem erect, 1.5–4 dm. high, usually sparingly villous; leaflets reddish or purplish, 8–20 mm. wide, sparingly strigillose or glabrous, ciliate on the margins; sepals 3–4 mm. long; petals bright yellow, 7–10 mm. long; capsule 7–9 mm. long, abruptly pointed. *O. rufa* Small [B]. Open places: Mass.—Minn.—Ga. *Canadian—Allegh.* My–Au.

5. **X. cymòsa** Small. Stem erect, 2–11 dm. high, loosely pubescent or glabrous; leaflets 10–17 mm. broad, glabrate or sparingly pubescent, ciliate; sepals 3–4 mm. long; petals deep yellow, 7–10 mm. long; capsule 10–16 mm. long, gradually acute. *O. cymosa* Small [B]. Open places: Ont.—N.D.—Neb.—Tex.—Fla. *E. Temp.* Ap–S.

A CATALOGUE OF SELECTED DOVER BOOKS IN ALL FIELDS OF INTEREST

A CATALOGUE OF SELECTED DOVER BOOKS IN ALL FIELDS OF INTEREST

AMERICA'S OLD MASTERS, James T. Flexner. Four men emerged unexpectedly from provincial 18th century America to leadership in European art: Benjamin West, J. S. Copley, C. R. Peale, Gilbert Stuart. Brilliant coverage of lives and contributions. Revised, 1967 edition. 69 plates. 365pp. of text.
21806-6 Paperbound $3.00

FIRST FLOWERS OF OUR WILDERNESS: AMERICAN PAINTING, THE COLONIAL PERIOD, James T. Flexner. Painters, and regional painting traditions from earliest Colonial times up to the emergence of Copley, West and Peale Sr., Foster, Gustavus Hesselius, Feke, John Smibert and many anonymous painters in the primitive manner. Engaging presentation, with 162 illustrations. xxii + 368pp.
22180-6 Paperbound $3.50

THE LIGHT OF DISTANT SKIES: AMERICAN PAINTING, 1760-1835, James T. Flexner. The great generation of early American painters goes to Europe to learn and to teach: West, Copley, Gilbert Stuart and others. Allston, Trumbull, Morse; also contemporary American painters—primitives, derivatives, academics—who remained in America. 102 illustrations. xiii + 306pp. 22179-2 Paperbound $3.00

A HISTORY OF THE RISE AND PROGRESS OF THE ARTS OF DESIGN IN THE UNITED STATES, William Dunlap. Much the richest mine of information on early American painters, sculptors, architects, engravers, miniaturists, etc. The only source of information for scores of artists, the major primary source for many others. Unabridged reprint of rare original 1834 edition, with new introduction by James T. Flexner, and 394 new illustrations. Edited by Rita Weiss. 6⅝ x 9⅝.
21695-0, 21696-9, 21697-7 Three volumes, Paperbound $13.50

EPOCHS OF CHINESE AND JAPANESE ART, Ernest F. Fenollosa. From primitive Chinese art to the 20th century, thorough history, explanation of every important art period and form, including Japanese woodcuts; main stress on China and Japan, but Tibet, Korea also included. Still unexcelled for its detailed, rich coverage of cultural background, aesthetic elements, diffusion studies, particularly of the historical period. 2nd, 1913 edition. 242 illustrations. lii + 439pp. of text.
20364-6, 20365-4 Two volumes, Paperbound $6.00

THE GENTLE ART OF MAKING ENEMIES, James A. M. Whistler. Greatest wit of his day deflates Oscar Wilde, Ruskin, Swinburne; strikes back at inane critics, exhibitions, art journalism; aesthetics of impressionist revolution in most striking form. Highly readable classic by great painter. Reproduction of edition designed by Whistler. Introduction by Alfred Werner. xxxvi + 334pp.
21875-9 Paperbound $2.50

VISUAL ILLUSIONS: THEIR CAUSES, CHARACTERISTICS, AND APPLICATIONS, Matthew Luckiesh. Thorough description and discussion of optical illusion, geometric and perspective, particularly; size and shape distortions, illusions of color, of motion; natural illusions; use of illusion in art and magic, industry, etc. Most useful today with op art, also for classical art. Scores of effects illustrated. Introduction by William H. Ittleson. 100 illustrations. xxi + 252pp.
21530-X Paperbound $2.00

A HANDBOOK OF ANATOMY FOR ART STUDENTS, Arthur Thomson. Thorough, virtually exhaustive coverage of skeletal structure, musculature, etc. Full text, supplemented by anatomical diagrams and drawings and by photographs of undraped figures. Unique in its comparison of male and female forms, pointing out differences of contour, texture, form. 211 figures, 40 drawings, 86 photographs. xx + 459pp. 5⅜ x 8⅜.
21163-0 Paperbound $3.50

150 MASTERPIECES OF DRAWING, Selected by Anthony Toney. Full page reproductions of drawings from the early 16th to the end of the 18th century, all beautifully reproduced: Rembrandt, Michelangelo, Dürer, Fragonard, Urs, Graf, Wouwerman, many others. First-rate browsing book, model book for artists. xviii + 150pp. 8⅜ x 11¼.
21032-4 Paperbound $2.50

THE LATER WORK OF AUBREY BEARDSLEY, Aubrey Beardsley. Exotic, erotic, ironic masterpieces in full maturity: Comedy Ballet, Venus and Tannhauser, Pierrot, Lysistrata, Rape of the Lock, Savoy material, Ali Baba, Volpone, etc. This material revolutionized the art world, and is still powerful, fresh, brilliant. With *The Early Work,* all Beardsley's finest work. 174 plates, 2 in color. xiv + 176pp. 8⅛ x 11.
21817-1 Paperbound $3.00

DRAWINGS OF REMBRANDT, Rembrandt van Rijn. Complete reproduction of fabulously rare edition by Lippmann and Hofstede de Groot, completely reedited, updated, improved by Prof. Seymour Slive, Fogg Museum. Portraits, Biblical sketches, landscapes, Oriental types, nudes, episodes from classical mythology—All Rembrandt's fertile genius. Also selection of drawings by his pupils and followers. "Stunning volumes," *Saturday Review.* 550 illustrations. lxxviii + 552pp. 9⅛ x 12¼.
21485-0, 21486-9 Two volumes, Paperbound $7.00

THE DISASTERS OF WAR, Francisco Goya. One of the masterpieces of Western civilization—83 etchings that record Goya's shattering, bitter reaction to the Napoleonic war that swept through Spain after the insurrection of 1808 and to war in general. Reprint of the first edition, with three additional plates from Boston's Museum of Fine Arts. All plates facsimile size. Introduction by Philip Hofer, Fogg Museum. v + 97pp. 9⅜ x 8¼.
21872-4 Paperbound $2.00

GRAPHIC WORKS OF ODILON REDON. Largest collection of Redon's graphic works ever assembled: 172 lithographs, 28 etchings and engravings, 9 drawings. These include some of his most famous works. All the plates from *Odilon Redon: oeuvre graphique complet,* plus additional plates. New introduction and caption translations by Alfred Werner. 209 illustrations. xxvii + 209pp. 9⅛ x 12¼.
21966-8 Paperbound $4.00

DESIGN BY ACCIDENT; A BOOK OF "ACCIDENTAL EFFECTS" FOR ARTISTS AND DESIGNERS, James F. O'Brien. Create your own unique, striking, imaginative effects by "controlled accident" interaction of materials: paints and lacquers, oil and water based paints, splatter, crackling materials, shatter, similar items. Everything you do will be different; first book on this limitless art, so useful to both fine artist and commercial artist. Full instructions. 192 plates showing "accidents," 8 in color. viii + 215pp. 8⅜ x 11¼. 21942-9 Paperbound $3.50

THE BOOK OF SIGNS, Rudolf Koch. Famed German type designer draws 493 beautiful symbols: religious, mystical, alchemical, imperial, property marks, runes, etc. Remarkable fusion of traditional and modern. Good for suggestions of timelessness, smartness, modernity. Text. vi + 104pp. 6⅛ x 9¼.
20162-7 Paperbound $1.25

HISTORY OF INDIAN AND INDONESIAN ART, Ananda K. Coomaraswamy. An unabridged republication of one of the finest books by a great scholar in Eastern art. Rich in descriptive material, history, social backgrounds; Sunga reliefs, Rajput paintings, Gupta temples, Burmese frescoes, textiles, jewelry, sculpture, etc. 400 photos. viii + 423pp. 6⅜ x 9¾. 21436-2 Paperbound $4.00

PRIMITIVE ART, Franz Boas. America's foremost anthropologist surveys textiles, ceramics, woodcarving, basketry, metalwork, etc.; patterns, technology, creation of symbols, style origins. All areas of world, but very full on Northwest Coast Indians. More than 350 illustrations of baskets, boxes, totem poles, weapons, etc. 378 pp.
20025-6 Paperbound $3.00

THE GENTLEMAN AND CABINET MAKER'S DIRECTOR, Thomas Chippendale. Full reprint (third edition, 1762) of most influential furniture book of all time, by master cabinetmaker. 200 plates, illustrating chairs, sofas, mirrors, tables, cabinets, plus 24 photographs of surviving pieces. Biographical introduction by N. Bienenstock. vi + 249pp. 9⅞ x 12¾. 21601-2 Paperbound $4.00

AMERICAN ANTIQUE FURNITURE, Edgar G. Miller, Jr. The basic coverage of all American furniture before 1840. Individual chapters cover type of furniture—clocks, tables, sideboards, etc.—chronologically, with inexhaustible wealth of data. More than 2100 photographs, all identified, commented on. Essential to all early American collectors. Introduction by H. E. Keyes. vi + 1106pp. 7⅞ x 10¾.
21599-7, 21600-4 Two volumes, Paperbound $10.00

PENNSYLVANIA DUTCH AMERICAN FOLK ART, Henry J. Kauffman. 279 photos, 28 drawings of tulipware, Fraktur script, painted tinware, toys, flowered furniture, quilts, samplers, hex signs, house interiors, etc. Full descriptive text. Excellent for tourist, rewarding for designer, collector. Map. 146pp. 7⅞ x 10¾.
21205-X Paperbound $2.50

EARLY NEW ENGLAND GRAVESTONE RUBBINGS, Edmund V. Gillon, Jr. 43 photographs, 226 carefully reproduced rubbings show heavily symbolic, sometimes macabre early gravestones, up to early 19th century. Remarkable early American primitive art, occasionally strikingly beautiful; always powerful. Text. xxvi + 207pp. 8⅜ x 11¼. 21380-3 Paperbound $3.50

ALPHABETS AND ORNAMENTS, Ernst Lehner. Well-known pictorial source for decorative alphabets, script examples, cartouches, frames, decorative title pages, calligraphic initials, borders, similar material. 14th to 19th century, mostly European. Useful in almost any graphic arts designing, varied styles. 750 illustrations. 256pp. 7 x 10. 21905-4 Paperbound $4.00

PAINTING: A CREATIVE APPROACH, Norman Colquhoun. For the beginner simple guide provides an instructive approach to painting: major stumbling blocks for beginner; overcoming them, technical points; paints and pigments; oil painting; watercolor and other media and color. New section on "plastic" paints. Glossary. Formerly *Paint Your Own Pictures*. 221pp. 22000-1 Paperbound $1.75

THE ENJOYMENT AND USE OF COLOR, Walter Sargent. Explanation of the relations between colors themselves and between colors in nature and art, including hundreds of little-known facts about color values, intensities, effects of high and low illumination, complementary colors. Many practical hints for painters, references to great masters. 7 color plates, 29 illustrations. x + 274pp. 20944-X Paperbound $2.50

THE NOTEBOOKS OF LEONARDO DA VINCI, compiled and edited by Jean Paul Richter. 1566 extracts from original manuscripts reveal the full range of Leonardo's versatile genius: all his writings on painting, sculpture, architecture, anatomy, astronomy, geography, topography, physiology, mining, music, etc., in both Italian and English, with 186 plates of manuscript pages and more than 500 additional drawings. Includes studies for the Last Supper, the lost Sforza monument, and other works. Total of xlvii + 866pp. $7\frac{7}{8}$ x $10\frac{3}{4}$. 22572-0, 22573-9 Two volumes, Paperbound $10.00

MONTGOMERY WARD CATALOGUE OF 1895. Tea gowns, yards of flannel and pillow-case lace, stereoscopes, books of gospel hymns, the New Improved Singer Sewing Machine, side saddles, milk skimmers, straight-edged razors, high-button shoes, spittoons, and on and on . . . listing some 25,000 items, practically all illustrated. Essential to the shoppers of the 1890's, it is our truest record of the spirit of the period. Unaltered reprint of Issue No. 57, Spring and Summer 1895. Introduction by Boris Emmet. Innumerable illustrations. xiii + 624pp. $8\frac{1}{2}$ x $11\frac{5}{8}$. 22377-9 Paperbound $6.95

THE CRYSTAL PALACE EXHIBITION ILLUSTRATED CATALOGUE (LONDON, 1851). One of the wonders of the modern world—the Crystal Palace Exhibition in which all the nations of the civilized world exhibited their achievements in the arts and sciences—presented in an equally important illustrated catalogue. More than 1700 items pictured with accompanying text—ceramics, textiles, cast-iron work, carpets, pianos, sleds, razors, wall-papers, billiard tables, beehives, silverware and hundreds of other artifacts—represent the focal point of Victorian culture in the Western World. Probably the largest collection of Victorian decorative art ever assembled—indispensable for antiquarians and designers. Unabridged republication of the Art-Journal Catalogue of the Great Exhibition of 1851, with all terminal essays. New introduction by John Gloag, F.S.A. xxxiv + 426pp. 9 x 12. 22503-8 Paperbound $4.50

A HISTORY OF COSTUME, Carl Köhler. Definitive history, based on surviving pieces of clothing primarily, and paintings, statues, etc. secondarily. Highly readable text, supplemented by 594 illustrations of costumes of the ancient Mediterranean peoples, Greece and Rome, the Teutonic prehistoric period; costumes of the Middle Ages, Renaissance, Baroque, 18th and 19th centuries. Clear, measured patterns are provided for many clothing articles. Approach is practical throughout. Enlarged by Emma von Sichart. 464pp. 21030-8 Paperbound $3.50

ORIENTAL RUGS, ANTIQUE AND MODERN, Walter A. Hawley. A complete and authoritative treatise on the Oriental rug—where they are made, by whom and how, designs and symbols, characteristics in detail of the six major groups, how to distinguish them and how to buy them. Detailed technical data is provided on periods, weaves, warps, wefts, textures, sides, ends and knots, although no technical background is required for an understanding. 11 color plates, 80 halftones, 4 maps. vi + 320pp. 6⅛ x 9⅛. 22366-3 Paperbound $5.00

TEN BOOKS ON ARCHITECTURE, Vitruvius. By any standards the most important book on architecture ever written. Early Roman discussion of aesthetics of building, construction methods, orders, sites, and every other aspect of architecture has inspired, instructed architecture for about 2,000 years. Stands behind Palladio, Michelangelo, Bramante, Wren, countless others. Definitive Morris H. Morgan translation. 68 illustrations. xii + 331pp. 20645-9 Paperbound $2.50

THE FOUR BOOKS OF ARCHITECTURE, Andrea Palladio. Translated into every major Western European language in the two centuries following its publication in 1570, this has been one of the most influential books in the history of architecture. Complete reprint of the 1738 Isaac Ware edition. New introduction by Adolf Placzek, Columbia Univ. 216 plates. xxii + 110pp. of text. 9½ x 12¾. 21308-0 Clothbound $10.00

STICKS AND STONES: A STUDY OF AMERICAN ARCHITECTURE AND CIVILIZATION, Lewis Mumford.One of the great classics of American cultural history. American architecture from the medieval-inspired earliest forms to the early 20th century; evolution of structure and style, and reciprocal influences on environment. 21 photographic illustrations. 238pp. 20202-X Paperbound $2.00

THE AMERICAN BUILDER'S COMPANION, Asher Benjamin. The most widely used early 19th century architectural style and source book, for colonial up into Greek Revival periods. Extensive development of geometry of carpentering, construction of sashes, frames, doors, stairs; plans and elevations of domestic and other buildings. Hundreds of thousands of houses were built according to this book, now invaluable to historians, architects, restorers, etc. 1827 edition. 59 plates. 114pp. 7⅞ x 10¾. 22236-5 Paperbound $3.00

DUTCH HOUSES IN THE HUDSON VALLEY BEFORE 1776, Helen Wilkinson Reynolds. The standard survey of the Dutch colonial house and outbuildings, with constructional features, decoration, and local history associated with individual homesteads. Introduction by Franklin D. Roosevelt. Map. 150 illustrations. 469pp. 6⅝ x 9¼. 21469-9 Paperbound $4.00

THE ARCHITECTURE OF COUNTRY HOUSES, Andrew J. Downing. Together with Vaux's *Villas and Cottages* this is the basic book for Hudson River Gothic architecture of the middle Victorian period. Full, sound discussions of general aspects of housing, architecture, style, decoration, furnishing, together with scores of detailed house plans, illustrations of specific buildings, accompanied by full text. Perhaps the most influential single American architectural book. 1850 edition. Introduction by J. Stewart Johnson. 321 figures, 34 architectural designs. xvi + 560pp.
22003-6 Paperbound $4.00

LOST EXAMPLES OF COLONIAL ARCHITECTURE, John Mead Howells. Full-page photographs of buildings that have disappeared or been so altered as to be denatured, including many designed by major early American architects. 245 plates. xvii + 248pp. 7⅞ x 10¾. 21143-6 Paperbound $3.50

DOMESTIC ARCHITECTURE OF THE AMERICAN COLONIES AND OF THE EARLY REPUBLIC, Fiske Kimball. Foremost architect and restorer of Williamsburg and Monticello covers nearly 200 homes between 1620-1825. Architectural details, construction, style features, special fixtures, floor plans, etc. Generally considered finest work in its area. 219 illustrations of houses, doorways, windows, capital mantels. xx + 314pp. 7⅞ x 10¾. 21743-4 Paperbound $4.00

EARLY AMERICAN ROOMS: 1650-1858, edited by Russell Hawes Kettell. Tour of 12 rooms, each representative of a different era in American history and each furnished, decorated, designed and occupied in the style of the era. 72 plans and elevations, 8-page color section, etc., show fabrics, wall papers, arrangements, etc. Full descriptive text. xvii + 200pp. of text. 8⅜ x 11¼.
21633-0 Paperbound $5.00

THE FITZWILLIAM VIRGINAL BOOK, edited by J. Fuller Maitland and W. B. Squire. Full modern printing of famous early 17th-century ms. volume of 300 works by Morley, Byrd, Bull, Gibbons, etc. For piano or other modern keyboard instrument; easy to read format. xxxvi + 938pp. 8⅜ x 11.
21068-5, 21069-3 Two volumes, Paperbound $10.00

KEYBOARD MUSIC, Johann Sebastian Bach. Bach Gesellschaft edition. A rich selection of Bach's masterpieces for the harpsichord: the six English Suites, six French Suites, the six Partitas (Clavierübung part I), the Goldberg Variations (Clavierübung part IV), the fifteen Two-Part Inventions and the fifteen Three-Part Sinfonias. Clearly reproduced on large sheets with ample margins; eminently playable. vi + 312pp. 8⅛ x 11. 22360-4 Paperbound $5.00

THE MUSIC OF BACH: AN INTRODUCTION, Charles Sanford Terry. A fine, nontechnical introduction to Bach's music, both instrumental and vocal. Covers organ music, chamber music, passion music, other types. Analyzes themes, developments, innovations. x + 114pp. 21075-8 Paperbound $1.25

BEETHOVEN AND HIS NINE SYMPHONIES, Sir George Grove. Noted British musicologist provides best history, analysis, commentary on symphonies. Very thorough, rigorously accurate; necessary to both advanced student and amateur music lover. 436 musical passages. vii + 407 pp. 20334-4 Paperbound $2.50

JOHANN SEBASTIAN BACH, Philipp Spitta. One of the great classics of musicology, this definitive analysis of Bach's music (and life) has never been surpassed. Lucid, nontechnical analyses of hundreds of pieces (30 pages devoted to St. Matthew Passion, 26 to B Minor Mass). Also includes major analysis of 18th-century music. 450 musical examples. 40-page musical supplement. Total of xx + 1799pp.
(EUK) 22278-0, 22279-9 Two volumes, Clothbound $15.00

MOZART AND HIS PIANO CONCERTOS, Cuthbert Girdlestone. The only full-length study of an important area of Mozart's creativity. Provides detailed analyses of all 23 concertos, traces inspirational sources. 417 musical examples. Second edition. 509pp. (USO) 21271-8 Paperbound $3.50

THE PERFECT WAGNERITE: A COMMENTARY ON THE NIBLUNG'S RING, George Bernard Shaw. Brilliant and still relevant criticism in remarkable essays on Wagner's Ring cycle, Shaw's ideas on political and social ideology behind the plots, role of Leitmotifs, vocal requisites, etc. Prefaces. xxi + 136pp.
21707-8 Paperbound $1.50

DON GIOVANNI, W. A. Mozart. Complete libretto, modern English translation; biographies of composer and librettist; accounts of early performances and critical reaction. Lavishly illustrated. All the material you need to understand and appreciate this great work. Dover Opera Guide and Libretto Series; translated and introduced by Ellen Bleiler. 92 illustrations. 209pp.
21134-7 Paperbound $1.50

HIGH FIDELITY SYSTEMS: A LAYMAN'S GUIDE, Roy F. Allison. All the basic information you need for setting up your own audio system: high fidelity and stereo record players, tape records, F.M. Connections, adjusting tone arm, cartridge, checking needle alignment, positioning speakers, phasing speakers, adjusting hums, trouble-shooting, maintenance, and similar topics. Enlarged 1965 edition. More than 50 charts, diagrams, photos. iv + 91pp. 21514-8 Paperbound $1.25

REPRODUCTION OF SOUND, Edgar Villchur. Thorough coverage for laymen of high fidelity systems, reproducing systems in general, needles, amplifiers, preamps, loudspeakers, feedback, explaining physical background. "A rare talent for making technicalities vividly comprehensible," R. Darrell, *High Fidelity*. 69 figures. iv + 92pp. 21515-6 Paperbound $1.00

HEAR ME TALKIN' TO YA: THE STORY OF JAZZ AS TOLD BY THE MEN WHO MADE IT, Nat Shapiro and Nat Hentoff. Louis Armstrong, Fats Waller, Jo Jones, Clarence Williams, Billy Holiday, Duke Ellington, Jelly Roll Morton and dozens of other jazz greats tell how it was in Chicago's South Side, New Orleans, depression Harlem and the modern West Coast as jazz was born and grew. xvi + 429pp.
21726-4 Paperbound $2.50

FABLES OF AESOP, translated by Sir Roger L'Estrange. A reproduction of the very rare 1931 Paris edition; a selection of the most interesting fables, together with 50 imaginative drawings by Alexander Calder. v + 128pp. 6½x9¼.
21780-9 Paperbound $1.50

AGAINST THE GRAIN (A REBOURS), Joris K. Huysmans. Filled with weird images, evidences of a bizarre imagination, exotic experiments with hallucinatory drugs, rich tastes and smells and the diversions of its sybarite hero Duc Jean des Esseintes, this classic novel pushed 19th-century literary decadence to its limits. Full unabridged edition. Do not confuse this with abridged editions generally sold. Introduction by Havelock Ellis. xlix + 206pp. 22190-3 Paperbound $2.00

VARIORUM SHAKESPEARE: HAMLET. Edited by Horace H. Furness; a landmark of American scholarship. Exhaustive footnotes and appendices treat all doubtful words and phrases, as well as suggested critical emendations throughout the play's history. First volume contains editor's own text, collated with all Quartos and Folios. Second volume contains full first Quarto, translations of Shakespeare's sources (Belleforest, and Saxo Grammaticus), Der Bestrafte Brudermord, and many essays on critical and historical points of interest by major authorities of past and present. Includes details of staging and costuming over the years. By far the best edition available for serious students of Shakespeare. Total of xx + 905pp.
21004-9, 21005-7, 2 volumes, Paperbound $5.50

A LIFE OF WILLIAM SHAKESPEARE, Sir Sidney Lee. This is the standard life of Shakespeare, summarizing everything known about Shakespeare and his plays. Incredibly rich in material, broad in coverage, clear and judicious, it has served thousands as the best introduction to Shakespeare. 1931 edition. 9 plates. xxix + 792pp. (USO) 21967-4 Paperbound $3.75

MASTERS OF THE DRAMA, John Gassner. Most comprehensive history of the drama in print, covering every tradition from Greeks to modern Europe and America, including India, Far East, etc. Covers more than 800 dramatists, 2000 plays, with biographical material, plot summaries, theatre history, criticism, etc. "Best of its kind in English," *New Republic.* 77 illustrations. xxii + 890pp.
20100-7 Clothbound $8.50

THE EVOLUTION OF THE ENGLISH LANGUAGE, George McKnight. The growth of English, from the 14th century to the present. Unusual, non-technical account presents basic information in very interesting form: sound shifts, change in grammar and syntax, vocabulary growth, similar topics. Abundantly illustrated with quotations. Formerly *Modern English in the Making.* xii + 590pp.
21932-1 Paperbound $3.50

AN ETYMOLOGICAL DICTIONARY OF MODERN ENGLISH, Ernest Weekley. Fullest, richest work of its sort, by foremost British lexicographer. Detailed word histories, including many colloquial and archaic words; extensive quotations. Do not confuse this with the Concise Etymological Dictionary, which is much abridged. Total of xxvii + 830pp. 6½ x 9¼.
21873-2, 21874-0 Two volumes, Paperbound $6.00

FLATLAND: A ROMANCE OF MANY DIMENSIONS, E. A. Abbott. Classic of science-fiction explores ramifications of life in a two-dimensional world, and what happens when a three-dimensional being intrudes. Amusing reading, but also useful as introduction to thought about hyperspace. Introduction by Banesh Hoffmann. 16 illustrations. xx + 103pp. 20001-9 Paperbound $1.00

POEMS OF ANNE BRADSTREET, edited with an introduction by Robert Hutchinson. A new selection of poems by America's first poet and perhaps the first significant woman poet in the English language. 48 poems display her development in works of considerable variety—love poems, domestic poems, religious meditations, formal elegies, "quaternions," etc. Notes, bibliography. viii + 222pp.
22160-1 Paperbound $2.00

THREE GOTHIC NOVELS: THE CASTLE OF OTRANTO BY HORACE WALPOLE; VATHEK BY WILLIAM BECKFORD; THE VAMPYRE BY JOHN POLIDORI, WITH FRAGMENT OF A NOVEL BY LORD BYRON, edited by E. F. Bleiler. The first Gothic novel, by Walpole; the finest Oriental tale in English, by Beckford; powerful Romantic supernatural story in versions by Polidori and Byron. All extremely important in history of literature; all still exciting, packed with supernatural thrills, ghosts, haunted castles, magic, etc. xl + 291pp.
21232-7 Paperbound $2.00

THE BEST TALES OF HOFFMANN, E. T. A. Hoffmann. 10 of Hoffmann's most important stories, in modern re-editings of standard translations: Nutcracker and the King of Mice, Signor Formica, Automata, The Sandman, Rath Krespel, The Golden Flowerpot, Master Martin the Cooper, The Mines of Falun, The King's Betrothed, A New Year's Eve Adventure. 7 illustrations by Hoffmann. Edited by E. F. Bleiler. xxxix + 419pp. 21793-0 Paperbound $2.50

GHOST AND HORROR STORIES OF AMBROSE BIERCE, Ambrose Bierce. 23 strikingly modern stories of the horrors latent in the human mind: The Eyes of the Panther, The Damned Thing, An Occurrence at Owl Creek Bridge, An Inhabitant of Carcosa, etc., plus the dream-essay, Visions of the Night. Edited by E. F. Bleiler. xxii + 199pp. 20767-6 Paperbound $1.50

BEST GHOST STORIES OF J. S. LEFANU, J. Sheridan LeFanu. Finest stories by Victorian master often considered greatest supernatural writer of all. Carmilla, Green Tea, The Haunted Baronet, The Familiar, and 12 others. Most never before available in the U. S. A. Edited by E. F. Bleiler. 8 illustrations from Victorian publications. xvii + 467pp. 20415-4 Paperbound $3.00

THE TIME STREAM, THE GREATEST ADVENTURE, AND THE PURPLE SAPPHIRE—THREE SCIENCE FICTION NOVELS, John Taine (Eric Temple Bell). Great American mathematician was also foremost science fiction novelist of the 1920's. *The Time Stream,* one of all-time classics, uses concepts of circular time; *The Greatest Adventure,* incredibly ancient biological experiments from Antarctica threaten to escape; The *Purple Sapphire,* superscience, lost races in Central Tibet, survivors of the Great Race. 4 illustrations by Frank R. Paul. v + 532pp.
21180-0 Paperbound $3.00

SEVEN SCIENCE FICTION NOVELS, H. G. Wells. The standard collection of the great novels. Complete, unabridged. *First Men in the Moon, Island of Dr. Moreau, War of the Worlds, Food of the Gods, Invisible Man, Time Machine, In the Days of the Comet.* Not only science fiction fans, but every educated person owes it to himself to read these novels. 1015pp. 20264-X Clothbound $5.00

LAST AND FIRST MEN AND STAR MAKER, TWO SCIENCE FICTION NOVELS, Olaf Stapledon. Greatest future histories in science fiction. In the first, human intelligence is the "hero," through strange paths of evolution, interplanetary invasions, incredible technologies, near extinctions and reemergences. Star Maker describes the quest of a band of star rovers for intelligence itself, through time and space: weird inhuman civilizations, crustacean minds, symbiotic worlds, etc. Complete, unabridged. v + 438pp. 21962-3 Paperbound $2.50

THREE PROPHETIC NOVELS, H. G. WELLS. Stages of a consistently planned future for mankind. *When the Sleeper Wakes,* and *A Story of the Days to Come,* anticipate *Brave New World* and *1984,* in the 21st Century; *The Time Machine,* only complete version in print, shows farther future and the end of mankind. All show Wells's greatest gifts as storyteller and novelist. Edited by E. F. Bleiler. x + 335pp. (USO) 20605-X Paperbound $2.25

THE DEVIL'S DICTIONARY, Ambrose Bierce. America's own Oscar Wilde—Ambrose Bierce—offers his barbed iconoclastic wisdom in over 1,000 definitions hailed by H. L. Mencken as "some of the most gorgeous witticisms in the English language." 145pp. 20487-1 Paperbound $1.25

MAX AND MORITZ, Wilhelm Busch. Great children's classic, father of comic strip, of two bad boys, Max and Moritz. Also Ker and Plunk (Plisch und Plumm), Cat and Mouse, Deceitful Henry, Ice-Peter, The Boy and the Pipe, and five other pieces. Original German, with English translation. Edited by H. Arthur Klein; translations by various hands and H. Arthur Klein. vi + 216pp. 20181-3 Paperbound $1.50

PIGS IS PIGS AND OTHER FAVORITES, Ellis Parker Butler. The title story is one of the best humor short stories, as Mike Flannery obfuscates biology and English. Also included, That Pup of Murchison's, The Great American Pie Company, and Perkins of Portland. 14 illustrations. v + 109pp. 21532-6 Paperbound $1.00

THE PETERKIN PAPERS, Lucretia P. Hale. It takes genius to be as stupidly mad as the Peterkins, as they decide to become wise, celebrate the "Fourth," keep a cow, and otherwise strain the resources of the Lady from Philadelphia. Basic book of American humor. 153 illustrations. 219pp. 20794-3 Paperbound $1.50

PERRAULT'S FAIRY TALES, translated by A. E. Johnson and S. R. Littlewood, with 34 full-page illustrations by Gustave Doré. All the original Perrault stories—Cinderella, Sleeping Beauty, Bluebeard, Little Red Riding Hood, Puss in Boots, Tom Thumb, etc.—with their witty verse morals and the magnificent illustrations of Doré. One of the five or six great books of European fairy tales. viii + 117pp. 8⅛ x 11. 22311-6 Paperbound $2.00

OLD HUNGARIAN FAIRY TALES, Baroness Orczy. Favorites translated and adapted by author of the *Scarlet Pimpernel.* Eight fairy tales include "The Suitors of Princess Fire-Fly," "The Twin Hunchbacks," "Mr. Cuttlefish's Love Story," and "The Enchanted Cat." This little volume of magic and adventure will captivate children as it has for generations. 90 drawings by Montagu Barstow. 96pp. (USO) 22293-4 Paperbound $1.95

THE RED FAIRY BOOK, Andrew Lang. Lang's color fairy books have long been children's favorites. This volume includes Rapunzel, Jack and the Bean-stalk and 35 other stories, familiar and unfamiliar. 4 plates, 93 illustrations x + 367pp.
21673-X Paperbound $2.00

THE BLUE FAIRY BOOK, Andrew Lang. Lang's tales come from all countries and all times. Here are 37 tales from Grimm, the Arabian Nights, Greek Mythology, and other fascinating sources. 8 plates, 130 illustrations. xi + 390pp.
21437-0 Paperbound $1.95

HOUSEHOLD STORIES BY THE BROTHERS GRIMM. Classic English-language edition of the well-known tales — Rumpelstiltskin, Snow White, Hansel and Gretel, The Twelve Brothers, Faithful John, Rapunzel, Tom Thumb (52 stories in all). Translated into simple, straightforward English by Lucy Crane. Ornamented with headpieces, vignettes, elaborate decorative initials and a dozen full-page illustrations by Walter Crane. x + 269pp. 21080-4 Paperbound $2.00

THE MERRY ADVENTURES OF ROBIN HOOD, Howard Pyle. The finest modern versions of the traditional ballads and tales about the great English outlaw. Howard Pyle's complete prose version, with every word, every illustration of the first edition. Do not confuse this facsimile of the original (1883) with modern editions that change text or illustrations. 23 plates plus many page decorations. xxii + 296pp.
22043-5 Paperbound $2.50

THE STORY OF KING ARTHUR AND HIS KNIGHTS, Howard Pyle. The finest children's version of the life of King Arthur; brilliantly retold by Pyle, with 48 of his most imaginative illustrations. xviii + 313pp. 6⅛ x 9¼.
21445-1 Paperbound $2.50

THE WONDERFUL WIZARD OF OZ, L. Frank Baum. America's finest children's book in facsimile of first edition with all Denslow illustrations in full color. The edition a child should have. Introduction by Martin Gardner. 23 color plates, scores of drawings. iv + 267pp. 20691-2 Paperbound $2.25

THE MARVELOUS LAND OF OZ, L. Frank Baum. The second Oz book, every bit as imaginative as the Wizard. The hero is a boy named Tip, but the Scarecrow and the Tin Woodman are back, as is the Oz magic. 16 color plates, 120 drawings by John R. Neill. 287pp. 20692-0 Paperbound $2.50

THE MAGICAL MONARCH OF MO, L. Frank Baum. Remarkable adventures in a land even stranger than Oz. The best of Baum's books not in the Oz series. 15 color plates and dozens of drawings by Frank Verbeck. xviii + 237pp.
21892-9 Paperbound $2.00

THE BAD CHILD'S BOOK OF BEASTS, MORE BEASTS FOR WORSE CHILDREN, A MORAL ALPHABET, Hilaire Belloc. Three complete humor classics in one volume. Be kind to the frog, and do not call him names . . . and 28 other whimsical animals. Familiar favorites and some not so well known. Illustrated by Basil Blackwell. 156pp. (USO) 20749-8 Paperbound $1.25

EAST O' THE SUN AND WEST O' THE MOON, George W. Dasent. Considered the best of all translations of these Norwegian folk tales, this collection has been enjoyed by generations of children (and folklorists too). Includes True and Untrue, Why the Sea is Salt, East O' the Sun and West O' the Moon, Why the Bear is Stumpy-Tailed, Boots and the Troll, The Cock and the Hen, Rich Peter the Pedlar, and 52 more. The only edition with all 59 tales. 77 illustrations by Erik Werenskiold and Theodor Kittelsen. xv + 418pp. 22521-6 Paperbound $3.00

GOOPS AND HOW TO BE THEM, Gelett Burgess. Classic of tongue-in-cheek humor, masquerading as etiquette book. 87 verses, twice as many cartoons, show mischievous Goops as they demonstrate to children virtues of table manners, neatness, courtesy, etc. Favorite for generations. viii + 88pp. 6½ x 9¼. 22233-0 Paperbound $1.25

ALICE'S ADVENTURES UNDER GROUND, Lewis Carroll. The first version, quite different from the final *Alice in Wonderland,* printed out by Carroll himself with his own illustrations. Complete facsimile of the "million dollar" manuscript Carroll gave to Alice Liddell in 1864. Introduction by Martin Gardner. viii + 96pp. Title and dedication pages in color. 21482-6 Paperbound $1.25

THE BROWNIES, THEIR BOOK, Palmer Cox. Small as mice, cunning as foxes, exuberant and full of mischief, the Brownies go to the zoo, toy shop, seashore, circus, etc., in 24 verse adventures and 266 illustrations. Long a favorite, since their first appearance in St. Nicholas Magazine. xi + 144pp. 6⅝ x 9¼. 21265-3 Paperbound $1.75

SONGS OF CHILDHOOD, Walter De La Mare. Published (under the pseudonym Walter Ramal) when De La Mare was only 29, this charming collection has long been a favorite children's book. A facsimile of the first edition in paper, the 47 poems capture the simplicity of the nursery rhyme and the ballad, including such lyrics as I Met Eve, Tartary, The Silver Penny. vii + 106pp. 21972-0 Paperbound $1.25

THE COMPLETE NONSENSE OF EDWARD LEAR, Edward Lear. The finest 19th-century humorist-cartoonist in full: all nonsense limericks, zany alphabets, Owl and Pussycat, songs, nonsense botany, and more than 500 illustrations by Lear himself. Edited by Holbrook Jackson. xxix + 287pp. (USO) 20167-8 Paperbound $2.00

BILLY WHISKERS: THE AUTOBIOGRAPHY OF A GOAT, Frances Trego Montgomery. A favorite of children since the early 20th century, here are the escapades of that rambunctious, irresistible and mischievous goat—Billy Whiskers. Much in the spirit of *Peck's Bad Boy,* this is a book that children never tire of reading or hearing. All the original familiar illustrations by W. H. Fry are included: 6 color plates, 18 black and white drawings. 159pp. 22345-0 Paperbound $2.00

MOTHER GOOSE MELODIES. Faithful republication of the fabulously rare Munroe and Francis "copyright 1833" Boston edition—the most important Mother Goose collection, usually referred to as the "original." Familiar rhymes plus many rare ones, with wonderful old woodcut illustrations. Edited by E. F. Bleiler. 128pp. 4½ x 6⅜. 22577-1 Paperbound $1.25

TWO LITTLE SAVAGES; BEING THE ADVENTURES OF TWO BOYS WHO LIVED AS INDIANS AND WHAT THEY LEARNED, Ernest Thompson Seton. Great classic of nature and boyhood provides a vast range of woodlore in most palatable form, a genuinely entertaining story. Two farm boys build a teepee in woods and live in it for a month, working out Indian solutions to living problems, star lore, birds and animals, plants, etc. 293 illustrations. vii + 286pp.
20985-7 Paperbound $2.50

PETER PIPER'S PRACTICAL PRINCIPLES OF PLAIN & PERFECT PRONUNCIATION. Alliterative jingles and tongue-twisters of surprising charm, that made their first appearance in America about 1830. Republished in full with the spirited woodcut illustrations from this earliest American edition. 32pp. 4½ x 6⅜.
22560-7 Paperbound $1.00

SCIENCE EXPERIMENTS AND AMUSEMENTS FOR CHILDREN, Charles Vivian. 73 easy experiments, requiring only materials found at home or easily available, such as candles, coins, steel wool, etc.; illustrate basic phenomena like vacuum, simple chemical reaction, etc. All safe. Modern, well-planned. Formerly *Science Games for Children.* 102 photos, numerous drawings. 96pp. 6⅛ x 9¼.
21856-2 Paperbound $1.25

AN INTRODUCTION TO CHESS MOVES AND TACTICS SIMPLY EXPLAINED, Leonard Barden. Informal intermediate introduction, quite strong in explaining reasons for moves. Covers basic material, tactics, important openings, traps, positional play in middle game, end game. Attempts to isolate patterns and recurrent configurations. Formerly *Chess.* 58 figures. 102pp. (USO) 21210-6 Paperbound $1.25

LASKER'S MANUAL OF CHESS, Dr. Emanuel Lasker. Lasker was not only one of the five great World Champions, he was also one of the ablest expositors, theorists, and analysts. In many ways, his Manual, permeated with his philosophy of battle, filled with keen insights, is one of the greatest works ever written on chess. Filled with analyzed games by the great players. A single-volume library that will profit almost any chess player, beginner or master. 308 diagrams. xli x 349pp.
20640-8 Paperbound $2.50

THE MASTER BOOK OF MATHEMATICAL RECREATIONS, Fred Schuh. In opinion of many the finest work ever prepared on mathematical puzzles, stunts, recreations; exhaustively thorough explanations of mathematics involved, analysis of effects, citation of puzzles and games. Mathematics involved is elementary. Translated by F. Göbel. 194 figures. xxiv + 430pp. 22134-2 Paperbound $3.00

MATHEMATICS, MAGIC AND MYSTERY, Martin Gardner. Puzzle editor for Scientific American explains mathematics behind various mystifying tricks: card tricks, stage "mind reading," coin and match tricks, counting out games, geometric dissections, etc. Probability sets, theory of numbers clearly explained. Also provides more than 400 tricks, guaranteed to work, that you can do. 135 illustrations. xii + 176pp.
20338-2 Paperbound $1.50

CATALOGUE OF DOVER BOOKS

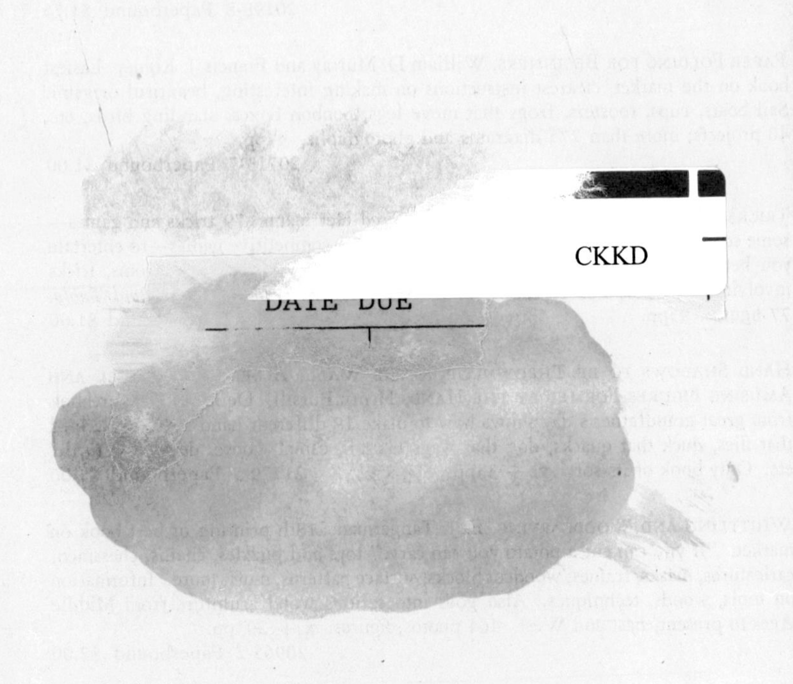